Lecture Notes in Computer Science   11266

Commenced Publication in 1973
Founding and Former Series Editors:
Gerhard Goos, Juris Hartmanis, and Jan van Leeuwen

Editorial Board

More information about this series at http://www.springer.com/series/7412

Yuxin Peng · Kai Yu
Jiwen Lu · Xingpeng Jiang (Eds.)

Intelligence Science and Big Data Engineering

8th International Conference, IScIDE 2018
Lanzhou, China, August 18–19, 2018
Revised Selected Papers

 Springer

Editors
Yuxin Peng
Peking University
Beijing, China

Kai Yu
Shanghai Jiao Tong University
Shanghai, China

Jiwen Lu
Tsinghua University
Beijing, China

Xingpeng Jiang
Central China Normal University
Wuhan, China

ISSN 0302-9743   ISSN 1611-3349   (electronic)
Lecture Notes in Computer Science
ISBN 978-3-030-02697-4   ISBN 978-3-030-02698-1   (eBook)
https://doi.org/10.1007/978-3-030-02698-1

Library of Congress Control Number: 2018958315

LNCS Sublibrary: SL6 – Image Processing, Computer Vision, Pattern Recognition, and Graphics

This Springer imprint is published by the registered company Springer Nature Switzerland AG
The registered company address is: Gewerbestrasse 11, 6330 Cham, Switzerland

Preface

IScIDE 2018, the International Conference on Intelligence Science and Big Data Engineering, took place in Lanzhou, China, during August 18–19 2018. As one of the annual events organized by the Chinese Golden Triangle ISIS (Information Science and Intelligence Science) Forum, this meeting was the eighth of a series of annual meetings promoting the academic exchange of research on various areas of intelligence science and big data engineering in China and abroad.

To keep the quality of the conference, every paper was reviewed by at least three reviewers. After that, we checked for plagiarism in each paper accepted for IScIDE 2018 and rejected 12 of them because of evidence of plagiarism. We received 121 full paper submissions for IScIDE 2018, and only 59 papers were finally accepted, for an acceptance rate of 48.8%.

We would like to thank all the reviewers for spending their precious time on reviewing papers and for providing valuable comments that helped significantly in the paper selection process.

We would like to express special thanks to the conference general co-chairs, Lei Xu, Dewen Hu, and Bin Hu, for their leadership, advice, and help on crucial matters concerning the conference. We would like to thank all Steering Committee members, Program Committee members, Invited Speaker Committee members, Organizing Committee members, and Publication Committee members for their hard work. We would like to thank Xinbo Gao, Dewen Hu, Jian Yang, Yuanqing Li, Dongliang Xie, and Dan Jiang for delivering the invited talks. Finally, we would like to thank all the authors of the submitted papers, whether accepted or not, for their contribution to the high quality of this conference. We count on your continued support of the ISIS community in the future.

August 2018

Yuxin Peng
Kai Yu
Jiwen Lu
Xingpeng Jiang

Organization

General Chairs

Lei Xu   Shanghai Jiao Tong University, China
Dewen Hu   National University of Defense Technology, China
Bin Hu   Lanzhou University, China

Program Chairs

Yuxin Peng   Peking University, China
Kai Yu   Shanghai Jiao Tong University, China
Jiwen Lu   Tsinghua University, China

Organization Chair

Zhixin Ma   Lanzhou University, China

Local Arrangements Chair

Songyang Liu   Lanzhou University, China

Keynote Chairs

Kun Zhang   Carnegie Mellon University, USA
Feiping Nie   Northwestern Polytechnical University, China
Shikui Tu   Shanghai Jiao Tong University, China

Special Issue Chairs

Yong Xia   Northwestern Polytechnical University, China
Weishi Zheng   Sun Yat-sen University, China
Yadong Mu   Peking University, China
Xingpeng Jiang   Central China Normal University, China
Yang Yu   Nanjing University, China

Publicity Chairs

Qiguang Miao   Xidian University, China
Huiguang He   Institute of Automation, Chinese Academy of Sciences, China

Special Session Chairs

Shiliang Sun   East China Normal University, China
Wankou Yang   Southeast University, China
Jinzhong Lin   Fudan University, China

Oral Session Chairs

Hanli Wang   Tongji University, China
Deyu Meng   Xi'an Jiaotong University, China
Cewu Lu   Shanghai Jiao Tong University, China

Poster Chairs

Xi Li   Zhejiang University, China
Di Huang   Beihang University, China

Registration Chairs

Rong Ma   Lanzhou University, China
Minqiang Yang   Lanzhou University, China

Demo Chairs

Bo Chen   Xidian University
Tao Zhou   Ningxia Medical University

Industrial Chairs

Mingkui Tan   South China University of Technology, China
Liang Lin   Sun Yat-sen University, China
Changdong Wang   Sun Yat-sen University, China

Program Committee

Akira Hirose   The University of Tokyo, Japan
Andrey Krylov   Lomonosov Moscow State University, Russia
Baoliang Lu   Chang'an University, China
Binbin Lin   University of Michigan, USA
Bingyao Yu   Tsinghua University, Beijing, China
Bo Chen   Xidian University, China
Caixun Wang   Tsinghua University, Beijing, China
Cewu Lu   Shanghai Jiao Tong University, China
Chang-Dong Wang   Sun Yat-sen University, Zhongshan, China
Changshui Zhang   Tsinghua University, Beijing, China
Changyin Sun   Southeast University, Nanjing, China

Chao Liu	Chongqing University, China
Cheng Ma	Tsinghua University, Beijing, China
Cheng-Yuan Liou	National Taiwan University, China
Chunhua Shen	University of Adelaide, Australia
Chunze Lin	École centrale de Nantes, French
Dacheng Tao	University of Technology, Sydney, Australia
Dajiang Lei	Chongqing University of Posts and Telecommunications, China
Daoqiang Zhang	Nanjing University of Aeronautics and Astronautics, Nanjing, China
Daoxiang Zhou	Chongqing University, China
Deng Cai	Zhejiang University, Hangzhou, China
Dewen Hu	National University of Defense Technology, Changsha, China
Deyu Meng	Xi'an Jiaotong University, China
Dezhong Yao	University of Electronic Science and Technology of China, China
Qingshan Liu	Nanjing University of Information Science & Technology, Nanjing, China
Qiongmin Zhang	Chongqing University of Technology, China
Qiuli Wang	Chongqing University, China
Ruojin Cai	Tsinghua University, Beijing, China
Seiichi Ozawa	Kobe University, Japan
Sheng Huang	Chongqing University, China
Shiguang Shan	Chinese Academy of Sciences, China
Shikui Tu	Shanghai Jiao Tong University, China
Shiliang Sun	East China Normal University, China
Shizheng Zhang	Zhengzhou University of Light Industry, China
Shutao Li	Hunan University, Changsha, China
Shuyan Li	Tsinghua University, Beijing, China
Si Wu	Beijing Normal University, China
Tao Zhou	Ningxia Medical University, China
Ting Xie	Chongqing University of Technology, China
Vincent S. Tseng	National Chiao Tung University, China
Wang Caixun	Tsinghua University, Beijing, China
Wanhua Li	Tsinghua University, Beijing, China
Wankou Yang	Southeast University, Nanjing, China
Wei-Shi Zheng	Sun Yat-sen University, China
Wei Wang	Chongqing University, China
Weixun Chen	Hunan University, China
Wencheng Zhu	Tsinghua University, Beijing, China
Wenying Wen	Jiangxi University of Finance and Economics, China
Di Huang	Beihang University, China
Fang Fang	Peking University, Beijing, China
Feiping Nie	Northwestern Polytechnical University, China
Feiyu Chen	Chongqing University, China
Gang Pan	Zhejiang University, China

Guangyi Chen	Nanjing University, China
Hai Nan	Chongqing University of Technology, China
Hanli Wang	Tongji University, China
Hao Liu	Tsinghua University, Beijing, China
Haomiao Sun	Tsinghua University, Beijing, China
Heikki Kalviainen	Lappeenranta University of Technology, Finland
Hiroyuki Iida	Japan Advanced Institute of Science and Technology, Japan
Hongbin Li	Chongqing University, China
Hongxing Wang	Chongqing University, China
Hongyi Sun	Tsinghua University, Beijing, China
Huchuan Lu	Dalian University of Technology, Dalian, China
Huiguang He	Institute of Automation, Chinese Academy of Sciences, China
Huimin Ma	Tsinghua University, Beijing, China
James Kwok	Hong Kong University of Science and Technology, Hong Kong, China
Jian-Huang Lai	Sun Yat-sen University, Zhongshan, China
Jian Yang	Nanjing University of Science and Technology, Nanjing, China
Jifei Han	Tsinghua University, Beijing, China
Jing Chen	Peking University, China
Jufu Feng	Peking University, Beijing, China
Kai Yu	Shanghai Jiao Tong University, China
Karl Ricanek	University of North Carolina Wilmington, USA
Kazushi Ikeda	Nara Institute of Science and Technology, Japan
Kun Zhang	Carnegie Mellon University, USA
Lei Chen	Tianjin University, China
Wen Zhang	Wuhan University, China
Wenzhao Zheng	Tsinghua University, Beijing, China
Xiaofei He	Zhejiang University, Hangzhou, China
Xiaojuan Cheng	Beihang University, China
Xiaorong Gao	Tsinghua University, Beijing, China
Xiaoxiang Zheng	Zhejiang University, China
Xi Li	Zhejiang University, China
Xinbo Gao	Xidian University, Xi'an, China
Xin Feng	Chongqing University of Technology, China
Xin Geng	Southeast University, Nanjing, China
Xingpeng Jiang	Central China Normal University, China
Xinpeng Zhang	Chongqing University, China
Xin Yuan	Lanzhou University, China
Xudong Lin	Tsinghua University, Beijing, China
Xuelong Li	Xi'an Optics and Fine Mechanics, Chinese Academy of Scienses, Xi'an, China
Xueping Wang	Hunan University, China
Yadong Mu	Peking University, China
Yahong Han	Tianjin University, China
Yang Yu	Nanjing University, China

Contents

Deep Neural Networks

Objects and Language

Classification and Clustering

Imaging

Biomedical Signal Processing

Robots and Intelligent System

Application of Ant Colony Algorithm Based on Monopoly and Competition Idea in QoS Routing

Yongsheng Li[1], Yong Huang[2([⊠])], Shibin Xuan[1], and Liangdong Qu[2]

[1] College of Information Science and Engineering,
Guangxi University for Nationalities, Nanning 530006, China
lkflys@163.com
[2] College of Software and Information Security,
Guangxi University for Nationalities, Nanning 530006, China
lyshlh@163.com

Abstract. Ant colony algorithm is easy to fall in local best and its convergent speed is slow in solving multiple QoS constrained unicast routing problems. Therefore, an ant colony algorithm based on monopoly and competition is proposed in this paper to solve the problems. In the choice of nodes, improves pheromone competition, avoids monopoly of pheromone prematurely, stimulates ants to attempt the paths which have less pheromone and improves the global search ability of ants. Stagnation behavior is judged by the monopoly extent of the pheromone on the excellent path. Moreover, the catastrophic is embedded in the global pheromone update operation. According to simulations, its global search is strong and it can range out of local best and it is fast convergence to the global optimum. The improved algorithm is feasible and effective.

Keywords: Ant colony algorithm · QoS routing · Monopoly · Competition
Catastrophic

1 Introduction

With the rapid development of internet, the function of the network is changed from transmitting the original data to transmitting lots of data, such as voice, image and so on. More and more business request the quality of service of the network. The management and control of the Qos network has become one of a hot research area by many researchers. Qos routing is the one of the core technologies in the Qos management and control, of which the key issue is to find a Qos routing, which meet multiple constraints, to achieve the optimal allocation of network resources. Wang et al. has proved the QoS routing problem with multiple constraints is NP-complete [1]. Commonly, the heuristic algorithms can solve the QoS routing problem, such as genetic algorithms [2], particle swarm optimization [3], neural networks [4], etc. and there have been a number of research results.

ACO algorithm is proposed by the Italian scholar Dorigo, who made it 90's of the last century, which is a new heuristic algorithm [5]. It does not depend on the specific

© Springer Nature Switzerland AG 2018
Y. Peng et al. (Eds.): IScIDE 2018, LNCS 11266, pp. 3–11, 2018.
https://doi.org/10.1007/978-3-030-02698-1_1

mathematical description and have the feature of positive feedback and distributed computing etc., which has been widely applied to solve the TSP [6], Job-shop scheduling [7], path planning [8] and the other combinatorial optimization problems. Lots of research results show that the ACO algorithm in solving complex optimization problems (in particular discrete optimization problem) has great advantages and potential applications. However, there are some shortcomings, which ACO algorithm is easy to fall into local minima and solve slowly.

Aiming at the shortcomings of the ACO algorithm and considering to the characteristics of the QoS unicast routing, this paper propose an ant colony algorithms based on monopoly and competition to solve QoS routing problems. In the algorithm of node selection strategy, the competition of pheromone is enhanced and the node selection strategy is dynamically adjusted, so that the randomness of algorithm node selection is strengthened, and the global search is extended. Stagnation behavior is judged by the monopoly extent of the pheromone on the excellent path and the catastrophic is embedded in the global pheromone update operation. So that solution can jump out of the local minimum interval to avoid falling into the local optimum Simulation experiments show that the improved algorithm can effectively solve the QoS routing problem.

2　Qos Unicast Routing Problem Description

To facilitate the analysis, a network can be represented as an undirected weighted graph $G(V, E)$, where $V = \{v_1, v_2 \ldots v_n\}$, which is the set of the network nodes. s is set to the source node, d is the target node. $p = (s, i, j \ldots d)$ is the path from the source node to the target node. For each $e \in E$, has four parameters, namely bandwidth function: bandwidth(e), delay function:delay(e), delay jitter function: delay-jitter(e), cost function:cost (e). Therefore, QoS unicast routing problem can be described as finding a transmission path p, which meet the indicators of the QoS and minimize resource consumption [9]:

(1)　bandwidth constraint: $B(p) \geq B$;
(2)　delay constraint: $D(p) \leq D$;
(3)　delay-jitter constraint: $J(p) \leq DJ$
(4)　minimum consumption of cost:Min(cost(p)).

Where bandwidth B is bottleneck bandwidth, that is $\min\{B_l, l \in p(s, d)\} \geq B$, the delay is the total delay of the path p, that is $D(p) = \sum_{l \in p} D_l$, the delay-jitter is the total delay-jitter of the path p, $DJ(p) = \sum_{l \in p} DJ_l$.

3　Basic Ant Colony System

The ant colony optimization algorithm (ACO) is proposed in the early 1990s. In 1996, ant colony system is proposed by Dorigo and Gmabardella [10], the performance of the ACO is effectively improved. In [10], they made three improvements as follows:

$$s_k = \begin{cases} \arg \max_{u \in allowed_k} \{[\tau(r,u)]^{\alpha}[\eta(r,u)^{\beta}]\}, & q \leq q_0 \\ S & q > q_0 \end{cases} \tag{1}$$

$$p_{ij}^k(t) = \begin{cases} \dfrac{\tau_{i,j}^{\alpha}(t)\eta_{i,j}^{\beta}(t)}{\sum\limits_{s \in allowed_k}(\tau_{i,s}^{\alpha}(t) \cdot \eta_{i,s}^{\beta}(t))}, & j \in allowed_k \\ 0 & otherwise \end{cases} \tag{2}$$

(1) A new selection strategy that combination of deterministic selection and random selection is adapted, which both can utilize the advantage of prior knowledge and can tendentiously explore. For an ant at node r to move to the next city s, the state transition rule is given by the following formula.

Where s_k is the next node of ant k, q is the random number draw from [0, 1], q_0 is a parameter $(0 \leq q_0 \leq 1)$. S is a random variable selected by the probability distribution given in Eq. (2). $allowed_k$ is a node set that these node can be selected for ant k in the next time, α is the pheromone heuristic factor, which reflects the effect of pheromone by ant accumulates when ants move to the other nodes. β is a heuristic factor, had reflects the extent of the heuristic information is focused when the ants select path. $\tau_{ij}(t)$ is the pheromone of path (i,j) at t time. η_{ij} is the visibility of path (i,j), which is corresponding with the inverse of the distance from node i to node j.

(2) Only the global optimal ant path performs global updating rule. After each iteration, the pheromone is enhanced only occur on the path walked by the best ant. For other pathway, the pheromone will be gradually reduced due to volatile mechanism, which can make the ant colony more inclined to select optimal path. Consequently, the convergence rate will be increased and the search efficiency will be enhanced. Global update rule is described as follows:

$$\tau(r,s) \leftarrow (1-\rho) \cdot \tau(r,s) + \rho \cdot \Delta\tau(r,s) \tag{3}$$

$$\Delta\tau(r,s) = \begin{cases} Q/L_{gb} & if \ (r,s) \in g \\ 0 & else \end{cases} \tag{4}$$

Here ρ is the pheromone volatile coefficient, $0 < \rho < 1$, L_{gb} is the current global optimal path. Q is a constant that indicate initial pheromone intensity between two nodes.

(3) Using the local update rule. The pheromone will be local updated when these ants build a path, which make the pheromone release by ants to reduce when they pass the path. The local update rule is used to decrease influence on other ants and make them search other edges. Therefore, the ants can avoid that they prematurely converge to a same solution.

The local update rule is represented by Eq. (5).

$$\tau(r, s) \leftarrow (1 - \rho_0) \cdot \tau(r, s) + \rho_0 \cdot \tau_0 \tag{5}$$

$$\tau_0 = (nL_{nn})^{-1} \tag{6}$$

Where n is the number of nodes. L_{nn} is a path length generated heuristically by the recent neighborhood.

4   Application of Ant Colony Algorithm Based on Monopoly and Competition Idea in QoS Routing

4.1   Improved Node Selection Strategy Based on the Idea of Monopoly and Competition

The search process of ant colony algorithm is the process of interaction between positive and negative feedback, as well as the process of pheromone monopoly and competition in each path. After a cycle, the ants leave more pheromones on the better path, and more pheromones attract more ants to choose the path, and gradually form the pheromone monopoly. In the end, all ants chose the path. However, if the pheromone monopoly is too strong, the algorithm will easily fall into premature and stagnation. By formula (1) and (2), we can see that the ant chooses the next node with a certain randomness. Even if there a great probability to select a better path, the ant is likely to choose another path. This is similar to the development of an industry in the society. In order to avoid the formation of complete monopoly in this industry, the government institutions always set up certain rules to maintain moderate competition. So that the industry can develop in a healthy way. Moreover, after a cycle, part of the pheromones on the path is forced to evaporate, which ensures that the pheromone will not accumulate endlessly and avoid complete monopoly. However, if the competition of pheromone is too strong, the randomness of search will be too large and the algorithm convergence time will be long. Therefore, the monopoly and competition of the pheromone are a pair of contradictions. It is because of their interaction that the ant colony algorithm can gradually converge to a better solution. In this paper, we enhancing the competition of pheromone when ants choose the next node in order to avoid premature formation of pheromone monopoly and stimulate the ants to try less pheromone path, so as to improve the global search ability of ants. The improved node selection strategy is:

$$p_{ij}^k(t) = \begin{cases} \dfrac{\tau_{i,j}^\alpha(t) \eta_{i,j}^\beta(t)}{\sum\limits_{s \in allowed_k} (\tau_{i,s}^\alpha(t) \cdot \eta_{i,s}^\beta(t)) \bullet \theta_{i,j}}, & j \in allowed_k \\ 0 & otherwise \end{cases} \tag{7}$$

$$\theta_{i,j} = \dfrac{N_u N}{N_u N + \dfrac{m_{ij} \bullet \eta_{i,j} \bullet \phi}{\max(\eta_{i,j})}} \tag{8}$$

In the formula, N is the number of ants, N_u is the current iteration number, and m_{ij} is the total number of ants passing through the path (i, j). When the iteration goes to a local optimum, although the pheromone of local optimal path is increasing, but the ants that passed through the path is also increasing, it lead to the decrease of $\theta_{i,j}$, inhibited the effects on state transition probability that comes from the growth of pheromone, which is conducive to improve the algorithm's global search ability. Because $m_{ij} \leq N_u \bullet N$, $\eta_{i,j}/\max(\eta_{i,j}) \leq 1$, so $1 \geq \theta_{i,j} \geq \theta_{\min} = 1/(1+\varphi)$. The parameter φ can adjust the intensity of the X, and the smaller the φ is, the larger the $\theta_{\min}$ is. The number of ants passing through the path has less influence on the rules of state transition.

4.2 Global Search Strategy for Embedding Pheromone Monopoly Judgment and Catastrophic

When solving the QoS routing problem, the ants start from the same source point, so compared with the TSP problem, the ant colony algorithm is more likely to form the pheromone monopoly and fall into the local optimum when solving the QoS routing problem. In order to solve this problem, the solution likely to fall into local optimum is subjected to catastrophic operations, in the global pheromone updating of the algorithm, catastrophic is introduced to kill the outstanding individual (path), so as to improve the competition ability of other individuals and improve the ability of jumping out of the local extreme value range.

The judgment of pheromone monopoly is the basis of premature treatment. Experiments show that both the premature convergence and global convergence lead to the occurrence of pheromone monopoly in the qualified path. Therefore, the monopoly extent of pheromone in each qualified path can be used as one of the judgment criteria of premature convergence. The extent of monopoly function of the pheromone in each qualified path can be defined as:

$$\sigma^2 = \sum_{i=1}^{g} \left[\frac{\tau_{iw} - \tau_{bw}}{f} \right]^2 \tag{9}$$

$$f = \begin{cases} \max |\tau_{iw} - \tau_{bw}|, & \max |\tau_{iw} - \tau_{bw}| > 1 \\ 1 & otherwise \end{cases} \tag{10}$$

Among them, g is the number of qualified paths found by m ants in the current iteration. τ_{iw} is the average pheromone on the qualified path found by the first i ants (τ_{iw} = the sum of the pheromone on the qualified path/the number of edges on the path). τ_{bw} is the average pheromone of the optimal path found by m ants. f is a normalized scaling factor and it can limits the size of σ^2. The formula (10) shows that σ^2 can reflect the extent of pheromone monopoly on the path. The smaller the σ^2 is, the more likely the pheromone monopoly is, when all the m ants choose the same path, $\sigma^2 = 0$. If in the consecutive Z iterations, the optimal path obtained by the algorithm is not significantly improved and $\sigma^2 < C$ (C is a given constant), at this time, the algorithm may have formed a pheromone monopoly and fall into the local optimum, at this time, algorithm can update the global pheromone embedded catastrophic, the rules are as follows:

$$\tau(r,s) \leftarrow (1 - \rho_0) \cdot \tau(r,s) + d \cdot \rho_0 \cdot \Delta\tau(r,s) \qquad (11)$$

In the formula above, d is the directional control factor, when the individual fitness is higher than the average value, the negative feedback mechanism is used to search, at this time, d = −1, and conversely, the positive feedback mechanism is used to search, at this time, d = 1. Thus, when it falls into the local optimum, the algorithm reduces the probability of selecting the sub interval of high fitness, increases probability of selecting low fitness interval, increases the diversity of the candidate solution and avoids the local convergence.

4.3 Algorithm Steps Description

Step 1: delete these nodes that do not meet the delay constraint and bandwidth constraint, obtain the new network topology, and select the rout based on the network topology.

Step 2: Initialize the pheromone intensity of all the links in the network. Set various parameters of the algorithm and the number m of the ants, DiedaiNum is the maximum number of loop. Initial the number of loop $l = 0$.

Step 3: set the current loop number $l = l + 1$, and set the increment of the pheromone of various links $\Delta\tau_{ij} = 0$, $t = 0$, m ants are deployed in the source node. Tabu table are generated for every ant, and source node is deployed in the tabu table.

Step 4: $t = t + 1$, for each ant k do not finish searching, according to the Eq. (7), select the next node j from the current node i, if node j does not exist, then note the ant has finished searching, else ant k is deployed in the tabu table, and if node j is target node, then note the ant has finished searching, else continue, use the Eq. (5) to update the local pheromone.

Step 5: repeat Step 4 until the m ants all have finished searching, record all the qualified paths and the optimum paths until the current loop, and record all the pheromone and fitness function value of all the qualified links.

Step 6: If in the consecutive Z iterations, the optimal path obtained by the algorithm is not significantly improved and $\sigma^2 < C$, then updating the pheromone of optimum paths according to the Eq. (11), otherwise updating the pheromone of optimum paths according to the Eq. (3). If $l < $ DiedaiNum, then go to step 3, else go to step 7.

Step 7: output the optimum path $Path_{best}$, end.

4.4 Simulation

The proposed improved ACO is implemented in MATLAB, In order to validate the validity of improved ACO we selected an example to experimentize. As shown in Fig. 1, the network includes 25 nodes. In our experiment, we only consider the constraints: link bandwidth, delay and cost, so, the link attribute can be described with a 3-tuple (bandwidth, delay, cost). Nevertheless, computation of multiplicative property can reference additivity.

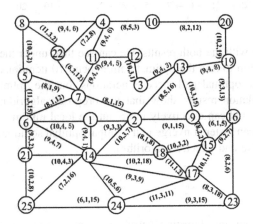

Fig. 1. Network model

Given there are service request of routing, the source node $vs = 25$, the target node $vd = 20$, the bandwidth $B = 8$, the delay $D = 48$, the parameters: $\alpha = 1$, $\beta = 2$, $\rho_0 = 0.2$, $\rho = 0.3$, $Q = 2$, $\varphi = 0.6$, **$C = 0.05$, $Z = N/5$,** assign an initial value is 1 to the pheromone of each links, the number of the ants deployed in the source node is $m = 10$, the number of iterations is 20.

Figure 2 shows the curves of the optimum path's cost and delay within 20 iterations using the improved algorithm and the basic ant colony algorithm (the vertical coordinate denotes the cost and delay, the unit of the delay is second, the horizontal ordinate denotes the iterations). Table 1 shows the comparison of the results about the parameters between the proposed algorithm and the basic ant algorithm.

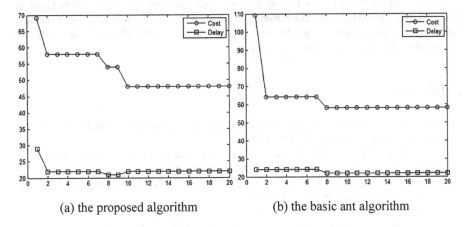

(a) the proposed algorithm   (b) the basic ant algorithm

Fig. 2. The curve of the results of the both algorithm

Table 1. The comparison of the results about the parameters between the both algorithms

The type	The optimum path	Cost	Delay	Iteration
The proposed algorithm	25 → 21 → 6 → 5 → 8 → 4 → 10 → 20	48	22	10
The basic ant algorithm	25 → 21 → 6 → 7 → 5 → 8 → 22 → 4 → 10 → 20	58	22	8

After comparing with the both results, we can see that, under the same conditions, the proposed algorithm to find the optimal path expend 10 iterations and the basic ant algorithm to find the optimal path need 8 iterations, but the optimal path is a local optimum. The simulation results show that the proposed algorithm have preferable global search ability and can effectively jump out the local minima, consequently, the proposed algorithm can faster converge to the global optimal solution. So, the proposed algorithm is a feasible and effective algorithm.

5 Conclusion

This paper presents an ant algorithm based on monopoly and competition idea for solving Qos routing problems. In nodes selection strategy, pheromone competition is enhanced to stimulate ants to attempt the paths which have less pheromone, avoid falling into local optima. Stagnation behavior is judged by the monopoly extent of the pheromone on the excellent path. Moreover, the catastrophic is embedded in the global pheromone update operation to make the algorithm successfully to jump out the local minima. The experimental results show that, this algorithm can effectively and quickly achieve QoS routing optimization. How to improve the algorithm, and apply it to the more complex QoS multicast routing problem. It is worth intensive study in the future.

Acknowledgment. This work is supported by the Application Research Program of 2016 the Guangxi province of China young and middle-aged teachers basic ability promotion project (No. KY2016YB133), the Research Program of 2014 Guagnxi University for Nationalities of China (No. 2014MDYB029), the Key project of science and technology research in Guangxi education (No. 2013ZD021), the innovation team project of xiangsihu youth scholars of Guangxi University For Nationalities, and the Research Program of 2014 Guagnxi University for Nationalities of China (No. 2014MDYB028).

References

1. Wang, Z., Corwcroft, J.: Quality-of-service routing for supporting multimedia applications. IEEE J. Sel. Areas Commun. **14**(7), 1228–1234 (1996)
2. Li, H.J., Jing, Y.Y., Liu, H.J.: Research on secure QoS routing algorithm for distributed fiber Bragg grating sensor networks. Laser J. **38**(6), 89–92 (2017)
3. Sun, X.X., Wang, X.W., Huang, M.: Adaptive harmony PSO based trusted QoS routing scheme. J. Syst. Simul. **28**(3), 741–748 (2016)
4. Liu, H.Y., Sun, F.C.: Satellite networks QoS routing algorithm based on an orthogonal polynomials neural network. J. Tsinghua Univ. (Sci. Technol.) **53**(4), 556–561(2013)

5. Colorni, A., Dorigo, M., Maniezzo, V., et al.: Distributed optimization by ant colonies. In: Varela, F., Bourgine, P. (eds.) Proceedings of the ECAL 1991, European Conference of Artificial Life, pp. 134–144. Elsevier, Paris (1991)
6. Xu, K., Lu, H., Cheng, B., Huang, Y.: Ant colony optimization algorithm based on improved pheromones double updating and local optimization for solving TSP. J. Comput. Appl. **37** (6), 1686–1691 (2017)
7. Zhang, H.G., Gong, X.: A Generalized ant colony algorithm for job-shop scheduling problem. J. Harbin Univ. Sci. Technol. **22**(1), 91–95, 102 (2017)
8. You, X.M., Liu, S., Lv, J.Q.: Ant colony algorithm based on dynamic search strategy and its application on path planning of robot. Control Decis. **32**(3), 552–556 (2017)
9. Gao, L.C.: QoS routing algorithm base on Q-learning and improved ant colony in mobile ad hoc networks. J. Jilin Univ. (Sci. Ed.). **53**(3), 483–488 (2015)
10. Dorigo, M., Maniezzo, V., Colorni, A.: Ant system: optimization by a colony of cooperating agents. IEEE Trans. System Man Cybern. Part B **26**(1), 29–41 (1996)

Graph Based RRT Optimization
for Autonomous Mobile Robots

Wilbert G. Aguilar[1,2,3]([✉]), David S. Sandoval[1], Jessica Caballeros[1],
Leandro G. Alvarez[1], Alex Limaico[1], Guillermo A. Rodríguez[1],
and Fernando J. Quisaguano[1]

[1] CICTE Research Center, Universidad de Las Fuerzas Armadas ESPE,
Sangolquí, Ecuador
wgaguilar@espe.edu.ec
[2] FIS Faculty, Escuela Politécnica Nacional, Quito, Ecuador
[3] GREC Research Group, Universitat Politècnica de Catalunya,
Barcelona, Spain

Abstract. In this article, we present the application of Graph Theory in the development of an algorithm of path planning for mobile robots. The proposed system evaluates a RRT algorithm based on the individual cost of nodes and the optimized reconnection of the final path based on Dijkstra and Floyd criteria. Our proposal includes the comparisons between different RRT* algorithms and the simulation of the environments in different platforms. The results identify that these criteria must be considered in all the variations of RRT to achieve a definitive algorithm in mobile robotics.

Keywords: RRT* · Mobile robots · Dijkstra algorithm · Floyd algorithm
Path planning

1 Introduction

An important goal in the development of an autonomous mobile robot is Path Planning, several techniques have been successfully applied and one of the most important is Rapidly Exploring Random Tree or RRT which is defined as a search in a known space, of a continuous pattern or trajectory that describes the connection between an initial state and a target state belonging to the space. This continuous pattern results from the random generation of a collision-free vertex tree which fills the known space, with the particularity that its growth is deter-mined towards free regions of greater space [1].

There are several variations of this tool such as RRT* which takes into account the step size between the vertices and the minimum separation between neighboring vertices which optimizes tree generation [2]. Although RRT* can optimize tree development without increasing computational cost, an exploration rate below RRT is shown. This occurs because it is scanned and optimized in several cases [3, 4]. When working with nodes and links for the generation of the RRT tree, we can consider them directly Standard graph algorithms. In [5] several optimization techniques are described to address these inefficiencies as slow convergence or high cost of communication or

© Springer Nature Switzerland AG 2018
Y. Peng et al. (Eds.): IScIDE 2018, LNCS 11266, pp. 12–21, 2018.
https://doi.org/10.1007/978-3-030-02698-1_2

computation. The most prominent application in Graph Theory is the development of Route Planning for transport networks [6–8]. The application of heuristic criteria and graph algorithms has been widely used in mobile robotics [9, 10], a star is an algorithm widely used in this field, several modifications are analyzed in [11]. To solve the same problem, but focused on mobile robots in static environments, it is proposed to combine algOrithms of path planning, resulting in hybrids as in [12], Where the algorithms are optimized; Dijkstra's shortest pattern, generation of PSO trajectories, genetic algorithms, evolutionary algorithms among others. This article will present the application of algorithms optimized as Floyd and in a mobile robot Dijkstra [13, 14] for interiors that bases its operation on Lidar [15].

2 Related Works

The variations that can be given to this code depends on the dynamic constraints applied in controlled simulations [16]. The variation depends on the dynamics of the mobile platform or robot in which it will be implemented. And if it is not able to modify its address instantaneously it will be necessary a modification called SRRT which considers curved trajectories in the tree [17]. This algorithm requires that the workspace be known and to have adequate information of the environment, there is the possibility of realizing the improvement proposed by [18]. In which objects are classified as valuable and worthless. Valuable objects are placed between the start and end points. They are considered the limits of them, to focus the growth of the tree. It is demonstrated that in this way the RRT chooses the most feasible way.

Simulated environments allow the development of new algorithms with a RRT* sampling approach [19], its contribution rests on the strategic reconnection that allows the movement of the agent without ruling out previous trajectories. Also, do not expect the tree to be complete in real time. The algorithm combines tree expansion and recoil in two modes; Dynamic root displacement and parches sampling. The downside is that working in real time requires large memory capacity and limited environment. It is necessary to consider that in RRT the tree is realized in the objective direction. In [20], they proposed an application on the path of a manipulator. The modification that arises in the algorithm is performed at the level of executing the sample function in the part of the space corresponding to where the target point exists. This space is known as useful space and can be identified as a circular function with center in the goal or other shapes as a square. In some cases, the sample space must be modified, allowing the sample space to behave dynamically. If the dimensional space in which sampling is performed increases the RRT something-rhythm reduces its performance. In [21] we propose a variation to RRT based on the incremental sampling called Fast Convergence Rapidly-exploring Random Tree (FCRRT). The major improvements are that exploration and optimization are performed separately to allow retention of the scanning force. And the use of Lazy-RRG which accelerates the rate of convergence. Lazy-RRG is to delay collision checks for some or all nodes. With this it is achieved that only a small group of nodes is involved final.

Applications that have this tool within the planning of movements as; The intra-operative planning of 3D movements in clinical operations carried out by robots [22] and fast anticipation of collisions [23], the evasion of obstacles which raises the guarantee of the RRT algorithm and translates it into a predictive algorithm [24]. Other algorithms for motion planning in autonomous vehicles consider; The angular variation between vertices [25], potentially hazardous areas [12], dynamic paths [26] or neural networks [26]. Achieving secure control under conditions of uncertainty is a problem that must be considered when placing autonomous robots in the real world. It is necessary to consider safety parameters proposed in [23] which performs evaluations based on a graphical model using a fast algorithm of variational inference, applied mainly to autonomous vehicles in dynamic environments. The framework of these programs develop it for the synthesis of controllers that use signals corresponding to functions of the robot state, the environment and other safety parameters. In the same article experiments were carried out, one of which was applied safety programs on a system that incorporated a RRT* planner used effectively in UAV capable of flying safely against unknown obstacles.

A limitation in several mobile robots is the field of vision, solved from the use of safe patterns for unknown dynamic environments [27]. Consequently, it is possible to experiment with the application of 3D sensors to increase the range of vision and generate a spatial trajectory that optimizes the performance of the movement baker [28]. By using long-range laser sensors LIDAR, it is possible to identify the environment better by sorting objects that could be traversed by the au-toner robot [29]. The final purpose of this applications is to cope with a UAV mobile robot for 3D planning, [30], rapid-flight applications such as military explorations require rapid trajectories that allow maneuvers [31]. The RRT algorithms are fully extrapolated to 3D environments for the detection of collisions for fixed-wing UAVs in complex environments [32], and land mobile applications [29, 33].

3 Our Approach

In our work, we apply as already mentioned Graph Theory, more detailed the algorithm of Floyd and the algorithm of Dijkstra for weighted graphs. RRT* [2, 4, 33] it benefits from the heuristic criterion that results from a new node or vertex generated in a random way in the RRT space, to evaluate if the neighboring nodes called neighbors have a lower Euclidean cost between the vertex generated and the near vertices. Total cost of the root to each neighboring vertex, with the purpose of connecting only vertices that imply lower cos-to. We consider that this evaluation can be optimized by taking into account more optimization criteria. So in this case we fusion the Floyd theory and implement it in the cost function. In a previous work, we analyze the behavior of RRT algorithms in mapping 3D environments [34], from there an analysis of the optimum of the algorithms was realized; RRT*, RRT-GD, and RRT-Limits. It was verified the reduction of the search time of the optimal path, the need to limit regions avoiding computational costs, necessary in robot navigation.

In the application of algorithms, it contemplates the use of terms according the subject like [35], we represent the problem of path planning with the following terms:

- X is a connected subset of where $d \in N, d \geq 2$.
- $G = (V, E)$ is a graph composed by a set of vertexes V and edges E.
- **Xobs** and **Xgoal**, subsets of **X**, are the obstacle region and the goal region respectively.
- The obstacle free space $X \backslash Xobs$ is denoted as **Xfree** and the initial state **Xinit** is an element of **Xfree**.
- A path in **Xfree** must be a collision-free path.

We must also consider traditional definitions of both Floyd and Dijkstra. Since the tree is interpreted as a graph, then this is a binary group constituted of a non-empty finite set $V(D)$ and a ordered pair set $A(D)$ made up of certain elements in $V(D)$, it can be denoted as $D = (V, A)$ [36]. Each pair of elements vi and vj $(1 \leq i, j \leq m)$ in $V(D)$, have an edge E.

Dijkstra [37] on direct form selects a node of the network, for this case will be the initial node, different to the root node co what we find the nearest node:

- Find $i \in P$ such that $D_j = \min D_j$ being $P = P \cup \{i\}$. P contains all nodes
- This case only has one dimension of nodes and the labels are updated by means of the minimum value and the accumulated cost is discarded.

The Floyd algorithm requires that the array D, preserve distance indicators between nodes, enabling all combinations associated with nearby nodes and their corresponding distance avoiding $i = j$. The algorithm RRT* allows to know which nodes are close to the objective node, in this idea includes Floyd with the following steps:

- Initialization with values ∞ or values that are considered as such, in space X.
- Iteration, for $n = 0, 1, \ldots, N + 1$ the matrix is updated with the minimum distances that relate a vertex i to another vertex j as follows:

$$D_{ij}^{(n+1)} = \min \left\{ D_{ij}^n, D_{i,n+1}^n + D_{n+1,i}^n \right\} \vee i \neq j \qquad (1)$$

In each iteration it is verified if it is better route than the previous one, or if it is due to use the initial node.

Reviewing the known RRT* algorithm we have a generality and starting point considering the following functions [34]:

- Function *Sample*: returns **Xrand**, an independent identically distributed sample from **Xfree**.
- Function *Nearest Neighbor*: returns the closest vertex **xnear** to the point $x \in$ **Xfree**, in terms of the Euclidean distance function.
- Function *Steer* returns a point **xnew** at a distance ε, from **xnear** in direction to **xrand**.
- Function *Obstacle Free*: Given two points $x, x' \in$ **Xfree**, the function returns true if the line segment between **xand x'** lies in **Xfree** and returns false otherwise.

- Function *Near Vertices*: returns a set V' of vertices that are close to the point $x \in Xfree$ within the closed ball of radius r centered at x.
- Function *Parent*: returns the parent vertex $xparent \in E$ of $x \in Xfree$.

The algorithm grows by adding random vertices and connecting to satisfy the neighborhood condition defined by the following algorithm for RRT*:

Algorithm: RRT* - Dijkstra Modification
1:    $V'\leftarrow V; E=E'$;
2:      xnearest←Nearest(G,x);
3:      xnew←Steer(xnearest,x);
4:      if ObstacleFree(xnearest,xnew) then
5:        $V'=V'\cup\{xnew\}$;
6:        xmin←xnearest;
7:        Xnear←**all(NearVertices(G, xnew))**;
8:        for all xnear∈Xnear do
9:        if ObstacleFree(xnear,xnew) then
10:   **c'←Cost(Line(xnear,xnew))**;
11:          if c'<**Cost(xnearest)** then
12:            xmin←xnear; //Choose new parent for xnew
13:            $E'=E'\cup\{(xmin,xnew)\}$;
14:      for all xnear∈Xnear\\{xmin\} do **minfuntion()**
15:      if ObstacleFree(xnear,xnew) then
16:      c'←Cost(xnew);
17:      if c'<Cost(xnear) then
18:      xparent←Parent(xnear); //Rewire
19:      $E'\leftarrow E'\backslash\{(xparent,xnear)\}$;
20:      $E'\leftarrow E'\cup\{(xnew,xnear)\}$;
21:      return $G'=(V',E')$

This initial algorithm was implemented using the repository developed by [38], Which implements the methods and functions necessary for development. In this case the all the near vertices are evaluated without the accumulated cost between the root and the last node.

Therefore, the application of Floyd algorithm requires to set all individual cost into a matrix between each near nodes, and a special function called global parent tract haves all possibilities to develop de Floyd iteration and solve the best path.

Algorithm: RRT* - Dijkstra Modification
1:   V'←V;E=E';
2:     xnearest←Nearest(G,x);
3:   xnew←Steer(xnearest,x);
4:   if ObstacleFree(xnearest,xnew) then
5:     V'=V'∪{xnew};
6:   xmin←xnearest;
7:   Xnear←**all(NearVertices(G, xnew));**
8:     for all xnear∈Xnear do
9:     if ObstacleFree(xnear,xnew) then
10:   A ←Cost((xnear_i ,xnear j)); - - < Also cost of xnew
11:   to the min_vertex(A) then
12:             xmin←xnear; //Choose new parent for xnew
13:             E'=E'∪{(xmin,xnew)};
14:     for all xnear∈Xnear\{xmin} do **minfuntion()**
15:   if ObstacleFree(xnear1_xnear2) then
16:   c'←Cost(xnearest);
18:       xparent←Rewire(xnearest);
19:       E'←E'\{(xparent,xnear)};
20:       E'←E'∪{(xnew,xnear)};
21:     return G'=(V',E')

Also no in all cases the new node is included into a new wired node, because if the node where so far, the algorithm discards them.

4   Results and Discussion

The following results are expressed in the way to present the best path into a max iteration of; 4000 and 9000 showing the fitness of the algorithm and improving the cost. The following figures represent the random tree in whose axes are positions in two-dimensional space (Figs. 1, 2 and 3).

A low number of iterations determines a weak path. The Floyd variation allows patterns similar to the original performance with the particularity of generating fewer nodes and rewriting, reducing the computational cost. Finally, Dijkstra being a function similar to the original cost, determines the same route, with the particularity of reduction of time in the moment to explore the neighboring nodes.

The metric of evaluation used to obtain the error on the cost is the percent error of measurements taken at different algorithms, defined as follows (Tables 1 and 2):

$$Performance\% = \frac{|RRT^* \; value - Variation \; value|}{Variation \; value} * 100\% \qquad (2)$$

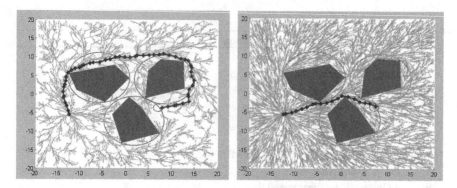

Fig. 1. Simple RRT*

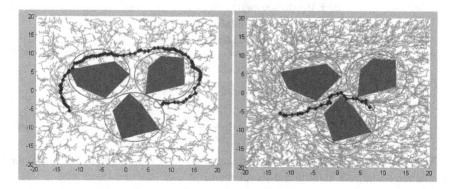

Fig. 2. RRT* Floyd variation

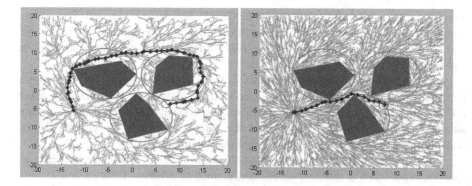

Fig. 3. RRT* Dijkstra variation

Table 1. Results analysis

Algorithm	Iterations	Global cost	Rewired nodes	Time
RRT*	4000	56.98	2800	22.50
	9000	21.60	14873	86.47
RRT – Floyd	4000	61.81	**2281**	**14.50**
	9000	25.32	**11953**	**38.46**
RRT - Dijkstra	4000	**55.98**	2800	15.14
	9000	**21.60**	14873	41.25

Table 2. Performance analysis

Algorithm	Performance cost %	Time performance %
RRT – Floyd	−7.81	**55.17**
	−14.69	**124.83**
RRT - Dijkstra	*1.79*	48.61
	0.00	109.62

There is a notable performance increment about the time cost, that case low computational cost. RRT – Floyd required less nodes to be rewired, then the time is reduced. The performance about the cost describes that Floyd is les performed because has near nodes and more travel but it's also a great path. Dijkstra has same path but in less time because it doesn't waste time in global cost evaluation.

5 Conclusions and Future Work

Our proposal for RRT* based on graph theory can accurately improves the basic algorithm, especially in the time cost and the computer cost. Also these algorithms would be applied into real time applications introducing robots into dynamical environments. We know about related work and planned to take all advantages into an autonomous planning.

In the next step we applied the algorithms in real time on a mobile robot. It can know its odometry by a LIDAR sensor and applies SLAM upgraded the basic RRT* Algorithm and a risk.4.

Acknowledgement. This work is part of the project MultiNavCar, 2016-PIC-025, from the Universidad de las Fuerzas Armadas ESPE, directed by Dr. Wilbert G. Aguilar.

References

1. Valle, S.M.L.: Rapidly exploring random trees: a new tool for path planning. Iowa State University, p. 4 (1998)
2. Islam, F.: RRT*-smart: rapid convergence implementation of RRT* towards optimal solution. In: Proceedings of IEEE International Conference on Mechatronics and Automation (ICMA), pp. 1651–1656 (2012)

3. Alterovitz, R., Patil, S., Derbakova, A.: Rapidly-exploring roadmaps: weighing exploration vs. refinement in optimal motion planning. In: Robotics and Automation (ICRA), pp. 3706–3712 (2011)
4. Rios-Martinez, J., Spalanzani, A., Laugier, C.: Probabilistic autonomous navigation using Risk-RRT approach and models of human interaction. In: CONACYT (2012)
5. Salihoglu, S., Widom, J.: Optimizing graph algorithms on pregel-like systems. In: Proceedings of the VLDB Endowment, vol. 7 (2014)
6. Bast, H., et al.: Route planning in transportation networks. In: Kliemann, L., Sanders, P. (eds.) Algorithm Engineering. LNCS, vol. 9220, pp. 19–80. Springer, Cham (2016). https://doi.org/10.1007/978-3-319-49487-6_2
7. Lopez, D., Lozano, A.: Transport network models for routing algorithms. In: Proceedings of 2016 IEEE 13th International Conference on Networking, Sensing, and Control (2016)
8. Pfetsch, M.E., Borndörfer, R.: Routing in line planning for public transport. Konrad-Zuse-Zentrum f'ur Informationstechnik Berlin (2004)
9. Pareekutty, N., James, F., Ravindran, B.: RRT-HX: RRT with heuristic extend operations for motion planning in robotic systems. In: Proceedings of the ASME 2016 International Design Engineering Technical Conferences (2016)
10. Penev, I., Karova, M.: Path planning algorithm for mobile robot. In: Conference: 15th International Conference on Applied Computer Science (2015)
11. Ducho, F., Babinec, A., Kajan, M.: Path planning with modified a star algorithm for a mobile robot. In: Modelling of Mechanical and Mechatronic Systems, pp. 59–69 (2014)
12. Purcaru, C., Precup, R.E., Iercan, D., Fedorovici, L.O., David, R.C.: Hybrid PSO-GSA robot path planning algorithm in static environments with danger zones. In: System Theory, Control and Computing (ICSTCC) (2013)
13. Liu, H., Stoll, N., Junginger, S., Thurow, K.: A Floyd-Dijkstra hybrid application for mobile robot path planning in life science automation. In: 8th IEEE International Conference on Automation Science and Engineering (2012)
14. Weiren, S., Kai, W.: Floyd algorithm for the shortest path planning of mobile robot. Chin. J. Sci. Instrum. **10**, 2088–2092 (2009)
15. Tomoiagă, T., Predoi, C., Coşereanu, L.: Indoor mapping using low cost LIDAR based systems. In: Applied Mechanics and Materials (2016)
16. Kim, J.: Motion planning of aerial robot using rapidly-exploring random trees with dynamic constraints. In: International Conference on Robotics & Automation (2003)
17. Yang, K., et al.: Spline-based RRT path planner for non-holonomic robots. J. Intell. Robot. Syst. **73**, 763–782 (2013)
18. Kuffner, J.J., LaValle, S.M.: RRT-connect: an efficient approach to single-query path planning. In: International Conference on Robotics & Automation (2000)
19. Naderi, K., Rajamaki, J., Hamalainen, P.: RT-RRT*: a real-time path planning algorithm based on RRT*. In: MIG (2015)
20. Ge, J., Sun, F., Liu, C.: RRT-GD: an efficient rapidly-exploring random tree approach with goal directionality for redundant manipulator path planning. In: International Conference on Robotics and Biomimetics (2016)
21. Kang, R., Liu, H., Wang, Z.: Fast convergence RRT for asymptotically-optimal motion planning. In: International Conference on Robotics and Biomimetics (2016)
22. Zhao, Y.J., et al.: 3D motion planning for robot-assisted active flexible needle based on rapidly-exploring random trees. J. Autom. Control Eng. **3**(5) (2015)
23. Huh, J., Lee, D.D.: Learning high-dimensional mixture models for fast collision detection in rapidly-exploring random trees. In: IEEE International Conference on Robotics and Automation (ICRA) (2016)

24. Lite, D.: Learning high-dimensional mixture models for collision deteccion. In: International Conference on Robotics and Biometrics (2016)
25. Kim, H., Kim, D., Shin, J.U., Kim, H., Myung, H.: Angular rate-constrained path planning algorithm for unmanned surface vehicles. Ocean Eng. **84**, 37–44 (2014)
26. Gautam, S.A., Verma, N.: Path planning for unmanned aerial vehicle based on genetic algorithm & artificial neural network in 3D. In: Data Mining and Intelligent Computing (ICDMIC) (2014)
27. Bouraine, S., Fraichard, T., Azouaoui, O.: Real-time safe path planning for robot navigation in unknown dynamic environments. In: Computing Systems and Applications (2016)
28. Xu, J., Duindam, V., Alterovitz, R., Goldberg, K.: Motion planning for steerable needles in 3D environments with obstacles using rapidly-exploring random trees and backchaining. In: Conference on Automation Science and Engineering, vol. 4 (2008)
29. Nguyen, T.H., Kim, D.H., Lee, C.H., Kim, H.K., Kim, S.B.: Mobile robot localization and path planning in a picking robot system using kinect camera in partially known environment. In: International Conference on Advanced Engineering Theory and Applications (2016)
30. Hebecker, T., Buchholz, R., Ortmeier, F.: Model-based local path planning for UAVs. J. Intell. Robot. Syst. **78**(1), 127–142 (2015)
31. Paranjape, A.A., Meier, K.C., Shi, X., Chung, S.J., Hutchinson, S.: Motion primitives and 3D path planning for fast flight through a forest. Int. J. Robot. Res. **34**(3), 357–377 (2015)
32. Farinella, J., Lay, C., Bhandari, S.: UAV collision avoidance using a predictive rapidly-exploring random Tree (RRT). In: American Institute of Aeronautics and Astronautics (2016)
33. Duchoň, F., et al.: Path planning with modified a star algorithm for a mobile robot. Procedia Eng. **96**, 59–69 (2014)
34. Aguilar, W.G., Morales, S.G.: 3D environment mapping using the kinect V2 and path planning based on RRT algorithms. Electronics **5**, 70 (2016)
35. Karaman, S., Frazzoli, E.: Incremental sampling-based algorithms for optimal motion planning. Robot. Sci. Syst. **104**, 2 (2010)
36. Wei, D.: An optimized Floyd algorithm for the shortest path problem. J. Netw. **5**, 1496 (2010)
37. Djojo, M.A., Karyono, K.: Computational load analysis of Dijkstra, A*, and Floyd-Warshall algorithms in mesh network. In: ROBIONETICS (2013)
38. Adiyatov, O., Varol, H.A.: Rapidly-exploring random tree based memory efficient motion planning. In: Mechatronics and Automation (ICMA), pp. 354–359 (2013)

Visual Based Autonomous Navigation
for Legged Robots

David Segarra[1($\boxtimes$)], Jessica Caballeros[1($\boxtimes$)],
and Wilbert G. Aguilar[1,2($\boxtimes$)]

[1] CICTE Research Center, Universidad de las Fuerzas Armadas ESPE,
Sangolquí, Ecuador
{desegarra, jacaballeros, wgaguilar}@espe.edu.ec
[2] GREC Research Group, Universitat Politècnica de Catalunya,
Barcelona, Spain

Abstract. This article will create the design and mathematical analysis of a four legged robot, underlining some of characteristics of the circuitry and physical design that it needs to have for being able to move from one point to another, focusing on the investigation of the kinematics of the quadruped robot using the Denavit Harteberg matrix, the analysis and processing of the images for the obstacle evasion with the purpose of avoiding collisions between the quadruped robot and static obstacles that have a defined form within the environment in which the robot is located. In addition, an Inertial Measurement Unit (IMU) was used with the implementation of the Kalman filter to detect the orientation estimation that the quadruped is in the plane of movement, this IMU called 9x3 is the same used in the MoMoPa3 project. For navigation a motion planning algorithm was implemented that is one of the main components of the artificial intelligence algorithms used to make the best route decisions. Using probabilistic techniques and partial information on the environment, RRT generates paths that are less artificial.

Keywords: Quadruped robot · Path planning · RRT · IMU · Kalman filter
Obstacle avoidance

1 Introduction

Today there are several applications that need the use of articulated robots capable of moving in irregular areas in a given environment [1], so one of the main problems to solve by the branch of Robotics, is the development of path planning methods, that allow the movement of the quadruped robot to be autonomous [2].

The search for anti-collision methods has become one of the most fascinating topics in mobile robotics, and to achieve its goal, route planning algorithms and obstacle prevention algorithms are needed to avoid the obstacles that arise along the way [3]. The main problem of the planning of movement is to try to find a trajectory free of collisions, taking as reference a state of departure towards an objective state. During the last 15 years these techniques based on random sampling have had a particular interest.

© Springer Nature Switzerland AG 2018
Y. Peng et al. (Eds.): IScIDE 2018, LNCS 11266, pp. 22–34, 2018.
https://doi.org/10.1007/978-3-030-02698-1_3

The RRT (Rapidly-exploring Random Tree) is based on the creation of branches of a tree in space, which iteratively samples new states and then directs the existing node that is closer to each sample to a new sample and thus forming a tree with ramifications [5].

For [4] the main purpose of road planning is to create algorithms that allow paths to be established considering restrictions in the movements of mobile robots. The Path Planning applications are oriented to different tasks that interact with the human being in different areas: (i) health: because it allows the support of robots in tasks for the elderly and people with paraplegia, (ii) military: because it is focused on the supervision of remote-controlled robots, autonomous and intelligent weapons, (iii) industrial: monitoring of robots with artificial intelligence through the use of mobile robots [5].

One key point to place the quadruped in space is being able to know its orientation to know where it needs to move, the determination of orientation of moving objects involved in numerous fields of science [6]. In order to get proper data to do the orientation estimate, there is necessary to process the raw signals received from the IMU, methods of signal processing are intensely researched to enhance the performance of the existing detection hardware [6, 7]. There are several methods to manipulate the data, in this article the Kalman filter was chosen to be implemented to obtain the estimation of the orientation of the Robot.

2 Related Works

The locomotion robots with legs can move with great ease in natural terrains, since they use discreet support points for each foot, in comparison with the robots with wheels, which require a continuous support surface. The main advantages of this type of robots are that they have great adaptability and maneuverability in terrains that are irregular [8]. Therefore, they are very good at moving on irregular terrains, thanks to two characteristics that they possess:

- Vary the configuration of your legs to adapt to the irregularities of a surface.
- The feet can make contact with the ground at the particular points according to the ground conditions.

For these reasons, the use of legs in robots is the best option for locomotion in irregular terrain. Although standing on four legs is relatively stable, the action of walking remains difficult because to remain stable, the center of gravity of the robot must move actively during the March.

One example of a quadruped robot developed at the Robotics Research Center, NTU, Singapore is LAVA [9]. The robot has two engines that are located in the hip section and the third is in the knee section. Each of the servomotors is coupled through a worm gear system to ensure a stable lock of the system that allows the engines to turn off when the vehicle with skid only needs to stop, which saves electrical energy

The human being can observe the world in three dimensions, each person can see different characteristics of their environment vividly with this three-dimensional perspective, so, with computer vision we try to simulate the vision that the human has by replacing the eye and the brain of the person with a camera and a processor. There are an infinity of applications in which you can use computer vision, one of them is the perceptual user interface where the computer is intended to have the capacity to detect and produce analogous signals from the human senses, such as allowing computers perceive sounds of their environment and produce a voice for it, also giving computers sense of touch and feedback force, and in this case, giving computers the ability to see, but these computer vision algorithms that claim to be part of a perceptual user interface must be fast and efficient [10].

There are different positions for the external camera, one of them is with a zenith perspective, that is, with the camera placed perpendicularly to the plane of the ground, to the plane of movement of the robot, this perspective was used by the work done by [11] in which the position of mobile robots and a ball in a game called robotsoccer was defined; Here, the color patterns were distinguished by the camera, but before that, the algorithm created first was a circle detection.

Finally, we turn our attention to the cognitive level of the robot. Cognition generally represents decision-making to achieve your objectives of higher order. Given a map and a target location, route planning involves identifying a path that will cause the robot to reach the target location when it is executed. Route planning is a strategic competence for problem solving since the robot must decide what to do in the long term to achieve its objectives [8], robots with RRT path planning are usually used in military applications because it is focused on the supervision [12–14] of remote-controlled robots, autonomous and intelligent weapons.

This work will control an autonomous robot, a tetrapod, using a camera with overhead perspective, which will use computer vision to navigate in chaotic environments, and to establish the orientation we will use an IMU in conjunction with a Kalman filter, that is, the angle in the plane of movement that the spider is in space. There are some methods for the recognition of the orientation, for example in [15] an online system was developed, due to the freezing of the gait (FOG), to detect falls and control tremors, a symptom present in the last stages of the disease of Parkinson. The system consists of a 3D camera sensor based on the Microsoft Kinect architecture, but for simplicity an IMU was used instead of this approach because it is a device that is effective, small and light because of this, it is used in several fields of science [6, 16].

3 Legged Robot

The physical design of the spider is somewhat simple, the mechanical part of SpiderRobot is composed of ABS plastic, the four legs of it are also made of this material. The axes and the form of movement of each joint are shown in Fig. 1.

In total, the robot has 12 motors, tree in each leg. The Robot's control unit is based on an ATMega328 same one used on the Arduino Nano's board. The PWM signals

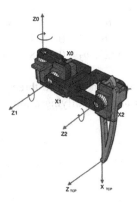

Fig. 1. Motion axes of each leg.

from the outputs of the ATMega328 board are transmitted over a connector board to servo motors. The block diagram of the system can be expressed as shown in Fig. 2. The camera sees the objects that surround the spider and where it is, and, with the help of the IMU its orientation, then the computer processes the RRT and finds a path that must be followed, calculates the direction that the robot must turn, and a command is transmitted over Bluetooth link to microcontroller. to the quadruped so that it is executed by the servos, and then the mechanism runs.

The design of the robot in 3D can be seen on the left of Fig. 3 and on its right the robot printed and armed with all its electronic components.

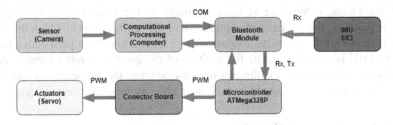

Fig. 2. Block diagram

The mathematical analysis was made using the Denavit Hartenberg method, focused on one of the four legs of the robot, resulting in the following Eq. 1

Fig. 3. Spider robot

$$N = \begin{bmatrix} \theta_1 & 0 & 3 & \pi/2 \\ \theta_2 & 0 & 5.4 & 0 \\ -\pi/2 + \theta_3 & 0 & 7.8 & 0 \end{bmatrix} \tag{1}$$

By using Matlab software, the leg motion and three joints is simulated by entering the values of the 3 angles, each angle modifying a joint, as shown in Fig. 4.

4 Mapping

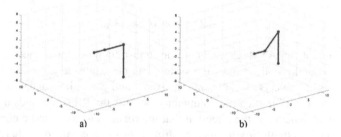

Fig. 4. (a) Leg simulation for: $\theta = 0$, $\theta = 0$, $\theta = 0$. (b) Leg simulation for: $\theta = 0$, $\theta1 = \pi/4$, $\theta3 = -\pi/4$

In this project the following considerations were taken; For the computational development the OpenCV Software is used, available in the Python tool, the decision making is based on the information captured from the environment in which the robot is located through the camera as a visual sensor, from a top-down perspective to the plane, the IMU is used to estimate the position of the quadruped robot located inside the camera's viewing area, for the evasion of obstacles, the properties of the static obstacles are known, which are identified using their morphology (rectangles).

4.1 Obstacle Avoidance Operation

The system has a camera connected directly to the computer by USB, image processing will be carried out on the Python platform that allows us to import artificial libraries, capture images from the screen in such a way that they remain in a range greater than 20 frames per second, enough to make the analysis of the obstacle evasion, according to the obtained data, Bluetooth commands will be sent to the robot to take the necessary measures (see Fig. 8).

4.2 Finding the Orientation of the Robot

To achieve a robust and reliable system in the determination of the orientation of the robot, artificial vision and inertial sensors that make up an Inertial Measurement Unit (IMU) have been used.

The IMU is a device used in several fields, since they are effective, small and light, the IMU used is the same as the work done in [16] developed at the Universitat Politècnica de Catalunya called 9 × 3 and has the objective of evaluating the symptoms of Parkinson disease (PD). The 9 × 3 Unit can be used 2 or 3 days working continuously and independently registers the signals of each of its sensors. The scheme shown in Fig. 5 shows the modules used for the execution of this investigation.

The signals used for the orientation estimation are obtained only from the inertial sensors integrated in LSM9DS0 [17] which provides raw data and is composed of a 9-axis system composed of a magnetometer, a triaxial accelerometer and a triaxial gyroscope. These signals are sent in real time to the STM32F415RG microcontroller from STMicroelectronics, which processes and filters using a Kalman filter, obtaining

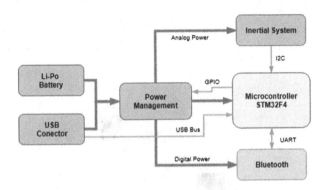

Fig. 5. 9§3 general structure used with main connections.

as a result the estimation of the orientation which is subsequently sent to the PC to calculate the angle to be rotated. The quadruped robot with respect to the trajectory given by the RRT.

Additionally, In the robot we have two colors to locate the orientation of the robot spider red in the back and green front, from this, two points are obtained with coordinates, and next to a third point that indicates the route that the robot should follow Form a scalene triangle (Fig. 6).

To find the angle of the robot with respect to the trajectory, we use the law of cosines and clearing the angle alpha obtaining Eq. 2 [17].

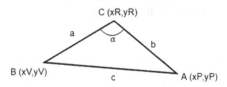

Fig. 6. Scalene triangle, to determine the angle of rotation of the robot.

$$\alpha = cos^{-1}\left(\frac{c^2 + b^2 - a^2}{2 \cdot c \cdot b}\right) \tag{2}$$

Because this formula only gives positive values of the angle, we also calculate the angle from the slopes of the lines that form it using Eq. 3.

$$\alpha_2 = \tan^{-1}\left(\frac{m - m'}{1 + m \cdot m'}\right) \tag{3}$$

An image of how the camera sees in overhead perspective can be observed in Fig. 7, in this you can notice an obstacle detected and the two colors placed on the spider.

Fig. 7. Quadruped robot and object

5   Rapidly-Exploring Random Trees

The RRT is one of the most suitable planners to solve several large route planning problems, since it shares many of the properties of the existing random planning techniques [18]. Below is the basic RRT expansion algorithm, taken from [18].

Algorithm 1 Basic RRT expansion method

For i=1 ... K do
    Xrand = random configuration
    Xnear = nearest neighbor in tree τ to Xrand
    Xnew = extend Xnear toward Xrand for step length
    If (Xnew can connect to Xnear along valid edge) then
        τ.AddVertex(Xnew). T.AddEdge(Xnew,Xnear)
    end if
  end for
  return τ

Figure 8 shows the flowchart of the trajectory tracking algorithm obtained from the RRT.

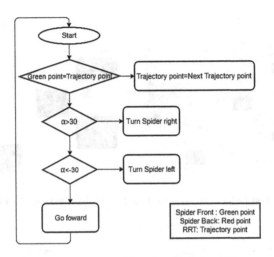

Fig. 8. Trajectory tracking algorithm flow diagram

6 Experimentation and Results

6.1 Optimal Delta for RRT

The objective of this experimentation is to get the fastest time to find the solution path with the RRT and have enough distance between each point that makes up the route, allowing the robot to move easily, because with many points, the robot would have to follow each one of them thus delaying the operation of moving from one point to another. For this, the delta constant of the RRT algorithm will vary between the following values: delta equal to 40, 30, 25, 20 and finally 10, the tests were performed in iterations of 10 times each and the average time was calculated.

In Fig. 9, we observe the results obtained for a Delta equal to 40 and the time in finding the route towards the objective, while decreasing the value of Delta as shown in Figs. 10 and 11, the time to find the route is considerably reduced, but as the Delta value continues to decrease, the mean time begins to elapse as shown in Figs. 12 and Fig. 13, so when increasing the time again it was seen that with a delta equal to 25 the best result was found.

Table 1 shows the results obtained by varying the variable to be able to determine the optimal Delta.

At the conclusion of the tests it was determined that the Delta equal to 25 is the most optimal to find the right path that the spider robot must follow, meeting the requirements of time and distance.

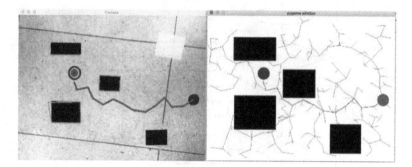

Fig. 9. Test Delta = 40, average time to find the route equal 19.45 s.

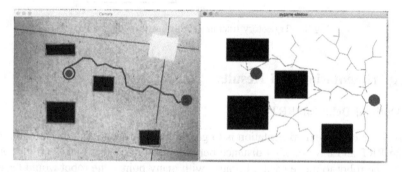

Fig. 10. Test Delta = 30, average time to find the route equal 15.25 s.

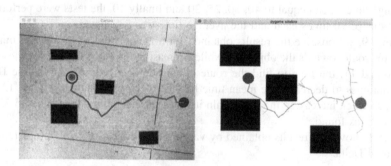

Fig. 11. Test Delta = 25, average time to find the route equal 5.4 s.

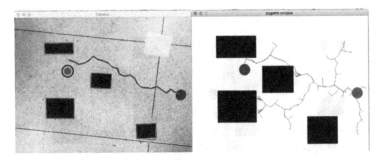

Fig. 12. Test Delta = 20, average time to find the route equal 12.74 s.

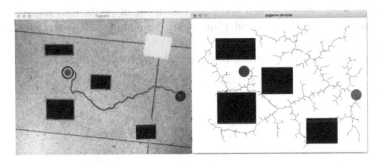

Fig. 13. Test Delta = 10, average time to find the route equal 21.41 s.

Table 1. Average time in each Delta value.

Delta (pixel)	Time (s)
40	19.45
30	15.25
25	5.40
20	12.74
10	21.41

6.2 Mean Squared Error Between the Real and Ideal Route

We chose the Mean Squared Error to evaluate the error, since we obtain only positive values, facilitating the appreciation of the error.

The MSE of an estimator measures the average of squared errors i.e. the difference between the estimator and what is estimated. The difference occurs because of the randomness or because the estimator does not take into account the information that could produce a more accurate estimation 1.

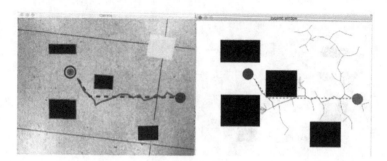

Fig. 14. Real rute vs ideal, comparison between 2 routes to obtain the MSE.

With the MSE, we evaluated the quality of a set of predictions regarding their variation. As shown in Figs. 14 and 15 in which the route traced by the RRT and the ideal route are visualized.

To obtain the average Quadratic Error, we use the Matlab software.

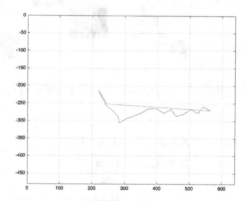

Fig. 15. Real rute vs ideal, comparison between 2 routes to obtain the MSE., using Matlab software.

The results obtained from the MSE in the X and Y axes of each route are shown in Table 2. MSE of the real vs ideal route.

Table 2. MSE of the real vs ideal route.

Axis	Error (pixel^2)
X	4.9651
Y	108.4149

7 Conclusions and Future Work

The method used in this work for the path planning is the RTT (rapidly exploring random tree) since it is able to determine relatively quickly the route that the quadruped robot must follow to reach its objective, allowing the autonomous movement of the robot in a certain area. With obstacles detected with morphological segmentation.

To estimate the orientation of the robot in the plane of movement, an IMU with the Kalman filter was used to combine the inertial data, and with the combination of 2 theorems the angle of rotation that the robot has to turn to follow the trajectory given by the RRT was used.

At the end of the performed tests to obtain the proper Delta value, we realized that the time it took to find the solution path increased with a relatively large or small Delta, therefore, we determined the optimal point of the value of Delta.

The optimum route is not the one that less distance traveled the robot but based on the obtained results it can be said that the best route to follow is the one that avoids the obstacles to a pertinent distance to not have collisions not planned by maneuver very close of the obstacle.

After carrying out the experimentation of the project, the level of illumination of the environment was established as a key point, since it is a determining factor for the optimal performance of the software, it was found that having a low level of lighting produces difficulties when recognizing colors and objects.

As a future work, other algorithms for path planning such as PRM, A*could be tested and implemented to see if they work better that the RRT described in this paper [19], and an object detection module with a path planner and deep learning methods.

References

1. Fernández, J.F., Barrientos, T.A.: Análisis, desarrollo y evaluación de modos de marcha para un robot hexápodo (2016)
2. Menéndez, C.: Navegación de robots autónomos en entronos dinámicos (2012)
3. Alvarez, P.Q., Antonio, J., Estrada, R., Fernández, A.A., Torres, J.G.R.: Técnicas para evasión de obstáculos en Robótica Móvil, p. 1 (2010)
4. Barraquand, J., Kavraki, L., Latombe, J.-C., Motwani, R., Li, T.-Y., Raghavan, P.: A random sampling scheme for planning. **16**, 759–774 (1997)
5. Ortiz, O., Santana, A.: Planificación de caminos para Robots en Realidad Virtual (2017)
6. Sabatini, A.M., Member, S.: Quaternion-based extended Kalman filter for determination orientation by inertial and magnetic sensing - Google Search. **53**, 1346–1356 (2006)
7. Sabatini, A.M.: Inertiel sensing in biomechanics : a survey of computational techniques bridging motion analysis and personal navigation. Comput. Intell. Mov. Sci. 70–100 (2006). https://doi.org/10.4018/978-1-59140-836-9
8. Siegwart, R., Nourbakhsh, I.R.: Introduction to Autonomous Mobile Robots. MIT Press, Cambridge (2004)
9. Zielinska, T., Heng, J.: Mechanical design of multifunctional quadruped. Mech. Mach. Theory **38**, 463–478 (2003). https://doi.org/10.1016/s0094-114x(03)00004-1
10. Bradski, G.R.: Computer vision face tracking for use in a perceptual user interface (cité 1923 fois). Intel Technol. J. **2**, 12–21 (1998). 10.1.1.14.7673

11. Antipov, V., Kokovkina, V., Kirnos, V., Priorov, A.: Computer vision system for recognition and detection of color patterns in real-time task of robot control. In: 2017 Systems of Signal Synchronization, Generating and Processing in Telecommunications SINKHROINFO (2017). https://doi.org/10.1109/sinkhroinfo.2017.7997496
12. Aguilar, W.G., Angulo, C.: Real-time model-based video stabilization for microaerial vehicles. Neural Process. Lett. **43**, 459–477 (2016). https://doi.org/10.1007/s11063-015-9439-0
13. Aguilar, W.G., Angulo, C.: Real-time video stabilization without phantom movements for micro aerial vehicles. EURASIP J. Image Video Process. **2014**, 46 (2014). https://doi.org/10.1186/1687-5281-2014-46
14. Aguilar, W.G., et al.: Pedestrian detection for UAVs using cascade classifiers and saliency maps. In: Rojas, I., Joya, G., Catala, A. (eds.) IWANN 2017. LNCS, vol. 10306, pp. 563–574. Springer, Cham (2017). https://doi.org/10.1007/978-3-319-59147-6_48
15. Amini Maghsoud Bigy, A., Banitsas, K., Badii, A., Cosmas, J.: Recognition of postures and Freezing of Gait in Parkinson's disease patients using Microsoft Kinect sensor. In: International IEEE/EMBS Conference on Neural Engineering, NER, pp. 731–734. IEEE (2015)
16. Rodríguez-Martín, D., et al.: A waist-worn inertial measurement unit for long-term monitoring of Parkinson's disease patients. Sensors **17**, 827 (2017). https://doi.org/10.3390/s17040827
17. Aguilar, W.G., Segarra, D., Jessica, C.: RRT path planning and morphological segmentation based navigation for a tetrapod robot (accepted)
18. LaValle, S.M.: Rapidly-exploring random trees: a new tool for path planning. **129**, 98–111 (1998). 10.1.1.35.1853
19. Aguilar, W.G., Morales, S., Ruiz, H., Abad, V.: RRT* GL based optimal path planning for real-time navigation of UAVs. In: Rojas, I., Joya, G., Catala, A. (eds.) IWANN 2017. LNCS, vol. 10306, pp. 585–595. Springer, Cham (2017). https://doi.org/10.1007/978-3-319-59147-6_50

An Environment-Adaptation Based Binaural Localization Method

Tao Song and Jing Chen[✉]

Department of Machine Intelligence, Speech and Hearing Research Center,
and Key Laboratory of Machine Perception, Ministry of Education,
Peking University, Beijing 100871, China
ist_songtao@pku.edu.cn, chenj@cis.pku.edu.cn

Abstract. The degrading effect of reverberation on automatic sound localization is a challenging problem for many intelligent applications. Motivated by the environment-adaption ability of human auditory system, we modified the previous model by introducing the phase of room classification. 4 room types representing reverberation time from 0.32 to 0.89 s were used to evaluate the performance of the new method, and the result showed the localization accuracy could be improve about 1%–9%, depending on the sound location. The limitation and the further work of the method is analyzed and discussed.

Keywords: Binaural localization · Environment-adaptation · Reverberation

1 Introduction

The ability to localize sound resources in adverse acoustic environments (e.g. reverberation and interfering sound) is necessary for many intelligent applications. Although extensive research has been done in the field of localization, the localization in adverse acoustic conditions remains a challenging task, especially when the acoustic environment changes and keeps unknown. For the typical localization technique using microphone-array, the performance depends on the array configuration and generally increases with the number of microphones [1]. Otherwise, the performance of human auditory system is very robust against adverse of acoustic conditions, apparently, they only rely on binaural input. Hence, it's important to develop the localization technique by integrating the property of auditory perception.

The sound localization of human listeners mainly relies on two binaural cues, interaural time difference (ITD) and interaural level difference (ILD). These two cues are not only determined by sound location but also altered by reflections from walls and objects in a given room [2, 3]. Several binaural models have been proposed to mimic the localization ability of auditory system in reverberant environments. Some works focused on extracting accurate binaural features (ITD and ILD) from the reverberant input. Based on the novel idea that binaural cues were robust when interaural coherence was high, Faller et al. screened binaural cues in time domain with interaural coherence before localization [4]. More directly, Pang et al. used reverberation weighting to separately suppress the early and late reverberation of the input signals, and then the sound location was estimated with ITDs and ILDs calculated from the de-reverberated

Y. Peng et al. (Eds.): IScIDE 2018, LNCS 11266, pp. 35–43, 2018.
https://doi.org/10.1007/978-3-030-02698-1_4

signal [5], however this method required the estimation of reverberation time and only one localization cue, ILD, was actually influenced by this de-reverberation process. Other works used binaural features directly. As binaural cues calculated within a short-term analysis window fluctuate following a certain distribution [6], May et al. modeled this fluctuation of binaural cues with a Gaussian mixture models (GMMs) and the direction of sound source was determined by a probability model [7]. In this work, localization model was trained only in a simulated room with fixed reverberation time of 0.5 s, and it was assumed to be generalized to other rooms. Youssef et al. used a three-layer neural network to estimate the sound location from binaural features in reverberant rooms without consideration of interference [8]. Although these works stated their methods were robust in reverberant environment, usually the reverberation time was limited in a range between 0–0.4 s, and the performance still be influenced by the relative long reverberations.

However, it has been reported that the degrading effect of reverberation on sound localization could be reduced for human listeners after a period of learning or adaptation [9, 10]. To reveal the underlying mechanism of this adaptation effect, many experiments have been conducted (see a review by [11]). In these experiments, binaural cues were manipulated by using ear blocks etc., and the adaptation was carried out either by sound exposure or training. It was found that the original localization ability of subjects was not interfered by adaptation [12–14] and the benefit of the adaptation to altered cues could be detected even months after adaptation experiments [15, 16]. With these findings, Mendonça hypothesized that auditory system adapted to altered cues by generating new cues-combination rules while kept previous rules unaltered, and the localization result was given by the cues-combination rule corresponding to current context [11]. In other words, listeners' could be adaptive to the acoustic environment first, and then the performance of sound localization could be improved.

Based on the hypothesis above, a room-adaptive binaural localization method was proposed and the performance was evaluated in this work. In this method, similar to the work of May et al. [7], a GMMs based localization model was used. To mimic the ability of environment adaptation, localization model was tuned to multiple rooms with different sets of model parameters. When locating sound source, a room classification model was introduced to estimate the listening room based on binaural features, the estimation result was further used to select the proper parameters for localization model, and finally the location result was given by the localization model. The framework of proposed method is given in Fig. 1. The evaluation was conducted by comparing three conditions: room-fixed, room-adapted, and room-matched. The room-fixed method corresponded to the baseline system that was proposed by May et al. [7] and the processing of room classification was not considered, the room-adapted represented the current method, and the room-matched method represented the ideal situation when the training room was the same as the test room.

The paper is organized as follows. Section 2 explains the details of proposed method. In Sect. 3, a localization experiment is described and the evaluation result is presented. Discussions are given in Sect. 4.

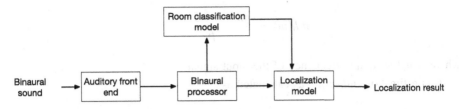

Fig. 1. The framework of environment-adaptation based binaural localization method

2 Method Description

The proposed method consists mainly of 4 parts: auditory front-end, binaural processor, room classification model and localization model, and they are described separately.

2.1 Auditory Front-End

In auditory front-end, 32-channel gammatone filterbank was used to simulated the frequency selectivity of human cochlear, the center frequencies were equally distributed in equivalent rectangular bandwidth (ERB) scale between 80 Hz and 5 kHz. In each channel, the neural transduction process in the inner hair cells was approximated by halfwave-rectification followed by a square-root compression. After all these processes, derived binaural auditory signal in frequency channel i was denoted as r_i and l_i.

2.2 Binaural Processor

In each frequency channel, ITD and ILD were estimated using a rectangular window of 20 ms, the overlap between successive frame was set to 50%. The time delay between left and right channel was calculated using cross-correlation function,

$$\hat{\tau}_i(m) = \max_{\tau} C_i(m, \tau)$$
$$= \max_{\tau} \sum_{n=0}^{W-1} l_i\left(m\frac{W}{2} - n\right) r_i\left(m\frac{W}{2} - n - \tau\right) \tag{1}$$

where m is frame index; W is window length in samples. Since the resolution of time delay calculated above was limited by the sample interval, exponential interpolation was applied to increase the accuracy of estimation result [17]. The fractional part δ_i was calculated as

$$\delta_i(m) = \frac{\log C_i(m, \hat{\tau}_i(m) + 1) - \log C_i(m, \hat{\tau}_i(m) - 1)}{4\log C_i(m, \hat{\tau}_i(m)) - 2\log C_i(m, \hat{\tau}_i(m) - 1) - 2\log C_i(m, \hat{\tau}_i(m) + 1)} \tag{2}$$

The final ITD estimation $\hat{itd}_i(m)$ was derived as

$$\hat{itd}_i(m) = \frac{(\hat{itd}_i(m) + \delta_i(m))}{fs} \tag{3}$$

where fs is the sample frequency of the input signal.

ILD estimation, $\hat{ild}_i(m)$, was calculated as

$$\hat{ild}_i(m) = 10\log \frac{\sum_{n=o}^{W-1} l_i(m\frac{W}{2} + n)^2}{\sum_{n=o}^{W-1} r_i(m\frac{W}{2} + n)^2} \tag{4}$$

2.3   Room Classification

In room classification model, the relationship between binaural cues and reverberant rooms was modeled by GMMs which was trained by iterative Expectation-Maximization (EM) algorithm [18]. For a given room R, the probability density function (pdf) of binaural features in frequency channel i was approximated by the weighted summation of K Gaussian components

$$p(X_i|R) = \sum_{j=1}^{K} \omega_j N\left(\mu_j, \Sigma_j\right) \tag{5}$$

where $X_i = \left[\hat{itd}_i, \hat{ild}_i\right]$, μ_j, Σ_j are the mean vector and covariance matrix of the j_{th} components.

For binaural signals from an unknown room, listening room can be estimated by maximizing likelihood function

$$\hat{R} = \max_R \sum_i p(X_i|R) \tag{6}$$

2.4   Localization

In each candidate reverberant room, the influence of sound location on binaural cues was also modeled by GMMs in each frequency channel. For a certain sound location S in room R, the pdf of binaural cues in frequency channel i was approximated by Q Gaussian components

$$p(X_i|S, R) = \sum_{j=1}^{Q} \omega_j N\left(\mu_j, \Sigma_j\right) \tag{7}$$

where $X_i = \left[\hat{itd}_i, \hat{ild}_i(m)\right]$, μ_j, Σ_j are the mean vector and covariance matrix of the j_{th} components.

For given binaural sound, the listening room was first estimated, based on room classification result $\widehat{R}$, source location was determined by maximizing the summation of likelihood function from all frequency channels

$$\widehat{S} = \max_{S} \sum_i p(X_i|S, \widehat{R}) \tag{8}$$

3   Localization Experiment

3.1   GMM Settings

The key parameter of GMMs is the component number, and it was set to 10 in localization model and 20 in room classification model. The stopping criteria of EM algorithm was set to 10^{-5} and the maximal iteration steps were limited within 2000.

3.2   Reverberant Rooms

Binaural room impulse responses (BRIRs) recorded in 4 rooms [19] were used. All 4 rooms vary both in geometry and reverberation time. BRIRs were recorded using a Cortex Intruments Mk.2 Head and Torso Simulator (HATS). The sound source was placed in the front horizontal plane of HATS with a distance of 1.5 m. The azimuth of sound source ranges from $-90°$ to $90°$ in step of $10°$. The reverberation times, T_{60}, are listed in Table 1.

Table 1. The reverberation time T_{60} of four rooms used in experiment

Room	A	B	C	D
T_{60} (s)	0.32	0.47	0.68	0.89

4   Training

Speech randomly selected from the training set of TIMIT database [20] was used as target. All speech were uttered by male speakers, and the total duration was about 30 s. A energy based voice activity detection (VAD) was applied to remove silent frames in which the energy was 40 dB lower than the global maximum. Binaural signals were then synthesized by convolving BRIRs with speech. A diffused white noise was used as interference signals. Three signal-to-noise-ratios (SNRs) 0, 10, 20 dB, were considered.

The time delay between left and right channel calculated from cross-correlation function could be as large as 20 ms, but it was beyond the range for real ITDs. So, in each frequency channel, frames with estimated ITD exceeds the range of $[-1, 1]$ ms were removed. To ensure that models were trained only with binaural features were associated with the target source, in the binaural processor, only frames with SNR exceeded 0 were used. The room classification model and localization model were trained with the same data.

4.1 Evaluation

Similar to the training session, speech uttered by male speakers randomly selected from the test set of TIMIT database was used as target, and they were different from the training set. The similar VAD method was applied to speech before convolving with BRIRs. Different diffused white noise was used as interference, and 3 SNR conditions (0, 10, 20 dB) were tested.

Three methods were evaluated, named room-fixed, room-adaptive and room-matched. In the room-fixed method, binaural signal from all rooms was localized by using the same localization model that was trained in room B (a room with moderate reverberation). The room-adaptive was corresponding to the proposed method. In the room-matched method, binaural signal was localized by the localization model trained in the same rooms, in other words, room-matched method could be treated as the ceiling performance of the proposed roome-adaptive method. Mean localization errors and localization accuracy were calculated as performance measurement. Theoretically, room-adaptive method was supposed to work better than room-fixed method due to the advantage of room adaptation, but the performance of the both methods was worse than the room-matched method. In room B, the performance of room-adaptive method might be worse than the other two methods since room-fixed method and room-matched method were equivalent in this condition.

The localization accuracy of each location in 4 rooms is shown in Fig. 2. Similar to the performance of human [21], the localization accuracy reaches maximum when sound source was located in the front, and it decreased as the sound source moved toward both flanks.

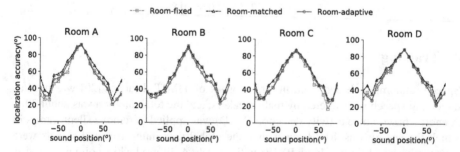

Fig. 2. The mean localization error of fixed room method, matched room method and proposed method in four reverberant rooms

To visualize the difference between these three methods, the benefit of introducing room adaptation was calculated by subtracting localization accuracy (%), or location errors in degree of the room-fixed method from the other two method. The decrease of

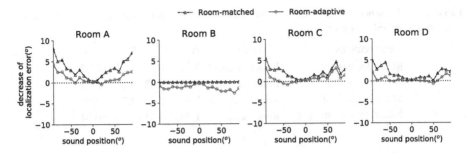

Fig. 3. The decrease of localization errors of matched room method and proposed method relative to the room-fixed method in 4 reverberant rooms

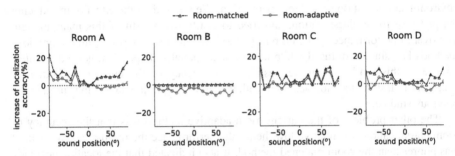

Fig. 4. The increase of localization accuracy of matched room method and proposed method relative to the room-fixed method in 4 reverberant rooms

mean localization error is shown in Fig. 3. The positive benefit could be obtained in all rooms except room B, which was consistent with the our assumption. In rooms A, C and D, the benefit increased as the sound source moved toward to flanks, where the localization performance was poor. The maximum benefit was about 5° for proposed method and 8° for matched room method. Similar pattern could be found in the increase of localization accuracy shown in Fig. 4. The maximum benefit was 9% for the room-adaptive method and 22% for the room-matched method.

Gains averaged in different SNRs and sound positions are listed in Table 2. As far as room A, C and D are concerned, for both matched room method and room-adaptive method, the benefit of room-adaptive processing was observed. The benefit decreased as the increase of reverberation level, which might be attributed to the similarity of binaural cues when reverberation time was long. Compared to the room matched method, the benefit of room-adaptive method was much smaller. It was probably caused by the poor accuracy of room classification model, which was only around 40%. Maybe, the current GMMs models was not sufficient to model the relation between cues and reverberation.

Table 2. Performance improvement of room-adaptive method and room-match method relative to room-fix method in 4 rooms

Measurements	Method	A	B	C	D
Mean decrease of localization error($^{\circ}$)	Room-matched	3.35	0	2.12	1.82
	Room-adaptive	1.15	−1.38	0.77	0.38
Mean increase of localization accuracy (%)	Room-matched	7.86	0	4.73	4.50
	Room-adaptive	1.79	−4.34	1.85	−0.04

5 Discussions

In this work, a room-adaptive binaural localization method was proposed, to mimic the environment-adaptation of auditory perception. We focused on the effect of introducing the phase of room-classification, and tried to testify the benefit of this manipulation. The overall performance of the current method was hard to compare with the previous methods [7], since a diffused white noise rather spatialized speech was used as interference in this work. To further testify the efficiency of the current method, it is necessary to test the location accuracy with the similar experimental settings used in the previous studied.

The other problem of the current room-adaptive method was that the accuracy of room classification was relative low, hence the performance the room-adaptive method was poorer than the room-matched method. It also indicated that the location accuracy could be improved if the room classification could work better.

In summary, a room-adaptive based binaural localization method was proposed to mimic the environment-adaptation of auditory perception. The GMMs model was trained in different reverberant room conditions and used to estimate the where the sound source was located. The major benefit of introducing room classification occurred in sound locations where the localization accuracy was relatively poor, and the maximum decrease of localization error was about 5° for this room-adaptive method.

Acknowledgments. The work was supported by the National Natural Science Foundation of China (No. 61473008, No. 61771023 and No. 11590773), and a Newton alumni funding by the Royal Society, UK.

References

1. Brandstein, M., Ward, D.: Microphone Arrays: Signal Processing Techniques and Applications. Springer, Berlin (2013). https://doi.org/10.1007/978-3-662-04619-7
2. Hartmann, W.M.: Localization of sound in rooms. J. Acoust. Soc. Am. **74**, 1380–1391 (1983)
3. Ihlefeld, A., Shinn-Cunningham, B.G.: Effect of source spectrum on sound localization in an everyday reverberant room. J. Acoust. Soc. Am. **130**, 324–333 (2011)
4. Faller, C., Merimaa, J.: Source localization in complex listening situations: selection of binaural cues based on interaural coherence. J. Acoust. Soc. Am. **116**, 3075–3089 (2004)

5. Pang, C., Liu, H., Zhang, J., Li, X.: Binaural sound localization based on reverberation weighting and generalized parametric mapping. IEEE/ACM Trans. Audio, Speech Lang. Process. **25**, 1618–1632 (2017)
6. Nix, J., Hohmann, V.: Sound source localization in real sound fields based on empirical statistics of interaural parameters. J. Acoust. Soc. Am. **119**, 463–479 (2006)
7. May, T., van de Par, S., Kohlrausch, A.: Based on a binaural auditory front-end. IEEE/ACM Trans. Audio, Speech Lang. Process. **19**, 1–13 (2011)
8. Youssef, K., Argentieri, S., Zarader, J.L.: A learning-based approach to robust binaural sound localization. In: IEEE/RSJ International Conference on Intelligent Robots and Systems (IROS), Tokyo, pp. 2927–2932 (2013)
9. Seeber, B.U., Müller, M., Menzer, F.: Does learning a room's reflections aid spatial hearing? In: Proceedings of the 22nd International Congress on Acoustics, Buenos Aires, pp. 1–8 (2016)
10. Shinn-Cunningham, B.: Learning reverberation: considerations for spatial auditory displays. In: International Conference on Auditory Displays, Atlanta, pp. 1–9 (2000)
11. Mendonça, C.: A review on auditory space adaptations to altered head-related cues. Front. Neurosci. **8**, 1–14 (2014)
12. Irving, S., Moore, D.R.: Training sound localization in normal hearing listeners with and without a unilateral ear plug. Hear. Res. **280**, 100–108 (2011)
13. Kumpik, D.P., Kacelnik, O., King, A.J.: Adaptive reweighting of auditory localization cues in response to chronic unilateral earplugging in humans. J. Neurosci. **30**, 4883–4894 (2010)
14. Carlile, S., Blackman, T.: Relearning auditory spectral cues for locations inside and outside the visual field. J. Assoc. Res. Otolaryngol. **15**, 249–263 (2014)
15. Butler, R.A.: An analysis of monaural displacement of sound in space. Percept. Psychophys. **41**, 1–7 (1987)
16. Zahorik, P., Bangayan, P., Sundareswaran, V., Wang, K., Tam, C.: Perceptual recalibration in human sound localization: learning to remediated front-back reversals. J. Acoust. Soc. Am. **120**, 343–359 (2006)
17. Zhang L, Wu X.: On cross correlation based discrete time delay estimation. In: 2005 IEEE International Conference on Acoustics, Speech, and Signal Processing, Philadelphia, pp. 981–984 (2005)
18. Dempster, A.P., Laird, N.M., Rubin, D.B.: Maximum likelihood from incomplete data via the EM algorithm. J. R. Stat. Soc. B **39**, 1–38 (1977)
19. Hummersone, C., Mason, R., Brookes, T.: Dynamic precedence effect modeling for source separation in reverberant environments. IEEE Trans. Audio Speech Lang. Process. **18**, 1867–1871 (2010)
20. Disc, N.S., et al.: TIMIT Acoustic-Phonetic Continuous Speech Corpus. NISTIR, Gaithersburg (1993)
21. Blauert, J.: Spatial Hearing: The Psychophysics of Human Sound Localization. MIT press, Cambridge (1997)

Towards Efficient Lesion Localization Based on Template Occlusion Strategy in Intelligent Diagnosis

Yan He[1], Kehua Guo[1(✉)], Ruifang Zhang[1], Jialun Li[1], and Wei Liu[2]

[1] School of Information Science and Engineering, Central South University,
Changsha 410083, China
guokehua@csu.edu.cn
[2] School of Informatics, Hunan University of Chinese Medicine,
Changsha 410208, China

Abstract. In recent years, Artificial Intelligence (AI) has made great achievements in medical field, and intelligent diagnosis becomes an important topic simultaneously. Recent research on the intelligent diagnosis of diseases is mainly focus on disease identification. However, how to detect lesion location is still a difficult problem which is of great significance. In medicine, lesion localization is required by the treatment of many diseases. Lesion location information can helps physicians make a further understanding of the disease, assists them in diagnosis and therapy, and increases the likelihood of the disease being cured. In this paper, we propose an efficient lesion localization method based on template occlusion (LLTO) for locating lesion areas in disease images. First, OpenCV cascade classifier is used to train candidate boxes representation model and the TensorFlow is used to train disease discrimination model. Second, to implement the lesion location task, the lesion candidate boxes are generated by the candidate boxes representation model in the test image. Third, the disease discrimination model is used to select true lesion areas from candidate boxes and integrate them to get the image marked with lesion locations. The experiment proves the efficiency and effectiveness of our method.

Keywords: Lesion localization · Template occlusion
Intelligent medical diagnosis · Target detection

1 Introduction

The development of artificial intelligence has brought great changes to our daily life. Especially in the medical field, there are great achievements realized through artificial intelligence [1]. For example, the Nature recently published a cancer related research paper which has realized skin cancer classification based on photographs by using the deep learning algorithm [2]. Iflytek also employed the deep learning method to the medical image-aided diagnosis system. Nevertheless, the current research on intelligent diagnosis of diseases mainly focuses on disease identification. It is truly important to identify the disease in the intelligent diagnosis of diseases, but it is also of great significance to be able to identify the specific location of the disease to make the

© Springer Nature Switzerland AG 2018
Y. Peng et al. (Eds.): IScIDE 2018, LNCS 11266, pp. 44–56, 2018.
https://doi.org/10.1007/978-3-030-02698-1_5

diagnosis more accurate. In medical field, lesion location information can help doctors make further diagnosis and therapy and improve the possibility of disease cure, and lesions localization is required by the treatment of many diseases. At present, the decision of the lesion localization mainly depends on the judgment of doctors, which is not only difficult to do but also greatly increase the workload of the doctor. Therefore, how to locate the lesion areas on the medical images has great practical significance.

Lesion location can also be understood as the target detection. There are many target detection algorithms for target positioning. The representative algorithms are YOLO (You Only Look Once: Unified, Real-Time Object Detection) and Faster RCNN (Regions with Convolutional Neural Network). The YOLO target detection algorithm turns the target detection problem into a regression problem. Weights are trained through training data and are directly called when testing [3]. The speed of this algorithm is fast, but it has a poor positioning accuracy and the detection accuracy on the small target is not that ideal [4]. The Faster RCNN generates candidate boxes by feature extraction and then makes a further refinement of the position of them [5]. The positioning accuracy of the Faster RCNN is improved comparing with the YOLO, yet the detection speed is slower. In general, both of the YOLO and the Faster RCNN need to take a lot of time to mark the true position of the object in training data. Moreover, YOLO and Faster RCNN are mostly used for detecting targets in daily life images but not commonly used for professional diagnosis of disease.

In order to solve the above problem, we propose LLTO, a lesion localization method based on template occlusion. Firstly, LLTO uses the OpenCV cascading classifier to train the candidate boxes representation model. When a test image is input, the candidate boxes of the image are generated through the lesion candidate boxes representation model. Then, we use the template occlusion method to process the generated lesion candidate boxes. Concretely, the remaining candidate boxes are masked by the template if one certain candidate box is discriminated by the disease discrimination which is trained already to determine whether the current image is a feature image. If the discrimination result of the above process is exactly true, so the discriminated candidate box contains lesion area and will be recorded in the image. Finally, all the candidate boxes recorded are marked on the test image, and the marked areas are the lesion locations. The innovations of this paper are as follows. (1) Combining OpenCV cascade classifier with deep learning to improve the accuracy of the lesion location. (2) There is no need to mark the true position of the object in training data manually comparing with YOLO and Faster RCNN.

The rest of this paper is organized as follows. Section 2 explains the related works of intelligent auxiliary medical diagnosis and target detection, Sect. 3 describes the concrete realization process of LLTO, and Sect. 4 provides experimental results. Section 5 concludes our work and points out the future work.

2   Related Work

In the past years, computer-aided intelligent medical diagnosis has played a great role in the medical diagnosis. The most famous research is computer aided diagnosis (CAD) study of breast and pulmonary nodule lesions, which represents current research status and the highest level of the CAD in medical imaging [6]. In addition, there are many other CAD related studies such as ROC study [7].

As the development of artificial intelligence, deep learning occupies a more and more important position in the medical diagnosis. In 2017, Cell has published a cover article which is about an system for diagnosing eye diseases and pneumonia based on deep learning whose accuracy can rival that of the top-level doctors in related specialties [8]. Reference [9] established CC-Cruiser, an artificial cataract intelligence platform using deep learning algorithm. Furthermore, many diagnostic systems have emerged and put into practical application, such as the Tencent miying platform [10], whose auxiliary diagnosis ability mainly includes diagnostic risk monitoring systems and medical records intelligent management systems. However, whether the research studies above has been developed and in use or is now being studied, they mainly concentrate upon the disease identification rather than the annotation of lesion locations.

There are many target detection technologies commonly used in research and practical environment. The traditional methods mainly include cascade classifier framework [11] and HoG+SVM [12]. Deep learning has also obtains some achievements in the target detection technology. YOLO and Faster RCNN are two most representative technologies.

In this paper, we combine traditional target detection methods with deep learning to implement the lesion location. We firstly use the OpenCV cascading classifier to train the candidate boxes representation model. The test images generate candidate boxes through the candidate boxes representation model, and then use the template occlusion method to process candidate boxes. Finally, the disease discrimination model is used to select true lesion areas and the candidate boxes to achieve the effect of lesion location.

3   Methodology

3.1   LLTO Overview

In LLTO, the discrimination model of disease and the representation model of candidate boxes are used to classify images, and the medical image that contains the candidate boxes is processed by the template occlusion method to obtain lesion locations. Google Inception v3 model is used to train data sets for the discrimination model of disease. We define the training data set of the discrimination model of disease as follows.

Definition 3.1. Training set of the discrimination model $(TSDM)$: $TSDM = \{FIS, NIS\}$, where FIS means feature image set and NIS means non-feature image set. FIS and NIS are satisfying:

$FIS = \{FI_1, FI_2, \ldots, FI_i, \ldots, FI_n\}$, where FI_i presents a disease image (or feature image, FI);

$NIS = \{NI_1, NI_2, \ldots, NI_i, \ldots NI_n\}$, where NI_i is a normal image (or non-feature image NI).

For the representation model of candidate boxes, we use the OpenCV cascade training classifier to train its training data set. The training set of representation model of candidate boxes is defined as follows:

Definition 3.2. Training set of the representation model ($TSRM$): $TSRM = \{PSS, NSS\}$, where PSS presents positive sample set and NSS presents negative sample set. PSS and NSS satisfy:

$PSS = \{PS_1, PS_2, \ldots, PS_i, \ldots, PS_n\}$, where PS_i is the image with the lesions (or positive sample, PS).

$NSS = \{NS_1, NS_2, \ldots, NS_i, \ldots, NS_n\}$, where NS_i is the image that does not contain lesions (or negative sample, NS).

If a test image is input into the representation model, a series of candidate boxes will be generated in the image. In this paper, we define the test image as the original image (OI), and the series of candidate boxes as the candidate box set. The image obtained through representation model of candidate boxes that containing the candidate box set is described as the candidate box representation image ($CBRI$).

Definition 3.3. Candidate box set (CBS): $CBS = \{CB_1, CB_2, \ldots, CB_i, \ldots, CB_n\}$, where $CB_i = (x, y, width, height)$ is an rectangular area (candidate box) in the image $CBRI$ where regarded as a candidate lesion location. x and y are respectively the abscissa and the ordinate of the first pixel in CB. $width$ and $height$ are respectively the width and the height of CB.

The example of CB is shown in Fig. 1, in this figure, each white box represents a CB.

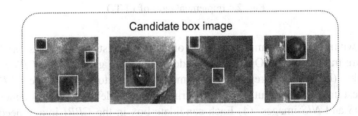

Fig. 1. Example of candidate box

The CBS representation model can find out candidate boxes in OI. However, due to the low accuracy of the CBS representation model of identifying lesion locations, it cannot correctly distinguish between true lesions and non-lesion areas with similar morphology to the lesion area. Although the lesion location can be partially found out, it may also mistakenly identify the non-lesion areas which are the suspicious lesion areas. Therefore, candidate boxes generated by the CBS representation model need to be further identified to remove the non-lesion areas in it. We use the template occlusion method to identify each candidate box in the CBS singly through the discrimination model to determine whether it is the true lesion location.

The whole implementation process of LLTO is shown in Fig. 2.

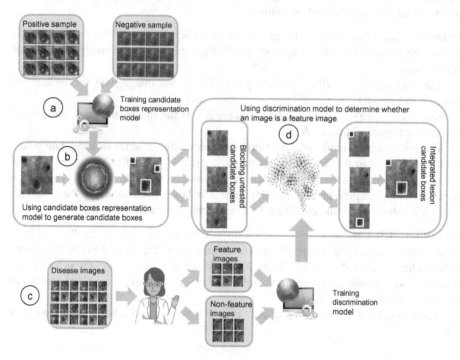

Fig. 2. Implementation of LLTO

As shown in Fig. 3, the implementation of LLTO consists of four parts. (a) The *PSS* and *NSS* are trained by the OpenCV cascade classifier to generate the *CBS* representation model. (b) The *CBS* representation model is used to generate the *CBRI* when there is a test image (*OI*) input. (3) The discrimination model of disease is generated by training *FIS* and *NIS* images. (4) Each candidate box in the *CBRI* image needs to be identified in the way of template occlusion method. In this measure, one tested image only contains one candidate box (candidate box test image, *CBTI*) to be identified and the other candidate boxes are hidden by the template. If the current *CBTI* is discriminated as *FI*, the candidate box in this *CBTI* is considered to contain a lesion area and is reserved. Otherwise, it doesn't contain a lesion area. After all the *CBTI* are identified, the reserved candidate boxes are integrated into the *OI* and the labeled areas are lesion locations.

3.2 The CBS Representation Model

We use the OpenCV cascade classifier to generate the CBS generation model. The OpenCV cascade classifier is a classifier to detect objects, which is suitable for detecting rigid objects but not for elastic objects. The core technology of the OpenCV cascaded classifier is feature selection and extraction. There are many methods for feature extraction at present. For example, the feature extraction method based on Cooperative representation [13]. In LLTO method, we choose the Haar feature for feature extraction. The Haar feature is a feature that can reflect the changes in gray level of the image. In the following contents, we will introduce the generation of the Haar feature [14].

Generating the Haar features. In LLTO, five kinds of basic Haar features are chosen to generate a set of Haar sub-feature. In this paper, we define any feature in the basic features as the BF, and we choose one of the five features, X3 characteristics, for illustration. The X3 feature is shown in the black and white square in the following Fig. 3:

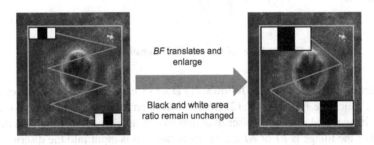

Fig. 3. Five basic Haar features

Many detection windows are generated during we extract Haar features of an image. We define these detection windows as the detection window set. The specific definition is as follows.

Definition 3.4. Detection window set (DWS): $DWS = \{DW_1, DW_2, \ldots, DW_i, \ldots, DW_n\}$, where $DW_i = (x, y, width, height)$ is a scanning window of feature extraction, where x and y respectively represents the abscissa and the ordinate of the first pixel in DW, $width$ and $height$ respectively represents the width and the height of DW. The red box in Fig. 3 represents a DW.

As shown in Fig. 3, the Haar sub-features are generated by the BF continuously scanning in the detection window. After a round of scanning, BF is enlarged and then the preceding process is repeated. The scanning operation stops when the size of BF is equal to the size of DW.

3.3 Training of Discrimination Model

Discrimination model is used to determine whether an image is the feature image. In this paper, we use the discrimination model to determine whether a $CBTI$ is FI. We denote the training data set of the discrimination model with $S = \{FIS, NIS\}$.

In intelligent diagnosis, in order to achieve the diagnosis of certain diseases, the true clinical medical image data obtained from hospitals are universally trained by deep learning method to generate a discrimination model. In LLTO, we use Google Inception v3 architecture to train the data set S and obtain the discrimination model of disease. The flow is represented as follows (Fig. 4):

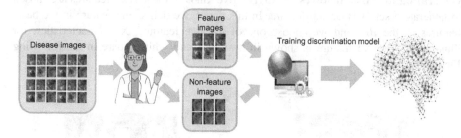

Fig. 4. Flow of training discrimination model

The detailed description is as follows. (1) Doctors classify images of a disease into two types, which are *FIS* and *NIS*. (2) Google Inception v3 model is applied to train the classified images and obtain the discrimination model suitable for identifying a disease. (3) Whether the image is *FI* or *NI* can be identified if it is input into the discrimination model. As a result, we get the discrimination model for disease identification, which can be used for candidate boxes selection in lesion location.

3.4   Lesion Localization Process

The *CBRI* is generated by the CBS representation model, in which contains two types of candidate boxes, one is the candidate boxes containing lesions, and the other is that the ones may not contain lesions. Therefore, we propose the template occlusion method to extract the real lesion areas from candidate boxes. The flow of template occlusion is shown in the D part in Fig. 2.

In order to avoid the interaction caused by each other, we singly identify the candidate boxes in *CBS* rather than in batch. When one candidate box is being identified, all the rest candidate boxes are hidden by the template frame and then the image becomes a *CBTI*, ensuring that there is only one candidate box for identification each time. The template frame used to hide the candidate box is not selected arbitrarily. The template frame should be selected from the background area of the image *OI* containing the current *CB* identified, to avoid the influence of the added template frame on the identification result.

The template occlusion method process can be described as Algorithm 1:

Algorithm 1. Template occlusion algorithm

1	Input: *CBRI*;
2	Output: the set of *CBTI*
3	List *list*;
4	*temp_image* = *CBRI*;
5	For each candidate box *i* in *CBRI* do
6	For each candidate box *j* in *CBRI* do
7	If *j* is not *i* then
8	Fill the *j* with the background in *temp_image*;
9	End if;
10	End for;
11	*l*.add (*temp_image*);
12	*temp_image* = *CBTI*; // reset *temp_image*
13	End for;
14	Return *l*;

Next, we will identify the set of *CBTI* through the discriminant model to select the real lesion locations.

Candidate boxes selection and integration. Candidate box selection is to discriminate and select real lesion locations from the set of *CBTI* by the discrimination model. If the *CBTI* is identified as *FI*, it means that the candidate box contained in the *CBTI* is the lesion area, so the position of the candidate box is recorded. Otherwise, the candidate box contained in the *CBTI* is the non-lesion area and the position of the candidate box will not be recorded. After all the candidate boxes in the *CBRI* are identified, the recorded ones will be marked on the *OI*, and the marked areas are the true lesion locations.

The process of locating lesion areas on the *OI* can be described as Algorithm 2.

Algorithm 2. Lesion localization algorithm

1	Input: the set of *CBTI*
2	Output: image // The image after the localization of the lesion
3	List *l*;
4	For each *CBTI* in *CBTIS* do
5	*CBTI* using the discrimination model
6	If *CBTI* is *FI* then
7	Record the position of the candidate box in the *CBTI* to the list
8	End if;
9	End for;
10	Label positions recorded in the *l* on *OI*;
11	Return image;

In Algorithm 2, we can get the result of whether a *CBTI* is the feature image, and the image marked by lesion locations will be output in the end.

4   Experimental Results

LLTO method is suitable for many diseases to locate lesion areas. In order to verify the effectiveness of the method, we chose a representative dermatosis disease to conduct experiments. The experimental data are the real clinical data from the Third Xiangya Hospital, Central South University.

4.1   The Lesion Localization Effect of LLTO

Whether a medical image is the feature image can be identified by LLTO, and for the feature image, the lesion locations will be marked. In this experiment, we use some medical images of the demodicosis disease to test the lesion localization effect of LLTO. We choose 15 images that the doctor concludes that there are demodicosis areas on them. Then these images are processed by our method to obtain the marked lesion locations on them. The experiment results are shown in Fig. 5.

Combining the judgment for these results of the doctors, we can get that LLTO method achieves a good result of locating the demodicosis areas. No matter there is one single or multiple lesion locations in the image, these areas can be marked out correctly by LLTO.

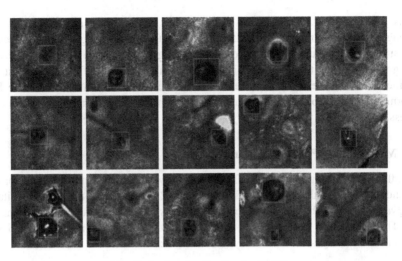

Fig. 5. Lesion localization by LLTO

4.2   Time Cost Comparison

This experiment is to prove the feasibility and the low time cost of locating lesion areas in LLTO. In the experiment of time cost comparison, we define three types of time cost, they are: (1) time cost of the test data classification $(T(c))$, (2) time cost of the data classification and the lesion area locating $(T(cl))$, and (3) time cost of candidate boxes generating $(T(g))$. Considering that the time cost of data test is not necessarily the same on machines of different performance, we adopt the time ratio to compare the three types of time cost (the time of one data classified is defined as the time unit). We set 5 groups of data and each contains 20 data. The results of the time cost comparison are as Fig. 6.

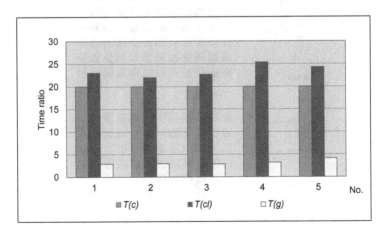

Fig. 6. Time cost comparison

As shown in Fig. 6, the process of the data classification and the lesion area locating costs approximately 1 to 1.25 times of the time comparing with the process of the test data classification. The speed of generating the candidate boxes is the fastest among them, and the time cost of it is about less than a quarter of that of the test data classification. As a result, the process of lesion localization can reach relative high performance and it does not cost much more time than the general data classification process.

4.3 Multi-clinical Patient Experiments

To further prove the accuracy of LLTO method, we conduct a multi-clinical patient trial and test using undiagnosed cases. We select 72 patients from the third Hospital of Xiangya, and one of the images from each patient's medical record are selected as test data. The doctor and LLTO should independently identify and mark out the lesion areas in the 72 images. The accuracy is defined as:

$$Accuracy = \frac{num_{correct}}{num_{total}} \times 100\%, \tag{1}$$

Where num_{total} is the total number of the test images, $num_{correct}$ is the number of images which the lesion areas are marked out correctly.

In Fig. 7, we can see the results of diagnosis of the doctor and LLTO. We assume that the doctor has the accuracy of 100% in locating the lesion areas. The yellow square marks the different result between the doctor and LLTO. It can also be interpreted as the error result of LLTO. Consequently, there are 12 errors in the 72 images of LLTO and the accuracy is 83.3%, indicating that LLTO can get a satisfied result of locating the lesion areas.

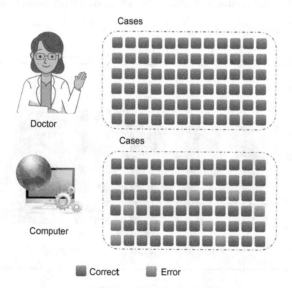

Fig. 7. The result of multi-clinical patient test

5 Conclusion

In recent years, artificial intelligence has made great achievements in medical field, and intelligent diagnosis becomes an important topic simultaneously. Recent research on the intelligent diagnosis of diseases is mainly focus on disease identification. However, how to detect lesion location is still a difficult problem which is of great significance. In medical field, lesion localization is required by the treatment of many diseases. Lesion location information helps physicians make a further understanding of the disease, assists them in diagnosis and therapy, and increases the likelihood of the disease being cured. Therefore, how to propose a lesion localization method has great practical sense. In LLTO, OpenCV cascade classifier is used to train candidate boxes representation model and the TensorFlow is used to train disease discrimination model. In order to implement the lesion location task, the lesion candidate boxes are firstly generated by the candidate boxes representation model in the test image, and then the disease discrimination model is used to select true lesion areas from candidate boxes and integrate them to get the image marked with lesion locations. In the following work, our next step is to validate whether LLTO can be extended to the identification and classification of other diseases in different clinical circumstances. Simultaneously, we will explore the feasibility of LLTO in clinical practice.

Acknowledgment. This work is supported the by Natural Science Foundation of China (61672535, 61472005, 61772254), the Key Laboratory of Information Processing and Intelligent Control of the Fujian Innovation Fund (MJUKF201735) and the Fundamental Research Funds for the Central Universities of Central South University (2018zzts585). The authors declare that they have no conflicts of interest.

References

1. Yang, W., Wang, Z., Zhang, B.: Face recognition using adaptive local ternary patterns method. Neurocomputing **213**, 183–190 (2016)
2. Esteva, A., Kuprel, B., Novoa, R.A., et al.: Dermatologist-level classification of skin cancer with deep neural networks. Nature **542**(7639), 115–118 (2017)
3. Redmon, J., Divvala, S., Girshick, R., et al.: You only look once: unified, real-time object detection. In: IEEE Conference on Computer Vision and Pattern Recognition, pp. 779–788. IEEE Computer Society (2016)
4. Chen, Z., Crandall, D., Templeman, R: Detecting Small, Densely Distributed Objects with Filter-Amplifier Networks and Loss Boosting (2018)
5. Ren, S., He, K., Girshick, R., et al.: Faster R-CNN: towards real-time object detection with region proposal networks. In: International Conference on Neural Information Processing Systems. pp. 91–99. MIT Press (2015)
6. Rd, A.S., Sensakovic, W.F.: Automated lung segmentation for thoracic CT impact on computer-aided diagnosis. Acad. Radiol. **11**(9), 1011–1021 (2004)
7. Chan, H.P., Sahiner, B., Helvie, M.A., et al.: Improvement of radiologists' characterization of mammographic masses by using computer-aided diagnosis: an ROC study. Radiology **212**(3), 817–827 (1999)

8. Kermany, D.S., Goldbaum, M., Cai, W., et al.: Identifying medical diagnoses and treatable diseases by image-based deep learning. Cell **172**(5), 1122–1131.e9 (2018)
9. Long, E., Lin, H., Liu, Z., et al.: An artificial intelligence platform for the multihospital collaborative management of congenital cataracts. Nat. Biomed. Eng. **1**(2), 0024 (2017)
10. Jia, K., Kenney, M., Mattila, J., et al.: The application of artificial intelligence at Chinese digital platform giants: Baidu, Alibaba and Tencent (2018)
11. Cuimei, L., Zhiliang, Q., Nan, J., et al.: Human face detection algorithm via Haar cascade classifier combined with three additional classifiers. In: 2017 13th IEEE International Conference on Electronic Measurement & Instruments (ICEMI), pp. 483–487. IEEE (2017)
12. Ouerhani, Y., Alfalou, A., Brosseau, C.: Road mark recognition using HOG-SVM and correlation. In: Optics and Photonics for Information Processing XI, vol. 10395, p. 103950Q. International Society for Optics and Photonics (2017)
13. Yang, W., Wang, Z., Sun, C.: A collaborative representation based projections method for feature extraction. Pattern Recognit. **48**(1), 20–27 (2015)
14. Viola, P., Jones, M.: Rapid object detection using a boosted cascade of simple features. In: Proceedings of the 2001 IEEE Computer Society Conference on Computer Vision and Pattern Recognition, 2001. CVPR 2001, vol. 1, pp. I-511–I-518. IEEE (2003)

A Service Agents Division Method Based on Semantic Negotiation of Concepts

Sheping Zhai, Zhaozhao Li[(⊠)], Hongyu Duan, and Shan Gao

School of Computer Science and Technology,
Xi'an University of Posts and Telecommunications, Xi'an 710121, China
leeezhaozhao@163.com

Abstract. In the process of Web service discovery, it is necessary to use ontology for negotiation and interaction between service participating agents, and the accuracy of service discovery is greatly influenced by the heterogeneity of ontology and the mistaken understanding of messages. The method based on concept semantic negotiation is proposed, which completed semantic negotiation through concept explanation, improved the concept understanding ability of service agent, and divided the agents which have similar functions and understanding ability into the same services community according to the result of negotiation. Finally, an experimental comparison between our method and other methods is presented, the results show that the semantic negotiation method has the least error rate.

Keywords: Services discovery · Ontology · Semantic Web
Semantic negotiation

1 Introduction

Booth et al. [1] argued that Semantic Web services were an abstract object which has lifecycle and can implement its diverse function through specific agents in different stages. Agents propose service requests and release service functions as service requester and provider, the service requester and the provider agents discover, select, compose, invoke, and execute services by sending and receiving messages [2]. The discovery and selection of Web services are the most important, because they relate to whether the results of the service composition, invocation, and execution can meet the functions and requirements of the service requests and service consumers.

Web service discovery refers to the way in which users find the services they want in many published Web services, combine them into their applications, or execute them directly to complete a computing task [3]. The service requester agent uses the abstract service descriptions to verbalize demands during services discovery, meanwhile, the service provider agent also expresses the function information by using abstract service conceptual relation. And during the stage of service selection, the abstract services produce specific services meeting the needs of the users by instantiating the parameters [4]. The choices of the parameters values relate to the pragmatic information of the service participants (the meaning of the service concept and the contexts' information used).

© Springer Nature Switzerland AG 2018
Y. Peng et al. (Eds.): IScIDE 2018, LNCS 11266, pp. 57–67, 2018.
https://doi.org/10.1007/978-3-030-02698-1_6

Since the service participant agent usually uses its own private ontology to label the specific services semantically, this causes there are problems in the agents' comprehension of the partial and heterogeneous service descriptions, also causes the failure of interaction process during the service discovery [5].

The method of concept semantics negotiation is proposed in order to improve the ability of understanding between service agents. By describing the models and concepts of service agent, the agents participated in the service can negotiate on the service concepts semantically, and the agents are partitioned in the same Service Community with similar functionality and understanding ability according to the results of the consultation, thereby to improve the understanding ability of the service agents in the same community.

2 Ontology Concept Understanding of Service Agents

In order to solve the problem of mutual understanding between service agents due to the heterogeneity of service description ontology, there are some methods that use ontology alignment, which not suitable for the agent modified constantly in an open environment, this is because the mapping between ontologies should be established before agents' interaction [6]. Moreover, grouping the service agents according to the different standards can effectively solve the problem of conceptual understanding between agents. For example, during the Web service discovery, the same group of service agents often have similar demands and higher concept understanding ability, which can increase the accuracy and efficiency of service discovery.

There is a method which developed a model using an alignment ontology framework and selected the most similar pairs, then suggested a protocol to support semantic negotiation [7]. However, the heterogeneous of ontology leads to the inexactness of similarity between ontologies. Meanwhile, each agent has specific domain knowledge of its own actions, it is impossible and unnecessary to adopt a common ontology.

For the division of service agents, a method based on objectives of agents divided the agents cooperated with the common target tasks into a group, and the similarity calculation method were used to compare the compatibility of objective of different agents. Some methods proposed an agent division method which divided the agents with the same service resources, beliefs and knowledge into the same service community by comparing the service belief knowledge and trusting relationship among the members of the group [8]. However, these methods are lacking of the consideration of the context knowledge used by service concept and cannot re-divide service agents according to the change of the concept semantic understanding during the process of service interaction.

A new method of using the conceptual semantic negotiation protocol between service agents is proposed. The problem of concept understanding discussed above can be solved effectively by dividing service community with the method of ontology concept interpretation. The definition of the agent model and the concept of service ontology are as follow.

Definition 1. Service Agent model.

The model with more than one service agents can be expressed as: $M = <S_a, S_A, S_c, Action, ST, CS>$.

Where S_a represents the set of service participant agents, S_A represents the common concept model of the abstract description of service agents, and S_c represents the set of the specific concept relationship and the explanation of service description. *Action* is the set of semantic negotiation external methods that can be called by service provider or requester agents. *ST* indicates the condition knowledge intention of service concept interpretation obtained by agents. *CS* represents the knowledge description of the service agents' private concepts, and reflects the service domain knowledge of the service participant agents.

In order to understand the message of other service agents in the case of incomplete knowledge of service, an effective way is to understand the conceptual meanings through semantic negotiation, so as to improve the success rate of information inter-action between service agents.

Definition 2. Service Ontology Concept Explanation.

A service ontology concept explanation e is a triple with form $ec = (E, A, ea)$, where ec is the concept being interpreted, E is the set of concepts consisting of the ontology concept equivalent to ec (only consider the equivalence relation between concepts in order to simplify validation), A is the set of agents understanding and agreeing with the explanation, ea illustrates which agent provides the explanation. For example, suppose agent a_2 has an explanation of negotiation ontology as follow: $Price = (\{Cost\}, <a_1, a_2, a_3, a_4>, a_1)$, which indicate that a_2 considers *Price* and *Cost* are the same semantic concept, a_1, a_2, a_3 and a_4 understand and agree with the explanation, and the initial explanations are given by a_1. As the negotiation process progressing, the content of this interpretation will change accordingly, for example, adding new interpretations and modifying the existed set of concept understanding agents.

3   Concept Semantic Negotiation Between Service Agents

The meanings of the concept of service description between service agents are understood by the sending and receiving of semantic negotiation messages. The semantic negotiation message is defined as follows.

Definition 3. Semantic negotiation message.

The semantic negotiation message is represented as a six-tuple with form $<i, j, S_{ia}, p, t, mc>$, S_a is used to represent the set of service participant agents, where $i, j \in S_a$ and represent the sender and receiver agents of the negotiation message. $S_{ia} \in S_a$ represents the set of all agents that involved in the current concept, p represents the behavior primitives of the messages, t represents the negotiation time determined by S_{ia} and mc represents the content of the negotiation messages.

3.1   Service Conceptual Relationship

Table 1 lists part of the service concept and composition of semantic negotiation. According to the level of service description, the service concept of negotiation can be divided into abstract service concept models and specific service concept explanations.

During the semantic negotiation, the abstract service concepts belong to the common knowledge of the service agents involved in semantic negotiation. For example, when describing a service, *ServiceProfile* is used to describe the functional and non-functional attributes of the services, while *ServiceName, Contact, Description, Category* and other subclasses are used to describe the service attributes.

Specific service concept explanations reflect the private domain knowledge of the service agents, expressed as the explanatory listing of the domain concepts, understanding and agreeing with the set of the agents and the source of the conceptual explanation. Such as $Area = (\{Size\}, <a1>, a1)$ reflect the understanding of Agent $a1$ on the private concept *Area*, indicate that *Area* can be interpreted by the use of the concept *Size*.

Table 1. Part of the relationship of service concept in semantic negotiation

Concept hierarchy	Concept name	Concept structure
Abstract service concepts model	ServiceProfile	{ServiceName, Contact, Description, Category}
	ServiceModel	{AtomicProcess, SimpleProcess, CompositeProcess}
	ServiceGrounding	{wsdlDocument, wsdlOperation, wsdlService}
	...	...
Specific service concepts explanation	Area	$(\{Size\}, <a1>, a1)$
	Department	$(\{Room\}, <a1, a2>, a2)$
	Price	$(\{Cost\}, <a1, a2, a3, a4>, a1)$
	Cheap	$(\{Inexpensive\}, <a1, a2, a3>, a3)$
	...	...

3.2   Semantic Negotiation Protocol

Communication between agents is done by exchanging messages, and each message uses behavior primitives to express the desired messages' result of message sender. On the basis of the original 22 behavioral primitives of FIPA-ACL, we designed 4 message behavior primitives shown in Table 2 for semantic negotiation interactions between agents. The behavior primitives are used to understand the implication of concept between the requester agents and the interpreter agents.

Table 2. Semantic negotiation behavior primitives and meanings

Negotiation behavior primitive	Meaning
CM-REQUEST	Agent i requests agent j to help understanding the meaning of some ontology concepts that compose the content of ACL messages, the list of concepts needed to be understood is contained in this speech behavior message
CM-RESPONSE	Agent j interprets the concept meaning of the requests after receiving the CM-REQUEST speech behavior message from agent i, the explanatory list of the corresponding request concept appears in this speech behavior message
CM-ACCEPT	Agent i informs j the list of concept explanations it accepts through this speech behavior message after receiving the explanations from agent j
CM-NOT-UNDERSTAND	For CM-REQUEST message from agent i, agent j informs i not to give explanation of the requested concept by this speech behavior message

Three different actions are completed by adding the negotiation speech behavior messages above.

(1) Request another service agent to help understanding the meaning of the concept that describe the service request or service advertisement.
(2) Give the concept interpretation results of the request.
(3) Return to a list of concepts which agents themselves can't understand.

Request Behavior. The process of semantic negotiation request is shown in Fig. 1. When a message received by Agt_1 contains a concept explanation that does not exist in its private ontology, then the semantic negotiation process is initiated, which is divided into the following two situations.

(1) During the process of interacting ACL messages with other agents, there are some incomprehensible ontology concepts contained in the message. Agt_1 can start the semantic negotiation process, search for concept semantic negotiation with the agent that is connected and active. For example, Agt_1 can construct a message: CM-REQUEST (Agt_1, Agt_2, C) and send it to Agt_2, which indicates that Agt_1 sends a request to Agt_2 to help understand the concept set C.
(2) If Agt_2 receives the concept understanding request from Agt_1 and can't explain the concept of the request, it will start the semantic negotiation process, request a concept explanation of an active agent that is connected to Agt_2, and return the final results to Agt_1.

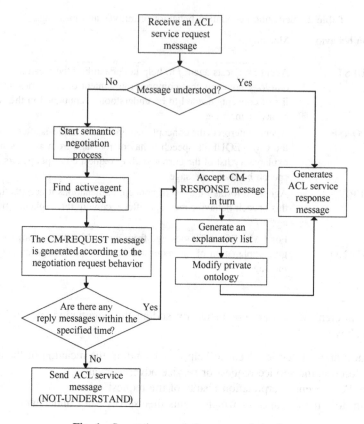

Fig. 1. Semantic negotiation request behavior

Response Behavior. Semantic negotiation response behavior process is shown in Fig. 2, Agt_2 receives a concept understanding request message (CM-REQUEST) from Agt_1, then response as follows:

(1) Agt2 can understand the meaning of the concept requested, and then responds with a CM-RESPONSE speech behavior message which contains an interpretation of the request concept.

(2) If Agt_2 can't understand the concept meaning of the request, it will ask an active agent for help (that is initiates a new negotiation). After that, if the concept is fully understood (or not understood), Agt_2 uses the CM-RESPONSE (or CM-NOT-UNDERSTAND) speech behavior message to respond to Agt_1.

3.3   The Division of Service Agents

Through the semantic negotiation of service concept, new concepts and context information are added into the concept interpretation lists of service participant agents (concept interpretation providers). The service provider agents then modify the service privacy ontology and establish the concept correspondence relation between privacy

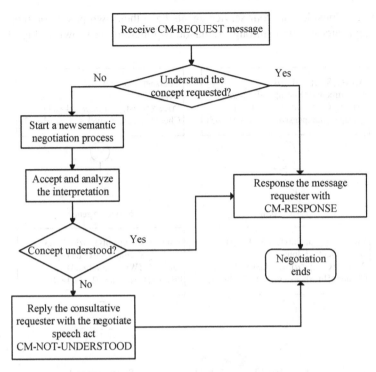

Fig. 2. Semantic negotiation response behavior

ontologies. By calculating the semantic similarity of the concept of privacy ontology, the agents are partitioned in the same service community with similar background knowledge.

Definition 4. The sets of service participant agents can be expressed as $S_a = \{a_1, a_2, \ldots, a_n\}$, where $a_n \in S_{pa} \cup S_{ra}$, O_i is the privacy ontology of service agent a_i ($i = 1, 2, \ldots, n$), and the ontology concept set which agent a_i obtained through semantic negotiation is expressed as $CS_i = \{C_{i1}, C_{i2}, \ldots, C_{ik}\}$.

Definition 5. Assuming that the ontology concepts that service agent a_i and a_j obtained through semantic negotiation are CS_i and CS_j, then the similarity σ_{ij} of the two service agents' ontology O_i and O_j is defined as follows:

$$\sigma_{ij} = \frac{CS_i \cap CS_j}{CS_i \cup CS_j} \tag{1}$$

Where $|CS_i \cap CS_j|$ indicates the number of the negotiation concepts that the two service agent ontology have, and $|CS_i \cup CS_j|$ represents the number of all negotiation concepts the two service agents owned, and $|CS_i \cup CS_j| = |CS_i| + |CS_j| - |CS_i \cap CS_j|$.

Example: Consider that four service agents have their own private ontologies, and their concept interpretations through semantic negotiation are shown in Fig. 3.

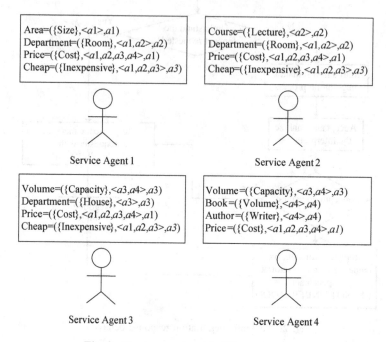

Fig. 3. Service agents with different concepts

The similarities between four service agents' ontologies are calculated as follows:

$$\sigma_{12} = \frac{|\{\text{Department, Price, Cheap}\}|}{5} = \frac{3}{5} = 0.6$$

$$\sigma_{13} = \frac{|\{\text{Department, Price, Cheap}\}|}{5} = \frac{3}{5} = 0.6$$

$$\sigma_{14} = \frac{|\{\text{Price}\}|}{7} = \frac{1}{7} = 0.14$$

$$\sigma_{23} = \frac{|\{\text{Department, Price, Cheap}\}|}{5} = \frac{3}{5} = 0.6$$

$$\sigma_{24} = \frac{|\{\text{Price}\}|}{7} = \frac{1}{7} = 0.14$$

$$\sigma_{34} = \frac{|\{\text{Volume, Price}\}|}{6} = \frac{2}{6} = 0.3$$

According to the calculation results, we can divide the service agents into two groups. Since a_1, a_2 and a_3 are divided into a group because they have higher concept understanding similarity. During service discovery, the same group of agents should interact previously. This can improve the accuracy of service discovery results effectively because they have higher service concept understanding ability and similar service functions.

4 Evaluation

We designed four service agents in the support of the multi-agents system development platform JADE, the four agents used the semantic negotiation messages previously designed for concept understanding.

In the initial stage, four agents bound different private ontology. Their structures changed from the inclusion relation between the simplest two concepts to the inclusion and equivalence relation between the complex eight concepts. And each ontology had a concept of 50 or so. In addition, the housing leasing ontology was used as the ontology registration of the other four service agents, which makes the four service agents both have their own ontology and can share the same ontology concept with other agents.

The method of concept semantic negotiation (CMN) proposed in this paper was compared with the method based on planning understanding (PB) mentioned in document [9] and the B-SDR method using the hierarchical clustering technique mentioned in document [10]. We chose one agent as the service concept understanding requester randomly, constructed the content of the negotiation message with its private ontology, and then started the concept semantic negotiation between the service agents.

We gradually increased the number of service requests ($1 \leqslant t_{rq} \leqslant 150$) and changed the number of ontology concepts contained in the request message content ($2 \leqslant nc \leqslant 8$, that is, the change from simple to complex conceptual inclusion relations). When service provider agent received the service request messages, we started the concept negotiation and counted up the total number which the result of the negotiation is CM-NOT-UNDERSTAND, recorded as Num(CMN), Num(PB) and Num(B-SDR). And we expressed the percentages of concepts that can't be understood as Per-num(CMN), Per-num(PB) and Per-num(B-SDR), record as error rate, which are calculated as follows:

$$Per-\text{num(CMN)} = \frac{\text{Num(CMN)}}{nc \times t_{rq}} \tag{2}$$

$$Per-\text{num(PB)} = \frac{\text{Num(PB)}}{nc \times t_{rq}} \tag{3}$$

$$Per-\text{num(B-SDR)} = \frac{\text{Num(B-SDR)}}{nc \times t_{rq}} \tag{4}$$

Figure 4 reflected the interaction of the negotiation messages between two concepts (which means $nc = 2$), the results showed that the semantic negotiation method is better than the other two. When the number of service request increases, the Per-num of concept understanding of the method of concept semantic negotiation is the lowest. As showed in Fig. 4 the Per-num of semantic negotiation method is 8.2% while Per-num (B-SDR) is 12.2% and Per-num(PB) is 9% for the condition of "t = 150".

Figure 5 showed that the error rate of the three methods were all increased when we changed the number of concepts contained in the negotiation message to 8($nc = 8$), the performance of the semantic negotiation method is better than other methods. As showed in picture, the Per-num keep rising when we increasing the number of requests. And the Per-num of semantic negotiation method is always the lowest, even if we add the number to 150 times.

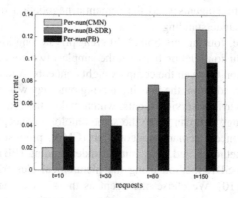

Fig. 4. Number of requests and understanding error rates (Simple messages)

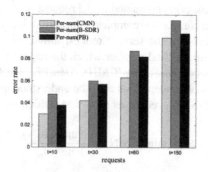

Fig. 5. Number of requests and understanding error rates (Complex messages)

5   Conclusion

The problem of concept understanding for heterogeneous private service ontologies between agents need to be solved during the service discovery, in this paper, the service agent model was described, the concept interpretation form was defined and a set of message behavior primitives of concept semantic negotiation was designed. On this basis, the understanding ability on service concept was improved through the negotiation and interaction between service agents. And then updated the private service knowledge ontology and recorded the context information of the service concept negotiation, so that the number of interactions of agents with similar function and higher ontology concept similarity can be increased. Experiments showed that through our concept semantic negotiation method, the concept understanding ability between agents divided into the same service community can be improved.

References

1. Booth, D., Haas, H.: Web services architecture, W3C working group. In: Concurrency & Computation Practice & Experience, pp. 72–81. Wiley (2004)
2. García-Sánchez, F., Guedea-Noriega, H.H.: Intelligent agents and semantic web services: friends or foes. In: Valencia-García, R., Lagos-Ortiz, K., Alcaraz-Mármol, G., Del Cioppo, J., Vera-Lucio, N., Bucaram-Leverone, M. (eds.) CITI 2017. CCIS, vol. 749, pp. 29–43. Springer, Cham (2017). https://doi.org/10.1007/978-3-319-67283-0_3
3. Le, D.N., Kanagasabai, R.: Semantic Web service discovery: state-of-the-art and research challenges. J. Pers. Ubiquit. Comput. **17**, 1741–1752 (2013)
4. Moghaddam, M., Davis, J.G.: Service selection in web service composition: a comparative review of existing approaches. In: Bouguettaya, A., Sheng, Q., Daniel, F. (eds.) Web Services Foundations, pp. 321–346. Springer, New York (2014). https://doi.org/10.1007/978-1-4614-7518-7_13
5. Rajendran, T., Balasubramanie, P.: An optimal agent-based architecture for dynamic web service discovery with QoS. In: International Conference on Computing Communication and NETWORKING Technologies, pp. 1–7. IEEE Press, Karur (2010)
6. Fellah, A., Malki, M., Elçi, A.: Web services matchmaking based on a partial ontology alignment. J. Inf. Technol. Comput. Sci. **8**, 9–20 (2016)
7. Noureddine, D.B., Gharbi, A., Ahmed, S.B.: Reflective multi-agent model using semantic similarity measure and negotiation protocol for solving heterogeneity. In: Ifsa-Scis, pp. 1–8. IEEE Press, Otsu (2017)
8. Wu, L., Su, J., Su, K., et al.: A concurrent dynamic logic of knowledge, belief and certainty for multi-agent systems. J. Knowl.-Based Syst. **23**, 162–168 (2010)
9. Sheping, Z., Hai, W., Juanli, W.: A pragmatics web service oriented approach to understanding the semantics of concepts. In: 2008 IEEE Asia-Pacific Conference on Services Computing (APSCC 2008), pp. 1069–1074. IEEE Press, Taiwan (2008)
10. Buccafurri, F., Rosaci, D., Sarnè, G.M.L., Ursino, D.: An agent-based hierarchical clustering approach for e-commerce environments. In: Bauknecht, K., Tjoa, A.M., Quirchmayr, G. (eds.) EC-Web 2002. LNCS, vol. 2455, pp. 109–118. Springer, Heidelberg (2002). https://doi.org/10.1007/3-540-45705-4_12

Network Traffic Generator Based on Distributed Agent for Large-Scale Network Emulation Environment

Xiao-hui Kuang, Jin Li$^{(\boxtimes)}$, and Fei Xu

National Key Laboratory of Science and Technology on Information System
Security, Beijing, China
tianyi198012@163.com

Abstract. Network traffic generation is one of the key technologies used to construct a network emulation environment for research into network protocols, applications and security mechanisms. In this paper, a new network traffic generator called NTGEDA is proposed that can generate network layer, application layer and user layer traffic on demand based on distributed agent architecture and the multi-level traffic model. The generator is evaluated by means of implementation. Experimental results illustrate that it is efficient in the network emulation environment for network protocols and security research.

Keywords: Network traffic generation · Distributed agent
End-user emulation · Behavior model

1 Introduction

An increasing number of researchers involved in the design of network services, protocols and security mechanisms have come to rely on results from emulation environments to evaluate correctness, performance and scalability [1–5]. To better understand the behavior of these protocols, devices and applications, as well as to predict their performance when deployed across the internet, the emulation environments must closely match real network characteristics, not just in terms of structure, nodes and links, but also network traffic.

However, achieving a deep understanding of network traffic characteristics has proven to be a challenging task due to the increase in the number of internet users, application types and network bandwidth. There are several network traffic generators available to the networking research community [6–18], many of which differ significantly with respect to the traffic model, protocol layer and application scenarios. Furthermore, existing network traffic generators fail to generate multi-level network traffic on demand and cannot reflect the behavioral characteristics of internet users; this is due to the fact that they focus primarily on traffic characteristics at a packet and flow level, or for a specific scenario such as web traffic or YouTube workload, rather than attempting to model user behavior model on the terminal node.

© Springer Nature Switzerland AG 2018
Y. Peng et al. (Eds.): IScIDE 2018, LNCS 11266, pp. 68–79, 2018.
https://doi.org/10.1007/978-3-030-02698-1_7

In this paper, we continue the direction of research into network traffic generation methods and present a novel approach for our network emulation environment called LSNEMUlab [19–21], which supports research into network protocols and security mechanism. Our approach combines classic network traffic generation methods in the field of network and application layer generation using distributed agent network traffic generation framework and models. Furthermore, we address the influence of internet users on network traffic and propose a user layer traffic model and behavior description.

We implement our approach and integrate it with LSNEMUlab, generating not only network layer and application layer traffic, but also traffic that reflects user behavior characteristics.

In summary, we make the following contributions to the problem of network traffic generation:

- **Network traffic generation framework.** We introduce a novel network traffic generation framework based on distributed agents that combines network layer, application layer and user layer traffic models. It can help researchers to implement the flexible generation of different types of network traffic.
- **User behavior model.** We introduce the user behavior model to elegantly characterize the impact of human behavior on the network traffic. This can improve the fidelity of the generated traffic better.
- **Efficient implementation.** We demonstrate that, by importing the network traffic generator into our LSNEMUlab, security experimentation can be efficiently executed in our environment.

The remainder of this paper is organized as follows. A survey of related work is presented in Sect. 2. In Sect. 3, we give a brief description of the large-scale network emulation experimental environment LSNEMUlab. The proposed network traffic generation framework is described in Sect. 4. Experimental setup and results are presented and discussed in Sect. 5. Finally, we draw some conclusions and outline directions for possible future work in Sect. 6.

2 Related Work

The network traffic generators can be classified into three different types, according to the layer on which they work [6]:

- **Application-level traffic generators:** Emulate the behavior of specific network applications in terms of the traffic they produce.
- **Flow-level traffic generators:** Used when the replication of realistic traffic is requested only at the flow level (e.g., number of packets and bytes transferred, flow duration).
- **Packet-level traffic generators:** With this term, we refer to generators based on a packet's inter-departure time (IDT) and packet size (PS). The size of each packet sent, as well as the time elapsed between subsequent packets, is chosen by the user, typically by setting a statistical distribution for both variables.

Bit-Twist [7] is a libpcap-based Ethernet packet generator that replicates traffic from captured PCAP files [8]. KUTE [9] is a Linux kernel-level packet generator that can be set up for any given packet sending rate, after which it calculates the corresponding IPT value and sends out packets to the network interface using active sleep between consecutive packets. D-ITG [10] is one of the leading application and packet-level network generators. This distributed and platform-independent tool is equipped with many features and has great future potential. The architecture of the NTG is inspired by the D-ITG in several ways. Scapy [11] and Karat Packet Builder [12] are packet-level traffic generators that are particularly suitable for firewall testing using a variety of network protocols. The packet-level traffic generator Ostinato [13] furnishes the user with a sophisticated GUI and detailed packet header customization. HAR-POON [14] is a traffic generator that can produce synthetic traffic based on various flow characteristics. Moreover, the tool can also analyze real measurements to extract such values, meaning that it can create artificial traffic with characteristics that are close to the original live measurement. SWING [15] is another high-level traffic generator that can generate traffic based on the characteristics of a real trace. However, many of these traffic generators only work on one layer, and all of them cannot emulate user behavior. Moreover, the fidelity of traffic cannot match experimental requirements for the performance of networks and network devices, security, quality of service and experience, new protocols, etc.

3   LSNEMUlab Network Experimental Environment

LSNEMUlab creates a large-scale network environment by accurately modeling networks, resources and terminals, thereby enabling researchers to study the performance and security of network protocols and distributed systems. In addition, real application software, network protocol stacks and hardware can be used unchanged in the network environment, which supports the study of complex applications, equipment and network protocols whose internal dynamics are difficult to model accurately. In short, LSNEMUlab provides a large-scale experimental network infrastructure with dynamic reconfiguration capability that enables scientific and systematic experimentation by supporting controllable, repeatable, and observable experiments.

LSNEMUlab is composed of an operation management network, experimental network, and a power and serial-port controlled network.

The management network is composed of the integrated management system, the image server and the control interface of the experimental node. The management system is responsible for the unified management, configuration and supervisory control of the experimental environment's soft hardware resources. The image server is used to store all image files required by the experimental environment, and also provides image making and reloading services.

The power and serial-port control network control the power switch of each node remotely and provide an interface for the centralized management of the experimental nodes in order to meet the experimental environment's manageability requirements under abnormal circumstances.

The experimental network is composed of various types of experimental nodes that are connected with the programmable intranet. The programmable intranet is made up of a plurality of high-speed, piled Ethernet switches. The type of the experimental node determines the scalability, flexibility and fidelity of the experimental environment. There are two types of experimental node in LSNEMUlab: the emulation experimental node and the real system/network node. Moreover, each type of emulation experimental node contains three subtypes. The network emulation node emulates a network on a single experimental node to meet the scalability requirements; the route emulation node realizes the route equipment functions through the soft route technique; finally, the terminal emulation node emulates a plurality of terminals and server nodes for different soft hardware platforms via the virtualization technique in order to meet the heterogeneous requirement.

In LSNEMUlab, the user first needs to define the topology he wishes to emulate using the integrated management system, then constructs the experimental environment according to the topology. Owing to physical resource limitations, LSNEMUlab develops a distributed network emulator based on NS2, which is the core of network emulation node. In order to solve the problem of interaction between the virtual network and the real system, we adopt the packet analysis and reconstruction technique, as well as the packet forward technique, to realize the correct switching and routing of the virtual/real network packet. After the construction, the user can verify the environment using a topology discover tool such as CAIDA [22].

The components and their relationship with LSNEMUlab are illustrated in Fig. 1. For more details, the interested reader is referred to [19–21].

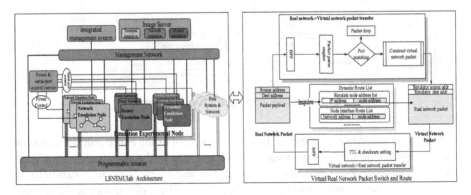

Fig. 1. LSNEMUlab network experimental environment

The topology for LSNEMUlab is required to incorporate not only network structure information and network properties (such as link bandwidth and node IP address), but also the network traffic. The use of manual network traffic generation is reasonable with respect to the scalability limit. Following recent efforts in designing LSNEMUlab, we identified the need for a network traffic generator that would further simplify the use of LSNEMUlab as a network testbed replacement and instrument for the creation of multi-layer network traffic.

4   Network Traffic Generation Framework

4.1   Network Traffic Generation Framework Based on Distributed Agent

Our network traffic generation framework, which is named NTGEDA, consists of two parts: the traffic generation manager and distributed agent. It utilizes network, application and user layer traffic models to implement the flexible generation of different traffic models and protocols (Fig. 2).

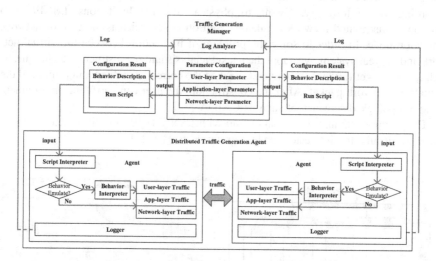

Fig. 2. Network traffic generation framework

The traffic generation manager is the control terminal of the entire traffic generation process, responsible for the parameter configuration, log analysis and other functional modules. It issues traffic scripts to distributed agents and analyzes log information from each agent in order to monitor the generation process. The traffic model in the scripts, which is the key component, contains the characteristic pattern and experimental demand. There are three traffic models in NTGEDA, each of which will be described in detail below.

The distributed agent, which is the executor of NTGEDA, uses multiple threads to interpret the scripts from the traffic generation manager. It can generate and receive traffic at the same time, as well as record logs during testing.

4.2 Network Layer Traffic Model

On the network layer, the main factors affecting the traffic characteristics are packet size and packet transmission interval. By specifying different distribution functions for these two parameters (e.g. constant, exponential, etc.), we can obtain network traffic with different characteristics. Packet data size also can be portrayed by the distribution function.

In order to generate the complex patterns of network traffic, NTGEDA utilizes hybrid traffic patterns in which the transmission period is a unit. Each unit could be a traffic pattern, and traffic patterns can be nested, as illustrated in Fig. 3.

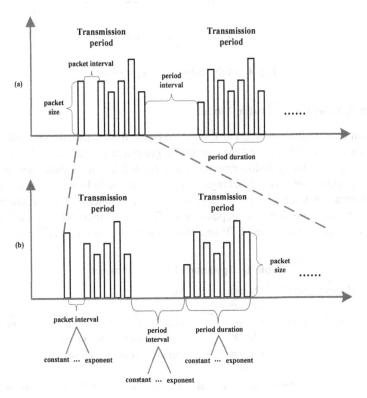

Fig. 3. Nested traffic patterns, superimposed

At the same time, by setting different start times and durations, NTGEDA can also transmit multiple data streams in order to create more complex network traffic, thus achieving a multi-data-stream aggregation of results (see Fig. 4).

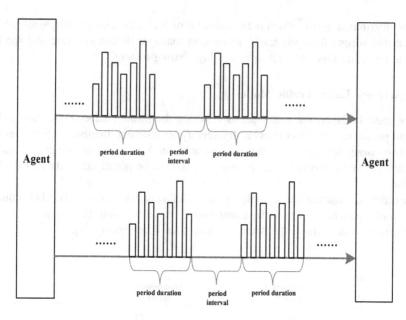

Fig. 4. Multi-data-stream aggregation

4.3 Application Layer Traffic Model

Application layer traffic can be divided into two types: the interactive type and the bulk transfer type. In the interactive type, the traffic generated by the two sides is approximately equal. The transmission rate is closely related to user habits such as typing speed and thinking time. Telnet belongs to this type. In the bulk transfer type, one side of communication mainly requests, and the one-way pattern of communication is obvious. The transmission rate depends on MTU and other network performance indexes between the two sides. FTP and Web belong to this type (Table 1).

Table 1. Application layer traffic model

Type of application layer traffic	Level	Characteristic parameters	Meaning
Interactive type	Flow	Session interval	Time between the beginning of the previous session and the current session
		Session duration	Duration time of each session
	Session	Packet interval	Interval between packet sending in each session
		Packet size	Packet size in each session
Bulk transfer type	Flow	Session interval	Time between the beginning of the previous session and the current session
	Session	Packet number	Packet number of each session
		Packet size	Packet size in each session

NTGEDA constructs the application layer traffic model from the flow and session levels. At each level, the interval, duration, data size and other characteristics are used to construct the model. The flow level contains all traffic between the two communication sides and each flow may contain one or more sessions.

4.4 User Layer Traffic Model

In order to characterize the impact of human behavior on the network traffic, we construct a user layer traffic model which is notated using time, location, operation and content.

- **Time attribute:** Describes the time point and duration of the actors that trigger network traffic. This attribute could be represented by traffic **start time** and traffic **duration**.
- **Location attribute:** Describes the IP address, port and function system used by the actors. Each function system has its own IP address and ports. This attribute could be represented by basic parameters such as **SrcIP/DstIP**, **SrcPort/DstPort**, etc.
- **Operation attribute:** Describes the way actors trigger network traffic, which is made up of the type of action and interaction patterns. The main types of actions include releasing command, browsing the web, etc. Every type of action is likely to include a subtype. Interaction patterns include C/S, B/S, P2P, etc. The type of actions could be represented by **protocol** (TCP/UDP), traffic **model** and data **content**, while interaction patterns could be represented by basic parameters such as **SrcIP/DstIP**, **SrcPort/DstPort**, etc.
- **Content attribute:** Describes the data characteristics and data size of network traffic. Different purposes of experimental scenario cause different forms of network traffic data. For example, a penetrator using buffer overflows will often send an overly long string. Data size is the key parameter for distinguishing user behavior. For example, although the duration of a DDoS attack is often very short, the amount of data that can be triggered is stunningly large. Data characteristics can be represented by data **content**, while data size can be represented by **duration** and data **length**.

To implement the user layer traffic model, NTGEDA uses behavior description to describe the user in the terminal. The basic behavior description is illustrated as follows:

```
<actionid>
    <id> IdNum </id>
        <percent> 1 </percent>
        <operation> IdNum, StartTime, Duration, Src, Dst,
Protocol, Model, Data
        </operation>
</actionid>
```

Every <actionid></actionid> flag corresponds to a response behavior. Every response behavior is mainly made up of the following parts:

-

This property is required and is used to match the actionid of events which the current agent receives. If matched, the operation below will be executed.

-

This property is optional; the default is 1.

If there is a percent property, when the IdNum received by the agent is equal to the id, the agent calculates a random number. If the value is larger than percent, the operation below will be executed.

-

This property is required and should number at least 1. If there are more than 1, they will be executed successively. Operation is the key property in behavior description by which the agent performs response behavior (Table 2).

Table 2. Elements of operation property

Element	Required	Description
IdNum	No	Current id, used to trigger response behavior
StartTime	Yes	The time response behavior will be executed. The default is 0
Duration	Yes	The amount of time it will take to execute the response The default is 1 s
SRC	Yes	Includes SrcIP and SrcPort, i.e. source IP address and source port
DST	Yes	Includes DstIP and DstPort, i.e. destination IP address and destination port
Protocol	Yes	Protocol type, such as UDP or TCP
Model	Yes	Traffic model
Data	No	Content in packets
Length	Yes	The size of packets in the traffic model; can specify a range, such as random distributed between M and N bytes

The traffic model is divided into two forms:

- Independent type
  {PeriodInterval, PeriodDuration, PacketSize, PacketInterval}
- Nested type B
  {PeriodInterval, {Independent type or Nested type}}; if this type includes a nested type, then it will continue to extend.

The user can implement a certain distribution function by specifying the values of parameters. NTGEDA supports constant, random, exponential and poisson distribution functions.

5 Experimental Analysis

Using LSNEMUlab, we constructed a large-scale network, and deployed our traffic generator agent in it, as illustrated in Fig. 5.

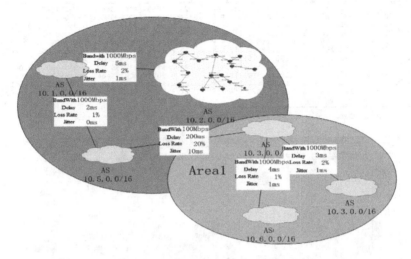

Fig. 5. Large-scale network environment for traffic generation

Three layers of traffic were generated, and the results were captured by Wireshark installed on a terminal node.

Figures 6 and 7 show that NTGEDA can generate network layer and application layer traffic on demand. Moreover, Fig. 8 shows the network traffic for a host that fell victim to DDoS. In this experimental scenario, the NTGEDA emulates three botnet hosts that send packets of 1Mbps one by one. As expected, the network traffic received by the victim host was increased.

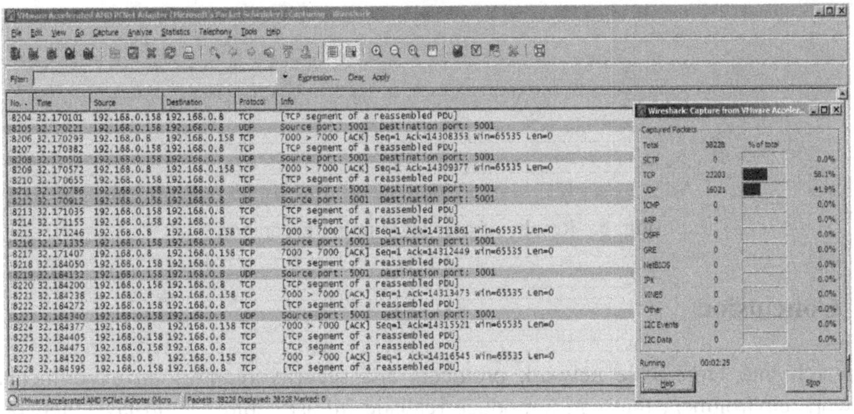

Fig. 6. Results of UDP traffic generation

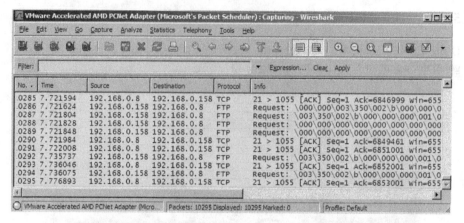

Fig. 7. Results of FTP traffic generation

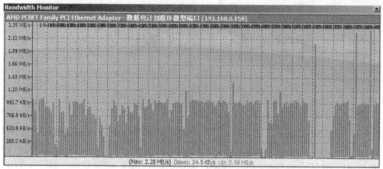

Fig. 8. Results of user layer traffic generation

6   Conclusion

Research into large-scale network protocols and security requires an experimental network environment that can reproduce realistic infrastructure. Multi-layer network traffic generation is therefore a requirement if the environment is to be used in an easy and effective way. In this paper, we have described a novel network traffic generation framework called NTGEDA for our network emulation environment LSNEMUlab. Our evaluation shows that NTGEDA can generate multi-layer network traffic on demand.

We will continue to improve the design of the network traffic generation framework by integrating the more representative hierarchical model into it.

References

1. Maier, S., Herrscher, D., Rothermel, K.: Experiences with node virtualization for scalable network emulation. J. Comput. Commun. **2007**(30), 943–956 (2007)
2. Paxson, V., Floyd, S.: Why we don't know how to simulate the internet. In: Proceedings of the 1997 Winter Simulation Conference, Atlanta, GA, January 1997
3. Simmonds, R., Unger, B.W.: Towards scalable network emulation. J. Comput. Commun. **26** (3), 264–277 (2003)
4. White, B., Lepreau, J., Stoller, L., et al.: An integrated experimental environment for distributed systems and networks. In: Proceedings of the Fifth Symposium on Operating Systems Design and Implementation, Boston, MA, USA, December 2002, pp. 255–270 (2002)
5. PlanetLab. http://www.planet-lab.org
6. Botta, A., Dainotti, A., Pescape, A.: A tool for the generation of realistic network workload foremerging networking scenarios. Comput. Netw. **56**(15), 3531–3547 (2012)
7. Bit-Twist. http://bittwist.sourceforge.net/
8. TCPDUMP/LIBPCAP public repository. http://www.tcpdump.org/
9. Sebastian Zander, G.A., Kennedy, D.: KUTE—a high performance Kernel-based UDP traffic engine. http://caia.swin.edu.au/reports/050118A/CAIA-TR-050118A.pdf
10. DIT-G:Distributed Internet Traffic Generator. http://traffic.comics.unina.it/software/ITG/documentation.php
11. Scapy. http://www.secdev.org/projects/scapy/
12. Karat Packet Builder. https://sites.google.com/site/catkaratpacketbuilder/
13. Ostinato. https://code.google.com/p/ostinato/
14. Sommers, J., Barford, P.: Self-configuring network traffic generation. In: The 4th ACM SIGCOMM Conference on Internet Measurement, pp. 68–81 (2004)
15. Vishwanath, K., Vahdat, A.: Swing: realistic and responsive network traffic generation. IEEE/ACM Trans. Netw. **17**(3), 712–725 (2009)
16. Srivastava, S., Anmulwar, S., Sapkal, A.M.: Comparative study of various traffic generator tools. In: Proceedings of RAECS UIET Panjab University Chandigarh, March, 2014
17. Mishra, S., Sonavane, S., Gupta, A.: Study of traffic generation tools. Int. J. Adv. Res. Comput. Commun. Eng. **4**(6), 4–7 (2015)
18. Reddy, V.R., Safwan, M., Deepamala, N., Shobha, G., Premkumar, S.J.: Network traffic simulator from real time captured packets. Int. J. Appl. Eng. Res. **12**(20), 10134–10137 (2017). ISSN 0973-4562
19. Xiaohui, K., Gang, Z., Hong, T.: A re-structure large scale network emulation environment. Comput. Sci. 4 (2009)
20. Gang, Z., Xiaohui, K., Weimin, Z.: An emulation environment for vulnerability analysis of large-scale distributed system. In: GCC 2009 (2009)
21. Kuang, X., Li, X., Zhao, J.: Architecture of network environment for high-risk security experimentation. In: AsiaARES 2013 (2013)
22. CAIDA. http://www.caida.org

Statistics and Learning

Optimization of Hypergraph Based News Recommendation by Binary Decision Tree

Wanrong Gu[1,2(✉)], Xianfen Xie[2,3(✉)], Yijun Mao[1], and Yichen He[1]

[1] South China Agricultural University,
Guangzhou 510642, People's Republic of China
guwanrong@scau.edu.cn

[2] Guangdong Key Laboratory of Big Data Analysis and Processing,
Guangzhou 510006, People's Republic of China
xiexianfen@163.com

[3] JiNan University, Guangzhou 510632, People's Republic of China

Abstract. News personalized recommendation has long been a favourite domain for recommender research. Traditional approaches strive to satisfy the users by constructing the users' preference profiles. Naturally, most of recent methods use users' reading history (content-based) or access pattern (collaborative filtering based) to recommend proper news articles to them. Besides, some researches encapsule the news content and access pattern in a recommender by vector space model. In this paper, we propose to use a hypergraph ranking for obtaining the preference rough, and then utilize the binary decision tree for eliminating the definition subjectivity of hypergraph. In this way, we can combine the content attributes on news content attributes, users and user's access pattern in a unified hypergraph and get more accuracy results, whereas we needn't to construct the user profile and select the possible important attributes empirically. Finally, we designed several experiments compared to the state-of-the-art methods on a real world dataset, and the results demonstrate that our approach significantly improves the accuracy, diversity, and coverage metrics in mass data.

1 Introduction

As the World Wide Web becomes the source and distribution channels of news, more and more people read news online or by mobile phone rather than buying newspaper or watching TV. However, massive news information online also brings information overload problem for the users. Personalized news recommender is valuable in many real world applications and has received increasing interest in recent years. Common personalized news recommendation researches can be roughly divided into two categories: (1) constructing user profiles by analyzing news article content from user's reading history (content-based filtering)

Supported by the open fund project of Guangdong Key Laboratory of Big Data Analysis and Processing (2017006).

Y. Peng et al. (Eds.): IScIDE 2018, LNCS 11266, pp. 83–101, 2018.
https://doi.org/10.1007/978-3-030-02698-1_8

[16]; and (2) analyzing users' behaviors and utilize the collaborative activities (collaborative filtering) [11]. Most of recent researches integrate the two kinds of methods above, called hybrid method [13,14]. Despite extensive advances, some critical issues in news personalized recommendation have not been well solved in previous researches, i.e., how to combine the news content and user access pattern completely, how to rank the candidate news articles reasonably and objectively.

For the tasks of news recommendation, the most common approach is to build the readers' preference profiles based on their reading history. In this way, user profiling is conducted by extracting representative words or phrases from the content of reading history or analyzing similar access patterns. However, news content may contain a gigantic amount of words or phrases. Moveover, many users tend to glance at news articles and are interested in some news named entities, e.g., the persons' names, where it happened, when the event happened, etc. Therefore, some important news content attributes are valuable to construct the users' preference profiles. In [14], named entities are used for user profile building as an entity vector. Such a representation might loss the news access data, for example, what entities is preferred by a given user.

In our work, to address the above issues, we propose a novel news recommendation algorithm by combing the user access pattern, news articles, news content attributes and mining the implicit relations among them. Motivated by [6], we utilize a unified hypergraph to model the objects in news personalization recommendation and the relationship among them firstly. A hypergraph is a set of a group of sub-sets, in which the sub-sets, called hyperedges, contain a group of vertexes [1]. After doing that, we further model the news personalized recommendation problem as a ranking problem on a unified hypergraph. And then, we select the final recommendation result through binary decision tree. The contribution of our paper is three-fold:

- **A hypergraph representation of news recommendation problem.** We explore the implicit relation-ships among users, articles, classes and news content attributes, and represent these objects and correlations as a unified hypergraph.
- **We propose to model high-order relations in news articles by hypergraph instead of traditional recommendation.** In this way, there is no information loss in representing various types of relations in news personalization recommendation.
- **We propose a binary decision tree based news selection algorithm.** In view of the fact that the definition of hyperedge and its weight is subjective. We propose to use binary decision tree algorithm to amend the subjective process and obtain the final recommendation results.

The remaining of this paper is organized as follows. Section 2 covers related work relevant to our study, including news personalized recommendation and hypergraph-based learning and ranking. In Sect. 3, we describes the data model and problem statement in this paper. Besides, we introduce the hypergraph based ranking algorithm and the recommendation optimization through binary decision tree in this section. Section 4 shows the experiments and results compared to the state-of-the-art approaches. Finally, we conclude this paper and discuss the future work in Sect. 5.

2 Related Work

In this paper, we combine content-based and collaborative filtering news recommendation methods to exploit correlations among readers and news content using hypergraph-based learning technology. In this section, we provide a brief review of related works about news personalized recommendations and hypergraph-based ranking.

2.1 Content-Based Methods

Content-based news recommenders construct user profile based on news content and recommend news articles similar to user profile [18]. In practice, news content is often represented as vector space model (VSM), and calculated the similar with user profile. For example, [4] utilized K-Nearest Neighbor method to recommend news to specific user. [17] employed the Naïve Bayesian method to classify web pages and construct user profile, and then recommend for user by the similar of web pages and user profiles. Liu et al. [16] (called ClickB in experimental Section) proposed a recommendation using news content based on click behavior. However, it should be noted that not all items are easy to express as VSM, such as audio, image and video news objects [19]. Another problem is that it might be insufficient to construct user profiles by a word bag for capturing the preference of users [14].

2.2 Collaborative Filtering

Collaborative filtering based news recommendations utilize the behaviors of user and analyze the collaborative among users and news articles. By this way, this method is content-free and can be roughly divided into two categories: Heuristic-based and Model-based. Heuristic-based method is inspired by the real-world phenomena [10]. Model-based methods train a model for predicting the utility of the current user u on item j, such as [5] and [11] (called Goo in experimental Section). Purchase recommendation and rating prediction are the most important applications in collaborative filtering recommendation research. In news recommender, the rating can be seen as binary, where a click on a piece of news can be rated as 1, and 0 otherwise [11]. The success of the collaborative filtering based recommendation system relies on the availability of lots of users and items. However, we can observe that the user-item matrix usually is a spare matrix that will lead to poor recommendation [2]. Another problem is that, with the rapid change of the newly-published news, it suffers from the well-known *cold-start* problem.

2.3 Hybrid Approaches

This method combines collaborative filtering, content-based methods and other factors [9]. Many news recommendation methods are hybrid, such as Bilinear [8], Bandit [15], SCENE [14] and TwoHy [13], which will be discussed and analysed in our experimental section. From the perspective of news recommendation, our work is similar to EMM News Explorer [3] and Newsjunike [12] in the use of news content and named entities for news recommendation. However, EMM News system did not provide personalized recommender and Newsjunike did not address the issues as we do in mining the correlations among users and news articles.

Although the above studies have achieved great success in news recommendation applications, the fail to make full use of the high-order relations in the news content and readers. In this work, we propose to use unified hypergraph, rather than the traditional model, to precisely capture the high-order correlations and hence enhance the performance of news recommendation.

2.4 Hypergraph-Based Learning and Ranking

Our study is also related to hpyergraph-based learning and ranking [6,20,21]. Bu et al. propose a unified hypergraph ranking method for music recommendation with combining social media information and music content [6]. [7] introduce a clustering algorithm in hypergraph to extract maximally coherent groups using high-order similarities. In [21], Zhou et al. study a general framework which is applicable to classification, clustering and embedding on hypergraph. The above studies only focus on the basic applications on hypergraphs. However, by modeling the multiple types of news objects and their relations as a hypergraph, we consider news recommendation as a hypergraph-based ranking problem.

3 Data Model and Recommendation Framework

In this section, we introduce our data model of news reading community, some basic notations and our exploration problem.

3.1 Preliminaries

We follow the definitions on [21] to describe this data model. Let V denote a finite set of objects, and E be a family of subsets e of V such that $\bigcup_{e \in E} = V$, in which e contains a subset of objects in V. We call $\mathbf{G} = (V, E)$ a *hypergraph* with the *vertex* set V and the *hyperedge* set E. The degree of a hyperedge e is defined by $\delta(e) = |e|$, i.e., the number of vertices in subset e. The degree of a vertex v is defined by $d(v) = \sum_{v \in e} w(e)$, where $w(e)$ denotes the weight of the hyperedge e. We say that there is a *hyperpath* between vertices v_1 and v_k when there is an alternative sequence of distinct vertices and hyperedges $v_1, e_1, v_2, e_2, \ldots, e_{k-1}, v_k$ such that $\{v_i, v_{i+1}\} \subseteq e_i$ for $1 \leq i \leq k - 1$. A hypergraph is *connected* if there is a path for every pair of vertices. Similar to the define of simple graph, a vertex-hyperedge *incidence matrix* $\mathbf{H} \in \Re^{|V| \times |E|}$ in which each entry $h(v, e)$ is 1 if $v \in e$ and 0 otherwise. Then we have:

$$\delta(e) = \sum_{v \in V} h(v, e), \tag{1}$$

$$d(v) = \sum_{e \in E} w(e) h(v, e). \tag{2}$$

Let $\mathbf{D}_e$ and $\mathbf{D}_v$ be two diagonal matrices containing the vertex and hyperedge degrees respectively. Let $\mathbf{W}$ be a $|E| \times |E|$ diagonal matrix containing hyperedge weights.

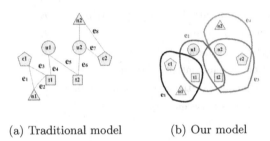

(a) Traditional model   (b) Our model

Fig. 1. An illustrative example of data model using traditional approach and unified hypergraph representation. Remark: u represents user, t denotes named entity, c denotes news class, n denotes news article.

In Fig. 1, a simple illustrative example of our proposed model is shown, in which traditional method can be only good at modeling the relationship between two objects, whereas our proposed approach could encapsule the news content and user access pattern with combining all meaningful information.

3.2 Data Model

In news recommendation application, multiple types of objects are involved, i.e., users, news articles, news content attributes, etc. In this paper, we use 4 types of objects for our proposed news personalized recommendation: user, news article, news class, news content attributes. Firstly, we collected the news articles read by at least 5 readers. Then we reduced the set of users by restricting each user has at least read 5 news articles. The objects in our data set is shown in Table 1.

Table 2 shows the count of our defined hyperedges. We formalize a hypergraph $\mathbf{G}$ that contains 8 different relations with different objects. In general, hyperedge can be generalized from a news article and all of its readers. However, the incidence matrix would become very density. Each hyperedge is explained as follows:

Table 1. Objects in our data set.

Objects	Notations
Users	$\mathcal{U}$
Classes	$\mathcal{C}$
News articles	$\mathcal{N}$
News content attributes	$\mathcal{A}$

Table 2. Relations in the data set.

Relations	Notations
User-article-content attribute	$E^{(1)}$
User-article-article	$E^{(2)}$
User-content attribute-content attribute	$E^{(3)}$
User-user-article	$E^{(4)}$
User-user-content attribute	$E^{(5)}$
Article-article-content attribute	$E^{(6)}$
Article-content attribute-content attribute	$E^{(7)}$
Similarities between classes	$E^{(8)}$

- $E^{(1)}$: A user reads a news article which contains a news content attribute. Typically, we assign the weight of this hyperedge to be 1 which we assume this user read the corresponding news article once.
- $E^{(2)}$: A user reads two news articles. Similar to $E^{(1)}$, we assign the weight to be 1 constant.
- $E^{(3)}$: A user connects two news content attributes. We set this weight to be 1.
- $E^{(4)}$: Two users read the same news article. We set this weight to be 1.
- $E^{(5)}$: Two users share the same news content attribute. This hyperedge will be divided into two categories:
  (1) User-User-News Class: Each user's reading history contains many classifications. Therefore, we set the weight as follows:

$$\omega(e^{(5)}_{u_i,u_j,c}) = |\{(u_i, u_j, c)|u_i \in \mathcal{U}, u_j \in \mathcal{U}, c \in \mathcal{C}\}|. \tag{3}$$

Normalization of this weight is as follows:

$$\omega(e^{(5)}_{u_i,u_j,c}) = \frac{\omega^*(e^{u_i,u_j,c})}{\sqrt{\sum_{l=1}^{|\mathcal{U}|} \omega^*(e^{u_l,u_j,c}) \sum_{k=1}^{|\mathcal{U}|} \omega^*(e^{u_i,u_k,c})}} \tag{4}$$

  (2) User-User-Other Content Attribute: We set this weight to be 1.
- $E^{(6)}$: Two news articles share the same news content attribute. We set this weight to be 1.

- $E^{(7)}$: A news article contains two different news content attributes. We set this weight to be 1.
- $E^{(8)}$: Each pair of classes might have different similarity. We define the weight as $\omega(e_{i,j}^{(8)})$ as follows:

$$\omega(e_{i,j}^{(8)}) = \alpha \frac{\omega(e_{i,j}^{(8)})}{\max(\omega(e^{(8)}))} \tag{5}$$

where $\max(\omega(e^{(8)}))$ denotes the maximum of the similarity between all news classes pairs and α denotes the weight of this factor.

Finally, we get the vertex-hyperedge incidence matrix $\mathbf{H}$, the degree matrix $\mathbf{D}_e$, $\mathbf{D}_v$ and the weight matrix $\mathbf{W}$.

3.3   Ranking on Unified Hypergraph

In this section, we discuss how to perform news ranking on unified hypergraph similar to [6]. We can define the cost function of $\mathbf{f}$ as follows:

$$Q(\mathbf{f}) = \frac{1}{2} \sum_{i,j=1}^{|V|} \sum_{e \in E} \frac{1}{\delta(e)} \sum_{\{v_i, v_j\} \subseteq e} \omega(e) \left\| \frac{f_i}{\sqrt{d(v_i)}} - \frac{f_j}{\sqrt{d(v_j)}} \right\|^2$$
$$+ \mu \sum_{i=1}^{|V|} \|f_i - y_i\|^2, \tag{6}$$

where $\mu > 0$ is the regularization factor. In order to obtain the optimal ranking result, we could minimize $Q(\mathbf{f})$ function:

$$\mathbf{f}^* = \arg\min_f Q(\mathbf{f}). \tag{7}$$

The right side of Eq. (6) is the smoothness constraint, which means that vertices contained by many common hyperedges should have similar ranking scores. For example, if two news articles or two news named entities have been associated by many common objects, the both articles or named entities will probably have similar ranking scores. Another example is the ranking of the users. If two readers join in many common interest objects, they may probably have similar ranking scores. The parameter μ controls the weight of the two terms in this cost function. Considering the hyperedges will have different sizes, that is, each hyperedge will have different number of vertices, we γ need to equally treat the hyperedges with different sizes by normalizing with its degree $\delta(e)$. Finally, we can get the equation as follows:

$$\mathbf{f}^* = (\mathbf{I} - \lambda\mathbf{A})^{-1}\mathbf{y} \tag{8}$$

Here, $\mathbf{A} = \mathbf{D}_v^{-\frac{1}{2}} \mathbf{H}\mathbf{W}\mathbf{D}_e^{-1}\mathbf{H}^T \mathbf{D}_v^{-\frac{1}{2}}$. Note that, the matrix $\mathbf{I} - \lambda\mathbf{A}$ is highly sparse and the computation of the inversion on this matrix is very efficient.

Our news recommendation algorithm has three phases: offline training, hypergraph based ranking and final selection. In the first phase, we construct the unified hypergraph as described and get the matrix $\mathbf{H}$, $\mathbf{D}_e$, $\mathbf{D}_v$ and $\mathbf{W}$, where the matrices $\mathbf{D}_e$ and $\mathbf{D}_v$ can be computed based on $\mathbf{H}$ and $\mathbf{W}$. In the second phase, we need to construct the query vector $\mathbf{y}$ firstly. After doing that, the ranking results can be calculated. At last, we use binary decision tree based news articles selection algorithm to generate the recommended result.

3.4 Selection Optimization

Ranking based on hypergraph model can get the news results. But the definition of hyperedges and their weights are subjective, which we can see in Experiments Section. Therefore, we propose a selection optimization to amend their subjective and select the final recommendation results for users.

The key stages are as follows:

(1) **Select several closer users and farther users.**
    After rankding calculation, we can get the other users' rank values. Therefore, we select k closer users $(u_i | i \in Z \bigcap i \in [i, k])$, and k farther users $(u_i' | i \in Z \bigcap i \in [i, k])$.
(2) **Sample and feature construction.**
    The decision tree is generally applicable to the less attribute values. In fact, there are many values in the attributes in news recommendation, such as reporters, key persons, etc. In this way, there may be many branches in some nodes. Therefore, we propose to utilize binary decision tree to solve this issue. Attribute values are mapped to the 0–1 binary space.

Take our corpus as an example. The attributes we used to construct the binary decision tree: reporter, area, genre, classification and key person. Example of state association table is shown in Table 3.

This table covers the training and the prediction set for decision tree learning, of which, the training set consists k nearest neighbor users and k farther users, and the prediction set contains the candidate news articles and needs to be accepted or not. The table contains other characteristics as follows:

I. The Accept values of k nearest neighbor users are 1.
II. The Accept values of k farther users are 0.
III. The Accept values of candidate news articles are unknown.
IV. In the training set, matrix elements are calculated by the coincidence of current user and the row user. The equation is as follows:

$$
\mathbf{D}_{tr|v,a} = \begin{cases} 0 & \dfrac{|a_v| \bigcap |a_u|}{|a_v| \bigcup |a_u|} < Threshold \\ 1 & \dfrac{|a_v| \bigcap |a_u|}{|a_v| \bigcup |a_u|} \geq Threshold \end{cases} \tag{9}
$$

Where v denotes the one of the given user's k nearest neighbor users or k farther users; a denotes the attribute (reporter, area, genre, key person or

Table 3. Example of state association table. Remark: the rows are k closer users, k farther users and candidate news articles. The columns are attributes in the news content. The value will be set to 1 if the row element has the column attribute, and 0 otherwise. Other symbols: R - reporter, A - area, G - genre, C - classification and K - key person, n_k denotes a candidate news article.

	R	A	G	C	K	Accept
u_1	1	1	0	1	0	1
u_2	0	0	0	1	1	1
...	...	...	...	...	...	...
u_k	1	1	0	1	0	1
u_1'	0	1	0	1	0	0
u_2'	0	0	1	0	1	0
...	...	...	...	...	...	...
u_k'	0	0	1	0	0	0
n_1	0	1	0	0	0	?
n_2	1	0	0	1	1	?
...	...	...	...	...	...	...
n_m	0	1	1	0	0	?

classification.); u denotes the given user. We set $Threshold = 0.5$ empirically.

V. The matrix element of candidate news articles is decided by the following equation:

$$\mathbf{D}_{tr|v,a} = \begin{cases} 1 & a \text{ appears in news article } n \text{ and} \\ & \text{the reading history of user } u. \\ 0 & \text{Otherwise.} \end{cases} \tag{10}$$

(3) **Binary decision tree building.**

After the above two stages, the third stage can be seen as a binary classification problem. In other words, we need to set the value of the candidate news articles as 1 or 0 to decide to accept or not. Each node of binary decision tree model has two out degrees. Each internal node (including root node) is an attribute, and the leaf nodes are classification tags. The branches are the value of the out-degree nodes, 0 or 1.

The key step of binary decision tree building is construction of tree nodes and their branches. The lower the level of the node means the more important of the corresponding attribute. In this paper, we use information gain ratio to measure the importance of the attribute node. The lower the level of the node means the larger of information gain ratio. It is easy to know that the root node is most important and its level is lowest. The information gain ratio is the ratio of the information gain of attribute and the empirical entropy of training data:

$$g_R(D, a) = \frac{g(D, a)}{H(D)} \tag{11}$$

Where $H(D)$ denotes the empirical entropy of training data, D is training set, $g(D, a)$ is the information gain of attribute a in training data set D. The empirical entropy of data set D can be calculated as follows:

$$H(D) = -\sum_{k=1}^{2} \frac{|c_k|}{|D|} log_2 \frac{|c_k|}{|D|} \tag{12}$$

Where c_k denotes the classification. After obtain the empirical entropy of training set, we need to calculate the empirical condition entropy $H(D|a)$ for the given attribute a:

$$H(D|a) = \sum_{i=1}^{n} \frac{|D_i|}{|D|} H(D_i) = -\sum_{i=1}^{2} \frac{|D_i|}{|D|} \sum_{k=1}^{2} \frac{|D_{ik}|}{|D_i|} log_2 \frac{|D_{ik}|}{|D_i|} \tag{13}$$

The information gain $g(D, a)$ is:

$$g(D, a) = H(D) - H(D|a) \tag{14}$$

It should be noted that the information gain ratio is a relative concept, namely when the collection object changes, the value of information gain needs to calculate again. In fact, the collection has changed when the construction of decision tree node is generating. Given the training set D (i.e., the $2k$ users in state association table) and its attribute set A (in this paper, $A = reporter, area, genre, key person, classification.$), the construction process of binary decision tree Tr of user u is shown as follows:

(1) If the records in training set all belong to a category (1 or 0), the tree Tr should be a single node binary tree. Set the root node as the category value, and return Tr.
(2) If the attribute set A is null set, set the root node as the value of majority of the category, and return Tr.
(3) Otherwise, calculate the rest information gain of the attributes $g_R(D, a)$, and select attribute a corresponding to the maximal information gain.
(4) Cut the attribute set into two non empty subsets D_0 and D_1 according to the possible value 1 and 0. Construct the sub nodes according to the majority of subset D_0 and D_1. Return the Tr with the sub nodes and their incidental sub sets.
(5) Use the non empty subsets D_0 and D_1 as the training sets and $A\backslash\{a\}$ as the new attribute set, and recursive call the step (1)–(4).

This algorithm is similar to C4.5. But in this paper, we don't need to set the threshold of the attributes. The time complexity of this algorithm is $O(k|n|log(|n|))$, here $n = |A|$. k denotes the number of category, $k = 2$. Therefore, it is easy to know that the majority complexity is in hypergraph based ranking step.

4 Experiments

4.1 Data Collection

To evaluate our algorithm, we gather the news data from South China Net[1], where the data collection ranges from Jan 1th, 2012 to May 24th, 2012. We preprocess the data by removing the cold news articles (i.e., the news article read by less than 5 users) and the non-active users (i.e., the users who read less than 5 news articles) for verify our recommendation performance. After preprocessing, $2,021$ users are stored with $31,395$ news articles. The data set includes the following information: News ID, Users, Publish Date, News title, Classification ID, News Content, Genre, Key Persons, Reporters, Area. The information in the data set is manual annotation.

4.2 Evaluation

To evaluate the performance of our algorithm and other baseline approaches, for each news reader, we select 20% news articles as test data for evaluation purpose, and the others as training data. We use F1-score, Diversity, Normalized Discount Cumulative Gain (NDCG) and Coverage as our evaluation metrics. Precision is the number of correctly recommended news articles divided by the total number of articles in recommended list. Recall is the number of correctly recommended news articles divided by the total number of articles which should be recommended. F1-score is the harmonic mean of Precision and Recall. Let $R(u)$ be a news recommended list of a user u, and the diversity of u can be computed as follows:

$$Diversity = 1 - \frac{\sum_{i,j\in R(u),i\neq j} sim(i,j)}{\frac{1}{2}|R(u)|(|R(u)|-1)}, \tag{15}$$

where i and j are two different news articles in recommendation list for user u, and $sim(i,j)$ denotes the news profile similarity between the news item i and j. We use NDCG to measure the ranking quality of the recommended list based on a user's actual accessing sequence. NDCG at position n is defined as:

$$NDCG@n = \frac{1}{IDCG} \times \sum_{i=1}^{n} \frac{2^{r_i}-1}{\log_2(i+1)}, \tag{16}$$

where r_i is the relevance rating of item at rank i. In news recommendation case, r_i is set to be 1 if the user has read this recommended news article and 0 otherwise. IDCG is chosen to ensure that the perfect ranking has a NDCG value of 1.

In recommender research, the users and items may appear the long-tail phenomenon. In order to insure that more items could be commended by the recommender, the coverage evaluation metric is proposed:

$$Coverage = \frac{|\bigcup_{u\in U} R(u)|}{|I|}, \tag{17}$$

---

[1] http://www.southcn.com.

where $|I|$ denotes the size of item set. This equation denotes that the ratio of all recommended items for all users on all items. It is noted that all items are recommended when $Coverage = 1$.

4.3 Parameter Tuning

In our method, there are two parameters for tuning, i.e., the weight α of each class in Eq. (5), the regularization factor λ in Eq. (8). We use F1-score as the evaluation metric for tuning this two parameters, and Fig. 2 shows the results.

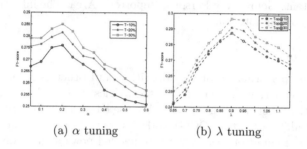

(a) α tuning   (b) λ tuning

Fig. 2. Parameter tuning.

From the results, we set the parameters α as 0.2 and λ as 0.9 for our experiments.

In the process of the optimization using binary decision tree, we need to set the parameters: (1) k parameter, i.e., k nearest neighbor users and k farther users. (2) m parameter, the number of candidate news articles. m is not less than the length of the recommendation list, i.e., in Top@10 recommendation list, it needs to insure that $m > 10$. Limited by the space of this paper, we set the $(m, k) = (30, 30), (m, k) = (40, 30)$ and $(m, k) = (60, 30)$, while the F1 score could reach the best results $0.3152, 0.3197$ and 0.3211 in Top@10, Top@20 and Top@30 respectively.

4.4 Construction of Hypergraph

In our proposed hypergarph method, we integrate 8 hyperedges into the construction of our model. Different hyperedges definition will lead to different recommendation results. Besides, the query vector **y** would affect the recommendation result with different initialized assignments.

Comparison of Different Hyperedge Combinations. The hypergraph model we proposed provides an elegant representation of the news data and users, encapsulating multiple relations among objects. To evaluate the effect of such model, we consider different combinations of hyperedges:

(1) Con-1: Use $E^{(1)}$ and $E^{(8)}$. Only consider news content, which is essentially a content-based approach.
(2) Con-2: Use $E^{(1)}-E^{(3)}$ and $E^{(8)}$: It is also essentially a content-based method. But in this condition, it highlights the area, key person, reporter, genre and other factors.
(3) Con-3: Use $E^{(4)}-E^{(5)}$: Each hyperedge includes two different users, which can be regarded as collaborative filtering.
(4) Con-4: Use $E^{(4)}-E^{(5)}$ and $E^{(8)}$: It is also a collaborative filtering. But the $E^{(8)}$ highlights the classification factor and has content-based weight.
(5) Con-5: Use all hyperedges. It is a hybrid recommender, which not only considers the content attributes, but also considers the collaborative in recommender.

The results of hypergraph ranking (HRB) and binary decision tree optimization (HRBopt) are shown in Fig. 3, in which Fig. 3(a) compares the F1-score using HRB, whereas Fig. 3(b) demonstrates the result of HRBopt.

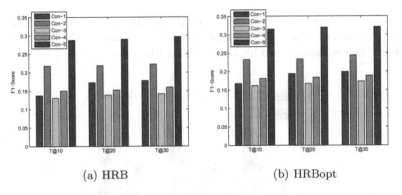

(a) HRB   (b) HRBopt

Fig. 3. Comparison of different hypergraph constructions with considering different hyperedges.

It can be seen from the experimental results: **I.** The results considered ALL hyperedges outperforms other constructions on both HRB and HRBopt. The straightforward reason may be that using all hypergraphs could coverage the correlations better needed in personalized news recommendation mining. **II.** HRBopt outperforms HRB in all conditions, especially in Con-1–4. This shows that the HRBopt algorithm can optimize the defect of the subjective definition on hypergraph model.

Comparison of Different Assignments of Query Vector y. The performance of our method will be also affected by the query vector **y**. There are three strategies to set the query vector **y** for our news recommendation:

I. Set the entry of **y** corresponding to the user u to be 1 and the others to be 0. For example, in information retrieval, the user supplies the query *computer*. But the system only considers the *computer* index and ignores the *PC*, *calculation machine* and other synonyms.

II. Set the entries of **y** corresponding to the user u with all the objects in hypergraph connected to this user by hyperedge to be 1, otherwise, to be 0. For example, the user wants to search information about computer and submit *computer* query, while the system set all synonyms of *computer* as the same initial weights.

III. Similar to **I**, set the entry of **y** corresponding to the user u to be 1. Besides, we define a parameter c_{uv} to measure the weight of u and v. Here, u is the given user, and v denotes the other point in hypergraph model. The value of c_{uv} can be set to:

$$c_{uv} = \begin{cases} 0 & when\ v \in \mathcal{U},\ and\ u \neq v \\ \frac{|a_i|}{|\mathcal{A}_i|} & when\ v \notin \mathcal{U},\ and\ v \notin \mathcal{N} \\ 1 & when\ v \in \mathcal{N} \end{cases} \tag{18}$$

If u connected to an object v, set the corresponding entry of v in **y** to be $a_{u,v}$. The $a_{u,v}$ denotes a measure of the relatedness between user u and the connected object v. The comparison results are shown in Fig. 4.

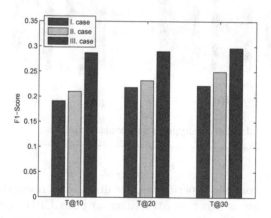

Fig. 4. Comparison result on different query vector.

From the comparison, we can observe that using the assignment of query vector **y** as **III**) case outperforms the other cases (**I** and **II**). The reason behind this is that: (1) In **I** case, it ignores the related objects in the hypergraph which may influence the user's preference. (2) In **II** case, it is essentially a hybrid approach integrates content-based and collaborative filtering based method. It is also not a good choice, because it fails to reflect the user's preference with different degrees.

4.5   Comparison with Other Methods

We also implement several state-of-the-art news recommendation methods: Goo [11], ClickB [16] and SCENE [14] for comparison. The details of these methods are described as follows:

- Goo: It is a collaborative filtering method, in which MinHash clustering, PLSI topic model and covisitation counts are taken into account for a recommendation.
- ClickB: The method is a content-based recommendation, in which the profiles of users are constructed based on the history click behavior. This approach utilizes a Bayesian framework for predicting users' news interests.
- SCENE: This method proposes a two stages recommendation. For the first stage, the recommender selects news clusters which the user may be interested, and then uses a greedy algorithm to model the news selection problem.
- HRB: Hypergraph Ranking Based recommendation, see Sect. 3.3 in this paper.
- HRBopt: Hypergraph Ranking Based recommendation with our proposed Optimization algorithm, see Sects. 3.3 and 3.4.

In order to ensure the fair comparison, the parameters of these baseline approaches are optimally tuned in our experiments.

Stability and Accuracy Evaluation. For the stability experiment, we select 50 users to provide news recommendation results for them in three recommender methods (ClickB, SCENE and ours). Figure 5 shows the comparison results as Top @10, Top @20 and Top @30 result lists for each user.

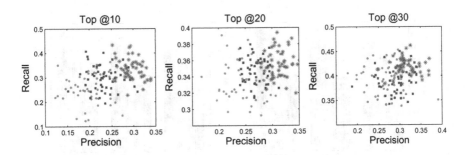

Fig. 5. Precision-Recall plot of different recommenders. Remark: ◯ (green) denotes the ClickB; □ (blue) denotes the HRB; and ∗ (red) represents HRBopt. (Color figure online)

In this sample evaluation, we can observe that besides the higher accuracy, the distribution of our approach is more stability than other methods. In accuracy experiment, we shows all users and cold users recommendation result in

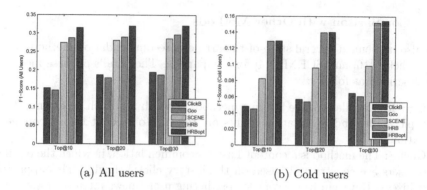

Fig. 6. Comparison on F1-score of different approaches with all users and cold users.

Fig. 6. For the cold user set building, we select the users who read less than 30 news articles, totally selected 562 users for evaluation.

Figure 6 shows the results of our proposed method compared with the other baseline approaches, in which Fig. 6(a) uses all users, whereas Fig. 6(b) demonstrates the recommendation result of cold users. From the experimental results we can observe that: (1) our proposed algorithm outperforms the other approaches. (2) our method is more stability than the other methods. The straightforward reason is that our model captures the important information of user's reading preferences and the access pattern among users.

Ranking Ability Evaluation. A good recommendation list has both good accuracy and ranking ability. The ranking ability evaluation result is shown in Fig. 7 as NDCG.

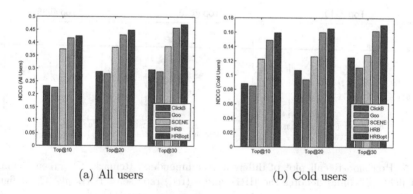

Fig. 7. Comparison on NDCG of different approaches with all users and cold users.

From the experimental result, our methods outperform the other baselines. The straightforward reason may be that our methods is ranked by the hypergraph ranking algorithm.

Diversity Evaluation. The recommendation news list of our method performs a great diversity. For this metric evaluation, we choose Goo [11] (a collaborative filtering based method), ClickB [16] (a content-based method) and SCENE [14] (a hybrid method using LSH for clustering and greedy algorithm for news selection) as the comparison baselines. Table 4 shows the result of the comparison with $|R(u)| = 10, |R(u)| = 20$ and $|R(u)| = 30$ which we define $T@10$ to represent $|R(u)| = 10$.

Table 4. Diversity evaluation on different recommendation lists. Remark: T@n-Recommended result with Top@n.

Methods	T@10	T@20	T@30
Goo	0.1914	0.1419	0.0904
ClickB	0.1541	0.1210	0.0776
SCENE	0.4510	0.3218	0.2879
HRB	0.4051	0.3074	0.1941
HRBopt	0.4151	0.3121	0.2014

From the comparison result, we can observe that our method outperforms the Goo and ClickB methods significantly. The straightforward reason may be that we diverse the news articles using the common hyperedges, not only relies on the center of the users' profiles. With the result list enlarge, the diversity decreases not very significantly on our method, whereas the other baseline approaches decrease very much because they rely on the user's profile too much. Unfortunately, the SCENE outperforms our methods in diversity because they use a greedy selection process to insure the diversity metric.

Coverage Evaluation. The coverage evaluation is shown in Fig. 8.

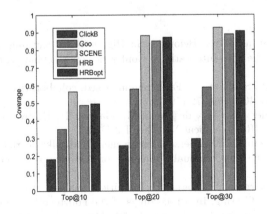

Fig. 8. Coverage comparison of different methods.

From the results we can observe that: (1) The coverage on Top@10 is not very good. The reason is that the length of recommendation list is too short and the coverage has a low upper limit. (2) The coverage on Top@30 is higher than Top@20, but the promotion is not very large. Therefore, recommend more news articles could not promote the coverage significantly. Generally, SCENE method obtains the good diversity and coverage, but needs to implement the extra selection process. Our methods outperform the other baselines in this metric.

5   Conclusion and Future Work

In this paper, we address the news personalized recommendation problem in hypergraph model, and focus on the combining various types of news content information and user-news relationship. We model this recommendation problem as a ranking problem on a unified hypergraph, which defines the multiple objects in the news recommendation as vertices, and the relations among these objects as hyperedges. In this way, the high-order relations in this hypergraph can be naturally captured. Besides, content-based and collaborative filtering based recommendation are combined in this framework. In order to eliminate the subjective definition of hypergraph, we propose a optimization algorithm through binary decision tree for the final news articles selection. The experimental results on a real world data set collected have demonstrated that our proposed approach significantly outperforms the traditional news recommenders on accuracy, stability and NDCG metric. Besides, our algorithms also good at diversity and coverage metrics.

For future work, to process mass newly-published news articles, we plan to deploy some components onto the Map-Reduce framework on our distributed system. Moreover, we also plan to integrate this hypergraph model into our news search engine due to the effectiveness in this work. Another remarkable point is the interest effectiveness of news named entities on news recommendation, which is able to insights on the exploration of news reading behaviors.

References

1. Agarwal, S., Branson, K., Belongie, S.: Higher order learning with graphs. In: Proceedings of the 23rd International Conference on Machine Learning, pp. 17–24. ACM (2006)
2. Balabanović, M., Shoham, Y.: Fab: content-based, collaborative recommendation. Commun. ACM **40**(3), 66–72 (1997)
3. Best, C., van der Goot, E., de Paola, M., Garcia, T., Horby, D.: Europe media monitor-EMM. JRC Technical Note No. I 2 (2002)
4. Billsus, D., Pazzani, M.J.: A personal news agent that talks, learns and explains. In: Proceedings of the Third Annual Conference on Autonomous Agents, pp. 268–275. ACM (1999)
5. Breese, J., Heckerman, D., Kadie, C.: Empirical analysis of predictive algorithms for collaborative filtering. In: Proceedings of the Fourteenth conference on Uncertainty in Artificial Intelligence, pp. 43–52. Morgan Kaufmann Publishers Inc. (1998)

6. Bu, J., et al.: Music recommendation by unified hypergraph: combining social media information and music content. In: Proceedings of the International Conference on Multimedia, pp. 391–400. ACM (2010)

7. Bulo, S.R., Pelillo, M.: A game-theoretic approach to hypergraph clustering. In: NIPS, vol. 22, pp. 1571–1579 (2009)

8. Chu, W., Park, S.T.: Personalized recommendation on dynamic content using predictive bilinear models. In: Proceedings of the 18th International Conference on World Wide Web, pp. 691–700. ACM (2009)

9. Claypool, M., Gokhale, A., Miranda, T., Murnikov, P., Netes, D., Sartin, M.: Combining content-based and collaborative filters in an online newspaper. In: Proceedings of ACM SIGIR Workshop on Recommender Systems, vol. 60. Citeseer (1999)

10. Cota, R., Ferreira, A., Nascimento, C., Gonçalves, M., Laender, A.: An unsupervised heuristic-based hierarchical method for name disambiguation in bibliographic citations. J. Am. Soc. Inf. Sci. Technol. **61**(9), 1853–1870 (2010)

11. Das, A.S., Datar, M., Garg, A., Rajaram, S.: Google news personalization: scalable online collaborative filtering. In: Proceedings of the 16th International Conference on World Wide Web, pp. 271–280. ACM (2007)

12. Gabrilovich, E., Dumais, S., Horvitz, E.: Newsjunkie: providing personalized newsfeeds via analysis of information novelty. In: Proceedings of the 13th International Conference on World Wide Web, pp. 482–490. ACM (2004)

13. Li, L., Li, T.: News recommendation via hypergraph learning: encapsulation of user behavior and news content. In: Proceedings of the Sixth ACM International Conference on Web Search and Data Mining, pp. 305–314. ACM (2013)

14. Li, L., Wang, D., Li, T., Knox, D., Padmanabhan, B.: SCENE: a scalable two-stage personalized news recommendation system. In: ACM Conference on Information Retrieval (SIGIR) (2011)

15. Li, L., Chu, W., Langford, J., Schapire, R.E.: A contextual-bandit approach to personalized news article recommendation. In: Proceedings of the 19th International Conference on World Wide Web, pp. 661–670. ACM (2010)

16. Liu, J., Dolan, P., Pedersen, E.R.: Personalized news recommendation based on click behavior. In: Proceedings of the 15th International Conference on Intelligent User Interfaces, pp. 31–40. ACM (2010)

17. Pazzani, M., Billsus, D.: Learning and revising user profiles: the identification of interesting web sites. Mach. Learn. **27**(3), 313–331 (1997)

18. Schafer, J., Konstan, J., Riedi, J.: Recommender systems in e-commerce. In: Proceedings of the 1st ACM Conference on Electronic Commerce, pp. 158–166. ACM (1999)

19. Shardanand, U., Maes, P.: Social information filtering: algorithms for automating "word of mouth". In: Proceedings of the SIGCHI Conference on Human factors in Computing Systems, pp. 210–217. ACM Press/Addison-Wesley Publishing Co. (1995)

20. Sun, L., Ji, S., Ye, J.: Hypergraph spectral learning for multi-label classification. In: Proceedings of the 14th ACM SIGKDD International Conference on Knowledge Discovery and Data Mining, pp. 668–676. ACM (2008)

21. Zhou, D., Huang, J., Schölkopf, B.: Learning with hypergraphs: clustering, classification, and embedding. Adv. Neural Inf. Process. Syst. **19**, 1601 (2007)

A Continuous Method for Graph Matching Based on Continuation

Xu Yang[1,2(✉)], Zhi-Yong Liu[1,2,3], and Hong Qiao[1,2,3]

[1] State Key Laboratory of Management and Control for Complex Systems,
Institute of Automation, Chinese Academy of Sciences,
Beijing 100190, People's Republic of China
{xu.yang,zhiyong.liu,hong.qiao}@ia.ac.cn
[2] DongGuan University of Technology, Dongguan, China
[3] Center for Excellence in Brain Science and Intelligence Technology,
Chinese Academy of Sciences, Shanghai 200031, People's Republic of China

Abstract. Graph matching has long been a fundamental problem in artificial intelligence and computer science. Because of the NP-complete nature, approximate methods are necessary for graph matching. As a type of approximate method, the continuous method is widely used in computer-vision-related graph matching tasks, which typically first relaxes the original discrete optimization problem to a continuous one and then projects the continuation solution back to the discrete domain. The continuation scheme usually provides a superior performance in finding a good continuous local solution within reasonable time, but it is limited to only the continuous optimization problem and therefore cannot be directly applied to graph matching. In this paper we propose a continuation scheme based algorithm directly targeting at the graph matching problem. Specifically, we first construct an unconstrained continuous optimization problem of which the objective function incorporates both the original objective function and the discrete constraints, and then the Gaussian smooth based continuation is applied to this problem. Experiments witness the effectiveness of the proposed method.

Keywords: Graph matching · Continuous method
Continuation method · Energy minimization
Combinatorial optimization

1 Introduction

Graph matching has long been a fundamental problem in computer science and artificial intelligence, especially computer vision and pattern analysis related tasks [1,2]. Graph matching aims at finding the assignments between vertices of two or more graphs, which term is generally associated with the second or high order constraints. Particularly, when using only the first order constraints, graph matching degenerates to a special case usually named by bipartite graph matching which only considers the vertex similarity [3]. Many attentions has

© Springer Nature Switzerland AG 2018
Y. Peng et al. (Eds.): IScIDE 2018, LNCS 11266, pp. 102–110, 2018.
https://doi.org/10.1007/978-3-030-02698-1_9

been devoted to this research direction due to its wide availability and ease of usage. From the algorithmic perspective, the first order problem is to some extent already solved. There exist many algorithms to find its global solution within polynomial time, such as the Hungarian algorithm. When further incorporating the second or high order constraints, graph matching utilizes not only the vertex information, but also the structural cues between vertices. In this paper, we only focus on the second order constraints, under which the pairwise relations between vertices should be maintained.

Almost all the graph matching tasks are NP-complete. A exceptional case is the classic graph isomorphism problem, which requires that the two graphs to be matched should be exactly the same, even including the attribute descriptions of all the vertices and edges. It is known to be a GI-hard problem, which means that it is neither solvable in polynomial time nor NP-complete. Almost all the other graph matching problems are known to be NP-complete, including subgraph isomorphism which is a generalized version of graph isomorphism to unequal sized graphs. Because of the NP-complete nature, getting global solutions of the graph matching problems is expensive. Except certain specific application like chemical structure matching [4], most tasks only need to get a suboptimal solution in reasonable time, and therefore approximate methods are necessary.

As a type of approximate method, the continuous method is widely used in computer-vision-related graph matching tasks. It typically first relaxes the discrete (combinatorial) graph matching problem to a continuous optimization problem, and then projects the continuous solution back to the discrete domain. The major concern of using the continuous method is about the balance between efficiency and efficacy. It would naturally lead to higher efficiency, to solve a continuous optimization problem by for example gradient based optimization techniques instead of solving a discrete optimization problem. Since the relaxed continuous optimization problem usually involves a nonconvex objective function, the gradient based methods may be trapped in bad local optimum. Therefore many researchers choose to further relax the original problem to a convex optimization problem, which could often avoid bad local optimum. However, when the original objective function is complex and the convex surrogate function is largely different from it, the performance would significantly deteriorate. In this situation, the continuation scheme would provide an interesting and useful choice to find a better local optimum within reasonable time [5].

The main idea of the continuation scheme is to first start from a convex optimization problem which is easy to solve, then gradually transform through a series of subproblems, and finally terminate at the original problem. By elaborately designing the initial convex optimization problem and the transformation process, the local optimal points of all the subproblems would form a sequence and follow a path to a usually better solution of the original problem. The reason for the better performance is that by such a graduated manner, in the front part the general profile of the original function could be more considered without being trapped by local changes of functions or bad initialization, and in the latter part more details of the original function are investigated without losing too much information of the original problem. However, the continuation method is only applicable to continuous optimization problem, while graph matching is essentially a combinatorial optimization problem. For the final discrete solution, a straightforward way is directly projecting the continuation solution back to the discrete domain, similar to most other continuous methods for graph matching. However, in this way the gap between the continuous optimization problem solved by the continuation method and the combinatorial nature of graph matching is to some extent omitted, which would result in the partial loss of efficacy [6,7].

In this paper, we propose a continuation scheme based continuous method directly targeting at the graph matching algorithm. Specifically, first a barrier function incorporating the original objective function and the discrete constraints is proposed, and thus the problem is changed to an unconstrained optimization problem. Then a continuation scheme based on Gaussian smooth is applied to the barrier function, in which the closed form of the objective function in each step of the continuation scheme is deduced. Below the proposed method is introduced in Sects. 2 and 3, and after the experimental assessment in Sect. 4, finally Sect. 5 concludes the paper.

2   Formulation

A graph $\mathcal{G} = \{V^G, E^G, A^G, W^G\}$ consists of a vertex set $V^G = \{1, 2, \cdots, N\}$, an edge set $E^G = \{e_1^G, e_2^G, \cdots, e_{|E^G|}^G\} \subseteq V^G \times V^G$, a vertex attribute set $A^G = \{a_1^G, a_2^G, \cdots, a_N^G\}$, and an edge weight set $W^G = \{w_1^G, w_2^G, \cdots, w_{|E^G|}^G\}$, where $|E^G|$ denotes the number of edges. The vertex attribute vector $a_i^G \in \mathbb{R}^{l_A \times 1}$ is used to describe the own characteristics of the vertex i, where l_A denotes the number of attribute vector dimension. For example, when using a graph to represent a local feature set in an image, the attribute could be an appearance descriptor around a feature point, such as the 128 dimensional SIFT descriptor [8]. The edge weight vector $w_k^G \in \mathbb{R}^{l_W^G \times 1}$, or denoted by $w_{\{i,j\}}^G$, is used to describe the edge $e^k = \{i, j\}$, or say the relations between the vertices i and j. For example, the length and orientation of the edge e^k can be treated as its weights, and the weight dimension is $l_W^G = 2$.

Given two graphs $\mathcal{G}$ and $\mathcal{H} = V^H, E^H, A^G, W^G$, the assignments between their vertices can be represented by an assignment matrix $\mathbf{X} \in \{0,1\}^{N \times N}$. When $\mathbf{X}_{iu} = 1$ it means that the vertex i in $\mathcal{G}$ is assigned to the vertex u in $\mathcal{H}$. Under the common used one-to-one matching constraints, the assignment matrix is actually a permutation matrix. The similarity measures of vertices and edges between $\mathcal{G}$ and $\mathcal{H}$ are pre-computed and stored in a matrix $\mathbf{S} \in \mathbb{R}^{N^2 \times N^2}$ usually called affinity matrix, which is often used in graph matching algorithms [9]. It can be defined as follows:

$$\mathbf{S}_{i'j'} = \mathbf{S}_{((i-1)N+u)((j-1)N+v)}$$
$$= \begin{cases} S^V(a_i^G, a_u^H), & \text{if } i = j, u = v, \\ S^E(w_{\{i,j\}}^G, w_{\{u,v\}}^H), & \text{if the edges } \{i,j\} \in \mathcal{G} \text{ and } \{u,v\} \in \mathcal{H} \text{ exist}, \\ 0, & \text{otherwise.} \end{cases} \tag{1}$$

In the above definition, $S^V(a_i^G, a_u^H)$ denotes the similarity measure between the vertices i in $\mathcal{G}$ and u in $\mathcal{H}$, and $S^E(w_{\{i,j\}}^G, w_{\{u,v\}}^H)$ denotes the similarity measure between the edges $\{i, j\}$ in $\mathcal{G}$ and $\{u, v\}$ in $\mathcal{H}$. Note in realistic applications, the two types of similarity measures should be normalized or be given weights according to prior information. Based on the affinity matrix $\mathbf{S}$, the graph matching problem can usually be formulated by the following quadratic assignment problem (QAP) [10]:

$$\mathbf{X} = \arg_{\mathbf{X}} \max \, \mathrm{vec}(\mathbf{X})^T \mathbf{S} \mathrm{vec}(\mathbf{X}), \tag{2}$$

$$\text{s.t.} \sum_{u=1}^{N} \mathbf{X}_{iu} = 1, \sum_{i=1}^{N} \mathbf{X}_{iu} = 1, \mathbf{X}_{iu} = \{0, 1\}, \forall i, u.$$

In this formulation, $\mathrm{vec}(\mathbf{X})$ denotes the row-wise replica of $\mathbf{X}$, and the constraints mean that $\mathbf{X}$ is a permutation matrix.

As explained in Sect. 1, the continuation scheme is only applicable to the continuous optimization problem, and directly applying it to the object function (2) in the relaxed continuous domain would result in the ignorance of the combinatorial nature of graph matching. Therefore a barrier function $B(\mathbf{X})$ is constructed to incorporate both the objective function (2) and the constraints, and the original discrete optimization problem is replaced by the following unconstrained continuous optimization problem:

$$\mathbf{X} = \arg_{\mathbf{X}} \max B(\mathbf{X}) = \arg_{\mathbf{X}} \max F(\mathbf{X}) - \zeta(C_1(\mathbf{X}) + C_2(\mathbf{X}) + C_3(\mathbf{X})), \tag{3}$$

where

$$F(\mathbf{X}) = \mathrm{vec}(\mathbf{X})^T \mathbf{S} \mathrm{vec}(\mathbf{X}), \tag{4}$$

$$C_1(\mathbf{X}) = \sum_{i=1}^{N} (\sum_{u=1}^{N} \mathbf{X}_{iu} - 1)^2, \tag{5}$$

$$C_2(\mathbf{X}) = \sum_{u=1}^{N} (\sum_{i=1}^{N} \mathbf{X}_{iu} - 1)^2, \tag{6}$$

and

$$C_3(\mathbf{X}) = \sum_{i=1}^{N} \sum_{u=1}^{N} (\mathbf{X}_{iu}^2 (1 - \mathbf{X}_{iu})^2). \tag{7}$$

In the above optimization problem (3), the regularization terms $C_1(\mathbf{X})$, $C_2(\mathbf{X})$, and $C_3(\mathbf{X})$ respectively correspond the constraints $\sum_{u=1}^{N} \mathbf{X}_{iu} = 1$, $\sum_{i=1}^{N} \mathbf{X}_{iu} = 1$, and $\mathbf{X}_{iu} = \{0, 1\}$. The weight parameter ζ is used to balance the original objective function $F(\mathbf{X})$ and these regularized terms. By properly setting ζ, maximizing the barrier function $B(\mathbf{X})$ would automatically lead to a discrete solution or a continuous solution close to the discrete domain.

3   Continuation Scheme Based Optimization

Since the barrier function is usually highly nonconvex, the continuation scheme is used to maximize it. As mentioned in Sect. 1, the continuous scheme starts at a convex optimization problem, and gradually transforms to the original objective function, i.e. the barrier function here. For a maximization problem, a convex optimization problem indicates a concave objective function. Generally, the optimal concave relaxation function should be the concave envelope of the function to be maximized, which however would need expensive computation. Gaussian smooth provides an alternative method, which has been proved to be a best approximation of the concave envelope under certain conditions [11][1].

For the proposed barrier function, the initial concave relaxation function, together with the subsequent objective functions in the continuation process, can be obtained by Gaussian smooth, i.e. by convolving the barrier function with an isotropic Gaussian kernel with varying bandwidth. Specifically, the convolution takes the following form by using a scalar variate x as an example:

$$\mathcal{B}(x, \sigma) = B(x) * k(x, \sigma) = \int_{-\infty}^{\infty} B(x)k(y - x, \sigma)dy, \tag{8}$$

where the Gaussian kernel $k(x, \sigma)$ with bandwidth σ is defined by

$$k(x, \sigma) = \frac{1}{\sqrt{2\pi}\sigma} e^{\frac{\|x\|^2}{2\sigma^2}}. \tag{9}$$

Since the four terms $F(\mathbf{X})$, $C_1(\mathbf{X})$, $C_2(\mathbf{X})$, and $C_3(\mathbf{X})$ all have polynomial forms, closed forms exist for their convolved functions [12], which are shown below respectively. For $F(\mathbf{X})$, the convolved function is

$$\mathcal{F}(\mathbf{X}, \sigma) = F(\mathbf{X}) * k(\mathbf{X}, \sigma)$$
$$= \text{vec}(\mathbf{X})^T \mathbf{S}\text{vec}(\mathbf{X}) + \sum_{i=1}^{N}\sum_{u=1}^{N} \mathbf{S}_{((i-1)N+u)((i-1)N+u)}\sigma^2. \tag{10}$$

The second term in (10) is actually σ^2 multiplied by the sum of the diagonal entries in $\mathbf{S}$, i.e. the sum of all $S^V(a_i^G, a_u^H), \forall i, u$. Since in each step of the continuation process the bandwidth σ is fixed, the second term is a constant with respect to the variate $\mathbf{X}$. Therefore the second term can be removed, and for simplicity $\mathcal{F}(\mathbf{X})$ is directly written as

$$\mathcal{F}(\mathbf{X}, \sigma) = \text{vec}(\mathbf{X})^T \mathbf{S}\text{vec}(\mathbf{X}), \tag{11}$$

which is exactly $F(\mathbf{X})$. For $C_1(\mathbf{X})$, the convolved function is

$$\mathcal{C}_1(\mathbf{X}, \sigma) = C_1(\mathbf{X}) * k(\mathbf{X}, \sigma)$$
$$= \sum_{i=1}^{N}\left(\sum_{u=1}^{N} \mathbf{X}_{iu} - 1\right)^2 + N\sigma^2. \tag{12}$$

---

[1] It uses convex envelope in the reference, which is the same in principle.

Since the second term is also a constant, $\mathcal{C}_1(\mathbf{X})$ could be directly given by

$$C_1(\mathbf{X}, \sigma) = \sum_{i=1}^{N}(\sum_{u=1}^{N}\mathbf{X}_{iu} - 1)^2,\tag{13}$$

which is also equal to $C_1(\mathbf{X})$. Similarly, for $C_2(\mathbf{X})$ the convolved function is directly written as

$$C_2(\mathbf{X}, \sigma) = \sum_{u=1}^{N}(\sum_{i=1}^{N}\mathbf{X}_{iu} - 1)^2.\tag{14}$$

For $C_3(\mathbf{X})$, the convolved function is

$$\begin{aligned}
C_3(\mathbf{X}, \sigma) &= C_3(\mathbf{X}) * k(\mathbf{X}, \sigma)\\
&= \sum_{i=1}^{N}\sum_{u=1}^{N}(\mathbf{X}_{iu}^2(1 - \mathbf{X}_{iu})^2) + N\sigma^2 + 3N\sigma^4\\
&\quad -6\sigma^2\sum_{i=1}^{N}\sum_{u=1}^{N}(\mathbf{X}_{iu}(1 - \mathbf{X}_{iu})).
\end{aligned}\tag{15}$$

For the similar reason, $C_3(\mathbf{X})$ is directly written as

$$C_3(\mathbf{X}, \sigma) = \sum_{i=1}^{N}\sum_{u=1}^{N}(\mathbf{X}_{iu}^2(1 - \mathbf{X}_{iu})^2) - 6\sigma^2\sum_{i=1}^{N}\sum_{u=1}^{N}(\mathbf{X}_{iu}(1 - \mathbf{X}_{iu})).\tag{16}$$

Therefore, for $B(\mathbf{X})$ the convolved function $\mathcal{B}(\mathbf{X}, \sigma)$ is

$$\begin{aligned}
\mathcal{B}(\mathbf{X}, \sigma) &= B(\mathbf{X}) * k(\mathbf{X}, \sigma)\\
&= \mathcal{F}(\mathbf{X}, \sigma) + \mathcal{C}_1(\mathbf{X}, \sigma) + \mathcal{C}_2(\mathbf{X}, \sigma) + \mathcal{C}_3(\mathbf{X}, \sigma)\\
&= F(\mathbf{X}) - \zeta(C_1(\mathbf{X}) + C_2(\mathbf{X}) + C_3(\mathbf{X}) - 6\sigma^2\sum_{i=1}^{N}\sum_{u=1}^{N}(\mathbf{X}_{iu}(1 - \mathbf{X}_{iu})))\\
&= B(\mathbf{X}) + 6\zeta\sigma^2\sum_{i=1}^{N}\sum_{u=1}^{N}(\mathbf{X}_{iu}(1 - \mathbf{X}_{iu}))
\end{aligned}\tag{17}$$

When $\sigma \to \infty$, the Hessian matrix of $\mathcal{B}(\mathbf{X}, \sigma)$ is negative definite because its non-diagonal entries are zeros and diagonal entries are all negative. Thus $\mathcal{B}(\mathbf{X}, \sigma)$ can be used as the objective function in the initial convex optimization problem, for which the closed-form solution exist:

$$\mathbf{X}_{iu} = \frac{1}{2}, \forall i, u.\tag{18}$$

Then by gradually decreasing σ from ∞ to 0, the objective function to be maximized would finally reach the barrier function $B(\mathbf{X})$. Practically, by initializing

$\mathbf{X}$ as (18), the bandwidth σ only needs to start from a relative large positive rather than ∞. The optimization scheme is summarized in Algorithm 1.

Algorithm 1. Continuation Scheme Based Optimization

Input: Two graphs $\mathcal{G}$ and $\mathcal{H}$
Initialize the continuous assignment matrix $\mathbf{X}^0$ by (18)
Initialize the bandwidth σ^0
Construct the affinity matrix $\mathbf{S}$ by (1)
repeat
  Solve $\mathbf{X}^{t+1} = \max \mathcal{B}(\mathbf{X}^t, \sigma^t)$
  $\sigma^{t+1} = \sigma^t - \delta\sigma$
  $t = t + 1$
until $\mathbf{X}^{t+1} = \mathbf{X}^t$ or $\sigma^t = 0$
Project $\mathbf{X}^{t+1}$ to the discrete domain
Output: A discrete assignment matrix $\mathbf{X}$

---

4   Experimental Results

In this section, the proposed method is evaluated by comparing it with some state-of-the-art graph matching algorithms. The algorithms for comparison include the spectral matching (SM) algorithm [9], and the probabilistic graph matching (PGM) algorithm [13].

The experiments are carried out on the synthetic point correspondence task, in which a graph is used to represent a point set. First two spatial point sets are sampled from a 2 dimensional surface as follows: Generate N points $G_i \in \mathbb{R}^{1\times2}, i = 1, 2, \cdots, N$ by sampling from a 2 dimensional uniform distribution $[0, 1]^{2\times1}$; Generate a ground truth permutation matrix $\mathbf{X} \in \{0, 1\}^{N\times N}$; Get the second point sets $H_u \in \mathbb{R}^{1\times2}$ by the following transformation:

$$H_u = G_i + \eta, \text{if } \mathbf{X}_{iu} = 1, \tag{19}$$

where $\eta \in \mathcal{N}(0, \sigma_\eta)^{1\times2}$ is the additional 2 dimensional Gaussian noise. For either point set, a point is represented by a graph vertex, and the connections between vertices, i.e. the graph edges, are randomly built considering the structure density. The two graphs share a similar graph structure, but the structure is disturbed by the Gaussian noise η, by randomly adding and removing $\frac{1}{2}|E^H|$ edges in the second graph $\mathcal{H}$ where $|E^H|$ denotes its edges number.

Three comparisons are performed respectively with respect to the noise level, problem scale, and structure density. The noise level σ_η ranges from 0 to 0.2 by a step size of 0.02, the problem scale N ranges from 20 to 40 by a step size of 2, and the structure density ρ ranges from 0.1 to 1 by a step size of 0.1. Following the law of single variable, in each comparison only one factor changes and the other two are set to be their median. For example, in the comparison with respect to the noise level, it is set that $N = 30$ and $\rho = 0.5$.

The results for the three comparisons are shown in Fig. 1. It can be observed that generally the proposed method denoted by 'OUR' outperforms the competitors in all the three comparisons, and it even achieves perfect performance in the comparison with respect to the noise level. Note that in the third comparison, when the edge density ρ is near 1, which indicates full connection, the proposed method is outperformed by the other two algorithms. A possible reason may be that the full connection introduces too much noise, which results in the inconsistency between the optimum of the objective function and the ground truth assignments.

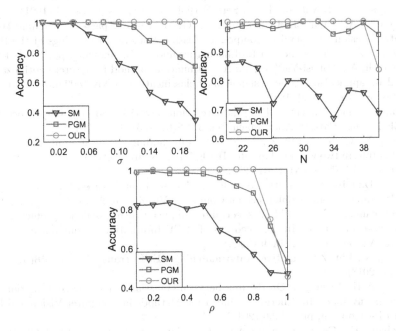

Fig. 1. Results on synthetic point matching. Upper-left: comparison results w.r.t. noise level. Upper-right: comparison results w.r.t. problem scale. Lower: comparison results w.r.t. edge density.

5   Conclusion

This paper proposes a novel graph matching algorithm based on the continuation scheme, by first transforming the original quadratic assignment problem to an unconstrained continuous optimization problem and then solving it by Gaussian smooth based continuation. The proposed method has been validated by experiments with respect to different factors.

Acknowledgment. This work is supported partly by the National Natural Science Foundation (NSFC) of China (grants 61503383, 61633009, U1613213, 61627808, 61502494, and U1713201), partly by the National Key Research and Development Plan of China (grant 2016YFC0300801 and 2017YFB1300202), and partly by the Development of Science and Technology of Guangdong Province Special Fund project (grant 2016B090910001).

References

1. Conte, D., Foggia, P., Sansone, C., Vento, M.: Thirty years of graph matching in pattern recognition. Int. J. Pattern Recogn. Artif. Intell. **18**(03), 265–298 (2004)
2. Zhao, C., Guo, L.: PID controller design for second order nonlinear uncertain systems. Sci. China Inf. Sci. **60**(2), 022201 (2017)
3. Scott, G., Longuet-Higgins, H.: An algorithm for associating the features of two images. Proc. Royal Soc. Lond. Seri. B Biol. Sci. **244**(1309), 21–26 (1991)
4. Smalter, A., Huan, J., Lushington, G.: GPM: a graph pattern matching kernel with diffusion for chemical compound classification. In: Proceedings of the IEEE International Conference on BioInformatics and BioEngineering, pp. 1–6 (2008)
5. Shameli, A., Abbasi-Yadkori, Y.: A continuation method for discrete optimization and its application to nearest neighbor classification. arXiv:1802.03482 (https://arxiv.org/abs/1802.03482)
6. Liu, Z., Qiao, H.: GNCCP - graduated nonconvexity and concavity procedure. IEEE Trans. Pattern Anal. Mach. Intell. **36**(6), 1258–1267 (2014)
7. Yang, X., Qiao, H., Liu, Z.: An algorithm for finding the most similar given sized subgraphs in two weighted graphs. IEEE Trans. Neural Netw. Learn. Syst. https://doi.org/10.1109/TNNLS.2017.2712794
8. Lowe, D.: Object recognition from local scale-invariant features. In: Proceedings of IEEE International Conference on Computer Vision, vol. 2, pp. 1150–1157 (1999)
9. Leordeanu, M., Hebert, M.: A spectral technique for correspondence problems using pairwise constraints. In: Proceedings of IEEE International Conference on Computer Vision, pp. 1482–1489 (2005)
10. Yang, X., Liu, Z.: Adaptive graph matching. IEEE Trans. Cybern. **48**(5), 1432–1445 (2018)
11. Mobahi, H., Fisher, J.: On the link between Gaussian homotopy continuation and convex envelopes. In: Energy Minimization Methods in Computer Vision and Pattern Recognition, pp. 43–56 (2015)
12. Mobahi, H.: Closed form for some Gaussian convolutions. arXiv:1602.05610v2 (https://arxiv.org/abs/1602.05610)
13. Egozi, A., Keller, Y., Guterman, H.: A probabilistic approach to spectral graph matching. IEEE Trans. Pattern Anal. Mach. Intell. **35**(1), 18–27 (2013)

Finding Overlapping Communities by Increasing the Determinacy of SLPA in Complex Networks

Jingyi Zhang[✉], Zhixin Ma, Qijuan Sun, Jun Yan, Xiao Zhang, and Mengjia Shen

School of Information Science and Engineering,
Lanzhou University, Lanzhou, China
jyzhang2016@lzu.edu.cn

Abstract. Community detection algorithms help to analyze the real structure of networks. The advantages of overlapping community detection algorithms are evident in discovering high-quality communities. SLPA is a classical algorithm of overlapping community detection, which is updated by random sequence in the phase of label updated. At the same time, it randomly selects one of the labels to update when a node receives multiple labels that appear most frequently in the phase of label propagation. Aiming at the uncertainty of randomness in the algorithm, an improved algorithm of SLPA based on nodes PageRank values and nodes similarity index is proposed in this paper. A large number of experiments have been carried out in benchmark data sets and real data sets. The results show that the proposed algorithm has excellent adaptability and robustness

Keywords: Overlapping community detection · Label propagation Determinacy

1 Introduction

Community detection of complex networks has become one of the hot spots in the field of big data research. It has important theoretical and behavioral significance for topology analysis, function analysis and behavior prediction of complex networks [1]. In 2002, the concept of community structure proposed by Newman believed that the nodes within the community are closely connected and the connections between the communities are loose [2]. For example, a community is composed of a series of specific interest enthusiasts in a social network. Individuals with the same interests are closely connected, and individuals with different interests are relatively sparsely connected.

Community detection algorithms are divided into non-overlapping and overlapping community detection algorithms. Non-overlapping community detection algorithms can detect several independent communities. Each node uniquely belongs to a community. However, the overlapping communities is more common in real large-scale networks [3]. Therefore, overlapping community detection algorithms have more

Y. Peng et al. (Eds.): IScIDE 2018, LNCS 11266, pp. 111–122, 2018.
https://doi.org/10.1007/978-3-030-02698-1_10

practical significance. Firstly, overlapping communities are more in line with the real network structure. Secondly, overlapping nodes are usually the key nodes in the networks. Therefore, using overlapping community detection algorithms to find the community structure in the real networks has become the focus of community detection, which is also the core of this paper.

In this paper, we propose an algorithm based on node PageRank values and node similarity index (DSLPA, deterministic speaker-listener label propagation algorithm) for the defects brought by randomness in SLPA. The algorithm improves the node updated order and node label selection. It increases the determinism and accuracy of the original algorithm to some extent.

2 Related Work

We review the state of the art and find existing algorithms are mainly based on network topology structure, such as graph segmentation, spectral analysis, hierarchical method, random walk, seed expansion, modularity optimization and dynamic. The early community detection algorithms are mainly graph segmentation. The representative algorithm K-L [4] uses the greedy strategy to divide a network into two sub-networks. The spectral analysis [5] divides the network into two by extracting the eigenvector corresponding to the second eigenvalue based on the Laplacian matrix. Hierarchical methods include split and condensed methods. The former splits the entire network from top to bottom until a single node is considered as a community. For example, GN [2]. The latter regards a single node as a community through a single link until it is merged into a community from bottom to top. For example, FastQ [6]. Random walk jump from the current node to the neighbor nodes in a certain probability, and the representative algorithm is MCL [7]. Seed expansion finds the optimal value of fitness function. For example, LFM [8] uses the fitness function to expand the randomly selected seed until the value of the objective function reaches the maximum. Modularity optimization finds the modularity as the maximum value of the objective function to extract the optimal community structure. The representative algorithm is Louvain [9].

In 2007, Raghavan et al. [10] applied the idea of label propagation to community detection for the first time and proposed LPA. The basic idea is to use label information of labeled nodes to predict label information of unlabeled nodes. The time complexity of LPA is linear, but it can only detect non-overlapping communities. To find overlapping communities, COPRA proposed by Gregory [11] in 2010 introduces a new label structure for each node based on LPA so that each node can carry multiple labels and detect overlapping communities. But the algorithm limits the number of communities to which overlapping nodes belong. On the basis of COPRA, BMLPA proposed by Wu et al. [12] eliminates the restriction of the number of communities nodes belong to. In 2013, SLPA proposed by Xie et al. [13] also expanded the LPA. Each node in the algorithm has a memory to storage labels. The labels are processed by postprocessing thresholds. Nodes with multiple labels are called overlapping nodes. So LPA to SLPA also completed the leap from non-overlapping community detection algorithms to overlapping community detection algorithms. Wang et al. [14] based on SLPA's improved HLPA creatively uses a hybrid update strategy to improve the efficiency of label propagation and avoid the swaying phenomenon that occurs when the label is propagated in a bipartite graph or an approximate bipartite graph. In 2016, Liu et al. [15] proposed the ELPA by combining the natural advantages of the link community with the efficiency of the LPA. In 2017, Hou et al. [16] proposed the SSCLPA detect the initial community based on the similarity score of the Sørensen-Dice index, using different update strategies for allocated and unallocated nodes and constraint conditions so as to obtain the community structure in the networks.

In this paper, while summarizing the above algorithms, according to the node PageRank value and node similarity, the DSLPA is proposed based on SLPA, which improves the stability of the original algorithm. The third section describes the basic idea, specific steps and time complexity of the DSLPA. The forth section validates the algorithm through simulation experiments and compares it with SLPA and some classical algorithms. The fifth section summarizes the work of the paper and gives the future research directions.

3 DSLPA

3.1 Related Definition

The study of complex networks originated from graph theory. The network can be represented by a graph, which is specifically expressed as Definition 1.

Definition 1. Network. A network can be abstracted as a graph $G = (V, E)$ consisting of a set of points V and edges E. The number of nodes is $n = |V|$ and the number of edges is $m = |E|$. Each edge in E corresponds to a pair of nodes in V.

The edge with a direction is called directed graph. Otherwise, it is an undirected graph. The graph with the weight of edges is called a weighted graph. Otherwise, it is called unweighted graph. The unweighted graph can also be seen as a weighted graph with the weight of 1. The following work in this paper is based on an undirected and unweighted graph.

To measure the importance of the nodes in the networks, this paper uses the PR value obtained by the PageRank algorithm. This algorithm is a key technology for Google to judge the importance of web pages. The larger the PR value is, the more important the web pages are.

Definition 2. PageRank algorithm. All pages on the web are used as nodes and link relationships are used as edges. The entire web is represented as a directed network composed of nodes and edges. The PR value of the web page is calculated as shown in formula (1).

$$PR(p_i) = \frac{1 - \alpha}{n} + \alpha \sum_{p_j \in M(p_i)} \frac{PR(p_j)}{L(p_j)} \tag{1}$$

$PR(pi)$ denotes the PageRank value of the vertex p_i; n denotes the total number of web pages. It denotes the total number of nodes in the networks in this paper; $L(p_j)$ denotes the out-degree of the web page p_j; $M(p_i)$ denotes the set of web pages connected to the out-degree of the web page p_i. This paper shows the set of neighbor nodes of the node p_i; α is a scaling constant, the closer to 1 the algorithm converges more slowly, and the closer to 0 the algorithm converges faster. It usually takes a value of 0.85.

In an undirected network, the improved iterative equation is shown in formula (2).

$$PR(p_i) = \frac{1-\alpha}{n} + \alpha \sum_{p_j \in M(p_i)} \frac{PR(p_j)}{d(p_j)} \tag{2}$$

$d(p_j)$ represents the degree of node p_j. Then all the nodes in the network are arranged according to their PR values from small to large, and the sequence of vertices obtained is the order of updating the list of labels of nodes in each iteration. The initial PR values for each node in the network are set to $1/n$.

Definition 3. Jaccard index. It is also known as Jaccard's similarity index which is used to compare similarities among individuals. The larger the index value is, the higher the degree of similarity between individuals is. It is defined as formula (3).

$$sim(u, v) = \frac{|\Gamma(u) \cap \Gamma(v)|}{|\Gamma(u) \cup \Gamma(v)|} \tag{3}$$

$\Gamma(u)$ is the neighbor nodes of the node u, which does not include the u itself; $\Gamma(v)$ is the neighbor nodes of the node v, which does not include the v itself. In this definition, $\Gamma(u) \cap \Gamma(v)$ does not include node u and node v, but $\Gamma(u) \cup \Gamma(v)$ contains node u and node v. In order to make the value of formula $sim(u, v)$ within the range of (0, 1], the molecular part on the right of the equation plus 2. In this paper, we use formula (4) to calculate the similarity of nodes.

$$sim(u, v) = \frac{|\Gamma(u) \cap \Gamma(v)| + 2}{|\Gamma(u) \cup \Gamma(v)|} \tag{4}$$

Definition 4. Modularity. To assess the quality of community division, the modularity function Q proposed by Newman and Girvan [17] has been widely used in traditional community detection algorithms. The extended modularity function EQ proposed by Shen [18] can measure the quality of overlapping communities. The definition is shown in formula (5).

$$EQ = \frac{1}{2m} \sum_{c \in C} \sum_{v,u} \frac{1}{o_v o_u} \left(A_{vu} - \frac{k_v k_u}{2m} \right) \sigma_{vc} \sigma_{uc} \tag{5}$$

m is the number of edges in the network; C is the community obtained after the algorithm is executed; o_v is the number of communities to which node v belongs; A_{vu} represents whether there is a connecting edge between node v and node u. If node v and node u connected, A_{vu} equals 1, otherwise, A_{vu} equals 0; k_v is the degree of node v in the network; σ_{vc} represents whether node v belongs to community c, the value of 1 means it belongs to, the value of 0 means it does not belong to.

The range of EQ is $[0, 1)$. The lager the value, the more obvious the structure of the community is. Normally, a significant community structure can be observed for values between 0.3 and 0.7.

3.2 Algorithm Ideas

The idea of DSLPA proposed in this paper mainly includes the following three aspects.

(1) The PageRank algorithm is used to calculate the PR values of all nodes in the network, and the nodes are arranged in ascending order of their PR values. The resulting sequence of nodes is the order of the list of node labels updates during each subsequent iteration of the algorithm.
(2) Using the index of similarity of nodes. When a node receives more than one most frequent labels, it calculates the similarity between the nodes of these labels and the current node. Then updating the label with the highest similarity node to the label of the current node list.
(3) The algorithm uses an extended modularity function EQ as a measure of the quality of overlapping communities.

3.3 Algorithm Steps

The specific steps for implementing the DSLPA are as follows.

The first step, initialization. The relevant parameters are initialized. Including the adjacency matrix net, the maximum number of iterations T and the post-processing threshold r. The adjacency matrix net shows the structure of the current required processing network, consisting of an n-order matrix, as shown in formula (6).

$$net = \begin{bmatrix} A_{11} & A_{12} & \cdots & A_{1n} \\ A_{21} & A_{22} & \cdots & A_{2n} \\ \vdots & \vdots & \vdots & \vdots \\ A_{n1} & A_{n2} & \cdots & A_{nn} \end{bmatrix} \tag{6}$$

Taking any arbitrary A_{kl}, $k \in [1, n]$, $l \in [1, n]$. It defined as formula (7).

$$A_{kl} = \begin{cases} 1, & \text{There is an edge between } k \text{ and } l \\ 0, & \text{There is no edge between } k \text{ and } l \end{cases} \tag{7}$$

The maximum number of iterations T has nothing to do with the size and structure of the network and T is usually setting up as 1000. Post-processing threshold $r \in [0, 0.5]$. The algorithm outputs non-overlapping communities when $r > 0.5$. At the same time, each label list of nodes is initialized with a unique label. The corresponding PR value of each node is calculated using formula (2). The order of updating the node label list is obtained by PR value from small to large.

The second step, label propagation. It updates the label list of each node iteratively according to the nodes update sequence generated during the phase of initialization. A node that currently needs to update the label list is given and performed corresponding update operations according to the labels of its neighbor nodes. Each neighbor node randomly selects a label whose probability is proportional to the frequency of occurrence of the label in its list and sends the selected label to the current node. The most frequently received label is unique and the label is added to the current node label list. When the label with the highest number of occurrences is not unique, the current node and the node that has passed the most occurrence label are calculated according to formula (4). The label with the highest similarity value corresponding to the node is added to the current node label list. Until the number of iterations of the loop reaches the user-defined maximum number of iterations T, this step ends.

The third step, post-processing and output. Calculating the probability of occurrence of mutually exclusive labels in each node label list. Comparing the obtained results with the post-processing threshold r. Deleting all labels with a probability less than r. The remaining labels are the community collections to which the node belongs. The output is the final community division of the network.

3.4   Time Complexity Analysis

Setting the network with n nodes and m edges. k represents the average degree of nodes in the network. T is the maximum number of iterations that user defined.

(1) Initializing the node label list $O(n)$.
(2) Calculating the PR value of node $O(nk)$.
(3) In the label propagation process, the outer loop is the maximum iteration number T. The intermediate loop is the number of nodes in the network. And the inner loop is the number of neighbors per node, that is the network average degree k. So $O(Tnk)$ is needed.
(4) $O(Tn)$ is required for post-processing.
(5) Therefore, the total time complexity of the algorithm is $O(Tnk)$. It can be found in the experiments of artificial networks and real networks that the algorithm can converge quickly.

4   Experiments and Analysis

In this paper, the algorithm runs on the MATLAB R2016a platform and draws in Pajek. In order to verify the adaptability and robustness of the DSLPA, test experiments were performed on three sets of artificial data sets and three sets of real data sets.

4.1   Artificial Data Sets

In the LFR [19] artificial data sets, three sets of data are generated using different parameters. The experimental parameters are shown in Table 1.

Table 1. Related parameters of LFR benchmark networks.

Data sets	n	k	maxk	minc	maxc	on	om	μ
R1	128	10	30	20	50	10	2	0.1
R2	500	20	40	20	50	10	2	0.1
R3	1000	50	50	20	50	20	3	0.1

n is the total number of nodes in the network; k is the average degree of nodes; maxk is the maximum degree of nodes; minc is the number of nodes included in the smallest community; maxc is the number of nodes included in the largest community; on is the number of overlapping nodes; om is the number of communities to which the overlapping nodes belong to; μ is the blending parameters.

The experimental results of DSLPA on three sets of artificial data sets are shown in Table 2. Compared with the real division results of network, the algorithm can identify the key overlapping nodes in the networks. DSLPA has obvious advantages in community detection of large network data sets by observing EQ values. Therefore, the feasibility and effectiveness of DSLPA in artificial data sets are verified.

Table 2. Experimental results of three sets of artificial data sets.

Data sets	Running time (s)	Overlapping nodes	EQ
R1	9.4197	{59, 83, 87, 109}	0.5901
R2	89.9573	{280, 347, 354, 394, 399}	0.8127
R3	397.6154	{138, 412, 733, 924}	0.8122

4.2   Real Data Sets

In this paper, we choose three data sets that represent the real network commonly used in testing community detection algorithms: Zachary's Karate Club Network, Dolphins Social Network and Books about US Politics. The related parameters are shown in Table 3.

Table 3. Experimental results of three sets of artificial data sets.

Data sets	n	m	k	Description
Zachary	34	78	4.5	Zachary's karate club network [20]
Dolphins	62	159	5.1	Dolphins social network [21]
Polbooks	105	441	8.4	Books about US politics [22]

The DSLPA has been experimentally observed in the Zachary data set after several experiments that the network is divided into three communities. The results of the community division are shown in Table 4.

Table 4. Experimental results of three sets of artificial data sets.

Community sets	Node sets
1	1, 2, 3, 4, 8, **9**, 10, 12, 13, 14, 18, 20, 22, **31**
2	**9**, 15, 16, 19, 21, 23, 24, 25, 26, 27, 28, 29, 30, **31**, 32, 33, 34
3	5, 6, 7, 11, 17

The red nodes (node 9 and node 31) in Fig. 1 are overlapping nodes which belong to community 1 and 2. From the perspective of community structure, the division result obtained by the DSLPA is reasonable and effective.

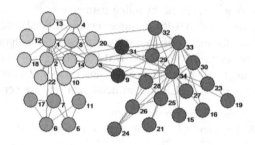

Fig. 1. Community division results of Zachary. (Color figure online)

The DSLPA has been experimentally observed in the Dolphins data set after several experiments that the network is divided into five communities. The results of the community division are shown in Table 5.

Table 5. Experimental results of three sets of artificial data sets.

Community sets	Node sets
1	2, 20, 49, 6, 33, 61, 57, 32, 23, 27, 26, 55, 40, 8, 58, 10, 7, 18, 28, 14, 42
2	5, 36, 30, 25, 12, 56, 60, 24, 52, **9**, 19, 22, 16, 46
3	4, 37, 41, 21, 51, 13, 45, 35, 50, 53, **9**, 17, 44, 39, 38, 34
4	47, 50
5	3, 11, 43, 31, 29, 62, 54, 1, 48

The red node (node 9) in Fig. 2 is an overlapping node which belongs to communities 2 and 3. Because the nodes 47 and 50 are located at the edge of the network and are loosely linked to the rest of the other nodes in the network, they are divided into a single community. From the perspective of community structure, the community division results obtained by the DSLPA are reasonable and effective.

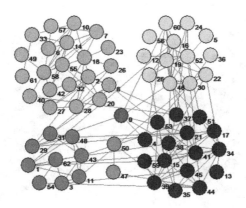

Fig. 2. Community division results of Dolphins. (Color figure online)

The DSLPA has been experimentally observed in the Polbooks data set after several experiments that the network is divided into four communities. The results of the community division are shown in Table 6.

Table 6. Community division results of DSLPA in Polbooks.

Community sets	Node sets
1	1, 2, 3, 5, 6, **7**, 8, **30**
2	103, 62, 60, 64, 63, 61, 90, 101, 99, 82, 89, 88, 92, 96, 102, 97, 93, 100, 87, 90, 85, 75, 84, 80, 83, 76, 67, 74, 29, 81, 94, 95, 31, 78, 32, 77, 73, 72, 91, 71, 79
3	70, 86, 65, 66, 53, 104, 52, 59, 105, 69, 68
4	17, 16, 22, 24, 33, 38, 39, 35, 55, 37, 36, 44, 46, 51, 58, 57, 41, 14, 13, 28, 11, 34, 18, 26, 19, 56, 4, 23, 25, 9, 48, 27, 42, 49, **7**, 43, 21, 10, 12, 20, **30**, 15, 47, 54, 50, 45, 40

The red nodes (node 7 and node 30) in Fig. 3 are overlapping nodes which belong to communities 1 and 4. From the perspective of community structure, the division result obtained by the DSLPA is reasonable and effective.

Table 7 shows the modularity values, execution times and overlapping nodes of DSLPA on three real networks. It can be seen that the DSLPA can identify the key overlapping nodes in the network.

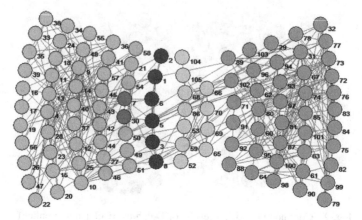

Fig. 3. Community division results of Polbooks.

Table 7. Experimental results of three sets of real data sets.

	Zachary	Dolphins	Polbooks
EQ	0.4424	0.5325	0.5329
Running time(s)	0.5859	2.6668	9.8261
Overlapping nodes	{9, 31}	{9}	{7, 30}

Table 8. The *EQ* of DSLPA and other algorithms.

EQ	DSLPA	SLPA	COPRA	BMLPA	LFM
Zachary	**0.4424**	0.3472	0.3239	0.3478	0.2146
Dolphins	**0.5325**	0.3879	0.4206	0.4355	0.2374
Polbooks	**0.5329**	0.4568	0.4586	0.5011	0.3476

Table 8 shows the results of comparing the *EQ* values obtained by the DSLPA with the algorithms in [8, 12, 13] and [14] for three real networks. The comparison results show that the DSLPA can identify high-quality overlapping community structures on the real networks.

Figure 4 shows the *EQ* comparison results for each algorithm in the three real data sets. It can be seen that the *EQ* values obtained by DSLPA in most networks are significantly better than other algorithms. Therefore, the feasibility and effectiveness of DSLPA in real network data sets are verified.

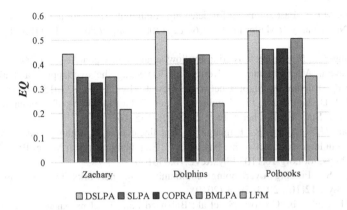

Fig. 4. *EQ* of five algorithms on three real-world networks.

5   Conclusion

An improved algorithm for SLPA proposed in this paper based on the importance of nodes in the network and the size of the influence between nodes. It integrates the PageRank value of nodes and the similarity between nodes to achieve the effective propagation of the label. The purpose of this paper to achieve DSLPA is as follows. (1) Improving the instability of SLPA caused by the random sequence updating the label list and random selection of labels when the most frequent labels are not unique. (2) Improving the execution efficiency of SLPA to some extent. Experiments were conducted on artificial data sets and real data sets. The results show that the DSLPA not only increases the stability of the original algorithm, but also has higher execution efficiency.

The complex networks in real life is a dynamically changing network. The appearance or disappearance of some nodes in the network also plays a key role in the evolution of the community. Therefore, the effective application of community detection algorithms to dynamic complex networks and identifying key nodes that affect community division will be the focus of our future research work.

References

1. Di, J., Dayou, L., Bo, Y., et al.: Fast and complex networks clustering algorithm based on local probing. J. Electron. **39**(11), 2540–2546 (2011)
2. Girvan, M., Newman, M.E.J.: Community structure in social and biological networks. Proc. National Acad. Sci. U.S. Am. **99**(12), 7821 (2002)
3. Palla, G., Derényi, I., Farkas, I., et al.: Uncovering the overlapping community structure of complex networks in nature and society. Nature **435**(7043), 814 (2005)
4. Kernighan, B.W., Lin, S.: An efficient heuristic procedure for partitioning graphs. Bell Syst. Tech. J. **49**(2), 291–307 (1970)
5. Fiedler, M.: Algebraic connectivity of graphs. Czech. Math. J. **23**(23), 298–305 (1973)

6. Newman, M.E.J.: Fast algorithm for detecting community structure in networks. Phys. Rev. E Stat. Nonlinear Soft Matter Phys. **69**(6 Pt 2) (2004). https://doi.org/10.1103/PhysRevE.69.066133
7. Van Dongen, S.M.: Graph clustering by flow simulation. Open Access (2000)
8. Lancichinetti, A., Fortunato, S., Kertész, J.: Detecting the overlapping and hierarchical community structure in complex networks. New J. Phys. **11**(3), 033015 (2009)
9. Blondel, V.D., Guillaume, J.L., Lambiotte, R., et al.: Fast unfolding of communities in large networks. J. Stat. Mech. Theory Exp. **2008**(10), 155–168 (2008)
10. Raghavan, U.N., Albert, R., Kumara, S.: Near linear time algorithm to detect community structures in large-scale networks. Phys. Rev. E Stat. Nonlinear Soft Matter Phys. **76**(3 Pt 2) (2007). https://doi.org/10.1103/PhysRevE.76.036106
11. Gregory, S.: Finding overlapping communities in networks by label propagation. New J. Phys. **12**(10), 2011–2024 (2010)
12. Wu, Z.H., Lin, Y.F., Gregory, S., et al.: Balanced multi-label propagation for overlapping community detection in social networks. J. Comput. Sci. Technol. **27**(3), 468–479 (2012)
13. Xie, J., Szymanski, B.K.: Towards linear time overlapping community detection in social networks. In: Tan, P.-N., Chawla, S., Ho, C.K., Bailey, J. (eds.) PAKDD 2012. LNCS (LNAI), vol. 7302, pp. 25–36. Springer, Heidelberg (2012). https://doi.org/10.1007/978-3-642-30220-6_3
14. Wang, T., Qian, X., Wang, X.: HLPA: a hybrid label propagation algorithm to find communities in large-scale networks. In: IEEE International Conference on Awareness Science and Technology, pp. 135–140. IEEE (2015)
15. Liu, W., Jiang, X., Pellegrini, M., et al.: Discovering communities in complex networks by edge label propagation. Sci. Rep. **6** (2016). Article number: 22470
16. Hou, C.J., Ratnavelu, K.: A semi-synchronous label propagation algorithm with constraints for community detection in complex networks. Sci. Rep. **7** (2017). Article number: 45836
17. Newman, M.E.J., Girvan, M.: Finding and evaluating community structure in networks. Phys. Rev. E **69**(2) (2004). https://doi.org/10.1103/physreve.69.026113
18. Shen, H., Cheng, X., Cai, K., et al.: Detect overlapping and hierarchical community structure in networks. Physica A: Stat. Mech. Appl. **388**(8), 1706–1712 (2009)
19. Lancichinetti, A., Fortunato, S., Radicchi, F.: Benchmark graphs for testing community detection algorithms. Phys. Rev. E Stat. Nonlinear Soft Matter Phys. **78**(2) (2008). https://doi.org/10.1103/physreve.78.046110
20. Zachary, W.W.: An information flow model for conflict and fission in small groups. J. Anthropol. Res. **33**(4), 452–473 (1977)
21. Lusseau, D., Schneider, K., Boisseau, O.J., et al.: The bottlenose dolphin community of Doubtful Sound features a large proportion of long-lasting associations. Behav. Ecol. Sociobiol. **54**(4), 396–405 (2003)
22. Newman, M.E.: Modularity and community structure in networks. In: APS March Meeting. pp. 8577–8582. American Physical Society (2006)

Learning Semantic Double-Autoencoder with Attribute Constraint for Zero-Shot Recognition

Kun Wang, Songsong Wu$^{(\boxtimes)}$, Yufeng Qiu, Fei Wu, and Xiaoyuan Jing

School of Automation, Nanjing University of Posts and Telecommunications, Nanjing, China
kunw0221@163.com, sswu@njupt.edu.cn

Abstract. Existing zero-shot recognition (ZSR) approaches generally learn a projection function from the labelled training (source) dataset. However, applying the learned projection function without adaptation to the test (target) dataset is prone to the domain shift problem. In this paper, we propose a semantic double-autoencoder with attribute constraint (SDAWAC) mechanism to overcome the problem effectively. Specifically, we take the semantic encoder-decoder paradigm to learn a projection function in the source and target domains simultaneously. In addition, we introduce one constraint on source domain attributes into this work to improve the performance of our model. The experimental results on three benchmark datasets demonstrate the efficacy of our proposed method.

Keywords: Zero-shot recognition · Projection function
Domain shift · Semantic double-autoencoder · Attribute constraint

1 Introduction

Traditional approaches to classification are usually based on supervised learning [21]. Specifically, a object classifier [19] is learned from a large number of annotated training (source) data to classify unseen (target) classes. However, providing large quantities of labelled training data for each target class is a huge and expensive task, especially we even have to recognise a new object class without ever seeing its any visual instance before. Encouraged by this problem, some scholars propose the extreme zero-shot recognition (ZSR) [1,5,14,25] and have achieved some inspiring breakthroughs in this field. It is necessary to point out that source and target classes must be disjoint in ZSR, which differs from conventional supervised learning.

Since the target domain has no labelled training data, existing ZSR approaches typically suppose that source and target classes are related in a semantic embedding space [5,17]. Such a semantic embedding space can be an attribute space [4,7,8,11,26] and each class is represented as a binary attribute

© Springer Nature Switzerland AG 2018
Y. Peng et al. (Eds.): IScIDE 2018, LNCS 11266, pp. 123–134, 2018.
https://doi.org/10.1007/978-3-030-02698-1_11

vector. Another commonly adopted space is a semantic word vector space [10] which describes each class label by a high-dimensional word vector learned from large-scale textual [25,26] databases like Wikipedia. We adopt the former as the semantic embedding space in this paper. Formally, ZSR first learns a projection function from a visual feature space to an attribute space using the source dataset. Then it uses the learned projection function to project test data into the attribute space. Finally, it gets the class label of each test image by comparing its attribute with target class prototypes.

General ZSR methods based on attribute learning suffer from the domain shift problem [2,12,25,28], as is shown in Fig. 1. First of all, we need to emphasize that the source class prototypes are surrounded by attributes of their instances tightly. Specifically, these approaches mainly employ the learned projection function to predict target classes directly. Such a projection function is only concerned with the labelled source dataset where relevant information of the unlabelled target dataset is missing. When applied without adaptation to the target, the estimated attributes of target data and their class prototypes will be separated, resulting in the domain shift.

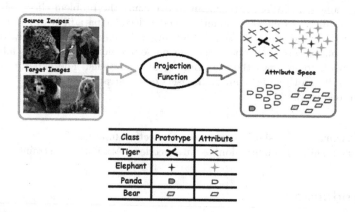

Fig. 1. An illustration of the domain shift problem in zero-shot recognition.

In this paper, we propose to solve the domain shift problem by developing a semantic double-autoencoder with attribute constraint [7,8,12] (SDAWAC) model, which is illustrated in Fig. 2. More Specifically, we first use an encoder [13, 22] (projection function) to project the source dataset into the semantic attribute space. Then we adopt a decoder [13] to project the achieved attributes back into the feature space and get novel feature representations about the source. Finally, we obtain the best encoder through minimizing the reconstruction error between the original and novel feature data. We assume the learned projection function is also suitable for test data and achieve another relevant information about the semantic encoder-decoder in the target domain. We thus get one better projection function through combining the source and target domains.

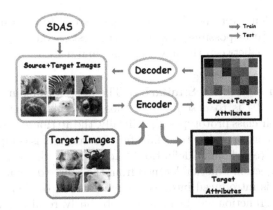

Fig. 2. The overall framework of the proposed semantic double-autoencoder with attribute constraint method.

Meanwhile, we introduce an attribute constraint into our model: Source Domain Attribute Similarity (SDAS) [21]. This constraint is to ensure that the attributes from the given source data instead of predefining are near to their faithful values as far as possible.

In summary, our paper makes several contributions below: (1) We take the semantic encoder-decoder paradigm which takes both the source and target domains into account to learn a projection function. (2) We introduce an additional attribute constraint SDAS to our model to reduce the effect of attribute noise in the source domain. (3) We conduct extensive experiments on three benchmark datasets and the experimental results show that our proposed SDAWAC is superior to existing ZSR methods.

2   Related Work

Attribute-Based ZSR. The semantic attribute space is considered as the bridge for knowledge transfer from the labelled training dataset to test data because attributes are shared among classes (both source and target). In Direct/Indirect Attribute Prediction [6], the attribute predictor is learned from the source domain data and then transferred to obtain attributes of target data. Finally, it achieves the class label of each target image by comparing its attribute with target class prototypes. In Attribute Embedding Learning [3], it learns an embedding space by maximizing the compatibility between the visual feature and class label spaces to classify novel unseen classes.

Semantic Autoencoder. The semantic autoencoder [13,22] has received increasing interests from academia in recent years. A simplest semantic autoencoder contains a encoder, a decoder and one semantic hidden layer shared by the encoder and decoder. Specifically, the encoder aims to project the input data into the hidden layer and the decoder is designed to reconstruct the original input data as far as possible. It achieves the best encoder by minimizing the reconstruction error between the novel and original data.

Source Domain Attribute Similarity. The Source Domain Attribute Similarity (SDAS) [7,21] is to describe the "closeness" of semantic attributes of source data to their class prototypes. Specifically, the attributes from the labelled source data can't correspond to their faithful representations completely because of various factors (e.g. human definition). Accordingly, this condition leads to an inaccurate projection function learned from the source domain dataset. If the SDAS constraint is introduced into the source domain, the performance of the learned projection function will be enhanced obviously, resulting in the accuracy improvement.

3 Methodology

3.1 Problem Definition

Let $Z_s = [z_1^s, \ldots, z_{c_s}^s]$ denotes a set of c_s source class labels and $Z_t = [z_1^t, \ldots, z_{c_t}^t]$ a set of c_t target class labels with $Z_s \cap Z_t = \varnothing$. Each class label corresponds to an m-dimensional vector describing absence/presence of attributes. We thus have $P_s = [p_1^s, \ldots, p_{c_s}^s] \in R^{m \times c_s}$ and $P_t = [p_1^t, \ldots, p_{c_t}^t] \in R^{m \times c_t}$ as source and target class prototypes respectively. Each image is represented using a d-dimensional visual feature vector, so we denote $X_s = [x_1^s, \ldots, x_{n_s}^s] \in R^{d \times n_s}$ as the given n_s source images and $Y_s = [y_1^s, \ldots, y_{n_s}^s] \in R^{m \times n_s}$ as the corresponding semantic attributes. In addition, suppose there are n_t target images $X_t = [x_1^t, \ldots, x_{n_t}^t] \in R^{d \times n_t}$. The goal of zero-shot recognition is to estimate unknown target attributes $Y_t = [y_1^t, \ldots, y_{n_t}^t] \in R^{m \times n_t}$ using the given source data. We summarize some notations emerging in the work and the corresponding descriptions in Table 1.

Table 1. Notations and the corresponding descriptions in the paper.

Notation	Description	Notation	Description
X_s	Source images	X_t	Target images
Y_s	Source attributes	Y_t	Target attributes
P_s	Source class prototypes	P_t	Target class prototypes
D	Encoder	D_1	Decoder

3.2 Model Formulation

We first learn one encoder $D \in R^{m \times d}$ to project the given source features $X_s \in R^{d \times n_s}$ into $Y_s \in R^{m \times n_s}$ by $DX_s = Y_s$. The Y_s is then projected back into the feature space with a decoder $D_1 \in R^{d \times m}$, resulting in novel feature representations $X_s^* \in R^{d \times n_s}$ by $X_s^* = D_1 Y_s$. We wish that the obtained X_s^* are as similar as possible to the original input data X_s and write the objective as:

$$\min_{D,D_1} \|X_s - D_1 D X_s\|_F^2 \ s.t. \ D X_s = Y_s \tag{1}$$

To further simplify the objective, we assume $D_1 = D^T$ and the learned objective thus becomes:

$$\min_{D} \|X_s - D^T Y_s\|_F^2 + \lambda \|D X_s - Y_s\|_F^2 \tag{2}$$

where λ is a weighting coefficient that controls the importance of first and second terms, which correspond to the decoder and encoder respectively. In the previous work, we presume that the source and target domains share the same encoder and get another objective about D in the target domain:

$$\min_{D,Y_t} \|X_t - D^T Y_t\|_F^2 + \lambda_1 \|D X_t - Y_t\|_F^2 \tag{3}$$

In order to express the information of D completely, we decide to combine Eq. (2) with Eq. (3):

$$\min_{D,Y_t} \|X_s - D^T Y_s\|_F^2 + \lambda \|D X_s - Y_s\|_F^2 + \|X_t - D^T Y_t\|_F^2 + \lambda_1 \|D X_t - Y_t\|_F^2 \tag{4}$$

Now, we introduce the attribute constraint: Source Domain Attribute Similarity (SDAS) into Eq. (4) to ensure that the Y_s got from the labelled source data are near to their faithful values. Specifically, that is the critical regularisation term $\sum_{i=1}^{n_s} \sum_{j=1}^{c_s} m_{i,j}^s \|y_i^s - p_j^s\|_2^2$, where p_j^s is the j-th source class prototype and y_i^s is the given semantic attribute of a test image x_i^s. Because the class label of x_i^s is known, $m_{i,j}^s$ can be estimated simply, $m_{i,j}^s = 1$ if x_i^s belongs to the j-th source class, otherwise $m_{i,j}^s = 0$.

$$\min_{D,Y_t,Y_s} \|X_s - D^T Y_s\|_F^2 + \lambda \|D X_s - Y_s\|_F^2 + \|X_t - D^T Y_t\|_F^2$$
$$+ \lambda_1 \|D X_t - Y_t\|_F^2 + \lambda_2 \sum_{i=1}^{n_s} \sum_{j=1}^{c_s} m_{i,j}^s \|y_i^s - p_j^s\|_2^2 \tag{5}$$

3.3   Model Optimisation

It is important to point out that Eq. (5) is concerned with D, Y_s and Y_t, so we consider using an alternating optimisation to resolve it. Specifically, we turn this learned objective into three sub objectives as follows:

(1) Fix D, Y_t, updata Y_s:

$$Y_s^* = \arg\min_{Y_s} \|X_s - D^T Y_s\|_F^2 + \lambda \|D X_s - Y_s\|_F^2 + \lambda_2 \sum_{i=1}^{n_s} \sum_{j=1}^{c_s} m_{i,j}^s \|y_i^s - p_j^s\|_2^2 \tag{6}$$

We take a derivative of Eq. (6) and set it zero:

$$Y_s^* = (DX_s + \lambda DX_s + \frac{\lambda_2}{2}R_s^T)(I_{n_s} + \lambda I_{n_s} + \lambda_2 I_{n_s})^{-1} \tag{7}$$

where I is the identity matrix and the unknown R_s is written as follows:

$$R_s = \begin{bmatrix} (\sum_{j=1}^{c_s} 2m_{1,j}^s p_j^s)^T \\ \cdots \\ (\sum_{j=1}^{c_s} 2m_{n_s,j}^s p_j^s)^T \end{bmatrix} \tag{8}$$

(2) Fix D, Y_s, updata Y_t:

$$Y_t^* = \arg\min_{Y_t} \|X_t - D^T Y_t\|_F^2 + \lambda_1 \|DX_t - Y_t\|_F^2 \tag{9}$$

It is a standard least squares problem and we achieve the closed form solution:

$$Y_t^* = (DD^T + \lambda_1 I_{n_t})^{-1}(DX_t + \lambda_1 DX_t) \tag{10}$$

(3) Fix Y_t, Y_s, updata D:

$$D^* = \arg\min_{D} \|X_s - D^T Y_s\|_F^2 + \lambda\|DX_s - Y_s\|_F^2 + \|X_t - D^T Y_t\|_F^2 + \lambda_1\|DX_t - Y_t\|_F^2 \tag{11}$$

We obtain the derivative of this equation and set it zero as follows:

$$(Y_s Y_s^T + Y_t Y_t^T)D + D(\lambda X_s X_s^T + \lambda_1 X_t X_t^T) = Y_s X_s^T + Y_t X_t^T + \lambda Y_s X_s^T + \lambda_1 Y_t X_t^T \tag{12}$$

We denote $A = Y_s Y_s^T + Y_t Y_t^T$, $B = \lambda X_s X_s^T + \lambda_1 X_t X_t^T$, $C = Y_s X_s^T + Y_t X_t^T + \lambda Y_s X_s^T + \lambda_1 Y_t X_t^T$ and have the following form:

$$AD + DB = C \tag{13}$$

Equation (13) is a famous Sylvester equation and can be solved simply by $D^* = sylvester(A, B, C)$.

Finally, we adopt the following convergence condition to get the best encoder D:

$$\|D_k - D_{k-1}\|_2^2 \leq \alpha \tag{14}$$

where value $\alpha = 20$ is given at training stage. Our model algorithm is summarised in Algorithm 1.

3.4   Classification

Once the encoder D is estimated by Algorithm 1, zero-shot recognition can be performed simply. We embed a new test image x_i^s to the semantic attribute space by $y_i^s = Dx_i^s$. The class label of x_i^s is achieved by comparing the similarity between the estimated y_i^s and target class prototypes P_s:

$$f(y_i^s) = \arg\min_{j} \|y_i^s - p_j^s\|_F^2 \tag{15}$$

where p_j^s is the j-th target class prototype and f is a similarity function which returns the class label of image.

Algorithm 1 Semantic Double-Autoencoder with Attribute Constraint for ZSR

Input:
    X_s: source images
    X_t: target images
    Y_s: source attributes
    P_s: source class prototypes
    P_t: target class prototypes
    $\lambda, \lambda_1, \lambda_2, \alpha$: free parameters
Output:
    Y_t: the coefficients for target images
    D: the encoder
1: Initialize
     Y_s by P_s,
     $m_{i,j}^s = 1$ if x_i^s belongs to the j-th source class, otherwise $m_{i,j}^s = 0$,
     D by Eq. (2).
2: While not converge do
3:    Update Y_t by Eq. (10);
4:    Update Y_s by Eq. (7);
5:    Update D by Eq. (13);
6:    check the convergence condition:
      $\|D_k - D_{k-1}\|_2^2 \leq \alpha$
7: end

---

4 Experiment

4.1 Experimental Setup

Datasets and Settings. We conduct a set of experiments on three benchmark datasets. (a) Animals with Attributes (AwA): It consists of 30475 images from 50 coarse-grained animals and each class label is provided with a 85D attribute vector. We adopt 40 classes as source classes and the other 10 classes as target classes. (b) Caltech UCSD Birds (CUB): It is a fine-grained bird dataset containing 200 different bird species, with 11788 images in total. Each image has been annotated with 312 binary attributes. 150 classes and 50 classes are chosen as the source and target domain data respectively. (c) aPascal-aYahoo (aPY): It contains two datasets: a PASCAL VOC 2008 dataset of 12695 images and 2644 images collected with Yahoo image search engine. The object images are given with 64 semantic attributes. For aPY, the PASCAL (20 classes) part serves the source domain and the Yahoo (12 classes) as test data. Table 2 summarizes the statistics in three datasets. Our model has three free parameters to determine and we evaluate the best values by continuous debuggings: $\lambda = 150000$, $\lambda_1 = 0.01$, $\lambda_2 = 1.2$.

Features. All recent ZSR approaches use deep features extracted from 2 popular Deep Convolutional Neural Network (CNN) architectures, i.e. VGG and GoogLeNet. In our experiments, we use VGG which uses the 4096-dim activations of last layer but one as features for all datasets.

Table 2. A brief summary of three benchmark datasets in our experiments. Notations - "A-D": the dimension of attribute; "F-D": the dimension of feature.

Dataset	Images	A-D	F-D	Source/target classes
AwA	30475	85	4096	40/10
CUB	11788	312	4096	150/50
aPY	15339	64	4096	20/12

4.2   Experimental Results

Classification Accuracy. We compare the classification accuracy of the proposed SDAWAC with five ZSR approaches RKT, UDA, ESZSL, GSE and MFMR which are among the more efficient existing ZSR approaches. Table 3 shows that our SDAWAC can get a better result than other five competitors on three benchmark datasets. For the AwA dataset, we obtain the promising accuracy 83.1%, even higher than the best baseline approach MFMR. For the CUB dataset, our accuracy is not as good as in AwA and reaches 58.3%, but the performance improvement is still inspiring. For aPY, our model improve the optimal method GSE 56.3% by 1.6%.

Table 3. Comparison in terms of classification accuracy (%) on three benchmark datasets using deep VGG features. Notations - "F": features; "SI"; side information; "A": attribute space; "V": VGG features.

Method	F	SI	AwA	CUB	aPY
UDA [21]	V	A	73.2	39.5	38.9
RKT [19]	V	A	75.3	38.6	40.6
MFMR [31]	V	A	79.8	47.7	48.2
ESZSL [2]	V	A	75.2	44.5	31.4
GSE [32]	V	A	78.4	57.4	56.3
Ours	V	A	83.1	58.3	57.9

Table 4. Our computational cost (in second) in comparison with four ZSR models in AwA dataset.

Method	Training	Testing
RKT [19]	49.71	10.65
UDA [21]	71.13	15.85
SMS [17]	38.63	7.32
GSE [32]	16.63	3.65
Ours	3.15	1.05

Computational Cost. The computational cost of our proposed SDAWAC in AwA in comparison with four popular ZSR methods: UDA, RKT, SMS and GSE, is summarized in Table 4. We can observe that our method is at least 5 times faster than the best baseline approach GSE for training. At testing stage, ours is still the fastest one.

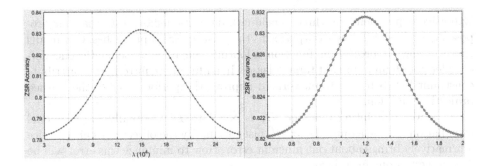

Fig. 3. The effects of parameters λ and λ_2.

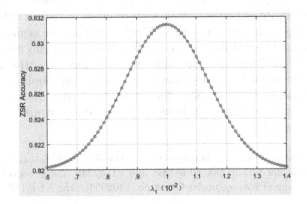

Fig. 4. The effect of parameter λ_1.

The Effects of Parameters. It is necessary to point out that the performance of our model is also concerned with three free parameters: λ, λ_1 and λ_2. Figures 3 and 4 show how the recognition accuracy in AwA dataset is affected by the above parameters. In the experiments, we study the effect of one parameter by fixing the other two. We observe that the three images all take on a Gaussian distribution. Further, the recognition accuracy is more sensitive to λ_1 than λ and λ_2. It is of no difficulty to explain this phenomenon. Parameter λ_1 is the weighting coefficient that controls the decoder and encoder in the target domain. Our model is to learn an efficient projection function in the source and target domains simultaneously, so λ_1 has a greater impact on the performance. On the other hand, it also proves the validity of our proposed approach.

4.3   Experimental Analysis

Extensive experiments on three benchmark datasets demonstrate that our proposed method significantly outperforms the existing ZSR approaches. Our model tries to learn a projection function adopting the popular semantic encoder-decoder paradigm to predict target classes. In addition, we introduce an attribute constraint SDAS to our framework to improve the performance of the learned projection function.

It is worth pointing out that D_1 is equal to D^{-1} actually in the model formulation and we assume $D_1 = D^T$ just to simplify the objective. This practice affirmatively brings about one inherent error. How to minimize this error will be encouraged in our future work.

5   Conclusion

In this paper, we propose a novel model called semantic double-autoencoder with attribute constraint to settle the domain shift problem in ZSR. It differs from the existing ZSR approaches in three ways. Firstly, we take the semantic encoder-decoder paradigm to learn one projection function considering the source and target domains simultaneously. Secondly, we introduce one strong constraint SDAS into our framework to reduce the effect of source attribute noise. Thirdly, our model is superior to the competitors in terms of computational cost. The experimental results on three benchmark datasets demonstrate the validity of our proposed method.

Acknowledgments. This work is supported by the National Natural Science Foundation of China under Grant nos. 61402238 and 61502245, the Postdoctoral Science Foundation of Jiangsu Province under Grant no. 1302054C, the NUPTSF under Grant no. NY212029.

References

1. Zhang, Z., Saligrama, V.: Zero-shot learning via joint latent similarity embedding. In: Proceedings of the IEEE Conference on Computer Vision and Pattern Recognition, pp. 6034–6042 (2016)
2. Romera-Paredes, B., Torr, P.H.: An embarrassingly simple approach to zero-shot learning. In: 32th IEEE International Conference on Machine Learning, pp. 2152–2561 (2015)
3. Akata, Z., Perronnin, F., Harchaoui, Z, Schmid, C.: Label embedding for attribute-based classification. In: Proceedings of the IEEE Conference on Computer Vision and Pattern Recognition, pp. 819–826 (2013)

4. Wang, X., Ji, Q.: A unified probabilistic approach modelling relationships between attributes and objects. In: Proceedings of the IEEE International Conference on Computer Vision, pp. 2120–2127 (2013)
5. Bucher, M., Herbin, S., Jurie, F.: Improving semantic embedding consistency by metric learning for zero-shot classiffication. In: Leibe, B., Matas, J., Sebe, N., Welling, M. (eds.) ECCV 2016. LNCS, vol. 9909, pp. 730–746. Springer, Cham (2016). https://doi.org/10.1007/978-3-319-46454-1_44
6. Lampert, C.H., Nickisch, H., Harmeling, S.: Attribute-based classification for zero-shot visual object categorization. IEEE Trans. Pattern Anal. Mach. Intell. **36**, 453–465 (2014)
7. Yu, J., Wu, S.: Robust zero-shot learning with source attributes noise. In: Proceedings of the 2016 International Conference on Progress in Informatics and Computing, pp. 205–209 (2016)
8. Jayaraman, D., Grauman, K.: Zero-shot recognition with unreliable attributes. In: Advances in Neural Information Processing Systems, pp. 3464–3472 (2014)
9. Zhang, Z., Saligrama, V.: Zero-shot learning via semantic similarity embedding. In: Proceedings of the IEEE International Conference on Computer Vision, pp. 4166–4174 (2015)
10. Mikolov, T., Sutskever, I., Chen, K., Corrado, G.S., Dean, J.: Distributed representations of words and phrases and their compositionality. In: Advances in Neural Information Processing Systems, pp. 3111–3119 (2013)
11. Lampert, C.H., Nickisch, H., Harmeling, S.: Learning to detect unseen object classes by between-class attribute transfer. In: CVPR, pp. 951–958 (2009)
12. Guo, Y., Ding, G., Jin, X., Wang, J.: Learning predictable and discriminative attributes for visual recognition. In: 29th AAAI Conference on Artificial Intelligence, pp. 3783–3789 (2015)
13. Kodirov, E., Xiang, T., Gong, S.: Semantic autoencoder for zero-shot learning. In: Proceedings of the IEEE Conference on Computer Vision and Pattern Recognition (2017)
14. Changpinyo, S., Chao, W., Gong, B., Sha, F.: Synthesized classifiers for zero-shot learning. In: CVPR, pp. 580–587 (2016)
15. Kodirov, E., Xiang, T., Fu, Z., Gong, S.: Learning robust graph regularisation for subspace clustering. In: British Machine Vision Conference (2016)
16. Patel, V.M., Gopalan, R., Li, R., Chellappa, R.: Visual domain adaptation: a survey of recent advances. IEEE Sig. Process. Mag. **32**(3), 53–69 (2015)
17. Guo, Y., Ding, G., Jin, X., Wang, J.: Transductive zero-shot recognition via shared model space learning. In: 30th AAAI Conference on Artificial Intelligence (2016)
18. Shigeto, Y., Suzuki, I., Hara, K., Shimbo, M., Matsumoto, Y.: Ridge regression, hubness, and zero-shot learning. In: Appice, A., Rodrigues, P.P., Santos Costa, V., Soares, C., Gama, J., Jorge, A. (eds.) ECML PKDD 2015. LNCS (LNAI), vol. 9284, pp. 135–151. Springer, Cham (2015). https://doi.org/10.1007/978-3-319-23528-8_9
19. Wang, D., Li, Y., Lin, Y., Zhuang, Y.: Relational knowledge transfer for zero-shot learning. In: 30th AAAI Conference on Artificial Intelligence, pp. 2145–2151 (2016)
20. Krizhevsky, A., Sutskever, I., Hinton, G.E.: Imagenet classification with deep convolutional neural networks. In: Advances in Neural Information Processing Systems, pp. 1097–1105 (2012)
21. Kodirov, E., Xiang, T., Fu, Z., Gong, S.: Unsupervised domain adaptation for zero-shot learning. In: ICCV, pp. 2452–2460 (2015)
22. Rifai, S., Vincent, P., Muller, X., Glorot, X., Bengio, Y.: Contractive auto-encoders: explicit invariance during feature extraction. In: 28th International Conference on Machine Learning, pp. 833–840 (2011)

23. Berg, T.L., Berg, A.C., Shih, J.: Automatic attribute discovery and characterization from noisy web data. In: Daniilidis, K., Maragos, P., Paragios, N. (eds.) ECCV 2010. LNCS, vol. 6311, pp. 663–676. Springer, Heidelberg (2010). https://doi.org/10.1007/978-3-642-15549-9_48

24. Yu, F.X., Cao, L., Feris, R.S., Smith, J.R., Chang, S.: Designing category-level attributes for discriminative visual recognition. In: CVPR (2013)

25. Lei Ba, J., Swersky, K., Fidler, S.: Predicting deep zero-shot convolutional neural networks using textual descriptions. In: ICCV, pp. 4247–4255 (2013)

26. Elhoseiny, M., Saleh, B., Elgammal, A.: Write a classifier: zero-shot learning using purely textual descriptions. In: ICCV (2013)

27. Rohrbach, M., Stark, M., Schiele, B.: Evaluating knowledge transfer and zero-shot learning in a large-scale setting. In: Computer Vision and Pattern Recognition, pp. 1641–1648 (2011)

28. Yu, X., Aloimonos, Y.: Attribute-based transfer learning for object categorization with zero/one training example. In: Daniilidis, K., Maragos, P., Paragios, N. (eds.) ECCV 2010. LNCS, vol. 6315, pp. 127–140. Springer, Heidelberg (2010). https://doi.org/10.1007/978-3-642-15555-0_10

29. Perrot, M., Habrard, A.: Regressive virtual metric learning. In: Advances in Neural Information Processing Systems, pp. 1810–1818 (2015)

30. Reed, S., Akata, Z., Yan, X., Logeswaran, L., Schiele, B., Lee, H.: Generative adversarial text-to-image synthesis. In: 33th International Conference on Machine Learning (2016)

31. Xu, X., Shen, F., Yang, Y., Zhang, D., Shen, H.T., Song, J.: Matrix tri-factorization with manifold regularizations for zero-shot learning. In: 2017 IEEE Conference on Computer Vision and Pattern Recognition, pp. 2007–2016 (2017)

32. Yang, L., Ling, S.: Describing unseen classes by exemplars: zero-shot learning using grouped simile ensemble. Unsupervised deep embedding for clustering analysis. In: 2017 IEEE Winter Conference on Applications of Computer Vision (2017)

Zero-Shot Leaning with Manifold Embedding

Yun-long Yu, Zhong Ji[✉], and Yan-wei Pang

School of Electrical and Information Engineering, Tianjin University, Tianjin, China
jizhong@tju.edu.cn

Abstract. Zero-Shot Learning (ZSL) has gained its popularity recently owing to its promising characteristic that requires no training data to recognize new visual classes. One key technique is to transfer knowledge from the seen classes to the new unseen classes in an intermediate embedding space for both visual and textual modalities. Therefore, the construction of the embedding space is extremely important. Manifold embedding is able to well capture the intrinsic structure of the embedding space. To this end, with the assumption that the distribution of the semantic categories in the word vector space has an intrinsic manifold structure, this paper proposes a Manifold Embedding based ZSL (ME-ZSL) approach by formulating the manifold structure for the visual to textual embedding with the intra-class compactness, the inter-class separability, and the locality preservation. The linear, closed-form solution makes ME-ZSL efficient to compute. Extensive experiments on the popular AwA and CUB datasets validate the effectiveness of ME-ZSL.

Keywords: Zero-Shot Learning · Image classification
Manifold embedding · Word vector

1 Introduction

Recent years, machines have been significantly promoted their performance to recognize objects due to the prosperous progresses of the machine learning and feature learning techniques [7,12,18,22,31,32]. Substantial results have even been reported surpassing human-level performance on the 1000-class ImageNet dataset [8]. Compared with the success of object recognition techniques, however, machines are still in an infant stage to imitate humans' strong inferential capability. Considering this capability from the visual aspect, humans achieve it partly because that they obtain massive prior knowledge from various sources, e.g., TV, Internet, textbook, etc. The prior knowledge enables them to contextualize how unknown objects might look and recognize these objects when they first look at them. For example, imaging that a boy has never seen a panda, however, he has the prior knowledge that a panda is a bear with large, distinctive black patches around its eyes and across its round body. When he sees a panda in the first time, he may recognize the panda with his own inferential capability.

© Springer Nature Switzerland AG 2018
Y. Peng et al. (Eds.): IScIDE 2018, LNCS 11266, pp. 135–147, 2018.
https://doi.org/10.1007/978-3-030-02698-1_12

Zero-Shot Learning (ZSL) is one way of attempting to enable computer vision systems to possess the visually inferential capability. It is an effective technique to cope with the dramatic expanse of data categories and the lack of labeled instances. ZSL aims at recognizing instances from unseen or unknown classes without a single training example [1,3–6]. The prior knowledge in ZSL is the seen classes for which many labeled instances exist, and the descriptions or names for the unseen classes. Accordingly, the challenging problem is how to transfer the prior knowledge from the seen classes (i.e., classes with the labeled data for training) to unseen classes (i.e., classes with no labeled training data). Generally, the key to implement this transfer is to construct an embedding space for the modalities of both visual and textual descriptions. Then, the similarity between the seen classes and unseen classes can be measured directly in this sharing space.

Attribute is an excellent intermediate semantics to construct the embedding space and have led to many successful ZSL methods [2,14,16,17]. It defines a few properties of an object, such as shape, color and the presence or absence of a certain body part. They are shared across both seen and unseen categories, which makes the knowledge transfer possible. In this way, both visual features and class names of the seen classes, as well as the unseen classes' names are embedded into the attribute space at the training stage. At the testing stage, the attribute values of a given test image are first obtained in the attribute space, and then its class name is inferred via the linkage between the attribute and the class name. Word vector is another promising embedding space for ZSL. Based on a distributed language representation, the word vector techniques represent words as vectors. Popular methods include Word2Vec [19] and Glove [23]. The word vector space is constructed with a linguistic knowledge base, e.g., Wikipedia, and then the names of all the classes are projected onto it. The word-vector-based approaches utilize the seen classes to teach a general mapping function from the visual space to word vector space, and then utilize this function to map the visual features of the unseen classes to the word vector space, finally retrieve the nearest neighbor words as their labels. In this paper, we focus on the word-vector-based approaches.

Although the existing ZSL approaches differ themselves in the embedding spaces and the embedding methods, many of them are based on Euclidean space. [6] demonstrates that the distribution of the semantic categories in the word vector space has a rich intrinsic manifold structure. Thus, they propose a semantic manifold distance to replace the conventional distance metric by a novel absorbing Markov chain process. However, they do not formulate the objective function with manifold embedding. Actually, it is reasonable to further exploit the manifold more directly in the embedding space. To this end, we present a novel Manifold Embedding based ZSL (ME-ZSL) approach in this paper, its flowchart is illustrated in Fig. 1.

It is worthy highlighting the contributions of this work:

(1) A novel ME-ZSL approach is proposed. Specifically, it simultaneously formulates the manifold structure for the visual to textual embedding in three

aspects, i.e., the intra-class compactness, the inter-class separability, and the locality preservation.

(2) The proposed ME-ZSL approach has a linear, closed-form solution, which is very efficient to compute. The global optimal solution ensures its effectiveness.

(3) Experiments on the popular AwA and CUB datasets demonstrate its compelling performance to both the state-of-the-art word-vector-based approaches and the attribute-based approaches.

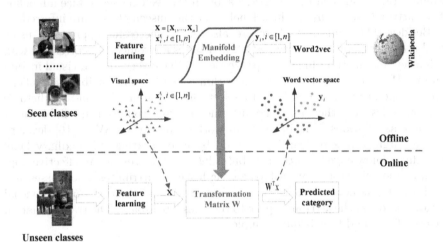

Fig. 1. The flowchart of the proposed Manifold Embedding Based ZSL (ME-ZSL) approach. In the offline training stage, the images and the words are firstly represented with feature vectors via feature learning and word2vec modules, respectively. Then, both the visual features and the word vectors of the seen classes are employed to model the proposed ME-ZSL, and outputs the cross-modal embedding matrix $\mathbf{W}$. In the testing stage, the visual feature of the unseen image $\mathbf{x}$ is embedded into the word vector space, and its category is predicted by finding the nearest word vector to $\mathbf{W}^T\mathbf{x}$ among the names of the unseen classes.

2   Related Work

From the perspective of embedding space, ZSL approaches fall into the following two categories: attribute-based approaches and word-vector-based approaches.

2.1   Attribute-Based Approaches

Attribute-based approaches build an attribute space for the seen and unseen classes, enabling learning unseen classes with their descriptions. In this way, the attribute is actually an intermediate space transferring the prior knowledge from the seen classes to the unseen classes. Among these studies, DAP (Directed Attribute Prediction) [14] and IAP (Indirected Attribute Prediction) [14] are pioneering work. DAP employs the attributes as an intermediate space between the visual features and the labels. In this case, a probabilistic classifier is learned for each attribute at the training stage, and the unseen classes are then inferred with the learned estimators. In IAP, the attributes make a connection between two layers of labels, the lower one for seen classes and the higher one for unseen classes. In this case, a probabilistic classifier is learned for each seen class at the training stage, and then the predictions for all the seen classes induce a labeling of the attribute layer, from which labels over the unseen classes are inferred.

Besides DAP and IAP, there are also many successful approaches using attributes. For example, observing that attributes are sometimes correlated with each other and abstract linguistic properties can have very diverse visual instantiations, [10] uses a random forest algorithm to exploit the attribute classifiers' receiver operating characteristics to select the discriminative and predictable decision nodes. And they also generalize the idea to account for unreliable class-attribute associations. Within the framework of multi-class SVM, [16] develops a max-margin zero-shot learning method by incorporating the auxiliary label relatedness knowledge from the attributes into it. [24] presents an effective approach by modeling the relationships among features, attributes, and classes as a two linear layers network to learn an embedding transformation from the visual feature space and the attribute space to the class label space. The transformation is closed-form, and thus is quite simple.

2.2   Word-Vector-Based Approaches

Recently, word-vector-based approaches have been proven to be another promising direction to ZSL with the rapid progress of computational linguistics techniques. Many neural language models, such as Bilinear language Model [20], Word2Vec [19], and GloVe [23] have been proposed to represent a word with a continuous vector. In this way, all the categories' textual names are embedded into the vector space. Generally, the words are closer in the space if they co-occur in similar contexts. In this case, the key of zero-shot learning is how to effectively embed the visual features into the word vector space. Since no human labeling is required, word-vector-based approaches avoid most of the issues in attribute-based approaches.

Several attempts have been made in this direction, of which [3] and [26] are among the first researchers. [3] proposes a method called DeViSE, which trains a linear mapping to link the image representation with the word vector using a hinge loss function. [26] employs a regression model to achieve the visual to word vector mapping. By searching the k relevant label embedding vectors with KNN, [21] embeds images into the word vector space via a convex combination of the top k classes' label embedding vectors. In [15], the authors compare 4 alternative learning algorithms, i.e., Linear Regression (LR), Canonical Correlation Analysis (CCA), Singular Value Decomposition (SVD), and Neural Network (NN).

They also show that ZSL scales well to a large and noisy dataset. Recently, SJE (Structured Joint Embedding) [1] relates the input embedding and output embedding through a compatibility function, and ZSL is implemented by finding the label responsible for the highest joint compatibility score. Further, LatEm [29] employs a bilinear compatibility model to learn a collection of maps as latent variables for the current image-class pair. The model is trained with a ranking based objective function that penalizes incorrect rankings of the true class for a given image. [5] shows that word vector can be used effectively to initialize the semantic space and develop a semi-supervised vocabulary-informed learning approach to push the zero-shot learning to a more open setting.

In addition, considering that the attribute and word vector may provide complementary information, there are some studies paying attention to exploiting both of them to enhance the performance of ZSL [1,4,13]. For example, [1] comprehensively evaluates the combinations with different output embedding approaches obtained from attribute and word vectors. They show that the attribute-based approaches usually benefit from the word vector embedding, and the combination performs equivalently to or even better than the individuals alone. [4] presents a novel heterogeneous multi-view hypergraph label propagation method in a transductive embedding space, which effectively exploits the complementary information offered by both attribute and word vectors.

3   The Proposed Method

The proposed ME-ZSL approach relies on the cross-modal embedding from a visual space to a word vector space. The embedding is performed with the assumption that the word vectors of all the categories lie in a manifold structure [6]. We first introduce notations, followed by the details of constructing the object functions to map the visual features into the word vector space.

3.1   Problem Definition

Let $S = \{s_1, \cdots, s_i, \cdots, s_n\}$ denote a set of n seen class labels and $U = \{u_1, \cdots, u_m\}$ a set of m unseen class labels, and $S \cap U = \emptyset$. The visual feature matrix for the labeled training dataset is $\mathbf{X} = [\mathbf{X}_1, \cdots, \mathbf{X}_i, \cdots, \mathbf{X}_n]$, where $\mathbf{X}_i = [\mathbf{x}_i^1, \cdots, \mathbf{x}_i^{t_i}]$ $(i \in [1, n])$ is a matrix for the i-th seen class, t_i is the training sample number of the i-th seen class, and $\mathbf{x}_i^j \in \mathbb{R}^p$ is the visual feature vector representing the j-th labeled sample from the i-th seen class. Let $\mathbf{Y} = [\mathbf{y}_1, \cdots, \mathbf{y}_i, \cdots, \mathbf{y}_n] \in \mathbb{R}^{q \times n}$ as the set of the corresponding word vectors for S. The aim of ME-ZSL is to learn an embedding function $F : \mathbb{R}^p \to \mathbb{R}^q$ with the training set $\Psi_S = \{(X_i, s_i), 1 \leq i \leq n\}$. Then given a test sample $\mathbf{x}$, its class name can be predicted by finding the nearest word vector to $F(x)$ among the labels of unseen classes. Note that the function can be linear or nonlinear. In this paper, we only focus on the linear one, such that the embedding function is $\mathbf{y}_t = \mathbf{W}^T \mathbf{x}$, where $\mathbf{W}$ is a cross-modal embedding matrix.

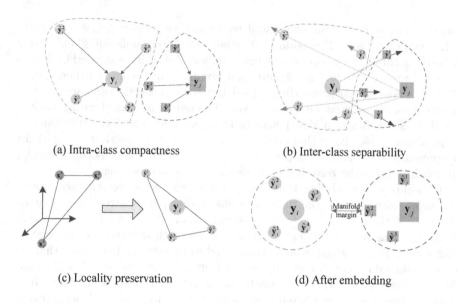

(a) Intra-class compactness   (b) Inter-class separability

(c) Locality preservation   (d) After embedding

Fig. 2. Illustration of the objective function composition of ME-ZSL. There are two seen classes i and j, and their word vectors $\mathbf{y}_i$ and $\mathbf{y}_j$. The examples after embedding in the semantic space are denoted as $\hat{\mathbf{y}}_i^k$ and $\hat{\mathbf{y}}_j^k$, respectively.

3.2   Manifold Embedding-Based ZSL (ME-ZSL)

Manifold builds a more accurate intrinsic structure of data and has been widely used in many domains [9,11,30]. ME-ZSL fomathbfulates the manifold structure of the embedding data from three aspects, as illustrated in Fig. 2. Specifically, Fig. 2(a) is to model the intra-class compactness between the word vectors and the embedding outputs from the same class. Figure 2(b) is to represent the inter-class separability between the word vector in one class and the embedding outputs from the other classes. Both the ideas of the intra-class compactness and the inter-class separability are borrowed from the graph embedding framework in [30]. The idea behind them is the maximum margin criterion. Note [16] also adopts the maximum margin idea to ZSL, however, it is in the multi-class SVM framework while ours is in the manifold learning. The third one, Fig. 2(c), is to model the locality preservation property developed in [9], which aims at preserving the local structure of the visual space in the embedding linguistic space. Correspondingly, the objective function J is composed of the three parts: J_1, J_2 and J_3. We discuss them in detail below.

Assuming that we have obtained the linear embedding matrix $\mathbf{W}$, we attempt to capture the embedding vector $\hat{\mathbf{y}}_i^k$ for a given visual sample $\mathbf{x}_i^k$, that is, $\hat{\mathbf{y}}_i^k = \mathbf{W}^T\mathbf{x}_i^k$. One natural idea is that $\hat{\mathbf{y}}_i^k$ should be as close as the word vector $\mathbf{y}_i$ in the embedding space as they represent the same word. To this end, we minimize the objective function J_1 to indicate this intra-class compactness:

$$J_1 = \sum_{i=1}^{n} \sum_{j=1}^{t_i} \left\| \hat{\mathbf{y}}_i^j - \mathbf{y}_i \right\|_2^2 \tag{1}$$

$$= \mathbf{W}^T \mathbf{X} \mathbf{X}^T \mathbf{W} + \mathbf{Y} \mathbf{Y}^T - 2\mathbf{W}^T \mathbf{X} \mathbf{Y}^T$$

Actually, it is the linear regression model widely used in the word-vector-based ZSL approaches [15, 26].

In contrast to the objective function J_1, the embedding outputs in one class should be as far as possible to the word vectors in the other classes. Therefore, we maximize this inter-class separability with:

$$J_2 = \sum_{\substack{i,j=1 \\ i \neq j}}^{n} \sum_{k=1}^{t_i} \left\| \hat{\mathbf{y}}_j^k - \mathbf{y}_i \right\|_2^2 \tag{2}$$

$$= \mathbf{W}^T \mathbf{X} \mathbf{X}^T \mathbf{W} + \mathbf{Y} \mathbf{Y}^T - 2\mathbf{W}^T \sum_{\substack{i,j=1 \\ i \neq j}}^{n} \mathbf{X}_i \mathbf{Y}_j^T$$

Besides minimizing the intra-class compactness and maximizing the inter-class separability, the intrinsic geometry of the data should be preserved. To this end, we model this objective by minimizing the locality preserving function as follows:

$$J_3 = \sum_{i=1}^{n} \sum_{j,k=1}^{t_i} s_i^{jk} \left\| \hat{\mathbf{y}}_i^j - \hat{\mathbf{y}}_i^k \right\|_2^2$$

$$= \sum_{i=1}^{n} \mathbf{W}^T \mathbf{X}_i \left(\mathbf{D}_i - \mathbf{S}_i \right) \mathbf{X}_i^T \mathbf{W} \tag{3}$$

$$= \sum_{i=1}^{n} \mathbf{W}^T \mathbf{X}_i \mathbf{L}_i \mathbf{X}_i^T \mathbf{W}$$

$$= \mathbf{W}^T \mathbf{X} L X \mathbf{W},$$

where $\mathbf{S}_i = \sum_{j,k=1}^{t_i} s_i^{jk}$ is a similarity matrix, $s_i^{jk} = e^{-\frac{\left\| \mathbf{x}_i^j - \mathbf{x}_i^k \right\|_2^2}{2\sigma^2}}$ measures the visual similarity between $\mathbf{x}_i^j$ and $\mathbf{x}_i^k$ with heat kernel, $\mathbf{D}_i = \sum_{j=1}^{t_i} s_i^{jk}$ is a diagonal matrix, $\mathbf{L}_i = \mathbf{D}_i - \mathbf{S}_i$ is a Laplacian matrix, $\mathbf{L} = diag\{\mathbf{L}_1, \ldots, \mathbf{L}_n\}$. In this case, objective function J_3 can be considered as a supervised locality preserving projection from visual space to linguistic space. Minimizing it is an attempt to ensure that if $\mathbf{x}_i^j$ and $\mathbf{x}_i^k$ are close to each other, their embeddings $\hat{\mathbf{y}}_i^j$ and $\hat{\mathbf{y}}_i^k$ are close as well.

As a result, we have the final objective function by minimizing the combination of the three parts as follows:

$$J = J_1 - \alpha J_2 + \beta J_3 + \lambda \|\mathbf{W}\|_2^2, \tag{4}$$

where the last term $\|\mathbf{W}\|_2^2$ is a regularizer, α, β and λ are parameters to control the weight of each term. By the simple algebra, the solution to the final objective function is written in the closed form:

Algorithm 1. The Procedure of ME-ZSL

Input: Training examples: visual features $\mathbf{X} = \{\mathbf{X}_i\}_{i=1}^n$, where $\mathbf{X}_i = [\mathbf{x}_i^1, \cdots, \mathbf{x}_i^{t_i}]$, and their corresponding semantic vectors $\mathbf{Y} = [\mathbf{y}_1, \cdots, \mathbf{y}_n] \in \mathbb{R}^{q \times n}$. Weight parameters: α, β and λ.

Output: The class name $\mathbf{y}_t$ of the testing example $\mathbf{x}_t \in \mathbb{R}^p$.

Method:

Step 1: Compute the diagonal matrix and similarity weight matrix of each class with $\mathbf{D}_i = \sum_{j=1}^{t_i} s_i^{jk}$ and $\mathbf{S}_i = \sum_{j,k=1}^{t_i} s_i^{jk}$, respectively. Their corresponding Laplacian matrix is $\mathbf{L}_i = \mathbf{D}_i - \mathbf{S}_i$;

Step 2: Construct the overall Laplacian matrix $\mathbf{L} = diag\{\mathbf{L}_1, \ldots, \mathbf{L}_n\}$;

Step 3: Compute the cross-modal embedding matrix $\mathbf{W} \in \mathbb{R}^{p \times q}$ using Eq. (5).

Step 4: Map the visual feature of the testing example $\mathbf{x}_t$ into the word vector space using the transfomathbf function $\mathbf{y}_t = \mathbf{W}^T \mathbf{x}$, then find the nearest semantic vector to $\mathbf{y}_t$ as the testing example's semantic vector.

$$\mathbf{W} = [(1-\alpha)\mathbf{X}X^T + \beta\mathbf{X}LX^T + \lambda\mathbf{I}]^{-1}(\mathbf{X}Y^T - \alpha\sum_{\substack{i,j=1 \\ i \neq j}}^{n} \mathbf{X}_i\mathbf{Y}_j^T), \qquad (5)$$

where $\mathbf{I}$ is a unit matrix. It is seen that the embedding matrix W is only determined by the visual features and word vectors of the training data. Consequently, ZSL is implemented with $\mathbf{y}_t = \mathbf{W}^T \mathbf{x}$. The main steps of the ME-ZSL algorithm are summarized in Algorithm 1.

Mathematically, ME-ZSL is a linear mapping approach and has an explicit solution, which makes it effective and efficient. Besides, the computational complexity for the three parts in Eq. (4) are $o(nt_i)$, $o(n^2t_i)$ and $o(nt_i^2)$, respectively. Therefore, the computational cost of the whole function is $o(nt_i + n^2t_i + nt_i^2)$.

4   Experiments

4.1   Datasets and Settings

We evaluate ME-ZSL on two popular datasets, AWA and CUB. Specifically, AWA [14] provides 50 classes of animal images (30,475 images), and 85 associated class-level attributes. CUB (Caltech-UCSD Birds-200-2011) [28] is more challenging, containing 11,788 images of 200 bird species with each class annotated in 312 attributes derived from a bird field guide website. For AWA, we use the standard ZSL setup in which 40 classes are used for seen classes and 10 classes are used for unseen classes. For CUB, we use the same ZSL split as that in [1] with 150 classes as the seen classes and 50 disjoint classes as the unseen classes. For each dataset, the train, validation, and test sets belong to mutually exclusive classes. And we employ training and validation, i.e., disjoint subsets of training set, for cross-validation. The average per-class top-1 accuracy on the test sets [1] is reported over 10 trials.

Convolutional Neural Networks (CNN) features are employed as the visual features in both datasets, since they have demonstrated the amazing power in visual recognition applications in recent years [3,25,27]. On both datasets, we use VGGNet features [25], which are 4,096-dim features from the fc7 layer of very deep 19-layer CNN pretrained on ILSVRC2014. We do not perform any task-specific pre-processing, such as cropping foreground objects or detecting irrelevant parts. Besides, Word2Vec [19] is used to generate the word vectors in ME-ZSL approach, which is trained on the English-language Wikipedia to represent word vectors on both datasets.

In the proposed ME-ZSL method, the parameters are set as follows: the inter-class separability parameter α is set to be 0.7, the locality preserving parameter β is set to be 2, and the regularizer parameter λ is set to be 100. In addition, both the state-of-the-art word-vector-based methods and attribute-based methods are chosen for the comparison study.

4.2   Evaluations on AwA

Five state-of-the-art word-vector-based methods (W), CCA [15], LR [15], DeViSE [3], SJE-W [1], and LatEm [29] are elected for comparison. Moreover, the famous attribute-based method (A), DAP [16], is also chosen for comparison. To make a fair comparison, except for SJE-W [1] and LatEm [29], all the comparative methods adopt the same visual features and word vectors as those in ME-ZSL, and we report their best performance after fine tune their parameters. As for SJE-W and LatEm, both utilizes GoogLeNet [27] as the visual feature, which is taken from the results from [1] and [29], respectively. Note that it is a fair comparison since the classification accuracy of GoogLeNet is generally better than that of VGGNet. Table 1 shows the comparison results. It is seen that ME-ZSL achieves the best performance among the competing methods. Specifically, it outperforms the best competing word-vector-based methods, DeViSE and LatEm, both in 3.0 absolute percentage points, and outperforms the other competing methods, CCA, LR, SJE-W, and DAP in 3.6, 7.6, 12.9, and 6.6 absolute percentage points, respectively. This is a very encouraging result.

We then evaluate the contribution of each part in ME-ZSL, as reported in Table 2. The average accuracy is 56.5% when only using the intra-class compactness constraint J_1 with the regularizer. Improvements of 5.4% and 4.0% are achieved if the inter-class separability constraint J_2 and the locality preserving property J_3 are added, respectively. The performance rises to 64.1% when all the constraints in Eq. (4) are considered. These demonstrate that the manifold assumption in ME-ZSL is reasonable and valuable.

In addition, we evaluate the impacts of parameters α, β, and λ, as documented in Fig. 3. The search ranges are set from 10^{-3} to 10^{3}. With extensive experiments, we find that the performance deteriorates rapidly when α is higher than 1 and β is higher than 5. Thus, we narrow down the search range from 0 to 1 for α and from 0 to 5 for β, respectively. As illustrated in Fig. 3(a), the performance reaches its peak when $\alpha = 0.7$, and then decreases rapidly. From Fig. 3(b), we observe that the performance is fairly good when β is between 1.5 and 2.5, and the performance achieves its peak when $\beta = 2$. These demonstrate that the performance benefits from the manifold embedding in ZSL. In addition, Fig. 3(c) shows that the best performance is obtained when $\lambda = 100$. We also evaluate the performance for each unseen class. Among them, "humpback whale" and "chimpanzee" classes are of a high accuracy of 99.72% and 96.72%, respectively. In contrast, the performance for the "seal" class is the worst one, which is only 5.30%. We find that most of the "seal" images are classified into the category of "humpback whale". There are two main reasons. One is that both "seal" and "humpback whale" are marine animals and have similar visual characteristics. The other reason is due to that there is a "blue whale" in the seen classes, which makes the marine animals in the unseen classes prone to be classified into the whale-related class.

Table 1. Comparison results on AwA. Different semantic embedding spaces are used for comparison, including attribute (A) and word vector (W).

Approach	Embedding space	Accuracy (%)
CCA [24]	W	60.5
LR [24]	W	56.5
DeViSE [9]	W	61.1
SJE-W [8]	W	51.2
LatEm [25]	W	61.1
DAP [15]	A	57.5
ME-ZSL (ours)	W	**64.1**

Table 2. Ablation study on AwA. Different semantic embedding spaces are used for comparison, including attribute (A) and word vector (W).

	J1+ regularizer	J1+J2+ regularizer	J1+J3+ regularizer	J1+J2+J3+ regularizer
Accuracy (%)	56.5	61.9	60.5	64.1

Finally, we report the running times of the training and testing stage, respectively. Our implementation is based on unoptimized Matlab code. On our computer with i5 4590 CPU and 12G memory, the training time takes about 11 min, and the testing time for all the 6180 images is only 0.45 s. This means that it only requires 0.073 millisecond to classify a test image. In contrast, it takes for DeViSE about 15 h for training and 0.52 s for testing. Therefore, ME-ZSL is quite efficient.

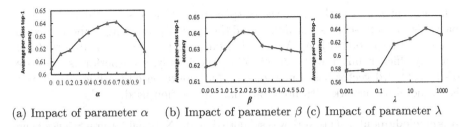

(a) Impact of parameter α (b) Impact of parameter β (c) Impact of parameter λ

Fig. 3. The influences of different λ on AwA and CUB datasets with attributes and Word2Vec, and $\mathcal{A}$ and $\mathcal{W}$ are short for attributes and Word2Vec, respectively.

4.3   Evaluations on CUB

The same six state-of-the-art methods are also elected for comparison in CUB. Similar comparison results are obtained to those in the AwA dataset, as documented in Table 3.

It is seen again that ME-ZSL beats all the comparative methods in performance. Specifically, it outperforms CCA, LR, SJE-W, DeViSE, LatEm and DAP in 2.9%, 4.4%, 4.9%, 5.1%, 1.5%, and 7.1% gains, respectively. We can notice that the overall performance in CUB is inferior to those in AwA, the main reasons are that CUB is a fine-grained dataset and there are more categories, which make it more challenging than AwA.

We then report the computational times for CUB. Under the same platform, the training time takes about 6 min, and the testing time for all the test images is only 0.08 s. Again, the promising efficiency of ME-ZSL is well demonstrated. As for the impact of parameters, similar results are observed to those in AWA.

Table 3. Comparison results on CUB. Different semantic embedding spaces are used for comparison, including attribute (A) and word vector (W).

Approach	Embedding space	Accuracy (%)
CCA [24]	W	30.4
LR [24]	W	28.9
DeViSE [9]	W	28.2
SJE-W [8]	W	28.4
LatEm [25]	W	31.8
DAP [15]	A	26.2
ME-ZSL (ours)	W	**33.3**

5   Conclusions

A novel Manifold Embedding based (ME-ZSL) approach to ZSL with word vector is presented in this paper. It employs the intrinsic manifold structure of the semantic categories in the word vector space from 3 aspects: the intra-class compactness, the inter-class separability, and the locality preserving property. Extensive experiments on both AwA and CUB datasets have demonstrated that it is not only efficient but also effective in ZSL. Specifically, it outperforms the state-of-the-art word-vector-based methods, and attribute-based methods. Besides, it is a general framework with no additional information used.

Acknowledgments. This work was supported by the National Natural Science Foundation of China under Grants 61771329 and 61632018.

References

1. Akata, Z., Reed, S., Walter, D., Lee, H., Schiele, B.: Evaluation of output embeddings for fine-grained image classification. In: Proceedings of the IEEE Conference on Computer Vision and Pattern Recognition (CVPR), pp. 2927–2936 (2015)
2. Deng, J., et al.: Large-scale object classification using label relation graphs. In: Fleet, D., Pajdla, T., Schiele, B., Tuytelaars, T. (eds.) ECCV 2014. LNCS, vol. 8689, pp. 48–64. Springer, Cham (2014). https://doi.org/10.1007/978-3-319-10590-1_4
3. Frome, A., et al.: Devise: a deep visual-semantic embedding model. In: Advances in Neural Information Processing Systems (NIPS), pp. 2121–2129 (2013)
4. Fu, Y., Hospedales, T.M., Xiang, T., Gong, S.: Transductive multi-view zero-shot learning. IEEE Trans. Pattern Anal. Mach. Intell. (PAMI) **37**(11), 2332–2345 (2015)
5. Fu, Y., Sigal, L.: Semi-supervised vocabulary-informed learning. In: Proceedings of the IEEE Conference on Computer Vision and Pattern Recognition (CVPR), pp. 5337–5346 (2016)
6. Fu, Z., Xiang, T., Kodirov, E., Gong, S.: Zero-shot object recognition by semantic manifold distance. In: Proceedings of the IEEE Conference on Computer Vision and Pattern Recognition (CVPR), pp. 2635–2644 (2015)
7. Girshick, R.: Fast R-CNN. In: Proceedings of the IEEE International Conference on Computer Vision, pp. 1440–1448 (2015)
8. He, K., Zhang, X., Ren, S., Sun, J.: Delving deep into rectifiers: surpassing human-level performance on ImageNet classification. In: Proceedings of the IEEE International Conference on Computer Vision, pp. 1026–1034 (2015)
9. He, X., Yan, S., Hu, Y., Niyogi, P., Zhang, H.J.: Face recognition using laplacianfaces. IEEE Trans. Pattern Anal. Mach. Intell. (PAMI) **27**(3), 328–340 (2005)
10. Jayaraman, D., Grauman, K.: Zero-shot recognition with unreliable attributes. In: Advances in Neural Information Processing Systems (NIPS), pp. 3464–3472 (2014)
11. Ji, Z., Pang, Y., He, Y., Zhang, H.: Semi-supervised LPP algorithms for learning-to-rank-based visual search reranking. Inf. Sci. **302**, 83–93 (2015)
12. Ji, Z., Wang, J., Su, Y., Song, Z., Xing, S.: Balance between object and background: object-enhanced features for scene image classification. Neurocomputing **120**, 15–23 (2013)
13. Ji, Z., Xie, Y., Pang, Y., Chen, L., Zhang, Z.: Zero-shot learning with multi-battery factor analysis. Sig. Process. **138**, 265–272 (2017)
14. Lampert, C.H., Nickisch, H., Harmeling, S.: Attribute-based classification for zero-shot visual object categorization. IEEE Trans. Pattern Anal. Mach. Intell. (PAMI) **36**(3), 453–465 (2014)

15. Lazaridou, A., Bruni, E., Baroni, M.: Is this a Wampimuk? Cross-modal mapping between distributional semantics and the visual world. In: Proceedings of the 52nd Annual Meeting of the Association for Computational Linguistics, vol. 1, pp. 1403–1414 (2014)

16. Li, X., Guo, Y.: Max-margin zero-shot learning for multi-class classification. In: Artificial Intelligence and Statistics, pp. 626–634 (2015)

17. Liu, M., Zhang, D., Chen, S.: Attribute relation learning for zero-shot classification. Neurocomputing **139**, 34–46 (2014)

18. Lu, J., Hu, J., Zhou, J.: Deep metric learning for visual understanding: an overview of recent advances. IEEE Sig. Process. Mag. **34**(6), 76–84 (2017)

19. Mikolov, T., Sutskever, I., Chen, K., Corrado, G.S., Dean, J.: Distributed representations of words and phrases and their compositionality. In: Advances in Neural Information Processing Systems (NIPS), pp. 3111–3119 (2013)

20. Mnih, A., Hinton, G.: Three new graphical models for statistical language modelling. In: International Conference on Machine Learning (ICML), pp. 641–648. ACM (2007)

21. Norouzi, M., et al.: Zero-shot learning by convex combination of semantic embeddings. arXiv preprint arXiv:1312.5650 (2013)

22. Peng, Y., He, X., Zhao, J.: Object-part attention model for fine-grained image classification. IEEE Trans. Image Process. **27**(3), 1487–1500 (2018)

23. Pennington, J., Socher, R., Manning, C.: Glove: global vectors for word representation. In: Proceedings of the 2014 conference on Empirical Methods in Natural Language Processing (EMNLP), pp. 1532–1543 (2014)

24. Romera-Paredes, B., Torr, P.: An embarrassingly simple approach to zero-shot learning. In: International Conference on Machine Learning (ICML), pp. 2152–2161 (2015)

25. Simonyan, K., Zisserman, A.: Very deep convolutional networks for large-scale image recognition. In: Computer Science (2014)

26. Socher, R., Ganjoo, M., Manning, C.D., Ng, A.: Zero-shot learning through cross-modal transfer. In: Advances in Neural Information Processing Systems (NIPS), pp. 935–943 (2013)

27. Szegedy, C., et al.: Going deeper with convolutions. In: Proceedings of the IEEE Conference on Computer Vision and Pattern Recognition (CVPR), pp. 1–9 (2015)

28. Wah, C., Branson, S., Perona, P., Belongie, S.: Multiclass recognition and part localization with humans in the loop. In: Proceedings of the IEEE International Conference on Computer Vision (ICCV), pp. 2524–2531. IEEE (2011)

29. Xian, Y., Akata, Z., Sharma, G., Nguyen, Q., Hein, M., Schiele, B.: Latent embeddings for zero-shot classification. In: Proceedings of the IEEE Conference on Computer Vision and Pattern Recognition (CVPR), pp. 69–77 (2016)

30. Yan, S., Xu, D., Zhang, B., Zhang, H.J., Yang, Q., Lin, S.: Graph embedding and extensions: a general framework for dimensionality reduction. IEEE Trans. Pattern Anal. Mach. Intell. (PAMI) **29**(1), 40–51 (2007)

31. Yang, W., Li, J., Zheng, H., Da, X., Xu, R.Y.: A nuclear norm based matrix regression based projections method for feature extraction. IEEE Access **6**, 7445–7451 (2018)

32. Yu, Y., Ji, Z., Guo, J., Pang, Y.: Transductive zero-shot learning with adaptive structural embedding. IEEE Trans. Neural Netw. Learn. Syst. **29**, 4116–4127 (2017)

Attribute Value Matching
with Limited Budget

Fengfeng Fan[1,2($\boxtimes$)], Zhanhuai Li[1,2], and Qun Chen[1,2]

[1] School of Computer Science, Northwestern Polytechnical University, Xi'an, China
fanfengfeng@mail.nwpu.edu.cn
[2] Key Laboratory of Big Data Storage and Management,
Northwestern Polytechnical University,
Ministry of Industry and Information Technology, Xi'an, China

Abstract. Equivalent but non-identical attribute values, e.g., synonyms or abbreviations, often result in data inconsistencies, and identifying the equivalents is an essential task in data cleaning or data integration. Since the inconsistencies between *hot* data tend to bring more distortion to data analysis, we prefer to match hot equivalent values with limited budget. Based on the matching probability of value pairs and the data hotness, we propose an heuristic strategy which prefers to resolve those *hot* equivalents such that maximal benefit to data consistency can be achieved with the limited budget. Experimental evaluations show the effectiveness of our approach.

Keywords: Attribute value matching · Hot data · Entity resolution
Data cleaning · Big data

1 Introduction

Due to typographical errors, aliases and abbreviations [1,2], the same real-world entities may take multiple distinct representations across data sources. Since such inconsistency may severely distort the results of data analysis, it is necessary to match and merge those equivalent values by a process called Attribute Value Matching or AVM [8], before delving into further data analysis. As a finer-granularity entity resolution, AVM identifies the equivalent attribute values that refer to the same entities.

Example 1. Table 1 lists several records for research papers, and Table 2 lists some extracted journal attribute values and their mappings to underlying entities. It can be observed that some entities (e_3 or e_4) take multiple representations and such inconsistencies need to resolved by the AVM process. The final result by resolving the inconsistencies is presented as follows:

$$\{e_1 = [y_1], e_2 = [y_2], e_3 = [y_3, y_4], e_4 = [y_5, y_6]\}$$

Y. Peng et al. (Eds.): IScIDE 2018, LNCS 11266, pp. 148–157, 2018.
https://doi.org/10.1007/978-3-030-02698-1_13

Table 1. Examples for research publications

pid	Title	Author	Journal	Year
p_1	Energy management in industrial plants	D. Bruneo, A. Cucinotta, A.L. Minnolo, A. Puliafito, M. Scarpa	Computers	2012
p_2	Beyond bits: the future of quantum information processing	A.M. Steane, E.G. Rieffel	Computer	2000
p_3	Priority assignment in real-time active databases	R.M. Sivasankaran, J.A. Stankovic, D. Towsley, B. Purimetla, K. Ramamritham	Journal on Very Large Data Bases	2003
p_4	A taxonomy of correctness criteria in database applications	K. Ramamritham, P.K. Chrysanthis	VLDB J	2002
p_5	Scalable package queries in relational database systems	M Brucato, JF Beltran, A Abouzied, A Meliou	VLDB J	2016
p_6	In defense of nearest-neighbor based image classification	O Boiman, E Shechtman, M Irani	CVPR	2008
p_7	Blind motion deblurring from a single image using sparse approximation	JF Cai, H Ji, C Liu, Z Shen	IEEE Conference on Computer Vision and Pattern Recognition	2009

Table 2. Mapping between `journal` values and entities

jid	Journal	Freq	Entity
y_1	Computers	1	e_1
y_2	Computer	1	e_2
y_3	Journal on Very Large Data Bases	1	e_3
y_4	VLDB J	2	e_3
y_5	CVPR	1	e_4
y_6	IEEE Conference on Computer Vision and Pattern Recognition	1	e_4

In data cleansing practice, the large data size and limited budget always prevent us from matching *all* equivalents within big data. With limited budget, it is a much more practical to employ a greedy approach, e.g. pay-as-you-go approach proposed in [15], and iteratively match the equivalent values such that maximal improvement to data consistency can be achieved.

There are two interesting observations motivating our proposal. The first observation is that all of the non-identical attribute values correspond to distinct entities implicitly, and only successful matching of equivalent values will improve the data consistency. Thus the limited budget ought to be spent to match those value pairs with highest matching probabilities.

The second observation is that NOT all attribute values are of equal importance in data analysis. Intuitively, those frequently accessed *hot* values are more important than those rarely accessed *cold* values, thus should receive more attentions. Since inconsistencies over *hot* values often result in much more distortion to data analysis, resolving the inconsistencies over *hot* values tends to bring more improvement to data consistency.

In summary, to improve the data consistency maximally, we integrate both the *matching probability* and the *data hotness* into benefit estimation, and propose an algorithm to greedily resolve the inconsistencies over *hot* values. Our contributions can be summarized as follows:

1. A *benefit* metric is devised based on the *matching relationship* and the *data hotness*.
2. The task of AVM with limited budget (AVM-LB) is modeled as a *benefit maximization* problem.
3. An algorithm is proposed to greedily revolve the hot value inconsistencies on-the-fly such that maximal benefit to data quality improvement can be achieved with limited budget.
4. Experimental evaluations show the effectiveness of our approach.

The rest of paper is organized as follows: Sect. 2 discusses the related work, Sect. 3 formalizes the problem, Sect. 4 presents the necessary algorithms, Sect. 5 shows the experimental evaluations, and Sect. 6 concludes our discussion.

2   Related Work

As a special case of entity resolution, the problem of Attribute Value Matching has received great attentions. The state of arts for matching attribute values can be roughly categorized into string-based [3–6], constraint-based [7], and probabilistic approaches [8,9].

The assumption that more similar values are more likely to be equivalent, lays the foundation of string-based approaches. However, as pointed out by [10], the effectiveness of string-based approaches, are highly dependent on the choice of string similarity metric and threshold.

The constraint-based approach, e.g., a functional dependency fd_1 : title $\rightarrow$ journal, can also be used for matching equivalent values. The constraint fd_1 states that two records should have equivalent journal values given they share the common value in title. However, the effectiveness of constraint-based approach heavily rely on the reliability of constraints (either discovered by automatic algorithm [11,12] or specified by domain expert) and the number of records captured by the matching pattern.

The authors of [8] proposed a probabilistic metric, *value correlation analysis* (VCA), to estimate the evidential supports from correlated attributes (e.g., author and title). By combining VCA and string similarity between journal values, considerable improvement has been achieved over string-based approach.

Other closely related work includes CrowdER [13] which employs the crowdsourcing in entity resolution, subgraph-cohesion-based approach [14] which employs the subgraph cohesion to entity resolution, pay-as-you-go approach [15] which greedily resolves the entities with high matching probability, and QDA [16] which proposes an query-driven-approach for on-line entity resolution.

Although the data hotness has played an important part in big data analytics and data storage community [17], it has been ignored for a long time in traditional data cleansing practice, which motivates our study in this paper.

3 Problem Statement

In this section, we will first introduce some preliminaries, and then formalize the problem of AVM with Limited Budget, or AVM-LB.

3.1 Matching Probability

As above mentioned, only successful matching of equivalents can bring data quality improvements, thus the principal consideration is the matching probability, which is defined as follows:

Definition 1. *For any attribute value pair* $\langle y_i, y_j \rangle_{i \neq j} \in \mathbf{Y} \times \mathbf{Y}$*, an indicator function will reply to the query whether* $\langle y_i, y_j \rangle_{i \neq j} \in \mathbf{Y} \times \mathbf{Y}$ *matches or not:*

$$I(y_i, y_j) = \begin{cases} 1 & y_i \cong y_j \\ 0 & otherwise \end{cases} \tag{1}$$

where $\cong$ *denotes the matching relationship, we assume that each call to the indicator function will cost one budget.*

Definition 2. *For any attribute value pair* $\langle y_i, y_j \rangle_{i \neq j}$*, the matching probability* $P(y_i, y_j)$ *is defined as the probability of positive reply from indicator function:*

$$P(y_i \cong y_j) = P(I(y_i, y_j) = 1) \tag{2}$$

Intuitively, more similar values are more likely to be equivalent. We approximate the matching probability $P(y_i \cong y_j)$ by a similarity function $\mathsf{sim}(y_i, y_j)$, which may either be a simple string similarity or some sophisticated metric, like [9].

3.2 Hotness

Hotness often reveals the attribute value's weight in data analysis, and it may be a function of the timeliness, occurrences, or access frequencies. For simplicity, we estimate the *hotness* of attribute values by their frequencies.

Definition 3. *The hotness for any attribute value pair* $\langle y_i, y_j \rangle$ *is defined:*

$$\mathsf{hot}(y_i, y_j) = \mathsf{freq}([y_i]) \cdot \mathsf{freq}([y_j]) \tag{3}$$

where $\mathsf{freq}(\cdot)$ *records the frequencies of attribute values, and the equivalent class* $[y_i]$*, denotes the set of attribute values co-referring to the same entity with* y_i*.*

Example 2. As shown in Table 2, with $\mathsf{freq}(y_3) = 1$ and $\mathsf{freq}(y_4) = 2$, we have $\mathsf{hot}(y_3, y_4) = 2$ by Definition 3. Similarly, we also have $\mathsf{hot}(y_1, y_2) = 1$ and $\mathsf{hot}(y_5, y_6) = 1$.

Example 3. After matching the value pair $\langle y_3, y_4 \rangle$, we will have

$$[y_3] = [y_4] = \{y_3, y_4\}$$

and

$$\mathsf{freq}([y_3]) = \mathsf{freq}([y_4]) = \mathsf{freq}(y_3) + \mathsf{freq}(y_4) = 3$$

3.3   Benefit

Benefit is defined to quantify the improvement to data consistency by resolving a pair of attribute values.

Definition 4. *The benefit by resolving a pair $\langle y_i, y_j \rangle$ can be defined by Eq. 4:*

$$\mathsf{B}(y_i, y_j) = \mathsf{I}(y_i, y_j) \cdot \mathsf{hot}(y_i, y_j) \tag{4}$$

Example 4. Continuing the examples shown in Table 2, by resolving the value pairs, $\langle y_1, y_2 \rangle$, $\langle y_3, y_4 \rangle$ and $\langle y_5, y_6 \rangle$, we will obtain the benefit by 0, 2 and 1 respectively.

3.4   Problem Definition

Based on above definitions, we model the task of AVM-LB as an optimal edge generation process in graph: for a initial graph $G(V, E)$ with each vertex $v \in V$ representing an attribute value and the initial edge set being an empty set: $E \leftarrow \emptyset$, and an edge $e_{i,j}$ can be added into the edge set E by resolving any attribute value pair $\langle y_i, y_j \rangle$. Then our objective is to construct K edges (or resolve K value pairs), such that the maximal benefit can be achieved:

$$E_K^* = \mathrm{argmax}(\sum_{e \in E_K} \mathsf{B}(e)) \tag{5}$$

where the edge set E_K is composed of K edges.

4   Algorithm

In this section, we introduce some essential building blocks and then present the algorithm for constructing the optimal edges.

4.1 Ranking Function

As Eq. 4 indicates, the benefit relies on the positive reply of indicator function and the hotness. To maximize the benefit, we define a Rank function to fine-tune the order of value pairs for resolving:

Definition 5. *Rank function takes two parameters as inputs:* matching probability *and* hotness:

$$\text{Rank}(y_i, y_j) = \text{P}(y_i \cong y_j) \cdot \text{sigmoid}(\alpha \cdot \text{hot}(y_i, y_j) + \beta) \tag{6}$$

where $\text{P}(y_i \cong y_j)$ is the estimated matching probability for attribute value pair $\langle y_i, y_j \rangle$. For simplicity, the conversion from hotnesses of value pairs into weights or preferences (between 0 and 1) is accomplished by a widely used $\text{sigmoid}(\cdot)$ transformation, in which $\alpha \geq 0$ and β are two tuning hyper-parameters.

4.2 Benefit Maximization

To efficiently generate the optimal edges (value pairs), it is necessary to maintain several auxiliary data structures: **Filter matrix** maintaining current matching relationships between attribute values, **Hotness matrix** maintaining the hotness between attribute values and **Matching probability matrix** maintaining the matching probabilities between attribute values.

Filter Matrix. The filter matrix maintains the current matching states between attribute values: "1" for match, "−1" for non-match, "0" for unknown thus need to be resolved, and "*" for those pairs that can be deduced by symmetry. The filter matrix **F** was initialized with upper triangular part being "0" and other part being "*". A snapshot of filter matrix of journal values in Table 2 is shown Eq. 7:

$$\mathbf{F} = \begin{bmatrix} * & 0 & 0 & 0 & 0 & 0 \\ * & * & 0 & 0 & 0 & 0 \\ * & * & * & 1 & 0 & 0 \\ * & * & * & * & -1 & -1 \\ * & * & * & * & * & 1 \\ * & * & * & * & * & * \end{bmatrix} \tag{7}$$

As the matching process goes on, more 0-labeled cells will be substituted either with "1" or "−1", depending on the replies of indicator function, until no more 0-cells is available or all the budget runs out.

Hotness Matrix. Hotness matrix **H**, with $\mathbf{H}[i, j] \leftarrow \text{hot}(y_i, y_j)$, maintains the hotnesses of attribute value pairs. The hotness matrix **H** will be updated by the successful matchings of equivalent attribute value pairs, seen in Example 3.

Matching Probability Matrix. Matching probability matrix $\mathbf{P}$, with $\mathbf{P}[i,j] \leftarrow P(y_i, y_j)$, maintains the matching probabilities between attribute values. The computation of matching probability is sketched in Algorithm 1.1, which was proposed in [9]. By taking into consideration the similarity over target attribute (e.g., Journal) and the similarities over correlated attributes (e.g., Author and Title), it has been verified to have better performance.

Algorithm 1.1. Matching Probability Matrix

    Input:
    Target attribute values $\mathbf{Y}$;
    Correlated attribute values $\mathbf{X}$;
    Output: matching probability matrix $\mathbf{P}$;
1 $cv \leftarrow$ CountVectorizer();
2 cv.fit($\mathbf{X}$);
3 $\mathbf{V} \leftarrow cv$.transform($\mathbf{X}$);
4 $\mathbf{V}' \leftarrow \mathbf{V}$.groupBy($Y$).agg(sum) ;
5 $\mathbf{U} \leftarrow$ normalize($\mathbf{V}', axis =$ "row", $norm =$ "L2") ;
6 $\mathbf{P}_X \leftarrow \mathbf{U} \cdot \mathbf{U}^T$;
7 $\mathbf{P}_Y \leftarrow \mathbf{Y} \times \mathbf{Y}$;
8 $\mathbf{P} \leftarrow \mathbf{P}_X \bigoplus \mathbf{P}_X$;
9 **return** $\mathbf{P}$;

In Algorithm 1.1, CountVectorizer in Line [1–3] is used to transform the correlated values in attribute $\mathbf{X}$ into vector representations. In line [4], word vectors are grouped by $\mathbf{Y}$ values and aggregated by sum operator. In line [5–6], word vector representations are normalized to unity-scale by "L2 norm", over which cosine similarities are estimated. Line [7] computes the string similarities between $\mathbf{Y}$ attribute values. Line [8] combines two measurements by evidential reasoning.

Greedily Resolving. Algorithm 1.2 greedily resolves the value inconsistencies based on the current matching states, matching probability and hotness, such that maximal benefit can be achieved. Line [1] checks the availability of budget. Line [2] gets the filter matrix such that the next optimal value pair is chosen from the **unknown** cells. Line [3] ranks the value pairs by Rank function defined in Eq. 6. Line [4] identifies the index for the optimal query q^*. Line [5–6] will break the process if no quality improvement can be made. Line [7–9] resolve the optimal query and update current states based on the received reply.

Algorithm 1.2 will keep running until no more quality improvement can be made or all available budget runs out.

5   Experimental Evaluation

In this section, we first illustrate the sigmoid transformations, and then empirically evaluate the benefits to data consistency by matching equivalent Journal

Algorithm 1.2. Greedily Resolving

Input: Filter matrix $\mathbf{F}$;
Hotness matrix $\mathbf{H}$;
Matching probability matrix $\mathbf{P}$;
Budget K;
Output: None;

1 **while** $K \geq 0$ **do**
2     $\bar{\mathbf{F}} \leftarrow (\mathbf{F} == 0)?1 : 0$;
3     $\mathbf{Rank} \leftarrow \frac{1}{1+\exp(-(\alpha+\beta\mathbf{H}))} * \mathbf{P} * \bar{\mathbf{F}}$;
4     $(i, j) \leftarrow \mathsf{argmax}(\mathbf{Rank})$;
5   **if** $\mathbf{Rand}[i, j] == 0$ **then**
6        $\lfloor$ **return**;
7     $ans \leftarrow \mathsf{I}(y_i, y_j)$;
8     $\mathsf{update}(\mathbf{M}, \mathbf{H}, \mathbf{F}, ans)$;
9     $\lfloor$ $K \leftarrow K - 1$;

values across two public available bibliographic datasets, DBLP[1] and CiteSeer[2], in which $1,666$ and $3,833$ distinct Journal values are extracted from $1,636,497$ and $45,783$ available tuples respectively.

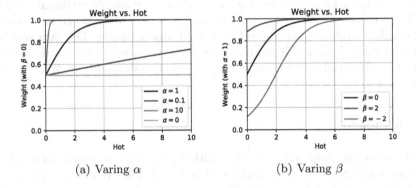

(a) Varing α (b) Varing β

Fig. 1. sigmoid transformation

The comparison of sigmoid transformations is plotted in Fig. 1. In Fig. 1(a), with $\beta = 0$ fixed, we vary the value of α. It can be observed that only small value of α, e.g. $\alpha = 0.1$, will make Weight sensitive to large range of hot values, while large value of α, e.g. $\alpha = 10$, will make Weight insensitive to large hot values. Additionally, $\alpha = 0$ makes Weight independent of hot values. In Fig. 1(b), with $\alpha = 1$ fixed, we vary the values of β. It can be observed that the values of β will

---

[1] http://dblp.uni-trier.de/.
[2] https://www.cs.purdue.edu/commugrate/data/citeseer/.

determine the lower bound of sigmoid transformation, e.g., $\beta = -2$, $\beta = 0$ and $\beta = 2$ will result in the lower bound values 0.12, 0.5 and 0.88 respectively.

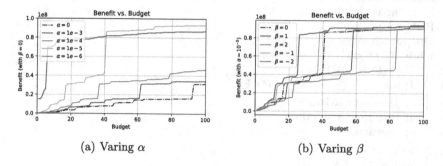

(a) Varing α (b) Varing β

Fig. 2. Benefit vs Budget

The parameters, α and β, determine sigmoid transformation, which further defines the Rank function. The comparison of benefit with different Rank is plotted in Fig. 2. The matching probabilities are estimated by the approach proposed in [9]. In Fig. 2(a) we vary the value of α while keeping $\beta = 0$ fixed. It can be observed that small α values tend to produce large benefit to data consistency, and the contribution of hotness can be eliminated by simply setting $\alpha = 0$. In Fig. 2(b) we vary the value of β while keeping $\alpha = 10^{-5}$ fixed. It can be observed that smaller β value tends to produce overall higher benefit. The rationale lies in that smaller β value will produce larger interval between the lower and the upper bounds, which will further enlarge the contribution of hotness in ranking value pairs to be resolved.

6   Discussion

In this paper, we proposed a greedy solution preferring to resolve inconsistencies over *hot* attribute values with limited budget. Unlike traditional approaches that take only the matching probability into consideration and ignore the weights of attribute values, we take both of them into benefit estimation, and greedily resolve those inconsistencies over *hot* attribute values, such that *maximal benefit* to data consistency can be achieved with limited budgets. Although the optimization for hyper-parameters is still under development, we hope that this study will make some contribution to data cleansing over big data.

Acknowledgments. The work was supported by the Ministry of Science and Technology of China, National Key Research and Development Program (Project Number 2016YFB1000703), the National Natural Science Foundation of China under No. 61732014 No. 61332006, No. 61472321, No. 61502390 and No. 61672432.

References

1. Batini, C., Scannapieco, M.: Data Quality: Concepts, Methodologies and Techniques. Springer, Heidelberg (2010)
2. Naumann, F., Herschel, M.: An introduction to duplicate detection. Synth. Lect. Data Manag. **2**, 1–87 (2010)
3. Navarro, G.: A guided tour to approximate string matching. ACM Comput. Surv. **33**, 31–88 (2000)
4. Jaro, M.A: Unimatch: A Record Linkage System: Users Manual. Bureau of the Census, Suitland (1978)
5. Gravano, L., Ipeirotis, P.G., Koudas, N., Srivastava, D.: Text joins in an RDBMS for web data integration. In: World Wide Web Conference Series, pp. 90–101 (2003)
6. Gong, J., Wang, L., Oard, D.W.: Matching person names through name transformation. In: ACM Conference on Information and Knowledge Management, Hong Kong, pp. 1875–1878 (2009)
7. Fan, W., Li, J., Ma, S., Tang, N., Yu, W.: Towards certain fixes with editing rules and master data. VLDB J. **21**, 213–238 (2012)
8. Fan, F., Li, Z., Chen, Q., Chen, L.: Reasoning about attribute value equivalence in relational data. Inf. Syst. **75**, 1–12 (2018)
9. Fan, F., Li, Z., Wang, Y.: Cohesion based attribute value matching: In 10th International Congress on Image and Signal Processing, BioMedical Engineering and Informatics, pp. 2257–2261 (2018). https://ieeexplore.ieee.org/document/8302315/
10. Yu, M., Li, G., Deng, D., Feng, J.: String similarity search and join: a survey. Front. Comput. Sci. **10**, 399–417 (2016)
11. Fan, W., Geerts, F., Li, J., Xiong, M.: Discovering conditional functional dependencies. IEEE Trans. Knowl. Data Eng. **23**, 683–698 (2011)
12. Diallo, T., Novelli, N., Petit, J.M.: Discovering (frequent) constant conditional functional dependencies. Int. J. Data Min. Modell. Manag. **120**, 205–223 (2016)
13. Wang, J., Kraska, T., Franklin, M.J., Feng, J.: CrowdER: crowdsourcing entity resolution. Proc. VLDB Endow. **5**, 1483–1494 (2012)
14. Wang, H., Li, J., Gao, H.: Efficient entity resolution based on subgraph cohesion. Knowl. Inf. Syst. **46**, 285–314 (2016)
15. Whang, S.E., Marmaros, D., Garcia-Molina, H.: Pay-as-you-go entity resolution. IEEE Trans. Knowl. Data Eng. **25**, 1111–1124 (2013)
16. Altwaijry, H., Kalashnikov, D.V., Mehrotra, S.: Query-driven approach to entity resolution. VLDB Endow. **6**, 1846–1857 (2013)
17. Hsieh, J.W., Kuo, T.W., Chang, L.P.: Efficient identification of hot data for flash memory storage systems. ACM Trans. Storage **2**, 22–40 (2006)

Moth-Flame Optimization Algorithm Based on Adaptive Weight and Simulated Annealing

Qiang Zhang, Li Liu[✉], Chengfei Li, and Fan Jiang

School of Information Science and Engineering, Lanzhou University,
Lanzhou, China
liulnd@lzu.edu.cn

Abstract. Moth-flame optimization algorithm has the demerit of being easily trapped in local optimum. To solve this problem, an improved algorithm ASMFO is proposed in this paper. Adaptive weight can be automatically changed so that the algorithm can get a greater search scope in the early stage and the precision of the optimal solution can be increased in the later stage of the algorithm. Moreover, the simulated annealing method is employed to accept new solutions with a certain probability, which can further alleviate the problem that MFO is easy to fall into local optimum and will also enhance the global search ability of MFO algorithm. The experimental results show that the improved algorithm is superior to other optimization algorithms in the convergence precision and the stability.

Keywords: Moth-flame optimization algorithm · Adaptive weight Simulated annealing

1 Introduction

Swarm intelligence algorithms generally originate from the imitation of some biological groups in the nature. Because they have the advantages that traditional optimization methods do not have, they have developed into a research hotspot in optimization problems, such as particle swarm optimization [1], genetic algorithms [2] and so on. Moth-flame optimization algorithm is a novel swarm intelligence algorithm. In 2015, Australian scholar Mirjalili was inspired by the flight mode of the moth and proposed a new heuristic intelligent optimization algorithm, that is, moth-flame optimization algorithm (MFO) [3]. The main idea of this algorithm is to solve the problem by simulating the behavior of moths. The algorithm has good stability and better search capability. Therefore, it has received extensive attention from scholars and has been applied in many fields, such as power system [4], image processing [5, 6], multilayer perceptron [7], power load forecasting [8] and so on.

MFO algorithm has better search capability, and it has the advantages of simple structure, less parameters and higher efficiency, whereas the algorithm still has the demerit of being easily trapped in local optimum, slow convergence speed, low convergence precision and other problems. In this paper, moth-flame optimization algorithm based on adaptive weight and simulated annealing (ASMFO) is proposed to alleviate the problems. Adaptive weight method is used to improve the local search

© Springer Nature Switzerland AG 2018
Y. Peng et al. (Eds.): IScIDE 2018, LNCS 11266, pp. 158–167, 2018.
https://doi.org/10.1007/978-3-030-02698-1_14

ability of moth-flame optimization algorithm and improve the convergence accuracy, and the simulated annealing algorithm is introduced to change the position of the moth, which improves the global search ability and avoids the local optimum. Consequently ASMFO algorithm can effectively improve other algorithms with the characteristics of a certain probability to accept new states, thus avoiding the local optimum.

2 Moth-Flame Optimization Algorithm

In MFO algorithm, the candidate solution is a moth, and the positions of the moths in the search space are the problem's variables. Moths can fly in any space by changing their position vectors. In MFO algorithm, the matrix M represents a set of moths, and the array OM is used to store corresponding fitness values. Another core component of the algorithm is flame. The flame matrix is represented by F, and the dimensions of array M and F are equal. Array OF is used to store corresponding fitness values. In MFO algorithm both the moths and flames are considered to be the solutions. The main difference between the moths and the flames is the process in which they are updated in each iteration. The actual search agents are the moths which move around the search space and the best positions obtained by the moths are represented by flames. The flames can be treated as flags where each moth searches around a flag and updates its position when it finds a better solution. With this method, a moth never loses its track from the best solution.

MFO algorithm is solved by three steps, which can achieve an approximate global optimum. The three parts of the algorithm are as follows:

$$MFO = (I, P, T). \tag{1}$$

I generates a random moth population and corresponding fitness values.

$$I : \varphi \rightarrow \{M, OM\}. \tag{2}$$

P is the main function by which moths move around the search space. P receives the matrix of M and returns its update.

$$P : M \rightarrow M. \tag{3}$$

T function returns true if the termination criteria are satisfied and return false if termination criteria are not satisfied.

$$T : M \rightarrow \{ture, false\}. \tag{4}$$

The next step after initialization is to run the P function iteratively until the T function is reached. The P function is the main function of moths to move around the search space and update their location relative to the flame.

$$M_i = S(M_i, F_j). \tag{5}$$

where M_i is the ith moth, and F_j is the jth flame. S is the logarithmic spiral function. The logarithmic spiral function defined by MFO algorithm is as follows:

$$S(M_i, F_j) = R_i \cdot e^{bn} \cdot \cos(2\pi n) + F_j. \tag{6}$$

where R_i indicates the distance between the ith moth and the jth flame, b is a constant to define the shape of the logarithmic spiral, and n is a random number between -1 and 1.
R is defined as follows:

$$R_i = |F_j - M_i|. \tag{7}$$

where M_i is the ith moth and F_j is the jth flame. R_i is the distance between an ith moth and jth flame.

The flight path of the moth is simulated by formula (6), and the next position of the moth relative to the flame is also determined. In this formula, b represents the degree of distance of the moth relative to the next position of the flame. When b = 1, the next position of the moth is farthest from the flame, and b = -1 represents the position closest to the flame. Formula (6) only defines the moth's flight to the flame, which makes it very easy for MFO algorithm to fall into a local optimum. The position of the moth is updated with respect to the flame as per (6). Each time the list of flames is updated, the flames are sorted according to their fitness value, after which the moths are updated. The first moth is updated with respect to the best flame, while the last moth updates the list with respect to the worst flame position. Such an update mechanism helps MFO avoid falling into a local optimum.

In order to improve the search efficiency of the optimal solution, an adaptive mechanism suitable for the number of flames can be used to make the number of flames adaptively reduce during the iteration process. The formula is as follows:

$$flame\ no = round(N - m * \frac{N - 1}{L}). \tag{8}$$

Where m is the current number of iterations, N is the maximum number of flames, and L is defined as the maximum number of iterations.

Equation (8) indicates that the number of flames is N at the beginning. However, the moths update their positions only with respect to the best flame in the final steps of iterations. The gradual decrement in number of flames balances exploration and exploitation of the search space.

3   Improvement of MFO Algorithm

3.1   Adaptive Weight

Since MFO algorithm uses a logarithmic spiral function to update the position of moth, this function simply defines the moth to fly to the flame, which makes the moth easily fall into a local optimum and has certain deficiencies in the global optimization. This paper adopts an adaptive weight method. When the moth approaches the flame to find the optimal solution, the value of the adaptive weight decreases, so that the moth's local optimum ability will be improved. The formula is as follows:

$$w = \sin(\frac{\pi \times l}{2 \times L} + \pi) + 1. \tag{9}$$

Where l represents the current number of iterations, and L represents the maximum number of iterations.

Apply adaptive weights to the formula for moth updates:

$$S(M_i, F_j) = R_i \cdot e^{an} \cdot \cos(2\pi n) + wF_j. \tag{10}$$

The adaptive weight w changes the value of F_j, and w is adaptively reduced from 1 to 0 with increasing number of iterations. When the moth approaches the flame, the value will be smaller, increasing the local search ability of the moth. Avoid moths missing the optimal solution.

3.2   Simulated Annealing Algorithm

Simulated annealing algorithm [9] can make the model jumping out of local low point and then we find the optimization with the control of annealing scheme. Simulated annealing algorithm has the advantage of accepting new states with a certain probability can effectively improve other algorithms, so as to avoid falling into the local optimum.

MFO algorithm based on adaptive weights not only improves the global search ability of the moth, but also improves the local detection performance greatly. However, further improvement is needed to make up for the local optimal problem. Therefore, this paper introduces the simulated annealing algorithm again, and proposes ASMFO algorithm. Simulated annealing algorithm has the property of accepting new solutions with a certain probability, and further makes up for the defect that the MFO algorithm is prone to fall into local optimum. ASMFO algorithm also ensures that the moth population searches in a better direction.

3.3 ASMFO Algorithm

ASMFO algorithm description as shown in Algorithm 3.1.

Algorithm 3.1:MFO algorithm based on adaptive weight and simulated annealing

Input: the value of the range of the objective function variable
Output: the approximate optimal value of the corresponding function

1. MFO algorithm initialization, set the population size n, the number of iterations L and the logarithm spiral shape constant b and other parameters;
2. Given the initial temperature $t_0 = f_{max} - f_{min}$, f_{max} and f_{min} are the maximum and minimum fitness of the initial moth population, respectively;
3. Calculate the fitness value of the moth position and sort it, then assign the value to the flame to form the initial spatial position of the flame;
4. Moth adaptive weight search, which uses the formula (5) and formula (10) to update the position of the moth;
5. Calculate the fitness value of moths and flames and sort them, and use the position with better fitness value as the position of the next-generation flame;
6. Reduce the number of flames using the adaptive mechanism of formula (8);
7. A new spatial position is randomly generated for the moth, and then the difference Δd from the fitness value of the previous position is calculated. If $\Delta d < 0$, the moth enters a new spatial position; otherwise, it proceeds to the next step;
8. Randomly generate a number r between $(0,1)$, if satisfied $r < \min[1, \exp(-\Delta d / t)]$, the moth enters the new spatial location and executes 4; otherwise, it proceeds to the next step;
9. Temperature lower setting, $t = 0.9 \times t$;
10. Determine whether to reach the maximum number of iterations. If it is satisfied, continue execution. Otherwise, skip to step 4;
11. Output the optimal result and the program ends.

---

4 Experimental Results and Analysis

In the experiment, nine test functions are used to verify the performance of ASMFO algorithm and particle swarm algorithm (PSO), genetic algorithm (GA), and basic MFO algorithm. The algorithm parameters in the experiment are set as follows: the population size was 30 individuals, the maximum number of iterations was 1000, the function dimension was set to 30 dimensions, 30 runs independently, and the final test result was the average of 30 independent optimal running values.

4.1 Test Function

Nine benchmark test functions were selected from the literature [10] as the test function of this experiment, as shown in Table 1. These functions are often used mathematical functions to verify the performance of group intelligent algorithms.

Table 1. Test function and related values

Function	f_{min}	Range	Dimension
$F_1(x) = \sum_{i=1}^{n} x_i^2$	0	[−100, 100]	30
$F_2(x) = \sum_{i=1}^{n} |x_i| + \prod_{i=1}^{n} |x_i|$	0	[−10, 10]	30
$F_3(x) = \sum_{i=1}^{n} \left(\sum_{j=1}^{i} x_j\right)^2$	0	[−100, 100]	30
$F_4(x) = \max_{1 \le i \le n} |x_i|$	0	[−100, 100]	30
$F_5(x) = \sum_{i=1}^{n} (x_i^2) - 10\cos(2\pi x_i) + 10$	0	[−5.12, 5.12]	30
$F_6(x) = -20\exp\left(-0.2\sqrt{\frac{1}{n}\sum_{i=1}^{n} x_i^2}\right) + e - \exp\left(\frac{1}{n}\sum_{i=1}^{n}\cos(2\pi x_i)\right) + 20$	0	[−32, 32]	30
$F_7(x) = \frac{1}{4000}\sum_{i=1}^{n} x_i^2 + 1 - \prod_{i=1}^{n}\cos(\frac{x_i}{\sqrt{i}})$	0	[−600, 600]	30
$F_8 = \left[\frac{1}{500} + \sum_{j=1}^{25} 1/(j + \sum_{i=1}^{2}(x_i - a_{ij})^6)\right]^{-1}$	0.998	[−65.56, 65.56]	2
$F_9(x) = [1 + (x_1 + x_2 + 1)^2)(19 - 14x_1 + 3x_1^2 - 14x_2 + 6x_1x_2 + 3x_2^2)]$ $\times [30 + (2x_1 - 3x_2)^2(18 - 32x_1 + 12x_1^2 + 48x_2 - 36x_1x_2 + 27x_2^2)]$	2.999	[−2, 2]	2

In order to perform a full-scale performance test on ASMFO algorithm, three types of test functions with different characteristics were selected. F_1–F_4 are unimodal test functions, F_5–F_7 are multimodal test functions, F_8 and F_9 are fixed-dimensional test functions.

4.2   Experimental Results and Analysis

This experiment compares ASMFO algorithm with MFO algorithm, particle swarm algorithm and genetic algorithm. The experimental results are shown in Tables 2, 3, and 4 respectively. The best results are shown in bold in the tables.

Table 2. Experimental results of unimodal test function

Function	Algorithm	Best	Worst	Mean	Std
F_1	PSO	6.44247027	28.6047372	14.14777878	4.87779115
	GA	5.64219159	44.11568688	24.79062017	10.14476297
	MFO	9.5928E−07	4.6093E−05	7.5581E−06	8.4415E−06
	ASMFO	**3.1132E−129**	**4.6136E−108**	**1.9274E−109**	**8.4816E−109**
F_2	PSO	6.44247027	28.60473722	4.390173383	1.41556419
	GA	1.19075277	3.15822457	1.74116166	0.44238885
	MFO	1.9063E−05	0.00012642	6.0213E−05	2.9402E−05
	ASMFO	**1.5794E−73**	**6.9312E−63**	**2.3729E−64**	**1.2647E−64**
F_3	PSO	437.6699399	1875.254823	888.886347	358.870893
	GA	11266.902016	29006.61557	18829.53855	4732.971926
	MFO	6.35739510	95.28625713	34.78855963	21.71567269
	ASMFO	**6.9681E−95**	**3.9857E−73**	**1.3325E−74**	**7.2761E−74**
F_4	PSO	6.34348413	16.34033120	12.13094671	2.76007705
	GA	10.77237783	19.13520829	14.74685534	2.679924803
	MFO	1.58532909	6.26086336	3.53795984	1.37849376
	ASMFO	**2.0928E−50**	**6.0737E−44**	**2.1195E−46**	**1.1079E−44**

Table 3. Experimental results of multimodal test function

Function	Algorithm	Best	Worst	Mean	Std
F_5	PSO	23.0315797	66.11898974	38.79050312	10.87197952
	GA	3.84974637	11.50592842	7.42886514	2.12945710
	MFO	17.90947950	46.76301511	30.61165432	8.68428996
	ASMFO	**0**	**0**	**0**	**0**
F_6	PSO	3.26415544	6.58335030	4.47039321	0.77858490
	GA	1.61784193	2.88877334	2.30607425	0.29967599
	MFO	0.00014982	0.00105265	0.00050489	0.00027118
	ASMFO	**8.8817E−16**	**8.8817E−16**	**8.8817E−16**	**0**
F_7	PSO	1.05234042	1.23343077	1.11898271	0.04124786
	GA	1.11481434	1.40197065	1.21738119	0.07820754
	MFO	3.4319E−06	0.03693290	0.01116933	0.01079854
	ASMFO	**0**	**0**	**0**	**0**

Table 4. Experimental results of fixed-dimensional test function

Function	Algorithm	Best	Worst	Mean	Std
F_8	PSO	0.9980038	11.71869956	4.11189598	3.22095081
	GA	0.99800384	0.99800453	0.99800387	1.29777131
	MFO	0.99800383	13.6186089	1.71637253	2.35261736
	ASMFO	**0.99800383**	**0.99802077**	**0.99800363**	**3.0832E−06**
F_9	PSO	3	3.00001671	3.00000311	4.1459E−06
	GA	3.00000368	30.0449059	6.60396389	9.34298666
	MFO	**2.99999992**	**2.99999992**	**2.99999992**	**2.0599E−15**
	ASMFO	3.00000023	3.00002472	3.00000820	7.2941E−06

The following 9 figures are the experimental results of the four algorithms respectively, they are repeated 1000 times and independently run 30 times the result of the last iteration. The best fitness values and convergence rate of the four algorithms can be seen from the figures.

From Tables 2, 3, and 4 and the convergence curves, it can be seen that ASMFO algorithm converges to the optimal value in functions F_5 and F_7. From the standard deviation, the stability of ASMFO algorithm is also significantly higher than the other three algorithms. Although F_1–F_4, F_6 did not converge to the optimal value after 1000 iterations, compared with the other three algorithms, the convergence speed, convergence accuracy, and stability were significantly improved. ASMFO algorithm and genetic algorithm have similar effects in the function F_8. ASMFO algorithm and MFO algorithm and particle swarm algorithm have similar effects in the function F_9. In general, ASMFO algorithm has improved convergence accuracy and stability, especially for unimodal functions and multimodal functions, convergence accuracy and stability have been improved. From Figs. 1, 2, 3, 4, 5, 6, 7, 8 and 9, the results of the optimization can be visually displayed.

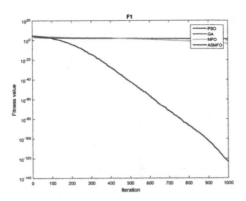

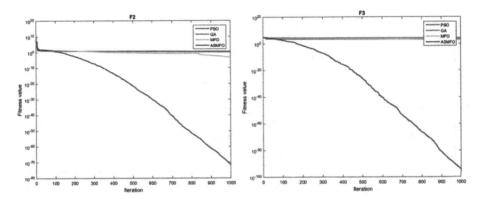

Fig. 1. F_1 fitness value iteration curve

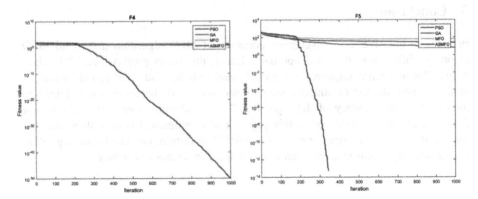

Fig. 2. F_2 fitness value iteration curve   **Fig. 3.** F_3 fitness value iteration curve

Fig. 4. F_4 fitness value iteration curve   **Fig. 5.** F_5 fitness value iteration curve

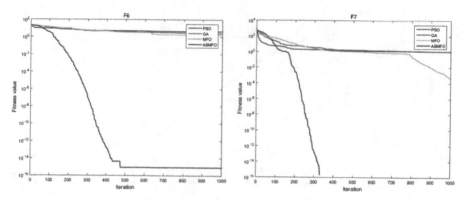

Fig. 6. F_6 fitness value iteration curve   **Fig. 7.** F_7 fitness value iteration curve

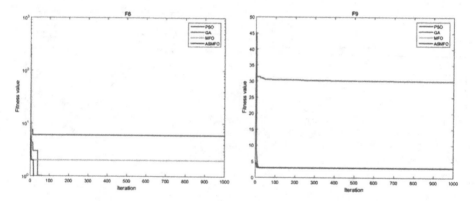

Fig. 8. F_8 fitness value iteration curve   **Fig. 9.** F_9 fitness value iteration curve

5 Conclusion

In order to improve the convergence accuracy of MFO algorithm and avoid MFO algorithm falling into the local optimal solution, this paper proposes ASMFO algorithm. The algorithm improves the global search ability and local search ability of moths through the use of adaptive weights and simulated annealing, which improves the convergence accuracy of the algorithm. The simulation experiments of nine test functions are also carried out in this paper. The experimental results show that the improved algorithm is superior to the basic MFO algorithm, particle swarm algorithm and genetic algorithm in convergence speed and convergence accuracy.

References

1. Qin, Z., Yu, F., Shi, Z., Wang, Y.: Adaptive inertia weight particle swarm optimization. In: Rutkowski, L., Tadeusiewicz, R., Zadeh, L.A., Żurada, J.M. (eds.) ICAISC 2006. LNCS (LNAI), vol. 4029, pp. 450–459. Springer, Heidelberg (2006). https://doi.org/10.1007/11785231_48
2. Bergey, P.K., Ragsdale, C.T., Hoskote, M.: A simulated annealing genetic algorithm for the electrical power districting problem. Ann. Oper. Res. **121**(1–4), 33–55 (2003)
3. Mirjalili, S.: Moth-flame optimization algorithm: a novel nature-inspired heuristic paradigm. Knowl.-Based Syst. **89**, 228–249 (2015)
4. Parmar, S.A., Pandya, M.H., Bhoye, M., et al.: Optimal active and reactive power dispatch problem solution using moth-flame optimizer algorithm. In: International Conference on Energy Efficient Technologies for Sustainability. IEEE (2016)
5. Aziz, M.A.E., Ewees, A.A., Hassanien, A.E.: Whale optimization algorithm and moth-flame optimization for multilevel thresholding image segmentation. Expert Syst. Appl. **83**, 242–256 (2017)
6. Muangkote, N., Sunat, K., Chiewchanwattana, S.: Multilevel thresholding for satellite image segmentation with moth-flame based optimization. In: International Joint Conference on Computer Science and Software Engineering, pp. 1–6. IEEE (2016)
7. Yamany, W., Fawzy, M., Tharwat, A., et al.: Moth-flame optimization for training multi-layer perceptrons. In: International Computer Engineering Conference, pp. 267–272. IEEE (2015)
8. Li, C., Li, S., Liu, Y.: A least squares support vector machine model optimized by moth-flame optimization algorithm for annual power load forecasting. Appl. Intell. **45**(4), 1–13 (2016)
9. Chen, H.G., Wu, J.S., Wang, J.L., et al.: Mechanism study of simulated annealing algorithm. J. Tongji Univ. **32**(6), 802–805 (2004)
10. Yang, X.S.: Appendix A: Test problems in optimization. In: Engineering Optimization, pp. 261–266. Wiley, Hoboken (2010)

An Improved DNA Genetic Algorithm Based on Cell-Like P System with Dynamic Membrane Structure

Wenqian Zhang and Wenke Zang[(⊠)]

Shandong Normal University, Jinan, China
wink@sdnu.edu.cn

Abstract. Inspired by the P system with dynamic membrane structure and the mechanism of DNA genetic information, an optimization algorithm based on improved DNA Genetic Algorithm (DNA-GA) and the cell-like P system with dynamic membrane structure called DNA-DMS, is proposed. By merging membrane computing and DNA genetic operation, DNA-DMS not only avoids the disadvantage of easily getting to local optional solutions, but also converges rapidly. At the same time, compared with the traditional DNA algorithm, our algorithm improves the ability of parallelism. We propose a novel DNA genetic operation, splicing operation that can enhance the variety of the population and promote fast convergence. The performance of DNA-DMS is tested through computational experiments and compared with those of standard DNA genetic algorithm, DE and PSO algorithm using 6 typical benchmark functions. The experimental results demonstrate that the DNA-DMS can overcome premature convergence and yield the global optimum with high efficiency.

Keywords: Membrane computing · P system · DNA genetic algorithm
Intelligent algorithm optimization

1 Introduction

Membrane computing, initiated by Păun [1], is inspired from the structure and function as well as interaction of living cells in tissues and organs. Membrane computing investigates distributed parallel computing models, also known as P systems [2]. Inspired by membrane computing and evolutionary algorithms, the membrane evolutionary algorithm was developed by using the hierarchical structures of the membrane and the search methodologies of the evolutionary algorithm [1]. Liu et al. has done a lot of research in the field of membrane computing applications, such as clustering methods [3]. Recently, P system with dynamic membrane structure has become a research hotspot. Liu et al. designed an evolutionary algorithm that employed a dynamical membrane structure and several reaction rules to solve the uncertain optimization problems [4–8]. Peng et al. proposed a method that employed membrane system with active membranes and differential evolution (DE) to solve automatic fuzzy clustering problems [9]. And Xiao et al. proposed an improved dynamic membrane evolutionary algorithm based on particle swarm optimization (PSO) and differential

© Springer Nature Switzerland AG 2018
Y. Peng et al. (Eds.): IScIDE 2018, LNCS 11266, pp. 168–177, 2018.
https://doi.org/10.1007/978-3-030-02698-1_15

evolution (DE) algorithms to solve constrained engineering design problems [10]. The combination of P system with dynamic membrane structure and intelligent algorithms is a trend. DNA genetic algorithm that also inspired by biological manipulation have attracted many researchers' attention in recent years and have been successfully applied to solve many difficult optimization problems. There are few articles that combined P system with dynamic membrane structure with DNA algorithms. Maroosi and Muniyandi have proposed a membrane computing-inspired genetic algorithm [11]. Zang et al. proposed an mDNA-GA method that used biological membrane structure without using any membrane evolution rules [12]. In this paper, we combine the cell-like P system with dynamic membrane structure that possess membrane evolution and communication rules with an improved DNA genetic algorithm and propose an optimization algorithm, DNA-DMS.

This paper is organized as follows. Section 2 presents some background material on DNA genetic algorithm. In Sect. 3, we introduce the cell-like P system with dynamic membrane structure. In Sect. 4, we provide the proposed algorithm DNA-DMS. Section 5 presents experimental results on 6 test functions along with their analysis. Section 6 provides some conclusions and future research directions.

2 DNA Genetic Algorithm

Since Adleman first developed DNA-based biological computing method for solving a computationally hard problem of the directed Hamiltonian path problem [13], researchers started to devote to the research and applications of DNA and RNA genetic algorithms [14]. Owing to advantages of DNA genetic algorithms, in this work, DNA–GA is used to optimize the performance of membrane computing.

2.1 DNA Encoding and Decoding

During the process of evolution, DNA genetic algorithm using base-based coding scheme has the ability to expressing rich genetic information, with a stronger global search capabilities. Because it has the advantages of genetic algorithm and biological DNA information, Not only can the coding way be more flexible, but also make the genetic algorithm process easier, overcoming the shortcomings of single genetic algorithm premature convergence and coding complexity and inflexibility.

A DNA sequence contains four types of nucleotides: Adenine (A), Guanine (G), Cytosine (C), and Thymine (T). DNA strings can be used as decoding information to manipulate DNA at the molecular level. Mathematically, a string composed of the four symbols A, G, C and T is used to encode a parameter of an optimization problem. This can be denoted by the four decimal codes: 0, 1, 2, and 3. There are two complementary pairs such as A–T and C–G, which corresponds to 0–3 and 1–2 after encoding. Through this correspondence, a sequence of DNA molecules that stores genetic information can be converted into a string of numbers that can be identified and computed by the computer.

2.2   DNA Genetic Operations

The Selection Operation

Selection operation determines which individual can be kept to next generation and how many individuals are replicated to offspring [15]. In the natural evolution, the species with a higher degree of adaptability to the living environment are inherited to the next generation more likely. To simulate this process, this paper uses a tournament selection method to generate a new generation of population. The basic idea is to choose one of the two individuals with a better fitness to the next generation, thus ensuring the efficiency and convergence of the algorithm.

The Crossover Operation

The crossover operation is mainly responsible for the global searching ability of the algorithm. In the process of transferring, genes far apart from one another can be combined together, and then new genetic materials can be generated.

The Mutation Operation

Mutation operation can generate new genetic material and maintain the population diversity. Take advantage of it can let search process avoid falling into local optimum and obtain better optimization performance. Therefore, mutation operation plays an important part in evolutionary algorithm [15]. The deletion and insertion operators derived from biologic DNA evolution are actually the operations of mutation [16]. The deletion mutation means that a fragment of bases is removed from a sequence and the insertion mutation means that a fragment of bases is inserted into a sequence. The purpose is to replace those bad individuals with better ones at the early stage of evolutionary process.

The Splicing Operation

A restriction enzyme can cut a double stranded DNA molecule efficiently. If DNA molecules and restriction enzymes simultaneously exist, then these enzymes can cut the DNA. If also ligase is present in the tube, then 'new' DNA molecules that not initially present can be formed by recombination. This biological process is called splicing. In this paper, we reference the Păun definition of splicing operation [17], which is widely used.

Inspired by the biologic DNA splicing operation and the splicing of RNA in the research field of biomedical, we propose a novel DNA genetic operation in this section.

External Splicing Operator

The variation of the left part of the chromosome is more likely to cause the change of fitness than the variation in the right part, so the first half of the chromosome is defined as 'hot spots', and the latter part is defined as 'cold spots' in a DNA sequence [18]. Based on this, we define an external splicing operator, which plays a role between two different chromosomes (individuals). And in biology, a DNA strand is a sequence of nucleotide bases, a subsequence of three consecutive nucleotides represents an amino acid. According to this biological fact, three bases of two different chromosomes head are swapped according to a certain probability, which can further increase population diversity.

For example, two new different double-strands DNA molecules generation processes are shown below 0 (Fig. 1).

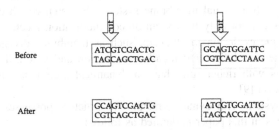

Fig. 1. An example of external splicing operator

Internal Splicing-Inversion Operator
This operator takes place inside the two individuals. Firstly, two parent individuals are randomly selected and named as $C^1 = C_1^1 C_2^1 \ldots C_L^1$, $C^2 = C_1^2 C_2^2 \ldots C_L^2$. Then randomly select a sequence $C_R^1 = C_i^1 \ldots C_{i+m1}^1$ from C^1, where, C_i^1 is a position in C^1 randomly selected, m1 is a positive integer and $1 \leq m_1 + i \leq L$. Analogously, $C_R^2 = C_j^2 \ldots C_{j+m2}^2$ $(1 \leq m_2 + j \leq L)$. Secondly, cut these two pieces down and reverse these two fragments. Finally, two new individuals are generated $C'^1 = C_1^1 \ldots C_{j+m2}^2 \ldots C_j^2 \ldots C_L^1$ and $C'^2 = C_1^2 \ldots C_{i+m1}^1 \ldots C_i^1 \ldots C_L^2$. (see as 0) (Fig. 2).

Fig. 2. An example of internal splicing-inversion operator

3   Cell-Like P System with Dynamic Membrane Structure

P systems could be divided into three groups: cell-like P systems, tissue-like P systems and neural-like P systems. The structure of cell-like P systems is the basic structure of other kinds of P systems. The membrane structure of a cell-like P system consists of several membranes arranged in a hierarchical structure inside a main membrane, called the skin membrane. A membrane without any other membrane inside is said to be elementary. A space delimited by one membrane and the membrane immediately below is called a region, and the region of an elementary membrane is the space it delimits. Each region can contain multiple sets of objects and evolution rules as well as communication rules [10].

There are two types of rules in usual P systems: evolution rules and communication rules. Evolution rules are used to evolve the objects in the regions, while communication rules are used to exchange the objects between internal and external regions of a membrane. Different from usual membrane systems, P systems with dynamic membrane structure possess not only the evolution-communication mechanism of objects but also a membrane evolution mechanism, such as membrane division, membrane merging, membrane dissolution and membrane creation and so on. Therefore, these kinds of P systems with richer rules have a dynamical membrane structure during evolution/computation [9].

A cell-like P system with dynamic membrane structure possesses evolution and communication rules in this paper is defined as follow:

$$\Pi = \left(O, u, \omega_1 \ldots \omega_q, R, i_0\right),$$

where

(1) o is a finite alphabet of objects;
(2) u is the initial membrane structure consisting of q membranes, $q \geq 1$ is the initial degree of the system;
(3) $\omega_1 \ldots \omega_q$ are multiple sets of objects placed in the regions of the q membranes;
(4) R is a finite set of rules
   (a) $[a \rightarrow b]_i$, for $i \in \{1, 2, \ldots, n\}$, $a, b \in O^*$ (evolution rules);
   (b) $[a]_i \rightarrow b$, for $i \in \{1, 2, \ldots, n\}$, $a \in O, b \in O \cup \{\lambda\}$ (membrane dissolution rules);
   (c) $[a \rightarrow [b]_{i1}]_{i2}$, for $i1, i2 \in \{1, 2, \ldots, n\}$, $a, b \in O$ (membrane creation rules);
   (d) $a[]_i \rightarrow [b]_i$, for $i \in \{1, 2, \ldots, n\}$, $a, b \in O$ (In communication rules);
   (e) $[a]_i \rightarrow []_i b$, for $i \in \{1, 2, \ldots, n\}$, $a, b \in O$ (Out communication rules).
(5) i_0 represents the storage area of the system.

In this paper, the improved DNA-GA worked as the objects evolution rules to evaluate the population in the DNA soup. After a certain degree of evolution, the population similarity is high and the solution space cannot continue to search, which will lead to the premature convergence of the algorithm. So after the iterative evolution process in the elementary membrane finished, we used the In communication rules to transport local best solutions from the elementary membrane into the middle membrane to form a new population. Besides, those objects of different middle membrane can interact and exchange information using In/Out communication rules according to a certain proportion in order to further enhance the diversity of the population.

In our work, only two types of membrane rules are introduced in cell-like P system with dynamic membrane structure: membrane dissolution rules and membrane creation rules. The three middle membranes can be regarded as three existing subsystems in the beginning, while during the process of computation, the cell-like P system with dynamic membrane structure can generate more subsystems that contain elementary membranes by membrane creation mechanism, and then generates m objects that form an initial population for each elementary membrane respectively, which can solve another problem according to the actual demands. It is fully demonstrated the powerful parallel computing ability of P system. The membrane dissolution rule is executed after the operation stops in the elementary membranes and the candidate solutions are transport to the middle membranes.

As shown in the 0 below, the cell-like P system with dynamic membrane structure has been set up. Membrane 0 is the skin membrane and membranes 1, 2 and 3 are the nonelementary membranes which are considered as the middle membrane, while the elementary membranes play a role as the DNA soup in our work. These arrows indicate communication rules. With the use of communication rules, objects can communicate with others in the surrounding regions to facilitate information exchange and material transmission. So, with the calculation progresses step by step, the membrane structure is changing continuously: the objects in the elementary membranes are evolved generation after generation, some of those objects communicate among different regions; some membranes complete the iterative calculation process and then are dissolved, and some new membranes are created in order to complete a new computing task. All membranes work in parallel in the cell-like P system with dynamic membrane structure. The calculation process will be repeated until the termination condition is satisfied. Normally set a maximum number of iteration as a termination condition, thus, the computation will stop after a predefined iteration. When the calculation halts, it produces a final result in the output region (Fig. 3).

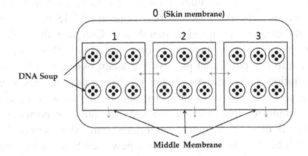

Fig. 3. Initial membrane structure

4   Proposed Algorithm: DNA-DMS

The DNA-DMS algorithm starts from the elementary membranes. The termination condition is that the number of generations reaches the maximum. If this condition is met, the algorithm will stop and the current best individuals will be output to the membranes in the next layer. Obviously, the number of individuals from all of the elementary membranes output to each of the middle membranes is N. The selection, crossover, mutation and splicing operations are executed with certain probabilities at each generation. A random number is generated first. The operation is performed only if the corresponding probability is larger than this random number. The steps of the DNA-DMS algorithm are described as follows:

Step 1. Determining the parameters N (the size of initial population), Pc (the probability of crossover operation), Pm (the probability of mutation operation), Ps (the probability of splicing operation) and setting the communication probability between the three middle membranes. Initialize the population of candidate solutions by randomly generating N individuals in each of the elementary membranes.

Step 2. The DNA-DMS algorithm in the elementary membranes starts to be executed. Perform selection operation by using the tournament method, and next, crossover, mutation and splicing operation is performed with the predetermined probability successively. The improved DNA genetic operations are repeated until the termination condition is met for each of the elementary membrane.

Step 3. Compute the fitness values of the solutions. Extract and transport the best candidate solutions Be (best value in elementary membranes) from each of the elementary membranes to the middle membranes and form the initial population. And all the elementary membranes dissolved.

Step 4. Communication. The DNA-DMS in the three middle membranes began functioning immediately. The three middle membranes will communicate and exchange the objects information according to the predetermined communication probability. And then the selection, crossover, mutation and splicing operations are executed and check until the termination condition is met.

Step 5. Compute the fitness values of the solutions and sort the solutions in descending order of their fitness values in each of the middle membranes. Extract and transport the best candidate solutions Bm (best value in middle membranes) from each of the middle membranes to the skin membrane. Compare these three candidate solutions and output the best of them Bg (global best value), then Stop.

During the entire computational process, we adopt two types of membrane rules: membrane dissolution rules and membrane creation rules. In this work, after the evolution of the object in the elementary membrane, the mission of the elementary membranes is completed, the local best objects Be are transported to the middle membranes, and all these elementary membranes will be dissolved by membrane dissolution rules. The P system can generate a new membrane using membrane creation rules along with these objects. This is also the embodiment of powerful parallelism in membrane computing. In fact, DNA-DMS is an integrated optimization algorithm, which not only takes advantage of the parallel computing of membrane computing and communication rules between different regions, but also takes advantage of the global optimization ability of DNA genetic algorithm.

5 Experimental Results and Analysis

In order to verify the performance of our approach, in this section, DNA-DMS is applied to 6 scalable benchmark functions. The specific description of these functions is shown in Table 1. These benchmark functions are widely adopted in benchmarking global optimization algorithms. All these benchmark functions are evaluated as the minimization problems.

Table 1. Test functions

Name	Test functions formula	Optimal value
Schaffer	$\min f_1(\mathbf{x}) = 0.5 + \frac{\sin^2\sqrt{x_1^2+x_2^2}-0.5}{(1+0.001(x_1^2+x_2^2))^2}$	0
Rastrigin	$\min f_2(\mathbf{x}) = (x_1^2 - 10 \times \cos(2\pi x_1) + 10)$ $+ (x_2^2 - 10 \times \cos(2\pi x_2) + 10)$	0
Rana	$\min f_3(\mathbf{x}) = x_1 \sin(\sqrt{|x_2+1-x_1|}) \times \cos(\sqrt{|x_1+x_2+1|})$ $+ 511.7329 + (x_2+1)\cos(\sqrt{|x_2+1-x_1|})$ $\times \sin(\sqrt{|x_1+x_2+1|})$	-511.708
Griewank	$\min f_4(\mathbf{x}) = 1 + \left(\frac{(x_1-100)^2}{4000} + \frac{(x_2-100)^2}{4000}\right)$ $- \cos\frac{x_1-100}{\sqrt{1}} \times \cos\frac{x_2-100}{\sqrt{2}}$	0
Matyas	$\min f_5(\mathbf{x}) = 0.26(x_1^2 + x_2^2) - 0.48 x_1 x_2$	0
Six-hump camel back	$\min f_6(\mathbf{x}) = (4 - 2.1x_1^2 + x_1^4/3)x_1^2 + x_1 x_2 + (-4 + 4x_2^2)x_2^2$	-1.0316

Functions from f1(x) to f4(x) are unimodal. Unimodal test functions are continuous, convex and have one global optimum. They are suitable to checkout the exploitation ability of algorithms. Functions from f5(x) to f6(x) are multimodal. For the six-hump camel back function, within the bounded region it owns six local minimum, two of them are global ones. In contrast to the unimodal functions, multi-modal functions have a global optimum as well as many local optima with the number increasing exponentially with dimension, which makes them suitable for testing the exploration ability of an algorithm.

To compare the results of different algorithms, standard DNA-GA, PSO and DE, each function is optimized over 100 independent runs. To be fair, we use the same set of initial random populations to evaluate different algorithms, and all of the compared algorithms start from the same initial population in each out of 50 runs. The population size is taken as 100. The largest iteration is taken as 100. The obtained best results are highlighted in italic (see Tables 2 and 3). The evolutionary iteration is averaged over several independent computations.

Table 2. Comparison of the optimal values of DNA-GA, PSO and DNA-DMS (run 50 times)

Function	DNA-GA	PSO	DE	DNA-DMS
$f_1(\mathbf{x})$	1.11e−07	6.15e−10	1.45e−08	*1.45e−12*
$f_2(\mathbf{x})$	5.49e−06	4.18e−06	1.41e−05	*6.19e−07*
$f_3(\mathbf{x})$	−510.159	−508.145	−506.068	*−511.708*
$f_4(\mathbf{x})$	0.01585	0.00986	0.00875	*1.31e−09*
$f_5(\mathbf{x})$	7.23e−09	7.10e−06	6.00e−06	*3.76e−09*
$f_6(\mathbf{x})$	−1.03155	−1.03156	*−1.03160*	−1.03160

Table 3. Comparison of the mean optimal values of DNA-GA, PSO, DE and DNA-DMS

Function	DNA-GA	PSO	DE	DNA-DMS
$f_1(\mathbf{x})$	8.67e−03	5.96e−02	3.42e−03	*5.27e−05*
$f_2(\mathbf{x})$	0.57430	0.31951	0.08019	*0.00114*
$f_3(\mathbf{x})$	−508.67454	−506.47353	−505.065	*−511.708*
$f_4(\mathbf{x})$	0.23633	1.01507	0.2787	*0.00021*
$f_5(\mathbf{x})$	0.00289	0.00484	0.09456	*0.00093*
$f_6(\mathbf{x})$	−1.0314	−1.0276	−0.7546	*−1.0316*

The DNA-DMS combines the genetic evolution mechanism of the population and the evolutionary-communication mechanisms of cell-like P system with dynamic membrane structure organically. From these results, we can see that our algorithm outperforms those other algorithms in all benchmark problems in both the average and the optimal cases. For f6(x), both our algorithm and DE algorithm can find the best value. But from the Table 3, we find that the result of the DNA-DMS algorithm is more stable and accurate than DE. This clearly indicates that our algorithm can find solutions that are much closer to the global optimal solution than other comparison algorithms. The algorithm overcomes such drawbacks in traditional genetic algorithm as "no memory" and cross-convex operation, which makes the algorithm search on a wider range and improves the quality of optimization.

6   Conclusions and Further Work

To accelerate the evolutionary process and increase the probability to find the optimal solution, we present a novel optimization algorithm based on an improved DNA-GA and the cell-like P system with dynamic membrane structure, DNA-DMS. On the one hand, we make full use of the parallel computing power of P system with dynamic membrane structure. On the other hand, we introduce a new splicing operation that including two kinds of operator. The traditional genetic operation of standard DNA genetic algorithm has been improved further that the algorithm can quickly converge.

We compare our algorithm with standard DNA-GA, PSO and DE using 6 optimization functions. The experimental results show that our algorithm can converge much closer to the global optimal solutions within much less number of iterations than the existing comparison algorithms. It is promising to use the new operation and new population evolution mechanism to improve the performance of traditional DNA genetic algorithms. The future work includes parametric analysis and improvement of algorithm, and other applications in engineering optimization.

References

1. Păun, G.: Computing with membranes. J. Comput. Syst. Sci. **61**, 108–143 (2000)
2. Peng, H., Zhang, J., Jiang, Y., Huang, X., et al.: DE-MC: a membrane clustering algorithm based on differential evolution mechanism. Rom. J. Inf. Sci. Technol. **17**, 76–88 (2014)
3. Liu, X., Xue, J.: Spatial cluster analysis by the bin-packing problem and DNA computing technique. Discret. Dyn. Nat. Soc. **2013**, 845–850 (2013)
4. Liu, X., Liu, H., Duan, H.: Particle swarm optimization based on dynamic niche technology with applications to conceptual design. Comput. Sci. **38**, 668–676 (2006)
5. Liu, X., Xue, J.: A cluster splitting technique by hopfield networks and P systems on simplices. Neural Process. Lett. **46**, 1–24 (2017)
6. Liu, X., Zhao, Y., Sun, M.: An improved apriori algorithm based on an evolution-communication tissue-like P system with promoters and inhibitors. Discret. Dyn. Nat. Soc. **2017**, 1–11 (2017)
7. Zhao, Y., Liu, X., Wang, W.: Spiking neural P systems with neuron division and dissolution. Plos One **11**, e0162882 (2016)
8. Liu, C., Fan, L.: Evolutionary algorithm based on dynamical structure of membrane systems in uncertain environments. Int. J. Biomath. **9**, 1650017 (2016)
9. Peng, H., Wang, J., Shi, P., Pérez-Jiménez, M.J., et al.: An extended membrane system with active membranes to solve automatic fuzzy clustering problems. Int. J. Neural Syst. **26**, 1650004 (2016)
10. Xiao, J., He, J.J., Chen, P., Niu, Y.Y.: An improved dynamic membrane evolutionary algorithm for constrained engineering design problems. Nat. Comput. **15**, 1–11 (2016)
11. Maroosi, A., Muniyandi, R.C.: Membrane computing inspired genetic algorithm on multi-core processors. J. Comput. Sci. **9**, 264–270 (2013)
12. Zang, W., Sun, M., Jiang, Z.: A DNA genetic algorithm inspired by biological membrane structure. J. Comput. Theor. Nanosci. **13**, 3763–3772 (2016)
13. Adleman, L.M.: Molecular computation of solutions to combinatorial problems. Science **266**, 1021–1024 (1994)
14. Zang, W., Zhang, W., Zhang, W., Liu, X.: A genetic algorithm using triplet nucleotide encoding and DNA reproduction operations for unconstrained optimization problems. Algorithms **10**, 76 (2017)
15. Dai, K., Wang, N.: A hybrid DNA based genetic algorithm for parameter estimation of dynamic systems. Chem. Eng. Res. Des. **90**, 2235–2246 (2012)
16. Li, Y., Lei, J.: A feasible solution to the beam-angle-optimization problem in radiotherapy planning with a DNA-based genetic algorithm. IEEE Trans. Bio-Med. Eng. **57**, 499–508 (2010)
17. Rogozhin, Y., Verlan, S.: Computational models based on splicing. In: Adamatzky, A. (ed.) Automata, Universality, Computation. ECC, vol. 12, pp. 237–257. Springer, Cham (2015). https://doi.org/10.1007/978-3-319-09039-9_11
18. Neuhauser, C., Krone, S.M.: The genealogy of samples in models with selection. Genetics **145**, 519–534 (1997)

Efficiency Ranking via Combining DEA Evaluation and Bayesian Prediction for Logistics Enterprises

Chenglong Cao[1,2]([⊠]) and Xiaoling Zhu[3]

[1] Anhui Finance and Trade Vocational College, Hefei 230601, China
chenglongcao@sina.cn
[2] Research Center of Logistics Engineering,
University of Science and Technology of China, Hefei 230026, China
[3] School of Computer and Information, Hefei University of Technology,
Hefei 230009, China
zhuxl@hfut.edu.cn

Abstract. Data envelopment analysis (DEA) has been widely used in economic development evaluation and enterprise performance analysis. When a conventional DEA is used directly for ranking, there are some issues. It requires the investment and income data, and the latter are after-the-fact data, so DEA is ex post analysis. In addition, coarse granular DEA results may cause parallel ranking at high frequencies. Therefore, combining DEA and Bayes, we propose an efficiency prediction approach that does not require income data. The approach can predict efficiency levels under different investment combinations, and help logistics enterprises to rationally allocate limited resources in their decision-making. Furthermore, an efficiency ranking algorithm is designed by incorporating the overall probability distribution of data set and then is applied to evaluate fourteen A-level logistics enterprises in Anhui, China. Empirical results show that the DEA-Bayes approach has good discrimination for efficiency ranking. Unlike expert scoring, our evaluation process is based on logistics enterprise data and easy to operate.

Keywords: Efficiency evaluation · Ranking · DEA · Bayesian approach

1 Introduction

Logistics is a basic industry in the national economy. It has a strong correlation with other industries and directly affects the development of the related industry chain [1]. Scientific evaluation of efficiency levels of logistics enterprises is helpful to promote the sound development of the logistics industry. Some evaluation methods were proposed, such as grey analytic hierarchy process, fuzzy comprehensive evaluation, and data envelopment analysis (DEA).

DEA has been widely used since it avoids subjective factors and has simple algorithm [2]. Ni et al. [3] evaluated logistics efficiency of Jiangxi province from 2005 to 2013 by a DEA model, and they analyzed the influencing factors using a Tobit regression model. Results show that the general level of logistics efficiency in Jiangxi is

© Springer Nature Switzerland AG 2018
Y. Peng et al. (Eds.): IScIDE 2018, LNCS 11266, pp. 178–188, 2018.
https://doi.org/10.1007/978-3-030-02698-1_16

high, but the pure technical efficiency is low, which leads to the waste of resources. Sherrie et al. [4] extended the research to the outside of the enterprise and found that an external market is also an important factor in determining logistics efficiency. An extended DEA model [5] with different input constraints is defined, and then the energy utilization efficiency for logistics enterprises in Hefei is evaluated. Markovits et al. [6] presented a method combining DEA with analytic hierarchy process, and the efficiency of logistics enterprises in 29 European countries was studied. Yang et al. [7] provided the cross-efficiency DEA and IAHP method. The former adopts an assessment mechanism; in the latter, an entropy method is used to introduce the decision makers' preference and determine the weights. Based on factor analysis, a rule of thumb, and Pastor Method, Pan et al. [8] constructed three sets of evaluation index. Using them, research efficiency of Chinese universities is analyzed. Using DEA CCR and BCC models, Deng et al. [9] conducted an empirical analysis on production efficiency and scale efficiency for 55 logistics enterprises in the stock markets of Shanghai and Shenzhen.

In the above research works [2–9], most focus on industrial analysis from a macro perspective. Unlike them, we make an empirical analysis on individual enterprises from a micro view. There are some issues that need to be further explored. (1) As we know, a conventional DEA method requires input-output indications. The output indications are after-the-fact data, and DEA has the function of ex post analysis. So DEA is difficult to give some objective suggestions before an enterprise makes a decision. (2) DEA results include comprehensive technical efficiency, purely technical efficiency, scale efficiency, scale reward, and classification results. They are coarser and difficult to achieve effective ranking. For example, classification includes three results: efficient, weak efficient, non-efficient. If ranking is based on classification, some companies will have the same ranking. It means that the parallel ranking will occur at high frequencies. (3) On the other hand, DEA algorithm usually aims at independent decision making unit (DMU), and it does not consider the overall probability distribution of data set. But, in fact, the ranking is related to data distribution.

In terms of other ranking methods, Storto et al. [10] combined DEA cross-efficiency and Shannon's entropy to calculate the ecological efficiency of a city. The method includes the calculation of efficiency scores, the calculation of cross-efficiency scores, and the combination of the scores by computing the Shannon's entropy index. In [11], DEA is enhanced with fuzzy analytic hierarchy process (FAHP), where each decision maker makes a dual comparison of decision criteria and qualities, and assigns a relative score. In [12], socio-economic ranking of the cities of Turkey was presented using DEA and linear discriminant analysis (LDA), and the cities were compared according to the socio-economic development efficiency scores.

In order to achieve effective ranking, the references [10–12] all try to introduce a new method into DEA such as entropy [10], FAHP [11] and LDA [12]. In this paper, we combine DEA and Bayes theory to construct a new efficiency evaluation model. The main contributions are: (1) the model predicts enterprise efficiency only according to the investment data. It will help some enterprises to make strategic deployments in advance and rationally allocate limited resources; (2) based on the Bayesian prediction results, we corporate the overall probability distribution of data set, design a ranking algorithm and conduct empirical research. Empirical analysis shows that our approach has good discrimination.

The remainder parts of this paper are organized as follows. Firstly, we choose evaluation indicators and then present an efficiency evaluation model in Sect. 2. Based on the investment data, we use Bayes method to predict efficiency classification, and design an ranking algorithm for logistics enterprises in Sect. 3. Then, the empirical analysis results are shown in Sect. 4. Finally, conclusions are presented in Sect. 5.

2 Efficiency Evaluation Based on DEA

DEA was put forward by Charnes, Cooper and Rhodes [2] in 1978. It is a nonparametric method in operations research. Using DEA, one can compare different organizations, called decision making units (DMUs). In general, some organizations are efficient, while others are non-efficient.

In this paper, we choose the investment data, such as staff, main business and warehouse area, as the DEA input variables. And further choose the income data, such as transport, warehouse and cargo handling, as the DEA output variables. Meanwhile, considering the uncertain marginal revenue of the current fast developing logistics industry, we adopt the BCC-DEA model [13] under the hypothesis of variable scale returns (Table 1).

Table 1. Evaluation indicators.

Types	Items	Units	Descriptions
Investment (Input)	Staff	People	Human resources situation
	Main business	Million Yuan	Enterprise operation situation
	Warehouse area	Square meter	Fixed assets and scales
Income (Output)	Transportation	Million Yuan	Transportation profitability
	Warehouse	Million Yuan	Warehousing profitability
	Loading & handling	Million Yuan	Cargo handling profitability

3 Efficiency Ranking Model Using DEA and Bayes Theory

Because prediction requires prior knowledge, we first input the investment and income data, run the DEA algorithm, obtain efficiency level, and build a priori knowledge base with the data including features, properties and efficiency level. Then, input the investment data, run the Bayesian prediction algorithm, obtain the efficiency level and the related probability distribution. Next, calculate the efficiency evaluation value (EV). At last, sort logistics enterprises according to their EVs. Finally, output a ranking list. The efficiency evaluation process is illustrated in Fig. 1.

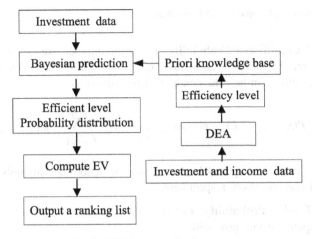

Fig. 1. Efficiency evaluation process

3.1 Efficiency Level Prediction

For the convenience of prediction, we quantify the DEA results as 1, 0 and −1, which correspond to DEA classification, i.e., efficient, weak efficient and non-efficient.

Feature Quantization. In our model, the investment indicators (staff, main business and warehouse area) are used as the features. But, they are continuous. In order to facilitate the use of Bayes, we discretize the continuous features. The mapping function for feature quantization is defined from floating point to integer as follows.

$$x_i = \left[(x'_i - min)(np - 1)/(max - min) + 1 \right] \tag{1}$$

Here, min and max are Minimum and Maximum value of the feature before mapping, respectively. np is Maximum value of the feature after mapping. The symbol [ ] represents the rounding operation. x'_i and x_i are the i-th feature before and after mapping, respectively.

Bayesian Classification. For an enterprise to be predicted, assuming that its feature vector is $x = (x_1, \ldots, x_n)$ and its classification is y, the conditional probability of Bayesian classification is $P(y|x_1, \ldots, x_n) = \frac{P(y) \prod_{i=1}^{n} P(x_i|y)}{P(x_1, \ldots, x_n)}$. For any given input x, $P(x_1, \ldots, x_n)$ is a constant, and thus the classification output is $\arg\max_y P(y|x_1, \ldots, x_n) = \arg\max_y P(y) \prod_{i=1}^{n} P(x_i|y)$.

3.2   Evaluation and Ranking Algorithm

Calculation of Conditional Probability. For a training set, we define the variable $count_{i,j,k}$ is the number of data items satisfying that the i-th feature is the property j in the k-th classification. Further, the conditional probability is

$$P(x_i = j | y = k) = pcount_{i,j,k} = count_{i,j,k} \left/ \sum_{j=1}^{np} count_{i,j,k} \right. \tag{2}$$

Here, $1 \leq i \leq nf, 1 \leq j \leq np, 1 \leq k \leq nc$. nf, np, nc denote the amounts of features, properties, and classifications, respectively.

Calculation of Joint Probability. For the DMU $(x_1, x_2, \ldots, x_{nf})$ that needs to be predicted, compute the join probability

$$P(y) \prod_{i=1}^{nf} P(x_i | y) \tag{3}$$

Where $P(y)$ is the probability of the classification y in the dataset.

Output of Classification Results. In all $y(1 \leq y \leq nc)$, return the value c that maximizes the following probability

$$c = \arg\max_y P(y) \prod_{i=1}^{nf} P(x_i | y) \tag{4}$$

Here, c is the predicted efficiency level of $(x_1, x_2, \ldots, x_{nf})$.

Calculation of Efficiency Evaluation. Define the efficiency value as

$$EV = c + k \sum_{y=1}^{nc} y P(y | x_1, \ldots, x_{nf}) \left/ \sum_{i=1}^{nf} x_i' \right. . \text{ Since } P(y | x_1, \ldots, x_n) \sim P(y) \prod_{i=1}^{n} P(x_i | y),$$

$$EV \sim c + k \sum_{y=1}^{nc} y P(y) \prod_{i=1}^{nf} P(x_i | y) \left/ \sum_{i=1}^{n} x_i' \right. \tag{5}$$

Here, k is a constant. EV gives priority to the DEA classification. Meanwhile, it takes into account the probability distribution of the classification and the original investment situations.

Output of Efficiency Ranking. We output the enterprises in the order of the EVs. Correspondingly, an efficiency ranking algorithm is given as follows.

Algorithm 1 Efficiency evaluation and ranking algorithm

Input: Feature vectors *TX*, DEA classification *TY*, and the
size m_1 of training set, the feature vectors *CX*, and the
size m_2 of test set; the number of classifications, features and properties are *nc*, *nf*, and *np*, respectively.
Output: Predicted classifications, efficiency values and
the ranking list.
1:   for *t*=1,2,..., m_1
2:       *x*=*TX*(*t*); *y*=*TY*(*t*);
3:       for *i*=1,2,..., *nf*
4:           count(*i*,*x*(*i*),*y*)=count(*i*,*x*(*i*),*y*)+1;
5:       end for
6:   end for
7:   for *k*=1,2,..., *nc*
8:       for *i*=1,2,..., *nf*
9:           Compute $pcount_{i,j,k}$ using Eq.(2)
10:         end for
11: end for
12: for *i*=1,2,..., m_2
13:      *x*=*CX*(i,:);
14:      for *y*=1,2,..., *nc*
15:          Search for $pcount_{i,x_i,y}$ and compute joint probability
             using Eq.(3)
16:      end for
17:      Compute and return *c* using Eq.(4)
18:      Compute and return *EV* using Eq.(5)
19: end for
20: Make a descending sort according to *EV*s.

---

4 Empirical Analysis of Logistics Enterprises

4.1 Data Source

We collected the investment and income data of 14 A-level logistics enterprises in
Anhui province in 2016. Considering that some contents involve privacy issues, the
names of all enterprises are replaced with the serial numbers from 1 to 14.

4.2 DEA Evaluation

The indicators, i.e., comprehensive efficiency, pure technical efficiency, scale efficiency, scale reward and efficiency level, are obtained by the BCC-DEA model. Scale

reward reflects the relationship between the scale and the reward, and it includes three types: increasing return to scale (IRS), constant return to scale (CRS), and decreasing return to scale (DRS).

Among the fourteen DMUs of Table 2, nine are efficient, two are weak efficient, and three are non-efficient. It reflects that most A-level logistics enterprises have the high level of efficiency in Anhui province, and a small number of enterprises need to improve the logistics efficiency urgently. In term of scale, nine DMUs are CRS, indicating that they achieve a good combination of input and output under the current technology and management level. Among the remaining DMUs, the pure technical efficiency values of DMU8 and DMU12 are one, which shows that the technical levels of the two enterprises have achieved the expected results. The DMU8 needs to increase investment scale. On the contrary, DMU12 needs to decrease the investment scale and improve scale rewards. For non-efficient DMU4, DMU7 and DMU9, their pure technical efficiency and scale efficiency are less than one, indicating that their technical capabilities and investment scale are both inadequate.

Table 2. DEA evaluation of fourteen logistics enterprises

DMUs	Comprehensive efficiency	Pure technical efficiency	Scale efficiency	Scale reward	Efficiency level
1	1	1	1	CRS	Efficient
2	1	1	1	CRS	Efficient
3	1	1	1	CRS	Efficient
4	0.955	0.959	0.996	IRS	Non-efficient
5	1	1	1	CRS	Efficient
6	1	1	1	CRS	Efficient
7	0.611	0.619	0.987	IRS	Non-efficient
8	0.313	1	0.313	IRS	Weak
9	0.359	0.782	0.459	IRS	Non-efficient
10	1	1	1	CRS	Efficient
11	1	1	1	CRS	Efficient
12	0.899	1	0.899	DRS	Weak
13	1	1	1	CRS	Efficient
14	1	1	1	CRS	Efficient

4.3   Efficiency Level Prediction

Efficiency classification can be predicted by the algorithm in Sect. 3.2. Table 3 shows that there are two errors among fourteen DMUs, which means that the prediction correct rate reaches 85.7%. In the table, staff, main business and warehouse area are the quantified results using Eq. (1). Then, they are as the input indicators of DEA.

Table 3. Efficiency prediction

DMUs	Staff	Main business	Warehouse area	DEA level	Predicted level
1	2	1	1	3	3
2	2	4	8	3	3
3	1	1	2	3	2
4	2	5	2	1	1
5	2	1	2	3	1
6	3	7	1	3	3
7	3	2	3	1	1
8	1	1	2	2	2
9	2	1	2	1	1
10	1	2	1	3	3
11	2	6	1	3	3
12	8	8	3	2	2
13	3	8	6	3	3
14	2	2	5	3	3

4.4   Efficiency Ranking

Using Eqs. (3), (4) and (5), the EV values and the ranking results are obtained. As can be seen from Fig. 2, the EVs of efficient DMUs are greater than the EVs of weak efficient and non-efficient DMUs. For example, when DMUs = 1, 2, 6, 10, 11, 13 and 14, they have high EV. Similarly, when DMUs = 4, 5, 7 and 9, they are non-efficient and their EV are low. For the DMUs with the same DEA level, the probability distribution is taken into account. The greater the average efficiency over various classifications is, the greater the EV. Similarly, the less the original investment quantity is, the greater the EV. Therefore, when DMUs = 1, 10, their EV values are higher than other efficient DMUs. Table 4 shows the final ranking results.

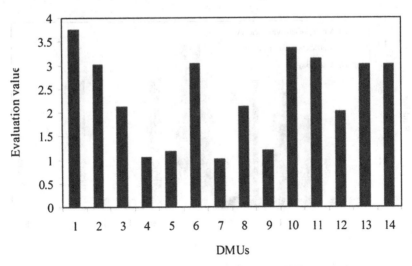

Fig. 2. Efficiency evaluation values of DMUs

Table 4. Efficiency ranking of logistics enterprises

DMU	Investment indicators			DEA evaluation					Rank
	Staff	Main business	Ware-house	Non-efficient	Weak efficient	Efficient	Efficiency level	EV	
1	139	1012.00	20000	0	0	0.0529	Efficient	3.75	1
2	96	18684.94	270000	0	0	0.0044	Efficient	3.00	6
3	51	84.00	52082	0	0.0179	0.0106	Efficient	2.13	8
4	181	19449.00	52000	0.0317	0	0	Non	1.04	13
5	87	759.00	62362	0.0317	0	0.0265	Efficient	1.18	12
6	307	33668.72	25060	0	0	0.0071	Efficient	3.06	4
7	213	3508.00	86000	0.0079	0	0	Non	1.01	14
8	45	103.29	55000	0	0.0179	0.0106	Weak	2.12	9
9	85	401.64	58000	0.0317	0	0.0265	Non	1.19	11
10	24	3205.42	8300	0	0	0.0141	Efficient	3.37	2
11	141	24867.01	16000	0	0	0.0176	Efficient	3.13	3
12	843	38530.89	82929	0	0.0179	0	Weak	2.03	10
13	234	37376.00	185500	0	0	0.0018	Efficient	3.00	7
14	171	5357.10	158012	0	0	0.0089	Efficient	3.02	5

4.5   Ranking Comparison

We used three methods to rank the efficiency of logistics enterprises. They are DEA comprehensive efficiency, DEA classification, and DEA-Bayes approach. As shown in Fig. 3, the red and blue bars are the results using the first two methods. When DMUs = 1, 2, 6, 10, 11, 13 and 14, their comprehensive efficiency values are equal to one and their classification results are all efficient. And they cannot be distinguished. Therefore, it is difficult to achieve an effective ranking only using DEA.

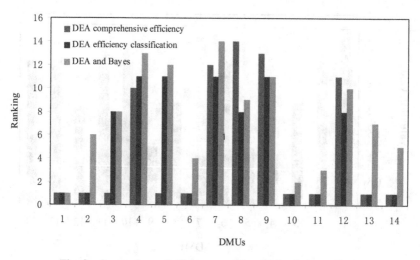

Fig. 3. Comparison of efficiency ranking (Color figure online)

In this paper, we introduce Bayes method and consider the overall statistical characteristics of the data. The ranking results achieve good discrimination.

5 Conclusions

Constructing an efficiency evaluation system is helpful for the healthy development of the logistics industry. Since the DEA ranking lacks significant distinction, we introduce Bayes approach. At first, we make a DEA evaluation and obtain efficiency level of an enterprise. Through feature quantization and Bayes classification, we predict the efficiency of an enterprise only using investment data. By integrating the probability distributions over the whole data set into the classification results, we present a new efficiency ranking method based on the EV. Then, we apply the DEA-Bayes method to evaluate fourteen A-level logistics enterprises in Anhui, China. Empirical results show that the method has good discrimination for efficiency ranking. Therefore, unlike expert scoring, our evaluation process is derived from enterprise data and easy to operate. Meanwhile, it can predict the efficiency level under different investment combinations, and provide objective proposals for enterprises decision-making.

In this work, the computational complexity of our ranking algorithm will increase as the number of the features and properties increase. For future work, we plan to study a feature reduction method to solve the time-consuming issue for efficiency ranking of logistics enterprises with a large number of features.

Acknowledgments. This work was supported by the Research Centre Project of Logistics Engineering in University of Science and Technology of China, the Anhui Provincial Major Research Project for Social Science Innovation and Development under Grant 2017ZD005, the Anhui Provincial Planned Text-book Project under Grant 2017ghjc384, and the Anhui Provincial Natural Science Foundation under Grant 1608085MF141.

References

1. Yi, S., Xie, J.: A study on the dynamic comparison of logistics industry's correlation effects in China. China Financ. Econ. Rev. **5**(1), 15 (2017)
2. Charnes, A., Cooper, W., Rhodes, E.: Measuring the efficiency of decision making units. Eur. J. Oper. Res. **2**(6), 429–444 (1978)
3. Ni, M., He, C., Yang, S.: Empirical research on logistics efficiency of Jiangxi and its influencing factors. J. East China Jiaotong Univ. **32**(4), 65–72 (2015)
4. Sherrie, D., Min, H., Joo, S.: Evaluating the comparative managerial efficiency of leading third party logistics providers in North America. Benchmarking **20**(1), 62–78 (2013)
5. Jiang, B., Fu, D., Liang, C.: Study on energy efficiency evaluation of logistics industry based on extended DEA. Chin. J. Manag. Sci. **23**, 518–524 (2015)
6. Markovits-Somogyi, R., Bokor, Z.: Assessing the logistics efficiency of European countries by using the DEA-PC methodology. Transport **29**(2), 137–145 (2014)
7. Yang, D., Xue, Y.: Study on logistics performance evaluation based on cross-efficiency DEA and entropy IAHP method. Oper. Res. Manag. Sci. **24**(3), 172–178 (2015)

8. Pan, J., Zong, X.: Study on evaluation index system of research efficiency in universities. Tsinghua J. Educ. **37**(9), 102–110 (2016)
9. Deng, X., Wang, X., Ada, S.F.N.: Productive efficiency and scale efficiency of Chinese logistics companies. Syst. Eng.- Theor. Pract. **29**(4), 34–42 (2009)
10. Storto, C.L.: Ecological efficiency based ranking of cities: a combined DEA cross-efficiency and Shannon's entropy method. Sustainability **8**(2), 1–29 (2016)
11. Rouyendegh, B.D., Oztekin, A., Ekong, J., Dag, A.: Measuring the efficiency of hospitals: a fully-ranking DEA–FAHP approach. Ann. Oper. Res. 1–18 (2016)
12. Ünsal, M.G., Nazman, E.: Investigating socio-economic ranking of cities in Turkey using data envelopment analysis (DEA) and linear discriminant analysis (LDA). Ann. Oper. Res. 1–15 (2018)
13. Banker, R.D., Charnes, A., Cooper, W.W.: Some models for estimating technical and scale inefficiencies in data envelopment analysis. Manag. Sci. **30**(9), 1078–1092 (1984)

A Brief Survey of Dimension Reduction

Li Song[1], Hongbin Ma[2,3(✉)], Mei Wu[2], Zilong Zhou[1],
and Mengyin Fu[3]

[1] School of Xuteli, Beijing Institute of Technology, Beijing, China
songli@bit.edu.cn
[2] School of Automation, Beijing Institute of Technology, Beijing, China
mathmhb@bit.edu.cn
[3] State Key Laboratory of Intelligent Control and Decision of Complex System,
Beijing Institute of Technology, Beijing, China

Abstract. Dimension reduction problem is a big concern which can reduce the scale of a database and keep the main features of these data simultaneously. This paper aims at reviewing and comparing different dimension reduction algorithms. Mainly, the performances of four basic algorithms (PCA, LDA, LLE and LE), their improved methods and deep learning methods are compared by reviewing the previous work. Their recognition accuracy and running time are carefully analyzed. We conclude that PCA and LDA are used more frequently in related fields. Combined methods usually perform better than original methods. Besides, deep learning method is also an approach developed in recent years, which outperforms existing traditional algorithms, though there are many barriers at present, such as obtaining huge labeled database, the computing and power limitation of different systems etc. Future research should focus on the processing of larger database. Finally, some new applications of dimension reduction are reviewed.

Keywords: Dimension reduction (DR) · PCA · LDA · Deep learning

1 Introduction

Unmanned control involves many key problems, such as motion control, image processing and so on, all of which include modeling. With the development of complexity, there are more and more parameters in these models. Therefore, a huge amount of data is required to find out the value of these parameters by solving equations. For example, multi-variable linear models can describe most dynamic systems [1]. Each of these parameters can influence the performance and robustness. When calculating these parameters directly, a huge scale of linear equations must be solved. The amount of calculation will explode when the number of these parameters increased, which is called "Curse of Dimensionality". Sometimes, it is even impossible to find out mathematical models between these parameters and physical quantities. They can only be estimated by analyzing data.

Therefore, sometimes it is quite difficult to identify these parameters directly because of the huge amount of calculations and the limitation of the computer

© Springer Nature Switzerland AG 2018
Y. Peng et al. (Eds.): IScIDE 2018, LNCS 11266, pp. 189–200, 2018.
https://doi.org/10.1007/978-3-030-02698-1_17

hardware. There is a crying need of reducing the dimension of the original data. Besides, retaining the main differences between different objects is also a concern.

In unmanned control field, dimension reduction (DR) was first used to solve parameter-dominated problems, which aim at decreasing the correlation among different dimensions by throwing away useless data. This paper provides a brief review as well as a comprehensive comparison among different existing methods of DR problem. Mainly four classical algorithms have been developed to solve DR problem. Principal component analysis (PCA) is a statistical method which is widely used in DR problem [2], though it is less efficient when processing high-dimensional data. Linear discriminant analysis (LDA) is a linear method which performs better on big data. However, it performs worse on small database [3]. Locally linear embedding (LLE) is a typical nonlinear algorithm which deals with high-dimensional data [4]. However, lower dimensional space also has useless data, and samples' category information has been removed [5]. Laplacian eigenmaps (LE) is a locally nonlinear solution for DR problem, it is unstable to noises and parameters yet [6]. Deep learning is a new method for dimension reduction with more hidden layers than classic machine learning.

Rest of the paper is organized as follows. Next part is mainly focused on the improved methods of the four basic algorithms and deep learning. At the same time, their performances are compared with the original performances. Then some disadvantages of the present methods are proposed. Finally, some applications of dimension reduction algorithms in unmanned aerial vehicle (UAV) systems are reviewed and summarized.

2 Statement of Dimension Reduction Problem

Dimension reduction is a process which can find low-dimensional data to represent an original high-dimensional data without losing its data structure. For example, the size of a gray-scale pixel face image is 200×200. It can be described by a 200×200 matrix whose elements are values of pixels.

$$A = \begin{bmatrix} a_{1,1} & a_{1,2} & \cdots & a_{1,200} \\ a_{2,1} & a_{2,2} & \cdots & a_{2,200} \\ \vdots & \vdots & \ddots & \vdots \\ a_{200,1} & a_{200,2} & \cdots & a_{200,200} \end{bmatrix} \tag{1}$$

After rearranging its elements into a column from left to right, the original data can be stored into a column vector

$$P = [p_1 \quad p_2 \quad \cdots \quad p_N]^T \tag{2}$$

where $N = 40000$ is dimension. Other approaches can also be used to extract features, whose number of dimensions is probably not 40000. When comparing the feature matrix mentioned above directly with another feature matrix, lots of calculations are needed, which is quite inefficient.

As to the feature vector P represented above, the aim of dimension reduction is finding a $m \times N$ matrix D. Then new column vector DP can be used to replace original data and it has m numbers which is fewer than before, and DP has similar data structure with A.

3 Methods

To survey the effectiveness of different kinds of methods, we compare the results of different papers. Generally, we can classify these papers into different classes based on their basic methods. We review these classes of methods in detail.

3.1 Principal Component Analysis

PCA is a basic method in DR problem. It is based on the covariance matrix of the original data. For example, in the field of face recognition, in case that there is one picture for one person, we use

$$C = \begin{bmatrix} c_1 & c_2 & \cdots & c_s \end{bmatrix} \tag{3}$$

to represent the whole face database, where c_i is the face of the ith person and s is the number of all people. The covariance matrix of C is calculated as

$$COV = \begin{bmatrix} \mathrm{cov}(c_1,c_1) & \mathrm{cov}(c_1,c_2) & \cdots & \mathrm{cov}(c_1,c_s) \\ \mathrm{cov}(c_2,c_1) & \mathrm{cov}(c_2,c_2) & \cdots & \mathrm{cov}(c_2,c_s) \\ \vdots & \vdots & \ddots & \vdots \\ \mathrm{cov}(c_s,c_1) & \mathrm{cov}(c_s,c_2) & \cdots & \mathrm{cov}(c_s,c_s) \end{bmatrix} \tag{4}$$

The eigenvalues and eigenvectors of COV are $\lambda_1, \lambda_2, \cdots, \lambda_s$ and $n_1, n_2, \cdots, n_s$, respectively. If we choose the first l eigenvectors, then the DR matrix is

$$DR_{PCA} = [n_1, n_2, \cdots, n_l]^T \tag{5}$$

Kim et al. [7] proposed a nonlinear improvement of PCA called kernel principal component analysis (KPCA) who combined PCA and polynomial kernel then used this method in face recognition process with an error rate of 2.5%. Since they use the tool of classical machine learning, more photos are needed per person, which bring additional calculations. Meng and Ke [8] found two new methods for eigenvalue extraction based on PCA using Kaiser criterion and eigenvalue curve. Their calculation times of 400 photos (40×10) are 26.714 s (Kaiser criterion) and 29.781 s (curve method). However, the correction rates are only 91.5% and 94%. Xu and Goodacre [9] used multi-block PCA (MPCA) for DR problem with multiple influential factors, whose cross-validated accuracy of data analysis can reach 99%. Nevertheless, they focused on the mathematical model rather than practical problems. Al-Arashi et al. [2] associated genetic algorithm (GA) to PCA in order to enhance the classification performance for face recognition, who tested 165 grayscale photos containing 15 people (11 photos for each person). To get an accuracy of 100%, they use 70 training and 10 eigenvectors in GAPCA, whose time is 0.0075 s. While in PCA, 15 eigenvectors are used, and the time is 0.0184 s. Definitely, GAPCA is better than PCA in their experiments, they did not do any experiments on big face database yet, which is more useful in practice. As GA is used, GAPCA is not able to solve one face problem in face recognition field, which means only one photo is used for one person. Kokiopoulou et al. [10] used polynomial filtering in PCA to avoid calculating eigenvalues, whose result is quite similar to the original PCA, though the computational and storage cost is reduced.

Many unmanned systems adopt PCA as a method to reduce dimensions and extract features. DuPont et al. [11] combined PCA with their classification algorithm on mobile robots, who developed a manifold curve to interpolate unknown coefficients with known coefficients produced by PCA. The classification results are more robust. Ma et al. [12] adopted PCABP neutral network to recognize objects in unmanned surface vehicles, whose recognition accuracy is above 85%. The above results are summarized in Table 1.

Table 1. Results of PCA-based DR methods

Methods	Accuracy	Running time
KPCA [7]	97.5%	N/A
Kaiser PCA [8]	91.5% (400 photos)	26.714 s
Eigenvalue PCA [8]	94% (400 photos)	29.781 s
MPCA [9]	99%	N/A
GAPCA [2]	100% (165 photos)	0.0075 s
Polynominal PCA [10]	Similar as PCA	Faster than PCA
PCA with manifold curve [11]	More robust than PCA	N/A
PCABP [12]	85%	N/A

3.2 Linear Discriminant Analysis

LDA, also called Fisher linear discriminant (FLD), is a classical algorithm in pattern recognition. The aim of this method is to increase the distances among different points in different classes, while decrease the distances in one class. LDA is a kind of linear classifier. When solving a k-classification problem, there are k linear functions

$$Y_b(x) = w_b^T x + w_{k_0} \quad b = 1, 2, \cdots, k \tag{6}$$

For b^*, if $Y_{b^*}(x_0)$ is the biggest one for a specific x_0, then x_0 belongs to class b^*. The error of a classification can be defined as

$$e = \frac{num}{N} \tag{7}$$

where *num* represents the number of samples which are classified wrongly, and N is the number of all samples. Then some classic methods such as Lagrange multiplier can be used to find a matrix $W = [w_1, w_2, \cdots, w_k]$ whose error rate is the smallest one.

Much previous work added some other methods to LDA. Ames and Hong [13] created a heuristic LDA method to solve DR problem, called sparse zero-variance discriminant analysis (SZVD). SZVD increases the accuracy of the result, its calculation time is longer than before yet. Mahmoudi and Duman [14] combined LDA and modified fisher analysis called modified fisher discriminant analysis (MFDA). MFDA has been used to detect credit card in a picture. Only from the detection result, its profit rate is 90.79%, which is quite higher than other methods mentioned in the paper. However, MFDA has many false positives. Dai and Yuen [15] improved the regularization scheme of LDA and created a new method called regularized discriminant analysis (RDA) for DR problem in face recognition. They trained 10 photos per person with 50 people, the recognition rate is 98.55%. Since they used machine learning method, at least 2 photos are used for one person, and its computation load is high because of the calculation of regulation.

Previous work also improved part of the processes in LDA. Wang et al. [16] used L_1-norm rather than L_2-norm in the process of LDA. Its performance is good when there is much noise in pictures, though its maximum recognition rates are almost the same with LDA-L2 method when there is less noise. Bose et al. [17] used nonparametric classification called generalized quadratic discriminant analysis (GQDA) rather than linear parametric method in LDA. In each database, GQDA is better than most of the other methods, and it is the fastest method. However, this method has only been tested in small database, whether it is useful in large database remains unsolved. The above results related to LDA are summarized in Table 2.

3.3 Locally Linear Embedding

LLE aims at keeping the data structure in DR problem. LLE can be realized in three steps. Firstly, each sample's k-nearest samples can be found by calculating 2-norms. Secondly, local weight matrix can be constructed based on k-nearest results. Thirdly, local weight matrix can be used as global matrix, which means local features are used to replace global features approximately. LLE is quite fast, which is widely used in nonlinear dimension reduction problem. For example, there is a curved surface in 3D space. LLE can make it flat and become a 2D plane. As a result, not only dimension has been reduced, but also the basic structure on the curved surface has been held.

Table 2. Results of LDA-based DR methods

Methods	Accuracy	Running time
SZVD [13]	Higher than LDA	Slower than LDA
MFDA [14]	90.79%	N/A
RDA [15]	98.55%	N/A
LDA-L1 [16]	Similar as LDA(less noise)	N/A
GQDA [17]	Higher than LDA	Faster than LDA

There are a few previous researches which focused on LLE. Bai et al. [18] used result based LLE for model identification. When damping coefficient is more than 0.1, LLE can work well while PCA can't work. However, they just use some simulation to validate rather than actual data. Zhang [19] presented a new method called enhanced supervised locally linear embedding (ESLLE) for face recognition. However, the result is not satisfying. Even using 20 train samples, the error rate is more than 30%, though its performance is better than other LLE method. A guided locally linear embedding method (GLLE) is presented by Alipanahi and Ghodsi [20]. In Balance dataset, GLLE can reach 94.4%, while LLE is only 69.2%. However, the number of these datasets is quite small. In Balance dataset, there are only 5 dimensions. Deng et al. [21] proposed an uncorrelated locally linear embedding (ULLE) for face recognition. ULLE has a good performance when the number of features is small, which can reach an accuracy of nearly 90%. However, when there are much more features, the recognition rate keeps an average of 92%, which is not very accurate in practice. Pang et al. [22] proposed a class-label locally linear Embedding (cLLE) method to keep the class-specific information. cLLE obtained an average error rate of 12.6026% using just 50 features, which is better than PCA (30.0194%, 150) and LLE (26.2318%, 100). As the features become more, PCA is more robust than cLLE. The above LLE-based results are summarized in Table 3.

3.4   Laplace Eigenmaps

The aim of LE can be described as follow:

$$min \sum_{i,j} (W_{ij} \times \|y_i - y_j\|^2) \tag{8}$$

where y_i and y_j are similar objects. That means if object i and object j are similar, then in low dimension space, they should be close. LE can also be realized by three steps. Firstly, a graph can be constructed using k-nearest methods or KNN methods. Secondly, the weight W_{ij} between object i and object j can be decided using thermal function. Thirdly, Laplace matrix can be constructed as

$$L = U - W \tag{9}$$

where U is a diagonal matrix whose diagonal elements are defined as

Table 3. Results of LLE-based DR methods

Methods	Accuracy	Running time
LLE [18] (simulation)	N/A	N/A
ESLLE [19]	<70%	N/A
GLLE [20]	94.4% (Balance dataset)	N/A
LLE [20]	69.2% (Balance dataset)	N/A
ULLE [21]	92%	N/A
cLLE [22]	87.3974%	N/A

$$U_{ii} = \sum_j w_{ij} \qquad (10)$$

and W is the weight matrix.

Some previous works were focused on machine learning in the process of LE. Zhao et al. [23] used modified algorithm to increase the speed of LE, whose learning speed is 9.99, which is similar to the original algorithm. However, the classification error rate is 7.13%, which is much lower. Tompkins and Wolfe [24] included regulation in LE for image analysis. The error rates are 43% (digit) and 52% (fingerprint), which are lower than linear method (55% and 60%). More work should focus on large database. Belkin and Niyogi [25] proposed a geometrically motivated algorithm which is computationally efficient, who just use some examples rather than experiments to test their method yet. Zhong et al. [26] proposed Relative Distance-based Laplacian Eigenmaps (RDLE) method to deal with DR problem in toy classification. Results showed that when the number of dimension is small, the error rate of this method is approximately 1% lower than LE. However, when the number of dimension is larger than 140, the two methods' performances are almost the same. Jafari and Almasganj [27] combined LE and latent variable models together for speech recognition. The phoneme recognition rates are 71.4% (LELVM) and 65.9% (PCA). The above results are summarized in Table 4.

3.5   Some Combinations

Of course, several of these basic methods can be combined to create multiple methods. There are some previous papers which combine LDA and PCA. Qi and Zhang [28] focused on 2D images and considered rows and cols of a image separately. They used 2DPCA for row-direction and 2DLDA for col-direction in DR problem. The recognition accuracy can reach 92.1% when selecting 8 training samples per class, and its running time is 3.5 s. However, the testing database is also quite small. Ge et al. [29] used a combination of PCA and LDA for speaker recognition. When there are 200 speakers in a database, the performance of this method can reach an accuracy of 98.5%, which is higher than PCA and LDA.

Table 4. Results of LE-based DR methods

Methods	Accuracy	Running time
Modified LE [23]	92.87%	9.99 s (training time)
Regulated LE [24]	57% (digit)	N/A
	48% (fingerprint)	N/A
GMLE [25] (only examples)	N/A	N/A
RDLE [26]	99%	Similar as LE
LELVM [27]	71.4%	N/A

Hou et al. [6] created a uniform framework for LLE and LE to solve eigenproblems in DR. The recognition accuracy is higher than before. Tang et al. [3] combined LDA and LE to solve number recognition problem. The recognition rate is 92.79%, which is higher than LDA (91.82%). The results of these combinations are summarized in Table 5.

Table 5. Results of combinations

Methods	Accuracy	Running time
2DPCA-2DLDA [28]	92.1%	3.5 s
PCA-LDA [29]	98.5%	N/A
LLE-LE [6]	Higher than LLE and LE	N/A
LDA-LE [3]	92.79%	N/A

3.6 Deep Learning

Deep learning is an efficient method in machine learning field. It was proposed by Hinton [35] in 2006 to solve dimension reduction problem. The concept of deep learning comes from artificial neutral network (ANN) [45]. ANN has one input layer, one hidden layer and one output layer. Deep learning network has more hidden layers than ANN which means it can be used to approximate nonlinear functions more precisely and more easily. Deep learning network can extract objects' features layer by layer like human's brain. As the number of hidden layers becomes larger, deep learning network's learning ability becomes stronger [36, 37]. An important model of deep learning is deep belief network (DBN) [36]. DBN is composed of several restricted Boltzmann machines (RBM) [44] and semi-supervised learning method is often adopted in DBN [38]. Another important model is deep convolutional neural network (CNN) [46] which is also a classic model in machine learning.

Shen et al. [39] proposed a Chinese web text classification model based on deep learning to solve DR problem in text classification, whose highest accuracy was 83.7% in their experiments. However, they did not do any experiments of classic dimension reduction methods on their database. Sun et al. [40] used modified marginalized stacked denoising autoencoders in deep learning to solve DR problem of sentiment classification, the accuracy of which is 81.08%, 8% higher than baseline. Li et al. [41] combined stacked denoising autoencoder with softmax for braille recognition whose highest accuracy is 92%, 27% higher than the result of original softmax algorithm. Yi et al. [42] constructed a deep-learning vocabulary network based on DBN and single-pass algorithm for DR problem of text clustering. For military category, the adoption of deep learning improved the accuracy by 0.1 according to their experiments, and running time is 10.84 s for 3000 features. Ahmad et al. [43] used a bilinear CNN architecture for face recognition, whose accuracy is 88.6% on Oxford 105K. The above results based on deep learning are summarized in Table 6.

Table 6. Results of deep learning methods

Methods	Accuracy	Running time
Autoencoders [39]	83.7%	N/A
Modified Autoencoders [40]	81.08%	N/A
Stacked Autoencoders [41]	92%	N/A
DBN [42]	N/A	10.84 s (3000 features)
Bilinear CNN [43]	88.6%(Oxford 105 K)	N/A

4 Conclusions

This paper aims at comparing the performances of different improved algorithms in DR problem. After the analysis of results, we conclude that PCA and LDA are used more often, while LLE and LE remain to be improved. What's more, most papers are focused on small database. Further research should test the performance of their algorithms in bigger database which is more suitable for practical applications. From the result of combinations, we can see that diversification methods usually perform better than original methods as they can take the advantage of different methods. Deep learning can improve the performance of original algorithms, though a huge database is needed.

Some papers adopted machine learning as an auxiliary tools [2, 15, 19, 28, 30]. Usually, the result of this method is better. However, this method needs much more data, which cannot be used in many microprocessors. Furthermore, it cannot be used to solve small sample recognition problems in pattern recognition, such as one face recognition problem. This kind of problems is a new topic which remains to be solved and improved.

5 Discussions

Dimension reduction is a big problem in unmanned systems. Solutions of this problem are widely used in some related fields, such as face recognition, fingerprint identification, information searching, computer vision, data mining, machine learning and so on. For example, with the increasing use of UAV, more and more data can be obtained by UAV, so dimension reduction methods are widely used in UAV systems to find out the internal relation of data. Zhang et al. [30] used feature dimension reduction data to train SVM for UAV in complex environment. Chebyshev fitting algorithm was used by He et al. [31] to recognize flight mode of UAV when cutting dimension in telemetric data. An et al. [32] adopted sparse projection for dimension reduction on original image captured by multi-rotor UAV in power inspection. Ren and Jiang [33] proposed subspace reliability analysis method for solving DR problem by removing unreliable feature dimensions directly, which was used to analyze UAV detection's micro-Doppler signature.

Of course, there are many other dimension reduction algorithms which do not belong to any of the four basic methods. When combining dimension reduction with models, DR problem is also called model reduction problem. If we consider parameters as dimensions of a system, then model reduction problem is also a DR problem. Qun et al. [34] did LPV model reduction about lateral system on UAV based on balanced truncation method, without losing its accuracy. This is a new application of dimension reduction, which could be quite efficient in simplifying the models in unmanned systems.

So far, deep learning has begun to be used in unmanned systems. However, there are many barriers to be dealt with. It is quite difficult to collect a huge set of labelled data in many applications. Besides, unmanned systems have many limitations. The high computational costs and large power required by deep learning may not be affordable on unmanned systems. Therefore, successful applications of deep learning techniques on unmanned systems demand further development in techniques such as model compression or powerful embedded computing chips. Moreover, the results of deep learning networks in practical problems are usually unexplainable, which will bring security issues, reliability problems and difficulties in fault diagnosis.

Acknowledgment. This work is partially supported by National Natural Science Foundation (NSFC) under Grants 61473038 and 91648117. And this work is also partially supported by Beijing Natural Science Foundation (BJNSF) under Grant 4172055.

References

1. Idan, M., Shaviv, G.E.: Robust control design strategy with parameter-dominated uncertainty. J. Guid. Control Dyn. **19**(3), 605–611 (1996)
2. Al-Arashi, W., Ibrahim, H., Suandi, S.: Optimizing principal component analysis performance for face recognition using genetic algorithm. Neurocomputing **128**, 415–420 (2014)
3. Tang, H., Fang, T., Shi, P.F.: Laplacian linear discriminant analysis. Pattern Recogn. **39**, 136–139 (2006)
4. Chen, J., Liu, Y.: Locally linear embedding: a survey. Artif. Intell. Rev. **36**, 29–48 (2011)
5. Deng, T., Deng, Y., Shi, Y., Zhou, X.: Research on improved locally linear embedding algorithm. In: Pan, L., Păun, G., Pérez-Jiménez, M.J., Song, T. (eds.) BIC-TA 2014. CCIS, vol. 472, pp. 88–92. Springer, Heidelberg (2014). https://doi.org/10.1007/978-3-662-45049-9_15
6. Hou, C.P., Zhang, C.S., Wu, Y., Jiao, Y.Y.: Stable local dimensionality reduction approaches. Pattern Recogn. **42**, 2054–2066 (2009)
7. Kim, K.I., Jung, K., Kim, H.J.: Face recognition using kernel principal component analysis. IEEE Sig. Process. Lett. **9**(2), 40–42 (2002)
8. Meng, H., Ke, X.: Further research on principal component analysis method of face recognition. In: Proceedings of 2008 IEEE International Conference on Mechatronics and Automation, pp. 421–425 (2008)
9. Xu, Y., Goodacre, R.: Multiblock principal component analysis: an efficient tool for analyzing metabolomics data which contain two influential factors. Metabolomics **8**, 37–51 (2012)
10. Kokiopoulou, E., Saad, Y.: PCA without eigenvalue calculations: a case study on face recognition. University of Minnesota (2005)

11. DuPont, E.M., Moore, C.A., Roberts, R.G.: Terrain classification for mobile robots traveling at various speeds: an eigenspace manifold approach. In: 2008 IEEE International Conference on Robotics and Automation, pp. 3284–3289 (2008)

12. Ma, Z.L., Wen, J., Liang, X.M., et al.: Extraction and recognition of features from multi-types of surface targets for visual systems in unmanned surface vehicle. J. Xi'an Jiaotong Univ. **48**(8), 60–66 (2014)

13. Ames, B.P.W., Hong, M.Y.: Alternating direction method of multipliers for penalized zero-variance discriminant analysis. Comput. Optim. Appl. **64**, 725–754 (2016)

14. Mahmoudi, N., Duman, E.: Detecting credit card fraud by modified fisher discriminant analysis. Expert Syst. Appl. **42**, 2510–2516 (2015)

15. Dai, D., Yuen, P.C.: Face recognition by regularized discriminant analysis. IEEE Trans. Syst. Man Cybern. Part B: Cybern. **37**(4), 1080–1085 (2007)

16. Wang, H.X., Lu, X.S., Hu, Z.L., Zheng, W.M.: Fisher discriminant analysis with L1-norm. IEEE Trans. Cybern. **44**(6), 828–842 (2014)

17. Bose, S., Pal, A., SahaRay, R., Nayak, J.: Generalized quadratic discriminant analysis. Pattern Recogn. **48**, 2676–2684 (2015)

18. Bai, J.Q., Yan, G., Wang, C.: Modal identification method following locally linear embedding. J. Xi'an Jiaotong Univ. **47**(1), 85–100 (2013)

19. Zhang, S.: Enhanced supervised locally linear embedding. Pattern Recogn. Lett. **30**, 1208–1218 (2009)

20. Alipanahi, B., Ghodsi, A.: Guided locally linear embedding. Pattern Recogn. Lett. **32**, 1029–1035 (2011)

21. Deng, T., Deng, Y.N., Shi, Y., Zhou, X.Q.: Research on improved locally linear embedding algorithm. Commun. Comput. Inf. Sci. **472**, 88–92 (2014)

22. Pang, Y.H., Teoh, A.B.J., Wong, E.K., Abas, F.S.: Supervised locally linear embedding in face recognition. In: International Symposium on Biometrics and Security Technologies, pp. 1–6 (2008)

23. Zhao, Z.Q., Li, J.Z., Gao, J., Wu, X.D.: A modified semi-supervised learning algorithm on Laplacian eigenmaps. Neural Process Lett. **32**, 75–82 (2010)

24. Tompkins, F., Wolfe, P.J.: Image analysis with regularized Laplacian eigenmaps. In: Proceedings of 2010 IEEE 17th International Conference on Image Processing, pp. 1913–1916 (2010)

25. Belkin, M., Niyogi, P.: Laplacian eigenmaps for dimensionality reduction and data representation. Neural Comput. **15**, 1373–1396 (2003)

26. Zhong, G.Q., Hou, X.W., Liu, C.L.: Relative distance-based Laplacian eigenmaps. In: IEEE Explore, pp. 1–5 (2009)

27. Jafari, A., Almasganj, F.: Using Laplacian eigenmaps latent variable model and manifold learning to improve speech recognition accuracy. Speech Commun. **52**, 725–735 (2010)

28. Qi, Y.F., Zhang, J.: (2D)2 PCALDA: an efficient approach for face recognition. Appl. Math. Comput. **213**, 1–7 (2009)

29. Ge, Z.H., Sharma, S.R., Smith, M.J.T.: PCA/LDA approach for text-independent speaker recognition. arXiv (2016)

30. Zhang, W.W., Ding, W.R., Liu, C.H.: Prediction of interference effect on UAV data link in complex environment. Syst. Eng. Electron. **28**(4), 760–766 (2016)

31. He, S.J., Liu, D.T., Yu, P.: Flight mode recognition method of the unmanned aerial vehicle based on telemetric data. Chin. J. Sci. Instrum. **37**(9), 2004–2013 (2016)

32. An, N., Yan, B., Xiong, J.: A multi-scale insulator tracking algorithm based on compressive sensing. Transducer Microsyst. Technol. **35**(2), 140–143 (2016)

33. Ren, J., Jiang, X.: Regularized 2-D complex-log spectral analysis and subspace reliability analysis of micro-Doppler signature for UAV detection. Pattern Recogn. **69**, 225–237 (2017)

34. Zong, Q., Ji, Y.H., Dou, L.Q., Zeng, F.L.: LPV model reduction of UAV lateral system. Control Decis. **25**(6), 948–952 (2010)
35. Hinton, G.E., Salakhutdinov, R.R.: Reducing the dimensionality of data with neural networks. Science **313**(5786), 504–507 (2006)
36. Hinton, G.E., Osindero, S., Teh, Y.W.: A fast learning algorithm for deep belief nets. IEEE Trans. Neutral Comput. **18**(7), 1527–1554 (2006)
37. Sutskever, I., Hinton, G.E.: Deep narrow sigmoid belief networks are universal approximators. IEEE Trans. Neutral Comput. **20**(11), 2629–2636 (2008)
38. Gao, Q., Ma, Y.M.: Research and application of the level of the deep belief network. Sci. Technol. Eng. **16**(23), 234–238 (2016)
39. Shen, F., Luo, X., Chen, Y.: Text classification dimension reduction algorithm for Chinese web page based on deep learning. In: International Conference on Cyberspace Technology, pp. 451–456 (2013)
40. Sun, M., Tan, Q., Ding, R.W., Liu, H.: Cross-domain sentiment classification using deep learning approach. In: IEEE International Conference on Cloud Computing & Intelligence Systems, pp. 60–64 (2014)
41. Li, T., Zeng, X.Q., Xu, S.J.: A deep learning method for Braille recognition. In: 2014 International Conference on Computational Intelligence and Communication Networks, pp. 1092–1095 (2014)
42. Yi, J.K., Zhang, Y.C., Zhao, X.H., Wan, J.: A novel text clustering approach using deep-learning vocabulary network. Math. Probl. Eng. **2017**(1), 1–13 (2017)
43. Ahmad, A., Abbes, A., Naeem, R.: Content-based image retrieval with compact deep convolutional features. Neurocomputing **249**, 95–105 (2017)
44. Ackley, D.H., Hinton, G.E., Sejnowski, T.J.: A learning algorithm for Boltzmann machines. Cogn. Sci. **9**(1), 147–169 (1985)
45. Vanhulle, M.M., Orban, G.A.: Entropy driven artificial neuronal networks and sensorial representation - a proposal. J. Parallel Distrib. Comput. **6**(2), 264–290 (1989)
46. Fukushima, K.: Neocognitron - a self-organizing neural network model for a mechanism of pattern-recognition unaffected by shift in position. Biol. Cybern. **36**(4), 193–202 (1980)

Deep Neural Networks

Regularized Extreme Learning Machine Ensemble Using Bagging for Tropical Cyclone Tracks Prediction

Jun Zhang and Jian Jin[✉]

Department of Computer Science and Technology, East China Normal University,
3363 North Zhongshan Road, Shanghai 200062, China
zj734264766@163.com, jjin@cs.ecnu.edu.cn

Abstract. This paper aims to improve the prediction accuracy of Tropical Cyclone Tracks (TCTs) over the South China Sea (SCS) and its coastal regions with 24 h lead time. The model proposed in this paper is a regularized extreme learning machine (ELM) ensemble using bagging. A new method is proposed in this paper to solve lasso and elastic net problem in ELM, which turns the original problem into familiar quadratic programming (QP) problem. The forecast error of TCTs data set is the distance between real position and forecast position. Compared with the stepwise regression method widely used in TCTs, 16.49 km accuracy improvement is obtained by our model. Results show that the regularized ELM ensemble using bagging has a better generalization capactity on TCTs data set.

Keywords: Regularized extreme learning machine · Bagging
Quadratic programming · Tropical Cyclone Tracks

1 Introduction

The Typhoon Cyclone (TC) ranks first in the top ten natural disasters of the world, which often causes heavy casualties and property losses. So, delivering the accurate Tropical Cyclone Tracks (TCTs) forecast is extremely significant to reduce the loss of disasters.

In the past, The forecasting techniques about TCTs can be divided into the following three major categories: (1) statistical; (2) dynamical and numberial; (3) statistical-dynamical [1]. In recent years, Artificial Neural Networks (ANNs) based techniques have been developed, which can produce forecast with good accuracy in TCTs [2–5]. However, we should spend as less time as possible for the prediction of the short term TCTs. The learning speed of ANNs is in general far slower than required. By contrast, Extreme Leaning Machine (ELM) is a new learning algorithm with fast learning speed and good generalization ability proposed by Huang et al., which has been successfully applied to a number of real-word applications [6–8]. Further, regularized ELM for regression problems

© Springer Nature Switzerland AG 2018
Y. Peng et al. (Eds.): IScIDE 2018, LNCS 11266, pp. 203–215, 2018.
https://doi.org/10.1007/978-3-030-02698-1_18

is proposed by Escandell-Montero et al., which can obtain the most appropriate architecture of the network [9]. The regularized regression methods include l_2 penalty (ridge), l_1 penalty (lasso) and mixtures of the two (elastic net). Huang et al. derived the closed-form expression for the ridge problem in ELM [11]. We usually solve the lasso and elastic net problem in ELM with the coordinate descent method due to its high speed [12].

Inspired by Elad which explained how to solve the lasso problem of Linear Systems, we propose a new method from a new perspective in this paper, which turns the lasso and elastic net problem in ELM into a familiar quadratic programming (QP) problem with expanding the scale of Independent variable in l_1 norm [10]. Quadratic programming problems can be solved with many methods in convex optimization [13]. Encouraging results have been achieved on TCTs data set with ensemble methods [14–17]. The performance of a single neural network can be expected to improve using an ensemble of neural networks with a plurality consensus scheme [18]. Bagging is one of the popular ensemble machine-learning methods. There are two key advantages on bagging algorithm: (1) reducing the learning time greatly with the network trained in parallel. (2) the out-of-bag examples produced by bootstrap sampling, which can be used to estimate the base learner [19,20]. Here we applied the regularized ELM ensemble using bagging to the short term forecasting of TCTs recorded in the 20-year period (1984–2003). The forecasting accuracy of our proposed model will be analysed in comparison with that obtained by stepwise regression.

2   Methodology

2.1   Extreme Learning Machine

ELM was proposed by Huang et al. for the single hidden layer feedforward neural networks (SLFNs) [6,7]. The structure of the network is shown in Fig. 1. The parameters of the hidden layer can be initialized randomly without tuning. Thus, the SLFNs can be simply considered as a linear system. The parameters we need to optimize are just the weights of output layer, which is a least-squares solution of linear system. we can work out it by means of the Moore-Penrose Generlized Inverse [21]. Thus, a simple learning method for ELM can be summarized in Algorithm 1.

2.2   Regularized ELM

The ELM algorithm is based on empirical risk minimization principle which is known as lack of robustness. In order to overcome the drawback, ridge ELM is proposed by Huang et al. [11], lasso ELM and elastic net ELM is proposed by Escandell-Montero et al. [9].

Ridge regression adds the l_2 norm of the output layer's weights on the loss function, which makes all the weights of output layer tend to small. That coincide with Bartlwtt's theory, the smaller the norms of weights are, the better

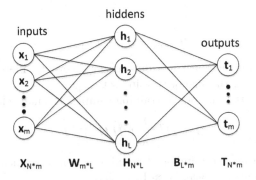

Fig. 1. ELM architecture.

Algorithm 1. Basic ELM

---

input: Given training set $D = \{(\mathbf{x_i},\mathbf{t_i}),\ \mathbf{x_i} \in \mathbf{R^n},\ \mathbf{t_i} \in \mathbf{R^m},\ i=1,...,N\}$;
    activation function $G(\mathbf{x})$ in hidden layer;
    the number of hidden nodes: L;
output: weights of input layer:W, bias of hidden layer: b, weights of output layer:B.
 1: Randomly generate the parameters $(\mathbf{w_i}, \mathbf{b_i})$, i=1,...,L.
 2: Calculate the hidden layer output matrix $\mathbf{H}$:

$$\mathbf{H} = \begin{bmatrix} G(\mathbf{w_1},\mathbf{b_1},\mathbf{x_1}) & \cdots & G(\mathbf{w_L},\mathbf{b_L},\mathbf{x_1}) \\ \vdots & \vdots & \vdots \\ G(\mathbf{w_1},\mathbf{b_1},\mathbf{x_N}) & \cdots & G(\mathbf{w_L},\mathbf{b_L},\mathbf{x_N}) \end{bmatrix} \tag{1}$$

 3: Calculate the output weight matrix B :

$$B = \mathbf{H}^\dagger \mathbf{T}. \tag{2}$$

    where $\mathbf{H}^\dagger$ is the Moore-Penrose inverse of matrix H.
 4: return W, b, B.

---

generalization performance the feedforward neural networks tend to have [22]. The mathematical model is

$$Min : \frac{1}{2}\sum_{i=1}^{m} \boldsymbol{\xi}_i^T \boldsymbol{\xi}_i + \lambda \frac{1}{2}\|B\|_2^2$$
$$s.t : \boldsymbol{o}_i = \boldsymbol{t}_i - \boldsymbol{\xi}_i, i = 1,...,m \tag{3}$$

where

$$\boldsymbol{B}_{L*m} = \begin{bmatrix} \beta_1 \cdots \beta_m \end{bmatrix} \tag{4}$$

$$\boldsymbol{o}_i = H\beta_i \tag{5}$$

and

$$T_{N*m} = \begin{bmatrix} t_1 \cdots t_m \end{bmatrix} \tag{6}$$

$\lambda \in (0, \infty)$. Based on the Karush-Kuhu-Tucker theorem, different solutions can be obtained according the size of training set [11].

In the case that the number of training samples is small relatively, we have

$$B = H^T(\frac{I}{\lambda} + HH^T)^{-1}T \tag{7}$$

In the other case that the number of training samples is much larger than the dimension of the feature space. N ≫ L, we have

$$B = (\frac{I}{\lambda} + H^T H)^{-1}H^T T \tag{8}$$

In theory, the above solutions can be used in any size of applications. The generalization performance of ELM is not sensitive to the dimension of the feature space (L) [11]. Thus, one may prefer to apply Eq. 8 in order to reduce computational costs. The detailed algorithm of ridge ELM is shown in Algorithm 2.

Algorithm 2. Ridge ELM

---

input: Given training set D = $\{(x_i,t_i),\ x_i \in R^n,\ t_i \in R^m,\ i=1,...,N\}$;
    activation function $G(x)$ in hidden layer;
    the number of hidden nodes: L;
    hyperparameters: λ.
output: weights of input layer:W, bias of hidden layer: b, weights of output layer:B.
 1: Randomly generate the parameters (w_i, b_i), i=1,...,L.
 2: Calculate the hidden layer output matrix **H**:
 3: Calculate the output weight matrix B :

$$B = H^T(\frac{I}{\lambda} + HH^T)^{-1}T \tag{9}$$

or

$$B = (\frac{I}{\lambda} + H^T H)^{-1}H^T T \tag{10}$$

4: return W, b, B.

The lasso method has a tendency to prefer sparse solutions. we can obtain more zero values about the weights of output layer. The structure of neural network tends to be more compatible with lasso method. The loss function of lasso problem in ELM can be written as

$$Min : \frac{1}{2} \sum_{i=1}^{m} \xi_i^T \xi_i + \lambda\frac{1}{2}\|B\|_1^1 \tag{11}$$

because

$$B_{L*m} = [\beta_1 \cdots \beta_m] \tag{12}$$

Thus

$$\|B\|_1^1 = \|\beta_1\|_1^1 + \dots + \|\beta_m\|_1^1 \tag{13}$$

where $\lambda \in (0, \infty)$. Bring Eq. 13 into Eq. 11, the lasso function can be re-written as

$$Min : \frac{1}{2} \sum_{i=1}^{m} \xi_i^T \xi_i + \lambda \frac{1}{2} (\|\beta_1\|_1^1 + \dots + \|\beta_m\|_1^1) \tag{14}$$

The Eq. 14 can be divided into m equations to solve

$$Min : \frac{1}{2} \xi_1^T \xi_1 + \lambda \frac{1}{2} \|\beta_1\|_1^1$$

$$\vdots$$

$$Min : \frac{1}{2} \xi_m^T \xi_m + \lambda \frac{1}{2} \|\beta_m\|_1^1 \tag{15}$$

We can obtain B by solving the above m equations separately. Each of the m equations can be solved in the same way because they have the same form. In the next, by taking one of the m equations as example, we illustrate how to solve the lasso problem in ELM with a new trick.

$$Min : \frac{1}{2} \xi_1^T \xi_1 + \lambda \frac{1}{2} \|\beta_1\|_1^1$$

$$s.t : o_1 = t_1 - \xi_1 \tag{16}$$

We make $\beta_1 = u - v$ in Eq. 16, $u, v \in R^m$. $u^T v = 0$ when we add non-negative constraints on u, v, if we assume the k-th entry in both u and v is non-zero(and positive), $u_k > v_k$, then by replacing these two entries by $u_k' = u_k - v_k, v_k' = 0$, we will eventually choose u_k', v_k' as the final solution in order to minimize the loss function. Thus

$$Min : \|\beta\|_1^1 \quad is \quad equivalent \quad to \quad Min : 1^T u + 1^T v \tag{17}$$

Using this trick, Eq. 16 can be represented by

$$Min : \frac{1}{2} \xi_1^T \xi_1 + \lambda \frac{1}{2} (1^T u + 1^T v)$$

$$s.t \quad o_1 = t_1 - \xi_1$$

$$u, v \geq 0 \tag{18}$$

Elastic net is a mixture of ridge and lasso. The loss function can be described as

$$Min : \frac{1}{2} \sum_{i=1}^{N} \xi_i^T \xi_i + \lambda \frac{1}{2} \|B\|_1^1 + \eta \frac{1}{2} \|B\|_2^2 \tag{19}$$

Similar to the lasso problem solving, we can also decompose this equation into the same m sub-equations. one of the m sub-equations can be written as

$$Min : \frac{1}{2}\xi_1^T\xi_1 + \lambda\frac{1}{2}\|\beta_1\|_1^1 + \eta\frac{1}{2}\|\beta_1\|_2^2$$
$$s.t : \quad o_1 = t_1 - \xi_1 \tag{20}$$

$\lambda \in (0,\infty)$, $\eta \in (0,\infty)$, Using the same trick as Eqs. 18, 20 can be expressed as

$$Min : \frac{1}{2}\xi_1^T\xi_1 + \lambda\frac{1}{2}(1^Tu + 1^Tv) + \eta\frac{1}{2}(u-v)^T(u-v)$$
$$s.t : \quad o_1 = t_1 - \xi_1$$
$$u, v \geq 0 \tag{21}$$

The loss function of Eqs. 18 and 21 is convex function. Obviously, both Eqs. 18 and 21 are QP problems. The QP problem [13] can be described as

$$Min : \frac{1}{2}x^T Px + q^T + r$$
$$s.t : \quad Gx \leq h$$
$$Ax = b \tag{22}$$

and we can use quadprod function in MATLAB's Optimization Toolbox to solve Eq. 22.

The detailed algorithm of lasso and elastic net ELM is shown in Algorithm 3.

Algorithm 3. Lasso and Elastic net ELM

---

input: Given training set D = $\{(x_i,t_i), x_i \in R^n, t_i \in R^m, i=1,...,N\}$;
   activation function $G(x)$ in hidden layer;
   the number of hidden nodes: L;
   hyperparameters λ;
   hyperparameters η is only for elastic net ELM.
output: weights of input layer:W, bias of hidden layer: b, weights of output layer:B.
 1: Randomly generate the parameters (w_i, b_i), i=1,...,L.
 2: Calculate the hidden layer output matrix **H**:
 3: Calculate the output weight matrix B by solving m sub-problems:
   for i = 1:m
      (i) β_i replaced by $u - v, u, v \geq 0$. The initial ith sub-problem becomes a QP problem.
      (ii) obtain β_i by solving QP problem.
   endfor
 4: return W, b, B.

2.3   Bagging

The two key components of bagging are bootstrap and aggregation. Bagging adopts bootstrap sampling to generate different base learners, that is, the training set is divided into a training set and a validation set which is different for each base learner by putting back the samples, and then the model for each base learner can be trained with base learning algorithm. Bagging adopts the most popular strategies for aggregating the model's outputs of each base learner, in other words, voting for classification and averaging for regression [19,20]. The bagging algorithm is summarized in Algorithm 4.

Algorithm 4. Bagging

!h

input: Given training set $D = \{(x_i, t_i), x_i \in R^n, t_i \in R^m, i=1,...,N\}$;
    Base Learning Algorithm Φ;
    The number of base learners: E.
output: Aggregation of C base learners.
  1: Process:
    for i = 1:E
        (i) obtain new training set and new validation set through bootstrap sample .
        (ii) train base learner with Φ in new training set and new validation set
    endfor
  2: return the model of E base learners.

---

3   Experiments and Results

3.1   Data Set

The TCTs data set used in this experiments is published by China Meteorological Administration, which form in or move into SCS in July, August and September of 1984–2003 and last for at least 48 h. The TCTs data is recorded every 12 h from the moment TC appears at SCS.

A TCTs and its changes are associated with its intensity, accumulation and replenish of energy, and various nonlinear changes in its enviroment flow field, which are referred to as variable in this paper. The variables can be divided into two categories: (1) the climatology and persistence factors representing changes of TC itself, such as changes in the latitude, longitude, and intensity of a TC at 12 and 24 h before prediction time, and (2) the physical variables calculated from the NCEP/NCAR global reanalysis data, representing the ambient flow field of the TC center [23]. The observations of TCTs is a position on the surface of our planet, which are latitude and longitude of the earth. The predicators (v1 to v16) and observations (Lat.t and Lon.t) of TCTs data set are listed in Table 1. Our object is to predict the latitude and longitude based on the predictors in the next 24 h.

Table 1. Variable information of the TCTs data set

Variable	Min.	Max.	Mean	Std.	Meaning
v1	105.3	123.0	114.7	4.2	Initial Lon.(E)
v2	−39.8	23.1	−9.1	10.3	Zonal motion at −12 h
v3	−36.1	20.8	−9.3	9.7	Zonal motion at −24 h
v4	106	130	116.8	4.9	Lon.(E) at −12 h
v5	−43.9	20.8	−9.6	10.3	Zonal motion from −24 to −12 h
v6	−4.5	2.7	−1.0	1.2	Lon differ between 0 and −12 h
v7	106.0	124.6	115.2	4.3	Lon.(E) at −6 h
v8	106.0	128.3	116.3	4.7	Lon.(E) at −24 h
v9	10.8	23.5	18.7	2.4	Initial Lat.(N)
v10	−17.4	26.6	4.2	6.5	Meridional motion at −12 h
v11	0.0	525.2	50.2	62.0	Squared zonal motion at −24
v12	10.0	23.7	18.2	2.4	Lat.(N) at −12 h
v13	−8.2	4.8	−2.1	2.2	Lon differ between 0 and −24 h
v14	10.3	23.8	18.5	2.4	Lat.(N) at −6 h
v15	9.6	24.0	18.0	2.5	Lat.(N) at −24 h
v16	10	60	23.5	9.9	Max surface wind at −6 h
Lat.t	101.4	128.3	112.8	4.7	Current Lon.(E)
Lon.t	12.2	30.0	19.8	2.6	Current Lat.(N)

The total number of TCTs dataset is 750. Researchers usually use the first 720 samples as the training set and the last 30 samples as the testing set.

3.2   Experiment Set up

From the above description of the data set, obviously the problem in this paper is a regression problem. Thus, the output of our model is the average of all the base learners' output. All base learners should be different and produce good forecasting accuracy in order to obtain better generalization capability. There are two differences between all base learners: (i) the training set and the validation set, and (ii) hyperparameters in the base learning algorithm. The model of each base learner use same process training.

Next, we elaborate the training process about a single base learner. Firstly, by sampling with replacement 720 times for 720 training samples DATA, we use the sampled data as a training set DATA_TRA. Validation data DATA_VAL, known as out-of-bag estimation, is the difference between DATA and DATA_TRA. The above mentioned sampling step is known as bootstrap sampling. Then, we come to learn the parameters in the base learning algorithm based on the DATA_TRA and DATA_VAL. The hyperparameters which have great influence on the performance of the model need to be set before model training. Usually, there are more than one hyperparameter, and the points sampled by sobol series [24, 25] are better uniform in multidimensional space. The hyperparameters of the model are determined by grid search combined with sobol series in this paper. In order to choose a reasonable value about the hyperparameters in the model, we set the initial ranges of each hyperparameter firstly, and then, many pairs of points generated by sobol series in the initial hyperparameters space, which are used as hyperparameters inputs to the model; next, the first k best-performing models in DATA_VAL will be selected among trained models, the minimum and maximum values for each hyperparameter in the first k models will be used as the final hyperparameter ranges. These operations can be view as scale intervals which performance is better. From the perspective of probability, as long as the final hyperparameter is select from enough candidates which generated in the final hyperparameter ranges, the final model performance must tends to be better. The process of hyperparameters selection is summarized as Algorithm 5.

Algorithm 5. hyperparameters selection

---

input: Given original training set D;
Base Learning Algorithm Φ;
The initial ranges of hyperparameters RH;
The number of model's candidates in hyperparamters selection: C;
The number of best-performing candidates: k.
output: The final ranges of hyperparameters, DATA_TRA, DATA_VAL.
1: obtain DATA_TRA and DATA_VAL by bootstrap sampling in D.
2: generate C points in RH by sampling with sobol.
3: obtain C models by using each point as the hyperparameters in Φ.
4: train C models in DATA_TRA.
5: choose the best performing k models in DATA_VAL.
6: the mininum and maximum value are obtained for each hyperparemeter in k models and used as the final ranges of hyperparameters.
7: return the final ranges of hyperparameters, DATA_TRA, DATA_VAL.

Finally, taking one base learner of ridge ELM as an example, we explain the choice about the specific ranges of each hyperparameter and parameters in the experiment. The hyperparameter λ in ridge ELM which is the trade-off between $\frac{1}{2}\sum_{i=1}^{m}\xi_i^T\xi_i$ and $\frac{1}{2}\|B\|_2^2$ and the number of hidden nodes is L. The initial ranges of λ and L are $(0, 1000)$ and $(1, 1000)$. Obviously, the effects of $\lambda \in (0, 1)$ and $\lambda \in (1, 1000)$ on the experiment results are two different situations. It is unreasonable and a waste of computing resources if we perform hyperparameters selection on $\lambda \in (0, 1000)$. That's because, the length of $\lambda \in (0, 1)$ and $\lambda \in (1, 1000)$ vary greatly. 200 models are trained separately in both cases in order to determine whether $\lambda \in (0, 1)$ or $\lambda \in (1, 1000)$. The initial ranges of λ and L to $(0, 1)$ and $(1,1000)$ are set respectively based on the experiment results. $C = 1000$, $k = 5$ are set in the algorithm of hyperparameters selection empirically. The final hyperparameters and parameters is determined based on the best performing candidate among 40 candiates which hyperparameters generated in the final hyperparameters ranges. A final base learner can be obtained based on the above algorithm. The number of base learners is 10 in this experiment. The final ensemble model is achieved after 10 base learners are trained. The lasso and elastic net ELM ensemble model are also trained based on the above mentioned idea.

3.3   Result and Analysis

The number of testing set DATA_TES which consist of four TC is 30. 30 prediction points of latitude and longitude are obtained after making prediction on DATA_TES with final ensemble model. The mean distance errors (MDE) Δd evaluate the model more intuitively in general, which is calculated by

$$\Delta d = \sqrt{\overline{\Delta x}^2 + \overline{\Delta y}^2} \times 110 (\text{km}) \qquad (23)$$

where Δx and Δy are mean absolute errors (MAE) between the predictions and observations of longitude and latitude on the DATA_TES.

Stepwise regression which has good performance in meteorological data is a kind of multivariable regression, specifically focusing on selecting variables. The results of step regression are often used as benchmark in meteorological area.

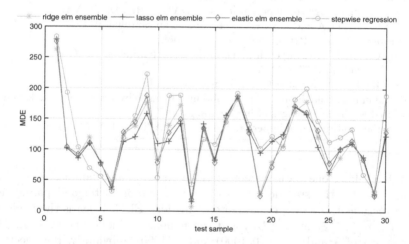

Fig. 2. The MDE of regularized ELM ensemble and stepwise regression on DATA_TST

The MDE of regularized ELM ensemble using bagging and stepwise regression are shown in Fig. 2. we can see that the MDE of the regularized ELM ensemble using bagging is lower than the MDE of the stepwise regression from Fig. 2. The MDE between ridge ELM ensemble, lasso ELM ensemble and elastic net ELM emsemble have small difference, but the gap between the structure of the network is obvious. Table 2 shows the average of number of hidden nodes in regularized

ELM ensemble models. The results shows that the lasso ELM ensemble is able to find more efficient networks (similar accuracy and more compact structure).The MDE of the first DATA_TST is nearly 275 Km in all predication models, which is an unacceptable forecast. we think the first DATA_TST is a singular TC which is difficult to make accurate predictions.

Table 2. Comparsion of the average of hidden node number in regularized ELM ensemble which include ridge,lasso and elastic net.

Method	Mean hidden nodes number
Ridge	179
Lasso	68.8
Elastic net	202

The Table 3 shows the MAE of longitude and latitude and the MDE in the ensemble model and the base learner in ensemble model, which indicates that ensemble model improve model's performance significantly compared with a single base learner.

Table 3. Comparsion of the MAE in the ridge, lasso and elastic net ELM ensemble and its base learner.

Method	MAE (Lon.)	MAE (Lat.)	MDE
Ridge	0.7859	0.5360	104.64
Base learner in ridge	0.8310	0.5551	109.93
Lasso	0.7746	0.5570	104.95
Base learner in lasso	0.8037	0.5768	108.82
Elastic net	0.7925	0.5396	105.46
Base learner in elastic net	0.8246	0.5471	108.85

Table 4 shows that experimental results of regularized ELM ensemble, stepwise regression and Bayesian neural network (BNN) respectively. Compared with stepwise regression, regularized ELM ensemble improved significantly in all evaluation indiactors. Regularized ELM ensemble is more accurate than the stepwise regression, the MDE dropped 16.49 km. A decrease of 8.12 Km in MDE is obtained by our proposed model compared to the BNN model, which is proposed by Zhu et al. in 2016 [5].

Table 4. Experimental results of the regularized ELM ensemble, stepwise regression and BNN.

Method	MAE (Lon.)	MAE (Lat.)	MDE
Stepwise regression	0.9471	0.5618	121.13
BNN	0.8649	0.5503	112.76
Elastic net ELM ensemble	0.7925	0.5396	105.46
Lasso ELM ensemble	0.7746	0.5570	104.95
Ridge ELM ensemble	0.7859	0.5360	104.64

4   Conclusion

In this study, a novel algorithm for solving lasso ELM and elastic net ELM. If we hope the network is compact, lasso ELM is a good choice. Compared to lasso ELM and elastic ELM, rigde ELM only a little training time because it does not need to solve the QP problem. The regularized ELM ensemble using bagging has been applied to the forecasting of TCTs over the SCS. Stepwise regression analysis is then used for comparison. The comparison results show that the model in this paper is superior to the stepwise regression model in forecast accuracy. For the data in TCTs data set which is train together in the model, due to we can not recognize whether it is a non-singular TC. In the further work, we will try to cluster the DATA first effectively to avoid the influence between singular TC and non-singular TC.

References

1. Roy, C., Kovordányi, R.: Tropical cyclone track forecasting techniques - a review. Atmos. Res. **104–105**(1), 40–69 (2012)
2. Ali, M.M., Kishtawal, C.M., Jain, S.: Predicting cyclone tracks in the north Indian Ocean: an artificial neural network approach. Geophys. Res. Lett. **34**(4), 545–559 (2007)
3. Wang, Y., Zhang, W., Fu, W.: Back Propogation(BP)-neural network for tropical cyclone track forecast, pp. 1–4 (2011)
4. Chaudhuri, S., Dutta, D., Goswami, S., Middey, A.: Track and intensity forecast of tropical cyclones over the North Indian Ocean with multilayer feed forward neural nets. Meteorol. Appl. **22**(3), 563–575 (2015)
5. Zhu, L., Jin, J., Cannon, A.J., Hsieh, W.W.: Bayesian neural networks based bootstrap aggregating for tropical cyclone tracks prediction in South China sea. In: Hirose, A., Ozawa, S., Doya, K., Ikeda, K., Lee, M., Liu, D. (eds.) ICONIP 2016. LNCS, vol. 9949, pp. 475–482. Springer, Cham (2016). https://doi.org/10.1007/978-3-319-46675-0_52
6. Huang, G.B., Zhu, Q.Y., Siew, C.K.: Extreme learning machine: a new learning scheme of feedforward neural networks. Proceedings of International Joint Conference on Neural Networks, vol. 2, pp. 985–990 (2004)
7. Huang, G.B., Zhu, Q.Y., Siew, C.K.: Extreme learning machine: theory and applications. Neurocomputing **70**(1–3), 489–501 (2006)

8. Huang, G.B., Wang, D.H., Lan, Y.: Extreme learning machines: a survey. Int. J. Mach. Learn. Cybern. **2**(2), 107–122 (2011)
9. Escandell-Montero, P., Soria-Olivas, E., Magdalena-Benedito, R.: Letters: regularized extreme learning machine for regression problems. Neurocomputing **74**(17), 3716–3721 (2011)
10. Elad, M.: Sparse and Redundant Representations, pp. 3–14. Springer, New York (2010). https://doi.org/10.1007/978-1-4419-7011-4
11. Huang, G.B., Zhou, H., Ding, X., Zhang, R.: Extreme learning machine for regression and multiclass classification. IEEE Trans. Syst. Man Cybern. Part B **42**(2), 513 (2012)
12. Friedman, J., Hastie, T., Tibshirani, R.: Regularization paths for generalized linear models via coordinate descent. J. Stat. Softw. **33**(1), 1 (2010)
13. Boyd, S., Vandenberghe, L., Faybusovich, L.: Convex optimization. IEEE Trans. Autom. Control **51**(11), 1859–1859 (2006)
14. Lee, T.C., Wong, M.S.: The use of multiple-model ensemble techniques for tropical cyclone track forecast at the Hong Kong Observatory. In: WMO Commission for Basic Systems Technical Conference on Data Processing and Forecasting Systems, pp. 554–565(12) (2002)
15. Wang Q: The study on ensemble prediction of typhoon track. J. Meteorol. Sci. (2012)
16. Goerss, J.S.: Tropical cyclone track forecasts using an ensemble of dynamical models. Mon. Weather Rev. **128**(4), 1187 (2000)
17. Huang, X., Jin, L., Shi, X.: A nonlinear artificial intelligence ensemble prediction model based on EOF for typhoon track. In: International Joint Conference on Computational Sciences & Optimization, pp. 1329–1333. IEEE (2011)
18. Hansen, L.K.: Neural network ensemble. IEEE Trans. Pattern Anal. Mach. Intell. **12**, 993–1001 (1990)
19. Breiman, L.: Bagging predictors. Mach. Learn. **24**, 123–140 (1996)
20. Zhou, Z.H.: Ensemble Methods: Foundations and Algorithms. Taylor & Francis, Abingdon (2012)
21. Banerjee, K.S.: Generalized inverse of matrices and its applications. Technometrics **15**(1), 197–197 (1971)
22. Bartlett, P.L.: The sample complexity of pattern classification with neural networks: the size of the weights is more important than the size of the network. IEEE Trans. Inf. Theory **44**(2), 525–536 (1998)
23. Jin, L., Huang, X., Shi, X.: A study on influence of predictor multicollinearity on performance of the stepwise regression prediction equation. Acta Meteorologica Sinica **24**, 593–601 (2010)
24. Sobol, I.M.: On the distribution of points in a cube and the approximate evaluation of integrals. USSR Comput. Math. Math. Phys. **7**, 86–112 (1967)
25. Joe, S., Kuo, F.Y.: Remark on algorithm 659: implementing Sobol's quasirandom sequence generator. ACM Trans. Math. Softw. (TOMS) **29**, 49–57 (2003)

The Day-Ahead Electricity Price Forecasting Based on Stacked CNN and LSTM

Xiaolong Xie[(⊠)], Wei Xu, and Hongzhi Tan

Shanghai Electric Group Co. Ltd. Central Academy, 960, Zhongxing Road,
Shanghai, China
xiaolongxie88@163.com,
{xuwei5, tanhzh}@shanghai-electric.com

Abstract. In the competitive electricity market, achieving accurate electricity price forecasting is important to the participants. To improve the electricity price forecasting accuracy, the stacked CNN and LSTM model is proposed in this paper. Periodic patterns exist in the electricity price time series, i.e., dependency between different timestamps exists, and the features are selected based on the patterns to forecast day-ahead electricity price. Then, the CNN model is designed and the original time series is transformed into image-like samples based on the periodic patterns, which will help CNN to learn the data more effectively. Next, LSTM model is designed based on the selected features. Last, the stacking method, which is an ensemble learning strategy, is adopted to achieve better accuracy by fusing the forecasted values of CNN and LSTM models. The proposed model is validated on the Pennsylvania - New Jersey - Maryland market data, and the results show that the proposed model can indeed improve the forecasting accuracy.

Keywords: Day-ahead · Electricity price forecasting
Convolutional neural network · Long short-term memory network
Stacking

1 Introduction

The electricity industry starts to transform from traditional vertically integrated electric utility structure to competitive market scheme since early 1990s [1, 2]. In the competitive electricity markets, participants, which are generally producers and consumers, need to optimize their production schedule or bidding strategies according to the status of the market to maximize their benefits, and the electricity price forecasting can provide such information. Therefore, forecasting the electricity price accurately is important to these participants.

The Pennsylvania - New Jersey - Maryland (PJM) market is a widely known power market in US and is one of the successful models [3]. PJM market uses locational marginal pricing (LMP), which is a popular energy pricing mode [4]. LMP is the price at a node in the grid and can vary at different nodes [5]. LMP is composed of three parts, which are the energy cost, the transmission congestion part, and the marginal loss, i.e., it reveals important information, particularly the congestion information,

© Springer Nature Switzerland AG 2018
Y. Peng et al. (Eds.): IScIDE 2018, LNCS 11266, pp. 216–230, 2018.
https://doi.org/10.1007/978-3-030-02698-1_19

which can influence the strategies of different participants; thus, achieving accurate LMP forecasting is crucial to all the participants.

Most of the relative literatures focus on day-ahead electricity price forecasting because the majority part of the total energy is traded in day-ahead market, and other short-term markets, which include real-time market, reserves market, etc., are generally executed to make the balance between power supply and demand [6]. Therefore, the day-ahead market is important, and we also focus on day-ahead electricity price forecasting in this paper.

The day-ahead market does not allow for continuous trading, for instance, as for generator companies, they should submit their bids for the electricity price at each timestamp, which is one hour in PJM market, of the next day (day d) before the closing time on day $(d - 1)$. After closing time, the market will verify the constraints and the electricity prices can be established. Therefore, all the hourly electricity prices of day d are determined at the same time on day $(d - 1)$, and we can only use the available information before closing time to forecast 24 hourly day-ahead electricity prices. This is quite different from conventional time series forecasting. This process is illustrated in Fig. 1.

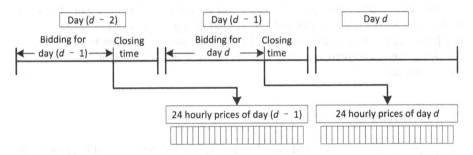

Fig. 1. The day-ahead bidding process of PJM market [2].

Figure 1 shows the day-ahead bidding process, and the day-ahead electricity price forecasting should be executed before the closing time on previous day, and the 24 hourly prices should be forecasted simultaneously.

Many researchers have proposed different approaches for the day-ahead electricity price forecasting, such as neural network [7], extreme learning machine [8], fuzzy model [4], cascade neural network [9], and some hybrid model [10]. Researchers have made different attempts to achieve more accurate forecasting.

Thus, the objective of this paper is also trying to improve the forecasting accuracy. Electricity price forecasting is a time series forecasting problem, and periodic patterns exist in the price time series, i.e., dependency exists between the prices of different timestamps. However, most of the above mentioned models are feed-forward models, i.e., these models treat prices of different timestamps as independent variables, which violates the characteristics of price time series, although these model perform well. Therefore, to improve the accuracy further, the dependency of prices at different timestamps will be modeled in the proposed model in this paper.

The recurrent neural network (RNN) is a natural choice because it considers the dependency between features of different timestamps and has achieved great performance in time series forecasting. Besides RNN model, the convolutional neural network (CNN) can also consider the correlation between different input elements. Moreover, CNN and long short-term memory (LSTM) network, which is one of the most advanced RNN, are two popular deep learning algorithms, which have shown great capabilities to solve complicated problems, and outperform traditional machine learning algorithms in many fields. Therefore, these deep learning algorithms are adopted, designed based on the features of electricity price time series, and then applied to the day-ahead electricity price forecasting task, hoping the deep learning model can improve the accuracy. Both CNN and LSTM models are adopted to forecast day-ahead electricity prices, and how to improve the performance based on these two models is another problem. Stacking is a meta-algorithm for ensemble learning, which can combine several machine learning models into one model and improve the performance. Therefore, the stacking strategy is adopted to fuse the forecasted values of CNN and LSTM models, aiming to improve the forecasting accuracy further, which can support the participants of the electricity market more effectively.

The remainder of this paper is organized as follows. Section 2 introduces the background knowledge of CNN and LSTM, and then gives a brief introduction to the day-head electricity price time series of PJM market and analyzes its features; Sect. 3 details the proposed approach; Sect. 4 shows the experiment results; and Sect. 5 draws the conclusion.

2   The Background

The proposed approach is based on CNN and LSTM models; thus, this section will introduce these two models briefly. Then, the background of PJM market, which is adopted as the case study in this paper, will be presented, and the data of PJM market will be analyzed to find the features for electricity price forecasting.

2.1   Convolutional Neural Networks

The convolutional neural network (CNN), which was proposed by LeCun et al. [11], can mimic the natural visual perception mechanism of human being and thus can extract the representations of the original images effectively. Although there are massive variants of CNN architectures, the basic components are similar [12], which are convolutional layer, pooling layer and fully-connected layer generally.

The convolutional layer is the most important part of CNN. Actually, CNNs can be seen as neural networks that use convolution in place of general matrix multiplication in at least one of their layers, i.e., CNN is an modification of conventional artificial neural networks, which adopts convolution layer instead of conventional layer. In convolutional layer, the input is convoluted with several filters, where each filter is a smaller matrix, and corresponding feature maps can be obtained after the convolution operation.

Another specific layer for CNN is the pooling layer, which replaces the output of the net at a certain location with a summary statistic of the nearby outputs. The most popular pooling layers are max-pooling, which outputs the maximum of a rectangular neighborhood, and average-pooling, which uses the average of the rectangular neighborhood.

The convolutional and pooling layers are generally used to extract features, and then one or more fully-connected layers are usually adopted after one or more groups of convolutional and pooling layers. The fully-connected layer can put the information from feature maps together, and then output them to latter layers.

CNN is still a kind of feed-forward neural network, which is similar to conventional artificial neural networks, but it uses convolutional layer to extract features from inputs, and the feature extraction is very effective, particularly for images. For images, the nearby pixels have strong correlations, and similarly for time series, some periodic timestamps also have strong correlations. Although CNN is a feed-forward network, its convolutional layer can extract the relations of nearby and correlated input elements, and this characteristic can be used to model the periodic dependency between different timestamps in time series. Thus, CNN is adopted to forecast the electricity price in this paper, hoping that the convolutional layer can improve the feature extraction in electricity price forecasting. Besides, the two dimensional convolution operation in CNN is designed to process images, which is quite different from time series; therefore, samples for CNN should be designed based the features of time series, such that their shape is similar to images.

2.2   Long Short-Term Memory Networks

The LSTM, which is one of the most advanced RNN model, was originally proposed by Hochreiter and Schmidhuber [13], and has shown remarkable performance in dealing with time series forecasting, because it has the ability to capture the long-term dependencies existing in the time series [14]. To illustrate the mechanism of LSTM clearly, the architecture of RNN is introduced first.

RNN is suitable for processing time series. When conventional artificial neural network is applied to time series forecasting, it ignores the dependency among the inputs and outputs of different timestamps, while this characteristic exists in time series. In contrary, RNN considers this dependency, because the information extracted from previous timestamps is used as the input of latter timestamps. RNN consists of several RNN cells, which have the same structure, as shown in Fig. 2.

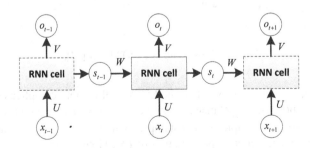

Fig. 2. The structure of RNN cell.

The mechanism of RNN cell can be represented by the following equations.

$$s_t = \tanh(W \cdot s_{t-1} + U \cdot x_t + b_s) \tag{1}$$

$$o_t = Vh_t \tag{2}$$

Where s_t and s_{t-1} are the hidden states of RNN cells of timestamps t and $(t-1)$, x_t is the input of timestamp t, o_t is the output of timestamp t, W is the weight matrix between hidden states of timestamps t and $(t-1)$, U and V are the weight matrices from input to hidden state and from hidden state to output. Those three weight matrices need to be learned during the training phase of RNN.

In RNN, the cells are organized according to the timestamps. According to Fig. 2. and Eqs. (1–2), it can be observed that the information extracted from previous timestamps is stored in hidden state, and the information can be used as part of the inputs of latter timestamps, i.e., RNN has some kind of memory; thus, it is effective in solving time series forecasting.

Although RNN can consider previous information when forecasting future values, it still has a weakness, which is the vanishing gradient problem. The training algorithm for RNN is called backpropagation through time (BPTT) algorithm, and too many backpropagations over a long period may cause this problem because every back-propagation will bring a multiply operation, i.e., RNN seems to forget the long-term information.

LSTM can handle this vanishing gradient problem because its cell's special structure, which is shown in Fig. 3.

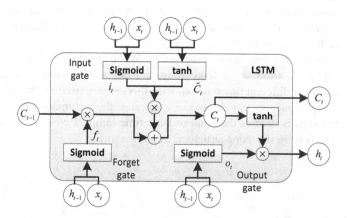

Fig. 3. The structure of LSTM cell [14].

From Fig. 3, it can be observed that the LSTM cell is more complicated than RNN cell, and it is composed of forget gate f_t, input gate i_t, output gate o_t and candidate value $\tilde{C}_t$. The x_t represents the input at timestamp t, and h_{t-1} is the output at timestamp $(t-1)$, then the mathematical representation of LSTM cell is shown as follows.

$$f_t = \sigma(W_f[h_{t-1}, x_t] + b_f) \tag{3}$$

$$i_t = \sigma(W_i[h_{t-1}, x_t] + b_i) \tag{4}$$

$$\tilde{C}_t = \tanh(W_C[h_{t-1}, x_t] + b_C) \tag{5}$$

$$C_t = f_t * C_{t-1} + i_t * \tilde{C}_t \tag{6}$$

$$o_t = \sigma(W_o[h_{t-1}, x_t] + b_o) \tag{7}$$

$$h_t = o_t * \tanh(C_t) \tag{8}$$

Where the $*$ represents Hadamard product operation. In LSTM cell, the forget gate f_t determines how much of the previous cell states are forgotten, and the output gate o_t determines how much to output [15].

2.3 The Electricity Price Time Series in PJM Market

The dataset adopted in this paper is the electricity price time series in PJM market. The data can be downloaded from the repository, and the dataset includes historical electricity price data and the load forecasting data, i.e., for generator companies, before the day-ahead bidding, the forecasted load of each hour of the next day will be available; thus, we can use the load forecasting data of the next day to forecast the corresponding hourly day-ahead electricity prices. We collected the LMP and corresponding load forecasting data from Apr. 1, 2012 to Feb. 28, 2017, which is almost five years. Everyday has 24 hourly electricity prices, and the forecasting of 24 hourly prices of day d should be executed on day $(d - 1)$. In this section, we will analyze the data and try to find some important features for day-ahead electricity price forecasting.

The correlation coefficient can reflect the correlation between two features. At the first glance of the data, it can be observed that some periodic patterns exist in the time series; thus, we need to extract the exact pattern, which means we need to find which timestamp provides the most valuable information in predicting the electricity price of day d. It is supposed that there exist two kinds of periodic patterns, which are daily and weekly patterns. Therefore, to investigate these assumed patterns and select important features simultaneously, the correlation coefficients between the hourly electricity prices of day d and corresponding hourly prices of each day before day d in two weeks are calculated, i.e. the hourly prices from day $(d - 14)$ to $(d - 1)$ are used to calculate the correlation coefficients with corresponding hourly prices of day d. The coefficients can show how strong the correlation between these two days, and important days can be selected based on these coefficients. There are two popular correlation coefficients, which are Pearson and Spearman coefficients. The Pearson coefficient is only effective in measuring the linear relationship between two features; thus, the Spearman coefficient is adopted to select important features. The results are shown in Fig. 4.

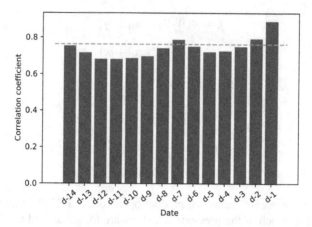

Fig. 4. The correlation coefficients of different days in previous two weeks (*bar chart*) and the threshold for feature selection (*dotted line*).

From the figure, we can observe that the most correlated feature is the corresponding hourly electricity prices of day $(d-1)$, which has the largest coefficient (0.8899), and this makes sense in time series. The most recent timestamp is generally the most important feature in predicting future timestamp. The next two features are corresponding hourly prices of day $(d-2)$ and day $(d-7)$, and their coefficients are similar, which are larger than other days in two weeks before day d. Thus, the corresponding hourly price of day $(d-7)$ is also adopted as an important feature in forecasting hourly price of day d. Besides, the large correlation coefficient of day $(d-7)$ means that the electricity price time series really has the weekly periodic pattern.

Besides, the correlation coefficient between the electricity price and corresponding forecasted load is 0.747394, which is only a little smaller than the electricity price of day $(d-7)$; thus, the forecasted load is also adopted to generate sample inputs.

Based on these periodic patterns, the important features are selected, and then, the CNN and LSTM models should be designed based on these patterns, such that these models can capture the dynamic mechanism of day-ahead electricity price time series and make accurate forecasting.

3   The Methodology

The proposed approach is the stacked CNN and LSTM model, meaning that the CNN and LSTM models are first adopted to forecast the electricity price individually, and then the stacking strategy is used to fuse the forecasted values of CNN and LSTM models to obtain a more accurate forecasting. This section will detail the proposed approach, and it consists of three parts: (i) the training of CNN model; (ii) the training of LSTM model and (iii) the stacking process. The details are show as following.

3.1 The CNN Model for Electricity Price Forecasting

As mentioned above, although CNN is perfectly suitable for image processing, it also can be applied to time series forecasting, because it can extract the neighborhood information of nearby input elements, which are nearby pixels in images, and nearby pixels generally has strong correlations. The electricity price time series has periodic patterns, which can also be seen as some kind of nearby input elements because they also have strong correlations. However, time series is a one dimensional vector, while the images are two dimensional matrices; thus, the electricity price time series should be preprocessed such that its shape is similar to images, and nearby input elements should also have strong correlations. Only in this way, can the two dimensional convolutional layer in CNN model extract the information from the electricity price time series effectively.

Based on the analysis in Sect. 2.3, the electricity price of day d has strong correlations with corresponding hourly electricity prices of day $(d-1)$ and day $(d-7)$; thus, these prices are used as input elements. Besides, the forecasted load information is also available; thus, they are also used as input elements. Besides corresponding hourly electricity prices, the consecutive hourly electricity prices also have some influences on future electricity prices. Therefore, all these features are adopted as input elements, and are transformed into two dimensional matrices based on the correlations, such that the inputs are similar to images. The generated samples and the structure of CNN are illustrated in Fig. 5.

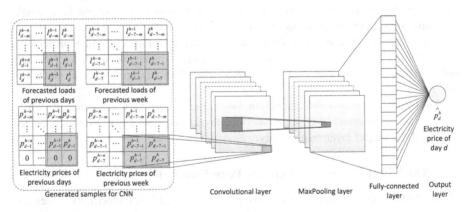

Fig. 5. The format of samples for CNN model (*l represents forecasted load and p represents the electricity price*) and the architecture of CNN model.

From Fig. 5, it can be observed that the vector-like time series is transformed into four two dimensional matrices, which can be used as four channels of the input tensor of one sample for CNN, and the format is similar to images. Four matrices represent the forecasted loads and electricity prices of previous days and previous week. For each matrix, there are two variables, which are the number of days ahead m and the number of hours ahead per day n. Take the top left matrix for instance, it represents the forecasted loads of previous m days. Every row represents one day, and every column is one hour. The l_{d-i}^{h-j} represents the forecasted load of hour $(h - j)$ in day $(d - i)$, where $0 \leq i \leq m$, $0 \leq j \leq n$. The output of the sample is the electricity price of hour h in day d, p_d^h. The first row indicates the forecasted load of the $(n + 1)$ hours of day $(d - m)$, which are $\{l_{d-m}^{h-n}, l_{d-m}^{h-n+1}, \cdots, l_{d-m}^{h-1}, l_{d-m}^{h}\}$. Besides the previous m days, the forecasted loads of corresponding hours in day d are also included in the last row because they are assumed available when predicting day-ahead electricity price. The other three matrices, which represent the forecasted loads of corresponding days in previous week, the electricity prices in previous m days and previous week, are generated in the similar way. One important thing needs to be noticed is that the all the electricity prices of day d are unknown; thus; the last row of right bottom matrix, which contains the electricity prices of day d, is set zero vector. The original forecasted load and electricity price time series are converted to image-like matrices by using this transformation, i.e., the input and output of one sample for CNN model can be represented as

$$\left. \begin{cases} l_{d-m}^{h-n}, \cdots, l_{d-m}^{h}; \cdots; l_{d-1}^{h-n}, \cdots, l_{d-1}^{h}; l_d^{h-n}, \cdots, l_d^{h}; \\ p_{d-m}^{h-n}, \cdots, p_{d-m}^{h}; \cdots; p_{d-1}^{h-n}, \cdots, p_{d-1}^{h}; 0, \cdots, 0; \\ l_{d-7-m}^{h-n}, \cdots, l_{d-7-m}^{h}; \cdots; l_{d-6}^{h-n}, \cdots, l_{d-6}^{h}; l_{d-7}^{h-n}, \cdots, l_{d-7}^{h}; \\ p_{d-7-m}^{h-n}, \cdots, p_{d-7-m}^{h}; \cdots; p_{d-6}^{h-n}, \cdots, p_{d-6}^{h}; p_{d-7}^{h-n}, \cdots, p_{d-7}^{h}; \end{cases} \right\} \rightarrow p_d^h.$$

CNN is effective in solving images because the convolutional layer can extract neighborhood information from images, and nearby pixels in images generally have strong correlation. Besides the similar format, the elements in input matrices should also have strong correlations. As mentioned above, the day $(d - 1)$ and day $(d - 7)$ both have strong correlations with day d; thus, the four matrices can represent the correlation embedded in the time series. Therefore, it makes sense that CNN may perform well by using this kind of samples when forecasting day-ahead electricity prices.

The architecture of CNN model in this paper is not very complicated, which contains a convolutional layer, a max-pooling layer, a dropout layer, a fully-connected layer and the output layer sequentially.

3.2   The LSTM Model for Electricity Price Forecasting

The LSTM model is designed to use historical forecasted loads, electricity prices and the forecasted loads of day d to forecast the electricity prices of day d. As for historical data, the data of day d and day $(d - 7)$ are both considered, and the samples for LSTM model is designed based on this assumption. The samples for LSTM and the architecture of LSTM model are shown in Fig. 6.

From Fig. 6, it can be observed that there are two LSTM layers and each layer has three timestamps in the model, which are information of day $(d - 7)$, day $(d - 1)$, and day d, respectively. As for each timestamp, the forecasted loads and electricity prices of corresponding hour are used as inputs, and one important thing is that, the forecasted

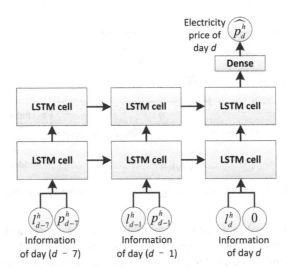

Fig. 6. The format of samples for LSTM and the architecture of LSTM model.

load of hour h in day d is available, while the electricity price, which is the output, is unknown. Therefore, the second input element of the third timestamp is set zero. The output is the electricity price of hour h in day d, i.e., the input and output of samples for LSTM model can be represented as the following format, i.e., $\{l^h_{d-7}, p^h_{d-7}, l^h_{d-1}, p^h_{d-1}, l^h_d, 0\} \rightarrow p^h_d$.

The number of input elements of LSTM model is smaller than CNN model, and only the most relevant information is considered in the inputs, because too many input elements may make the model too complicated and difficult to be trained.

3.3 The Stacked CNN and LSTM Model

The above two sections use CNN and LSTM model to forecast future electricity price, respectively, and the next step is fusing their forecasted values into a more accurate forecasting. The stacking strategy is adopted in this paper. Stacking is a popular ensemble learning strategy, while it needs a modification in this problem. The conventional stacking strategy uses cross validation on the training data, while it is not suitable for time series forecasting. The cross validation uses every part of the training data as validation data and other parts to train the model, i.e., it may use future timestamps to train the model and validate the model using previous timestamps. This is not acceptable for time series forecasting because this is kind of data leakage during the training phase, and then the validation results are not reliable, which will decrease the generalization capability of the model. Therefore, the stacking strategy for CNN and LSTM in electricity price forecasting is modified, which is shown in Fig. 7.

From Fig. 7, it can be observed that all the training dataset is split into two parts, which are training and validation data, and the split is based on timestamp, not the cross validation in conventional stacking. The previous part is the training data and the latter part is used as validation data.

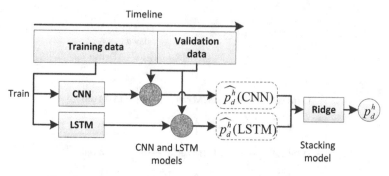

Fig. 7. The stacking strategy for electricity price forecasting.

First, the training data is adopted to train the CNN model and LSTM model according to the approach in the above two sections, and two models can be obtained. Then, these two models are used to forecast the samples in validation dataset, and their forecasted values of the electricity price of hour h in day d are represented as $\widehat{p_d^h}(CNN)$ and $\widehat{p_d^h}(LSTM)$. Next, the stacking model, which is a ridge regression model, is adopted to fuse the predictions of CNN and LSTM models. Its inputs are the forecasted values of these two models, and its output is the actual electricity price of hour h in day d. Last, based on the validation dataset, the ridge model is trained, and the CNN and LSTM model can be fused.

In summary, the CNN and LSTM models are trained on the training dataset, then their generalization capabilities are validated on the validation dataset, and the stacking model is trained on the validation dataset. Next, the stacked CNN and LSTM model can be used to forecast future electricity prices. The validation dataset is extracted based on timestamps, making the future forecasting reliable.

4 Experimental Results

The proposed approach is applied to the PJM dataset to validate its performance. The data is the electricity prices and corresponding forecasted loads of PJM market from Apr. 1, 2012 to Feb. 28, 2017. Based on the stacking strategy, the dataset should be split into three parts, where the training dataset is from Apr. 1, 2012 to Dec. 31, 2015; the validation dataset contains one year data, which is from Jan. 1, 2016 to Dec. 31, 2016; and the test dataset is from Jan. 1, 2017 to Feb. 28, 2017. The training data is used to train CNN and LSTM model, the validation data is used to train the ridge model, and the test data is used to test the performance of the stacked CNN and LSTM model.

Some hyper-parameters need to be determined. As for CNN model, the number of filters in convolutional layer is 32, the kernel size is 2, the zero-padding is adopted to make sure that the shape of the feature map is the same as input matrices, and the activation function is hyperbolic tangent function. In the max-pooling layer, the pool size is 2 multiply 2. The dropout layer is adopted and the rate is set 0.2. The number of neurons in fully connected layer is 2000, and the activation function is also hyperbolic tangent function. During the training phase, the Adam algorithm is adopted, the batch size is set to 100, and the number of epochs is set to 50. As for LSTM model, the output units of the first LSTM layer is 50, and then a dropout layer is adopted and the rate is 0.2; then the second LSTM layer with units equal 50 and a dropout layer is adopted; finally, the output layer has one output. The loss function is the mean squared error, the Adam algorithm is adopted, the batch size is set to 500, and the number of epochs is set to 50. The alpha of ridge model is set to 1.0.

To validate the accuracy of the proposed approach, the conventional artificial neural network (ANN) is adopted to do the comparison. By comparing with the performance of ANN, the effectiveness of the proposed approach can be validated. Because ANN model is the feed-forward network, which does not consider the dependency between different timestamps, the samples for ANN are extracted directly, i.e., one sample has five input elements, which are the forecasted loads and electricity prices of corresponding hours in day $(d - 7)$ and day $(d - 1)$, and the forecasted load of hour h in day d; the output is the electricity price of hour h in day d. Therefore, the samples for ANN can be represented as $\{l_{d-7}^h, p_{d-7}^h, l_{d-1}^h, p_{d-1}^h, l_d^h\} \rightarrow p_d^h$, which is similar to LSTM, but ANN does not have the memory mechanism. The ANN model has one hidden layer, and the number of hidden neurons is set to 2000.

Two error measurements are adopted to evaluate the performance of these models. One is root mean squared error (RMSE), which is $\text{RMSE} = \sqrt{\sum_{i=1}^{N} (y_i - o_i)^2 \big/ N}$, and the other one is mean absolute percentage error (MAPE), which is $\text{MAPE} = (100/N) \cdot \sum_{i=1}^{N} |(y_i - o_i)/y_i|$, where y_i and o_i represent the actual and forecasted values, respectively.

These models have some stochastic characteristics during the training processes; thus, to decrease the influence of the variations, every model is executed 10 times, and the average value of these two error measurements are calculated to represent the performance of these models. The results of these models are presented in Table 1. There are four models in the table, where ANN is the benchmark model, CNN and LSTM models are the base models of the proposed stacked model, and their performances are also listed to compare with other models.

Table 1. The accuracy of different models

	RMSE	MAPE
ANN model	3.408578	9.862251%
CNN model	3.023677	8.891508%
LSTM model	2.530097	6.949292%
Proposed model	2.296742	6.182146%

From Table 1, it can be observed that: (i) the proposed model outperforms other models, indicating that the proposed approach can indeed achieve better accuracy in electricity price forecasting; (ii) the performance of the proposed model is better than

CNN model and LSTM model, meaning that the adopted stacking strategy is effective in improving the forecasting accuracy; (iii) the LSTM model performs better than CNN model, and they both outperform conventional ANN model, meaning that the proposed architecture of CNN and LSTM, which both consider the time dependencies of different timestamps, are effective in time series forecasting. To illustrate the results, the error bar charts of RMSEs and MAPEs are shown in Figs. 8 and 9, respectively.

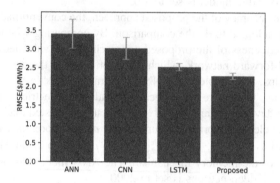

Fig. 8. The error bar chart of RMSEs of four models

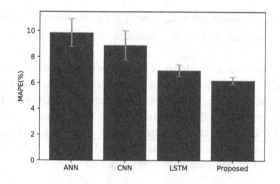

Fig. 9. The error bar chart of MAPEs of four models

From the error bar charts, not only the average errors of four models can be observed, but also the standard deviations, which represent the stability, are presented. It can be observed that the proposed stacked CNN and LSTM model is also the most stable model. The stability of LSTM model is better than ANN and CNN models. Maybe because many parameters need to be tuned for CNN model, its performance is not that stable.

In summary, the proposed CNN model and LSTM model, which are designed for electricity price forecasting, are effective, and they perform better than conventional models by considering the correlation between different timestamps in the time series; by using the stacking strategy, the proposed ensemble model can improve the electricity price forecasting further. Therefore, the proposed stacked CNN and LSTM model is effective and can achieve more accurate electricity price forecasting, which can support the participants of electricity market better.

5 Conclusion

Achieving more accurate day-ahead electricity price forecasting is crucial to the electricity market participants; thus, to improve the electricity price forecasting accuracy, the stacked CNN and LSTM model is proposed in this paper.

The CNN model is designed aiming to capture the dependency between different timestamps; thus, the periodic patterns of electricity price time series is analyzed, and correlated timestamps are extracted and reshaped to image-like two dimensional matrices, such that the two dimensional convolutional operation in CNN is suitable to handle the time series. This idea can be applied to other scenarios, and CNN can consider the correlations between different elements. To ensemble CNN and LSTM models, the stacking strategy is modified such that it can be applied to time series, and the stacking strategy is effective.

In summary, the contributions of this paper are three-folds: (i) the proposed CNN model can be adopted to consider the dependencies of different timestamps in time series forecasting; (ii) the proposed CNN and LSTM model can improve the electricity price forecasting accuracy; (iii) by using the stacking strategy, the proposed model can improve the accuracy further, and the stability is also better.

Besides the selected features, some other features also have influences on day-ahead electricity price, such as the feature indicating weekday or weekend. Therefore, more features should be extracted and adopted in the model to improve the electricity price forecasting accuracy in the future.

References

1. Liu, J.D., Lie, T.T., Lo, K.L.: An empirical method of dynamic oligopoly behavior analysis in electricity markets. IEEE Trans. Pow. Syst. **21**(2), 499–506 (2006)
2. Weron, R.: Electricity price forecasting: a review of the state-of-the-art with a look into the future. Int. J. Forecast. **30**(4), 1030–1081 (2014)
3. Lin, W.M., Gow, H.J., Tsai, M.T.: Electricity price forecasting using enhanced probability neural network. Energy Convers. Manag. **51**(12), 2707–2714 (2010)
4. Hong, Y.Y., Lee, C.F.: A neuro-fuzzy price forecasting approach in deregulated electricity markets. Electr. Pow. Syst. Res. **73**(2), 151–157 (2005)
5. Hong, Y.Y., Weng, M.T.: Investigation of nodal prices in a deregulated competitive market-case studies. In: IEEE International Conference on Electric Power Engineering, p. 161. IEEE Press, Powertech Budapest (1999)
6. Amjady, N.: Day-ahead price forecasting of electricity markets by a new fuzzy neural network. IEEE Trans. Pow. Syst. **21**(2), 887–896 (2006)
7. Vahidinasab, V., Jadid, S., Kazemi, A.: Day-ahead price forecasting in restructured power systems using artificial neural networks. Electr. Power Syst. Res. **78**(8), 1332–1342 (2008)

8. Zhang, Y., Li, C., Li, L., et al.: Electricity price forecasting by a hybrid model, combining wavelet transform, ARMA and kernel-based extreme learning machine methods. Appl. Energy **190**, 291–305 (2017)
9. Amjady, N., Daraeepour, A.: Design of input vector for day-ahead price forecasting of electricity markets. Expert Syst. Appl. **36**(10), 12281–12294 (2009)
10. Amjady, N., Keynia, F.: Day ahead price forecasting of electricity markets by a mixed data model and hybrid forecast method. Int. J. Electr. Power Energy Syst. **30**(9), 533–546 (2008)
11. LeCun, Y., Boser, B., Denker, J.S., et al.: Handwritten digit recognition with a back-propagation network. Adv. Neural. Inf. Process. Syst. **2**(2), 396–404 (1990)
12. Gu, J., Wang, Z., Kuen, J., et al.: Recent advances in convolutional neural networks. Pattern Recogn. **77**, 354–377 (2015)
13. Hochreiter, S., Schmidhuber, J.: Long short-term memory. Neural Comput. **9**(8), 1735–1780 (1997)
14. Qing, X., Niu, Y.: Hourly day-ahead solar irradiance prediction using weather forecasts by LSTM **148**, 461–468 (2018)
15. Yang, J., Kim, J.: An accident diagnosis algorithm using LSTM. Nucl. Eng. Technol. (2018, in press)

Covariance Based Deep Feature
for Text-Dependent Speaker Verification

Shuai Wang, Heinrich Dinkel, Yanmin Qian, and Kai Yu[✉]

Key Laboratory of Shanghai Education Commission for Intelligent Interaction
and Cognitive Engineering, SpeechLab, Department of Computer Science
and Engineering, Brain Science and Technology Research Center,
Shanghai Jiao Tong University, Shanghai, China
kai.yu@sjtu.edu.cn

Abstract. *d*-vector approach achieved impressive results in speaker verification. Representation is obtained at utterance level by calculating the mean of the frame level outputs of a hidden layer of the DNN. Although mean based speaker identity representation has achieved good performance, it ignores the variability of frames across the whole utterance, which consequently leads to information loss. This is particularly serious for *text-dependent* speaker verification, where within-utterance feature variability better reflects text variability than the mean. To address this issue, a new covariance based speaker representation is proposed in this paper. Here, covariance of the frame level outputs is calculated and incorporated into the speaker identity representation. The proposed approach is investigated within a joint multi-task learning framework for *text-dependent* speaker verification. Experiments on RSR2015 and RedDots showed that, covariance based deep feature can significantly improve the performance compared to the traditional mean based deep features.

Keywords: Deep features · Text-dependent speaker verification
Speaker recognition · *d*-vector · *j*-vector · Covariance discrimination

1 Introduction

Speaker verification (SV) is the task of verifying the identity of a certain person by means of his voice. Considering the restriction on the spoken text, speaker verification can be classified into two categories, text-dependent and text-independent. Moreover the duration of the model registration is detrimental to the user experience, although a tough challenge, short enrollment utterances are preferred over long ones. In short utterance environments, the additional

This work has been supported by the National Key Research and Development Program of China under Grant No. 2017YFB1002102 and the China NSFC projects (No. U1736202 and No. 61603252). Experiments have been carried out on the PI supercomputer at Shanghai Jiao Tong University.

Y. Peng et al. (Eds.): IScIDE 2018, LNCS 11266, pp. 231–242, 2018.
https://doi.org/10.1007/978-3-030-02698-1_20

information provided by the text gives the text-dependent SV an edge over the text-independent SV, while simultaneously being less convenient for the user. In recent research, deep neural network (DNN) was applied to speaker verification [1–5] and was critically acclaimed.

After a fast deep neural network training algorithm was published in [6,7], progressively more researchers turned their focus to deep learning. Motivated by the powerful non-linear learning ability, researchers started using DNN as a feature extractor, e.g. as a bottleneck feature in speech recognition [8,9] and language identification [10]. Moreover, bottleneck features can be used together with traditional features in a tandem manner, which can be seen in [11,12].

Other than using DNN as a feature extractor, some researchers extract model representations directly out of a DNN. In google's work [13], d-vector is proposed to produce speaker model and test utterance representations, subsequently cosine distance is used to calculate a score between model and test utterance d-vectors. Motivated by google's work, j-vector based on a multi-task learning framework was proposed in [14] and achieved significant improvements. In this paper, we introduce and survey the usage of covariance based representations into the frameworks of both d-vector and j-vector. Different scoring methods are then applied to the covariance based representations. Experiments manifest that the proposed covariance based representations surpass mean based approaches on RSR2015 [15] and RedDots [16] data-sets respectively.

The remainder of the paper is organized as follows, Sect. 2 briefly introduces previous works. Section 3 introduces our own v-vector. Section 4 showcases our experiment design and result analysis. Finally Sect. 5 concludes this paper.

2   Speaker Representation Using DNN

In [13], Google proposed to use a neural network to extract frame level vectors from commonly used cepstral features (PLP, FBANK, MFCC). The outputs of the last hidden layer are derived and averaged to get utterance-level representations (d-vector).

Based on Google's work, a multi-task framework which learns both speaker identity and text information is proposed in [14]. In this framework, the output nodes consist of both speakers and texts, where two types of multi-task joint training can be considered: speaker + phrase, speaker + phone. The architecture is shown in Fig. 1 (speaker + phrase as an example).

3   Covariance Based Deep Feature

Although mean based speaker identity representation has achieved good performance, it ignores the variability of frames across the whole utterance, which consequently leads to information loss. This is particularly serious for text-dependent speaker verification, where within-utterance feature variability better reflects text variability than the mean. To address this issue, a new covariance based speaker representation is proposed in this paper.

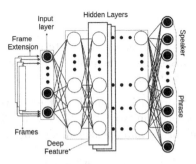

Fig. 1. j-vector approach

In [17] it was shown that covariance is a key factor to cluster narrow-band, wide-band and background (noise, music, silence) speech into the respective categories. Furthermore, in another related field of research, speaker anti-spoofing [18,19], it can be seen that covariance based discrimination for binary classification outperforms conventional mean based features. Moreover, covariance has been used in computer vision for human detection tasks to represent human descriptors as an important feature [20,21].

3.1 v-vector

In this paper we introduce covariance based vectors within the d/j-vector framework. After extracting the utterance-level representation $\mathbf{u}$ containing N frames with dimension D from speaker s, different types of vectors can be extracted by calculating the mean vector $(\boldsymbol{m})$ and covariance matrix $(\boldsymbol{\Sigma})$.

$$\boldsymbol{m} = \frac{1}{N} \sum_{i=1}^{N} \mathbf{u}_i \tag{1}$$

$$\boldsymbol{\Sigma} = cov(\mathbf{u}) \tag{2}$$

$$\boldsymbol{v}_f = \left[\boldsymbol{\Sigma}_{11}, \boldsymbol{\Sigma}_{12}, \ldots, \boldsymbol{\Sigma}_{1D}, \boldsymbol{\Sigma}_{22}, \boldsymbol{\Sigma}_{23}, \ldots, \boldsymbol{\Sigma}_{DD}\right]^{\top} \tag{3}$$

$$\boldsymbol{v}_d = \left[\boldsymbol{\Sigma}_{11}, \boldsymbol{\Sigma}_{22}, \boldsymbol{\Sigma}_{33}, \ldots, \boldsymbol{\Sigma}_{DD}\right]^{\top} \tag{4}$$

Since *covariance* matrices are symmetric, only the upper triangular part contains information, thus we obtain the $\boldsymbol{v}_f$ (full covariance vector) as a concatenation of the upper triangular part according to Eq. 3, the dimension of $\boldsymbol{v}_f$ is $\frac{D(D+1)}{2}$. $\boldsymbol{v}_d$ (diagonal covariance vector) is defined as the diagonal of $\boldsymbol{\Sigma}$ according to Eq. 4 and has a dimension of D. In terms of the methods of extracting vectors, we annotate the mean vector $\boldsymbol{m}$ as m-vector, $\boldsymbol{v}_f$ and $\boldsymbol{v}_d$ as v-vector. Both vectors can be extracted in d-vector or j-vector framework.

It should be stressed that we denote d/j vector in terms of the network types (multi-task or not), while m/v-vector in terms of the method of generating utterance-level representations from frame-level.

3.2   Scoring Methods

Having obtained the speaker model and test-utterance representations from the respective neural network, scoring leads to obtain a classification metric to either accept or reject a certain testcase.

CDS (Cosine Distance Scoring) is a simple but effective scoring method successfully used in the i-vector framework [22–24]. Cosine distance scoring is a dot product between test vector, $\mathbf{w}$ and speaker model mean, μ_m

$$score_w^m = \frac{\mathbf{w}^T \mu_m}{\|\mathbf{w}\| \, \|\mu_m\|} \tag{5}$$

GC (Gaussian Classifier) is a classical classifier following a generative manner, speaker identity vector representations(eg. i-vector) are modeled by a Gaussian distribution, where full covariance matrix is shared across all speakers. For an test vector representation $\mathbf{w}$, we evaluate the log likelihood score against the target m,

$$ln(\mathbf{w}|m) = \mathbf{w}^T \mathbf{\Sigma}^{-1} \mu_m - \frac{1}{2}(\mathbf{w}^T \mathbf{\Sigma}^{-1} \mathbf{w} + \mu_m^T \mathbf{\Sigma}^{-1} \mu_m) + const \tag{6}$$

where μ_m is the mean vector for speaker m, $\mathbf{\Sigma}$ is the common covariance matrix and $const$ is a speaker- and test vector-independent constant. Furthermore, the speaker-independent part and constant can be neglected and we get

$$ln(\mathbf{w}|m) = \mathbf{w}^T \mathbf{\Sigma}^{-1} \mu_m - \frac{1}{2}\mu_m^T \mathbf{\Sigma}^{-1} \mu_m \tag{7}$$

PLDA (Probabilistic Linear Discriminant Analysis) [25] uses a generative approach to score utterances [26,27]. A PLDA estimator is trained on the background data, which will be used to transform and score the test utterances against the target models. situation [14].

4   Experiments

4.1   Experiments on RSR2015

Experimental Setup and Baseline. RSR2015 Part 1 consists of overall 300 speakers, whereas 143 are females and 157 are males. The whole set is divided into background (*bkg*), development (*dev*) and evaluation (*eval*) subsets (Table 1).

Bkg and *dev* data is merged to obtain an extended training data set consisting of 194 speakers and 52244 utterances. The evaluation part is split into enrolement part and test part. The test set encompasses 1568008 tests, which is divided into 19052 true speaker tests and 1548956 impostor tests.

Table 1. Subset definition of RSR2015 part 1

Subset	# Female speaker	# Male speaker	# Total
bkg	47	50	97
dev	47	50	97
eval	49	57	106

The neural network was trained using 39-dimensional PLP features, which were extended by a frame window of 5 at left and right. Network initialization was done using a 6 hidden-layer, 1024 neuron RBM network. Sigmoid was used as the activation function.

The complete DNN comprises of 8 layers, 1 input layer with 429 (11×39) neurons, 6 hidden layers with 1024 neurons and a single output layer having 194 output neurons (one for each speaker) using the d-vector approach and 224 (194 speakers + 30 phrases) using the j-vector approach.

During the enrolement and evaluation phase, we feed forward one sample at a time into the network to acquire a 1024-dimensional representation, while the bottleneck features were extracted with a 45 dimensional representation. We used the 2nd, 4th and 7th layer (corresponding to the 1st, 3rd, 6th hidden layer) outputs as valid representations in our experiments.

Finally while scoring the output vectors, we first normalize the mean and variances against the before applying cosine distance. PLDA is trained for 20 iterations with a within-covariance smoothing factor[1] of 0.5. It's notable that PLDA here doesn't reduce the vector dimension. Moreover we apply z-norm [28] on the PLDA scores to further enhance the performance.

The GMM-UBM baseline (Table 2) follows a gender-independent approach [29], where 39-dimensional PLP features were used as input. A DNN based VAD was applied to all the features to filter silent segments out. Finally z-norm [28] was utilized on the scores, using 300 impostor utterances.

A GMM based i-vector baseline (Table 2) is also provided, in which *bkg* data is used to train the T matrix and PLDA classifier. 400-dim i-vectors are used.

Table 2. Baselines for RSR2015, in % EER

Method	EER
GMM-UBM	1.10
i-vector	1.39

---

[1] In order to get a good estimate of the within-class covariance, the product of this parameter and between-class covariance is adding to the within-class covariance.

The Comparison of Deep Features. Following google's work [1], we first extracted our features from the last hidden layer (7th layer) and indeed, the result outperforms the GMM-UBM baseline. We denote v_d as the diagonal covariance vector, m as the common d-vector baseline and $m \oplus v_d$ denotes the score fusion of m and v_d.

Table 3. 7th layer results, in % EER

Deep feature	d-vector			j-vector		
	GC	PLDA	CDS	GC	PLDA	CDS
m	0.79	2.90	15.95	0.12	1.15	9.35
v_d	0.81	1.95	9.28	**0.07**	0.71	4.74
$m \oplus v_d$	0.63	1.99	9.62	0.07	0.73	4.51

As we can see (Table 3), cosine distance is largely outperformed by GC and PLDA, hence we decided to exclude cosine distance out of the future experiments. Furthermore we see that the covariance based vector consistently surpasses the traditional mean based method.

Moreover, performance of the full covariance vector v_f inside the j-vector framework was investigated. To make the vector length comparable, bottleneck features of 45 dimensions were extracted from the last hidden layer to get 1035-dim (v_f) according to Eq. 3 (it's also hard to use 1024-dim feature for v_f). We denote D as the extracted feature dimension and v-dim as the dimension of the covariance vector. The comparison between the full and diagonal vectors can be found in Table 4. The full covariance approach (v_f) performs poorly inside the j-vector framework, so further research was discontinued.

Table 4. Full v.s. diagonal covariance vector, 7th layer j-vector, in % EER

Covariance	D	v-dim	GC	PLDA
Full(v_f)	45	1035	2.62	3.90
Diagonal(v_d)	1024	1024	**0.07**	**0.71**

Layer Comparison. The covariance based deep feature is investigated with the different positions of the DNN, i.e. the deep features are extracted from the different hidden layers (the 2nd, 4th and 7th layer in this paper). As we can observe in Table 5, the best results are achieved in layer 4.

Table 5. Layer-wise comparison, in % EER

Deep feature	Layer	d-vector		j-vector	
		GC	PLDA	GC	PLDA
m	2	0.20	0.19	0.15	1.07
	4	0.14	1.18	**0.08**	0.95
	7	0.79	2.90	0.12	1.15
v_d	2	0.10	0.85	0.07	0.64
	4	0.11	0.82	**0.05**	0.55
	7	0.81	1.95	0.07	0.71
$m \oplus v_d$	2	0.13	0.75	0.09	0.59
	4	0.12	0.83	**0.05**	0.56
	7	0.63	1.99	0.06	0.72

Open-Set Condition Analysis. Despite GC being the best result throughout, this is partly due to the test only having a closed set of speakers. Thus, by removing 1/4 enrolement speakers and corresponding test-cases (which means there are 1/4 speakers present in test set are not present in the enrolement set), we simulate real-life conditions to accurately estimate GC's performance (Table 6). We can see that in open set cases, PLDA relatively looses 10% accuracy, while GC looses 540%, compared to closed set cases.

Table 6. 4th layer deep feature results after removing 1/4 enrolement speakers (gender balanced), in % EER

Deep feature	d-vector		j-vector	
	GC	PLDA	GC	PLDA
m	0.61	1.31	0.47	1.16
v_d	0.54	0.95	**0.32**	0.72

Error Analysis. To further figure out where the performance gain comes from, an analysis on error types is given in Table 7. Here, we chose the best result, j-vector of the 4th layer to analyse the error pattern.

In text-dependent tasks there exist three kinds of impostors:

1. The enrolled speaker speaks a wrong utterance
2. An impostor speaks a correct utterance
3. An impostor speaks a wrong utterance

In real applications, error type 3 occurs the most and fortunately is the easiest one to detect. In fact in the test sets defined by RSR2015, this kind of error occupies nearly 90% of all test cases, we completely neglect these test cases because otherwise the EER will be extremely low.

Table 7. Err. distribution of false accepts (4th layer j-vector)

Error type	Deep feature	# Trials	# Err	Err. rate (%)
Speaker	m	996448	13562	1.36
	v_d		8943	0.89
Text	m	552508	1235	2.2
	v_d		123	0.22

From the Table 7, we can observe that the error rate (PLDA) is reduced by relatively 90% for text and 35% for speaker[2], respectively. We clarify the assumption that covariance does relate to text much more closely than mean. Since the improvement on speaker is not as significant, we can infer that this approach should also work on text-independent speaker verification tasks, but the improvement will not be as perceivable.

4.2   Experiments on RedDots Database

The RedDots project was initiated, with collaboration from multiple sites, as a follow-up to a special session during INTERSPEECH 2014 [16]. In this section, experiments on RedDots 2015 Quarter 4, part 1 will be discussed.

Experimental Setup and Baseline. RedDots only provided the the enrolement (1133 utterances) and test data (4726 utterances), thus we used the identical deep feature extractor as in Sect. 4.1, which leads to both channel and context mismatch. After applying VAD, the dataset was truncated, resulting in 1131 enrolement utterances and 4680 test utterances. The truncated test set encompasses 1275424 tests including 3850 true speaker and 1271574 impostor tests.

The baseline (Table 8) was run using the same configuration as in the corresponding RSR experiments, except that for i-vector, we use GC instead of PLDA since PLDA gives much worse performance than the GMM-UBM baseline. (*Bkg* data is borrowed from RSR2015 database, Sect. 4.1, leading to both channel and context mismatch.) Following the same metric with experiments on RSR2015, we didn't separate three types of error explicitly and only compute the overall EER.

---

[2] Speaker errors happen when an impostor speaker utters the correct text, is accepted, while text errors happen when an enrolled speaker utters the wrong text is accepted.

Table 8. Baselines for RedDots

Method	EER (%)
GMM-UBM	2.45
i-vector	3.30

The pattern of different layers on RedDots is the same as that on RSR2015, thus only the best results achieved in the 4th layer are presented in the following sections.

Results and Analysis. As can be perceived in Table 9, the d/j-vector(using GC) does not show commensurate performance on RedDots, which is a totally mismatched corpus (Besides the channel and context mismatch between RSR2015 *bkg* and RedDots data, RedDots also contains channel mismatch between enrolement and test data). However, it can still be observed that the newly proposed v-vector within our previous j-vector framework significantly outperforms the traditional mean based vector, which demonstrates the superiority and robustness of the new method and the best system is also slightly better than the baseline, even in a totally mismatched scenario.

Table 9. 4th layer results on RedDots, in % EER

Deep feature	d-vector	j-vector
m	6.91	6.03
v_d	5.04	**2.36**

We also give an analysis on false accept error distribution on RedDots, as can be observed in Table 10, error rate is reduced by more than 60% on text and 27% on speaker, which agrees with the observation on RSR2015 in Sect. 4.1. Another interesting observation, which also exists in the RSR2015 experiments, is that covariance based deep feature works better for j-vector, which is consistent with that j-vector framework takes more text information into consideration.

Table 10. Err. distribution of false accepts (4th layer j-vector)

Error type	Deep feature	# Trials	# Err	Err. rate (%)
Speaker	m	123703	16941	13.7
	v_d		12446	10.06
Text	m	34658	2506	7.23
	v_d		997	2.87

Additionally, the utterance lengths in RedDots vary from 75 to 375 frames(after VAD), most of them are shorter than those in RSR2015. To further explore the impact of the utterance length, an analysis of the false reject error rate is given in Fig. 2. We can conclude that errors mainly happen when the utterances are rather short (less than 200 frames). Moreover, v-vector performs better than i-vector on utterances longer than 150 frames.

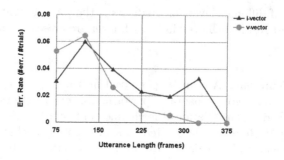

Fig. 2. False reject Err. rate w.r.t Utt length

Short utterances contain less text information and covariance matrices can not be estimated accurately with only a few frames, which explains why errors are more tending to occur when the utterances are shorter than 200 frames.

5   Conclusion

We proposed two kinds of covariance based approaches for deep feature extraction. While the diagonal covariance vector beats the mean based deep feature on both the RSR2015 and RedDots *text-dependent* speaker verification tasks, the full covariance vector is not yet applicable and needs some further research. We show that covariance based deep features are more capable of capturing text variability and perform better when incorporated into the joint multi-task learning framework. However, this approach's performance degrades when it comes to utterances shorter than 200 frames. For such short utterances, we will try to use some adaptation techniques rather than directly estimate the representation in the future work. Furthermore, other fusion techniques can be investigated.

References

1. Chen, K., Salman, A.: Learning speaker-specific characteristics with a deep neural architecture. IEEE Trans. Neural Netw. **22**(11), 1744–1756 (2011)
2. Heigold, G., Moreno, I., Bengio, S., Shazeer, N.: End-to-end text-dependent speaker verification. arXiv preprint arXiv:1509.08062 (2015)
3. Chen, Y.-H., Lopez-Moreno, I., Sainath, T.N., Visontai, M., Alvarez, R., Parada, C.: Locally-connected and convolutional neural networks for small footprint speaker recognition. In: INTERSPEECH (2015)

4. Lei, Y., Ferrer, L., McLaren, M., et al.: A novel scheme for speaker recognition using a phonetically-aware deep neural network. In: IEEE International Conference on Acoustics, Speech and Signal Processing (ICASSP), pp. 1695–1699. IEEE (2014)
5. Liu, Y., Qian, Y., Chen, N., Fu, T., Zhang, Y., Yu, K.: Deep feature for text-dependent speaker verification. Speech Commun. **73**, 1–13 (2015)
6. Hinton, G.E., Salakhutdinov, R.R.: Reducing the dimensionality of data with neural networks. Science **313**(5786), 504–507 (2006)
7. Hinton, G.E., Osindero, S., Teh, Y.-W.: A fast learning algorithm for deep belief nets. Neural Comput. **18**(7), 1527–1554 (2006)
8. Yu, D., Seltzer, M.L.: Improved bottleneck features using pretrained deep neural networks. In: INTERSPEECH, vol. 237, p. 240 (2011)
9. Grézl, F., Karafiát, M., Kontár, S., Cernocky, J.: Probabilistic and bottle-neck features for lvcsr of meetings. In: IEEE International Conference on Acoustics, Speech and Signal Processing (ICASSP), vol. 4, pp. IV–757. IEEE (2007)
10. Matejka, P., et al.: Neural network bottleneck features for language identification. In: Proceedings of IEEE Odyssey, pp. 299–304 (2014)
11. Fu, T., Qian, Y., Liu, Y., Yu, K.: Tandem deep features for text-dependent speaker verification. In: INTERSPEECH, pp. 1327–1331 (2014)
12. Richardson, F., Reynolds, D., Dehak, N.: Deep neural network approaches to speaker and language recognition. IEEE Sig. Process. Lett. **22**(10), 1671–1675 (2015)
13. Variani, E., Lei, X., McDermott, E., Lopez Moreno, I., Gonzalez-Dominguez, J.: Deep neural networks for small footprint text-dependent speaker verification. In: IEEE International Conference on Acoustics, Speech and Signal Processing (ICASSP), pp. 4052–4056. IEEE (2014)
14. Chen, N., Qian, Y., Yu, K.: Multi-task learning for text-dependent speaker verification. In: INTERSPEECH (2015)
15. Larcher, A., Lee, K.A., Ma, B., Li, H.: Text-dependent speaker verification: classifiers, databases and RSR2015. Speech Commun. **60**, 56–77 (2014)
16. Lee, K.A., et al.: The RedDots data collection for speaker recognition. In: INTERSPEECH (2015)
17. Hain, T., Johnson, S., Tuerk, A., Woodland, P., Young, S.: Segment generation and clustering in the HTK broadcast news transcription system. In: Proceedings of 1998 DARPA Broadcast News Transcription and Understanding Workshop, pp. 133–137 (1998)
18. De Leon, P.L., Pucher, M., Yamagishi, J., Hernaez, I., Saratxaga, I.: Evaluation of speaker verification security and detection of hmm-based synthetic speech. IEEE Trans. Audio Speech Lang. Process. **20**(8), 2280–2290 (2012)
19. Chen, L.-W., Guo, W., Dai, L.-R.: Speaker verification against synthetic speech. In: 7th International Symposium on Chinese Spoken Language Processing (ISCSLP), pp. 309–312. IEEE (2010)
20. Tuzel, O., Porikli, F., Meer, P.: Human detection via classification on Riemannian manifolds. In: IEEE Conference on Computer Vision and Pattern Recognition, CVPR 2007, pp. 1–8. IEEE (2007)
21. Yao, J., Odobez, J.-M.: Fast human detection from videos using covariance features. Technical report, Idiap (2007)
22. Dehak, N., Kenny, P., Dehak, R., Dumouchel, P., Ouellet, P.: Front-end factor analysis for speaker verification. IEEE Trans. Audio Speech Lang. Process. **19**(4), 788–798 (2011)
23. Kenny, P., Boulianne, G., Dumouchel, P.: Eigenvoice modeling with sparse training data. IEEE Trans. Speech Audio Process. **13**(3), 345–354 (2005)

24. Kenny, P.: A small footprint i-vector extractor. In: Odyssey, pp. 1–6 (2012)
25. Prince, S.J., Elder, J.H.: Probabilistic linear discriminant analysis for inferences about identity. In: IEEE 11th International Conference on Computer Vision, ICCV 2007, pp. 1–8. IEEE (2007)
26. Kenny, P., Stafylakis, T., Ouellet, P., Alam, M.J., Dumouchel, P.: PLDA for speaker verification with utterances of arbitrary duration. In: IEEE International Conference on Acoustics, Speech and Signal Processing (ICASSP), pp. 7649–7653. IEEE (2013)
27. Matějka, P., et al.: Full-covariance UBM and heavy-tailed PLDA in i-vector speaker verification. In: IEEE International Conference on Acoustics, Speech and Signal Processing (ICASSP), pp. 4828–4831. IEEE (2011)
28. Auckenthaler, R., Carey, M., Lloyd-Thomas, H.: Score normalization for text-independent speaker verification systems. Digit. Sig. Process. **10**(1), 42–54 (2000)
29. Reynolds, D.A., Quatieri, T.F., Dunn, R.B.: Speaker verification using adapted gaussian mixture models. Digit. Sig. Process. **10**(1), 19–41 (2000)

Radar HRRP Target Recognition with Recurrent Convolutional Neural Networks

Mengqi Shen[✉] and Bo Chen

National Laboratory of Radar Signal Processing, Xidian University, Xian, China
mengqishen_xidian@163.com, bchen@mail.xidian.edu.cn

Abstract. Conventional radar automatic target recognition (RATR) methods using High-Resolution Range Profile (HRRP) sequences require carefully designed feature extraction techniques and plenty of HRRP waveforms, which result in insufficient recognition rate and limit in real-time recognition. To address these issues a modified end-to-end architecture consisting of a convolutional neural network (CNN) followed by a recurrent neural network (RNN) is proposed. In this model the local features of HRRPs extracted by a CNN are passed to a RNN, which avoids manual feature extraction and takes advantage of its shared parameters mechanism which enables single HRRP recognition in real-time. The effectiveness of this model is shown in this paper with numerical results.

Keywords: Radar automatic target recognition (RATR)
High-resolution range profile (HRRP)
Convolutional neural network (CNN)
Recurrent neural network (RNN)

1 Introduction

A high-resolution range profile (HRRP) denotes the coherent summation of projection vectors of complex echoes from target scatters along the radar line-of-sight (LOS). It is a strong function of the target-radar aspect angle and contains abundant informative target structure signatures, e.g. target size, scatters distribution, etc. Compared to a SAR or ISAR image which requires complex preprocessing procedure and substantial amount of calculation [4,17,24,25], an HRRP is easy to obtain, store and process. In recent years, high-resolution radar automatic target recognition (ATR) has received considerable attention.

Several approaches have been proposed to achieve HRR-ATR [4,14,19,20,22, 24,34]. These approaches can be roughly divided into two categories according to the treatment of HRRP signatures. One uses single-look HRRP data and the other utilizes multi-look HRRP data. Lots of research [30] indicated that the recognition rate from multi-look HRRP can be higher than that from single-look HRRP because the utilization of multiple of HRRPs brings more information about target, which enlightens us to make use of the multi angle information.

© Springer Nature Switzerland AG 2018
Y. Peng et al. (Eds.): IScIDE 2018, LNCS 11266, pp. 243–251, 2018.
https://doi.org/10.1007/978-3-030-02698-1_21

Previous works about multi-look HRR-ATR mainly include two parts: feature extraction and classifier design [12, 23, 27]. For example, [9, 16, 26, 33] employed a feature set consisting of the (location, amplitude) pairs of fifteen principal wave-fronts selected from the single-look HRR signature by using the RELAX algorithm and fed them to a Hidden Markov Model (HMM) to deal with sequential information. Nevertheless, in these approaches the feature extraction and classifier design are two irrelevant parts which prevent the model automatically from learning the internal representations of the target. Both parts require carefully design and are hard to optimize.

But the development of deep learning has equipped us with different kinds of deep neural networks that are usually designed aiming at specific tasks with an end-to-end manner, which means that the model is trusted to learn the transformations between input and output directly from data. The end-to-end model learns mathematically optimal representations of data without interference by human's knowledge and has achieved state-of-the-art results in plenty of areas. For example, [2] used an end-to-end model to train self-driving cars, [29] provided another novel end-to-end network to generate image captions. In this paper, we utilize an end-to-end architecture consisting of a convolutional neural network (CNN) followed by a recurrent neural network (RNN), which avoids manual feature extraction and enables sequence learning.

It has been convincingly shown that CNNs function well when facing local feature extraction problems due to their shift-invariant attributes [18, 28]. As a result they are fit for HRRP recognition because HRRP signatures in the same category share local similarities between sequences. As for RNNs, it passes the extracted sequential features across sequence steps and uses iterative function loops to store temporal information, which enables the learning of time dependencies on multiple scales. Hence we choose CNNs to extract features of single HRRP and then use RNNs to acquire sequential information. In practice this design can be rather useful for that the number of radar echoes reflected from targets of interest are usually uncertain, especially for non-cooperative targets.

There are five sections in the remainder of the paper. Section 2 focuses on a briefly description about CNNs and RNNs. Then we apply the recurrent CNN architecture for HRRP target recognition in Sect. 3. The detailed experiments conducted with the proposed model are provided in Sect. 4. Conclusions come at the end of the paper in Sect. 5.

2   Preliminaries

In this section we will generally review the concepts of convolutional neural network (CNN) and recurrent neural network (RNN).

Convolutional Neural Networks. A CNN is a special case of feedforward neural networks. It consists of one or more convolutional layers, often with a down-sampling layer, which are followed by some fully connected layers in the standard neural network. Each feature map of a convolutional layer receives inputs from a set of features located in a small neighborhood in the previous layer known as local receptive fields.

The convolution operation and weight sharing mechanisms solve "the curse of dimensionality" problem as they reduce the number of parameters, allowing the network to be deeper with fewer parameters. Thus, with deeper layers and different filters CNNs are able to build low level features up to more abstract concepts through a series of convolutional layers.

Recurrent Neural Networks. While some recognition problems can be easily solved by CNNs, there are plenty of temporal data like frames from video, words from sentences which need sequential model to capture their time dependency. Recurrent neural networks utilize the hidden states to pass information across sequence steps, processing sequential data one element at a time. Therefore RNNs can map the entire "history" of previous inputs to each output and take advantage of hidden states to preserve sequential information [21].

A basic RNN is shown in Fig. 1. x denotes an input sequence where each data point x_i is a real-valued vector at time step i. h_i is called a hidden state at time step i which is a non-linear function of the input at the same time step x_i and the hidden state of the previous time step h_{i-1}:

$$h_i = \Phi(Wh_{i-1} + Ux_i). \tag{1}$$

where Φ denotes a non-linear function, and W and U are two different weights matrices. o_i is the output at time step i, which can be calculated by:

$$o_i = \Theta(Vh_i). \tag{2}$$

where Θ is another non-linear mapping function and V is the output weight matrix.

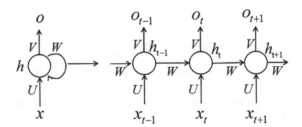

Fig. 1. A basic RNN model in both folded (*left*) and unfolded (*right*) ways

3 Recurrent Convolutional Neural Network for Radar HRRP Target Recognition

Conventional approaches used for radar HRRP target recognition usually ignored the sequential information and depended on carefully designed feature extraction methods. In order to resolve these issues, we apply the Recurrent CNN model in the end-to-end fashion for this recognition problem, which can not only automatically learn the mathematically optimal representation of HRRP data but also make use of multi-aspect HRRP signatures. The architecture of this model is illustrated in Fig. 2.

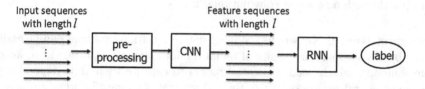

Fig. 2. Our proposed model for HRRP target recognition

3.1 Preprocessing

Before using a CNN for feature extraction, preprocessing procedure for HRRP is conducted due to its complex characteristics about scattering motion through range cells (MTRC) mentioned in [31]. First l_2 normalization is used to normalize the amplitude scale of HRRPs. Then we adopt centroid alignment [1] as a time-shift compensation technique to avoid the relative position shift in HRRP. Figure 3 illustrates examples of preprocessed HRRP data.

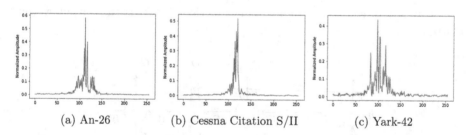

| (a) An-26 | (b) Cessna Citation S/II | (c) Yark-42 |

Fig. 3. Examples of preprocessed HRRP data in each categories. (a) An-26. (b) Cessna Citation S/II. (c) Yark-42.

3.2 Recurrent CNN Architecture

After preprocessing the normalized and aligned HRRPs are acquired. Let $X_{training} = \{x_1, x_2, ..., x_m\}$ be the training dataset, where x_i is the i-th HRRP data in the training sequence, m is the number of training samples. First we choose a fixed number l, which denotes the number of sequences used in training. l should neither be too large nor too small. Specifically, if l is too large it's difficult for a radar to keep tracking a non-cooperate target so the designed model is not practical. On the contrary if l is too small the model may not use multi-aspect information. At the beginning we simply select $l = 10$ and will discuss the influence of l later in the next section.

During training, a series of HRRP sequences $\{x_1, x_2, ..., x_l\}$ in the same category are input into a multi-layer CNN to obtain feature sequences $\{f_1, f_2, ..., f_l\}$, which are fed in a followed RNN to predict a label for these successive HRRP signatures. Equations (3)–(5) demonstrate the mathematical details.

$$f_i = H(x_i). \tag{3}$$

$$s_i = \Phi(W f_i + U s_{i-1}). \tag{4}$$

$$label = softmax(\Theta(V s_l)). \tag{5}$$

The non-linear transformation $H(\cdot)$ in CNN can be a composite function of operations such as Batch Normalization (BN) [15], rectified linear units (ReLU) [18], Convolution (Conv) or Pooling [18]. f_i is the feature vector of each HRRP sequence corresponding to x_i. s_i indicates the hidden state at each time step i. Only the last hidden state s_l is used to get a label for these input sequences.

As for testing we apply the well-trained model to recognize single HRRP data in order to achieve real-time recognition. Experiment results are listed in the next section.

4   Experimental Results and Analysis

4.1   Data Description

The results presented in this section are based on measured HRRP data from three real airplanes, which are extensively used in [3,5–8,10,11,31,32]. The basic parameters of the targets and radar are shown in Table 1 and the projections of target trajectories onto the ground plane are shown in Fig. 4.

Table 1. Parameters of planes and radar in the ISAR experiment

Radar parameters	Center frequency	5520 MHz	
	Bandwidth	400 MHz	
Aircraft	Length (m)	Width (m)	Height (m)
Yark-42	36.38	34.88	9.83
Cessna Citation S/II	14.40	15.90	4.57
An-26	23.80	29.20	9.83

When partitioning training datasets and testing datasets there are two things should be taken into account. (1) the training dataset is supposed to cover all possible targets aspect theoretically in order to obtain a well-trained model. (2) the elevation angles of targets in the testing dataset are different from those in the training dataset. Hence we choose the second and the fifth segments of Yark-42, the sixth and the seventh segments of Cessna Citation S/II and the fifth and the sixth segments of An-26 as training samples, and take other data segments as testing samples. All of them are successive in the same segment. The whole training dataset has 140000 HRRP samples and the testing dataset has 7800 HRRP samples in three categories. Each example has 256 range cells, i.e. is a 256-dimensional vector.

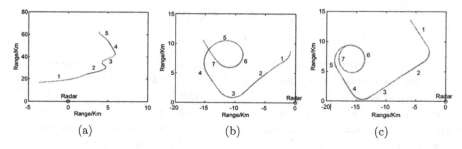

Fig. 4. Projections of target trajectories onto the ground plane. (a) Yak-42. (b) Cessna Citation S/II. (c) An-26.

4.2   Architectures and Performance

To evaluate the proposed model we have conducted a series of experiments on the dataset described above. In our model, a three-layer CNN with the same size of filters and a RNN with one hidden layer of 16-dimensional LSTM-cells are used [13]. Attempts to optimize the parameters of the model, such as the number of layers and the size of filters in CNN, the dimension of hidden state in LSTM and so on, have been made but these parameters have little effect on the final results. For the output layer the *softmax* activation function which is standard for 1 out of K classification tasks is utilized. The input sequence has a fixed length l. The recognition rates with different l are listed below in Table 2. In order to prove the model's ability and to compare to other methods which also used HRRP sequences to do recognition task. We run some experiments according to [12], whose basic idea is the combination of a feature extraction algorithm and a HMM classifier. The recognition results with the same CNN in our model to extract features and a HMM classifier using l HRRP data is also presented in Table 2. From Table 2 we can tell that recognition rates become better when l increases, which is apparent because the model can use more angular information.

It is necessary to point out that the HMM based model can achieve better result when dealing with Cessna but behave poor in An-26 so the average

Table 2. Summary of obtained results (recognition rate) with CNN-RNN model and CNN-HMM model

Model	$l = 3$	$l = 5$	$l = 8$	$l = 10$	$l = 20$	$l = 30$
CNN + LSTM-RNN	0.925	0.925	0.943	0.947	0.947	0.952
CNN + HMM	0.824	0.845	0.877	0.889	0.890	0.892

recognition rate is far behind the proposed model's. Meanwhile, [23,26,27] all have mentioned a vital procedure, down-sampling, to avoid mutual correlation problem, which requires carefully choosing of sampling rate and weakens the real-time recognition ability. In practice if more than one HRRP signatures can be obtained, the model is also able to utilize the multi-angle information due to its shared parameters. The results in Table 2 show that the recognition accuracy is remarkably improved if successive HRRP sequences are used in the proposed model.

5 Conclusions and Future Work

In this paper, we employ a compact end-to-end Recurrent CNN model for multi-aspect HRRP target recognition task. The model makes use of sequential information on extracted feature of HRRP waveforms and achieves better performance than other sequence model. The contrast experiments using HMM model on the same dataset are also provided. However, the RNN-based model requires same intervals between HRRP sequences and may become infeasible if this condition cannot be guaranteed. In the future we will try to loosen the "time interval" requirement and improve the model's practicability.

References

1. Chen, B., Liu, H.W., Bao, Z.: Analysis of three kinds of classification based on different absolute alignment methods. Mod. Radar **28**(3), 58–62 (2006)
2. Bojarski, M., et al.: End to end learning for self-driving cars (2016)
3. Chen, B., Liu, H., Chai, J., Bao, Z.: Large margin feature weighting method via linear programming. IEEE Trans. Knowl. Data Eng. **21**(10), 1475–1488 (2009)
4. Chiang, H.C., Moses, R.L., Potter, L.C.: Model-based classification of radar images. IEEE Trans. Inf. Theory **46**(5), 1842–1854 (2000)
5. Du, L., Liu, H., Bao, Z., Zhang, J.: Radar automatic target recognition using complex high-resolution range profiles. IET Radar Sonar Navig. **1**(1), 18–26 (2007)
6. Du, L., Liu, H., Bao, Z.: Radar HRRP statistical recognition: parametric model and model selection. IEEE Trans. Signal Process. **56**(5), 1931–1944 (2008)
7. Du, L., Liu, H., Bao, Z., Xing, M.: Radar HRRP target recognition based on higher order spectra. IEEE Trans. Signal Process. **53**(7), 2359–2368 (2005)
8. Du, L., Liu, H., Wang, P., Feng, B., Pan, M., Bao, Z.: Noise robust radar HRRP target recognition based on multitask factor analysis with small training data size. IEEE Trans. Signal Process. **60**(7), 3546–3559 (2012)

9. Du, L., Wang, P., Liu, H., Pan, M., Bao, Z.: Radar HRRP target recognition based on dynamic multi-task hidden markov model. In: Radar Conference, pp. 253–255 (2011)

10. Du, L., Wang, P., Liu, H., Pan, M., Chen, F., Bao, Z.: Bayesian spatiotemporal multitask learning for radar HRRP target recognition. IEEE Trans. Signal Process. **59**(7), 3182–3196 (2011)

11. Feng, B., Du, L., Liu, H.W., Li, F.: Radar HRRP target recognition based on K-SVD algorithm. In: IEEE CIE International Conference on Radar, pp. 642–645 (2012)

12. Fielding, K.H.: Spatiotemporal pattern recognition using hidden markov models. IEEE Trans. Aerosp. Electron. Syst. **31**(4), 1292–1300 (1995)

13. Graves, A.: Long short-term memory. In: Graves, A. (ed.) Supervised Sequence Labelling with Recurrent Neural Networks. SCI, vol. 385, pp. 37–45. Springer, Berlin (2012). https://doi.org/10.1007/978-3-642-24797-2_4

14. Hudson, S., Psaltis, D.: Correlation filters for aircraft identification from radar range profiles. IEEE Trans. Aerosp. Electron. Syst. **29**(3), 741–748 (2002)

15. Ioffe, S., Szegedy, C.: Batch normalization: accelerating deep network training by reducing internal covariate shift, pp. 448–456 (2015)

16. Ji, S., Liao, X., Carin, L.: Adaptive multiaspect target classification and detection with hidden markov models. IEEE Sens. J. **5**(5), 1035–1042 (2005)

17. Jones, G., Bhanu, B.: Recognizing occluded objects in SAR images. IEEE Trans. Aerosp. Electron. Syst. **37**(1), 316–328 (2001)

18. Krizhevsky, A., Sutskever, I, Hinton, G.E.: ImageNet classification with deep convolutional neural networks. In: International Conference on Neural Information Processing Systems, pp. 1097–1105 (2012)

19. Li, H.J., Yang, S.H.: Using range profiles as feature vectors to identify aerospace objects. IEEE Trans. Antennas Propag. **41**(3), 261–268 (1993)

20. Liao, X., Bao, Z., Xing, M.: On the aspect sensitivity of high resolution range profiles and its reduction methods. In: The Record of the IEEE 2000 International Radar Conference, pp. 310–315 (2000)

21. Mikolov, T., Karafit, M., Burget, L., Cernocky, J., Khudanpur, S.: Recurrent neural network based language model. In: INTERSPEECH 2010, Conference of the International Speech Communication Association, Makuhari, Chiba, Japan, September, pp. 1045–1048 (2010)

22. Mitchell, R.A., Westerkamp, J.J.: Robust statistical feature based aircraft identification. IEEE Trans. Aerosp. Electron. Syst. **35**(3), 1077–1094 (1999)

23. Nilubol, C., Pham, Q.H., Mersereau, R.M., Smith, M.J.T., Clements, M.A.: Hidden markov modelling for SAR automatic target recognition. In: IEEE International Conference on Acoustics, Speech and Signal Processing, vol. 2, pp. 1061–1064 (1998)

24. Novak, L.M.: State-of-the-art of SAR automatic target recognition. In: The Record of the IEEE 2000 International Radar Conference, pp. 836–843 (2000)

25. O'Sullivan, J.A., Devore, M.D., Kedia, V., Miller, M.I.: SAR ATR performance using a conditionally Gaussian model. IEEE Trans. Aerosp. Electron. Syst. **37**(1), 91–108 (2001)

26. Pei, B., Bao, Z.: Multi-aspect radar target recognition method based on scattering centers and HMMs classifiers. Acta Electronica Sinica **41**(3), 1067–1074 (2003)

27. Runkle, P., Nguyen, L.H., Mcclellan, J.H., Carin, L.: Multi-aspect target detection for SAR imagery using hidden markov models. IEEE Trans. Geosci. Remote. Sens. **39**(1), 46–55 (2001)

28. Szegedy, C., et al. Going deeper with convolutions. In: IEEE Conference on Computer Vision and Pattern Recognition, pp. 1–9 (2015)
29. Vinyals, O., Toshev, A, Bengio, S., Erhan, D.: Show and tell: a neural image caption generator, pp. 3156–3164 (2014)
30. Williams, R., Westerkamp, J., Gross, D., Palomino, A.: Automatic target recognition of time critical moving targets using 1D high range resolution (HRR) radar. IEEE Aerosp. Electron. Syst. Mag. 15(4), 37–43 (2000)
31. Xing, M., Bao, Z., Pei, B.: Properties of high-resolution range profiles. Opt. Eng. 41(2), 493–504 (2002)
32. Da Zhang, X., Shi, Y., Bao, Z.: A new feature vector using selected bispectra for signal classification with application in radar target recognition. IEEE Trans. Signal Process. 49(9), 1875–1885 (2001)
33. Zhu, F., Da Zhang, X., Hu, Y.F., Xie, D.: Nonstationary hidden Markov models for multiaspect discriminative feature extraction from radar targets. IEEE Trans. Signal Process. 55(5), 2203–2214 (2007)
34. Zyweck, A., Bogner, R.E.: Radar target classification of commercial aircraft. IEEE Trans. Aerosp. Electron. Syst. 32(2), 598–606 (1996)

DiffusionNet: Establish Convolutional Networks with Nitric Oxide Diffusion Model

Kai Gao, Hui Shen, Jianpo Su, and Dewen Hu[(✉)]

College of Artificial Intelligence, National University of Defense Technology,
Changsha, Hunan, China
dwhu@nudt.edu.cn

Abstract. Skip connections are used in DenseNets recently and have significantly improved network performance. In this paper, we compare skip connections with the diffusion process of endogenous Nitric Oxide (NO) between neurons, and propose DiffusionNets by replacing skip connections with NO diffusion model. Each layer is considered as a point spreading signal to space as well as receiving signal from space. The whole network transmits information with a diffusing way. DiffusionNets have several advantages: (1) generate more discriminative features. (2) more similar to neural information transmission. (3) higher classification accuracy. DiffusionNets were evaluated on CIFAR10 and CIFAR100 and outperform the original DenseNets.

Keywords: Convolutional neural network · Skip connection
NO diffusion model · DiffusionNet

1 Introduction

Convolutional networks have become the most important method in computer vision since AlexNet won the ImageNet title in 2012. Many variants such as VGGNet and InceptionNet are proposed to improve performance of convolutional networks [1]. However, it is still hard to train deeper networks considering the vanishing gradient problem. Using skip connections is an effective solution to the problem. Highway networks [2] proposed the conception of skip connections initially and successfully trained a network with depth of 150+. Following Highway network, ResNets [3, 14] optimized the network structure and increased the network depth up to 1000+. Further, DenseNets [4] proposed adding skip connections between any two layers and using the concatenation of all the features from front layers instead of the summation method in ResNets. But in DenseNets, skip connections are set with same weights ignoring the distance between layers.

It is well known that the generation of neural networks is closely related to biological neurotransmission. But there are many other kinds of information transmission methods in human body, such as Nitric Oxide (NO for short) diffusion transmission [6]. NO is a kind of non-locally diffused neurotransmitters [5, 11]. Unlike general neurotransmitters, its signaling is not limited to synapses. Considering the differences between skip connections and other kinds of connections, we try to compare skip connections with NO diffusion transmission. Surprisingly, they have similar properties:

© Springer Nature Switzerland AG 2018
Y. Peng et al. (Eds.): IScIDE 2018, LNCS 11266, pp. 252–261, 2018.
https://doi.org/10.1007/978-3-030-02698-1_22

(1) NO can transmit information to any neuron around. Similarly, skip connections can pass features to any back layer; (2) there is no material conversion during the process of NO diffusion transmission. And for skip connections, the features don't change either. But a difference is also noticed: NO diffusion transmission is a kind of attenuated transmission, which means that the concentration decreases as the distance between target and source increases. While in DenseNets, features are passed with same weights between layers through skip connections no matter how far the layers are.

In this paper, we study the relationship between skip connections and NO diffusion transmission and propose a novel network named DiffusionNet by introducing the diffusion mechanism of NO into DenseNets. The whole DenseNet can be separated into two parts: skip connections and the other connections. And we believe that skip connections play the role of NO diffusion transmission in deep neural networks. The main difference between DiffusionNets and DenseNets is that the weights of skip connections are calculated by NO diffusion model.

In traditional neural networks, space location doesn't make sense. But NO diffusion mechanism is related to spatial locations, so it is necessary to define the spatial position and size of the layers. Simply, each layer is considered as a diffusion source without volume, and the distance between layers can be represented by Euler distance. To simulate the diffusion process of NO, a signal transmitter and a signal receiver are added to each layer. The transmitter spreads signal to the entire external space while the receiver receives signal from other layers by perceiving the strength of information of the corresponding layer. To ensure the correct flow direction of information, we specify that the receivers can only receive signals from the front layers. Figure 1 shows the structural differences of DiffusionNet and DenseNet.

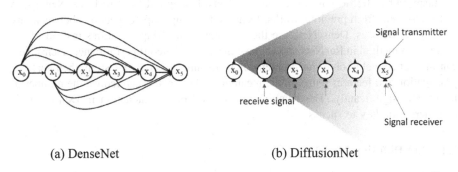

(a) DenseNet   (b) DiffusionNet

Fig. 1. Comparison of DenseNet and DiffusionNet structures. (a) DenseNet with 6 layers. (b) DiffusionNet with 6 layers. Each layer is equipped with a signal transmitter and signal receiver which can transmit information between layers. In (b), we show the information transmitting process of the first layer. Signal transmitter of first layer transmit signal to the space, and the receivers of the back layers receive signals of the first layer. It should be noted that receivers can only receive signals from front layers.

With experiments on CIFAR10 and CIFAR100 [13], DiffusionNets achieve better results than DenseNets. Two other diffusion models were also tested to verify the effects of different diffusion functions. Besides, a parameter efficient variant of DiffusionNet is also proposed to reduce the number of parameters.

The contributions of our article are mainly summarized as follows:

1. We compare skip connections with NO diffusion transmission and provide a new perspective to understand skip connection.
2. We propose DiffusionNet by introducing NO diffusion mechanism and improve the performance of network.
3. Another information transmission method is used to model artificial neural networks which makes the networks more similar to biological neural networks.

2   Related Work

2.1   Skip Connection

With the development of neural networks, the networks become deeper. Although deeper networks can improve the accuracy of classification, they are hard to train considering the vanishing gradient problem [1]. Many new techniques are proposed to deal with the problem, and skip connection is one of the most effective methods. Skip connections were originally proposed in Highway networks [2], which solve the problem that with the deepening of networks, gradient information is hard to pass back.

ResNets [3] were proposed following Highway Network in 2016. The structure of ResNet is quite similar to Highway Network. ResNets solve the same problem as Highway networks do with better performance. With the success of ResNets, many variants such as Wide ResNets and ResNet in ResNet appear [1]. These networks improved the structure of ResNets and achieve better performance.

Inspired by Highway networks and ResNets, Huang put forward DenseNets [4] in 2017. Different from previous works, DenseNets set up skip connections between any two layers. Besides, DenseNets use the concatenation of input features instead of the summation method in ResNets and achieve the state-of-the-art performance. DenseNets can also effectively solve the problem of vanishing gradients by strengthen features propagation and features reuse. In the meanwhile, DenseNets can reduce the number of parameters significantly. The combination method of concatenation in DenseNets is considered the key to success.

2.2   NO Diffusion Model

It has been confirmed that endogenous NO can serve as a signal transmission carrier in human body and it plays a very crucial role in information transmission process [7]. The discovery of NO breaks many traditional opinions in neural signal processing area. Traditional neurotransmission can only deliver signal from the presynaptic space to the postsynaptic space. But the diffusing characteristics of NO help to transmit information between any two neurons whether they are connected or not. Plenty of excellent researches about information transmission processing model and neural computing powers of NO are published. Krekelberg and Taylor have deeply studied the application of NO's diffusivity in neural computation [7]. Philippides combined NO diffusion principle with evolutionary algorithm and successfully established NO diffusion model for target recognition of robot [8]. Chen combined the NO diffusion model with Self-Organization Mapping (SOM), and proposed DGSOM algorithm, which has greatly improved the performance of the original SOM [9, 10, 12]. All these research results demonstrate that the NO diffusion model has a great potential for neural computing, which also provide support for our work.

3   DiffusionNet

3.1   ResNet and DenseNet

Assuming the outputs of layers to be $x_0, x_1, \ldots, x_l$, where x_0 represents the output of the initial convolutional layer with the input image x. We define $H_l(\cdot)$ to be a composite function of operations such as Batch Normalization (BN), rectified linear units (ReLU), Convolution (Conv) or Pooling.

ResNets. Traditional convolutional networks only use the output of the upper layer as the input of the next layer, which produces the transfer function $x_l = H_l(x_{l-1})$. ResNets add a skip connection from the input layer to the output, and the output of the lth layer is:

$$x_l = H_l(x_{l-1}) + x_{l-1} \tag{1}$$

DenseNets. In DesneNets, the input of the lth layer is the concatenations of the outputs of all front layers, so the output of the lth layer is:

$$x_l = H_l([x_0, x_1, \ldots, x_{l-1}]) \tag{2}$$

Where $[x_0, x_1, \ldots, x_{l-1}]$ refers to the concatenation of the front $l - 1$ layers.

3.2   NO Diffusion Model

Point source diffusion is commonly used to model the process of NO diffusion:

$$f_{(r,t)} = \frac{S_{t=0}}{8(\pi Dt)^{\frac{3}{2}}} \exp\left(\frac{-r^2}{4Dt} - \lambda t\right) \tag{3}$$

where $f_{(r,t)}$ is the density at time t, and r refers to the distance to the signal source. $S_{t=0}$ indicates the initial density of NO. D is the diffusion coefficient and λ is the delay speed related to half-period of NO. For simplicity, we set t to 1, λ to 0 and make $S_{t=0} = 8(\pi Dt)^{\frac{3}{2}}$, then Eq. 3 becomes a function only related to distance and can be written as:

$$f_{(r)} = \exp\left(\frac{-r^2}{4D}\right) \tag{4}$$

To further simplify, we approximate the NO point source diffusion model with commonly used functions, such as linear model:

$$f_{(r)} = 1 - \alpha r \tag{5}$$

and exponential model:

$$f_{(r)} = \beta^r \tag{6}$$

where α and β are hyper parameters to adjust the decay rate.

3.3 DiffusionNet

To represent the concept of distance in deep neural networks, we represent the distance between the ith layer and the jth layer with their Euler distance $r_{i,j}$. Moreover, we introduce the concept of signal strength to measure how much information is transmitted between layers. The signal strength between layer i and layer j can be represented as $f_{i,j}$. Then the output of lth layer can be written as:

$$x_l = H_l\left(\left[f_{l,0} \cdot x_0, f_{l,1} \cdot x_1, \ldots, f_{l,l-1} \cdot x_{l-1}\right]\right) \tag{7}$$

where x_i represents the output of the ith layer. Following ResNets and DenseNets, function $H_l(\cdot)$ includes three operations: batch normalization (BN), ReLU and a 3 * 3 Conv (2 * 2 pooling for transition layer). We use L to represent the depth of network and k to represent the number of output channels, which is also known as growth rate. Following the diffusion model of NO, we propose a new network named DiffusionNet using NO diffusion model. In DiffusionNets, different networks can be established simply by setting different diffusion function f. Actually, DiffusionNet and DenseNet are exactly the same with diffusion function $f = 1$. In order to verify the influence of different diffusion functions on network performance, two variants of DiffusionNet were designed by setting diffusion functions with linear and exponential functions. Furthermore, a parameter efficiency DiffusionNet is proposed to reduce parameter size. Figure 2 displays the differences of the networks.

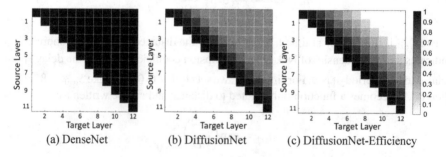

(a) DenseNet   (b) DiffusionNet   (c) DiffusionNet-Efficiency

Fig. 2. Signal strength between layers of (a) DenseNet, (b) DiffusionNet and (c) DiffusionNet-Efficiency. (i, j) represents the signal strength from source layer i to target layer j.

DiffusionNet-N. Following the diffusion model of NO, we propose DenseNet-N with diffusion function $f_{(i,j)} = \exp\left(\frac{-r_{i,j}^2}{4D}\right)$, which represents the signal strength between the ith layer and the jth layer. In order to compare with the original DenseNets, all layers are set on a straight line. And the distance between two adjacent layers is set to be 1. Considering the problem that too small features may affect the learning efficiency, the lower threshold of signal strength is set to be 0.5.

DiffusionNet-L. Linear model is introduced to approximate the NO diffusion process. Linear model is the most common model and is easy to understand. The network structure is same with DiffusionNet-N except that the diffusion function of DiffusionNet-L is $f_{(i,j)} = 1 - \alpha \cdot r_{i,j}$.

DiffusionNet-E. Another common model, exponential model, is also used in simulating the diffusion process. The diffusion function of DiffusionNet-E can be written as $f_{(i,j)} = \beta^{r_{i,j}}$. And other parameters are consistent with DenseNets-N and DenseNets-L. By introducing linear model and exponential models, we can test the influences of different diffusion model.

DiffusionNet-Efficiency. The main approach is to limit the spread distance of signal transmitters to reduce inefficient signal transmission, which is equivalent to limiting the minimum of signal strength. To do this, each receiver is equipped with a gating switch. The receivers can receive signal from front layers only if the signal strength is strong enough to open the gating switch. The threshold of the gating switch is denoted by *thr* ($0 < thr < 1$). And the output of lth layer can be written as:

$$x_l = H_l\left(\left[s_{l,0} \cdot f_{l,0} \cdot x_0, s_{l,1} \cdot f_{l,1} \cdot x_1, \ldots, s_{l,l-1} \cdot f_{l,l-1} \cdot x_{l-1}\right]\right) \tag{8}$$

in which $s_{i,j} \in \{0, 1\}$ is the gating switch. The gating switch will open when signal strength is larger than *thr*. The diffusion function of Diffusion-Efficiency is same with DenseNets-N.

4 Experiments

4.1 Dataset

Experiments were carried on commonly used datasets CIFAR10 and CIFAR100. Each dataset consist 50000 training images and 10000 testing images with size of 32 * 32 pixels. CIFAR10 contains 10 classes and CIFAR100 contains 100 classes. During data processing, we normalize the data by the channel means and standard deviation. And 5000 pictures were taken out as the validation set during training phase. In the last epoch, all training images were used for training.

4.2 Implement Detail

The implementation of this paper is based on the Tensorflow framework. Following DenseNet [4], DiffusionNet is divided into three blocks containing same number of layers. The stochastic gradient descent (SGD) is used for optimization. Minibatch size is set to 64 (unless GPU memory exceeded). Models are trained 300 epochs both on CIFAR10 and CIFAR100. Initial learning rate is set to 0.1, and is lowered by 10 at 50% and 75% of the total number of training epochs. We set weight decay to 10^{-4} with a Nesterov momentum of 0.9. Both datasets are not augmented to avoid the impact of different data augmentation methods and make a fair comparison. Dropout rate is set to 0.2 following DenseNets. D, α, β and *thr* are set to 5, 0.1, 0.8 and 0.25 respectively. Each experiment is conducted only once. All experiments were computed on a server with K80 GPUs.

We conducted four experiments using the original DenseNets and DiffusionNets (NO model, Linear model and Exponential model) with configurations $\{L = 40, k = 12\}$, $\{L = 40, k = 18\}$, $\{L = 40, k = 24\}$ and $\{L = 100, k = 12\}$ respectively both on CIFAR10 and CIFAR100. DiffusionNet-Efficiency were only tested on CIFAR100.

Table 1. Top1 error rates (%) on CIFAR10 and CIFAR100. k denotes the growth rate. * denote results of our implementation. The best results of same configurations are **bold** and overall best results are **blue**.

Method	Depth	Params	C10	C100
Network in Network	-	-	10.41	35.68
All-CNN	-	-	9.08	-
DSN	-	-	9.69	-
FractalNet	21	38.6M	10.18	35.34
ResNet[3]	110	1.7M	13.63	44.74
Stochastic Depth ResNet	110	1.7M	11.66	37.80
DenseNet(k=12)[4]	40	1.1M	7.00/6.72*	27.55/27.70*
DenseNet(k=18)	40	2.4M	6.20*	25.74*
DenseNet(k=24)	40	4.3M	5.81*	24.34*
DenseNet(k=12)	100	7.2M	5.77/5.48*	23.79/23.94*
DiffusionNet-N(k=12)	40	1.1M	6.70	27.54
DiffusionNet-N(k=18)	40	2.4M	**6.01**	**24.92**
DiffusionNet-N(k=24)	40	4.3M	5.60	24.57
DiffusionNet-N(k=12)	100	7.2M	5.52	23.50
DiffusionNet-L (k=12)	40	1.1M	6.57	27.02
DiffusionNet-L (k=18)	40	2.4M	6.07	25.03
DiffusionNet-L (k=24)	40	4.3M	5.56	**24.07**
DiffusionNet-L (k=12)	100	7.2M	**5.18**	23.58
DiffusionNet-E(k=12)	40	1.1M	**6.41**	**26.94**
DiffusionNet-E(k=18)	40	2.4M	6.13	25.47
DiffusionNet-E(k=24)	40	4.3M	**5.40**	24.28
DiffusionNet-E(k=12)	100	7.2M	5.20	**23.08**

4.3   Results

Accuracy. We trained with different depths and growth rates on CIFAR10 and CIFAR100 respectively without data augmentation. Main results of DiffusionNet-NO, DiffusionNet-Linear and DiffusionNet-Exponential can be seen in Table 1 and Fig. 1. Firstly, DiffusionNets use equivalent number of parameters as DenseNets with same depth L and growth rate k, which means no additional extra computation is needed in our methods. Secondly, DiffusionNets outperforms DenseNets on all different configurations. DiffusionNets achieved lowest error rates of 5.18 on CIFAR10 and 23.08 on CIFAR100 (Fig. 3).

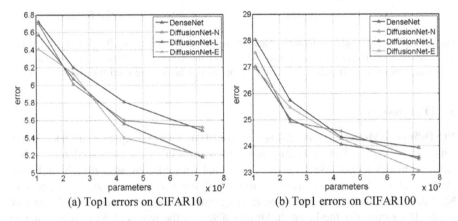

(a) Top1 errors on CIFAR10   (b) Top1 errors on CIFAR100

Fig. 3. Results of DenseNet and variants of DiffusionNet on CIFAR10 (a) and CIFAR100 (b).

Parameter Efficient DiffusionNet. As the distance between layers increases, signal strength decreases. Signal with too small strength can't provide sufficient information but consume the same amount of calculation. DiffusionNet-Efficiency is proposed to solve the problem. The network is tested on CIFAR100, and the results are shown in Fig. 4(a). DiffusionNet-Efficiency with L = 40 and k = 12 achieve considerable accuracy with only half parameters. And the best result of DiffusionNet-Efficiency outperforms that of DenseNets with only 70% parameters.

Single Layer Performance. A possible reason that DiffusionNets outperform DenseNets may be that DiffusionNets can produce more discriminative features on each single layer. A simple experiment was carried out to verify our guess. We set a network with only one block with different depths. The architectures of the networks are same with DenseNets and DiffusionNet-E except that only the output of the last layer is used for classification. Results are displayed in Fig. 4(b). Obviously, DiffusionNet performs better than DenseNet on single-layer classification experiment.

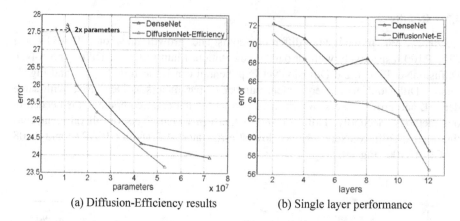

Fig. 4. (a) Testing results of DenseNet and DiffusionNet-Efficiency on CIFAR100. (b) Results of single layer classification. Horizontal axis shows the layers of networks and ordinate axis shows top1 errors on CIFAR100. Only the output of last layer were used for classification.

5 Discussion

NO Diffusion Model. In DenseNets, the front layer transmit signal to all back layers with same weights. But in DiffusionNet, skip connections are replaced by the NO diffusion model so that layers can transmit signal with different strength. NO diffusion model is a typical attenuation transmission model. It is very similar with the principle in NLP that the closer word can provide more information compared with the farther words. If we consider the layers in DiffusionNets as the words in NLP, it is easier to understand why our method achieve better results.

Parameter Efficiency. A parameter efficient variant of DiffusionNets is proposed to help reduce parameters. As is mentioned above, signal strength decreases with distance increasing. And signal with too small strength contribute little but cost the same amount of calculation. In DiffusionNet-Efficiency, layers can only transmit signal within a limited distance. Information with small strength is limited so that only efficient parameters are kept. And that's the reason why it works.

Generate Discriminative Features. As is shown in Fig. 4(b), DiffusionNets can produce more discriminative features compared with DenseNets. One possible reason is that the input of layers are more different in DiffuisionNets. In DenseNets, the input of each layer is the combination of the outputs from the front layers. So it is easy to infer that half of the input features of the second layer are same with the first layer, and 90% of the input features of the 10th layer are same with the 9th layer. Similar input features may result in similar output features. But in DiffusionNets, layers transmit information through diffusion mechanism and receive information with different signal strengths. Even though the input features of the adjacent layers are same, they have different signal strengths. In this way, we can make the input of the adjacent layers more different which helps to produce more discriminative features.

6 Conclusion

We replaced the skip connections in DenseNet with the NO diffusion model and proposed DiffusionNet, which outperforms DenseNet on both CIFAR10 and CIFAR100 datasets. The experimental results demonstrate that DiffusionNet can produce more discriminate features. The proposed parameters efficient variant of DiffusionNet can achieve considerable accuracy with only half parameters of DenseNet.

DiffusionNet is a network containing two different biological information transmission methods, neural transmission and NO diffusion transmission. That makes DiffusionNet more similar to the signal transmission process in human body. Neural network is inspired by biological neural networks initially. And this paper shows that it is possible to learn more from the mechanisms of biological information transmission.

Acknowledgments. This work was supported by the National Science Foundation of China (61420106001 91420302 and 61773391).

References

1. Alom, M.Z., Taha, T.M.: The history began from AlexNet: a comprehensive survey on deep learning approaches. arXiv preprint arXiv:1803.01164 (2018)
2. Srivastava, R.K., Greff, K.: Training very deep networks. In: NIPS, pp. 2377–2385 (2015)
3. He, K.: Deep residual learning for image recognition. In: CVPR, pp. 770–778 (2016)
4. Huang, G.: Densely connected convolutional networks. In: CVPR, pp. 2269–2361 (2016)
5. Jia, P., Yin, J.: Retrograde adaptive resonance theory based on the role of nitric oxide in long-term potentiation. J. Comput. Neurosci. **23**(1), 129–141 (2007)
6. Bredt, D.S.: Nitric oxide, a novel neuronal messenger. Neuron **8**(1), 3–11 (1992)
7. Krekelberg, B., Taylor, J.G.: Nitric oxide: what can it compute? Netw. Comput. Neural Syst. **8**(1), 1–16 (2009)
8. Philippides, A., Husbands, P.: Four-dimensional neuronal signaling by nitric oxide: a computational analysis. J. Neurosci. **20**(3), 1199–1207 (2000)
9. Chen, S., Zhou, Z., Hu, D.: Diffusion and growing self-organizing map: a nitric oxide based neural model. In: Yin, F.-L., Wang, J., Guo, C. (eds.) ISNN 2004. LNCS, vol. 3173, pp. 199–204. Springer, Heidelberg (2004). https://doi.org/10.1007/978-3-540-28647-9_34
10. Yin, J., Hu, D., Chen, S.: DSOM: a novel self-organizing model based on NO dynamic diffusing mechanism. Sci. China **48**(2), 247–262 (2005)
11. Sweeney, Y., Clopath, C.: Emergent spatial synaptic structure from diffusive plasticity. Eur. J. Neurosci. **45**(8), 1057–1067 (2017)
12. Palomo, E.J., López-Rubio, E.: The growing hierarchical neural gas self-organizing neural network. IEEE Trans. Neural Netw. learn. Syst. **28**(9), 2000–2009 (2017)
13. Krizhevsky, A.: Learning multiple layers of features from tiny images. Technical report (2009)
14. He, K., Zhang, X., Ren, S., Sun, J.: Identity mappings in deep residual networks. In: Leibe, B., Matas, J., Sebe, N., Welling, M. (eds.) ECCV 2016. LNCS, vol. 9908, pp. 630–645. Springer, Cham (2016). https://doi.org/10.1007/978-3-319-46493-0_38

Wavelet Autoencoder for Radar HRRP Target Recognition with Recurrent Neural Network

Mengjiao Zhang[✉] and Bo Chen

National Laboratory of Radar Signal Processing, Xidian University, Xi'an, China
mengjiao_123@hotmail.com, bchen@mail.xidian.edu.cn

Abstract. A Wavelet Autoencoder model with Recurrent Neural Network (WaveletAE with RNN) is developed for radar automatic target recognition (RATR), with an encoder-decoder layer, in which the weights of decoder are fixed as a set of overcomplete bases derived from mother wavelet. Imposing an sparsity constraint on the hidden units of encoder-decoder layer, interesting structure in the data is discovered, and superior recognition performance is achieved on the measured High Resolution Range Profiles (HRRP) data, showing the effectiveness of the proposed model. Specific results are represented in our experiments.

Keywords: High resolution range profiles (HRRP)
Recurrent neural network (RNN)
Radar automatic target recognition (RATR) · Overcomplete bases
Joint training · Wavelet · Autoencoder (AE)

1 Introduction

An HRRP is the amplitude of coherent summations of the complex time returning from target scatterers in each range cell, which represents as a one-dimensional signature along the radar line of sight (LOS). HRRP infers to some target structure signatures, such as target size, scatterer distribution, etc. Compared with inverse synthetic aperture radar (ISAR) and synthetic aperture radar (SAR) images, HRRP can be acquired more easily and processed more efficiently. Thus, radar HRRP target recognition has received intensive attention [1–5].

One classical RATR approach is comprised of a feature extraction step that computes robust features from target data and a classification step that calculates classes based on features. Extracting proper features can not only reduce system complexity and processing time, but also improve system performance in many aspects [6].

The energy spectrum, bispectrum features and high-order spectra features extracted using Fourier transform and wavelet transform performed well [7–9]. These traditional systems have mainly relied on elaborate models incorporating

© Springer Nature Switzerland AG 2018
Y. Peng et al. (Eds.): IScIDE 2018, LNCS 11266, pp. 262–275, 2018.
https://doi.org/10.1007/978-3-030-02698-1_23

carefully hand engineer features or large amount of prior knowledge [9]. However, without sufficient prior knowledge for the application, those features could be weak and incomplete. Many statistical recognition methods aim at choosing an appropriate model that describes HRRP's statistical property accurately [10]. But most of those statistical models ignore the correlation and dependency between range cells, and the algorithm for updating parameters will cost much time when the scale of dataset is too large. In [11], a linear dynamic model was investigated and [12] presented a hidden Markov model (HMM) to model the time correlations of HRRP. However, both the models are under Gauss Markov Assumptions [13] which will suffer from limited capabilities of Gaussian distributions in each HMM state and it cannot take long-term dependency into account. Recently, nonlinear deep neural networks like Deep Belief Networks (DBN) and Stacked Denoising Autoencoders (SDAE) show superior performance in various real world tasks, and the hierarchical structure with non-linear activation is similar to human visual cortex [14,15]. Yet, they adapt the layer-wise unsupervised greedy learning method and ignore the correlation and dependency as well. Recurrent Neural Network (RNN) [16], another important branch of the deep neural networks family, are mainly designed for handling sequences. In [17], an attention-based Recurrent Neural Network Model is proposed, which is effective to HRRP target recognition.

In this paper, we not only take advantages of the correlation and dependency, but also the spectrum features of HRRP. Instead of a two-stage approach with feature extractor and classifier, a model that directly captures the objective of the ultimate recognition task is taken into consideration. Combining the bases derived from mother wavelet with the RNN model, we propose a wavelet autoencoder (waveletAE) model with RNN for HRRP target recognition, which capitalizes on spectrum features as well as the correlation and dependency between range cells. Unlike other networks, adapted by a layer-wise unsupervised greedy learning method, this kind of model is supervised and can be jointly trained. Detailed experiments have been done to examine and analyze the recognition performance as well as the learned features.

The remainder of the paper is organized as follows. Section 2 is a brief description of HRRP, Fourier Transform and Wavelet Transform. An introduction of RNN is also included. Later, preprocessing of the HRRP and the waveletAE with RNN for HRRP target recognition are discussed in Sect. 3. Several detailed experiments and results are presented in Sect. 4, with the conclusion and future work provided in Sect. 5.

2   Preliminary

2.1   Fourier Transform and Wavelet Transform for HRRP

Generally, the size of the targets or their components is much larger than the wavelength of the HRRP radar. As a result, those complex target such as an aircraft can be divided into many range *cells*. Within the same range cell, radar signature is the summation of the complex time return from target scatterers. An HRRP x can be represented as:

$$x = e^{j\phi}[x_1, x_2, \ldots, x_d]^T, \text{ with } x_n = \sum_{i=1}^{I_n} \alpha_{in} e^{j\psi_{in}}, n = 1, 2, \ldots, d \qquad (1)$$

where ϕ is the initial phase, d is the number of range cells, I_n is the total number of scatterers in the nth range cell, and α_{in} and ψ_{in} are the reflectivity and phase of the ith scatterer in the nth range cell. The frequency spectrum y of x is given as follow:

$$y = FFT(x) = e^{j\phi}[y_1, y_2, \ldots, y_d]^T, \text{ with } y_f = \sum_{n=1}^{d} x_n e^{-j2\pi f n/d}, f = 1, 2, \ldots, d$$

$$(2)$$

where $FFT(\cdot)$ denotes the fast Fourier transform (FFT) and f is the spectral component. There are some advantages of the frequency spectrum of HRRP for target recognition due to its following appealing qualities: 1. FFT is an information preserving procedure, so there is no information lost in the transformation process; 2. Frequency spectrum is translation invariant, and the feature extracted from it can still enjoy this property; 3. A signal represented in the frequency domain may show characteristics which are not normally observed in the time domain (such as stationary) [6]. According to the Wide Sense Stationary-Uncorrelated Scattering model [18], and some mathematical derivation, it is concluded that the correlation function of two frequencies depends only on their difference, i.e. the frequency spectrum is a wide sense stationary process. From Eq. 2, we can see the frequency spectrum still suffers from initial-phase sensitivity. So only the amplitude of frequency spectrum $z = |y|$ is considered here. It is simple to prove z is still a stationary series.

Affected by many factors, HRRPs are non-stationary signals. Since wavelet transform provides high time resolution and low frequency resolution for high frequencies and high frequency resolution and low time resolution for low frequencies [19,20], Wavelet Transform (WT) and more particular the Discrete Wavelet Transform (DWT) is multi-resolution decomposition used to analyze signals and images. The basis function used in wavelet transform are locally supported; they are nonzero only over part of the domain represented. As an alternative to the short time Fourier Transform (STFT), it can overcome problems related to its frequency and time resolution properties.

The DWT is a special case of WT that provides a compact representation of a signal in time and frequency that can be computed efficiently. The DWT coefficients for sequence $f(n)$ is defined by the following equation [21]:

$$W_\phi(j_0, k) = \frac{1}{\sqrt{M}} \sum_n f(k)\phi_{j_0,k}(n)$$

(3)

$$W_\psi(j, k) = \frac{1}{\sqrt{M}} \sum_n f(k)\psi_{j,k}(n), for j \geq j_0$$

where j_0 is an arbitrary starting scale. The $\phi_{j_0,k}(n)$ and $\psi_{j,k}(n)$ in the equations are sampled versions of basis function $\phi_{j_0,k}(x)$ and $\psi_{j,k}(x)$, where $\psi(x)$ is a time function with finite energy and fast decay called the mother wavelet. The complementary inverse DWT is

$$f(n) = \frac{1}{\sqrt{M}} \sum_k W_\phi(j_0, k)\phi_{j_0,k}(n) + \frac{1}{\sqrt{M}} \sum_{j=j_0}^{\infty} \sum_k W_\psi(j, k)\psi_{j,k}(n) \quad (4)$$

Normally, we let $j_0 = 0$ and select M to be a power of 2 (i.e., $M = 2^j$) so that the summation in Eqs. 3 through 4 are performed over $n = 0, 1, 2, \ldots, M-1, j = 0, 1, 2, \ldots, J-1$, and $k = 0, 1, 2, \ldots, 2^j - 1$. The DWT analysis can be performed using a fast, pyramidal algorithm related to multirate filterbanks [22].

2.2 Recurrent Neural Network for HRRP Target Recognition

Recurrent neural networks (RNNs) are powerful models for sequential data. RNNs are inherently deep in time, since their hidden state is a function of all previous hidden states [23]. The network has an input layer x, hidden layer h (also called state) and output layer o. An HRRP, a 256-dimension vector, can be divided into several segments as a sequence, for echoes of the adjacent range cells have a time difference. In order to transform a vector x into a sequence, we apply a window function on the HRRP sample with the window size $w = 32$ and the overlap $p = 16$, as shown in Fig. 1. $x^{(t)}$ is the input of RNN at time t in the sequence, which can be described as $x^{(t)} = [x(p(t-1)+1) : x(p(t-1)+w)]$. At every time, there is an output to determine the label of the input, and all outputs vote for the final label.

Given an input sequence $x = (x^{(1)}, \ldots, x^{(T)})$, a standard recurrent neural network (RNN) computes the hidden vector sequence $h = (h^{(1)}, \ldots, h^{(T)})$ and output vector sequence $o = (o^{(1)}, \ldots, o^{(T)})$ by iterating the following equations from $t = 1$ to T:

$$h^{(t)} = f\left(W_{xh}x^{(t)} + W_{hh}h^{(t-1)} + b_h\right)$$

(5)

$$o^{(t)} = g\left(W_{ho}h^{(t)} + b_o\right)$$

(6)

where the W terms denote weight matrices (e.g. W_{xh} is the input-hidden weight matrix), the b terms denote bias vectors (e.g. b_h is hidden bias vector) and f, g is the hidden layer function. Usually, $f(z)$ is selected as the *sigmoid* activation function, and $g(z)$ is *softmax* function [16]:

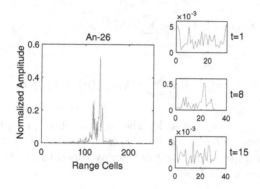

Fig. 1. Illustration of an HRRP sample from an aircraft target. (a) One-dimensional HRRP. (b)The samples of transformed HRRP sequence at time t.

$$f(z) = \frac{1}{1 + e^{-z}} \tag{7}$$

$$g(z_m) = \frac{e^{z_m}}{\sum_k e^{z_k}} \tag{8}$$

The model is trained using the backpropagation through time (BPTT) to maximize the correct probability of the predict output. A loss L measures how far each o is from the corresponding training target y, in which N represents the number of the input data:

$$L_1 = -\frac{1}{NT} \sum_{n=1}^{N} \sum_{t=1}^{T} \left(\sum_{i=1}^{K} y_{(ni)}^{(t)} log \left(o_{(ni)}^{(t)} \right) \right) \tag{9}$$

3   Wavelet Autoencoder for HRRP Target Recognition with Recurrent Neural Network

In Sect. 2, we have discussed about the RNN model for HRRP target recognition, which takes the correlation and dependency between range cells, with the one-dimensional input x transformed into a sequence. As mentioned before, spectrum features have several appealing qualities and HRRPs are non-stationary signals, so DWT is more suitable to analyze HRRPs compared to FFT. First at each time, we can take FFT or DWT on original input sequence and then take the transformed result as input of RNN. However, the spectrum transform is an extra process in this way, which is separated from RNN. What we want is a more integrated model—once the input of the model is given, the output is what we want.

As we all know, both FFT and DWT can be seen as the linear combination of a set of bases, and the signal can be reconstructed by another set of bases, which is similar to autoencoder (AE). In the encoder, the deterministic mapping f_θ transforms an input vector x into hidden representation s, and then decoder maps s back to $\hat{x}$, reconstruction of x, with function $g_{\theta'}$, $\theta = \{W, b\}$, and $\theta' = \{W', b'\}$. Instead of automatically learning the weights of AE, the weights of decoder can be set as the fixed overcomplete bases derived from the mother wavelet, W_{idwt}, and the reconstruction $\hat{x}^{(t)}$ at each time is a linear mapping from $s^{(t)}$:

$$s^{(t)} = f(W_{xs} x^{(t)} + b_s) \tag{10}$$

$$\hat{x}^{(t)} = W_{sx} s^{(t)} = W_{ifft(idwt)} s^{(t)} \tag{11}$$

This can be called wavelet bases autoencoder (WaveletAE). Theoretically, infinite bases can be obtained with a mother wavelet, and only several are involved for DWT. However, we can fix the weights of decoder W_{sx} as a set of overcomplete bases derived from mother wavelet to reconstruct the input as $\hat{x}$ and learn the transform matrix W_{xs} from x to s. The loss of WaveletAE over all HRRP samples $\{x_1, x_2, \ldots, x_N\}$ can be written as follow:

$$L_2 = -\frac{1}{NT} \sum_{n=1}^{N} \sum_{t=1}^{T} \|x_n^{(t)} - \hat{x}_n^{(t)}\|_2^2 \tag{12}$$

AE is forced to learn a *compressed* representation of the input. If there is structure in the data, for example, if some of the input features are correlated, then this algorithm will be able to discover some of those correlations. The argument above relied on the number of hidden units s being small, however, the bases are overcompelete in WaveletAE. But even when the number of hidden units is large (perhaps even greater than the number of input pixels), we can still discover interesting structure, by imposing other constraints on the network. In particular, if we impose a sparsity constraint on the hidden units, then the AE will still discover interesting structure in the data, even if the number of hidden units is large. We choose the following penalty term [24]:

$$penalty = \sum_{j=1} \rho \log \frac{\rho}{\hat{\rho}_j} + (1 - \rho) \log \frac{1 - \rho}{1 - \hat{\rho}_j} \tag{13}$$

$\hat{\rho}_j$ is the average activation of hidden unit j (averaged over the training set):

$$\hat{\rho}_j = \frac{1}{m} \sum_{i=1}^{m} s_i^{(t)} \tag{14}$$

and ρ is a sparsity parameter, typically a small value close to zero (say $\rho = 0.01$).

If we put s as the input of RNN, an integrated model, combining RNN with the wavelet bases autoencoder is derived. This model handles the time sequence and takes the spectrum features into account, called WaveletAE with RNN, as shown in Fig. 2. The goal of the model is to minimize reconstruction error to get the distinctive physical structure and maximize the correct probability of the predict output.

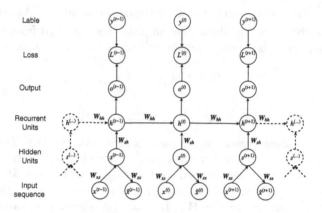

Fig. 2. WaveletAE with RNN

The calculation of recurrent units and hidden units are the same as RNN. But the loss function is a little bit different, which contains the loss of cross entropy L_1 and reconstruction L_2.

$$L = L_1 + L_2 + penalty$$
$$= -\frac{1}{NT} \sum_{n=1}^{N} \sum_{t=1}^{T} \left(\sum_{i=1}^{K} y_{(ni)}^{(t)} log \left(o_{(ni)}^{(t)} \right) + \| x_n^{(t)} - W_{sx} s_t^{(n)} \|_2^2 \right) + penalty \quad (15)$$

4   Experimental Results

In this section, we first study the methodology and introduce the measured HRRP data used in our experiments. And the detailed model settings and experiments to analyze the recognition performance of the model is presented and studied, respectively.

4.1   Measured HRRP Data

The results presented in this paper are based on measured data of three real airplanes. The center frequency and bandwidth of the radar are 5.52 GHz and 400 MHz, respectively. The projections of plane trajectories on to ground plane are segmented as displayed in Fig. 3.

The detailed size of each airplane and the parameters of the measured radar are listed in Table 1.

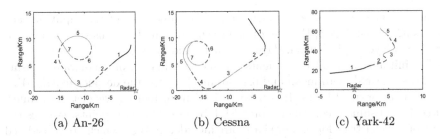

(a) An-26   (b) Cessna   (c) Yark-42

Fig. 3. The projections of three target trajectories onto the ground plane

Table 1. Parameters of radar and planes

Radar parameters	Center frequency		5.52 GHz
	Bandwidth		400 MHz
Airplane	Length (m)	Width (m)	Height (m)
A-26	23.80	29.20	9.83
Cessna citation S/II	14.40	15.90	4.57
Yark-42	36.38	34.88	9.83

According to the preconditions for choosing the training and testing dataset:
(a) The training dataset cover almost all of the target-aspect angles of the test
dataset; (b) The elevation angles of the test dataset are different from those of
the training dataset. Therefore, the fifth and sixth parts of An-26, the sixth and
seventh parts of Cessna and the second and fifth parts of Yark-42 are chosen as
training data with the remaining parts taken as testing data. There are 26000
HRRPs in each part except for the fifth part of Yark-42, which has 10000 HRRPs.
In addition, the dimension of HRRP is 256. In our experiment, there are 7800
HRRP samples for training and 2600 each class; the number of test samples is
5200.

As discussed in literature [25], it is a prerequisite for radar target recognition
to deal with the target-aspect, time-shift, and amplitude-scale sensitivity. Similar to the previous study [25,26], HRRP training samples should be aligned by
the time-shift compensation techniques in ISAR imaging to avoid the influence
of time-shift sensitivity. Each HRRP sample is normalized by L_2 normalization
algorithm to avoid the amplitude-scale sensitivity. In the following experiments,
all of the HRRPs are assumed to have been aligned and normalized. And considering the echoes from different range cells as a sequence, the HRRP sample is
separated into 15 time steps as mentioned in Sect. 2. So, for an HRRP sample,
there are 15 time steps, and at each time the dimension of input $x^{(t)}$ is 32.

4.2   Model Setting

The proposed model consists of the input layer, encoder-decoder layer, recurrent
layer, and the output layer. The number of the recurrent units in the model is

15, and weights are shared at each time. The dimension of the input at each time is 32 and output 3; the size of one recurrent unit $h^{(t)}$ is usually 4–30. And for hidden layer and recurrent layer, the activation is *sigmoid* function.

For initialization, all the weights are set to a matrix of small values, using truncated normal (zero mean and 0.1 variance) except the decoder matrix W_{sx}. W_{sx} is a set of fixed bases. It can be derived from mother wavelet with different scale and shift or consists of some FFT bases of different points. Besides the bases of wavelet transform, more bases are included in W_{sx}, which are considered overcomplete. Several mother wavelet bases are discussed in our experiment, for example, *biorNr.Nd* means biorthogonal wavelets, Nr, Nd are the reconstruction order and decomposition order respectively; dbN are the wavelets belong to Daubechies family, N means the order.

The model is trained in several epochs, in which all data from training set are sequentially presented. For optimization, we use the Adam optimizer [27], and reduce the value of the learning rate after several epochs to accelerate the convergence. The number of epoch is 300 and sparsity parameter ρ is 0.01.

4.3   Recognition Performance

To access the recognition performance of the proposed model on measured data, we carry out some experiments on other methods compared with our model. First we use the popular classification technique, Support Vector Machine (SVM), to classify the measured data directly; both linear SVM (LSVM) and kernel SVM (KSVM) are included. Then the dimension reduction methods are also applied to extract low-dimension features which contain most information of the data, and we take the Linear Discriminant Analysis (LDA), PCA into consideration, followed by a classifier SVM. Several statistic method, like Probabilistic principal component analysis (PPCA) to model the data with Gaussian latent variable and Hidden Markov Model (HMM) to model the time relations of the data, are also considered here to demonstrate our method's validity. At last, We measured the performance of some neural networks, e.g. Multilayer Perceptron (MLP), Deep Belief Network(DBN) and RNN. The superior recognition performance of our model is shown in Table 2.

Table 2. Classification performance of the proposed model with several traditional methods.

Methods	LSVM	KSVM (RBF)	LDA	PCA	PPCA	HMM	DBN	MLP	RNN with different inputs			WaveletAE with RNN
									$x^{(t)}$	FFT($x^{(t)}$)	DWT($x^{(t)}$) (bior2.4)	W_{xs} (bior.2.4)
ACRR	86.81	89.27	83.88	86.22	89.19	87.05	89.68	87.78	89.89	91.63	93.70	**94.42**

Several observations can be made from Table 2. First, the average correct recognition rates(ACRR) of LSVM here only serves as a simple baseline, thus

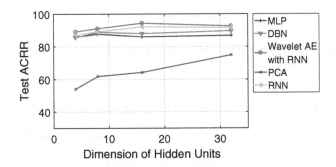

Fig. 4. The ACRRs with dimensionality of feature via different models.

it does not employ any feature extraction. KSVM is the SVM with kernel function which can used for non-linearly separate data. Using radial basis function kernel (RBF), SVM can achieve a higher accuracy. With single HRRP as input, the ACRR of PCA is 86.22%, of which the feature dimension is 77, containing 99% information of the input. While the performance of LDA is inferior, for the feature dimension is C-1, as C is the number of categories, and in our experiment, C is 3. A probabilistic formulation of PCA from a Gaussian latent variable model, called PPCA reaches a desirable result as well as the HMM. But LDA, PCA, and PPCA, which are feature exaction method, should be followed by a classifier. And when we use the SVM as the classifier, we should pick the relatively better one. The process of these methods consist of two stages, feature exaction and classifier selection. And the neural networks, DBN and MLP with a single hidden layer can also obtain effective performance, and the ACRR of RNN is better. However, our model is the most effective and efficient, which can directly predict object categories from input by a joint learning method.

We also plot the ACRRs of the methods varying with the dimensionality of extracted features in Fig. 4. From Fig. 4, we can see that networks works more stable and better than PCA, especially when the dimensionality is small, as the network can capture more discriminative features because of its nonlinear characteristic. And WaveletAE with RNN gives the leading performance, since the model makes use of the time dependency of a single HRRP data. What's more, the model also takes advantages of spectrum features with the decoder weights fixed as the overcomplete bases, which can be jointly trained.

We especially take the experiments on RNN and WaveletAE with RNN as following:

(1) $x^{(t)}$ as the input of the RNN model;
(2) the L_2 norm of FFT on $x^{(t)}$ as the input of RNN model;
(3) DWT with different mother wavelet (dbN, bior$n.d$) on $x^{(t)}$ as the input of RNN model;
(4) $x^{(t)}$ as the input of the proposed model shown in Fig. 2, and the W_{xs} is derived from the mother wavelet same as (3).

Table 3 depicts the ACRRs of the experiments mentioned above. To look insight into the performance of the methods on different targets, we also list the confusion matrix for RNN with different inputs and the proposed model with different W_{sx} (Table 4).

Table 3. Classification performance of the proposed model with RNN model

Methods	RNN with different inputs						WaveletAE with RNN			
	$x^{(t)}$	FFT($x^{(t)}$)	DWT($x^{(t)}$)				W_{xs}			
			bior1.5	bior2.4	db2	db3	bior1.5	bior2.4	db2	db3
Av.R	89.89	91.63	92.47	93.70	91.31	91.88	93.46	**94.42**	93.72	93.56

In Table 3, we show the superior recognition performance of the proposed model. Compared with RNN with different inputs, our method reaches the best result. Although take $x^{(t)}$ as the input of RNN model achieved a good performance 89.89%, FFT on $x^{(t)}$ has a better result, which shows the appealing qualities of FFT mentioned in Sect. 2. Moreover, substituting DFT for FFT increases the classification performance, while the best results achieved by the proposed model. Using the same mother wavelet, the performance of RNN with DWT on $x^{(t)}$ is inferior to the proposed model about $1\% \sim 2\%$.

Table 4. Confusion matrices of different methods.

Methods	RNN with different inputs									WaveletAE with RNN		
	$x^{(t)}$			FFT($x^{(t)}$)			DWT($x^{(t)}$)(bior2.4)			W_{xs}(bior2.4)		
	An-26	Cessna	Yark-42	An-26	Cessna	Yark-42	An-26	Cessna	Yark-42	An-26	Cessna	Yark-42
An-26	90.33	2.39	7.28	96.73	1.72	1.56	93.76	2.13	4.11	94.91	3.07	2.03
Cessna	18.65	81.00	0.35	16.6	80.50	2.90	9.50	89.60	0.90	8.05	91.10	0.85
Yark-42	1.25	0.42	98.33	2.00	0.33	97.67	1.67	0.58	97.75	1.92	0.83	97.25
Av.R	89.89			91.63			92.47			**94.42**		

For RNN, FFT and DWT on $x^{(t)}$ are extra processes of the model, which are two-stage methods, while for the proposed model, it can be jointly trained to predict the categories from input directly. The model has an encoder-decoder layer, in which the weights of decoder are fixed as a set of overcomplete bases derived from the mother wavelet, containing more bases than DWT. Imposing a sparsity constraint on hidden layer $s^{(t)}$, the model can find interesting structure in the data, even if the number of hidden units is large. Table 4 list the confusion matrices for different methods, and only mother wavelet *bior*2.4 is included, since it is the leading one for both RNN and proposed model. It can be concluded that An-26 and Cessna are more separable in the proposed model. Fig. 5 shows W_{sx} and the bases of inverse DWT. It is apparent that W_{sx} contains more bases than IDWT. Fig. 5 also demonstrates the hidden units in the proposed model and DWT on one of the HRRP samples $x^{(t)}$. Obviously, the hidden unites contains

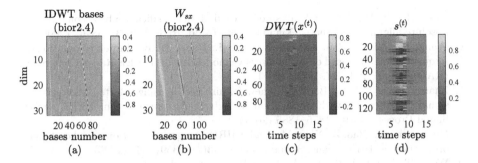

Fig. 5. (a) The inverse DWT bases of bior2.4. (b) The weights of decoder W_{xs}, which is derived from mother wavelet bior2.4. (c) The DWT on input in each time. (d) The hidden units of encoder-decoder layer in each time.

more information than DWT on $x^{(t)}$, for the reason that the bases of DWT is limited.

In summary, the proposed model is a more effective and efficient model than RNN, which takes a set of overcomplete bases into consideration, and obtain better recognition performance for radar HRRP target recognition. Besides, the model can be jointly trained with a specific objective function rather than takes two stages of extracting features and classification.

5 Conclusions

In this paper, a wavelet autoencoder with RNN is analyzed and utilized for radar HRRP target recognition, which extract features by a joint learning method, achieving better recognition performance than the RNN model with different inputs. We fixed the weights of decoder in our model as a set of overcomplete bases, and imposed an sparsity constraint on the hidden units, in order to discover interesting structure in the data, even if the number of hidden units is large. The experimental results on measured HRRP data show that our model achieves superior recognition than other methods.

References

1. Chen, B., Liu, H., Chai, J., Bao, Z.: Large margin feature weighting method via linear programming. IEEE Trans. Knowl. Data Eng. **21**(10), 1475–1488 (2009)
2. Jacobs, S.P.: Automatic target recognition using high-resolution radar range-profiles. Washington University (1997)
3. Feng, B., Du, L., Liu, H., Li, F.: Radar HRRP target recognition based on K-SVD algorithm. In: 2011 IEEE CIE International Conference on Radar (Radar), vol. 1, pp. 642–645. IEEE (2011)
4. Molchanov, P., Egiazarian, K., Astola, J., Totsky, A., Leshchenko, S., Jarabo-Amores, M.P.: Classification of aircraft using micro-doppler bicoherence-based features. IEEE Trans. Aerosp. Electron. Syst. **50**(2), 1455–1467 (2014)

5. Chen, B., Liu, H., Bao, Z.: Analysis of three kinds of classification based on different absolute alignment methods. Xiandai Leida (Mod. Radar) **28**(3), 58–62 (2006)

6. Wang, P., Dai, F., Pan, M., Du, L., Liu, H.: Radar HRRP target recognition in frequency domain based on autoregressive model. In: 2011 IEEE Radar Conference (RADAR), pp. 714–717. IEEE (2011)

7. Zhang, X.-D., Shi, Y., Bao, Z.: A new feature vector using selected bispectra for signal classification with application in radar target recognition. IIEEE Trans. Signal Process. **49**(9), 1875–1885 (2001)

8. Du, L., Liu, H., Bao, Z., Xing, M.: Radar HRRP target recognition based on higher order spectra. IEEE Trans. Signal Process. **53**(7), 2359–2368 (2005)

9. Wang, T., Wu, D.J., Coates, A., Ng, A.Y.: End-to-end text recognition with convolutional neural networks. In: 2012 21st International Conference on Pattern Recognition (ICPR), pp. 3304–3308. IEEE (2012)

10. Liu, H., Du, L., Wang, P., Pan, M., Bao, Z.: Radar HRRP automatic target recognition: algorithms and applications. In: 2011 IEEE CIE International Conference on Radar (Radar), vol. 1, pp. 14–17. IEEE (2011)

11. Penghui, W., Lan, D., Mian, P., Xuefeng, Z., Hongwei, L.: Radar HRRP target recognition based on linear dynamic model. In: 2011 IEEE CIE International Conference on Radar (Radar), vol. 1, pp. 662–665. IEEE (2011)

12. Pan, M., Lan, D., Wang, P., Liu, H., Bao, Z.: Multi-task hidden Markov modeling of spectrogram feature from radar high-resolution range profiles. EURASIP J. Adv. Signal Process. **2012**(1), 86 (2012)

13. Wang, X., Takaki, S., Yamagishi, J.: A comparative study of the performance of HMM, DNN, and RNN based speech synthesis systems trained on very large speaker-dependent corpora. In: 9th ISCA Speech Synthesis Workshop, vol. 9, pp. 125–128 (2016)

14. Bengio, Y., Courville, A., Vincent, P.: Representation learning: a review and new perspectives. IEEE Trans. Pattern Anal. Mach. Intell. **35**(8), 1798–1828 (2013)

15. Lee, H., Ekanadham, C., Ng, A.Y.: Sparse deep belief net model for visual area V2. In: Advances in Neural Information Processing Systems, pp. 873–880 (2008)

16. Mikolov, T., Karafiát, M., Burget, L., Černocký, J., Khudanpur, S.: Recurrent neural network based language model. In: Eleventh Annual Conference of the International Speech Communication Association (2010)

17. Liu, H., Xu, B., Chen, B., Jin, L.: Attention-based recurrent neural network model for radar high-resolution range profile target recognition. J. Electron. Inf. Technol., December 2016

18. Kim, K.-T., Seo, D.-K., Kim, H.-T.: Efficient radar target recognition using the MUSIC algorithm and invariant features. IEEE Trans. Antennas Propag. **50**(3), 325–337 (2002)

19. Tzanetakis, G., Essl, G., Cook, P.: Audio analysis using the discrete wavelet transform. In: Proceedings of the Conference in Acoustics and Music Theory Applications, vol. 66 (2001)

20. Jie, W., Jianjiang, Z., Jiehao, Z.: A radar target recognition method based on wavelet power spectrum and power offset. In: International Asia Conference on Informatics in Control, Automation and Robotics, pp. 130–134 (2010)

21. Gonzalez, R.C., Woods, R.E.: Digital Image Processing, 3rd edn. Prentice-Hall Inc., Upper Saddle River (2007)

22. Mallat, S.G.: A theory for multiresolution signal decomposition: the wavelet representation. IEEE Trans. Pattern Anal. Mach. Intell. **11**, 674–693 (1989)

23. Graves, A., Mohamed, A., Hinton, G.: Speech recognition with deep recurrent neural networks. In: 2013 IEEE International Conference on Acoustics, Speech and Signal Processing (ICASSP), pp. 6645–6649. IEEE (2013)
24. Ng, A.: Sparse autoencoder (2011)
25. Lan, D., Liu, H., Bao, Z., Zhang, J.: A two-distribution compounded statistical model for radar HRRP target recognition. IEEE Trans. Signal Process. **54**(6), 2226–2238 (2006)
26. Lan, D., Liu, H., Bao, Z.: Radar HRRP statistical recognition: parametric model and model selection. IEEE Trans. Signal Process. **56**(5), 1931–1944 (2008)
27. Kingma, D.P., Ba, J.: Adam: a method for stochastic optimization. arXiv preprint arXiv:1412.6980 (2014)

Stacked Discriminative Denoising Auto-encoder Based Recommender System

Kai Wang, Lei Xu, Ling Huang, Chang-Dong Wang[✉], and Jian-Huang Lai

School of Data and Computer Science, Sun Yat-sen University, Guangzhou, China
{wangk75,xulei28}@mail2.sysu.edu.cn, huanglinghl@hotmail.com,
changdongwang@hotmail.com, stsljh@mail.sysu.edu.cn

Abstract. Recommender systems are widely used in our life for automatically recommending items relevant to our preference. Matrix Factorization (MF) is one of the most successful methods in recommendation. However, the rating matrix utilized by the MF-based models is usually sparse, so it is of vital significance to integrate the side information to provide relatively effective knowledge for modeling the user or item features. The key problem lies how to extract representative features from the noisy side information. In this paper, we propose Stacked Discriminative Denoising Auto-encoder based Recommender System (SDDRS) by integrating deep learning model with MF based recommender system to effectively incorporate side information with rating information. Extensive top-N recommendation experiments conducted on three real-world datasets empirically demonstrate that SDDRS outperforms several state-of-the-art methods.

Keywords: Side information · Collaborative filtering
Matrix factorization · Stacked denoising auto-encoder

1 Introduction

The goal of recommender system is mainly in filtering the useless information based on users' preference, so as to help users to find items they are really interested in. Collaborative filtering (CF) [1–10] based recommender system is one of the most successful recommendation methods, which models the user and item feature based on the rating information, and assumes that users with similar feature will share similar preference to items. Since the Netflix Prize [11], matrix factorization (MF) techniques are widely used in collaborative filtering based recommender systems, like [1,12], etc. Unlike the traditional CF models like [2,13], which directly adopt the rating information to build the feature of recommendation entity, MF based model assumes that the feature of users and items can be represented by less factors in a latent space.

Rating information is of vital importance in CF based recommender systems. There are two kinds of rating information, which are **explicit rating** and

Y. Peng et al. (Eds.): IScIDE 2018, LNCS 11266, pp. 276–286, 2018.
https://doi.org/10.1007/978-3-030-02698-1_24

implicit rating [14]. The explicit ratings are multi-value variables, representing the preference of users or popularity of items, while the implicit ratings are binary, standing for whether users interested in items. Both the two kinds of rating information can provide useful semantic and sentimental knowledge when modeling the user and item feature. However, at most of time, each user can only interact with a small ratio of items, which usually causes the **data sparsity problem** [15]. In order to alleviate this problem, several works like [16,17] utilize the relative information from other sources which also called side information[1], to enrich the rating information. But the side information involves other kinds of knowledge except rating, which can be seen as noises in recommender system, and thus it's necessary to extract the effective feature inside side information. Deep learning do well in extracting denoised feature with rich semantic information from noisy data which makes increasingly more works focus on combining deep learning model with recommender system in recent, like [18–20]. However, most of those works only take into account the combination of side information and implicit ratings while seldom consider the explicit rating, which causes the acquired feature lose the explicit knowledge. Stacked Denoising Auto-encoder (SDAE) is a deep neural network with the parameters of each layer initialized by Denoising Auto-encoder (DAE) [21], which satisfies the **Local Unsupervised Criterion** [22] and **Denoising Criterion** [21], helping neural network keep good training and denoising properties. For the effectiveness of SDAE [23] in extracting denoised feature, [18,20,24] aims to adopt it as their deep model. However, those works mistake deep denoising auto-encoder for SDAE, causing deep models used in their recommender systems cannot keep the good properties of real SDAE.

In this paper, we integrate the SDAE based classifier with MF based recommender system to incorporate the above three kinds of information. Due to the discriminative property of classification model, our deep model is named Stacked Discriminative Denoising Auto-encoder (SDDAE), and our recommendation model is named Stacked Discriminative Denoising Auto-encoder based Recommender System (SDDRS).

To sum up, the contributions of this paper can be listed as follows:

- We present SDDAE which integrates the side information with explicit rating.
- We propose the SDDRS model by combining SDDAE with MF based recommender system, which tightly couples the side information with both explicit and implicit rating.
- Extensive experiments on three real-world datasets show that SDDRS outperform state-of-art recommender systems.

The organization of this paper is as follows. In Sect. 2, we briefly introduce SDAE and then describe the proposed SDDRS model in detail. In Sect. 3, we present experiment results and give an analysis on them. At last, we draw conclusions in Sect. 4.

---

[1] According to [18], for the reason of private, usually the useful user side information is hard to access, and thus we only consider the item side information in this paper.

2 Stacked Discriminative Denoising Auto-encoder for Recommender System

2.1 Preliminary

Stacked Denoising Auto-encoder. SDAE is a deep neural network which treats each layer as the hidden layer of Denoising Auto-encoder (DAE) [21] when initializing, for the reason that the DAE-based initialization satisfies both the **denoising criterion** [21], and the **local unsupervised criterion** [22]. The objective function of SDAE during initialization is shown as follows:

$$\sum_{i=1}^{k} \arg\min_{\mathbf{W}_i, \mathbf{b}_i} \|\mathbf{X}_i^o - \mathbf{X}_i\|_F^2 + \lambda_1(\|\mathbf{W}_i\|_F^2 + \|\mathbf{b}_i\|_2^2) \tag{1}$$

where $\mathbf{W}_i$, $\mathbf{b}_i$ represent the weighted matrices and bias vectors of the i-th DAE, and $\mathbf{X}_i^o$, $\mathbf{X}_i$ are the reconstructed output and the clean input of it.

2.2 Stacked Discriminative Denoising Auto-encoder Based Recommender System

The goal of SDDRS is to smoothly couple side information, explicit rating information and implicit rating information, so as to learn more representative features.

The establishment of our model is divided into three steps. Firstly, we propose SDDAE to connect explicit rating with side information. Then, we integrate SDAE with implicit rating matrix factorization to incorporate implicit rating with side information. At last, we fuse the above two models to construct SDDRS, which can couple three kinds of information to train features of recommender system.

Stacked Discriminative Denoising Auto-encoder. In this part, we propose SDDAE to connect explicit rating with side information.

Supposing the explicit rating matrix $\mathbf{R}$ and side information matrix $\mathbf{X}$ are observed, we can design the following process to construct the SDDAE model:

1. Set a rating threshold t and categorize items by comparing the average received rating of each item with t, and use $\mathbf{c} \in \mathbb{R}^m$ to denote the classification vector:

$$c_i = \begin{cases} 1, & \text{if } \operatorname{avg}(\mathbf{R}_{*i}) > t \\ 0, & \text{if } \operatorname{avg}(\mathbf{R}_{*i}) \leq t \end{cases} \tag{2}$$

where the $\operatorname{avg}(\mathbf{R}_{*i})$ represents the average rating of the i-th item.
2. Construct the $(k+1)$-layer SDDAE as follows:
   (a) Optimize the Eq. (1) to initialize parameters of each layer of SDDAE.
   (b) Put $\mathbf{X}$ into SDDAE, and treat the output of the k-th layer as the side feature $\mathbf{X}^f \in \mathbb{R}^{m \times d}$. Then, the $(k+1)$-th layer maps $\mathbf{X}^f$ to a classification probability vector $\mathbf{x}^p \in \mathbb{R}^m$.

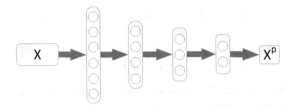

Fig. 1. 5-layer stacked discriminative denoising auto-encoder.

(c) The classification error can be calculated by following equation:

$$\mathcal{L}_c = \mathbf{c}\log(\mathbf{x}^p) + (\mathbf{1} - \mathbf{c})\log(\mathbf{1} - \mathbf{x}^p)$$

$$+\lambda_1\left(\sum_{l=1}^{k+1}\|\mathbf{W}_l\|_F^2 + \|\mathbf{b}_l\|_2^2\right) \tag{3}$$

where $\mathbf{1} \in \mathbb{R}^m$ represents the all-one vector. The parameters of SDDAE is fine-tuned by back-propagating the above error.

By ignoring the initialization phase, the model of SDDAE is shown in Fig. 1. SDDAE adopts the item classifications generated from explicit ratings to supervised train the side feature, which can help extract rating feature from side information, and therefore can establish the relationship between them.

Stacked Denoising Auto-encoder Based Implicit Rating Matrix Factorization. In this part, we integrate side information $\mathbf{X}$ with implicit rating $\bar{\mathbf{R}} \in \mathbb{R}^{m \times n}$ through combining SDAE with MF based recommender system. Similar to the previous part, we design following process to describe this model:

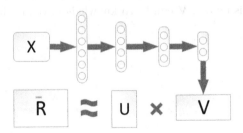

Fig. 2. Combination of 4-layers SDAE and implicit rating matrix factorization

1. Randomly initialize item feature matrix $\mathbf{V} \in \mathbb{R}^{m \times d}$ and user feature matrix $\mathbf{U} \in \mathbb{R}^{n \times d}$, and calculate the rating error by:

$$\|\bar{\mathbf{R}} - \mathbf{U}\mathbf{V}^T\| \odot \mathbf{I} \tag{4}$$

where $\odot$ is the Hadamard dot or element-wise dot, and the element in the indicator matrix $\mathbf{I}$ is

$$\mathbf{I}_{ij} = \begin{cases} \beta, & \text{if } \bar{\mathbf{R}}_{ij} > 0 \\ 1, & \text{if } \bar{\mathbf{R}}_{ij} = 0 \end{cases} \tag{5}$$

2. Construct a k-layer SDAE to connect side information with implicit rating matrix factorization as follows:
   (a) Optimize Eq. (1) to initialize the $\mathbf{W}_l$, $\mathbf{b}_l$ of each layer of SDAE and feed forward side information $\mathbf{X}$ to get the side feature $\mathbf{X}^f$.
   (b) Due to the inherent relevance between the rating and side information of the same item, there is a constraint between these two features:

$$\|\mathbf{X}^f - \mathbf{V}\|_F^2 \tag{6}$$

   (c) By integrating the above errors, the loss function establishes the relationship between side information and implicit rating, which is formulated as follows:

$$\mathcal{L}_r = \|(\bar{\mathbf{R}} - \mathbf{U}\mathbf{V}^T) \odot \mathbf{I}\|_F^2 + \lambda_2\|\mathbf{U}\|_F^2 + \lambda_3\|\mathbf{X}^f - \mathbf{V}\|_F^2$$
$$+ \lambda_1(\sum_{l=1}^{k} \|\mathbf{W}_l\|_F^2 + \|\mathbf{b}_l\|_2^2) \tag{7}$$

   where the λ_2, λ_3 and $\|\mathbf{U}\|_F^2$, $\|\mathbf{X}^f - \mathbf{V}\|_F^2$ are the user, item regular coefficients and regular terms respectively.
3. Train $\mathbf{U}$, $\mathbf{V}$ based on Eq. (7) and back-propagate this error to jointly train the parameters of SDAE.

Like the last part, by ignoring the initialization phase, Fig. 2 presents the above combination model. From the above process, the relationship between implicit rating and side information is bridged by item feature $\mathbf{V}$. In this case, when optimizing this model, $\mathbf{V}$ can learn knowledge from both side information and implicit rating.

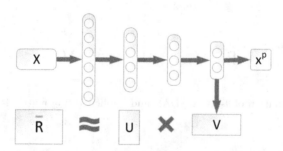

Fig. 3. Stacked discriminative denoising auto-encoder based recommender system

Stacked Discriminative Denoising Auto-encoder Based Recommender System. In this part, we propose SDDRS through integrating the above two models. For the reason that the construction process of SDDRS is similar to the above two models, we omit its description for saving space.

The goal of SDDRS is to smoothly integrate three kinds of information so as to learn more representative item features. As we mentioned before, the objective functions of the first model Eq. (3) integrates explicit rating with side information, while that of the second model Eq. (7) incorporates implicit rating with side information. Therefore, by combining the previous two models, SDDRS can couple these three kinds of information by using the side information to contact both of explicit and implicit rating. Based on this thought, the model of SDDRS is designed as Fig. 3, and its objective function is formalized by fusing the Eqs. (3) and the (7) as follows:

$$
\begin{aligned}
\mathcal{L} &= \alpha \mathcal{L}_c + (1-\alpha)\mathcal{L}_r \\
&= \alpha(\mathbf{c}^T \log(\mathbf{x}^p) + (1-\mathbf{c})^T \log(1-\mathbf{x}^p) + \lambda_1(\|\mathbf{W}_{k+1}\|_F^2 + \|\mathbf{b}_{k+1}\|_2^2)) \\
&\quad + (1-\alpha)(\|(\bar{\mathbf{R}} - \mathbf{U}\mathbf{V}^T) \odot \mathbf{I}\| + \lambda_2\|\mathbf{U}\|_F^2 + \lambda_3\|\mathbf{X}^f - \mathbf{V}\|_F^2) \\
&\quad + \lambda_1(\sum_{l=1}^{k} \|\mathbf{W}_l\|_F^2 + \|\mathbf{b}_l\|_2^2)
\end{aligned}
\tag{8}
$$

where α is the fusion parameter.

Based on the Eq. (8), the user feature $\mathbf{U}$ and item feature $\mathbf{V}$ are optimized by Stochastic Gradient Descending (SGD) while the parameters of SDDAE are jointly trained by back-propagating the loss.

After the Eq. (8) is optimized, we can get the trained user feature $\mathbf{U}$ and item feature $\mathbf{V}$. Then, the predicted implicit rating matrix is calculated by their inner product as follows:

$$
\bar{\mathbf{R}}^p = \mathbf{U}\mathbf{V}^T
\tag{9}
$$

3 Experiment

3.1 Dataset

Experiments are conducted on three datasets MovieLens100k, MovieLens1M, and MovieLens10M[2]. During experiment, for each dataset, we randomly choose 80% as training data, and the rest 20% as testing data. The above datasets mainly include rating information and movie genres, and the range of rating is 1–5. There are 20 kinds of movie genre, with one of them is named 'no genre', which is removed in advance and remains 19 kinds of genre. The statistic information of three datasets is shown as Table 1.

The range of each explicit rating is between 1 and 5, where the implicit ratings are got like [24] by firstly setting a threshold $t = 4$, and then binarize explicit ratings by comparing them with t, where ratings less than t are set to 0, and others

---

[2] https://grouplens.org/datasets/movielens/.

Table 1. Statistic information of three datasets

Dataset	Users	Items	Ratings	Sparsity
MovieLens100k	943	1682	100,000	93.7%
MovieLens1M	3882	6040	1,000,000	95.8%
MovieLens10M	71567	10681	10,000,000	98.7%

are set to 1. Besides, each movie has its own movie genres like 'Comedy', 'Action', and there are totally 19 kinds of genre. We use bag-of-word method to model them as side information during experiment.

3.2 Baseline Model and Experiment Settings

The compared algorithms used in this paper are Probabilistic Matrix Factorization (**PMF**) [25], Sparse Linear Method (**SLIM**) [26], additional Stacked Denoising Auto-encoder (**aSDAE**) [24], Sparse Linear Methods with Side Information (**SSLIM**) [27], Collaborative Denoising Auto-encoder (**CDAE**) [28]. Among the above algorithms, the result of aSDAE are directly extracted from the corresponding paper [24] for the reason that the code is not acceptable.

During the experiment, like [24], we adopt the root mean square error (RMSE) and recall@N as our evaluation metrics, where the previous metric reflects the general performance of the predicted scoring matrix and the latter metric directly represents the accuracy of the top-N recommendation.

As Eq. (8) shows, there are six parameters in SDDRS, which are fusion parameter α, regularization coefficients $\lambda_1, \lambda_2, \lambda_3, \beta$ of indicator matrix $\mathbf{I}$, epochs number epo, and learning rate η. Different parameters setting are utilized on three datasets as follows:

- MovieLens100k: $\alpha = 0.2$, $\lambda_1 = 0.3$, $\lambda_2 = 1$, $\lambda_3 = 70$, $\beta = 10$, $epo = 600$, and $\eta = 0.0002$
- MovieLens1M: $\alpha = 0.2$, $\lambda_1 = 0.3$, $\lambda_2 = 1$, $\lambda_3 = 20$, $\beta = 10$, $epo = 600$, and $\eta = 0.0002$
- MovieLens10M: $\alpha = 0.2$, $\lambda_1 = 0.3$, $\lambda_2 = 1$, $\lambda_3 = 50$, $\beta = 3$, $epo = 100$, and $\eta = 0.0002$.

3.3 Experimental Results

RMSE. The Table 2 shows the RMSE results of each algorithm on three dataset. In MovieLens10M dataset, the RMSE of both SLIM and SSLIM are denoted as N/A for the reason that this dataset is not suitable for these two models, and the aSDAE is '−' because its paper doesn't present the result on this dataset.

As we can see, the RMSE of SDDRS is lower than other algorithms on each dataset, which empirically demonstrates the efficiency of SDDRS in RMSE metric.

Table 2. RMSE of compared models on three datasets

Algorithm	PMF	SLIM	aSDAE	SSLIM	CDAE	SDDRS
MovieLens100k	0.5941	0.5863	0.5435	0.5735	0.5376	**0.5021**
MovieLens1M	0.5715	0.5213	0.5236	0.5163	0.5132	**0.5019**
MovieLens10M	0.6854	N/A	–	N/A	0.8146	**0.6501**

Recall@N. The recall@N represents the ratio of items user actually like in the recommended top-N list. Therefore, it is undoubtedly a very intuitive and important metric in the top-N recommendation task.

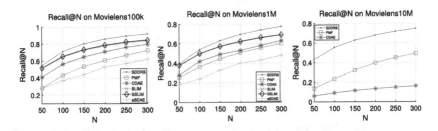

Fig. 4. Recall@N on each dataset

The three subgraphs in Fig. 4 present the recall@N performance of different algorithms on three datasets with $N = [50 : 50 : 300]$. And we can see that, SDDRS outperforms all of the other models, which empirically demonstrates the ability of top-N recommendation of our model. What's more, the sparser the dataset is, the more SDDRS outperforms other algorithms, which further proves the effectiveness of data utilization.

3.4 Parameters Analysis

In this section, we fix accessorial parameters $\eta = 0.0002$, $\lambda_1 = 0.3$, $\beta = 10$ and $d = 10$, and analyze the primary parameters of SDDRS. For speed up, those experiments are conducted on the smallest dataset MovieLens100k.

Epochs Number Analysis. Fix other parameters $\frac{\lambda_3}{\lambda_2} = 50$, $\alpha = 0.2$ and $d = 10$. Then, we set the epoch number $epo = [100 : 100 : 1000]$.

Figure 5 shows that, with the increase of epo, the performance of SDDRS is becoming better, and achieve convergence until $epo > 500$.

Regularization Ratio Analysis. Fix other parameters $epo = 600$, $\alpha = 0.2$ and $d = 10$. Then, we set the ratio between λ_3 and λ_2 as $[1, 10 : 10 : 100]$.

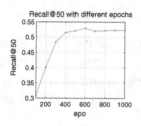
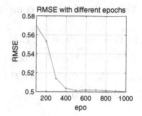

Fig. 5. Epochs analysis

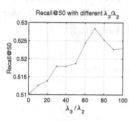
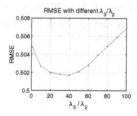

Fig. 6. Regularization ratio analysis

Figure 6 shows that, within a certain range, the item feature regularization coefficient λ_3 can positively influence the performance of SDDRS. However, when it becomes too large, this term will prevent item feature from learning rating information, which exerts a negative influence on SDDRS. As we can see, when the ratio in the range of 40–70, the RMSE is somewhat higher, while the recall@50 increases sharply.

Fusion Parameter Analysis. Fix other parameters $epo = 600$, $\frac{\lambda_3}{\lambda_2} = 50$ and $d = 10$. Then, we set the α as $[0 : 0.05 : 0.5]$.

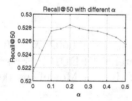
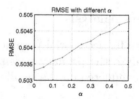

Fig. 7. Fusion parameter analysis

According to Fig. 7, we can find that within the range of 0–0.2, the increase of α enhances the performance of recall@50, because the explicit rating helps learn the popularity of items, which lead to a better rank. But the main information source is implicit rating matrix, causing the α need to be controlled in a lower level. RMSE is affected with the increase of α because it is calculated by comparing the predicted matrix with implicit rating matrix, which causes the explicit rating exerts a bad influence on it.

4 Conclusion

There are three kinds of information commonly used in recommender system, which are explicit rating, implicit rating, and side information, where both kinds of rating information are sparse but can directly represent the sentiment of recommendation entity, while the side information are rich but not so obvious as ratings. The goal of this paper is to smoothly couple these three kinds of information so as to help compensate each other. To this end, we propose a novel model named SDDRS by integrating SDDAE with MF based recommender system. The extensive top-N recommendation experiments conducted on three datasets show that SDDRS outperforms several state-of-the-art models, which empirically demonstrates its performance.

Acknowledgements. This work was supported by NSFC (61502543), Guangdong Natural Science Funds for Distinguished Young Scholar (2016A030306014), Tip-top Scientific and Technical Innovative Youth Talents of Guangdong special support program (2016TQ03X542), and the key research project of Guangzhou municipal colleges and universities "The Research and realization of 3D Game Engine based on Android" (No. 2012A164).

References

1. Li, M., Huang, D., Wei, B., Wang, C.D.: Event recommendation via collective matrix factorization with event-user neighborhood. In: IScIDE, pp. 676–686 (2017)
2. Balabanovic, M., Shoham, Y.: Content-based collaborative recommendation. Commun. ACM **40**(3), 66–72 (1997)
3. Hu, Q.Y., Zhao, Z.L., Wang, C.D., Lai, J.H.: An item orientated recommendation algorithm from the multi-view perspective. Neurocomputing **269**, 261–272 (2017)
4. Zhao, Z.L., Wang, C.D., Lai, J.H.: AUI&GIV: recommendation with asymmetric user influence and global importance value. PLoS One **11**(2), e0147944 (2016)
5. Zhao, Z.L., Wang, C.D., Wan, Y.Y., Lai, J.H., Huang, D.: FTMF: recommendation in social network with feature transfer and probabilistic matrix factorization. In: IJCNN, 847–854 (2016)
6. Zhao, Z.L., Wang, C.D., Wan, Y.Y., Lai, J.H.: Recommendation in feature space sphere. Electron. Commer. Res. Appl. **26**, 109–118 (2017)
7. Zhao, Z.-L., Huang, L., Wang, C.-D., Huang, D.: Low-rank and sparse cross-domain recommendation algorithm. In: Pei, J., Manolopoulos, Y., Sadiq, S., Li, J. (eds.) DASFAA 2018. LNCS, vol. 10827, pp. 150–157. Springer, Cham (2018). https://doi.org/10.1007/978-3-319-91452-7_10
8. Wang, C.D., Deng, Z.H., Lai, J.H., Yu, P.S.: Serendipitous recommendation in e-commerce using innovator-based collaborative filtering. IEEE Trans. Cybern. (2018, in press)

9. Zhao, Z.-L., Huang, L., Wang, C.-D., Lai, J.-H., Yu, P.S.: Low-rank and sparse matrix completion for recommendation. In: Liu, D., Xie, S., Li, Y., Zhao, D., El-Alfy, E.-S.M. (eds.) ICONIP 2017. LNCS, vol. 10638, pp. 3–13. Springer, Cham (2017). https://doi.org/10.1007/978-3-319-70139-4_1

10. Xu, Y.-N., Xu, L., Huang, L., Wang, C.-D.: Social and content based collaborative filtering for point-of-interest recommendations. In: Liu, D., Xie, S., Li, Y., Zhao, D., El-Alfy, E.-S.M. (eds.) ICONIP 2017. LNCS, vol. 10638, pp. 46–56. Springer, Cham (2017). https://doi.org/10.1007/978-3-319-70139-4_5

11. Koren, Y., Bell, R.M., Volinsky, C.: Matrix factorization techniques for recommender systems. IEEE Comput. **42**(8), 30–37 (2009)

12. Bengio, Y., Yao, L., Alain, G., Vincent, P.: Generalized denoising auto-encoders as generative models. In: NIPS, pp. 899–907 (2013)

13. Sarwar, B.M., Karypis, G., Konstan, J.A., Riedl, J.: Item-based collaborative filtering recommendation algorithms. In: WWW, pp. 285–295 (2001)

14. Xue, H.J., Dai, X., Zhang, J., Huang, S., Chen, J.: Deep matrix factorization models for recommender systems. IJCA **I**, 3203–3209 (2017)

15. Burke, R.D.: Hybrid recommender systems: survey and experiments. User Model. User-Adapt. Interact. **12**(4), 331–370 (2002)

16. Li, M.R., Huang, L., Wang, C.D.: Geographical and overlapping community modeling based on business circles for POI recommendation. In: IScIDE, pp. 665–675 (2017)

17. de Campos, L.M., Fernández-Luna, J.M., Huete, J.F., Rueda-Morales, M.A.: Combining content-based and collaborative recommendations: a hybrid approach based on bayesian networks. Int. J. Approx. Reason. **51**(7), 785–799 (2010)

18. Wang, H., Wang, N., Yeung, D.Y.: Collaborative deep learning for recommender systems. In: KDD, pp. 1235–1244 (2015)

19. Salakhutdinov, R., Mnih, A., Hinton, G.E.: Restricted Boltzmann machines for collaborative filtering. In: ICML, pp. 791–798 (2007)

20. Wei, J., He, J., Chen, K., Zhou, Y., Tang, Z.: Collaborative filtering and deep learning based recommendation system for cold start items. Expert Syst. Appl. **69**, 29–39 (2017)

21. Vincent, P., Larochelle, H., Bengio, Y., Manzagol, P.A.: Extracting and composing robust features with denoising autoencoders. In: ICML, pp. 1096–1103 (2008)

22. Erhan, D., Bengio, Y., Courville, A.C., Manzagol, P.A., Vincent, P., Bengio, S.: Why does unsupervised pre-training help deep learning? J. Mach. Learn. Res. **11**, 625–660 (2010)

23. Vincent, P., Larochelle, H., Lajoie, I., Bengio, Y., Manzagol, P.-A.: Stacked denoising autoencoders: learning useful representations in a deep network with a local denoising criterion. J. Mach. Learn. Res. **11**, 3371–3408 (2010)

24. Dong, X., Yu, L., Wu, Z., Sun, Y., Yuan, L., Zhang, F.: A hybrid collaborative filtering model with deep structure for recommender systems. AAA **I**, 1309–1315 (2017)

25. Salakhutdinov, R., Mnih, A.: Probabilistic matrix factorization. In: NIPS, pp. 1257–1264 (2007)

26. Ning, X., Karypis, G.: SLIM: sparse linear methods for top-n recommender systems. In: ICDM, pp. 497–506 (2011)

27. Ning, X., Karypis, G.: Sparse linear methods with side information for top-n recommendations. In: RecSys, pp. 155–162 (2012)

28. Wu, Y., DuBois, C., Zheng, A.X., Ester, M.: Collaborative denoising auto-encoders for top-n recommender systems. In: WSDM, pp. 153–162 (2016)

Deep Metric Learning for Software Change-Proneness Prediction

Yongxin Ge, Min Chen, Chao Liu$^{(\boxtimes)}$, Feiyi Chen, Sheng Huang,
and Hongxing Wang

School of Big Data and Software Engineering, Chongqing University,
Chongqing, China
`liu.chao@cqu.edu.cn`

Abstract. Software change-proneness prediction, which predicts whether or not class files in a project will be changed in their next release, can help software developers allocate resources more effectively and reduce software maintenance costs. Previous studies found that change-proneness prediction cannot work well with limited training data, especially for new projects. To address this issue, the cross-project change-proneness prediction is proposed, which builds a prediction model by using sufficient data form other projects, i.e. the source projects, and predicts the change-prone files in a target project. However, the cross-project prediction is unstable due to the large metric distinction between source projects, leading to a challenge for classifying change-prone files. To improve the cross-project prediction, we propose a Deep Metric Learning (DML) model to minimize such feature distinction before the file classification. Specifically, DML maps files in source projects into a particular space, where files from the same category, e.g. change-prone files, are getting closer while files from different categories are getting further. Besides, we also leverage an over-sampling approach to handle the highly imbalanced dataset for model training. We verify our model on 20 change-proneness datasets, and compare it with 5 cross-project change-proneness models. Results indicate that the proposed model can substantially improve the performance of change-proneness prediction.

Keywords: Change-proneness prediction · Cross-project prediction
Deep metric learning

1 Introduction

In a software life cycle, software development and maintenance always lead to massive code changing for adding new features [35], and repairing defects [1,31, 45]. To control the development cost and improve the product quality, researchers predict whether code files will be modified in their following release, i.e. the change-proneness prediction, because identifying change-prone files in advance can help development teams to wisely set a project schedule, and effectively allocate manpower and time [18,25,29,43].

© Springer Nature Switzerland AG 2018
Y. Peng et al. (Eds.): IScIDE 2018, LNCS 11266, pp. 287–300, 2018.
https://doi.org/10.1007/978-3-030-02698-1_25

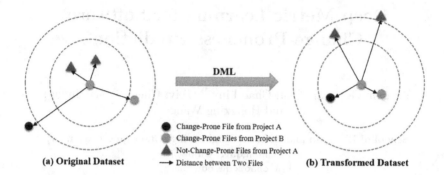

(a) Original Dataset   **(b) Transformed Dataset**

● Change-Prone File from Project A
● Change-Prone Files from Project B
▲ Not-Change-Prone Files from Project A
→ Distance between Two Files

Fig. 1. The theory of DML model. When our DML model is applied, DML maps file metrics from different source projects into a particular space, where change-prone files from the different (e.g., black and red circles) are getting closer while not change-prone files from different projects (e.g., blue triangles) are getting further. (Color figure online)

Over the past decade, researchers have spent significant efforts on software change-proneness prediction. They found many Object-Oriented (OO) metrics (e.g., the depth in inheritance tree) and internal quality attributes (e.g., the lines of code) correlate to the change-proneness. Among these metrics and attributes, statistical analysis shows that the lines of code and the CKJM (Chidamber and Kemerer Java Metric) suite[1] are the best combination to discriminate the change-prone files [25]. Afterward, based on these metrics, many Machine Learning (ML) models have been used to improve the prediction performance [29]. However, to predict the change-proneness of a project, such ML models are trained by historical labeled data within this project, which called the within-project prediction. And it hardly works for a new project because manually collecting labeled data in the new project is time-consuming and expensive [18,43].

To address the limitation of the within-project prediction, a better solution is to take advantage of the abundant labeled data from other projects, which called the cross-project prediction [6]. It trains a change-proneness model by using labeled data from other projects (i.e. source projects) and predicts the change-prone files in a target project. However, in some cases, the cross-project prediction is no better than the within-project prediction [27], and we assume that this phenomenon is caused by the large distinction in the metrics from different source projects. Specifically, as illustrated in Fig. 1(a), change-prone files from different projects (black and red circles) may be largely different in terms of metric distance, due to their distinctions in project scale, programming style, development process, and etc. Based on these differences, an ML model would wrongly predict the yellow circle to be not-change-prone for its far distance to other change-prone files.

---

[1] CKJM: https://www.spinellis.gr/sw/ckjm/.

To solve this metric problem, we apply a Deep Metric Learning (DML) model according to Hu et al. [17] on cross-project change-proneness prediction. However, the training data in a change-proneness dataset is highly imbalanced, which would lead to an over-sampling problem for our model training. To address this problem, we perform SMOTE (Synthetic Minority Oversampling Technique) on training data to balance its minority instances, and we can use the processed training data to build the DML model. Generally, the DML model aims to learn a non-linear transformation to minimize the metric distinction before the file classification. In specific, DML maps file metrics from different source projects into a particular space, where files from the same categories (e.g., change-prone files) are getting closer while files from different categories are getting further, as shown in Fig. 1(b). In this way, the files represented by the transformed metrics can be easily classified. For this advantage, the deep metric learning technique has been widely applied in other researches [37]. To the best of our knowledge, this technique has not been studied in the change-proneness prediction. The validity of the proposed model is verified by investigating 20 change-proneness datasets from 5 open source Java projects [39], and comparing it with 5 typical cross-project change-proneness models. Experimental results show that DML substantially outperforms 5 baseline models.

2   Related Work and Background

This section describes related work of the change-proneness prediction, and background of the metric learning.

Change-Proneness Prediction. Change-prone class files have been studied since two decades ago [24]. It is shown that larger sized classes are likely to be change-prone, in terms of the lines of code [21, 24]. Later, researchers found that some OO metrics are strongly correlated with the change-proneness, including coupling, cohesion, inheritance, complexity, and etc. [2, 4, 5] However, the highest values of those metrics do not always indicate the change-prone files [21]. For example, a large file in a newly initialized project tends to be change-prone, because developers are likely to make mistakes in that complex file. Meanwhile, a large file in a stable project would not cause modifications, as developers prefer to focus on other vulnerable code files. To mitigate this problem, researchers studied the combination of the code metrics and validated that the CKJM OO metric suite associated with the size metric is an effective combination [11, 28]. Following their researches, we adopt this metric combination as our model inputs.

Based on these software metrics, many ML models have been applied for change-proneness prediction, such as tree-based models [21, 22], artificial neural network [42], support vector machine [29], and ensemble methods [10, 13, 36]. However, such ML models are trained by historical labeled data within a project, namely the within-project prediction, and they cannot be applied to new projects because manually collecting labeled data in the new projects is time-consuming and expensive [18, 43].

To solve this shortcoming of within-project prediction, a better solution is to fully leverage the abundant labeled data from other projects, which is called the cross-project prediction [6,27]. It trains a change-proneness model by using labeled data from other projects, i.e. source projects, and predicts the change-prone files in a target project. However, in many cases, the cross-project prediction is not accurate for classification [27], and we assume that this phenomenon is caused by the large distinction in the metrics from different source projects, as referred to in Sect. 1.

Metric Learning. With the goal of learning better metric representation for classification, many metric learning algorithms have been proposed over the past decade [7,20]. Their common objective is to learn a transformation, where the distance between files from the same categories is reduced and that of different categories is enlarged as much as possible.

However, most existing metric learning methods only learn a linear transformation to map files into a new feature space, which is not powerful to reach the goal. To address this limitation, the kernel trick, a non-linear solution, is usually adopted, which first maps files into a high-dimensional feature space, and then learns a discriminative distance metric in the high-dimensional space [40,44]. However, the kernel trick based methods cannot process large-scale dataset efficiently, therefore a deep metric learning is developed to explicitly learn a non-linear transformation, which proves its substantial advantages over previous methods. Thus, we apply this deep learning based model to our research. Different from this DML model, we also leverage the SMOTE [8] method to handle our imbalanced dataset for better model training. Details on the proposed model are shown in the following section.

3   Methodology

This section describes the proposed model for the cross-project change-proneness prediction. We first present the main part of our DML model, then show the SMOTE method that processes training data for DML, and finally describe the classifier for change-proneness prediction.

3.1   Deep Metric Learning

DML aims to improve the cross-project prediction by minimizing metric distinction in different source projects. Generally, DML learns a multi-layer neural network to transform files represented by 7 software metrics (defined in Sect. 4) from different source projects, where within-category files (e.g., change-prone files) are getting closer while between-category files are getting larger. As the model performance is affected by many factors in the built network, such as its network structure, we thus provide the optimal model setting in this section and investigate the parameter impacts in Sect. 4.

Neural Network. The built network contains 5 layers with 3 hidden layers, 1 input and 1 output layer, where each layer has 7 nodes. Precisely, to transform metric values of n files, the network takes all these files $X \in \mathbb{R}^{n \times 7}$ for its input layer, and the output of the first layer is $H^{(1)} = s(W^{(1)}X + b^{(1)}) \in \mathbb{R}^{7 \times 7}$, then passes them through each layer from the first to the last. For the m-th layer, its output is:

$$H^{(m)} = s(W^{(m)}H^{(m-1)} + b^{(m)}) \in \mathbb{R}^{7 \times 7} \tag{1}$$

where $W^{(m)}$ and $b^{(m)}$ are the weights and biases of the m-th layer, respectively; s is a *tanh* activation function, $s(x) = (e^x - e^{-x})/(e^x + e^{-x})$. As weights and biases in network need to be initialized at first, we adopt a prevalent random initialization method as Hu et al. [17]. In specific, for m-th layer, all elements in the biases $b^{(m)}$ are set to zero, while the weights $W^{(m)}$ is initialized by an uniform random numbers ranging from $-\sqrt{6}/\sqrt{p^{(m)} + p^{(m-1)}}$ to $\sqrt{6}/\sqrt{p^{(m)} + p^{(m-1)}}$, where $p^{(m)}$ is the number of nodes in the m-th layer.

Objective Function. The purpose of DML is to learn a non-linear mapping f, i.e. the built network, that makes the distance between the files of the same category less than a threshold α, and enable the distance between the different categories greater than the threshold α. To optimize the built network, we adopt the objective function as below following [17]:

$$arg \min_f L = \frac{1}{n} \sum_{i,j}^{n/2} y_{i,j}D_{i,j}^2 + (1 - y_{i,j})g(\alpha - D_{i,j}^2)$$
$$+ \frac{\lambda}{2} \sum_{m=1}^{M} (||W^{(m)}||_F^2 + ||b^{(m)}||_2^2) \tag{2}$$

where x_i and x_j are the i-th and j-th files in training data; g is a logistic function, $g(x) = \frac{1}{\beta}log(1 + e^{\beta x})$; $D_{i,j}^2$ measures the difference between the outputs of two files $f(x_i)$ and $f(x_j)$, and $D_{i,j}^2 = ||f(x_i) - f(x_j)||_2^2$; $y_{i,j} = 1$ if two files x_i and x_j are from the same category, otherwise $y_{i,j} = 0$; λ is a regularization term to control the complexity of weights $W^{(m)}$ and biases $b^{(m)}$, and we $\lambda = 1$ in default as lower complexity is beneficial to the model performance; $\alpha = 0.5$ for its better performance in the experiment.

Gradient Descent Method for Network Optimization. By leveraging above objective function, we randomly select pairs of files as the model input, and iteratively optimize the weights and biases in the network by a gradient descent method, according to Hu et al. [17], it first computes the gradients of the loss function, and then updates the parameters $W^{(m)}, b^{(m)}$ in each layer and $m = 1, 2, ..., M$. The optimization stops when L converges to a small coefficient $\xi = 10^{-2}$, or all the training data has been used. Specifically, the gradients of the objective function L with respect to the parameters $W^{(m)}$ and $b^{(m)}$ can be computed as follows:

$$\frac{\delta L}{\delta W^{(m)}} = \frac{1}{n} \sum_{i,j}^{n/2} \left(\Delta_{ij}^{(m)} H_i^{(m-1)^T} + \Delta_{ji}^{(m)} H_j^{(m-1)^T} \right) + \lambda W^{(m)}$$

$$\frac{\delta L}{\delta b^{(m)}} = \frac{1}{n} \sum_{i,j}^{n/2} \left(\Delta_{ij}^{(m)} + \Delta_{ji}^{(m)} \right) + \lambda b^{(m)} \tag{3}$$

where the updating equations $\Delta_{ij}^{(m)}$ and $\Delta_{ji}^{(m)}$ are:

$$\Delta_{ij}^{(M)} = 2 \left(y_{ij} - (1 - y_{ij}) g'(\alpha - D_{i,j}^2) \right) \left(x_i^{(M)} - x_j^{(M)} \right) \odot g' \left(z_i^{(M)} \right)$$

$$\Delta_{ji}^{(M)} = 2 \left(y_{ij} - (1 - y_{ij}) g'(\alpha - D_{i,j}^2) \right) \left(x_j^{(M)} - x_i^{M)} \right) \odot g' \left(z_j^{(M)} \right)$$

$$\Delta_{ij}^{(m)} = \left(W^{(m+1)^T} \Delta_{ij}^{(m+1)} \right) \odot g' \left(z_i^{(m)} \right)$$

$$\Delta_{ji}^{(m)} = \left(W^{(m+1)^T} \Delta_{ji}^{(m+1)} \right) \odot g' \left(z_j^{(m)} \right) \tag{4}$$

Then, $W^{(m)}$ and $b^{(m)}$ can be updated by the following equations, where μ is the learning rate ($\mu = 10^{-3}$):

$$W^{(m)} = W^{(m)} - \mu \frac{\delta L}{\delta W^{(m)}}, b^{(m)} = b^{(m)} - \mu \frac{\delta J}{\delta b^{(m)}} \tag{5}$$

3.2 SMOTE

The training data in the change-proneness dataset is highly imbalanced [26], namely the percentage of change-prone files is varying in different source projects, as illustrated in Table 1. Highly imbalanced dataset would lead to an over-sampling problem to our model training, so that the trained model cannot adapt to the metric distinction problem with a high generalizability [19].

To overcome this issue, we leverage an approach called SMOTE (Synthetic Minority Oversampling Technique) [8], which creates minority instances (e.g., change-prone files) by assuming that they are lying near that category of real instances. Specifically, supposing we have ten change-prone files and fifty not-change-prone files, to create underlying change-prone instances we randomly pick two change-prone files in the space represented by their metric values, then synthesize a new instance along the line segments between the two selected files, where the distance proportion of the new instance to two files is generated by a uniform random number ranging from 0 to 1. In other study, we perform SMOTE on training data of change-proneness dataset to balance its minority instances and then use the processed training data to build the DML.

3.3 Classifier

In our change-proneness prediction, DML learns a better metric representation for files but the classifier finally determines the prediction accuracy. In the study, we use a decision tree algorithm named J48 [3] as the classifier, which is implemented by invoking the Weka tool with the default setting. This decision tree is based on an up-button strategy and the partition strategy of recursion [3]. It selects a metric to be placed in the root node, generates a branch for each possible metric value, and divides files into multiple subsets. Each subset corresponds to the branch of a root node, and recursively repeats the process on each branch. This process stops when all files have the same classification. We adopt this classifier due to its better performance in cross-project change-proneness prediction comparing with 4 other commonly used classifiers, as shown in Sect. 5.

4 Experiment

This section describes the studied dataset, dependent variable, independent variable, and evaluation criteria.

Dataset. To verify the validity of the proposed model, we investigate 20 change-proneness datasets from 5 open source projects, collected from a public dataset Qualitas Corpus [39]. Table 1 provides the number of files (#File) and percentage of change-prone files (%Change) in each project. This dataset can validate the model ability for a wide range, in terms of the file numbers and change-prone percentages.

Table 1. Studied 20 change-proneness datasets.

No.	Project	#Files	%Changed	No.	Project	#Files	%Changed
1	Argouml-0.24.0	1330	74.3%	11	Argouml-0.26.2	1504	31.1%
2	Argouml-0.30.2	1501	17.0%	12	Argouml-0.32.2	1505	40.0%
3	Freecol-0.9.0	491	31.4%	13	Freecol-0.9.2	493	54.6%
4	Freecol-0.9.3	484	04.7%	14	Freecol-0.9.4	484	03.9%
5	Jmeter-2.5.1	763	34.7%	15	Jmeter-2.6.0	808	20.2%
6	Jmeter-2.7.0	818	40.1%	16	Jmeter-2.8.0	830	58.1%
7	Jung-1.6.0	341	37.0%	17	Jung-1.7.2	464	07.1%
8	Jung-1.7.4	468	28.9%	18	Jung-1.7.5	467	03.0%
9	Weka-3.5.5	925	82.6%	19	Weka-3.5.6	969	12.1%
10	Weka-3.5.7	994	68.6%	20	Weka-3.5.8	1119	14.0%

Dependent Variable. The change-proneness of files refers to whether these files are changed in their next releases [9]. To measure the dependent variable, we first find the following release for each target change-proneness data in Table 1 from the Qualitas Corpus [39], then compare whether each file is changed via the version control system. In the comparison, a file is labeled as change-prone for any addition, deletion, or modification; otherwise, this file is labeled as not-change-prone [22].

Independent Variables. To capture the pattern of the change-proneness, we measure 7 software metrics that are strongly correlated to the change-proneness, following previous studies [25,27,38]. Table 2 lists the definitions of measured metrics, and these metrics are extracted by performing the CKJM[2] (Chidamber and Kemerer Java metric) tool [23] on obtained source code.

Table 2. Software metrics.

Metric	Description
DIT	Depth of Inheritance Tree
NOC	Number of Children
CBO	Coupling between Object Classes
WMC	Weighted Methods per Class
RFC	Response for a Class
LCOM	Lack of Cohesion in Methods
SLOC	Source Lines of Code

Training and Testing Data. With the collected 20 change-proneness datasets, We validate the proposed model by the leave-one-out cross-validation [34]. Specifically, for a prediction, one of these 20 projects is used as a testing data, and the rest of projects are used for model training. And we perform this kind of cross-project prediction in 20 times and report the mean and standard deviation of the prediction performance. Moreover, to reduce the effects of outliers in datasets, 7 measured metric values of files in a project are processed by the z-score normalization [30].

Evaluation Criteria. There are 4 types of prediction results: truly or falsely identifying change-prone files, respectively called TP (True Positive) and FP (False Positive); correctly or wrongly predicting not-change-prone files, respectively called TN (True Negative) and FN (False Negative) [32]. Precision and recall are usually used for model evaluation [33]. Precision indicates that the proportion of the changed files that are correctly predicted in the total number of changed files that is predicted by the model, $precision = TP/(TP+FP)$ [14]. Meanwhile, recall measures the proportion of the changed files that are correctly predicted in the total number of changed files, namely $recall = TP/(TP+FN)$ [41]. However, a model with a high precision usually leads to a low recall and vice versa. To avoid this dilemma, the F-measure, the harmonic mean of precision and recall, is widely used to measure the performance of change-proneness prediction [33].

---

[2] CKJM: https://www.spinellis.gr/sw/ckjm/.

5   Results and Discussion

This section investigates the effectiveness of the proposed model and the impacts of model parameters.

Prediction Performance. In the experiment, we compare DML with 5 typical cross-project change-proneness models [12,27] as listed in Table 3. These baseline models are re-implemented by invoking Weka tool [16] with default settings. The table also presents the performance comparison of the proposed model DML with these baselines in terms of mean ± standard deviation of F-measure across 20 change-proneness datasets.

Results show that the decision tree J48 is the best baseline for the cross-project change-proneness prediction, with mean F-measure 0.381. This is the reason why we choose this decision tree algorithm as our classifier (Sect. 3.3). Moreover, we can notice that DML achieves the best performance with mean F-measure 0.408, outperforming the second best by 7.01%. This result indicates that by applying the techniques of deep metric learning and SMOTE to the classifier J48, the performance of change-proneness prediction can be substantially improved, due to the reduction of the metric distinction between files from multiple source projects.

Table 3. Performance comparison of DML with 5 baseline models, where performance is represented by mean ± standard deviation of F-measure across 20 change-proneness datasets.

No.	Classifier	F-measure
1	Naive Bayes	0.312 ± 0.041
2	Support vector machine	0.353 ± 0.046
3	Logistic regression	0.379 ± 0.074
4	Decision table	0.369 ± 0.079
5	Decision tree (J48)	0.381 ± 0.065
6	DML	0.408 ± 0.089

Impacts of Model Parameters. As the performance of DML is largely influenced by the learning rate μ, the coefficient α, and the number of network hidden layers p, we investigate their impacts on the model respectively, namely adjusting one parameter at a time and keeping other parameters in default.

Table 4 shows the performance of DML set with different learning rate μ. We can observe that DML achieves the best performance when $\mu = 10^{-1}$ in terms of F-measure 0.408, due to its large precision 0.342 and recall 0.505. And decreasing learning rate value not only increases the time of model optimization but also reduces the model performance. Hence, setting the learning rate as 0.1 is optimal.

Table 4. Performance comparison of DML with different learning rate μ.

μ	Precision	Recall	F-measure
10^{-1}	**0.342**	**0.505**	**0.408**
10^{-2}	0.316	0.483	0.382
10^{-3}	0.289	0.462	0.356
10^{-4}	0.295	0.470	0.363
10^{-5}	0.296	0.475	0.365
10^{-6}	0.296	0.475	0.365
10^{-7}	0.296	0.475	0.365
10^{-8}	0.296	0.475	0.365

Table 5 compares the model performance configured with different coefficient α. Results indicate that when α equals 0.5, the model obtains the best values on F-measure 0.408 and recall 0.505. Although increasing the value of α can boost the prediction precision, the recall and F-measure values are substantially decreased. This condition implies that mapping same-category, e.g. change-prone, files more compact is advantageous to the reduction of metric distinction and beneficial to the cross-project change-proneness prediction.

Table 6 investigates the prediction performance of DML with different number of network hidden layers. We can notice that DML reaches the best performance with 3 hidden layers, in terms of recall and F-measure. Even though building more hidden layers may improve the model precision, the recall and F-measure cannot be enhanced simultaneously. This result suggests that a deeper network may miss the pattern of change-prone files, which may be caused by the disappearing gradients in optimizing the deeper network. Thus, it is worth to incorporate the deep residual network [15] in our model for further improvement in the near future.

Table 5. Performance comparison of DML with different coefficient α.

α	Precision	Recall	F-measure
0.5	0.342	**0.505**	**0.408**
1.0	0.328	0.490	0.393
1.5	0.322	0.471	0.383
2.0	0.356	0.452	0.380
2.5	0.344	0.418	0.378
3.0	0.337	0.439	0.382
3.5	0.354	0.391	0.372
4.0	0.356	0.390	0.373
4.5	0.373	0.384	0.379
5.0	**0.378**	0.392	0.385

Table 6. Performance comparison of DML with different number of network hidden layers p.

p	Precision	Recall	F-measure
3	0.342	**0.505**	**0.408**
4	0.345	0.493	0.406
5	0.350	0.498	0.404
6	**0.358**	0.466	0.405
7	0.340	0.487	0.400

6   Conclusion

Identifying change-prone files in the software development and maintenance has long been a challenging problem, especially building a model with abundant labeled data from other projects, i.e. the cross-project prediction. However, the cross-project prediction model has poor performance for the incapability of handling metric distinction of files from different source projects, so that the built model is not trained well. To solve this problem, we propose a Deep Metric Learning (DML) model to mitigate such metric distinction problem and leverage a SMOTE method to cope with the over-sampling issue in training data. The proposed model is verified by 20 change-proneness datasets, and compared with 5 typical cross-project change-proneness models. Experimental results indicate that DML outperforms baselines substantially.

Acknowledgement. The work described in this paper was partially supported by the Fundamental Research Funds for the Central Universities of China (No. 106112017CDJXSYY002), the National Natural Science Foundation of China (Grant no. 61402062, 61602068, 61602069, 61772093), Chongqing Research Program of Basic Science & Frontier Technology (Grant no. cstc2015jcyjA40037, cstc2016jcyjA0458, cstc2016jcyjA0468), and the Chongqing Major Theme Program (Grant No. cstc2017zdcy-zdzxX0002).

References

1. Anbalagan, P., Vouk, M.: On predicting the time taken to correct bug reports in open source projects. In: IEEE International Conference on Software Maintenance, ICSM 2009, pp. 523–526. IEEE (2009)
2. Arisholm, E., Briand, L.C., Foyen, A.: Dynamic coupling measurement for object-oriented software. IEEE Trans. Softw. Eng. **30**(8), 491–506 (2004)
3. Bhargava, N., Sharma, G., Bhargava, R., Mathuria, M.: Decision tree analysis on j48 algorithm for data mining. Proc. Int. J. Adv. Res. Comput. Sci. Softw. Eng. **3**(6) (2013)

4. Bieman, J.M., Andrews, A.A., Yang, H.J.: Understanding change-proneness in OO software through visualization. In: 2003 11th IEEE International Workshop on Program Comprehension, pp. 44–53. IEEE (2003)
5. Bieman, J.M., Jain, D., Yang, H.J.: OO design patterns, design structure, and program changes: an industrial case study. In: 2001 Proceedings of IEEE International Conference on Software Maintenance, pp. 580–589. IEEE (2001)
6. Briand, L.C., Melo, W.L., Wust, J.: Assessing the applicability of fault-proneness models across object-oriented software projects. IEEE Trans. Softw. Eng. **28**(7), 706–720 (2002)
7. Cai, X., Wang, C., Xiao, B., Chen, X., Zhou, J.: Deep nonlinear metric learning with independent subspace analysis for face verification. In: Proceedings of the 20th ACM International Conference on Multimedia, pp. 749–752. ACM (2012)
8. Chawla, N.V., Bowyer, K.W., Hall, L.O., Kegelmeyer, W.P.: Smote: synthetic minority over-sampling technique. J. Artif. Intell. Res. **16**, 321–357 (2002)
9. Elish, M.O., Al-Rahman Al-Khiaty, M.: A suite of metrics for quantifying historical changes to predict future change-prone classes in object-oriented software. J. Softw.: Evol. Process **25**(5), 407–437 (2013)
10. Elish, M.O., Aljamaan, H., Ahmad, I.: Three empirical studies on predicting software maintainability using ensemble methods. Soft Comput. **19**(9), 2511–2524 (2015)
11. Eski, S., Buzluca, F.: An empirical study on object-oriented metrics and software evolution in order to reduce testing costs by predicting change-prone classes. In: 2011 IEEE Fourth International Conference on Software Testing, Verification and Validation Workshops (ICSTW), pp. 566–571. IEEE (2011)
12. Fukushima, T., Kamei, Y., McIntosh, S., Yamashita, K., Ubayashi, N.: An empirical study of just-in-time defect prediction using cross-project models. In: Proceedings of the 11th Working Conference on Mining Software Repositories, pp. 172–181. ACM (2014)
13. Giger, E., Pinzger, M., Gall, H.C.: Can we predict types of code changes? An empirical analysis. In: 2012 9th IEEE Working Conference on Mining Software Repositories (MSR), pp. 217–226. IEEE (2012)
14. Goutte, C., Gaussier, E.: A probabilistic interpretation of precision, recall and F-score, with implication for evaluation. In: Losada, D.E., Fernández-Luna, J.M. (eds.) ECIR 2005. LNCS, vol. 3408, pp. 345–359. Springer, Heidelberg (2005). https://doi.org/10.1007/978-3-540-31865-1_25
15. He, K., Zhang, X., Ren, S., Sun, J.: Deep residual learning for image recognition. In: Proceedings of the IEEE Conference on Computer Vision and Pattern Recognition, pp. 770–778 (2016)
16. Holmes, G., Donkin, A., Witten, I.H.: Weka: a machine learning workbench. In: 1994 Proceedings of the 1994 Second Australian and New Zealand Conference on Intelligent Information Systems, pp. 357–361. IEEE (1994)
17. Hu, J., Lu, J., Tan, Y.P.: Discriminative deep metric learning for face verification in the wild. In: Proceedings of the IEEE Conference on Computer Vision and Pattern Recognition, pp. 1875–1882 (2014)
18. Huang, Y., Huang, Z., Wang, Y., BingWu, F.: Survey on data driven software defects prediction. Chin. J. Electron. **4**, 982–988 (2017)
19. Jeatrakul, P., Wong, K.W., Fung, C.C.: Classification of imbalanced data by combining the complementary neural network and SMOTE algorithm. In: Wong, K.W., Mendis, B.S.U., Bouzerdoum, A. (eds.) ICONIP 2010. LNCS, vol. 6444, pp. 152–159. Springer, Heidelberg (2010). https://doi.org/10.1007/978-3-642-17534-3_19

20. Koestinger, M., Hirzer, M., Wohlhart, P., Roth, P.M., Bischof, H.: Large scale metric learning from equivalence constraints. In: 2012 IEEE Conference on Computer Vision and Pattern Recognition (CVPR), pp. 2288–2295. IEEE (2012)

21. Koru, A.G., Tian, J.: Comparing High-Change Modules and Modules with the Highest Measurement Values in Two Large-Scale Open-Source Products. IEEE Press (2005)

22. Koru, A.G., Liu, H.: Identifying and characterizing change-prone classes in two large-scale open-source products. J. Syst. Softw. **80**(1), 63–73 (2007)

23. Kumar, L., Rath, S.K., Sureka, A.: Using source code metrics to predict change-prone web services: a case-study on ebay services. In: IEEE Workshop on Machine Learning Techniques for Software Quality Evaluation (MaLTeSQuE), pp. 1–7. IEEE (2017)

24. Lindvall, M.: Are large C++ classes change-prone? An empirical investigation. Softw.-Practice Exp. **28**(15), 1551–1558 (1998)

25. Lu, H., Zhou, Y., Xu, B., Leung, H., Chen, L.: The ability of object-oriented metrics to predict change-proneness: a meta-analysis. Empirical Softw. Eng. **17**(3), 200–242 (2012)

26. Lusa, L.: Smote for high-dimensional class-imbalanced data. BMC Bioinform. **14**(1), 106 (2013)

27. Malhotra, R., Bansal, A.J.: Cross project change prediction using open source projects. In: 2014 International Conference on Advances in Computing, Communications and Informatics (ICACCI), pp. 201–207. IEEE (2014)

28. Malhotra, R., Khanna, M.: Investigation of relationship between object-oriented metrics and change proneness. Int. J. Mach. Learn. Cybern. **4**(4), 273–286 (2013)

29. Malhotra, R., Khanna, M.: Examining the effectiveness of machine learning algorithms for prediction of change prone classes. In: 2014 International Conference on High Performance Computing & Simulation (HPCS), pp. 635–642. IEEE (2014)

30. Margulies, M., et al.: Genome sequencing in microfabricated high-density picolitre reactors. Nature **437**(7057), 376 (2005)

31. Mens, T., Tourwé, T.: A survey of software refactoring. IEEE Trans. Softw. Eng. **30**(2), 126–139 (2004)

32. Moser, R., Pedrycz, W., Succi, G.: A comparative analysis of the efficiency of change metrics and static code attributes for defect prediction. In: Proceedings of the 30th International Conference on Software Engineering, pp. 181–190. ACM (2008)

33. Powers, D.M.: Evaluation: from precision, recall and F-measure to ROC, informedness, markedness and correlation (2011)

34. Refaeilzadeh, P., Tang, L., Liu, H.: Cross-validation. In: Liu, L., Özsu, M.T. (eds.) Encyclopedia of Database Systems, pp. 532–538. Springer, Heidelberg (2009). https://doi.org/10.1007/978-0-387-39940-9_565

35. Riaz, M., Mendes, E., Tempero, E.: A systematic review of software maintainability prediction and metrics. In: Proceedings of the 2009 3rd International Symposium on Empirical Software Engineering and Measurement, pp. 367–377. IEEE Computer Society (2009)

36. Romano, D., Pinzger, M.: Using source code metrics to predict change-prone Java interfaces. In: 2011 27th IEEE International Conference on Software Maintenance (ICSM), pp. 303–312. IEEE (2011)

37. Song, H.O., Xiang, Y., Jegelka, S., Savarese, S.: Deep metric learning via lifted structured feature embedding. In: 2016 IEEE Conference on Computer Vision and Pattern Recognition (CVPR), pp. 4004–4012. IEEE (2016)

38. Spinellis, D.: ckjm chidamber and kemerer metrics software. Technical report, v 1.6. Technical report, Athens University of Economics and Business (2005). http:// www.spinellis.gr/sw/ckjm

39. Tempero, E., et al.: The qualitas corpus: a curated collection of Java code for empirical studies. In: 2010 17th Asia Pacific Software Engineering Conference (APSEC), pp. 336–345. IEEE (2010)

40. Tsang, I.W., Kwok, J.T., Bay, C., Kong, H.: Distance metric learning with kernels. In: Proceedings of the International Conference on Artificial Neural Networks, pp. 126–129. Citeseer (2003)

41. Tsuruoka, Y., Tsujii, J.: Boosting precision and recall of dictionary-based protein name recognition. In: Proceedings of the ACL 2003 Workshop on Natural Language Processing in Biomedicine, vol. 13, pp. 41–48. Association for Computational Linguistics (2003)

42. Van Koten, C., Gray, A.: An application of Bayesian network for predicting object-oriented software maintainability. Inf. Softw. Technol. **48**(1), 59–67 (2006)

43. Wang, D., Wang, Q.: Improving the performance of defect prediction based on evolution data. Chin. J. Softw. **27**(12), 3014–3029 (2016)

44. Yeung, D.Y., Chang, H.: A kernel approach for semisupervised metric learning. IEEE Trans. Neural Netw. **18**(1), 141–149 (2007)

45. Zhou, Y., Leung, H.: Predicting object-oriented software maintainability using multivariate adaptive regression splines. J. Syst. Softw. **80**(8), 1349–1361 (2007)

Infrared-Visible Image Fusion Based on Convolutional Neural Networks (CNN)

Xianyi Ren[1,2,3]([✉]), Fanyang Meng[1,2,3], Tao Hu[1,2,3], Zhijun Liu[1], and Changwei Wang[1,2,3]

[1] Shenzhen Institute of Information Technology, Shenzhen, China
renxianyi@tsinghua.org.cn, fymeng@pku.edu.cn,
happy.hut@163.com, zjlieu@126.com, wangcw@sziit.edu.cn
[2] Shenzhen Key Lab for Visual Media Processing and Streaming Media, Shenzhen, China
[3] Intelligent Vision Engineering Research Center of Guangdong Province, Guangzhou, China

Abstract. Image fusion is a process of combing multiple images of the same scene into a single image with the aim of preserving the full content information and retaining the important features from each of the original images. In this paper, a novel image fusion method based on Convolutional Neural Networks (CNN) and saliency detection is proposed. Here, we use the image representations derived from CNN Network optimized for infrared-visible image fusion. Since the lower layers of the network can seize the exact value of the original image, and the high layers of the network can capture the high-level content in terms of objects and their arrangement in the input image, we exploit more low-layer features of visible image and more high-layer features of infrared image in the fusion. And during the fusion procedure, the infrared target of an infrared image is effectively highlighted using saliency detection method and only the salient information of the infrared image will be fused. The method aimed to preserve the abundant detail information from visible image as much as possible, meanwhile preserve the salient information in the infrared image. Experimental results show that the proposed fusion method is rather promising.

Keywords: Image fusion · Convolutional Neural Networks (CNN) Saliency detection

1 Introduction

Image fusion is the combination of two or more different images to form a new image by using a certain algorithm. The combination of sensory data from multiple sensors can provide reliable and accurate information. For example, visible images have high spatial resolution and abundant texture details while they are limited by the light conditions and can't detect the hidden targets. While the infrared images have the extraordinary ability to detect the hidden targets or detect the person at night but they have poor image quality because of low contrast and details. The fusion of the visible image and infrared image combine the information of two images and is very useful to improve the ability of target detection and scene description. Image fusion has been

© Springer Nature Switzerland AG 2018
Y. Peng et al. (Eds.): IScIDE 2018, LNCS 11266, pp. 301–307, 2018.
https://doi.org/10.1007/978-3-030-02698-1_26

applied to many areas such as computer vision, automatic target recognition, remote sensing, robotics and medical image processing, etc.

Many fusion methods are implemented so far in literature. Among them multi-scale decomposition methods (pyramid, wavelet, contourlet, etc.) and data driven methods are the most successful methods. But these methods may introduce artifacts into the fused image. To overcome these problems optimization based fusion schemes are proposed. These methods take multiple iterations to find the optimal solution (fuse image). These optimization methods may oversmooth the fused image because of multiple iterations [1].

The recent advance of Deep Convolutional Neural Networks has produced powerful computer vision systems that learn to extract high-level semantic information from natural images and can do well in the task of object recognition [2]. It was shown that deep neural networks can extract image features which contain different information at each layer. In CVPR 2016, Gatys et al. proposed an image style transfer method based on CNN. [3] They use VGG-network [4] to extract deep features at different layers from the "content" image, "style" image and a generated image, respectively. The difference of deep features extracted from the generated image and source images is minimized by iteration. The generated image will contain the main object from the "content" image and texture features from the "style" image. Li proposed an infrared and visible image fusion using a deep learning framework. They decomposed the source images into base parts and detail content. The base parts are fused by weighted-averaging, and the deep learning network is used to extract the multi-layer features of the detail content and to get the fused detail content. The fused image is reconstructed by combing the fused base part and the detail content [5].

Inspired by the idea of them, we use the VGG network to extract the feature of the source images and fuse them by minimize the cost function which is constructed according to the task of the fusion. In this work, we show how the generic feature representations learned by high-performing CNN can be used to combine the information from visible image and infrared image. The fusion image is generated by incorporating different ratios of different layers features of the source images and only the salient regions in the infrared image take part in the fusion.

The rest of the paper is organized as follows: Sect. 2 describes the CNN and analyzes the image representation and fusion strategy of the fusion algorithm based on the CNN. Experimental results and the comparison with some other methods are shown in Sect. 3. In the Sect. 4, a conclusion is drawn.

2 Image Representations of the CNN

Many studies have shown that deep neural networks can extract different levels of image features. Here we use a widely used CNN network: VGG network, which was trained to perform object recognition and localization. [6] And we used the feature space provided by a normalized version of the 16 convolutional and 5 pooling layers of the 19-layer VGG network.

2.1   Content Representation of the CNN

As Gatys [4] pointed out, each layer in the network defines a non-linear filter bank whose complexity increases with the position of the layer in the network. Hence a given input image $\vec{x}$ is encoded in each layer of the CNN Network by the filter responses to that image. A layer with N_l distinct filters has N_l feature maps each of size M_l, where M_l is the height times the width of the feature map. So the responses in a layer l can be stored in a matrix $F^l \in R^{N_l \times M_l}$, where F_{ij}^l is the activation of the ith filter at position j in layer l.

To visualize the image information that is encoded at different layers of the hierarchy one can perform gradient descent on a white noise image to find another image that matches the feature responses of the original image. [6] Let $\vec{p}$ and $\vec{x}$ be the original image and the noise image that is generated separately, and P^l and F^l are their respective feature representations in layer l. We then define the squared-error loss between the two feature representations:

$$\gamma_{content}(\vec{p}, \vec{x}, l) = \frac{1}{2} \sum_{i,j} (F_{ij}^l - P_{ij}^l)^2 \tag{1}$$

The derivative of this loss with respect to the activations in layer l equals:

$$\frac{\partial \gamma_{content}}{\partial F_{i,j}^l} = \begin{cases} (F^l - P^l)_{i,j} & if \quad F_{i,j}^l > 0 \\ 0 & else \end{cases} \tag{2}$$

From which the gradient with respect to the image $\vec{x}$ can be computed using standard error back-propagation. Thus we can change the initially random image $\vec{x}$ until it generates the same responses in a certain layer of the CNN network as the original image $\vec{p}$.

When CNN Network is trained on object recognition, it develops a representation of the image that makes object information increasingly explicit along the processing hierarchy. [7] Therefore, along the processing hierarchy of the network, the input images is transformed into representations that are increasingly sensitive to the actual content of the image, but become relatively invariant to its precise appearance. Thus, higher layers in the network capture the high-level content in terms of objects and their arrangement in the input image but do not constrain the exact pixel values of the reconstruction very much. In contrast, reconstructions from the lower layers simply reproduce the exact pixel values of the original image.

2.2   The Visible and Infrared Image Fusion Based on CNN

Visible images have high spatial resolution and abundant texture details while they are limited by the light conditions and can't detect the hidden targets. While the infrared images have the extraordinary ability to detect the hidden targets or detect the person at night but they have poor image quality because of low contrast and details. We can easily draw the conclusion that the fused image should keep the abundant details and the high resolution of the visible image while preserve the object information in the infrared images.

Based on the conclusion of the last section, we can exploit more of the higher layer information of the salient infrared image as the salient object information and exploit more of the lower layer information of the visible image as the content information and then combine them to get the fused image.

To fuse the visible image $\vec{p}$ and infrared image $\vec{q}$, we synthesis a new image that simultaneously matches the content representation of $\vec{p}$ and the object representation of $\vec{q}$. Thus we jointly minimize the distance of the feature representations of a white noise image from the content representation of the visible image in several lower layers and the object representation of the infrared image on several layers of the VGG Network. The loss function we minimize is:

$$\gamma_{total}(\vec{p}, \vec{a}, \vec{x}) = \alpha\gamma_{content}(\vec{p}, \vec{a}, \vec{x}) + \beta\gamma_{object}(\vec{p}, \vec{a}, \vec{x}) \tag{3}$$

Where α and β are the weighting factors for content and object reconstruction, respectively. In the lower layers, the value of parameter α can be much larger than the value of parameter β, and on the contrast, the value of parameter α can be much smaller than the value of parameter β in the higher layers. And in all layers, only the salient regions in the infrared image take part the object information extraction. And due to the gradient with respect to the pixel values $\frac{\partial\gamma_{total}}{\partial\vec{x}}$ can be used as input for some numerical optimization strategy. Here we also use L-BFGS [8] which works best for image synthesis.

3 Experimental Results

In order to evaluate the effectiveness of the proposed algorithm, we test the algorithm on several different visible and infrared images and compare the results with some other fusion methods: the average algorithm, the contrast pyramid algorithm, the FSD pyramid algorithm, the dwt-based algorithm and the Morphological pyramid. The original visible and infrared images and the fused images using the four methods are shown in Fig. 1. Among which, Fig. 1(a), (b) are two source images, Fig. 1(c), (d), (e) and (f) are the fused image using the average algorithm, the contrast pyramid algorithm, the FSD pyramid algorithm, the dwt-based algorithm, the Morphological pyramid, and proposed algorithm, respectively.

As we can see from Fig. 1((a)–(h)), the fused image obtained by the proposed method can preserve more detail information in the source images and contains less artificial noise. Compared with other methods, the fused image obtain by proposed method contains most texture information in the visible image and the salient object in the infrared image. At the same time, the fused image contains much less artifacts and looks more natural.

On the other hand, we noticed that the brightness of the person in the fused image is a little weaker than the infrared image. The possible reason is that low-layer characters of the infrared image which represent the exact pixel values are less involved in the fusion and more low-layer characters of the visible image are involved.

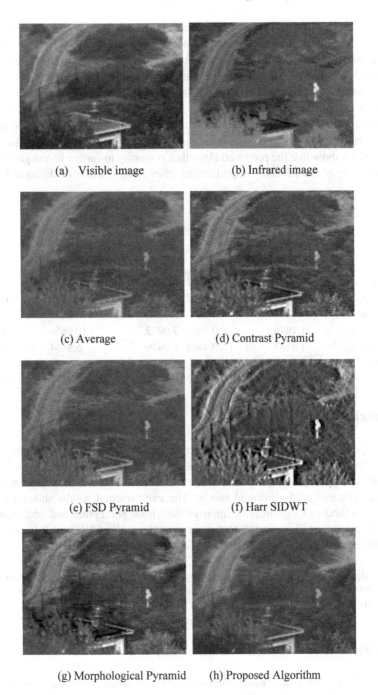

(a) Visible image   (b) Infrared image

(c) Average   (d) Contrast Pyramid

(e) FSD Pyramid   (f) Harr SIDWT

(g) Morphological Pyramid   (h) Proposed Algorithm

Fig. 1. The original visible and infrared images and the fused images

Other than the subjectively visible effects, an objective and quantitative evaluation of the effects of fused image is needed. We adopt five kinds of quantitative measures to evaluate the fusion results for above four fusion methods: mutual information, $Q_{AB/F}$ (Gradient based fusion performance metric), information entropy, average gradient, weighted structural similarity (MSSIM). Table 1 shows the quantitative analysis results of the four fusion methods. From the Table 1, we can see that the proposed algorithm outperform other methods in the mutual information and $Q_{AB/F}$ metrics, and in other three quantitative metrics, the proposed algorithm is at the middle level. The experimental results show that the proposed algorithm is worthy to further investigate and it is possible to apply the proposed algorithm to other image fusion of different types of images.

Table 1. Evaluation of the fusion results of different algorithm

Algorithm	Mutual information	$Q_{AB/F}$	Information entropy	Average gradient	WSSIM
Average	1.5794	0.3721	6.2178	3.2953	0.7922
Contrast pyramid	1.5004	0.3682	6.7163	6.6259	0.7604
FSD pyramid	1.5274	0.4138	6.3262	4.8801	0.7696
Harr SIDWT	1.1949	0.2851	7.0973	10.8866	0.5472
Morphological pyramid	1.4887	0.4404	6.9479	6.8864	0.6540
Proposed algorithm	2.4690	0.4653	6.5896	4.1812	0.7388

4   Conclusions

In this paper, an infrared and visible image fusion algorithm based on Convolutional Neural Networks (CNN) is proposed. In the algorithm, the visible and infrared images are combined by exploiting more low-layer features of the visible image and more high-layer features of the infrared image. The experimental results showed that the proposed method can significantly improve the visual perception and objective evaluation. In the future, it is worthy to further investigate the CNN and it is possible to apply the method to other kind image fusion of different types of images.

Acknowledgments. This paper is supported by the National Natural Science Foundation of China (NSFC. No. 61271420), Scientific Research Platform Cultivation Project of SZIIT (PT201704), Scientific Research Project of SZIIT (ZY201715).

References

1. Bavirisetti, D.P., Dhuli, R.: Fusion of infrared and visible sensor images based on anisotropic diffusion and Karhunen-Loeve transform. IEEE Sens. J. **16**(1), 203–209 (2016)
2. Krizhevsky, A., Sutskever, I., Hinton, G.E.: Imagenet classification with deep convolutional neural networks. In: International Conference on Neural Information Processing Systems, pp. 1097–1105 (2012)

3. Gatys, L.A., Ecker, A.S.: Image style transfer using convolutional neural networks. In: IEEE Conference on Computer Vision and Pattern Recognition (CVPR), pp. 2414–2423. IEEE (2016)
4. Simonyan, K., Zisserman, A.: Very deep convolutional networks for large-scale image recognition. Comput. Sci. (2014)
5. Li, H., Wu, X., Kittler, J.: Infrared and Visible Image Fusion using a Deep Learning Framework. arXiv:1804.06992 [cs.CV] (2018)
6. Mahendran, A., Vedaldi, A.: Understanding deep image representation by inverting them. In: IEEE Conference on Computer Vision and Pattern Recognition (CVPR), pp. 5188–5196. IEEE (2015)
7. Gatys, L.A., Ecker, A.S., Bethge, M.: Texture synthesis using convolutional neural networks. In: Advances in Neural Information Processing Systems, vol. 70, no. 1, pp. 262–270 (2015)
8. Zhu, C., Byrd, R.H., Lu, P., Nocedal, J.: Algorithm 778: L-BFGS-B: fortran subroutine for large-scale bound-constrained optimization. ACM Trans. Math. Softw. (TOMS) 23(4), 550–560 (1997)

Objects and Language

Joint Spoken Language Understanding and Domain Adaptive Language Modeling

Huifeng Zhang[1], Su Zhu[1], Shuai Fan[2], and Kai Yu[1,2](✉)

[1] Key Laboratory of Shanghai Education Commission for Intelligent Interaction
and Cognitive Engineering, SpeechLab,
Department of Computer Science and Engineering,
Brain Science and Technology Research Center, Shanghai Jiao Tong University,
Shanghai, China
{lloydzhang,paul2204,kai.yu}@sjtu.edu.cn
[2] AI Speech Co., Ltd., Suzhou, China
shuai.fan@aispeech.com

Abstract. *Spoken Language Understanding* (SLU) aims to extract structured information from speech recognized texts, which suffers from inaccurate automatic speech recognition (ASR) (especially in a specific dialogue domain). *Language Modeling* (LM) is significant to ASR for producing natural sentences. To improve the SLU performance in a specific domain, we try two ways: (1) *domain adaptive* language modeling which tends to recognize in-domain utterances; (2) joint modeling of SLU and LM which distills semantic information to build semantic-aware LM, and also helps SLU by *semi-supervised* learning (LM is unsupervised). To unify these two approaches, we propose a *multi-task* model (MTM) that jointly performs two SLU tasks (slot filling and intent detection), domain-specific LM and domain-free (general) LM. In the proposed multi-task architecture, the shared-private network is utilized to automatically learn which part of general data can be shared by the specific domain or not. We attempt to further improve the SLU and ASR performance in a specific domain with a few of labeled data in the specific domain and plenty of unlabeled data in general domain. The experiments show that the proposed MTM can obtain 4.06% absolute WER (Word Error Rate) reduction in a car navigation domain, compared to a general-domain LM. For language understanding, the MTM outperforms the baseline (especially slot filling task) on the manual transcript, ASR 1-best output. By exploiting the domain-adaptive LM to rescore ASR output, our proposed model achieves further improvement in SLU (7.08% absolute F1 increase of slot filling task).

Keywords: Spoken language understanding · Domain adaptive
Language modeling · Semi-supervised · Multi-task

This work has been supported by the National Key Research and Development Program of China under Grant No. 2017YFB1002102, the China NSFC project (No. 61573241) and the JiangSu NSF project (BK20161244). Experiments have been carried out on the PI supercomputer at Shanghai Jiao Tong University.

Y. Peng et al. (Eds.): IScIDE 2018, LNCS 11266, pp. 311–324, 2018.
https://doi.org/10.1007/978-3-030-02698-1_27

1   Introduction

As a critical component in a traditional spoken dialogue system, spoken language understanding (SLU) extracts semantic information from texts interpreted from automatic speech recognition (ASR). Meanwhile, language modeling (LM) generates natural sentences in ASR and can also be applied to rescore ASR n-best outputs which attempt to obtain more accurate recognized texts. Since SLU and LM are so related and share the same form of inputs (sentences), it is significative to combine two tasks, and in this paper, we aim to improve SLU performance on recognized texts by applying multi-task learning with LM which shares linguistic and semantic information to improve both tasks and domain adaptation schemes which introduce extra domain-free knowledge.

In a real end-to-end dialogue system, SLU which performs on ASR recognized texts requires high robustness, and the performance of SLU depends not only on the capability of SLU system but also on ASR accuracy which includes situations: (1) errors exist in semantic words; (2) correctly recognize semantic words. One method to reduce ASR error in the first situation is to rescore ASR n-best outputs for more accurate ASR results. Our proposed MTM applies domain adaptation by the shared-private network with a few of labeled data in the specific domain and plenty of unlabeled data in general domain to better model in-domain texts and improve SLU performance since the model learns extra domain-free knowledge from general domain data.

Slot filling and intent detection are two critical tasks in SLU. Slot filling can be treated as a sequence labeling task and can be applied many sequence models such as conditional random fields (CRFs) [5], recurrent neural networks (RNNs) [6], attention models [7] and focus models [8]. While intent detection can be treated as a text classification task and approaches include support vector machines (SVMs) [9], convolutional neural networks (CNNs) [10] and RNNs [2]. Figure 1 illustrates an example from Chinese navigation corpus which contains word labels following begin-in-out (BIO) annotation method and sentence labels which are unique to each sentence. For language modeling, the goal is to predict the current word conditioned on context words in the sentence. Popular models include N-grams [11], RNNs [12] and Hierarchical-RNNs [13]. The implementation of bi-directional language modeling achieves decent results on several tasks [3] and can be further improved by applying NCE training scheme [4].

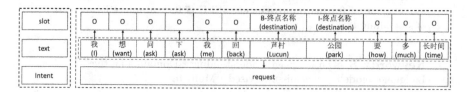

Fig. 1. An example from Chinese navigation corpus with aligned labels of slots and unaligned labels of intents.

Multi-task learning is derived by Caruana [28] and has efficiently applied in NLP problems [29] which includes sequence labeling [1] and text classification [2]. Meanwhile, adversarial shared-private model [20] is recently proposed for multi-task learning and achieves considerable improvements on several tasks [22,23]. The works that apply Multi-task for LM and SLU includes Liu's models which jointly trains SLU and LM which embeds SLU outputs of the previous time step, concatenates it with current word embedding and feeds into the RNN to improve language modeling. Moreover, Rei contributes his work of joint training several sequence labeling tasks with language modeling with the objective to let language modeling enhance the extracted features for sequence labeling tasks [1]. Such joint models promote information sharing between tasks and also simplify the SLU system since only one model needs to be trained.

Our work is based on previous research achievements, but few of them focus on the improve of SLU on recognized texts and apply adaptation for multi-task learning. Besides, most of the multi-task models apply multi-task learning to improve one task mainly. In this paper, our proposed model joint performs SLU and LM and utilizes domain adaptive schemes with a few of labeled data in the specific domain and plenty of unlabeled data in general domain. Meanwhile, it is noteworthy that we are the first one to apply the adversarial shared-private model for domain adaptation and find in practice that bi-directional language modeling achieves better performance on rescored WER than uni-directional language modeling. The experiments (in Tables 3 and 4) indicate the efficiency of multi-task learning which outperforms 0.28% absolute WER and increases 1.99% absolute slot F1. With the adaptive scheme, absolute WER eventually reduces 4.06% and absolute slot F1 increases 7.08%.

2   LSTM for Slot Filling, Intent Detection, and Language Modeling

Here we adopt recurrent neural network with long short-term memory (LSTM) as our slot filling, intent detection and language modeling model due to their superior performance in various NLP tasks [2,8,13].

2.1   LSTM

RNNs [14] can exploit context information in the time dimension. LSTM [15] is a variant of RNN and it performs better at exploiting long-range dependencies. While there are various types of LSTM, we use Jozefowicz's LSTM architecture [16] which is similar to Graves's LSTM architecture [17] but without peep-hole connections.

We denote x_t as input at each time step t and implement the LSTM by the following composition function:

$$i_t = \sigma \left(W_{xi} x_t + W_{hi} h_{t-1} + b_i \right) \tag{1}$$

$$f_t = \sigma \left(W_{xf} x_t + W_{hf} h_{t-1} + b_f \right) \tag{2}$$

$$g_t = \tanh\left(W_{xg}x_t + W_{hg}h_{t-1} + b_g\right) \tag{3}$$

$$o_t = \sigma\left(W_{xo}x_t + W_{ho}h_{t-1} + b_o\right) \tag{4}$$

$$c_t = f_t \odot c_{t-1} + g_t \odot i_t \tag{5}$$

$$h_t = o_t \odot \tanh\left(c_t\right) \tag{6}$$

where h_t is the hidden state at time step t, h_{t-1} is the hidden state of $t-1$ or the initial hidden state at time step 0, c_t is the cell state, i_t, f_t, i_t, i_t are the input, forget, cell and output gates respectively, W_{xi}, W_{xf}, W_{xg}, W_{xo} are the weight matrixes and b_i, b_f, b_g, b_o are the biases. tanh and sigmoid function σ are two activation functions.

We rewrite Eqs. 1–6 to a short form as following:

$$h_t = \text{LSTM}\left(x_t, h_{t-1}; \Theta\right) \tag{7}$$

where Θ represents all the parameters of LSTM. While applying bidirectional LSTM (BLSTM), at each time step t, two LSTM component: $\overrightarrow{h_t}$ and $\overleftarrow{h_t}$, moving forward and backward through the input, are the history and future context-dependent representations. Then the hidden representations from both directions are concatenated together as a context-specific representation for each input x_t which is conditioned on the whole input x. The representation can be write as follow:

$$\overrightarrow{h_t} = \overrightarrow{\text{LSTM}}\left(x_t, \overrightarrow{h_{t-1}}; \overrightarrow{\Theta}\right) \tag{8}$$

$$\overleftarrow{h_t} = \overleftarrow{\text{LSTM}}\left(x_t, \overleftarrow{h_{t+1}}; \overleftarrow{\Theta}\right) \tag{9}$$

$$h_t = \left[\overrightarrow{h_t}, \overleftarrow{h_t}\right] \tag{10}$$

2.2   Slot Filling and Intent Detection

Given a text sequence, we first project each word index to an embedding space $w = [w_0, w_1, ..., w_T]$ as the LSTM input.

For slot filling, the LSTM output h_t as a word representation is feed into a linear output layer, projected to tag label space. We apply softmax to the output and directly give the normalized distribution over all possible tag labels K for every word as the prediction.

$$\hat{P}_t^{\text{tag}} = \text{softmax}\left(W_{\text{tag}}h_t + b_{\text{tag}}\right) \tag{11}$$

where W_{tag} is the weight matrix and b_{tag} is the bias. Given a corpus with N training samples $\{(w_i; y_i)\}$, The model is optimized by minimizing the cross-entropy loss which is equivalent to minimize the negative log probability of the correct tag labels:

$$L_{\text{tag}} = -\sum_{i \in N}\sum_{t \in T_i} \log \hat{P}_{it}^{\text{tag}}\left[y_{it}\right] \tag{12}$$

where T_i is the length of i-th sentence and y is the ground-truth tag label.

For intent detection, we sum the LSTM's hidden states at each time step as h_{sum}, and feed it into a parallel linear output layer which projects the component to intent label space. Similar to slot filling, we achieve the normalized distribution over all possible intent labels after softmax.

$$h_{\text{sum}} = \sum_{t \in T} h_t \tag{13}$$

$$\hat{P}^{\text{intent}} = \text{softmax}\left(W_{\text{intent}} h_{\text{sum}} + b_{\text{intent}}\right) \tag{14}$$

and the corresponding cross-entropy loss is illustrated below:

$$L_{\text{intent}} = -\sum_{i \in N} \log \hat{P}_i^{\text{intent}}\left[y_i\right] \tag{15}$$

2.3   Language Modeling

In addition to language understanding, we propose a secondary objective of language modeling. This task traditionally attempts to predict the next word, while we aim to evaluate the word existence rationality in the whole sentence. Hence, we attempt to predict the target word probability conditioned on the whole sentence except the target word. Meanwhile, the task requires the model to learn more general patterns of semantic and syntactic composition, which can be helpful to language understanding.

Based on the LSTM output h_t as word representation, we add a language modeling parallel linear output to predict the word existence probability. For uni-directional model, the probability of each word shall be as follow:

$$\hat{P}_t^{\text{lm}} = \text{softmax}\left(W_{\text{lm}} \overrightarrow{h_{t-1}} + b_{\text{lm}}\right) \tag{16}$$

For bi-directional model [4], since the prediction shall be only conditioned on the whole sentence except the target word, we must design the loss objective so that only sections of the model that have not yet observed the target word are optimized to perform the prediction.

$$h_t' = \left[\overrightarrow{h_{t-1}}, \overleftarrow{h_{t+1}}\right] \tag{17}$$

$$\hat{P}_t^{\text{lm}} = \text{softmax}\left(W_{\text{lm}} h_t' + b_{\text{lm}}\right) \tag{18}$$

Finally, the language modeling objective loss is describe as follow:

$$L_{\text{lm}} = -\sum_{i \in N} \sum_{t \in T_i} \log \hat{P}_{it}^{\text{lm}}\left[y_{it}\right] \tag{19}$$

Figure 2 illustrates a diagram of unfolded BLSTM multi-task model (MTM) architecture with an example of sentence length 2. At each time step t, the model is optimized to predict current slot tag and word probability, and in the end, the model predicts sentence intent.

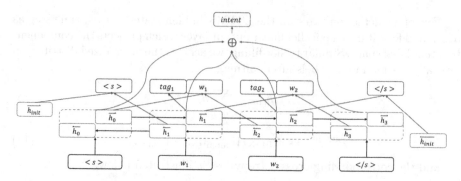

Fig. 2. Unfolded BLSTM multi-task model (MTM) architecture example of sentence length two where blue, red, green line and pane indicate language modeling, slot filling, intent detection outputs respectively. Here, we add $<s>$ and $</s>$ to indicate the start and end of the sentence, w_t denotes the input word at time step t. (Color figure online)

3 Adaptive Models

In comparison with labeled data in a specific domain, it is rather simple to achieve a large amount of unlabeled general domain data. Hence, traditional rescore commonly utilize a big language model trained by a large amount of unlabeled general domain data, while we attempt to utilize a few of data in the specific domain to adapt the language modeling.

There are numerous approaches to model adaptation such as output interpolation [18] and linear projection [19]. Since we also expect all the unlabeled data to feed into the model and help improve the robustness of language understanding, it is preferred to adapt the language modeling through the model. Therefore, the sharing schemes are applied to achieve the adaptation and we utilize three models: MTM, simple-shared-private MTM and shared-private MTM. We also introduce an adversarial loss to the shared-private model which is put forward by [20] to confine the shared model sections to extract domain independent feature and achieves significant improvement on several tasks [22] [23].

Here, we denote D_g, D_d as data in general and specific domain with sample amounts of N_g, N_d respectively.

3.1 Multi-task Model (MTM)

The most straightforward approach to train an adaptive model is to feed the data in general and specific domain together. In this model, two domains of data fully share their parameters and ignore the negative influence of information learned from general data. Figure 3a illustrates the structure of MTM and Eq. 20 describe the loss of joint training.

$$L = \lambda_{lm} * (L_{\mathrm{lm^g}} + L_{\mathrm{lm^d}}) + \lambda_{tag} * L_{\mathrm{tag}} + \lambda_{intent} * L_{\mathrm{intent}} \qquad (20)$$

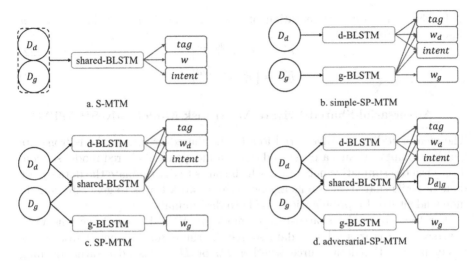

Fig. 3. Architectures of four adaptive multi-task models where D_g, D_d denotes data in general and specific domain, *tag*, *intent*, *w* are the outputs of slot filling, intent detection and language modeling, $D_{d|g}$ represents the prediction of data source whether D_d or D_g.

where g, d denote general and specific domain respectively and λ_{lm}, λ_{tag}, λ_{intent} are the weights of loss for each task which control their importance.

3.2   Simple-Shared-Private Multi-task Model (simple-SP-MTM)

As shown in Fig. 3b, the model provides private sections for specific domain data, and utilize general data as a feature extractor which provides extra word representation for specific domain tasks. We can compute concatenated specific domain word representation as follow:

$$h_t'^d = \left[h_t^d, h_t^g \right] \tag{21}$$

Meanwhile, there are two outputs of language modeling which are domain depended. Since we aim to achieve an adaptive specific domain language modeling, it is more rational to apply specific domain output while general output contributes to keep the general data information. The following models comply with the same scheme.

3.3   Shared-Private Multi-task Model (SP-MTM)

Based on simple-SP-MTM, SP-MTM further provides private model sections for general data and enables general data to share its information with reservation. The shared model sections supply feature separately for each task to utilize but also remains the knowledge learned from all data. SP-MTM is illustrated in Fig. 3c and word representation of data in general and specific domain is described in Eqs. 22–23.

$$h_t'^d = \left[h_t^d, h_t^{shared}\right] \tag{22}$$

$$h_t'^g = \left[h_t^g, h_t^{shared}\right] \tag{23}$$

3.4   Adversarial-Shared-Private Multi-task Model (adv-SP-MTM)

The idea of adding an adversarial loss to the shared-private model is from [20] which attempts to learn a purer and domain independent shared model section. There is a discriminative network D which learns to discriminate the data source domain while shared section as a generative network learns to generate domain independent word representations for D to discriminate.

For the adversarial shared-private model in Fig. 3d, shared sections work adversarially towards a linear data source discriminator, prevent it from accurately predict the data source whether D_d or D_g. This adversarial training encourages shared sections to be purer and ensure the word presentation of shared sections not to be domain depended.

We have tried several adversarial training schemes including (1) directly back propagating negative cross-entropy loss, (2) random data source labeling or (3) GAN (generative adversarial nets) [24] like training scheme (applied by Chen [21]), and in the end, scheme (1) achieves the best performance. Hence, the adversarial loss of discriminator we apply is negative cross-entropy loss which attempts to train the discriminator not to make the accurate prediction and the function is described in Eq. 24.

$$L_{\text{adv}} = -\sum_{i \in N_d \cup N_g} \log \hat{P}_i^{\text{adv}} \left[\{D_d | D_g\}_i\right] \tag{24}$$

Hence, we have the new loss of multi-task model with additional adversarial training.

$$L = \lambda_{lm} * (L_{\text{lm}^g} + L_{\text{lm}^d}) + \lambda_{tag} * L_{\text{tag}} + \lambda_{intent} * L_{\text{intent}} - \lambda_{adv} * L_{\text{adv}} \tag{25}$$

where λ_{adv} is the weight of adversarial loss.

4   Experiments

4.1   Setup

We prepare specific domain corpus of Chinese navigation and general domain corpus of Chinese SMS for the experiments which are all collected from real conversations.

The navigation corpus in Table 1 contains manual transcript texts with labels and corresponding recognized texts from ASR n-best results. We extract some manual transcript texts with labels as training set and a few of manual transcript texts with recognized n-best results as the testing set. Since the amount of training set is limited due to the difficulty of human labeling, we take training set as patterns and apply the slots database which contains slot values collected from internet and former corpus to expand training set. Since the final goal is to parse the manual transcript semantic information directly, the recognized texts apply the labels of manual transcript texts. The SMS corpus in Table 1 contains plenty of unlabeled data of daily conversations which provides extra linguistic knowledge.

Table 1. Statistics of the navigation, SMS corpus as specific and general domain data. The columns 1–3 indicate the amount of training, valid and testing sets. Column 4 represents the number of patterns and column 5 illustrates the sizes of vocabs.

Corpus	Train	Valid	Test	Pattern	Vocab
Navi (specific)	425k	75k	6598	12799	48038
SMS (general)	1000k	30k	-	-	45248

All the training set in each domain are further partitioned randomly into the training set and valid set and the detailed amounts of all the datasets are listed in Table 1. We apply google string tokenizer with a referred large vocab in both ASR and SLU phases to segment words and extract vocab from the training set of each domain with an appearance times filter of two which replaces words that appear less than two times in training set with <unk>.

Since BLSTM is one of the most widely used models on language understanding and language modeling tasks, and has achieved state-of-art results of certain tasks on several corpora such as [8], we apply it as the baseline with linear multi-task outputs which is similar to the work of Zhang [25].

4.2   Hyper Parameters

The parameters of the model are randomly initialized with the value from uniform distribution in $[-0.1, 0.1]$. We train the model with the mini-batch scheme in the size of 16. For all models, we set embedding size and hidden dimension of BLSTM as 100, apply dropout [26] of 0.5 to achieve a better training performance, clip all the gradients by the max norm of 5.0 to avoid the gradient explosion. Adam [27] is used as the optimizer and learning rate is set to 0.001. The loss weights of all tasks are directly set to 1.0 and the weight of adversarial loss is 0.4.

4.3   Evaluations

For language understanding, we evaluate the model slot F1 score through semantic tuples in the form of [slot : value] which are extracted from predicted slot tags. We also calculate the intent F1 score as the measure of model intent detection performance. And for language modeling, since the uni-directional and bi-directional model outputs conditioned on different context information, we directly use rescored WER to evaluate the language modeling capability.

Table 2. Results on navigation testing set of the uni-directional and bi-directional model include rescored word error rate and slot, intent F1 score on the manual transcript, ASR 1-best, rescored 1-best texts.

Data	Model	WER (%)	True		ASR 1-best		Rescored 1-best	
			Slot F1	Intent F1	Slot F1	Intent F1	Slot F1	Intent F1
Navi	uni-MTM	15.59	74.49	96.50	40.06	96.17	48.00	96.47
Navi	bi-MTM	**15.06**	**92.97**	**99.45**	**50.17**	**98.50**	**60.62**	**98.59**

We first evaluate the performance of bi-directional language modeling through the multi-task model (MTM) trained by navigation corpus and the results in Table 2 indicates that, in comparison with uni-directional model, the bi-directional model gives better performance not only on slot F1 and intent F1 but also rescored WER (from 15.34% to 15.06%). Hence, the models in further experiments are all bi-directional.

Secondly, we evaluate the performance of multi-task model (MTM) which is illustrated in Tables 3 and 4. Since the ASR 1-best and rescored 1-best texts contain the error from ASR, and the labels are directly from manual transcript texts which strictly requires the right predicted values of slots, the slot F1 score turns distinctly worse than the performance on manual transcript texts which also reveals the robustness problem encountered by the SLU systems in a real application. For the results on manual transcript texts, STM (LU) and MTM (trained by navigation data) models achieve slightly better performance on slot F1 (92.76% and 92.97%) and intent F1 (from 99.35% to 99.45%) and outperform on language modeling (from 15.34% to 15.06%). Moreover, from the results on ASR 1-best and rescored 1-best texts, we can observe significant increments on both slot and intent F1 which proves that the secondary language modeling objective help to improve the robustness of slot filling and intent detection. The slot F1 increases 1.75% from 48.42% to 50.17% and intent F1 increases 0.70% from 97.80% to 98.50% on ASR 1-best texts, while on rescored 1-best texts, slot F1 improves 1.99% from 58.63% to 60.62% and intent F1 improves 0.48% from 98.11% to 98.59%.

Then we apply shared-private models with a few of navigation data and plenty of SMS data to achieve domain adapted word representations. As described in Sect. 3.2, we apply specific domain language modeling output for rescoring since it adapts the word representation to the specific domain. The result of each model is illustrated in Tables 3 and 4 which indicates that the domain adapted word presentations efficiently improves the performance of language modeling. In the comparison of MTM and simple-SP-MTM, the improvement of slot F1 on rescored 1-best texts and WER expresses the benefits of leaving private parameters for specific domain data to adapt language modeling

Table 3. WER of LM rescoring results on navigation testing set where adv-SP-MTM achieves the best performance.

Data	Model	Task		WER (%)
		LU	LM	
SMS	STM (LM)	-	✓	17.12
Navi	STM (LM)	-	✓	15.34
Navi	MTM	✓	✓	**15.06**
Navi & SMS	MTM	✓	✓	14.85
Navi & SMS	simple-SP-MTM	✓	✓	13.38
Navi & SMS	SP-MTM	✓	✓	13.22
Navi & SMS	adv-SP-MTM	✓	✓	**13.06**

Table 4. Slot, intent F1 results on navigation manual transcript, ASR 1-best, rescored 1-best testing sets where adv-SP-MTM achieves best slot F1 on all testing sets, but unfortunately underperforms on intent F1 in comparison with MTM trained by navigation data. It is noteworthy that rescored 1-best texts are generated from corresponding model rescoring in Table 3 while the performance of STM (LU) on rescored 1-best texts is evaluated on the 1-best texts rescored by STM (LM)

Data	Model	True		ASR 1-best		Rescored 1-best	
		Slot F1	Intent F1	Slot F1	Intent F1	Slot F1	Intent F1
Navi	STM (LU)	92.76	99.35	48.42	97.80	58.63	98.11
Navi	MTM	**92.97**	**99.45**	**50.17**	**98.50**	**60.62**	**98.59**
Navi & SMS	MTM	93.56	98.92	50.89	97.15	60.44	97.42
Navi & SMS	simple-SP-MTM	93.10	99.06	50.08	98.00	64.65	97.83
Navi & SMS	SP-MTM	93.77	**99.48**	51.20	**98.44**	65.19	**98.26**
Navi & SMS	adv-SP-MTM	**93.90**	99.15	**51.38**	98.14	**65.71**	98.14

while decreasement on manual transcript and ASR 1-best texts reveals the risk of fully sharing information which possibly brings negative influences. In the comparison of simple-SP-MTM and SP-MTM, results indicate the effectiveness of private sections for general data, and it is useful for model to learn by itself which part of information should be shared and which should not. The adv-SP-MTM provides a sharing scheme and the increments emphasize the importance of restriction on sharing parameters. In the end, adv-SP-MTM reduces WER 4.06% from 17.12% to 13.06%. Meanwhile, SMS data helps the model to learn more linguistic information and further improve the robustness of language understanding performance. Slot F1 increases 1.14% from 92.76% to 93.90% on manual transcript texts and 2.96% from 48.42% to 51.38% on ASR 1-best texts. With the decreasement of WER and increasement of slot F1 on ASR 1-best texts, on rescored 1-best texts, slot F1 further outperforms 7.08% from 58.63% to 65.71%. It is noteworthy that, for intent F1, almost all the shared-private models slightly degrade in comparison with the non-adaptive MTM trained by

Table 5. WER of LM rescoring results on navigation testing set of adv-SP-MTM include interpolated rescored word error rate. Delta denotes the interpolation weights of scores from specific (d), general (g) domain language modeling outputs respectively.

Data	Model	Delta		WER (%)
		d	g	
Navi & SMS	adv-SP-MTM	1.0	0.0	**13.06**
		0.9	0.1	13.20
		0.7	0.3	15.13
		0.5	0.5	15.95
		0.0	1.0	22.48

navigation corpus (but imperceptibly outperforms STM), since the external word representation possibly introduce a few of misleading information that accumulates in the end to promote a wrong prediction for intent.

We also conduct additional experiments on rescoring interpolation of two outputs of specific and general domain in adv-SP-MTM in Table 5 where none of the interpolation weights we use outperforms specific domain output rescoring WER which verifies the analysis in Sect. 3.2.

In conclusion, with the multi-task learning and domain adaptive schemes, the model can achieve more accurate rescored texts and more robust language understanding capability. In the end, adv-SP-MTM achieves best rescoring WER with decreasement of 4.06% and slot F1 on rescored 1-best texts with increasement of 7.08% while almost equivalently intent F1 which imperceptibly increase 0.03%.

References

1. Rei, M.: Semi-supervised multitask learning for sequence labeling. arXiv preprint arXiv:1704.07156 (2017)
2. Liu, P., Qiu, X., Huang, X.: Recurrent neural network for text classification with multi-task learning. arXiv preprint arXiv:1605.05101 (2016)
3. Peris, A., Casacuberta, F.: A bidirectional recurrent neural language model for machine translation. Procesamiento del Lenguaje Nat. (55) (2015)
4. He, T., Zhang, Y., Droppo, J., et al.: On training bi-directional neural network language model with noise contrastive estimation. In: 2016 10th International Symposium on Chinese Spoken Language Processing (ISCSLP), pp. 1–5. IEEE (2016)
5. Raymond, C., Riccardi, G.: Generative and discriminative algorithms for spoken language understanding. In: Eighth Annual Conference of the International Speech Communication Association (2007)
6. Mesnil, G., Dauphin, Y., Yao, K.: Using recurrent neural networks for slot filling in spoken language understanding. IEEE/ACM Trans. Audio Speech Lang. Process. **23**(3), 530–539 (2015)
7. Liu, B., Lane, I.: Attention-based recurrent neural network models for joint intent detection and slot filling. arXiv preprint arXiv:1609.01454 (2016)

8. Zhu, S., Yu, K.: Encoder-decoder with focus-mechanism for sequence labelling based spoken language understanding. In: 2017 IEEE International Conference on Acoustics, Speech and Signal Processing (ICASSP), pp. 5675–5679. IEEE (2017)
9. Zhang, W., Yoshida, T., Tang, X.: Text classification based on multi-word with support vector machine. Knowl.-Based Syst. **21**(8), 879–886 (2008)
10. Kim, Y.: Convolutional neural networks for sentence classification. arXiv preprint arXiv:1408.5882 (2014)
11. Brown, P.F., Desouza, P.V., Mercer, R.L.: Class-based n-gram models of natural language. Comput. Linguist. **18**(4), 467–479 (1992)
12. Mikolov, T., Karafiát, M., Burget, L., et al.: Recurrent neural network based language model. In: Eleventh Annual Conference of the International Speech Communication Association (2010)
13. Morin, F., Bengio, Y.: Hierarchical probabilistic neural network language model. In: Aistats, vol. 5, pp. 246–252 (2005)
14. Elman, J.L.: Finding structure in time. Cogn. Sci. **14**(2), 179–211 (1990)
15. Hochreiter, S., Schmidhuber, J.: Long short-term memory. Neural Comput. **9**(8), 1735–1780 (1997)
16. Jozefowicz, R., Zaremba, W., Sutskever, I.: An empirical exploration of recurrent network architectures. In: International Conference on Machine Learning, pp. 2342–2350 (2015)
17. Graves, A.: Generating sequences with recurrent neural networks. arXiv preprint arXiv:1308.0850 (2013)
18. Tüske, Z., Irie, K., Schlüter, R., et al.: Investigation on log-linear interpolation of multi-domain neural network language model. In: 2016 IEEE International Conference on Acoustics, Speech and Signal Processing (ICASSP), pp. 6005–6009. IEEE (2016)
19. Ma, M., Nirschl, M., Biadsy, F., et al.: Approaches for neural-network language model adaptation. In: Proceedings of Interspeech 2017, pp. 259–263 (2017)
20. Liu, P., Qiu, X., Huang, X.: Adversarial multi-task learning for text classification. arXiv preprint arXiv:1704.05742 (2017)
21. Chen, X., Shi, Z., Qiu, X., et al.: Adversarial multi-criteria learning for Chinese word segmentation. arXiv preprint arXiv:1704.07556 (2017)
22. Zhu, S., Lan, O., Yu, K.: Robust spoken language understanding with unsupervised ASR-error adaptation. In: 2018 IEEE International Conference on Acoustics, Speech and Signal Processing (ICASSP), pp. 6179–6183. IEEE (2018)
23. Lan, O., Zhu, S., Yu, K.: Semi-supervised training using adversarial multi-task learning for spoken language understanding. In: 2018 IEEE International Conference on Acoustics, Speech and Signal Processing (ICASSP), pp. 6049–6053. IEEE (2018)
24. Goodfellow, I., Pouget-Abadie, J., Mirza, M., et al.: Generative adversarial nets. In: Advances in Neural Information Processing Systems, pp. 2672–2680 (2014)
25. Zhang, X., Wang, H.: A joint model of intent determination and slot filling for spoken language understanding. In: IJCAI, pp. 2993–2999 (2016)
26. Srivastava, N., Hinton, G., Krizhevsky, A.: Dropout: a simple way to prevent neural networks from overfitting. J. Mach. Learn. Res. **15**(1), 1929–1958 (2014)
27. Kingma, D.P., Ba, J.: Adam: a method for stochastic optimization. arXiv preprint arXiv:1412.6980 (2014)
28. Caruana, R.: Multitask learning. In: Thrun, S., Pratt, L. (eds.) Learning to Learn, pp. 95–133. Springer, Boston (1998). https://doi.org/10.1007/978-1-4615-5529-2_5

29. Collobert, R., Weston, J.: A unified architecture for natural language processing: deep neural networks with multitask learning. In: Proceedings of the 25th International Conference on Machine Learning, pp. 160–167. ACM (2008)

30. Liu, B., Lane, I.: Joint online spoken language understanding and language modeling with recurrent neural networks. arXiv preprint arXiv:1609.01462 (2016)

A Channel-Cascading Pedestrian Detection Network for Small-Size Pedestrians

Jiaojiao He[1,2], Ken Liu[2,3], Yongping Zhang[2(✉)], Tuozhong Yao[2],
Zhongjie Zhao[1], Jiangjian Xiao[4], and Chengbin Peng[4]

[1] Institute of Electronic Information and Control Engineering,
Chang'an University, Xi'an 713100, Shaanxi, China
[2] School of Electronic and Information Engineering,
Ningbo University of Technology, Ningbo 315211, Zhejiang, China
ypz@nbut.edu.cn
[3] Institute of Information, Chang'an University, Xi'an 713100, Shaanxi, China
[4] Ningbo Institution of Industrial Technology, Chinese Academy of Sciences,
Ningbo 315201, Zhejiang, China

Abstract. At present, there are several new challenges for multi-scale pedestrian detection in wide-angle field of view, especially small-size pedestrians. So the problem is how we can detect pedestrians efficiently and accurately with limited resources in wide-angle field of vision. In this work, we propose a Channel-Cascading pedestrian detection network for small-size pedestrians. In combination with the two-stage idea of Faster-RCNN in our detector, the optimized network was applied and the regional proposal network was improved. We propose a novel feature extraction network as optimized network, which we call the "Channel-Cascading Network" (CCN), that fuses information between channels by progressive cascading strategy and adapts our idea to other network designs. The experimental results show that our detector performs better for small-size pedestrians, it not only the precision of pedestrian detection in wide field of view is greatly improved especially small-size pedestrian, but also the speed is accelerated.

Keywords: Pedestrian detection · Wide-angle field of view
Channel-cascading network · Small-size

1 Introduction

Pedestrian detection, as special object detection, has gradually become a research hotspot in the field of computer vision, and is also an important technology to promote the development of smart city. It detects some specific pedestrian targets by extracting some features of the image, and provides the necessary technical basis for the higher level tasks such as behavior recognition and analysis, pedestrian attitude analysis and research [1]. Pedestrian detection has been widely used in intelligent video surveillance, vehicle assisted pedestrian protection system, intelligent traffic control, intelligent robot and so on. It has great commercial value. Pedestrians in the wide field of

© Springer Nature Switzerland AG 2018
Y. Peng et al. (Eds.): IScIDE 2018, LNCS 11266, pp. 325–338, 2018.
https://doi.org/10.1007/978-3-030-02698-1_28

view have more research value, such as large shopping centers, new entertainment places, railway stations, bus stations and other large sites. However, there are the following problems in the wide-angle visual field downlink detection: 1, the multidimensional pedestrian multiscale problem [2]; 2, the change span of the detection scene is large; 3, there are different degrees of occlusion between the pedestrians. In practical applications, besides the above problems, the camera's perspective and other environmental factors, such as light and brightness due to weather conditions, and different road conditions, are all required to interfere with the accuracy of pedestrian detection [3].

Recently a series of novel methods have emerged [4], extracting image features by neural network, showing considerable accuracy gains. In our works, we revisit impressive improvements in object detection by performing extensive experiments on the railway station pedestrian dataset. Due to the large span of pedestrian scale in the dataset, the detector is powerless for small-size pedestrians. After research find that the reason is that the resolution of the size pedestrians is too small, and information is lost more when extracting features. Our goal is to adapt the highly successful Faster R-CNN object detector to perform better in small-size pedestrians.

We combine the two-stage detection concept of Faster-RCNN and apply the optimized detection algorithm to specific pedestrian detection. There are numerous false detections and missed detections for small-size pedestrians. We study the relationship between image channels, and propose a novel method, which we call "Channel-Cascading Network" (CCN), This new method makes full use of different levels of image channel characteristics and adopts a cascade of progressive methods to further extend the advantages of convolutional neural network extraction features, and select the best search box combining with clustering algorithm, which makes the detection rate of pedestrians in the distance higher, and the phenomenon of false reporting and missing reporting in the whole pedestrian detection is significantly reduced, the speed has also been improved.

2 Related Work

At present, pedestrian detection methods can be broadly classified into two categories: traditional feature extraction methods and deep learning methods. Traditional methods of extracting features are obtained by manual design. The early scale invariant feature conversion (Sift) [5], directional gradient histogram (HOG) [6, 7] and linear support vector machine (SVM) [8] are used in pedestrian detection. Subsequently, the Deformable Part Model (DPM) [9] model and HOG combined with Local Binary Patterns (LBP) [10] as feature sets have all improved pedestrian detection results.

With the increasing popularity of deep neural networks, convolution neural network has become the mainstream for extracting image features and has been widely used in pedestrian detection [11]. At present, From RCNN [12], SPP-Net [13] and Fast-RCNN [14] to Faster-RCNN [15], the accuracy and speed of object detection have reached a new height. The two-stage Faster-RCNN has further improved the accuracy of common target detection due to the introduction of the Regional Proposal Network. However, there are still problems in the detection of small targets, because the default RPN in Faster R-CNN uses 9 search boxes to search for an area, scale is 128^2, 256^2, 512^2, and the ratio of length to width is 1:1, 1:2, and 2:1. Such search boxes are aimed at ImageNet and VOC data sets. For small objects, the corresponding search box is very small, so it is difficult to detect the feature map of small pedestrians in the wide angle field of view. Because of this selection of the size and ratio of anchors based on experience in Faster R-CNN, it has certain limitations: it is not effective for small-scale or remote pedestrian detection. If you can choose the right size search box at the beginning, it will certainly help the network to detect better. YOLO9000 [16], YOLOv3 [17], SSD [18] and so on have made some improvements in this respect, and all of them have selected suitable search boxes by clustering. However, as first stage detector, the accuracy of the detector is still not as accurate as that of the two stage detector. We select a reasonable search box based on clustering research, and show the importance of small size search box to distant pedestrians.

In order to improve the performance of object detection, a series of algorithms have been generated by improving the feature network extraction. He [19] uses the deeper network to extract image features, and the target detection accuracy is further improved, but it is useless in small sample datasets. In [20–24], based on the idea of layer jump, a more abundant image feature map is extracted. The results show that the layer jump operation is advantageous for feature extraction. For a limited set of data, we use the VGG [25] network as a benchmark. Different from the above algorithms, our algorithm does not use the idea of layer jump and additional information, instead of seeking the relationship between image channel information, adopts progressive cascading, it fully uses different layer image feature information, thus obtains aggregated image features.

3 Algorithm Description

Pedestrians have large span and slight distortion in a railway station pedestrian dataset, so small-size pedestrians that have low resolution are often missed. To address these challenges, two improvements are made to the standard Faster R-CNN object detector. First, optimizing the feature extraction network, and enhancing the image features through the cascade of image features. Second, improving the RPN search mechanism and introducing the clustering algorithm to automatically find a reasonable search box. Two modifications correspond to the blue dashed box and the green dashed box in Fig. 1 respectively. The overall architecture of the improved Faster R-CNN algorithm (Improved FRCNN) is shown in Fig. 1:

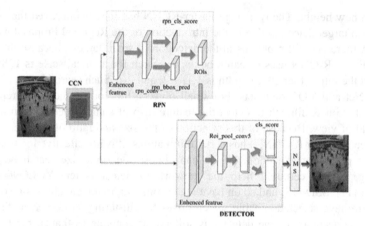

Fig. 1. Overall architecture diagram of the system (Color figure online)

3.1 Optimization of Feature Extraction Network

The effect of pedestrian detection is directly dependent on the selection of features. In general, if a better image feature can be extracted, even a simple classifier can also achieve a good detection effect. When the standard Faster-RCNN uses VGG16 to extract feature maps, it uses four times of pooling, so the feature map will be reduced to one-sixteenth, and only the last layer of feature information will be taken as the input of the subsequent network. Most of the details will be lost. For the villain in the distance, the information will be incomplete after reducing size by one-sixth, so the villain information cannot be reconstructed. For the convolution neural network, the low level feature semantic information is less, but the target location is accurate; the high level feature semantic information is rich, but the objective location is relatively rough. Therefore, in order to detect more small-size people, we propose a Channel-Cascading Network (CCN) to extract enhancement features. CCN adopts progressive cascading, makes full use of low-level channel feature information that facilitates detection of villain advantages, improves feature extraction network, and enhances features through channel information fusion.

Our optimized feature extraction network structure is as follows:

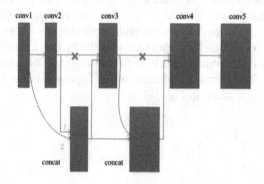

Fig. 2. CCN network architecture

The idea of constructing the network is that the feature map obtained by the low convolution layer has a small receptive field, and the response area of the same area responds to the small target more specifically, which is advantageous to detect the villain in the distance, the high convolution layer has passed several pooling layers. The resolution of the image is reduced, but the contour information is relatively abundant, which is conducive to the detection of pedestrians and intermediate pedestrians in the vicinity. As shown in Fig. 2, in order to detect multi-scale pedestrians with high precision, we discard the feature map information directly from the fifth layer, but concatenate the first layer after the pooling feature and the second layer feature, widen the channel, thus the feature map information is more abundant, and then the enhanced information is further sent to the third layers of volume. In the product network, the convolution output is then concatenate to the enhancement feature. The semantic information obtained by the method of layer by layer enhancement of semantic information is more beneficial to extract better features. In CNN, we use two dimensional plane diagrams. First, we can see the convolution feature map of different widths, and the wider the width, the more abundant the semantic information of the channel. Secondly, in the channel fusion process, we add 1 * 1 Convolution, not only for channel conversion, but also for reducing computation.

The construction of the optimized network is suitable for the connection of various convolution modules. Figure 3 gives the first progressive cascade structure diagram, in the figure, 1 and 2 correspond to 1, 2 in Fig. 2, respectively and the specific conversion formula is as follows:

$$F : X \to Y, X \in R^{H \times W \times C}, Y \in R^{H' \times W' \times C'}$$
$$Y = W * X = \sum_{n=1}^{c'} W_c^n * X^n \tag{1}$$

Where F denotes a convolution operation, X denotes the network input, and Y denotes the network output. In the convolution operation, $W = [W_1, W_2, ..., W_n]$ is recorded as a convolution weight, where W_n represents the weight of the n-th filter. The bias effect is ignored in Eq. (1).

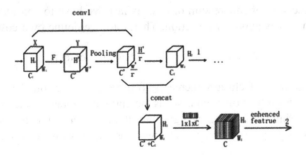

Fig. 3. Part cascading structure diagram

In order to achieve progressive cascading and enhance channel information, we use maximum pool size to compress the image size after every layer of convolution in the network. The r in Fig. 3 is the stride of the pool operation, ensuring the same size as the next convolution input size, here, $\frac{H'}{r} = H_2$, This operation reduces the size of the original information and minimizes the amount of calculation. After the output after each level of pooling is concatenated with convolution output, then they are channel-compressed with 1 * 1 Convolution to enrich channel information. It not only plays a role in dimensionality reduction, but also adds a nonlinear incentive on the previous level of learning by 1 * 1 Convolution, so as to enhance the expressive power of the network. After the enhancement feature is obtained, we use local response normalization to fuse the different features in the same space, highlight the image features, and use it as a next-level convolution input, followed by a progressive cascade to construct an optimized feature extraction network. Layer by layer cascading channel information, is conducive to the polymerization of more low-level useful information, reducing the loss of information in the feature delivery process.

3.2 Regional Proposal Network Optimization

For the two stage detection algorithm of Faster-RCNN, the first stage of RPN plays an important role in precision positioning. As a class of unknown detectors, RPN is a key step in pedestrian detection. Anchors selection is an important mechanism for obtaining target regions. The number and dimensions of search boxes in Faster-RCNN are manually set. Calculating the sizes of multi-scale pedestrians is invalidly and time-consuming. So we abandoned the idea of artificial setting at the beginning, adopted the idea of dimension clustering, and used the clustering algorithm to select a more suitable search box. In particular, the detection of pedestrians in the distance is further analyzed.

We use the k-means algorithm to cluster in the target box manually labeled in the training set, automatically find the statistical law of the target box, the number of the cluster number is set as the number of the selected search box, and the box corresponding to the cluster center is used as the search box.

After clustering analysis of the data set, we use hill-climbing algorithm to select the optimal number of search boxes suitable for detection.

Although the sum of square sum of error is fast, we have to consider whether the selected search box is good for detection. Therefore, we define own cost function as:

$$J(box, center) = 1 - IOU(box, center) \tag{2}$$

When the number of clusters increases to a certain value, our loss function will change slowly. This inflection point is set as the optimal number of search boxes. In the process of clustering using k-means algorithm, the loss function changes as shown in Fig. 4. When k > 9, the cost function changes very little, so we take k as 9.

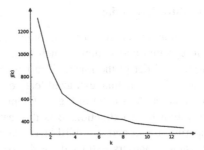

Fig. 4. Cost function curve

4 Experimental Evaluation

4.1 Dataset

In order to train and evaluate the pedestrian detector, a railway station pedestrian dataset was proposed. This dataset is sampled from the video of surveillance camera in 2016 train station, the image size is 960 × 1280, sampled in the daytime. Figure 5 shows some of our datasets. The scale of the pedestrian is large in the image, and it is difficult for the villain in the far distance to detect and there is a problem of serious occlusion. We randomly selected 6000 positive samples as the training set and another 1500 samples as the test set. The experimental machine CPU is i5, the memory is 6 GB, and the GPU is GTX1060. Network training and detection are based on TensorFlow.

Fig. 5. Wide angle field of view dataset enumeration

4.2   Valuation of the Feature Extraction

Generally, in the feature extraction process of convolutional neural networks, the activation degree of the foreground part is high, which makes the features more discernable and easy to classify and detect the image. The features extracted from each convolution layer are visualized and the final extracted features clearly show the global information of the detection target. As shown in Fig. 6. Graph (a) is the feature of the original algorithm's convolution layer extraction. B is the feature extracted from the optimized network. Comparing the second and third layer feature maps in (a) and (b), the details of the feature target contours extracted by the optimization network are clearer, the background is more pure, and the differences in the fourth layer feature maps are relatively large. In (b) is the effect of adding localized response normalization, which makes the relatively large value of the response relatively larger. Therefore, sending it to the fifth layer convolutional neural network makes the target information we need clearer. Figure 7 shows some of the visual comparison results, as shown in the figure, the first column is the original graph, the second one is the feature extracted by VGG, and the third column is the feature extracted by CCN.

(a) Extracted feature map byVGG

(b) Extracted feature map by CCN

Fig. 6. Visual contrast diagram of convolution neural network

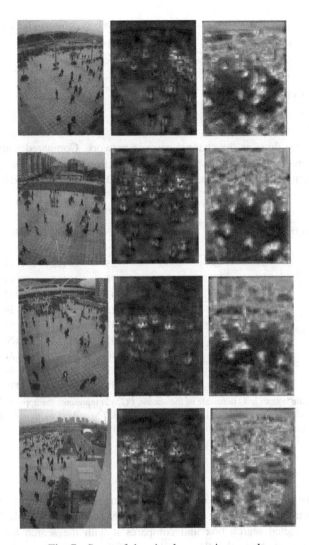

Fig. 7. Some of the visual comparison results

4.3   Analysis of Detection Results

The network parameters are set, the learning rate is 0.01, The maximum number of iterations is 40000 times. Railway station pedestrian dataset was adopted in the experiments.

First, we compared the influence of different network structures on the detection accuracy, and selected the most feature extraction network. In Table 1, the network structure is from top to bottom, respectively, structure one, structure two, structure three, structure four. 1, 2, 3, 4, 5 represent the number of layers of convolution respectively. The brackets represent a cascade of two levels. The table compares the influence of different layers of cascading information on pedestrian detection accuracy in the near, middle, and far distances. According to the results of the network structure comparison,

Table 1. Comparison of detection results at different levels of information at different levels

Network structure	AP(N_P)	AP(M_P)	AP(F_P)	MAP
1-2-3-4-5	66.1%	52.1%	29.4%	49.2%
1-(1,2)-3-(3,4)-5	75.7%	71.3%	39.5%	63.2%
1-(1,2)-(2,3)-4-5	77.9%	72.3%	43.6%	64.6%
1 -(1,2)-(2,3)-(3,4)- 5	77.6%	71.9%	30.3%	60.0%

we chose structure three as the channel cascade network. Compared with the original algorithm, the average detection speed of the optimized algorithm has increased by 25.2%, the villain's detection rate has increased by 30.3%, and the speed has improved.

Table 2. Anchor results comparison

No.	Anchor (Faster-RCNN)/pixel				Anchor (Improved Faster-RCNN)/pixel			
	x1	y1	x2	y2	x1	y1	x2	y2
1	−83	−39	100	56	−55	−112	71	127
2	−175	−87	192	104	−74	−132	89	147
3	−359	−183	376	200	−147	−181	162	196
4	−55	−55	72	72	−23	−54	39	69
5	−119	−119	136	136	−47	−90	62	104
6	−247	−247	264	264	−105	−138	120	152
7	−35	−79	52	96	−16	−37	31	52
8	−79	−167	96	184	−33	−74	48	89
9	−167	−343	184	360	−101	−181	116	196

(x1, y1) represents the bottom left coordinate of the anchor. (x2, y2) represents the upper right coordinate of the anchor. In the calculation process, the initial center point (7.5, 7.5) and the width and height are known, and the coordinates obtained from formula (2) contain negative numbers

$$X = X_{ctr} - 0.5 * (W - 1)$$
$$Y = Y_{ctr} - 0.5 * (H - 1)$$
$$(3)$$

Secondly, in the experiment, instead of selecting anchors by hand, I use the K-means algorithm on the training set, where the K value is 9, Table 1 shows the automatically selected search box and compares it with the search box selected in Faster-RCNN. Result shows that the overall size of the improved search box becomes smaller and there are multiple scales, indicating the selected search from the side. The box fits the characteristics of our dataset (Table 2).

Finally, combining two improved strategies, using improved Faster-RCNN for pedestrian detection, in Table 3, we compared the improved algorithm and the original algorithm, and the results show that the average detection speed of the optimized algorithm is increased by 25.2%. The detection rate increased by 30.3% and the speed increased.

Table 3. Comparison of test results from different algorithms

Algorithm	AP(N_P)	AP(M_P)	AP(F_P)	MAP	Run time
Faster-RCNN	66.1%	52.1%	29.4%	49.2%	0.294 s
Improved FRCNN	83.3%	80.1%	59.7%	74.4%	0.243 s

In order to further study the remote villain's detection, we use the control variable method. Firstly, the detection map is input, the optimized network is used to extract effective features, and then the improved RPN network positioning target is used to maintain the large-size search box in the RPN, and the influence of the small-size search box on the detection effect is studied. The test results are shown in Fig. 8. The results show that the small size of the search box we have automatically selected is more advantageous for detecting small objects.

Fig. 8. The result of F_P detection

Figure 9 shows the comparison results of the algorithm under some different viewing angles. Figure (a) shows the detection results of the Faster-RCNN at different viewing angles, and Figure (b) shows the results of the corresponding Improved FRCNN. It can be

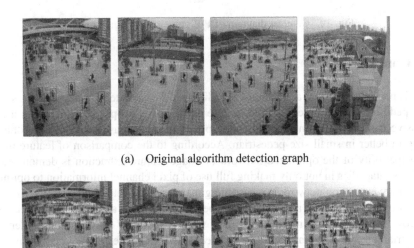

(a)   Original algorithm detection graph

(b)   Optimization algorithm detection graph

Fig. 9. Comparison of test results from different perspectives

seen from the figure that the improved algorithm is more suitable for detection under large field of view, while reducing the missed rate of remote villains and improving the overall detection rate.

In the PR curve of Fig. 10, the recall is the x-axis, and the precision is the y-axis. The recall rate and the precision rate are contradictory quantities. When the recall rate is high, the precision rate is often low. The PR curve reflects the trade-off between the accuracy and recall of the classifier. The red curve is the test result of the original classifier. The blue curve is the result of the optimized classifier. It can be seen from the graph that the blue curve completely surrounds the red curve, which means that the optimized classifier is superior to the original classifier. The result of curve comparison shows that our classifier has better performance.

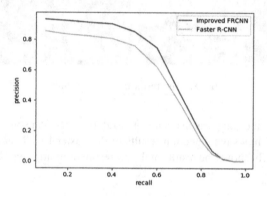

Fig. 10. Contrast diagram of P-R curve (Color figure online)

5   Conclusion

In this paper we propose a Channel-Cascading pedestrian detection network for small-size pedestrians, and a new idea of feature extraction is proposed, which uses a progressive cascade between channels and applies it to the Faster-RCNN algorithm to perform better in small-size pedestrian. According to the comparison of feature maps, the superiority of the optimized network in image feature extraction is demonstrated. Our advantage lies in not only making full use of pixel channel information to optimize the feature extraction network in a cascading manner in order to build a stronger feature extraction network, but also using unsupervised learning algorithms to effectively improve the RPN search mechanism, all of which effectively alleviate the problems of pedestrian detection that are difficult to detect small-size pedestrians. Finally, this idea can also be applied to other network structures.

Acknowledgements. This work is supported by the National Natural Science Foundation of China (NO. 61771270), the Natural Science Foundation of Zhejiang Province (No. 2017A610109) and (LQ15F020004), Key research and development plan of Zhejiang province (2018C01086).

References

1. Zhang, Q.: Research on pedestrian detection methods on still images. University of Science and Technology of China (2015). (In Chinese)
2. Lin, T.Y., et al.: Feature pyramid networks for object detection. In: IEEE Computer Society Conference on Computer Vision and Pattern Recognition, pp. 936–944 (2016)
3. Wang, B.: Pedestrian Detection Based on Deep Learning. Beijing Jiaotong University (2015). (In Chinese)
4. Krizhevsky, A., Sutskever, I., Hinton, G.E.: ImageNet classification with deep convolutional neural networks. In: International Conference on Neural Information Processing Systems, pp. 1097–1105 (2012)
5. Lowe, D.G.: Distinctive image features from scale-invariant keypoints. Int. J. Comput. Vis. 91–110 (2004)
6. Dalal, N., Triggs, B.: Histograms of oriented gradients for human detection. In: IEEE Computer Society Conference on Computer Vision and Pattern Recognition, pp. 886–893 (2005)
7. Zhu, Q., et al.: Fast human detection using a cascade of histograms of oriented gradients. In: IEEE Computer Society Conference on Computer Vision and Pattern Recognition, vol. 2, pp. 1491–1498 (2006)
8. Chen, P.H., Lin, C.J.: A Tutorial on -support vector machines. In: Applied Stochastic Models in Business & Industry, vol. 21, No. 2, pp. 111–136 (2005)
9. Felzenszwalb, P.F., et al.: Object detection with discriminatively trained part-based models. IEEE Trans. Pattern Anal. Mach. Intell. 32(9), 1627–1645 (2010)
10. Wang, X.: An HOG-LBP human detector with partial occlusion handling. In: Proceedings of IEEE International Conference on Computer Vision, September, Kyoto, Japan, pp. 32–39 (2009)
11. Kuo, W., Hariharan, B., Malik, J.: DeepBox: learning objectness with convolutional networks. In: IEEE International Conference on Computer Vision, pp. 2479–2487 (2015)
12. Girshick, R., Donahue, J., Darrell, T., et al.: Rich feature hierarchies for accurate object detection and semantic segmentation. In: IEEE Conference on Computer Vision and Pattern Recognition, pp. 580–587 (2014)
13. He, K., et al.: Spatial pyramid pooling in deep convolutional networks for visual recognition. In: IEEE Trans. Pattern Anal. Mach. Intell. 37(9), 1904–1916 (2015)
14. Girshick, R.: Fast R-CNN. Computer Science, pp. 1440–1448 (2015)
15. Ren, S., et al.: Faster R-CNN: towards real-time object detection with region proposal networks. In: International Conference on Neural Information Processing Systems, pp. 91–9 (2015)
16. Redmon, J., Farhadi, A.: YOLO9000: better, faster, stronger. In: IEEE Computer Society Conference on Computer Vision and Pattern Recognition, pp. 6517–6525 (2016)
17. Redmon, J., Farhadi, A.: YOLOv3: an incremental improvement. In: IEEE Computer Society Conference on Computer Vision and Pattern Recognition (2018)
18. Liu, W., et al.: SSD: single shot multibox detector. In: IEEE Computer Society Conference on Computer Vision and Pattern Recognition, pp. 21–37 (2016)
19. He, K., et al.: Deep residual learning for image recognition. In: IEEE Computer Society Conference on Computer Vision and Pattern Recognition, pp. 770–778 (2015)
20. Kong, T., et al.: HyperNet: towards accurate region proposal generation and joint object detection. In: Computer Vision and Pattern Recognition (2016)

21. Cai, Z., Fan, Q., Feris, Rogerio, S., Vasconcelos, N.: A unified multi-scale deep convolutional neural network for fast object detection. In: Leibe, B., Matas, J., Sebe, N., Welling, M. (eds.) ECCV 2016. LNCS, vol. 9908, pp. 354–370. Springer, Cham (2016). https://doi.org/10.1007/978-3-319-46493-0_22
22. Hariharan, B., et al.: Hypercolumns for object segmentation and fine-grained localization. In: IEEE Conference on Computer Vision and Pattern Recognition, pp. 447–456 (2014)
23. Long, J., Shelhamer, E., Darrell, T.: Fully convolutional networks for semantic segmentation. In: IEEE on Computer Vision and Pattern Recognition, pp. 3431–3440 (2015)
24. Sermanet, P., Kavukcuoglu, K., Chintala, S., et al.: Pedestrian detection with unsupervised multi-stage feature learning. In: IEEE Conference on Computer Vision and Pattern Recognition, pp. 3626–3633 (2013)
25. Simonyan, K., Zisserman, A.: Very Deep Convolutional Networks for Computer Science. pp. 730–734 (2014)

Collaborative Error Propagation for Single Sample Face Recognition

Jin Liu$^{(\boxtimes)}$, Langlang Li$^{(\boxtimes)}$, Qi Li, and Xue Wei

School of Electronic Engineering, Xidian University, Xi'an 710071, China
{jinliu,liqi}@xidian.edu.cn,
langzgy@163.com, wxyyxz@163.com

Abstract. Face recognition with single sample per person (SSPP) is a very challenging task because each class lacks sufficient training samples. To address this problem, this paper proposed a new face recognition method called collaborative error propagation (CEP) for single sample face recognition. First, we construct a facial variations dictionary through generic learning set, and rich collaborative representation dictionary. Then, we construct an error function by utilizing the global representation residual error of each testing sample, the error function as soft label influences the following patch classification. Later, partition the entire sample into many overlapping patch, obtain a new patch representation residual combined with the error function. Finally, using all the patch recognition results to get the voting result. Compared with the state of the art single sample face recognition methods, the experimental results demonstrate the efficacy of the proposed method, and shows more robust to complex facial variations, especially for disguise and uneven illumination.

Keywords: Single sample face recognition · Generic learning
Collaborative representation · Image patch

1 Introduction

Within the past three decades, face recognition is still attracting much attention because of its scientific challenges and its wide range of potential applications. Many recognition algorithms have been devised, such as principal component analysis (PCA) [1], linear discriminant analysis (LDA) [2] and LPP [3]. Recently, Wright et al. [4] propose a sparse representation based classification (SRC), and indicate that the test image can be approximated by a sparse linear combination of the training images, the choice of feature space is no longer critical. However, SRC has disadvantages of high computational complexity. Later, a collaborative representation based classification (CRC) scheme is given for face recognition [5, 6].

Furthermore, in many real-world FR applications (e.g., law enforcement, e-passport, driver license, etc.), we can obtain only a single sample per person (SSPP). We call face recognition in this scenario as single sample face recognition (SSFR). Most current face recognition techniques would suffer degraded performance or fail to work due to lack of sufficient samples, including SRC, or CRC will not get satisfactory performances. In order to overcome the lack of samples, and take full advantage of the

Y. Peng et al. (Eds.): IScIDE 2018, LNCS 11266, pp. 339–348, 2018.
https://doi.org/10.1007/978-3-030-02698-1_29

sample information, the idea of image patch is used for single sample face recognition. Base on different parts (e.g., eye, mouth, nose, cheek) of human faces have different importance in identifying the identity of a face. Many patch-based methods have been proposed, the final result is a vote of the results of all patches [7–9]. This makes the image patch method more robust to the disguise face recognition. Nevertheless, they do not solve the problem of lacking facial variations in the gallery set. To solve this problem, many researchers think that the variational features can be learned from an additional generic training set with multi-samples per person based on the assumption that the different individuals share similar variation information. Deng et al. [10] constructs an intra-class variation dictionary. Su et al. [11] proposed an adaptive generic learning method. Yang et al. [12] proposed a new sparse variation dictionary learning scheme for single sample face recognition. Ding et al. [13] proposed a variation feature representation(VFRC) method. Ji et al. [14] proposed a collaborative probabilistic labels (CPL) model for single-sample face recognition based on collaborative presentation and general learning method. Recently, because of the superiority of the cooperative representation speed, many single sample recognition algorithms base CRC combining image patch or general learning methods have been devised [15–17]. Although many improvements have been reported, CPL and other generic learning methods ignored the local information of the image certain extent. Motivated by this consideration, in this paper, we propose a new method called collaborative error propagation (CEP) for single sample face recognition. Compared with some related methods, the CEP has a number of advantages: (1) Constructing an error function by utilizing the global collaborative representation residual error, as a soft class label. (2) Obtain patch collaborative representation residuals, and then constructing a new function, get a new patch representation residuals combined with the error function. (3) Patch classification results are obtained by the new patch representation residuals, and it affected by global and local information. (4) The final identification result is obtained by all patch classification results voting. This combined with global and local information identification scheme make the result more robust. Experimental results show that the proposed methods not only outperforms than other related methods for SSPP problem, but also has good robust to facial variation caused by expression, illumination changes and disguise.

The rest of this paper is organized as follows. Section 2 reviews some related work. In Sect. 3, we describe the proposed method. The experimental evaluations are presented in Sect. 4. Finally, Sect. 5 concludes the paper.

2   Related Works

2.1   Normal Feature and Variational Feature of a Face Image

The face images we get from real life contain various variational information, including illumination, expression, disguise, and so on. The variational feature in each class can be obtained by subtracting the normal sample from other samples of the same class, denoted as:

$$V_i = I_i - N, \tag{1}$$

where $N \in R^{m \times n}$ is the normal face image, $I_i \in R^{m \times n}$ is the face image contain variational information, $V_i \in R^{m \times n}$ is denotes the face variation feature.

Because different individuals share the similar variation feature, the abundant face variations are useful to more accurately represent a testing face with unknown variations. In SSPP scenario, we can use a generic training set to get the variation dictionary. Suppose that we have collected a generic training set:

$$Q = [q_{11}, q_{12}, \cdots, q_{1K}, q_{21}, \cdots, q_{ij}, \cdots, q_{NK}] \in R^{d \times (N \times K)}, \tag{2}$$

where N is the number of individuals, and each person has K images that contain different variations(such as light, expression and disguise, etc.), $d = m \times n$ is the dimension of the image. d_{i1} is the normal image of i-th sample. So we can obtain the general variation dictionary:

$$V = [v_{11}, v_{12}, \cdots, v_{1(K-1)}, v_{21}, \cdots, v_{ik}, \cdots, v_{N(K-1)}] \in R^{d \times (N \times (K-1))}, \tag{3}$$

where $v_{ik} = q_{i(k+1)} - q_{i1}$.

2.2 General Collaborative Representation

Researches show that if we use only a few labeled samples as a dictionary to represent the unlabeled samples, the collaborative representation error can be big. Therefore, we add additional facial variations set to enrich collaborative representation dictionary.

Suppose that we have a generic training set: Q, through Eq. (3) we can obtain the general variation dictionary: $V \in R^{d \times (N \times (K-1))}$; gallery set: $G = [g_1, g_2, \cdots, g_C] \in R^{d \times C}$, It means that there are C people, and each person has only one training sample, $d = m \times n$ is the dimension of the image. We can construct a general collaborative representation dictionary:

$$X = [G, V] \in R^{d \times (C + N \times (K-1))}, \tag{4}$$

The testing sample is $y \in R^{d \times 1}$, application collaborative representation model:

$$\begin{bmatrix} \hat{\alpha} \\ \hat{\beta} \end{bmatrix} = \arg\min_{\alpha, \beta} \left\| y - [G, V] \begin{bmatrix} \alpha \\ \beta \end{bmatrix} \right\|_2^2 + \lambda_1 \left\| \begin{bmatrix} \alpha \\ \beta \end{bmatrix} \right\|_2^2, \tag{5}$$

where λ_1 is the regularization parameter. The solution of CRC with regularized least square in Eq. (5) can be easily and analytically derived as:

$$\begin{bmatrix} \hat{\alpha} \\ \hat{\beta} \end{bmatrix} = \{[G, V]^T [G, V] + \lambda_1 I\}^{-1} [G, V]^T y, \tag{6}$$

where I is an identity matrix. Let $P = ([G,V]^T[G,V] + \lambda_1 I)^{-1}[G,V]^T$. Clearly, P is independent of y so that it can be calculated as a projection matrix in advance. Once a testing sample y comes, we can just simply project y via P. This makes CRC very fast.

The regularized reconstruction error for each class is computed by:

$$r_c = \left\| y - g_c\hat{\alpha}_c - V\hat{\beta} \right\|_2^2 / \|\hat{\alpha}_c\|, \tag{7}$$

where g_c is the sub-dictionary associated with class c in the G, and $\hat{\alpha}_c$ is the coding coefficient associated with class c in the $\hat{\alpha}$.

According to the minimum residual error, we can judge the category of testing samples y:

$$identity(y) = \arg\min_c\{r_c\}, \tag{8}$$

3   Collaborative Error Propagation for Single Sample Face Recognition

3.1   Collaborative Error Propagation

In order to improve the recognition rate and robust of single sample face recognition under complex conditions such as uneven illumination and disguise, the global general collaborative representation residual error no direct use for classification, but for constructing an error function, as a class error label propagation to the following patch identification.

Using Eqs. (5), (6) and (7), the global regularized reconstruction error vector can be derived as:

$$R = [r_1, r_2, \cdots, r_c, \cdots, r_C], \tag{9}$$

The error function is constructed as:

$$P = [p_1, p_2, \cdots, p_c, \cdots, p_C], \tag{10}$$

where $p_c = e^{r_c/\min(R)}$, and min(R) is take the minimum element of the vector R.

Obviously, the larger p_c is, the greater the regularized reconstruction error of the sample g_c to the y, the probability of y belonging to the c-th class is smaller.

Different parts (e.g., eye, mouth, nose, cheek) of human faces contains different feature information, and they have different importance in identifying the identity of a face. Researches show that the disguises (i.e., scarf and sunglass) can be well dealt with by patch-based methods. We partition the testing sample y into S (overlapped) patches and denote these patches as: $\{y_1, y_2, \ldots, y_t, y_s\}$, the size of patch is $ps \times ps$ (overlapped pixels is pm). Correspondingly, the gallery set G and the generic variation set V can be partitioned as: $\{G_1, G_2, \cdots, G_t, \cdots, G_S\}$ and $\{V_1, V_2, \cdots, V_t, \cdots, V_S\}$.

For each local patch $y_t, t = 1, 2, \cdots, S$, its associated patch dictionary is: $[G_t, V_t]$, application collaborative representation model:

$$\begin{bmatrix} \hat{\alpha}_t \\ \hat{\beta}_t \end{bmatrix} = \{[G_t, V_t]^T [G_t, V_t] + \lambda_2 I\}^{-1} [G_t, V_t]^T y_t, \tag{11}$$

The regularized reconstruction error of local patch y_t for class c is computed by:

$$r_{tc} = ||y_t - g_{tc}\hat{\alpha}_{tc} - V_t \hat{\beta}_t||_2^2 / ||\hat{\alpha}_{tc}||, \tag{12}$$

Then, propagates the global collaborative reconstruction error to the local reconstruction error via the following function:

$$e_{tc} = \begin{cases} r_{tc} \times p_c, \min(R) < \min(r_{tc}) \\ r_{tc}, \min(R) > = \min(r_{tc}) \end{cases}, \tag{13}$$

where $\min(R)$ is take the minimum element of a vector R. e_{tc} is the new regularized reconstruction error combined global information with local information.

Obviously, if $\min(R) < \min(r_{tc})$, we believe that global recognition results are instructive than local, so we make further judgment the local reconstruction error by combining the global recognition error function; if $\min(R) > = \min(r_{tc})$, because local information is more robust to occlusion and uneven illumination, we do not change local reconstruction error.

3.2   Classifier

According to the minimum residual error, we can judge the class of testing patch samples y_t:

$$identity(y_t) = \underset{c}{\arg\min}\{e_{tc}\}, \tag{14}$$

Clearly, according to the Eq. (14), we can get classified results of local patch y_t, recorded as: $label_t$. And we partition all the sample into S (overlapped) patches, so we can get classified results of S patch: $\{label_1, label_2, \cdots, label_t \cdots, label_S\}$.

Through Eqs. (7) and (8) we can get classified results of global image, recorded as: $label_{S+1}$. Combining global and local recognition results, get a new class label:

$$Label = \{label_1, label_2, \cdots, label_t \cdots, label_S, label_{S+1}\}, \tag{15}$$

The final identification results are obtained according to the maximum voting model.

1	2	1
3	1	3
2	1	3

Fig. 1. The maximum voting model

The maximum voting model as shown in Fig. 1, if Label $= \{1, 2, 1, 3, 1, 3, 2, 1, 3\}$, get the number of votes to each class, four of the patch are identified by class 1, three of the patch are identified by class 2, two of the patch are identified by class 2. The final identification results according to the maximum voting model is class 1.

The main procedure of the proposed method is summarized in Table 1.

Table 1. The main procedure of CEP

Input:	Generic training set:Q , training set:G ,testing sample: y ,patch size: ps , overlapped pixels: pm ,regularization parameter:λ_1 , λ_2 .
Output:	The class label of the testing sample: y .
Step 1	Caculate general variation set V via Eq. (3);
Step 2	Caculate the global reconstruction error vector R via Eq.(9);
Step 3	Construct the error function P via Eq.(10)
Step 4	Partition G, V and y into S patches as: $\{G_1, G_2, \cdots, G_S\}$, $\{V_1, V_2, \cdots, V_S\}$,and $\{y_1, y_2, \cdots, y_S\}$ respectively;
Step 5	For $t = 1 : S$ do
	Construct the patch reconstruction error via Eq. (11) and (12);
	Caculate the new patch reconstruction error via Eq. (13).
	End for
Step 6	Caculate patch classified results via Eq. (14) and (15).
Step 7	Identify the testing sample y via the maximum voting model.

4 Experiments

In this section, the performance of the proposed method is evaluated and compared with PCRC [8], ESRC [10], VFRC [13] and CPL [14] under the same experiment condition on the AR [18] and Extended Yale B [19] databases. These databases with expression, illumination, and disguise variations. All the experiments are executed on PC with Core(TM) 3.4 GHz processor and 4.0 GB RAM, using Matlab R2015b software on WINDOWS 7 system.

4.1   Datasets

The AR database contains over 4000 color face images of 126 people, including frontal views of faces with different facial expressions, lighting conditions, and occlusions. In our experiments, we select 2600 face images from 100 individuals (50 men and 50 women). The pictures of 100 individuals are taken in two sessions (separated by two weeks) and each section contains 13 color images. The face image are resized to 80×80 pixels and converted to gray scale. The sample images of one person are shown in Fig. 2.

Extended Yale B face database includes 2,414 frontal-face images with 64 illumination conditions from 38 subjects. In our experiments, for each subject, we reserve the first 59 images for experiments. The face image are resized to 80×80 pixels and converted to gray scale. The sample images of one person are shown in Fig. 3.

Fig. 2. Sample images of one individual in the AR database

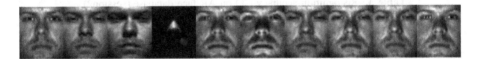

Fig. 3. Sample images of one individual in the extended Yale B database

4.2   Parameter Setting

Following the suggestions in the literatures [9, 13], in this paper, we set the regularization parameters as follows: $\lambda_1 = 0.01, \lambda_2 = 0.001..$ All the face image are resized to 80×80 pixels, the image patches are overlapped and the patch size is set as 20×20 (overlap is 10 pixels).

4.3   Experiments on the AR Database

We use the first 80 subjects from sessions 1 are used for the gallery and testing set while the other 20 subjects are used as the generic training set. We also use the face images from session 2 as the testing set to test the FR performance. Each person's image is divided into different subsets, including illumination, expression, and disguise (scarf and sunglass) in this experiment. The experimental results on session 1 and session 2 are shown in Tables 2 and 3, respectively.

Tables 2 and 3 tabulates the recognition accuracies of all methods on session 1 and session 2. The experiment results show that, our approach achieves the best performance on all the four different testing subsets. In PCRC method, it uses image patch multiple recognition, which leads to the face recognition performance not very bad, because the disguises (such as scarf and sunglass) can be well dealt with by patch-based methods, but it can't deal with the expression various better. The others methods ESRC, VFRC and CPL, learned a variation set from a generic training set, it makes a good performance for expression various; In our method, we construct a global reconstruction error function combined global and local information, and exploiting the advantages of generic variation information recognition, this made our method outperforms better than others under all different variations.

Table 2. Recognition accuracy (%) on AR face database (session1).

Method	Illumination	Expression	Disguise	All various
PCRC	95.0	86.7	89.4	81.3
ESRC	97.5	85.0	87.3	68.6
VFRC	84.1	73.9	72.8	75.8
CPL	95.7	88.3	71.6	83.6
CEP	99.6	91.3	92.9	93.7

Table 3. Recognition accuracy (%) on AR face database (session2).

Method	Illumination	Expression	Disguise	All various
PCRC	88.8	71.7	78.5	63.1
ESRC	85.8	71.2	65.6	45.0
VFRC	73.3	67.8	66.4	69.2
CPL	83.5	70.2	71.6	75.9
CEP	94.2	79.2	82.5	82.2

4.4   Experiments on the Extended Yale B Database

On the Extended Yale B database, we use the face images of the first 30 subjects to form the gallery and testing sets, and use the face images of the other 8 subjects as the generic training set. There are many illumination variations for every subject, so we choose the first sample for training and the following first 20, 30, 40, 50, 58 samples for testing, respectively. Table 4 lists the recognition rates by different methods. The experiment results show that, our proposed method outperforms better than others under all different amounts of illumination variations. In addition, the proposed combined global and local recognition mechanisms demonstrated good robustness to light changes and test sample quantities.

Table 4. Recognition accuracy (%) on Extended Yale B database.

Method	20	30	40	50	58
PCRC	88.3	82.8	82.7	83.4	77.8
ESRC	85.7	67.7	67.0	67.8	62.4
VFRC	91.7	79.0	76.1	78.7	74.4
CPL	92.8	89.4	85.7	86.5	84.3
CEP	93.5	92.7	90.8	88.5	85.8

5 Conclusion

In order for a more effective face recognition when the number of training samples per person is single sample and the testing sample has complex facial variations, in this paper we proposed a new face recognition method called collaborative error propagation for single sample face recognition. When face identification, we not merely using the global information of a single training sample, and using local information at the same time. By exploiting the advantages of global representation, patch representation and generic variation information, experiments demonstrate that the proposed CEP achieves the highest recognition accuracy, and shows more robust to complex facial variations compared with the state of the art methods, especially for disguise and uneven illumination.

Acknowledgments. We would like to thank the associate editor and all anonymous reviewers for their constructive comments and suggestions. This research was partially supported by the National Science Foundation of China (Grant No. 61101246) and the Fundamental Research Funds for the Central Universities (Grant No. JB150209).

References

1. Turk, M., Pentland, A.: Eigenfaces for recognition. J. Cogn. Neurosci. **3**(1), 71–86 (1991)
2. Wang, X., Tang, X.: Random sampling LDA for face recognition. In: IEEE Conference on Computer Vision and Pattern Recognition, CVPR 2004, vol. 2, pp. 259–265. IEEE Press (2004)
3. He, X., Yan, S., Hu, Y., et al.: Face recognition using laplacianfaces. IEEE Trans. Pattern Anal. Mach. Intell. **27**, 328–340 (2005)
4. Wright, J., Yang, A.Y., Ganesh, A., et al.: Robust face recognition via sparse representation. IEEE Trans. Pattern Anal. Mach. Intell. **31**(2), 210–227 (2009)
5. Zhang, L., Yang, M., Feng, X.: Sparse representation or collaborative representation: which helps face recognition? In: IEEE International Conference on Computer Vision (ICCV), pp. 471–478 (2011)
6. Zhang, L., Yang, M., Feng, X., et al.: Collaborative representation based classification for face recognition. arXiv preprint https://arxiv.org/arXiv:1204.2358 (2012)
7. Chen, S., Liu, J., Zhou, Z.: Making FLDA applicable to face recognition with one sample per person. Pattern Recogn. **37**(7), 1553–1555 (2004)

8. Zhu, P., Zhang, L., Hu, Q., Shiu, S.C.K.: Multi-scale patch based collaborative representation for face recognition with margin distribution optimization. In: Fitzgibbon, A., Lazebnik, S., Perona, P., Sato, Y., Schmid, C. (eds.) ECCV 2012. LNCS, vol. 7572, pp. 822–835. Springer, Heidelberg (2012). https://doi.org/10.1007/978-3-642-33718-5_59

9. Zhu, P., Yang, M., Zhang, L., Lee, I.-Y.: Local generic representation for face recognition with single sample per person. In: Cremers, D., Reid, I., Saito, H., Yang, M.-H. (eds.) ACCV 2014. LNCS, vol. 9005, pp. 34–50. Springer, Cham (2015). https://doi.org/10.1007/978-3-319-16811-1_3

10. Deng, W., Hu, J., Guo, J.: Extended SRC: undersampled face recognition via intraclass variant dictionary. IEEE Trans. Pattern Anal. Mach. Intell. 34(9), 1864 (2012)

11. Su, Y., Shan, S., Chen, X., Gao, W.: Adaptive generic learning for face recognition from a single sample per person. In: IEEE Conference on Computer Vision and Pattern Recognition, pp. 2699–2706 (2010)

12. Yang, M., Van, L., Zhang, L.: Sparse variation dictionary learning for face recognition with a single training sample per person. In: Proceedings of International Conference on Computer Vision, pp. 689–696 (2013)

13. Ding, R.X., Du, D.K., Huang, Z.H., et al.: Variational feature representation-based classification for face recognition with single sample per person. J. Vis. Commun. Image Represent. 30, 35–45 (2015)

14. Ji, H.K., Sun, Q.S., Ji, Z.X., et al.: Collaborative probabilistic labels for face recognition from single sample per person. Pattern Recogn. 62(C), 125–134 (2017)

15. Yu, Y.F., Dai, D.Q., Ren, C.X., et al.: Discriminative multi-scale sparse coding for single-sample face recognition with occlusion. Pattern Recogn. 66, 302–312 (2017)

16. Zhang, G., Sun, H., Ji, Z., et al.: Label propagation based on collaborative representation for face recognition. Neurocomputing 171(C), 1193–1204 (2016)

17. Liu, F., Tang, J., Song, Y., et al.: Local structure based multi-phase collaborative representation for face recognition with single sample per person. Inf. Sci. 346–347, 198–215 (2016)

18. Martinez, A.M., Benavente, R.: The AR face database. CVC Technical report 24, Barcelona, Spain (1998)

19. Georghiades, A.S., Belhumeur, P.N., Kriegman, D.J.: From few to many: illumination cone models for face recognition under variable lighting and pose. IEEE Trans. Pattern Anal. Mach. Intell. 23(6), 643–660 (2002)

Coarse Label Refined Knowledge Reasoning for Fine-Grained Visual Categorization

Xiangyu Zhao and Yuxin Peng[✉]

Institute of Computer Science and Technology, Peking University, Beijing, China
xiangyu.zhao.official@gmail.com, pengyuxin@pku.edu.cn

Abstract. Fine-grained visual categorization (FGVC) is a challenging task due to large intra-class variance and small inter-class variance, aiming at recognizing hundreds of subcategories which belong to the same basic-level category. Humans have the ability to acquire knowledge about the world and use the acquired knowledge to reason about entities. When identifying an image, humans first reason out candidate subcategories of the object based on their prior knowledge. Then discover the characteristics of the object, which can be used as assistance for comparisons between candidate subcategories. Finally humans can determine the ultimate prediction. Inspired by this behavior, we propose a Coarse Label Refined Knowledge Reasoning (CLRKR) approach for fine-grained visual categorization. Its main novelties and advantages are as follows: (1) **Knowledge Reasoning:** We first construct the knowledge graph based on subcategory-attribute correlations, then reason out candidate subcategories with corresponding probabilities based on the knowledge graph. (2) **Coarse Label Refinement:** We incorporate coarse classes, which are pre-defined groups of attributes, to get corresponding coarse labels. These coarse labels reveal the most characteristic features of the entity in the image, which help us to re-rank the sequence of candidate subcategories and acquire the foremost subcategory as the final prediction. Our CLRKR approach achieves the best performance according to the experimental results on the widely-used CUB-200-2011 dataset for fine-grained visual categorization comparing with the state-of-the-art methods.

Keywords: Fine-grained visual categorization
Knowledge reasoning · Knowledge graph · Coarse labels

1 Introduction

Conventional methods of fine-grained visual categorization mainly rely on discovering visual discriminations for categorization. Lin et al. [1] proposed an architecture that uses two separate CNN feature extractors to model the appearance due to where the parts are and what the parts look like. Zhang et al. [2] propose to select discriminative parts through exploiting useful information in part

© Springer Nature Switzerland AG 2018
Y. Peng et al. (Eds.): IScIDE 2018, LNCS 11266, pp. 349–359, 2018.
https://doi.org/10.1007/978-3-030-02698-1_30

clusters. Recently, researchers begin to focus on utilizing structured prior knowledge for image classification. Marino et al. [3] propose to use knowledge graphs as extra information to improve image classification. However, they merely use knowledge alone and do classification in basic-level categories. In fact, humans are more than just appearance-based classifiers; they combine prior knowledge with visual discriminations for categorization.

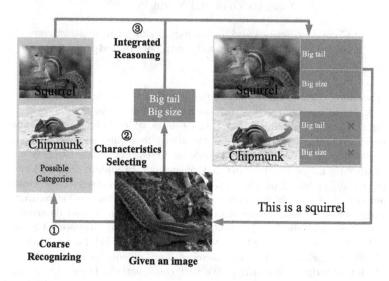

Fig. 1. Three steps of humans' categorization behaviors: coarse recognizing, characteristics selecting, and integrated reasoning.

Humans gain knowledge of the world from experiences in daily lives and use acquired knowledge to recognize objects [3]. When we see an object, we quickly recall what we have known, do analyses and make comparisons based on prior knowledge in our mind to get a final prediction. Generally, humans' categorization behaviors can be concluded into three steps: (1) *Coarse Recognizing:* given an object, we intuitively associate it with several candidate subcategories based on our experience and knowledge. (2) *Characteristics Selecting:* we carefully examine the object to find striking features, which can be used to distinguish between candidate subcategories. (3) *Integrated Reasoning:* we can finally reason out the most likely subcategory by making comparisons based on the presence of selected features in each subcategory. For instance, as shown in Fig. 1, when we see the image at the first glance, we can infer that this animal is either a squirrel or a chipmunk. Secondly, if we take a further look, we can find it relatively big in size and with a big tail. Combining these results with prior knowledge, we can conclude that it is a squirrel because "big tail" and "big size" are features belong to squirrels, rather than chipmunks'.

Inspired by this, we propose a **Coarse Label Refined Knowledge Reasoning (CLRKR)** approach for fine-grained visual categorization, which uses the prior knowledge of the correlations between subcategories and attributes to facilitate final subcategory prediction. The contributions of our proposed CLRKR approach can be summarized as follows:

1. **Knowledge Reasoning:** We first construct the knowledge graph based on subcategory-attribute correlations and train our model to acquire the underlying information via propagating through knowledge graph. Our model implicitly learns the correlations between subcategories and attributes, which are used as potential supervision for elevating the categorization performance.
2. **Coarse Label Refinement:** We introduce coarse classes, which are predefined groups of attributes, to mitigate intra-class variance and pay more attention on inter-class variance. We train the coarse classes to get proper coarse labels, which reveal the most characteristic features of the entity in the image. The introduction of coarse labels brings in additional supervised information, which helps to capture subtle differences between fine-grained subcategories and add strong restrictions to regulate categorization.

Comparing with more than ten state-of-the-art methods on the widely-used fine-grained visual categorization dataset CUB-200-2011 [4], our CLRKR approach achieves the best categorization accuracy, which verifies its effectiveness.

2   Our CLRKR Approach

In this section, we present the proposed Coarse Label Refined Knowledge Reasoning (CLRKR) approach for fine-grained visual categorization, which consists of two components: knowledge reasoning and coarse label refinement. Knowledge reasoning is to reason out the candidate subcategories with corresponding probability, and coarse label refinement is to acquire coarse labels and use these labels to re-rank candidate subcategories to get the final prediction. The framework is shown in Fig. 2. We first construct the knowledge graph based on the structured attribute information, and acquire knowledge about the fine-grained visual categorization task. Then we take the feature representation of the image as input, and reason out candidate subcategories via propagating through constructed knowledge graph. At the same time, the image propagates through several coarse class classifiers to get corresponding coarse labels. Finally we combine the coarse labels and subcategories based on the knowledge of subcategory-attribute relations to reason out the most likely subcategory as the final prediction. We present knowledge reasoning and coarse label refinement in the following subsections.

2.1   Knowledge Reasoning

Knowledge Graph Construction. We construct the knowledge graph based on the correlations between subcategories and attributes, where nodes denote the subcategories and attributes, and the edges denotes their correlations. Figure 3 shows the constructed knowledge graph on CUB-200-2011 dataset [4]. An edge only exists in the connection of a subcategory and an attribute, which indicates

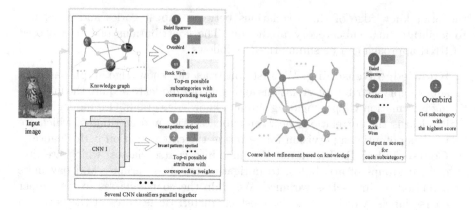

Fig. 2. An overall pipeline of our proposed CLRKR approach. This framework primarily consists of a knowledge reasoning process for subcategories prediction, several CNN classifiers for generating coarse labels, and a combination of acquired results to get the subcategory with the highest score.

that the subcategory possesses the attribute. For a fine-grained task, we have x subcategories, y attributes, and a correlation matrix which illustrates the correlations between subcategories and attributes. The correlated value $SA_{i,j}$ refers to the probability that the ith subcategory has the jth attribute. Given these information, we can construct the adjacency matrix $\bar{A}$ of the knowledge graph, which definition is as follows:

$$\bar{A}_{i,j} = \begin{cases} 1, & \text{if } SA_{i,j} > 0 \\ 0, & otherwise \end{cases} \tag{1}$$

Reasoning Through GGNN. In the reasoning process, we apply the constructed knowledge graph to Gated Graph Neural Networks (GGNN) [5], which is a kind of neural network for learning graph-structured data, to do reasoning and reason out the candidate subcategories.

We initialize the node referring to subcategory label i with a score s_i, which represents the probability that the given image belongs to this subcategory, while the nodes referring to each attribute with zero vector. The score vector $s = s_0, s_1, \ldots, s_{C-1}$ for all subcategories is acquired by a basic classifier, which will be illustrated in Sect. 3.1. For each node v in the knowledge graph, the input feature for the node can be represented as:

$$x_v = \begin{cases} [s_i, \mathbf{0}_{n-1}], & \text{if the node refers to the subcategory i} \\ [\mathbf{0}_n], & \text{if the node refers to an attribute} \end{cases} \tag{2}$$

where each input feature X_v is a vector with dimension n. Then we initialize the hidden state vector $h_v^{(1)}$ with the input feature x_v.

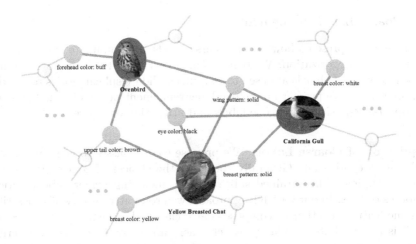

Fig. 3. An example of constructed knowledge graph for modeling subcategory-attribute correlations based on CUB-200-2011 dataset.

For the GGNN [5], the basic recurrent process is formulated as:

$$h_v^{(1)} = x_v^\top \tag{3}$$

$$a_v^t = A_v^\top [h_1^{(t-1)\top}, h_2^{(t-1)\top}, \ldots, h_{|v|}^{(t-1)\top}]^\top + b \tag{4}$$

$$z_v^t = \sigma(W^z a_v^{(t)} + U^z h_v^{(t-1)}) \tag{5}$$

$$r_v^{(t)} = \sigma(W^r a_v^{(t)} + U^r h_v^{(t-1)}) \tag{6}$$

$$\widetilde{h_v^{(t)}} = tanh(W a_v^{(t)} + U(r_v^{(t)} \odot h_v^{(t-1)})) \tag{7}$$

$$h_v^{(t)} = (1 - z_v^{(t)}) \odot h_v^{(t-1)} + z_v^{(t)} \odot \widetilde{h_v^{(t)}} \tag{8}$$

After T steps, we can obtain the final hidden state vectors for all nodes in the knowledge graph, which are utilized to output the final node scores.

Through the recurrent propagation of the GGNN [5], we output the node score for each node, and its definition is as follows:

$$o_v = g(h_v^\top, x_v) \tag{9}$$

We typically notice the output scores of the subcategory nodes, which indicate the probability that the image belongs to this subcategory. We re-rank the output scores of all the m subcategory nodes and get the top $(\alpha + 1)$ of them. We use these $(\alpha + 1)$ scores to get the top α subcategories and their corresponding weights. Let o_i denotes the output score of the ith number, the weight w_i of subcategory i means the probability of this subcategory to be the ground truth, which can be calculated as follows:

$$w_i = (o_i - o_{\alpha+1})/\Sigma_1^\alpha(o_i - o_{\alpha+1}) \tag{10}$$

By Eq. 10, we finally acquire top-α subcategories and their corresponding weights, which will be used in coarse label refinement.

2.2   Coarse Label Refinement

This subsection presents how we use coarse labels and candidate subcategories to refine the categorization. We define coarse classes and use them to acquire coarse labels. These labels are used to re-rank candidate subcategories to get the final prediction. We present our coarse label refinement possess in the following two aspects: acquisition of coarse labels and utilization of coarse labels.

Acquisition of Coarse Labels. We propose coarse classes to generate coarse labels for categorization. Generally speaking, the choices of coarse classes can be any grouping of fine-grained subcategories, including bigger subcategories, attributes or tags. In our CLRKR approach, we group attributes which describe the same feature together to compose coarse classes. For instance, "wing color: black" is an attribute; similar to which, there are other 14 colors to describe wing color. We can group these attributes that describe wing color together and get a coarse class "wing color". This type of coarse class typically describes the wing color of the bird in the image and has 15 attributes about wing colors. The chosen one of the 15 colors is called the coarse label of this coarse class. Compared with normal fine-grained labels, the introduction of coarse classes has these benefits: coarse classes are more natural and cheaper to obtain and they bring in additional supervised information, which can be used to capture subtle differences between fine-grained subcategories.

Given m types of coarse classes, where each type j contains k_j attributes, we generate m coarse class classifiers, each of which is trained to get the corresponding coarse label with great precision. For a given image, it corresponds to m coarse labels, which can be used as auxiliaries for categorization refinement. For each coarse class classifier, we augment the last fully-connected layer to let the output number be the same as that of attributes it contains. That is to say, for every image in a certain coarse class classifier j, we have k_j output scores. We re-rank the scores in a descending order to get the top-β of them and their corresponding attributes. Also, for each classifier j, we can get the first βth prediction accuracy respectively. These accuracies are designated as weights of the coarse label w_c, which indicates the fiducial probability of the βth prediction. For instance, let us suppose $\beta = 2$, given coarse class classifier j's top-1 accuracy is 70% and top-2's is 90%, we can have $w_{cj}^1 = 0.7$ and $w_{cj}^2 = 0.2$, which means that the first attribute has the probability of 0.7 to be the truth and that of the second attribute is 0.2. In conclusion, by introducing m coarse classes, we generate m classifiers. For each of the classifier, we can acquire β predictions with their corresponding weights, which utilization will be illustrated in the following section.

Utilization of Coarse Labels. Given an image, from Sect. 2.1, we can obtain α candidate subcategories with their corresponding weights w_i where $i = 1, \alpha$. And from section above, we get m classification results, which are relevant attributes. For each classifier j, the result is β attributes with their weights w_{cj}^k, where

$k = 1, \beta$. We propose to combine the candidate subcategories with the candidate attributes to re-sort the order of candidate subcategories. In addition to this, we utilize the subcategory attribute adjacency matrix to further facilitate our approach. For the ith subcategory and jth attribute, there is a value $p_{ij} \in [0, 1]$, which indicates the correlation degree of the two. The bigger the value, the stronger the correlation.

We manage to take all things above in consideration to calculate a score for each candidate subcategory. The subcategory with the highest score will be the final prediction. Let c_{jk} denotes the kth coarse label of the coarse class classifier j, where $j = 1, m$ and $k = 1, \beta$. The score s_i for each subcategory i is calculated as follows:

$$s_i = w_i \Sigma_1^m p_{ic_{jk}} \Sigma_1^\beta w_{cj}^k \qquad (11)$$

From Eq. 11, we can know that the calculation of score is relevant to subcategory's original weight, relevant attributes' certainty and the correlation degree between the subcategory and attributes. An increase in any factors above will raise the final score, which is strongly in line with our theorized expectations.

3 Experiments

This section evaluates our CLRKR's performance on the widely-used CUB-200-2011 dataset [4], and compares with more than ten state-of-the-art methods. We evaluate our CLRKR approach on the CUB-200-2011 dataset [4], which is the most widely-used benchmark for fine-grained visual categorization. It covers 200 subcategories and 312 attributes of birds, which contains 5,994 images for training and 5,794 images for testing. In our experiments, the image-level subcategory label, image attribute labels are used for knowledge graph construction and class attribute labels are utilized for assigning coarse labels for each attribute group.

3.1 Implementation Details

We present the implementation details in the following two aspects: knowledge reasoning and coarse label refinement.

Details of Knowledge Reasoning. We first adopt CUB-200-2011 dataset, which has $x = 200$ subcategories and $y = 312$ attributes, to construct our knowledge graph and put all $35,812$ edges in an edge list. We also utilize the 19-layer VGGNet [6] with Batch Normalization [7] as the basic CNN to produce scores to initialize the hidden states. The edge list and basic CNN output scores are used to initialize GGNN [5]. We adopt the "node selection" GGNN to output score vectors for each node. The dimension of annotation is set to 5 and that of the hidden state is set to 10. We set the propagation time T to 5. For the output of GGNN [5] of each image, we acquire the first $m = 200$ output scores and set $\alpha = 5$ to get top-6 scores and top-5 subcategories.

Details of Coarse Label Refinement. 312 attributes are divided into 28 coarse classes, and we select the best $m = 10$ of them based on their final classification accuracy. For each coarse class, we train a coarse class classifier and assign target labels for each image in subcategory level. Each coarse class classifier is implemented on the off-the-shelf Caffe [8] platform and use the GoogLeNet [9] framework. We modified the output number of the last fully connected layer to equal the number of attributes of each coarse class. In the training phases, each of them is trained with SGD and minibatch is set to 16. For the output of each classifier, we set $\beta = 2$ to get top-2 scores with their weights.

Table 1. Comparisons with state-of-the-art methods on CUB-200-2011 dataset.

Methods	Accuracy (%)
Our CLRKR approach	**85.7**
AGAFA [10]	80.4
VGG-BGLm [11]	75.9
RA-CNN [12]	85.3
Saliency-guided [13]	85.1
TSC [14]	84.7
FOAF [15]	84.6
PD [16]	84.5
STN [17]	84.1
Bilinear-CNN [1]	84.1
Multi-grained [18]	81.7
NAC [19]	81.0
PIR [2]	79.3
TL Atten [20]	77.9
MIL [21]	77.4

3.2 Comparisons with State-of-the-Art Methods

This subsection presents the experimental results and analyses of our CLRKR approach as well as the state-of-the-art methods on the widely-used CUB-200-2011 dataset, which verifies the effectiveness of our CLRKR approach. The results are shown in Table 1.

First, we compare our CLRKR approach with these methods that use attributes as aids for fine-grained visual categorization. It is noted that our CLRKR framework requires no ground truth part annotations and merely utilize the image-level subcategory label and the annotated attributes, but it achieves an accuracy of 85.7% that outperforms all previous state-of-the-art methods even those with part annotations and bounding box. AGAFA [10] proposes an attribute-guided attentive network to sequentially discover informative parts or regions and feed the regions into a recurrent neural network to yield the object-level representation. Its accuracy is 80.4% under the same setting and 84.1% with the bounding box, which is lower than us by 1.6%. VGG-BGLm [11] proposes to exploit the relationships through bipartite-graph labels and incorporate class bipartite graphs into CNN. Under the same setting, VGG-BGLm only achieves an accuracy of 75.9%. Even with bounding box, its accuracy is 80.4%, which is still lower than ours (85.7%). Compared with the all 28 coarse labels used in VGG-BGLm, we only used ten of them, which further indicate the efficiency and accuracy of our approach.

There are other methods that explore the fine-grained visual categorization task without using attributes. Most of them explicitly search discriminative regions and aggregate deep features of these regions for categorization. RA-CNN [12] recurrently discovers image regions over three scales and achieves an accuracy of 85.3%. However, its accuracy decreases to 84.7% when only two scales are used. As for the rest of the methods, their results are far lower than ours. Our CLRKR approach achieves better categorization accuracy than these methods since that we explore rich subcategory-attribute correlations information through the knowledge representation to learn more discriminative features and add strong restrictions to regulate categorization due to the introduction of coarse labels. These comparisons well demonstrate the effectiveness of our CLRKR approach over existing algorithms.

4 Conclusion

In this paper, we propose a Coarse Label Refined Knowledge Reasoning (CLRKR) approach for fine-grained visual categorization. We first construct the knowledge graph based on subcategory-attribute correlations. Then we take the feature representation of an image as input, and reason out candidate subcategories via propagating through constructed knowledge graph. At the same time, the image passes through several coarse class classifiers to get corresponding coarse labels. Finally we combine coarse labels with subcategories and incorporates subcategory-attribute relations to reason out the most likely subcategory as the final prediction. Compared with more than ten state-of-the-art methods, our CLRKR approach achieves the best accuracy, which verifies the effectiveness of our CLRKR approach.

There are several future directions to our work. First, we will apply knowledge reasoning with other visual tasks like Visual Questing Answering to get a more accurate prediction for candidate subcategories. Second, we will focus on how to use knowledge graph to do attribute classification in per-image level, which will improve the performance of fine-grained visual categorization especially when attribute labels are ambiguous in subcategory level.

Acknowledgment. This work was supported by the National Natural Science Foundation of China under Grant 61771025.

References

1. Lin, T.-Y., RoyChowdhury, A., Maji, S.: Bilinear CNN models for fine-grained visual recognition. In: Proceedings of the IEEE International Conference on Computer Vision, pp. 1449–1457 (2015)
2. Zhang, Y., et al.: Weakly supervised fine-grained categorization with part-based image representation. IEEE Trans. Image Process. **25**(4), 1713–1725 (2016)
3. Marino, K., Salakhutdinov, R., Gupta, A.: The more you know: using knowledge graphs for image classification. In: IEEE Conference on Computer Vision and Pattern Recognition (CVPR), July 2017
4. Wah, C., Branson, S., Welinder, P., Perona, P., Belongie, S.: The caltech-UCSD birds-200-2011 dataset (2011)
5. Li, Y., Tarlow, D., Brockschmidt, M., Zemel, R.: Gated graph sequence neural networks. In: International Conference on Learning Representations (ICLR) (2016). http://arxiv.org/abs/1511.05493
6. Simonyan, K., Zisserman, A.: Very deep convolutional networks for large-scale image recognition. arXiv preprint arXiv:1409.1556 (2014)
7. Ioffe, S., Szegedy, C.: Batch normalization: accelerating deep network training by reducing internal covariate shift. arXiv preprint arXiv:1502.03167 (2015)
8. Jia, Y., et al.: Caffe: convolutional architecture for fast feature embedding. In: ACM International Conference on Multimedia (ACM MM), pp. 675–678. ACM (2014)
9. Szegedy, C., et al.: Going deeper with convolutions. In: IEEE Conference on Computer Vision and Pattern Recognition (CVPR), pp. 1–9 (2015)
10. Yan, Y., Ni, B., Yang, X.: Fine-grained recognition via attribute-guided attentive feature aggregation. In: ACM International Conference on Multimedia (ACM MM), pp. 1032–1040. ACM (2017)
11. Zhou, F., Lin, Y.: Fine-grained image classification by exploring bipartite-graph labels. In: IEEE Conference on Computer Vision and Pattern Recognition (CVPR), pp. 1124–1133 (2016)
12. Fu, J., Zheng, H., Mei, T.: Look closer to see better: recurrent attention convolutional neural network for fine-grained image recognition. In: The IEEE Conference on Computer Vision and Pattern Recognition (CVPR), July 2017
13. He, X., Peng, Y., Zhao, J.: Fine-grained discriminative localization via saliency-guided faster R-CNN. In: ACM International Conference on Multimedia (ACM MM), pp. 627–635. ACM (2017)
14. He, X., Peng, Y.: Weakly supervised learning of part selection model with spatial constraints for fine-grained image classification. In: AAAI Conference on Artificial Intelligence (AAAI), pp. 4075–4081 (2017)
15. Zhang, X., Xiong, H., Zhou, W., Tian, Q.: Fused one-vs-all features with semantic alignments for fine-grained visual categorization. IEEE Trans. Image Process. **25**(2), 878–892 (2016)
16. Zhang, X., Xiong, H., Zhou, W., Lin, W., Tian, Q.: Picking deep filter responses for fine-grained image recognition. In: IEEE Conference on Computer Vision and Pattern Recognition (CVPR), pp. 1134–1142 (2016)
17. Jaderberg, M., Simonyan, K., Zisserman, A., et al.: Spatial transformer networks. In: Neural Information Processing Systems (NIPS), pp. 2017–2025 (2015)
18. Wang, D., Shen, Z., Shao, J., Zhang, W., Xue, X., Zhang, Z.: Multiple granularity descriptors for fine-grained categorization. In: International Conference on Computer Vision (ICCV), pp. 2399–2406 (2015)

19. Simon, M., Rodner, E.: Neural activation constellations: unsupervised part model discovery with convolutional networks. In: International Conference of Computer Vision (ICCV), pp. 1143–1151 (2015)
20. Xiao, T., Xu, Y., Yang, K., Zhang, J., Peng, Y., Zhang, Z.: The application of two-level attention models in deep convolutional neural network for fine-grained image classification. In: IEEE Conference on Computer Vision and Pattern Recognition (CVPR), pp. 842–850 (2015)
21. Xu, Z., Tao, D., Huang, S., Zhang, Y.: Friend or Foe: fine-grained categorization with weak supervision. IEEE Trans. Image Process. (TIP) 26(1), 135–146 (2017)

A Fast Positioning Algorithm Based on 3D Posture Recognition

Xianbing Xu[1,2], Chengbin Peng[2(✉)], Jiangjian Xiao[2], Huimin Jing[1],
and Xiaojie Wu[3]

[1] Institute of Mechanical Engineering and Mechanics, Ningbo University,
Ningbo 315201, Zhejiang, China
[2] Ningbo Institute of Industrial Technology, Chinese Academy of Sciences,
Ningbo 315201, Zhejiang, China
pengchengbin@nimte.ac.cn
[3] Ningbo Institute of Information Technology Applications, Chinese Academy
of Sciences, Ningbo 315201, Zhejiang, China

Abstract. Automatic workpiece grabbing on production line is important for improving the production efficiency in manufacturing industry. However, due to their irregular shapes, uncertain positions, and various posture changes, traditional edge detection, feature extraction and other methods are difficult to accurately identify and locate complex workpieces. In this paper, we propose an approach to recognize workpieces and determine their postures with deep learning. Based on object detection, input images containing workpieces are fed into an angle regression network which is used to determine the three-dimensional posture of workpieces. The classification, position and posture information are obtained from the final output. Experiments show that this method is more robust and can achieve higher accuracy than traditional feature extraction methods.

Keywords: Deep learning · Object detection · Angle regression
Posture determination

1 Introduction

In industrial production, the problem of automatic gripping of workpieces has plagued numerous manufacturers. Traditional mechanical arms work mostly based on manually predefined operations, and have poor adaptability and low anti-interference ability to complex tasks. Thus, its application is very limited. Nowadays, workpieces that manufacturers have to face are usually irregular in shape and structure with various postures and positions. In this case, when to handle workpieces, manufacturers often need a large number of laborers forces or specially designed machines. With the increase in labor costs and the development of modern manufacturing, the intelligence of traditional robot systems can no longer meet the needs for factory production. Recent advancements start to introduce machine vision technology into the production line, and traditional robot systems can now be equipped with more intelligent sensing devices. Such adoption can reduce labor cost and increase production efficiency.

© Springer Nature Switzerland AG 2018
Y. Peng et al. (Eds.): IScIDE 2018, LNCS 11266, pp. 360–370, 2018.
https://doi.org/10.1007/978-3-030-02698-1_31

1.1   Related Work

Machine vision is closely related to object detection technology. The object detection technique is to find specific objects from input images. Object detection technology is widely used and gradually becomes a research hotspot in the field of computer vision. Usually for object detection algorithms, people will evaluate the performance of object detection through time and accuracy of the detection. Higher accuracy and shorter detection time indicate the better performance. At present, the difficulty of achieving object detection is mainly due to the diversity of objects from different perspective, the diversity of different object categories, and the complexity of background environments. These factors are inevitable in the object detection for manufacturing, and can greatly undermine the accuracy of the object detection. The traditional object detection algorithm mainly adopts the sliding window method and extracts features of the input to detect. The classical object detection method based on Histogram of Oriented Gradient (HOG) is proposed by Navneet Dalal and Bill Triggs in 2005 [1]. After obtaining the HOG feature of the image, it is fed into an SVM classifier for classification. This approach is widely used in the object detection field, especially in pedestrian detection. In 2010, Felzenszwalb et al. put forward DPM [2] on the basis of HOG and achieved great success in pedestrian detection. Traditional object detection algorithms can achieve good results through step-by-step optimization of algorithms and design of various image features. However, in most object detections, accuracy and real-time performance are difficult to meet.

Deep learning has been applied to machine vision and object detection in recent years. It can be seen as a subset of machine learning. It is a method of automatically extracting the features of the original image. Its essence is the artificial neural network. Thus, deep learning originates from artificial neural networks and is a branch of artificial neural networks. In traditional machine learning, it is often necessary to manually design features to allow computers to learn, but the process of making features is cumbersome. In 1957, the famous artificial intelligence expert Frank Rosenblatt invented the Perceptron [3]. This is the earliest and simplest artificial neural network model. However, its classification ability is very limited, it can only deal with simple binary problems, and it cannot handle non-linear problems. By the 1970s, Paul Werbos proposed a back-propagation method to train artificial neural networks. In 1980, the algorithm was further adopted by Jeffrey Hinton, Yan Lecun, et al. to train neural networks with deep structure [2]. The back propagation method adjusts the weight of the network based on the classification error of the output layer of the neural network until the error cannot be reduced [4]. However, the back-propagation method based on gradient descent [5] can easily converge to local minimum when training network parameters and its learning performance is not good. There are also practical problems such as slow training speeds and large sample sizes. In 1998, Yan Lecun proposed a model for deep learning, the convolutional neural network [6]. The convolutional neural network proposed by Yan Lecun not only converges well in the training process, but also has strong generalization ability. In 2006, Jeffrey Hinton proposed the concept of deep learning and proposed one of the deep learning models, the deep belief network, along with an efficient semi-supervised algorithm: A layer-by-layer greedy algorithm to train deep beliefs the parameters of the network break the deadlock that has long been difficult to train in deep networks [7, 8].

Compared to the traditional machine vision and object detection algorithms, deep learning has more powerful computing performance, wider application scope, and better practical results. In 2012, Krizhevsky [9] achieved the best result in image classification challenge in Image Net, which reduced the image recognition error rate by 14%. In 2014, the region positioning network structure model based on the CNN feature map (Regions with CNN, R-CNN [10]) proposed by Ross Girshick et al. transformed the detection problem into a classification problem and achieved significant success in the image classification. It has provided a brand-new idea for the object detection problem. Since then, the deep learning method has once again set off an upsurge of research in the industry, and the method of object detection has also been continuously updated. As the CNN needs to input the fixed size of the picture, He et al. proposed SPPNet [11] and introduced the SPP layer to release the fixed size constraint. Subsequently, based on the R-CNN network structure, Girshick et al. proposed Fast R-CNN [12] and Faster R-CNN [13] which make the regional proposition, classification, and regression share the convolution feature together. Although the speed of detection has been further improved, it is still not fast enough. In 2016, Girshick et al. proposed a new algorithm called YOLO [14] to solve the object detection problem as a regression problem. The speed of detection has increased dramatically. In the same year, Google's DeepMind's [15] AlphaGo [16] defeated professional go-getter Shishi Li in the Korean Open Championship, once again arousing the interest of researchers to study deep learning. The application of deep learning has broad prospects. As long as there is a large enough data set, fast enough processing speed, and sufficient algorithm complexity, computers can accomplish many tasks that were previously completed by humans.

1.2 Contributions

Based on object detection algorithms, this study proposes a fast positioning algorithm for 3D pose recognition, which is mainly used for workpieces placed on the production line. In general, the contributions of our approach include the following aspects:

1. A network module for angle regression is proposed to train and predict the workpiece posture information.
2. Integration of the posture detection module along with the object detection model is proposed, so that the latter one can be applied to the workpiece positioning problem.
3. This approach is faster and with higher accuracy, comparing with traditional algorithms.

1.3   System Overview

The overall process of the system is shown in Fig. 1.

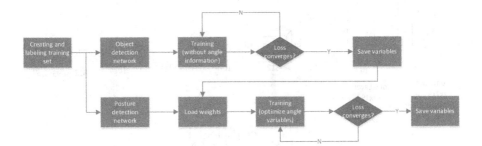

Fig. 1. System Overview. First, we use the object detection network to train. After training, the optimum parameters without angle information are finally obtained. After completing the first step of training and testing, the previous training parameters are loaded into the corresponding part of the posture detection network. Then, only the variables for the angle prediction module are trained and optimized at this step.

2   Network Structure and Loss Function

Our approach is based on a traditional object detection algorithm. We propose an angle regression module and the corresponding loss function to enhance the capability to detect the postures. We integrate this module with an object detection approach, and train with workpiece images.

2.1   Network Structure

Traditional deep learning algorithms (for example, YOLO) perform well in object detection, but lacks the posture information, which is important for grabbing workpieces. Therefore, it is necessary to improve on the basis of the original network to obtain the angle value. Without loss of generality, we consider the YOLO network as our base model, and in the following part, as an example, we demonstrate how to integrate posture estimations into this specific network. The network structure we proposed is shown in Fig. 2. The upper layer of the network contains 24 convolutional layers and 3 fully connected layers for calculating classification and location information. The network at the lower level is an angle regression layer. It is inserted when the picture becomes 28×28 after multi-layer convolution and pooling. It contains a convolutional layer and 3 full-connection layers for obtaining angle information. This layer of the network through the full connection layer, the final output size is $7 \times 7 \times 3$, where 3 is the output of the three angles.

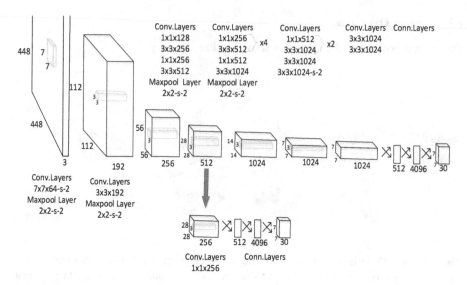

Fig. 2. The Architecture. One detection network has 25 convolutional layers followed by six fully connected layers. The top 24 convolutional layers and three full-connected layers are used to calculate the classification and location information. The bottom one convolutional layer and three fully connected layers are used to derive the angle information.

2.2 Loss Function

Due to the introduction of the angle regression network, based on the original loss function, we propose the angle error loss function. It can be formulated as follows:

$$L_{angle} = \sum_{i=0}^{S^2} I_i^{obj} \left[(Ax_i - \hat{A}x_i)^2 + (Ay_i - \hat{A}y_i)^2 + (Az_i - \hat{A}z_i)^2 \right] \qquad (1)$$

Where Ax, Ay and Az are network-predicted angles of rotation around the x, y and z axes respectively, $\hat{A}x$, $\hat{A}y$ and $\hat{A}z$ are corresponding label values, and I_i^{obj} denotes if object appears in cell i. We use the least square method to calculate the angular error.

It is important to note here that when the angular value of the angle is between $0°$ and $360°$, the difference between $1°$ and $359°$ reflected in the picture is very small, and they actually differ by only $2°$, but the difference in value is $358°$. So in the calculation, we introduce a minimization operation of $A - \hat{A}$:

$$A - \hat{A} = \min\{abs(A - \hat{A}), abs(A - \hat{A} - 360), abs(A - \hat{A} + 360)\} \qquad (2)$$

At the same time, we need to normalize the loss value so that its value is between 0 and 1. The formula is as follows:

$$L_{angle} = \sum_{i=0}^{S^2} I_i^{obj} \left[\frac{(Ax_i \times 360 - \hat{A}x_i)^2}{360^2} + \frac{(Ay_i \times 360 - \hat{A}y_i)^2}{360^2} + \frac{(Az_i \times 360 - \hat{A}z_i)^2}{360^2} \right]$$

$$(3)$$

In addition to the angle loss function, the loss function also includes three other parts [15]: coordinate error, IOU error, and classification error. The formulas are as follows:

$$L_{coord} = \lambda_{coord} \sum_{i=0}^{S^2} \sum_{j=0}^{B} I_{ij}^{obj} [(x_i - \hat{x}_i)^2 + (y_i - \hat{y}_i)^2]$$

$$+ \lambda_{coord} \sum_{i=0}^{S^2} \sum_{j=0}^{B} I_{ij}^{obj} [(\sqrt{w_i} - \sqrt{\hat{w}_i})^2 + (\sqrt{h_i} - \sqrt{\hat{h}_i})^2]$$

$$(4)$$

$$L_{iou} = \sum_{i=0}^{S^2} \sum_{j=0}^{B} I_{ij}^{obj} (C_i - \hat{C}_i)^2 + \lambda_{noobj} \sum_{i=0}^{S^2} \sum_{j=0}^{B} I_{ij}^{noobj} (C_i - \hat{C}_i)^2 \qquad (5)$$

$$L_{cls} = \sum_{i=0}^{S^2} I_i^{obj} \sum_{c \in classes} (p_i(c) - \hat{p}_i(c))^2 \qquad (6)$$

Where x, y, w, h, C, p are network prediction values, $\hat{x}, \hat{y}, \hat{w}, \hat{h}, \hat{C}, \hat{p}$ are all label values, I_i^{obj} denotes if object appears in cell i. I_{ij}^{obj} and I_{ij}^{noobj} denote that the jth bounding box predictor in cell i is "responsible" and "not responsible" for that prediction.

For the entire network, the total loss value is:

$$L = L_{coord} + L_{iou} + L_{cls} + L_{angle} \qquad (7)$$

3 Creating and Labeling the Training Set

The workpieces used in our training are of the three types shown in Fig. 3. They are all different in shape and size and do not have a high degree of symmetry. They are small in size, but not in micro parts [17]. They are larger than 2 cm in length, width and height, but no more than 5 cm. Because the original VOC2007 data set is used for classification of 20 types of objects such as cars, people and cats, it was necessary to create our own training set. Taking the first kind of workpieces as an example, in order to obtain the posture information, the 3D CAD model of the workpieces can be rotated around the x, y and z axes through OPENGL to obtain each posture image. We consider a certain posture as a reference, for example, (0°, 0°, 0°) for rotations around the x, y and z axes. If we capture images for every 1° differences, there are a total of 360^3 training images for a single type of workpiece. The training set leads to a huge amount of calculations and a lot of training time, and it is advisable to take part of the posture for training and testing, which is used to compare the test conditions of other postures. In our experiment, the rotation angle between the x axis and the y axis is $-15°$ to $14°$, and the rotation angle around the z axis is between $0°$ and $90°$. You can draw 81000 pictures and can accurately determine the rotation angle of each picture.

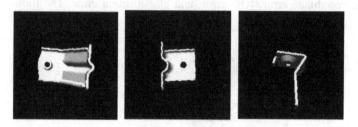

Fig. 3. Three kinds of workpieces.

After the training set image is generated, it is necessary to mark the picture. Similar to the training set of the YOLO network, we mark classification information of training pictures and also the bounding boxes of objects to be detected. The classification information is marked as 1, on behalf of the first category, and the bounding box of the workpieces can be obtained by the minimum bounding rectangle. In addition, we add the four values of Xmin, Ymin, Xmax and Ymax, which are written to the "annotation" file in the same format.

In addition, angle information needs to be added in the annotation file. When annotating angles, it should be noted that for any posture, there is only one picture in the training set corresponding to it. This is actually not enough. We also proved through experiments that it is difficult to train appropriate weights and get more accurate angle information. Here we think of a relatively simple and effective method that makes the number of pictures corresponding to each pose increase without increasing the training set. That is the dividing angle. We can find that when the angle difference between different postures is small, the naked eye is difficult to distinguish the difference of its posture. Therefore, when we take the rotation around the x and y axes, we take the median value every $5°$ to mark. Because the rotation around the z-axis can be seen as a rotation in the plane, the posture of the workpieces does not change much, so the intermediate value is marked every $10°$. For example, for a workpiece that rotates between $6°$ and $10°$ about the x and y axes, and a workpiece that rotates between the $11°$ and $20°$ around the z axis, the angle is uniformly labeled as (8, 8, 15). We believe that the workpiece in this angle range are all the same posture, and the rotation angles are (8, 8, 15). In the training set, there are 250 training images corresponding to one posture.

4   Training

In this part, we describe the training method, parameter settings, and the preservation of variables. Since the network contains an angle regression layer, if all the variables of the network are trained at the same time, it will cause the loss function to be difficult to converge. Here we take a two-step training approach.

First, we use the object detection network to train. Throughout training we use a batch size of 30, the initial learning rate is 0.01, the decay rate is 0.1 and decay steps are 30000. We train the network for about 37 epochs on the training and validation data sets from our training set. The gradient descent optimizer optimizes the variables. Through continuous training, the test result without angle information is finally obtained.

After completing the first step of training and testing, the training parameters is loaded into the corresponding part of the posture detection network. The variables for the angle prediction module are optimized, using the gradient descent optimizer. The batch size is still 30, the initial learning rate is 0.01, and it gradually changes from 0.01 to 0.0001, and the period is 37. The loss function is shown in Fig. 4.

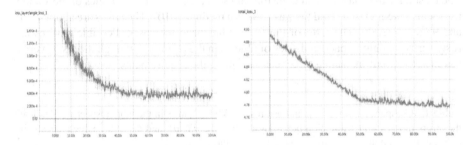

Fig. 4. Loss Function Graph. The left side is the angle loss function curve, and the right side is the total loss function curve. When the training period reaches 16 times, the function gradually converges.

5   Experiments

5.1   Test Result

For the test set, the images of workpieces are taken by the camera. The sizes of the workpieces are different from that of the training set. The surface of the workpieces affected by light, rust, etc. is also different from the training set image, but its angle range is still rotate about $-15°$ to $14°$ about the x-axis and y-axis, and $0°$ to $90°$ around the z-axis. There are 1600 pictures of each workpiece, and totally 4800 pieces for all the three workpieces.

We test the performance of our algorithm after training. The algorithm outputs detected classification, confidence, position information, and posture information. The average detection time is 0.043 s, the test results as shown in Fig. 5.

Fig. 5. Test Result Chart. The first row of the test results in the figure indicates the classification and the corresponding confidence level, and the second row indicates the angle of rotation around the *x*, *y* and *z* axes of the test. The green box indicates the position of the object. (Color figure online)

Comparison of training and test results is shown in Table 1. From the table, whether it is training or testing, the classification results are 100% correct; and the errors in the *x* and *y* direction bounding boxes are very low, and the error of the test set is about 1 mm. In the training set, when the rotation angle errors around the *x*, *y* and *z* axes averaged 2.371°, 3.667°, and 3.464° respectively, the loss value could not be reduced any more. The reason for the analysis is that the training set labels the rotation angles of the *x* and *y* axes with a median of 5°, and the *z* axis takes a median of 10°. The errors are about 4°, 4° and 9° respectively. The error of the test set has been increased. Possible reasons are rusty surfaces, light impact, and sizes of objects in the picture. Although the test set error has increased, but the actual effect can still meet the requirements.

Table 1. Comparison of training and test results. The *tx* and *ty* in the table respectively indicate the errors in the *x* and *y* direction bounding boxes. The *θX*, *θY* and *θZ* in the table respectively indicate the rotation angle errors around the *x*, *y* and *z* axes.

Workpieces	Training and test	Classification	tx (mm)	ty (mm)	θX (°)	θY (°)	θZ (°)
Class1	Training	100%	0.125	0.165	2.056	3.236	3.503
	Test	100%	1.013	1.036	5.160	6.745	10.725
Class2	Training	100%	0.112	0.098	2.489	3.869	3.432
	Test	100%	1.156	1.059	4.988	4.865	10.389
Class3	Training	100%	0.087	1.025	2.568	3.897	3.457
	Test	100%	1.073	1.089	5.046	6.088	10.149

5.2 Compared with YOLO

We compare it with the YOLO algorithm. The original YOLO [13] network divides the input image into an 7 × 7 grid and each grid cell predicts 2 bounding boxes. Each bounding box consists of 5 predictions represent the center and the shape of the box and confidence, and includes 20 class probabilities. Thus, the predictions are encoded in 7 × 7 × (2 × 5 + 20) = 7 × 7 × 30 tensor. If we do not change the core structure of the network, the angle information will be output together with the position and category information, as shown in Fig. 6. Compared with the results obtained by applying the algorithm of this paper, the training results are shown in Table 2.

Fig. 6. Original YOLO Net. It divides the image into an 7 × 7 grid and for each grid cell predicts 2 bounding boxes, confidence for those boxes, and 3 class probabilities and another 3 angle information. These predictions are encoded as an 7 × 7 × (2 × 5 + 3 + 3) = 7 × 7 × 16 tensor.

Table 2. Comparison of original YOLO net. The meaning of tx, ty, θX, θY and θZ in the table is as shown in Table 1.

Avg abs	Classification	tx (mm)	ty (mm)	θX (°)	θY (°)	θZ (°)
Ours	100%	0.108	0.429	4.038	4.334	8.464
Original net	100%	0.130	0.175	20.362	22.253	70.335

It can be seen from the table that when the angle information is output after the fully connected layers of the original network, the effect on the classification and positioning of the workpiece is not significant, but the posture information error is large, and the loss value of the angle is difficult to converge. Even if we have added penalty factors for angle loss in loss function for the original network, for example, multiplying angle loss by different parameters, the angle loss is still difficult to converge. The reason is that when the original network passes through the full connection layer, the picture size has been reduced to 7 × 7 through multi-layer convolutions and pooling, and the picture feature information has been discarded a lot. At this time, the angle feature is difficult to learn and the angle loss value does not converge, so the error is too large.

6 Conclusion

We propose a posture detection method which can be integrated into object detection methods to solve the problem of workpieces recognition and grabbing on the production line. By integrating posture recognition module into traditional object detection networks, our approach can not only obtain classification and positioning information, but also obtain relatively accurate posture information. Our implementation is fast detection speed compared to other object detection algorithms.

Acknowledgments. This work is supported by China Postdoctoral Science Foundation (2017M612047) and Qianjiang Talent Program (QJD1702031).

References

1. Dalal, N., Triggs, B.: Histograms of oriented gradients for human detection. In: 2005 IEEE Computer Society Conference on Computer Vision and Pattern Recognition, CVPR 2005, vol. 1, pp. 886–893. IEEE (2005)
2. Felzenszwalb, P., Girshick, R., McAllester, D., Ramanan, D.: Object detection with discriminatively trained part-based models. IEEE Trans. Pattern Anal. Mach. Intell. **32**(9), 1627–1645 (2010)
3. Rosenblatt, F.: Perceptron simulation experiments. J. Proc. IRE **48**(3), 301–309 (1960)
4. Dahl, G.E., Yu, D., Deng, L., Acero, A.: Context-dependent pre-trained deep neural networks for large-vocabulary speech recognition. IEEE Trans. Audio Speech Lang. Process. **20**(1), 30–42 (2012)
5. LeCun, Y., Bottou, L., Bengio, Y., Haffner, P.: Gradient-based learning applied to document recognition. Proc. IEEE **86**(11), 2278–2324 (1998)
6. LeCun, Y., Bengio, Y.: Convolutional networks for images, speech, and time series. In: The Handbook of Brain Theory and Neural Networks, vol. 3361, no. 10 (1995)
7. Hinton, G.E., Osindero, S., Teh, Y.W.: A fast learning algorithm for deep belief nets. J. Neural Comput. **18**(7), 1527–1554 (2006)
8. Ranzato, M., Boureau, Y.L., Chopra, S., Lecun, Y.: A unified energy-based framework for unsupervised learning. In: Proceedings of Conference on AI and Statistics, pp. 860–867 (2007)
9. Krizhevsky, A., Sutskever, I., Hinton, G.E.: ImageNet classification with deep convolutional neural networks. International Conference on Neural Information Processing Systems, vol. 60, pp. 1097–1105. Curran Associates Inc. (2012)
10. Girshick, R., Donahue, J., Darrell, T., Malik, J.: Rich feature hierarchies for accurate object detection and semantic segmentation. In: Proceedings of the IEEE Conference on Computer Vision and Pattern Recognition, pp. 580–587 (2014)
11. He, K., Zhang, X., Ren, S., Sun, J.: Spatial pyramid pooling in deep convolutional networks for visual recognition. In: Fleet, D., Pajdla, T., Schiele, B., Tuytelaars, T. (eds.) ECCV 2014. LNCS, vol. 8691, pp. 346–361. Springer, Cham (2014). https://doi.org/10.1007/978-3-319-10578-9_23
12. Girshick, R.: Fast R-CNN. Computer Science (2015)
13. Ren, S., He, K., Girshick, R., Sun, J.: Faster R-CNN: towards real-time object detection with region proposal networks. In: Advances in Neural Information Processing Systems, pp. 91–99 (2015)
14. Redmon, J., Divvala, S., Girshick, R., Farhadi, A.: You only look once: unified, real-time object detection. In: Proceedings of the IEEE Conference on Computer Vision and Pattern Recognition, pp. 779–788 (2016)
15. Beattie, C., Leibo, J.Z., Teplyashin, D.: DeepMind Lab (2016)
16. Yuan-Dong, T.: A simple analysis of AlphaGo. J. Acta Autom. Sin. **42**, 671–675 (2016)
17. Yu, X., Gao, G., He, W., Xu, J.: Machine vision based automatic micro-parts detection system. In: Sun, C., Fang, F., Zhou, Z.-H., Yang, W., Liu, Z.-Y. (eds.) IScIDE 2013. LNCS, vol. 8261, pp. 160–167. Springer, Heidelberg (2013). https://doi.org/10.1007/978-3-642-42057-3_21

An Adaptive Iterated Extended Kalman Filter for Target Tracking

Chunman Yan[1,2(✉)], Junsong Dong[1], Genyuan Lu[1],
Daoliang Zhang[1], Manli Chen[1], and Liang Cheng[1]

[1] College of Physics and Electronic Engineering, Northwest Normal University,
Lanzhou 730070, China
629588619@qq.com
[2] Engineering Research Center of Gansu Province for Intelligent Information
Technology and Application, Lanzhou 730070, China

Abstract. For the iterated extended Kalman filter (IEKF) in target tracking, the system model and noise estimation are always uncertain. In view of these problems, an improved adaptive iterated extended Kalman filter is proposed. The new algorithm is based on IEKF, combines the improved strong tracking filter to make it more fitting the maneuvering target tracking issue. Moreover, in our approach, the noise variance is adjusted in real-time by a noise parameter estimator which is based on the seasonable statistic characteristic of the noise. The estimator can availably reduce the influence of the time-varying noise. The simulation results indicate that the improved algorithm has higher estimation accuracy on the target position and speed for target tracking.

Keywords: Target tracking · Iterative extended Kalman filter
Strong tracking filter · Innovation

1 Introduction

With the rapid development of aerospace technology, maneuvering target tracking has become a hot and difficult point in the field of target tracking. Dealing with the problem that the system model is nonlinear, the extended Kalman filter (EKF) [1], unscented Kalman filter (UKF) [2], and cubature Kalman filter (CKF) [3] algorithm are proposed early or late. Among these algorithms, the EKF is a most widely used one at present [4]. EKF uses the first order Taylor series expansion of the nonlinear equation to linearize the system model. However, when the non-linearity is strong, the filter performance is declined [5]. In order to overcome the limitations of the EKF, an iterative extended Kalman filter (IEKF) [6] is proposed. The state estimate can be obtained by the iterative operation, which reduces the influence of linearization error and improves the filtering precision. However, because the algorithm is based on EKF, it still needs to know the accurate statistics characteristic of prior noise, which limits its application in target tracking. Whereupon, a strong tracking filter (STF) [7] is proposed. The sub-optimal fading factor is introduced into the EKF, which adjusts the prediction covariance matrix, and obtains the online filter gain, and enhances the tracking ability. Similarly, STF does not consider the filter effect of time-varying noise. Whereafter, an

© Springer Nature Switzerland AG 2018
Y. Peng et al. (Eds.): IScIDE 2018, LNCS 11266, pp. 371–377, 2018.
https://doi.org/10.1007/978-3-030-02698-1_32

adaptive extended Kalman filter (AEKF) [8] is proposed, which combines the adaptive Sage-Husa estimator with EKF. In the filtering process, the estimator is engaged to estimate the noise variance in real-time. However, with the high computationally cost.

This paper proposes an improved adaptive strong tracking IEKF (IAST-IEKF), which uses the framework of IEKF. An improved suboptimal fading factor is introduced into IEKF to adjust the one-step prediction covariance matrix, which can improve the filter efficiently. Moreover, an adaptive noise estimation method based on innovation statistics is derived, which can update the noise parameters, and then reduces the estimation error of the time-varying noise to the target state. Through the simulation experiment of target tracking, the validity of the improved algorithm is verified.

2   Improved Adaptive Strong Tracking IEKF

2.1   IEKF Algorithm

State and observation equation are as follows.

$$x(k) = f[k, x(k-1)] + W(k-1) \tag{1}$$

$$y(k) = h[k, x(k)] + V(k) \tag{2}$$

Equations (1) and (2) are nonlinear function, where k denotes discrete time. The mean of $W(k)$ and $V(k)$ is zero. Their variance matrices are Q and R.

The traditional IEKF algorithm is as follows [6].

$$\hat{x}(k|k-1) = f[\hat{x}(k-1|k-1)] \tag{3}$$

$$\hat{y}(k|k-1) = H(k|k-1)\hat{x}(k|k-1) \tag{4}$$

$$P(k|k-1) = F(k|k-1)P(k-1|k-1)F^T(k|k-1) + Q(k-1) \tag{5}$$

$$K(k) = P(k|k-1)H^T(k)(H(k)P(k|k-1)H^T(k) + R(k))^{-1} \tag{6}$$

$$\hat{x}(k|k) = \hat{x}(k|k-1) + K(k)(y(k) - \hat{y}(k|k-1) - H(k)(\hat{x}(k|k-1) - x(k-1|k-1))) \tag{7}$$

$$P(k|k) = P(k|k-1) - K(k)H(k)P(k|k-1) \tag{8}$$

Where $F(k) = \frac{\partial f}{\partial x}|x(k) = \hat{x}(k|k-1)$, and $H(k) = \frac{\partial h}{\partial x}|x(k) = \hat{x}(k|k-1)$.

2.2   Improved Strong Tracking Filter

The fading factor $\lambda(k)$ of traditional STF is as follows.

$$N(k) = \varepsilon(k) - H(k)Q(k-1)H^T(k) - R(k) \tag{9}$$

$$M(k) = H(k)F(k|k-1)P(k-1|k-1)F^T(k|k-1)H^T(k) \tag{10}$$

$$c(k) = \frac{tr[N(k)]}{tr[M(k)]} \tag{11}$$

$$\lambda(k) = \begin{cases} c(k) & c(k) > 1 \\ 1 & c(k) \leq 1 \end{cases} \tag{12}$$

Where $\varepsilon(k) = H(k)P(k|k-1)H^T(k) + R(k)$, $\varepsilon(k)$ is residual matrix, and $tr(\cdot)$ is matrix trace. Then, $N(k)$ can be described as follows.

$$N(k) = H(k)P(k|k-1)H^T(k) - H(k)Q(k-1)H^T(k) \tag{13}$$

Equation (5) can be transformed into (10) as follows.

$$M(k) = H(k)(P(k|k-1) - Q(k-1))H^T(k) \tag{14}$$

The filter divergence criterion is as follows [9].

$$\varepsilon(k)\varepsilon^T(k) \leq \kappa tr(E(\varepsilon(k)\varepsilon^T(k))) \tag{15}$$

Where κ is a reserve coefficient. When (15) is established, the filter is in normal working condition. The limiting factor $\beta(k)$ is as follows.

$$\beta(k) = \frac{\varepsilon(k)\varepsilon^T(k) - tr(H(k)Q(k)H^T(k) + R(k))}{tr(H(k)F(k|k-1)P(k-1|k-1)F^T(k|k-1)H^T(k))} \tag{16}$$

To sum up, the calculation of $M(k)$ is as follows.

$$M(k) = \beta(k)H(k)(P(k|k-1) - Q(k-1))H^T(k) \tag{17}$$

So, we put $\lambda(k)$ in (5), and then we have to one-step predict covariance as follows.

$$P(k|k-1) = \lambda(k)F(k|k-1)P(k-1|k-1)F^T(k|k-1) + Q(k-1). \tag{18}$$

2.3   Noise Parameter Estimator Based on the Innovation Statistics Characteristic

While tracking the maneuvering target, process noise covariance matrix and observation noise variance matrix tend to change over time. The prediction error of k time can be described with the innovation $e(k)$, which is defined as follows.

$$e(k) = y(k) - \hat{y}(k|k-1) = y(k) - H(k)\hat{x}(k|k-1) \tag{19}$$

The innovation variance matrix is as (20).

$$Ce(k) = E(e(k)e^T(k)) = H(k)P(k|k-1)H^T(k) + R(k) \tag{20}$$

Equation (6) can be rewritten as (21).

$$K(k) = P(k|k-1)H^T(k)Ce(k)^{-1} \tag{21}$$

Equation (21) is multiplied by $Ce(k)K^T(k)$ on both sides as follows

$$K(k)Ce(k)K^T(k) = P(k|k-1)H^T(k)K^T(k) \tag{22}$$

$P(k|k-1)$ is a symmetric matrix. The two sides of (22) are transposed as (23).

$$K(k)Ce^T(k)K^T(k) = K(k)H(k)P^T(k|k-1) = K(k)H(k)P(k|k-1) \tag{23}$$

Equation (23) is substituted into (8), then obtain the follows.

$$\begin{aligned} K(k)Ce^T(k)K^T(k) &= P(k|k-1) - P(k|k) \\ &= F(k|k-1)P(k-1|k-1)F^T(k|k-1) + Q(k-1) - P(k|k) \end{aligned} \tag{24}$$

Therefore, the estimation of process noise can be described as (25).

$$\hat{Q}(k-1) = K(k)Ce^T(k)K^T(k) + P(k|k) - F(k|k-1)P(k-1|k-1)F^T(k|k-1) \tag{25}$$

Equation (5) can be brought into (20), that is as follows.

$$\begin{aligned} R(k) &= Ce(k) - H(k)P(k|k-1)H^T(k) \\ &= Ce(k) - H(k)(F(k|k-1)P(k-1|k-1)F^T(k|k-1) + Q(k-1))H^T(k) \end{aligned} \tag{26}$$

Equation (24) can be rewritten as (27).

$$Q(k-1) + F(k|k-1)P(k-1|k-1)F^T(k|k-1) = K(k)Ce^T(k)K^T(k) + P(k|k) \tag{27}$$

Equation (27) is substituted into (26), and the estimated value of new observed noise is as follows.

$$\begin{aligned} \hat{R}(k) &= Ce(k) - H(k)(K(k)Ce^T(k)K^T(k) + P(k|k))H^T(k) \\ &= Ce(k) - H(k)K(k)Ce^T(k)K^T(k)H^T(k) - H(k)P(k|k)H^T(k) \end{aligned} \tag{28}$$

3 Simulation Results and Analysis

In order to evaluate the IAST-IEKF, it is compared with STF, AEKF and IEKF algorithm. The tracking precision is measured by the root mean square error (RMSE).

3.1 Simulation Model and Initial Value

The parameters of missile tracking maneuvering targets provided in [10] were used in the simulation experiment. At k moment, the vector of the position, velocity and acceleration of the target is $x(k) = [\, rx(k) \; ry(k) \; rz(k) \; vx(k) \; vy(k) \; vz(k) \; ax(k) \; ay(k) \; az(k)\,]^T$. The state equation and the observation equation are as follows.

$$x(k+1) = Fx(k) + \Gamma U(k) + W(k) \tag{29}$$

$$z(k) = h(x(k)) + V(k) \tag{30}$$

Where $F = \begin{bmatrix} I_{3\times3} & \Delta t I_{3\times3} & \frac{1}{\lambda^2}(e^{-\lambda\Delta t} + \lambda\Delta t - 1)I_{3\times3} \\ 0_{3\times3} & I_{3\times3} & \frac{1}{\lambda}(1 - e^{-\lambda\Delta t})I_{3\times3} \\ 0_{3\times3} & 0_{3\times3} & e^{-\lambda\Delta t}I_{3\times3} \end{bmatrix}$, $\Gamma = \begin{bmatrix} -(\Delta t^2/2)I_{3\times3} \\ -\Delta t I_{3\times3} \\ 0_{3\times3} \end{bmatrix}$,

Δt is measuring period. The maneuvering frequency λ is 1, the covariance of $W(k)$ is

$Q(0) = [0_{6\times6}, 0_{3\times6}; 0_{6\times3}, 0.1I_{3\times3}]$, $h(x(k)) = \left[\arctan\frac{ry(k)}{\sqrt{r^2x(k)+r^2z(k)}}, \quad \arctan\frac{-rx(k)}{rz(k)} \right]^T$, $R(k) = D^{-1}(k)a$

$(D^{-1}(k))^T$, $E[V(k)] = r(1) = 0_{2\times1}$, $D(k) = \begin{bmatrix} \sqrt{r^2x(k)+r^2y(k)+r^2z(k)} & 0 \\ 0 & \sqrt{r^2x(k)+r^2y(k)+r^2z(k)} \end{bmatrix}$,

$E[V(k)V^T(k)] = R(k)$, $R(k) = D^{-1}(k)a(D^{-1}(k))^T$, $a = 0.1I_{2\times2}$, $P(0) = [10^4 \times I_{6\times6}, 0_{6\times3};$ $0_{3\times6}, 10^2 \times I_{3\times3}]$.

3.2 Tracking Maneuvering Target in Noisy Environments

At 0–100 steps, the target is at (2500, 200, 2500), turning left at (−800, −210, −250) speed and (5, 5, 10) acceleration. At 100–500 steps, the target moves at a constant speed. $Q(k)$ and $R(k)$ are expanded 200 times, and other parameters are the same as those in [10]. The RMSE of position estimated is shown in Fig. 1. The RMSE of velocity estimated is shown in Fig. 2.

As seen in Figs. 1 and 2, the position and speed estimation accuracy of IAST-IEKF are better than those of the other three algorithms at 0–100 steps (turning maneuver occurs in the target), and it converges smoothly without fluctuation. This indicates that IAST-IEKF can track maneuvering targets well. In the whole tracking process, the position and velocity estimation accuracy of IAST-IEKF is better than that of AEKF, which indicates that IAST-IEKF has good adaptive ability to noise. When the target is moving at a constant speed, IAST-IEKF is not much different from STF and IEKF.

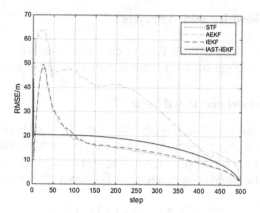

Fig. 1. The RMSE of the estimated position

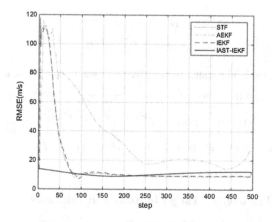

Fig. 2. The RMSE of the estimated velocity

4   Conclusion

An IAST-IEKF algorithm is proposed in this paper. Simulation results show that IAST-IEKF algorithm is higher than the accuracy of the IEKF, STF and AEKF tracking maneuvering target, the time-varying noise adaptive estimation precision is better than AEKF. IAST-IEKF can in larger maneuvering target tracking in noise environment. As IAST-IEKF adopts IEKF algorithm framework, the calculation amount needs to be further optimized. In addition, when the state dimension is high, it verifies whether the new noise estimation method will lose its positive character.

Acknowledgments. This work was supported by National Natural Science Foundation of China (61741119), the Fundamental Research Funds for the Universities of Gansu Province, and the Natural Science Foundation of Gansu Province (17JR5 RA074, 17JR5RA078). The authors would like to thank all editors and reviewers for their constructive comments and suggestions. Moreover, they would also want to thank professor Xiangyu Deng and coworkers for their helps for the preparations of our manuscript.

References

1. Ramadan, H.S., Becherif, M., Claude, F.: Extended Kalman filter for accurate state of charge estimation of lithium-based batteries: a comparative analysis. J. Sci. Direct. **42**, 29033–29046 (2017)
2. Mundla, N., Nayak, J., Marco, H.T., Sabat, S.L.: ARMA model based adaptive unscented fading Kalman filter for reducing drift of fiber optic gyroscope. J. Sens. Actuators A Phys. **251**, 42–51 (2016)
3. Zhao, X., Wang, S.C., Liao, S.Y., Ma, L., Liu, Z.G.: An ultra-tightly coupled tracking method based on robust adaptive cubature Kalman filter. J. Acta Automatica Sinica. **40**, 2530–2540 (2014)
4. Gu, F., Zhou, Y.J., Hu, Y.Q., Han, J.D.: Experimental investigation and comparison of nonlinear Kalman filters. J. Control Decision. **29**, 1387–1393 (2014)
5. Liu, Y.H., Li, T., Yang, Y.Y., Ji, X.W., Wu, J.: Estimation of tire-road friction coefficient based on combined APF-IEKF and iteration algorithm. J. Mech. Syst. Signal Process. **88**, 25–35 (2017)
6. Tian, Y., Chen, Z., Yin, F.L.: Distributed iterated extended Kalman filter for speaker tracking in microphone array networks. J. Acta Automatica Sinica. **40**, 2530–2540 (2014)
7. Zhou, D.H., Xi, Y.G., Zhang, Z.J.: A suboptimal multiple extended Kalman filter. Chin. J. Autom. **4**, 145–152 (1992)
8. Dai, L., Jin, G., Chen, T.: Application of adaptive extended Kalman filter in spacecraft attitude determination system. J. Jilin Univ. Eng. Technol. Ed. **38**, 466–470 (2008)
9. Ruan, X.G., Yu, M.M.: Modeling research of MEMS gyro drift based on Kalman filter. In: 26th control and decision conference, pp. 2949–2952. IEEE Press, Chang Sha (2014)
10. Huang, X.P., Wang, Y.: Kalman Filter Principle and Application. Publishing House of Electronics Industry, Beijing (2015)

Learning Siamese Network with Top-Down Modulation for Visual Tracking

Yingjie Yao[1], Xiaohe Wu[1(✉)], Wangmeng Zuo[1], and David Zhang[1,2]

[1] Vision Perception and Cognition, Harbin Institute of Technology, Harbin, China
yaoyoyogurt@gmail.com, xhwu.cpsl.hit@gmail.com, cswmzuo@gmail.com,
csdzhang@comp.polyu.edu.hk
[2] The Chinese University of Hong Kong (Shenzhen), Shenzhen, China

Abstract. The performance of visual object tracking depends largely on the target appearance model. Benefited from the success of CNN in feature extraction, recent studies have paid much attention to CNN representation learning and feature fusion model. However, the existing feature fusion models ignore the relation between the features of different layers. In this paper, we propose a deep feature fusion model based on the siamese network by considering the connection between feature maps of CNN. To tackle the limitation of different feature map sizes in CNN, we propose to fuse different resolution feature maps by introducing deconvolutional layers in the offline training stage. Specifically, a top-down modulation is adopted for feature fusion. In the tracking stage, a simple matching operation between the fused feature of the examplar and search region is conducted with the learned model, which can maintain the real-time tracking speed. Experimental results show that, the proposed method obtains favorable tracking accuracy against the state-of-the-art trackers with a real-time tracking speed.

Keywords: Visual tracking · Feature fusion · Siamese network

1 Introduction

Visual tracking is one of the fundamental and practical problems in computer vision. The applications of visual tracking range from visual surveillance, real-time systems, robotics, biometrics and medical interventions. Given an initial bounding box annotation of an object in the first frame, the task is to locate the target object in the future frames and estimate the scales. Although the object tracking has been studied for decades, accurate tracking of general objects in a dynamic environment remains a challenging task due to various appearance changes.

To tackle this issue, numerous approaches are motivated from the perspective of feature representation. Specially, the handcrafted features, e.g., HOG, Color Name, etc., have been extensively adopted due to its high computational

© Springer Nature Switzerland AG 2018
Y. Peng et al. (Eds.): IScIDE 2018, LNCS 11266, pp. 378–388, 2018.
https://doi.org/10.1007/978-3-030-02698-1_33

efficiency [1,7]. In recent years, benefited from the development of CNN, deep features have been successfully applied to resolve the object tracking problem due to its robust and discriminative representation [5,9,13]. However, the direct use of deep features extracted from the pre-trained model, such as VGG [15], which is trained on the large-scale image repository ImageNet [4] for object classification, limits the feature representation due to the gap between different tasks. It facilitates another branch of CNN-based trackers, which aims to train the CNN off-line with the large video dataset, whereas they only adopt the output of the last single layer for feature representation. The output of the bottom layers in CNN has higher resolution with low level information which is helpful to determine the location of the specific object. On the contrary, the output of top layers contains more semantic information which is more powerful to distinguish objects from background. Although some attempts have been made such as HCF [10] and HDT [12], their performance is far from state-of-the-art trackers in terms of accuracy and tracking speed. It can be explained by that they treat each layer's feature independently and ignore the intrinsic relationships between different layers.

In this paper, we propose a deep feature fusion tracker based on the convolutional neural network (CNN) for robust description of the tracking object inspired by the human visual pathway. The contributions of this paper are summarized as follows:

- We propose to fuse features from different layers of CNN with a top-down modulation in an off-line training manner. The bottom layer features are beneficial to target localization and the top layer features are helpful to distinguish the objects from background.
- Based on the Siamese network, we maintain the real-time tracking speed with improved tracking accuracy.
- Experimental results on the tracking benchmark demonstrate that the proposed method performs comparably against the state-of-the-art trackers and maintains a real-time tracking speed.

2 Related Work

2.1 Fully Convolutional Siamese Network

Siamese network based trackers lead the most recent trends in tracking due to its simple but flexible architecture [2,6,16]. In particular, the SiameseFC obtains special attention with high tracking speed and considerable tracking accuracy. It treats tracking as a matching task, which learns a function $f(\mathbf{x}, \mathbf{z})$ to compare the candidate images $\mathbf{z}$ with an exemplar image $\mathbf{x}$. The response is obtained by cross-correlation operator between the feature of exemplar and search region, as presented in Eq. (1),

$$f(\mathbf{x}, \mathbf{z}) = \psi(\mathbf{x}) \star \psi(\mathbf{z}) + b \cdot \mathbf{1}, \tag{1}$$

where ψ denotes the feature extractor, $\star$ indicates the cross-correlation operation and $b \cdot \mathbf{1}$ denotes a signal which takes value $b \in \mathcal{R}$ in every location. Instead of a single score, SiameseFC finally produces a score map. In tracking process, a search region centered at the previous position of the target is cropped and compared with the examplar to determine the displacement of the target, according to the position of the maximum score relative to the center of the score map.

SiameseFC learns feature representation and similarity function simultaneously in an end-to-end manner, which is implemented by introducing the full-convolutional layer for cross-correlation operation. Then the similarity function with an embedding feature representor is simply evaluated online during tracking, which facilitates a high real-time speed. However, the tracking accuracy is much far from the state-of-the-art methods. The mean reason is that SiameseFC only exploits the output of the last convolutional layer as feature representation for training and tracking, which is with the size of 17×17.

Table 1. The architecture of SiameseFC.

Layer	Kernel size	Stride	Channel	Exemplar size	Search size
			3	127×127	255×255
Conv1	11×11	2	96	59×59	123×123
Pool1	3×3	2	96	29×29	61×61
Conv2	5×5	1	256	25×25	57×57
Pool2	3×3	2	256	12×12	28×28
Conv3	3×3	1	192	10×10	26×26
Conv4	3×3	1	192	8×8	24×24
Conv5	3×3	1	128	6×6	22×22

2.2 Feature Fusion

In recent years, benefited from the development of CNN, deep features have been successfully applied to resolve the object tracking problem due to its robust and discriminative representation. As shown in [10], the output of the bottom layers in CNN has higher resolution with low level information which is helpful to determine the location of the specific object, while the output of top layers contains more semantic information which is more powerful to distinguish objects from background. To leverage the low level and semantic information for tracking, a branch of trackers use the hierarchical features in parallel. In particular, HCF [10] employs three different layer features independently for calculating three correlation filters, and produces three score maps with various sizes for the coarse-to-fine localization. While HDT [12] uses features from different CNN layers and uses an adaptive Hedge method to hedge several CNN trackers into a stronger one. Although these trackers attempt to utilize features from different layers, their performance is far from the state-of-the-art trackers in terms of tracking accuracy and speed, because they treat each layer's feature independently and ignore the intrinsic relationships between different layers.

3   Proposed Method

The SiameseFC off-line trains a Siamese network with a large video dataset in an end-to-end manner, which enables the weight adaption for video feature extraction. However, the feed-forward process only applies the output of the last layer as feature representation. To fully employ the low level and semantic information of CNN, we propose to utilize the multi-layer features by adopting a simple feature fusion method. Motivated by the recent work on top-down modulation [14], we introduce the advanced method of using hyper-column features based on the SiameseFC.

3.1   Hierarchical Feature Fusion

Deep features extracted from CNN have been extensively studied in various applications [10,19]. To demonstrate the effectiveness of using multi-layer deep features, we apply a simple hierarchical feature fusion strategy for the SiameseFC framework. Based on the AlexNet [9] architecture (Table 1), we aim to adopt the outputs of Conv3, Conv4 and Conv5 as feature representation to facilitate the tracking performance. Specifically, we extract features of these three convolutional layers for both the examplar and search region. Then we correlate the corresponding layer features of the examplar and search region to obtain three score maps individually. Finally, a coarse-to-fine localization is processed on these score maps. It is worth noting that, although the feature maps of different layers have different spatial sizes as described in Table 1, the obtained three score maps still have the same size due to the cross-correlation operation. However, the output of Conv3 actually has the higher resolution, which is beneficial to locate the target precisely. Therefore, we fuse the hierarchical features in a sequential manner for coarse-to-fine location. Experimental results provided in Sect. 4 demonstrate that the hierarchical feature fusion does improve the tracking accuracy with little sacrifice of tracking speed.

3.2   Top-Down Features Fusion Training Modulation

The above hierarchical feature based trackers suffer from the independent deep features. Generally, in CNN based trackers, different feature map sizes limit the feature fusion in an unit tracker in tracking process. To tackle this issue, we propose to fuse different resolution feature maps by introducing the top-down modulation in off-line training stage of SiameseFC. In detail, we develop a top-down way to fuse the different layers, namely, SiameseFC_TDM, motivated by the human visual pathway [14]. Figure 1 depicts the whole architecture of the proposed SiameseFC_TDM tracker.

For simplicity, we take one top-down modulation as an example for description and more modulations can be dealt by the same way. The whole network starts from the output of the last layer of the forward mode. It consists of two key modules, the **L** module and the **T** module. The lateral module **L** aims to learn the transformation of the low-level layer features. The **T** module is designed for

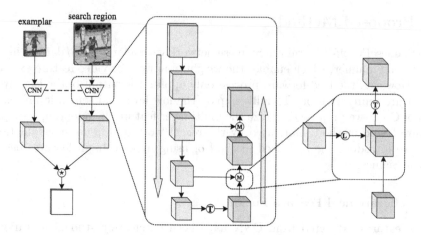

Fig. 1. The framework of the proposed SiameseFC_TDM. The left part is the overall architecture of SiameseFC and SiameseFC_TDM while the middle part is the pipeline of network equipped with the bottom-up and top-down pathway. There is an example of the TDM module in the right part. The bold **L**, **T** and **M** are the lateral connection, top-down fusion and top-down module, respectively.

the purpose of fusing the information from the top **T** module and the outputs of module **L**. The **T** module is expected to preserve the most useful information of the top **T** module and learn to guide the selection of features from module **L**.

As illustrated in Fig. 1, denote the i-th convolutional layer feature map as $\mathbf{x}_i \in \mathcal{R}^{W \times H \times C_i}$, where C_i represents the number of channels of the i-th convolutional layer. The lateral module $\mathbf{L}_i$ transforms the $\mathbf{x}_i$ to produce the lateral features $\mathbf{z}_i \in \mathcal{R}^{W \times H \times L_i}$, where L_i is the number of channels for the corresponding lateral output. Supposing the output of the top module **T** from the j-th convolutional layer is $\mathbf{y}_j \in \mathcal{R}^{W \times H \times T_j}$, where T_j denotes the number of channels for the last fused features. Then the concation of $\mathbf{y}_j$ and $\mathbf{z}_i$ is served as the input of the module $\mathbf{T}_i$, which generates the final output $\mathbf{y}_i \in \mathcal{R}^{W' \times H' \times T_i}$. $\mathbf{T}_i$ associates hierarchical deep features with different feature map size and optionally selects the most useful channels for more robust feature representation. We adopt the features from the last top-down module $\mathbf{T}_j^{out}$ for the tracking task. It can be found that the network is designed not only to encode the rich semantic information from higher layers but also keep the resolution same with low-level features, which is beneficial to locate the target accurately.

Specifically, the module **L** and **T** can be designed as a small sub-net or one or more layers. For implementation, we model the top-down module **T** as one de-convolutional layer except the module $\mathbf{T}_j^{out}$ is a 1×1 convolutional layer, and the lateral module **L** as a convolutional layer with 3×3 kernel size and padding 1. Here we describe the detailed feature fusion process of one top-down modulation. We assume that the outputs of the last two layers are $\mathbf{x}_5 \in \mathcal{R}^{H_5 \times W_5 \times C_5}$ and $\mathbf{x}_4 \in \mathcal{R}^{H_4 \times W_4 \times C_4}$, respectively. The top-down module $\mathbf{T}_5$ takes $\mathbf{x}_5$ as input and outputs the up-sampled top-down features $\mathbf{y}_5 \in \mathcal{R}^{H_4 \times W_4 \times C_5'}$. Note that $C_5' < C_5$ is kept to suppress the growth of channels. The input of the lateral module $\mathbf{L}_4$ is $\mathbf{x}_4$ and the output is $\mathbf{z}_4 \in \mathcal{R}^{H_4 \times W_4 \times C_4'}$. Then, the lateral feature $\mathbf{z}_4$ and top-down feature $\mathbf{y}_5$ are fused by the concat operation and are further dealt by next top-down modulation.

Both module $\mathbf{T}_i$ and module $\mathbf{L}_i$ is differentiable and can be implemented simply by the deep learning framework. The gradient updates from the loss function via top-down modules and lateral modules to the original AlexNet part. For the training progress, we gradually add one top-down modulation to our baseline tracker SiameseFC as described above. The training progress can be divided into two steps. Firstly, we initially the parameters of the added top-down modulation randomly. Parameters in SiameseFC are kept fixed while parameters in the TDM are learned through the SGD algorithm. Then, all the parameters in the whole network are trained end-to-end. When adding the second or more new top-down modulation, the parameters in the specific network which performs best in the validation set are used to initial the new net while the new added top-down module is initialized randomly. Then the same training strategy is applied to the network.

4   Experimental Results

In this section, we first describe the implementation details for training and test phases. Then we compare the proposed method with the representative trackers and state-of-the-arts, SiameseFC [2], CFNet [17], SINT [16], HCF [10], and Staple [1], on the object tracking benchmark (OTB50 [20], OTB100 [21]) and Visual Object Tracking 2016 Challenge (VOT2016) [8] to demonstrate the effectiveness of the proposed method. We also report the comparison of the tracking speed in the last part of this section. Our method is implemented on MatConvNet [18] library. All the experiments are run on a PC equipped with an Intel i7 CPU, 32 G RAM and a single NVIDIA 1080 GPU. The code and results will be publicly available.

4.1   Implementation Details

Training. The training set comes from the video detection of the ImageNet Large Scale Video Recognition Challenge (ILSVRC2015), which consists of more than 4400 videos and about one million of annotated frames. For the fair comparison, we pre-process the dataset for training following the original SiameseFC, with exemplar images with size 127×127 and search images with size 255×255 pixels. To evaluate the fusion of different layer features, we provide two variants of SiameseFC_TDM, i.e. the SiameseFC_TDM_1 with one TDM and the SiameseFC_TDM_2 with two TDMs. Hence, we apply the same strategy to train SiameseFC_TDM_2 based on SiameseFC_TDM_1, with its parameters fixed for the first 100 epochs and then train the whole network for the next 100 epochs. In training stage, we apply the SGD optimization with mini-batches of size 8.

The learning rate is annealed geometrically at each epoch from 10^{-2} to 10^{-5}, which is consistent with our baseline tracker SiameseFC.

Table 2. Evaluation results of the mean OP (in %) and tracking speed (fps) for different trackers. The best results are marked in red.

Tracker	SiameseFC	CFNet	SINT	HCF	Staple	SiameseFC_HFF	SiameseFC_TDM_1	SiameseFC_TDM_2
Mean OP	71.1	69.3	68.8	64.9	71.0	73.2	72.9	76.2
FPS	95.4	78.4	4.0	10.2	76.6	68.2	88.3	83.1

Tracking. In tracking, we first crop the target region in the initial frame according to the given ground-truth as the examplar, then it is resized to 127×127 pixels to feed to the network. For the subsequent frames, the search region centered on the position in the previous frame is cropped and resized to 255×255. After feature extraction, the examplar and the search region are correlated to generate the score map with size 17×17. Moreover, three search scales $1.01^{\{-1,0,1\}}$ are adapted to handle the scale variations, and the specific scale with the highest response is considered as the tracking result.

4.2   Comparison with the State-of-the-Arts

In this section, we perform comparison experiments on the Object Tracking Benchmark (OTB) and Visual Object Tracking 2016 Challenge (VOT2016). The OTB50 dataset is a famous tracking dataset which contains 50 fully annotated sequences and is a subset of OTB100, which has 100 sequences. The sequences in OTB dataset are also labeld with 11 different attributes, such as scale variation, occlusion, fast motion and in-plane rotation. On the OTB dataset, following the metric criteria in [20, 21], we adopt the one pass evaluation (OPE) with the success plot and its area-under-curve (AUC) score for comparison. The success plot shows the ratios of successful frames when the threshold varies from 0 to 1, where a successful frame means its overlap is larger than the given threshold. There are 60 sequences in VOT2016 and trackers are evaluated in terms of accuracy, robustness and expected average overlap(EAO). The accuracy measures the average overlap ratio between the predicted and ground truth bounding boxes while the robustness computes the average number of tracking failures over the entire sequence. The EAO metric averages the no-reset overlap of a tracker on several short-term sequences. **OTB50 and OTB100.** From Table 2, our simple hierarchical feature fusion method (named SiameseFC_HFF) achieves the mean OP of 73.2% on the OTB100 dataset, over the baseline SiameseFC tracker by 2.1%. By introducing the top-down feature fusion modulation, the SiameseFC_TDM_1 and SiameseFC_TDM_2 further improves the mean OP by 1.8% and 5.1%, respectively. For the comparison with the SiameseFC_HFF, one can note that, both variants of SiameseFC_TDM outperforms it, which demonstrates that the jointly training considering the relationship between different

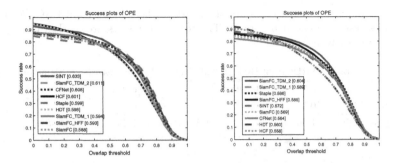

Fig. 2. Success plots of different trackers on the OTB50 (left) and OTB100 (right) dataset.

layer features helps to facilitate the tracking performance. Further from Fig. 2, the SiameseFC has an AUC of 56.9%, however, the SiameseFC_HFF obtains an AUC of 58.5%, and the variants of SiameseFC_TDM get an around AUC of 58.9% and 60.4%, which further demonstrates the effectiveness of our feature fusion modulation. To evaluate the number of top-down modulation, we find that SiameseFC_TDM_2 gets superior performance than SiameseFC_TDM_1 with a gain of 1.5% in AUC. Although SiameseFC_TDM_1 and SiameseFC_TDM_2 output feature maps of the same resolution, SiameseFC_TDM_2 fuses more layers and the feature maps contain more semantic information, thus it performs better than SiameseFC_TDM_1. Finally, we compare the proposed methods with the representative trackers and the state-of-the-arts in Fig. 2. Both the SiameseFC_TDM_2 and SiameseFC_TDM_1 perform favorably against the other trackers, especially the SiameseFC_TDM_2 with an AUC of 60.4%, outperforming the siamese net based and feature fusion based trackers largely. For comprehensive analysis, we present the AUC scores with different video attributes annotated in the OTB100 benchmark in Fig. 3. Our method is ranked among top two on 9 of 11 attributes, which can be explained by the efficient fusion of semantics and spatial details from the hyper-column features.

VOT2016. We also present the quantitative results on VOT2016 dataset in terms of EAO, accuracy and robustness in Table 3. The proposed method SiameseFC_TDM performs much better than the baseline tracker SiameseFC, in terms of EAO, accuracy and robustness. In particular, our method obtains the second best EAO with a value of 0.262 and has a 11% gain compared with the baseline tracker SiameseFC (0.235). While compared with other state-of-the-art trackers, SiameseFC_TDM ranks the second-best on EAO and the best on accuracy, which further demonstrate the effectiveness of the proposed siamese network with top-down modulation.

Qualitative Evaluation. In Fig. 4, we visualize tracking performance on several video sequences, e.g., Doll, DragonBaby, Girl2, Rubik and Skiing. Although almost all compared trackers predict the position accurately but at the same time they fail to estimate the scale correctly. Thanks to the top-down modulation in

Table 3. Comparison with the state-of-the-art trackers in terms of EAO, Accuracy and Robustness on VOT2016 dataset. The top three results are marked in red, blue and green respectively.

Tracker	Staple [1]	SRDCF [3]	HCF [10]	MDNet_N [11]	SiameseFC [2]	SiameseFC_TDM
EAO	0.295	0.247	0.220	0.257	0.235	0.262
Accuracy	0.54	0.52	0.47	0.53	0.50	0.54
Robustness	1.07	1.50	1.38	1.20	1.65	1.48

the siamese network, which enable the tracker to locate the object accurately and to have a robust performance when dealing with scale changes, which is consistent with Fig. 3. Our method also performs well on fast motion (i.e. Drag-onBaby) and occlusion (i.e. Girl) while other trackers either fail to track the object or fail estimate the scale accurately. The reason behind the improvement of tracking performance can be contributed to the top-down modulation, which utilizes features from both higher layers with more semantic information and lower layers with higher resolution.

4.3 Tracking Speed Comparison

It is well know that the SiameseFC is implemented in real-time due to the simple feed-forward tracking algorithm. It achieves a speed of 95.4 FPS as shown in Table 2. However, to enhance its tracking accuracy, we improve from the feature fusion perspective in the off-line training stage. In fact, the SiameseFC_TDM improves the tracking accuracy with little sacrifice of computation complexity due to the same simple feed-forward tracking algorithm. From Table 2, both two variants of SiameseFC_TDM obtain similar real-time tracking speed of 83.1 FPS and 88.3 FPS, respectively. Our hierarchical feature fusion model Siamese_HFF also gets the speed of 68.2 FPS in real-time.

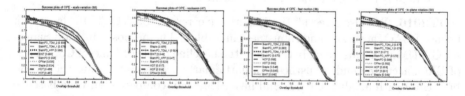

Fig. 3. Success plots on different attributes, e.g., scale variation, occlusion, fast motion and in-plane rotation in the OTB100 benchmark.

SINT   HCF   CFNet   SiameseFC   SiameseFC_TDM

Fig. 4. Visualization results compared with SiameseFC, CFNet, SINT and HCF on video sequences Doll, DragonBaby, Girl2, Rubik and Skiing.

5   Conclusion

Feature representation plays a critical role in visual object tracking. Benefited from the powerful representation ability of CNN feature, tracking accuracy has been improved in comparison with the handcrafted feature. Based on the observation that the lower layer feature maps of CNN contain more low level detail information, and the higher layers maintain more semantic information, we evaluate the utility of hierarchical features in parallel and propose to fuse different resolution feature maps considering their inner connection by introducing deconvolutional layer in the off-line training stage. Specifically, a top-down modulation is adopted for feature fusion. The experimental results demonstrate the favorable performance of our tracker both from the accuracy and efficiency.

References

1. Bertinetto, L., Valmadre, J., Golodetz, S., Miksik, O., Torr, P.H.: Staple: complementary learners for real-time tracking. In: CVPR, pp. 1401–1409 (2016)
2. Bertinetto, L., Valmadre, J., Henriques, J.F., Vedaldi, A., Torr, P.H.S.: Fully-convolutional siamese networks for object tracking. In: Hua, G., Jégou, H. (eds.) ECCV 2016. LNCS, vol. 9914, pp. 850–865. Springer, Cham (2016). https://doi.org/10.1007/978-3-319-48881-3_56
3. Danelljan, M., Hager, G., Shahbaz Khan, F., Felsberg, M.: Learning spatially regularized correlation filters for visual tracking. In: ICCV, pp. 4310–4318 (2015)
4. Deng, J., Dong, W., Socher, R., Li, L.J., Li, K., Fei-Fei, L.: ImageNet: a large-scale hierarchical image database. In: CVPR, pp. 248–255 (2009)
5. Dong, C., Loy, C.C., He, K., Tang, X.: Learning a deep convolutional network for image super-resolution. In: Fleet, D., Pajdla, T., Schiele, B., Tuytelaars, T. (eds.) ECCV 2014. LNCS, vol. 8692, pp. 184–199. Springer, Cham (2014). https://doi.org/10.1007/978-3-319-10593-2_13
6. Held, D., Thrun, S., Savarese, S.: Learning to track at 100 FPS with deep regression networks. In: Leibe, B., Matas, J., Sebe, N., Welling, M. (eds.) ECCV 2016. LNCS, vol. 9905, pp. 749–765. Springer, Cham (2016). https://doi.org/10.1007/978-3-319-46448-0_45
7. Henriques, J.F., Caseiro, R., Martins, P., Batista, J.: High-speed tracking with kernelized correlation filters. TPAMI 37(3), 583–596 (2015)
8. Kristan, M., et al.: The Visual Object Tracking VOT2016 Challenge Results, October 2016. http://www.springer.com/gp/book/9783319488806
9. Krizhevsky, A., Sutskever, I., Hinton, G.E.: ImageNet classification with deep convolutional neural networks. In: NIPS, pp. 1097–1105 (2012)
10. Ma, C., Huang, J.B., Yang, X., Yang, M.H.: Hierarchical convolutional features for visual tracking. In: ICCV, pp. 3074–3082 (2015)
11. Nam, H., Han, B.: Learning multi-domain convolutional neural networks for visual tracking. In: CVPR, pp. 4293–4302 (2015)
12. Qi, Y., et al.: Hedged deep tracking. In: CVPR, pp. 4303–4311 (2016)
13. Ren, S., He, K., Girshick, R., Sun, J.: Faster R-CNN: towards real-time object detection with region proposal networks. In: NIPS, pp. 91–99 (2015)
14. Shrivastava, A., Sukthankar, R., Malik, J., Gupta, A.: Beyond skip connections: top-down modulation for object detection. arXiv:1612.06851 (2016)
15. Simonyan, K., Zisserman, A.: Very deep convolutional networks for large-scale image recognition. arXiv:1409.1556 (2014)
16. Tao, R., Gavves, E., Smeulders, A.W.: Siamese instance search for tracking. In: CVPR, pp. 1420–1429 (2016)
17. Valmadre, J., Bertinetto, L., Henriques, J., Vedaldi, A., Torr, P.H.: End-to-End representation learning for correlation filter based tracking. In: CVPR, pp. 5000–5008 (2017)
18. Vedaldi, A., Lenc, K.: MatConvNet: convolutional neural networks for MATLAB. In: ICM, pp. 689–692. ACM (2015)
19. Wang, L., Ouyang, W., Wang, X., Lu, H.: Visual tracking with fully convolutional networks. In: ICCV, pp. 3119–3127 (2015)
20. Wu, Y., Lim, J., Yang, M.H.: Online object tracking: a benchmark. In: CVPR, pp. 2411–2418 (2013)
21. Wu, Y., Lim, J., Yang, M.H.: Object tracking benchmark. TPAMI 37(9), 1834–1848 (2015)

R-CASENet: A Multi-category Edge Detection Network

Yuan Shen, Houde Liu, and Zhenhua Guo$^{(\boxtimes)}$

Graduate School at Shenzhen, Tsinghua University, Shenzhen 518055, China
`zhenhua.guo@sz.tsinghua.edu.cn`

Abstract. Edge detection plays an important role in image processing. With the development of deep learning, the accuracy of edge detection has been greatly improved, and people have more and more requirements for edge detection tasks. Most edge detection algorithms are binary edge detection methods, but there are usually multiple categories of edges in an image. In this paper, we present an accurate multi-category edge detection network Richer-CASENet (R-CASENet). In order to make full use of CNN's powerful feature expression capabilities, we attempt to use more information from feature map for edge feature extraction and classification. Using the ResNet101 network as the backbone, firstly we merge the building blocks in different composite blocks and down-sample to obtain the feature map. Then we fuse the feature maps in different composite blocks to obtain the final fused classifier. Experiments show that we achieved better results on a public dataset.

Keywords: Edge detection · Convolutional neural network · ResNet

1 Introduction

Edge detection algorithms have been studied for several decades and are still a challenging research issue. With the development of the edge detection algorithm, its influence in the field of target detection, image segmentation, significance detection, and stereo modeling is also growing. With the development of deep learning, edge detection algorithms have also been ushered in new developments. We also have more and more requirements for edge detection results. For example, we want to get the edge of a specific target or detect multiple categories of edges in a single image. In some specific tasks, there will be cases where each pixel may belong to different categories of edges. There may also be edges with similar shape or nested relationships, which greatly increases the difficulty of detection.

Traditional edge detection algorithm usually detect all edges in one image because they simply used simple low-level features in the image such as gradients and could not get edges on a particular target specially. Later, algorithms such as Pb [18] and gPb [1] that use well-designed complex artificial features to extract edges have emerged, and significant progress has been made in public

© Springer Nature Switzerland AG 2018
Y. Peng et al. (Eds.): IScIDE 2018, LNCS 11266, pp. 389–400, 2018.
https://doi.org/10.1007/978-3-030-02698-1_34

data sets. Even so, these methods cannot detect the edges of a specific target, because the feature extraction method of traditional methods can not obtain high-level semantic information. In addition, the traditional methods are not general enough, and a large number of parameters need to be tuned for different circumstances.

The emergence of convolutional neural networks (CNN) improve this situation. Benefiting from the powerful expression capabilities of CNN, we have avoided the problems of tuning various parameters. The network itself can extract high-level semantic information and can therefore be used to detect the edges on a particular target. A notable algorithm using deep learning for edge detection is HED [21]. Edge detection is performed by fusing feature maps from different stages in VGG16 [20]. What's more, HED is an end-to-end network, its output is an edge probability map. HED is a binary classification network and cannot handle situations where multiple categories of edges appear in one image. The traditional edge detection algorithm can obtain multi-edge detection results after performing preprocessing and clustering on the results of binary edge detection. By using CNN, we can make an end-to-end multi-classification edge detection network, such as CASENet [22], see Fig. 1. CASENet uses ResNet101 [11] as backbone. In this network, multi-channels are used to store the feature maps of different categories of edges, combining low-level feature extraction and high-level edge classification to achieve state-of-the art results on public dataset.

However, compared to existing methods, the improvement by CASENet is not obvious enough. There are few multi-category edge detection networks, the room for improvement in accuracy is still large. For example, how to make better use of features to improve accuracy? Is it possible to combine blocks to get an optimal combination? In summary, our major contributions in this paper are:

- We propose a new architecture uses more information from layers in ResNet, making full better of features from each composite blocks to get higher accuracy.
- We try various combination of composite blocks in ResNet and find out the best one.
- We can outperform other state-of-art methods on SBD dataset [10].

2   Related Work

As one of the typical problems in the field of image processing, researches on edge detection started early. We classify edge detection algorithms into five categories: traditional edge detection algorithms, algorithms with complex artificial design features, edges are obtained by image segmentation, binary edge detection algorithms based on deep learning, and multi-category edge detection based on deep learning. Then we briefly introduce these algorithms by category.

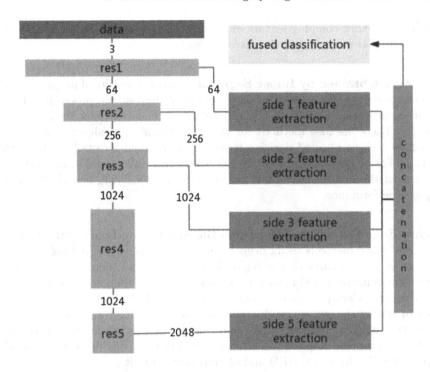

Fig. 1. An overview of the CASENet.

Traditional Edge Detection Algorithm: Early edge detection work mainly extracts some simple features in one image, then a threshold is set to obtain the edge after processing. The Sobel [12] operator uses the color gradient of the pixels in a small area to determine whether there is an edge. The calculation of the gradient is achieved using a convolution operation. The Canny [4] algorithm, which is still widely used today, uses non-maximum suppression to ensure that the edges in the result are sufficiently thick. The Canny operator interpolates pixels before calculating gradients to reach higher accuracy. In order to further reduce the number of edge points, Canny firstly uses strong filtering rules to obtain preliminary results, and then connects some edges to get the final result.

Algorithms with Complex Artificial Design Features: The expression capabilities of simple features such as gradients and textures are not powerful enough. Some subsequent works attempt to use artificially designed complex features, various parameters in their model are learned. Konishi et al. [13] firstly used data-driven parameters. The Pb [18] algorithm models the changes in brightness, color, and texture in the image. Using a large number of labeled edge images, its edge detection results no longer contain edges from the object's internal texture. Based on the Pb algorithm, Arbelaez et al. [1] proposed gPb, which has a more complex feature extraction process and is considered as the best traditional method, but it runs very slowly.

Edges are Obtained by Image Segmentation: the results of image segmentation can be used as a result of edge detection after simple processing. Early image segmentation algorithms also rely on simple image features, such as Meanshift [6]. There are also methods for classifying artificial design features using classifiers such as boosted decision trees [7] and random forests [14]. With the development of deep learning, FCN [17] firstly uses full convolutional network to segment the image. After that, DeepLab [5], PSPNet [23], etc. continuously improve performance.

Binary Edge Detection Algorithms Based on Deep Learning: There are more and more methods using deep learning to detect edges. N^4-field [9] uses CNN to extract features from a region in the image, and then connects similar image blocks in the global image. To obtain the edge of the image, deep-edge [2] firstly uses the Canny operator to obtain the potential edge points, and then uses CNN to extract the feature results to further filter the final result. By combining the feature maps at different scales and calculating the loss for each pixel, the HED [21] algorithm implements an end-to-end binary edge detection network. Later, RCF [16] improves HED and obtains better results.

Multi-category Edge Detection Based on Deep Learning: The edge of the binary classification does not meet current requirements. Deep-Contour [19] classifies the edge information into multiple categories by training a model for each category and extends edge detection from the binary problem to multiple categories. CASENet [22] is currently one of the best algorithms for multi-category edge detection networks and has achieved much better results than traditional methods on SBD dataset. CASENet uses the network's low-level layer to extract features and high-level layers to classify edges. DSN [22] network uses all classifiers from 5 composite blocks in ResNet101 and achieves similar results. CASENet and DSN have achieved good results in multi-category edge detection tasks, but they have not fully utilized the information in the composite blocks of ResNet. To solve this problem, we propose a new network architecture to combine building blocks in composite blocks of ResNet.

3   Approach

Inspired by CASENet, we have designed a new network architecture by modifying the ResNet101 network. The ResNet101 network consists of 100 convolutional layers, a global average pooling layer, and a fully connected layer. It achieved state-of-the-art results in the areas of image classification and target detection. ResNet101 has 5 composite blocks, in which the first composite block has only one convolutional layer, and the last four composite blocks have 3, 4, 23, and

3 sub-blocks. There are two convolutional layers with kernel size 1×1 and a convolutional layer with kernel size 3×3. In the backbone of the network, we follow the structure in CASENet, removing the original average pooling layer and the full connection layer. Since too small feature images can result in the loss of edge detection details, we change the stride of the first and fifth residual blocks to 1 and use hole convolution to ensure that the receptive field of vision is consistent after the stride changes.

3.1 Multi-label Loss Function

There are K categories of annotation images $\{\overline{\mathbf{Y}}_1, \ldots, \overline{\mathbf{Y}}_k\}$ in image I. These annotation images are binarized and a pixel may belong to more than one categories. $\overline{\mathbf{Y}}_k$ means the annotation of the k-th category edge. Considering the number of edge points in the image is much less than the number of non-edge points, we use another way to calculate the loss function. We have improved the multi-label loss function, using λ to balance edge points and non-edge points. The improved loss function is:

$$
\mathcal{L}(\mathbf{W}) = \sum_k \mathcal{L}_k(\mathbf{W})
$$

$$
= \sum_k \sum_\mathbf{p} \{ -\gamma_k \overline{\mathbf{Y}}_k(\mathbf{p}) log \mathbf{Y}_k(\mathbf{p} \mid \mathbf{I}; \mathbf{W}) \tag{1}
$$

$$
- (1 - \gamma_k)(1 - \overline{\mathbf{Y}}_k(\mathbf{p})) log(1 - \mathbf{Y}_k(\mathbf{p} \mid \mathbf{I}; \mathbf{W})) \}
$$

$$
\gamma_k = \frac{\lambda \mathbf{Y}_k^+}{\mathbf{Y}_k^+ + \mathbf{Y}_k^-} \tag{2}
$$

Where $\mathbf{W}$ is the parameters of the network, $\mathbf{p}$ is a pixel of image, $\mathbf{Y}_k^+$ and $\mathbf{Y}_k^-$ denote edge pixel and non-edge pixel, respectively. All the parameters in the model are learned.

3.2 R-CASENet Architecture

Compared with the traditional edge detection methods, CNN has a strong performance in high-level semantic classification, and it reaches high accuracy while determining the approximate position of a category edge. However, after several convolutional and pooling layers, a pixel of the high-level feature map cover a relatively large area on the original image. Therefore, the details of the edge detection result are implemented by the low-level layers. Further improvement of detection accuracy depends on the low-level layers of network. We fuse the building blocks in a composite block, and then get the low-level feature extractor and high-level edge classifiers. As show in Fig. 2 except some building blocks in res3 and res4, each building block in a composite block is connected to a convolutional layer with kernel size 1×1, the depth of channel is 25. We use the Eltwise layer in Caffe to directly add down-sampled features from each composite block to obtain a fused feature map. We use the 1×1 convolution kernel to convolve

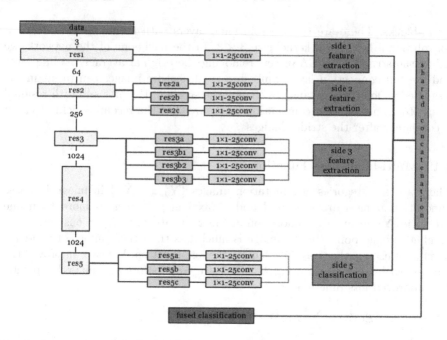

Fig. 2. An overview of the R-CASENet. The yellow rectangles represent the composite blocks of ResNet101, the green rectangles represent the building blocks, the blue lines represent the upsample operators. (Color figure online)

the fused feature map in each composite block, reduce the number of channels to 1, and use the deconvolution layer to transform the feature maps in different composite blocks into same size. Then we get the final feature map of this composite block. The final fused loss function is obtained by fusing feature maps from three low-level feature extraction layers and a high-level edge classification layer. For a K-category edge classification task, our high-level edge classification layer will get K feature maps, and the low-level feature extraction layer has only one Feature maps. While fusing, each classification layer is concatenated with the feature maps of the three low-level extraction layers to obtain the hybrid classification layer.

4   Experiment

4.1   Dataset

We use the public dataset SBD to evaluate the algorithms. The SBD dataset contains 11355 images in the PASCAL VOC 2011 [8] train-val set, divided into 8498 training images and 2857 test images. The data set contains 20 categories: aero, bike, bird, boat, bottle, bus, car, cat, chair, cow, table, dog, horse, mbike, person, plant, sheep, sofa, train and tv. Since the number of samples in each

Table 1. Results on the SBD benchmark. All MF scores are measured by %.

Backbone	Method	aero	bike	bird	boat	bottle	bus	car	cat	chair	cow	table	dog	horse	mbike	person	plant	sheep	sofa	train	tv	mean
-	InvDet	41.5	46.7	15.6	17.1	36.5	42.6	40.3	22.7	18.9	26.9	12.5	18.2	35.4	29.4	48.2	13.9	26.9	11.1	21.9	31.4	27.9
	HFL-FC8	71.6	59.6	68.0	54.1	57.2	68.0	58.8	69.3	43.3	65.8	33.3	67.9	67.5	62.2	69.0	43.8	68.5	33.9	57.7	54.8	58.7
	HFL-CRF	73.9	61.4	74.6	57.2	58.8	70.4	61.6	71.9	46.5	72.3	36.2	71.1	73.0	68.1	70.3	44.4	73.2	42.6	62.4	60.1	62.5
VGG16	CASENet	72.5	61.5	63.8	54.5	52.3	65.4	62.6	67.2	42.6	51.8	31.4	62.0	61.9	62.8	75.4	41.7	59.8	35.8	59.7	50.7	56.8
	R-CASENet	72.1	61.5	64.6	55.1	52.6	66.7	63.0	67.3	43.1	51.6	34.0	64.1	62.5	63.2	77.2	41.9	60.1	35.6	60.7	51.3	57.4
	DSN	81.6	75.6	78.4	61.3	67.6	82.3	74.6	82.6	52.4	71.9	45.9	79.2	78.3	76.2	80.1	51.9	74.9	48.0	**76.5**	66.8	70.3
ResNet101	CASENet	**83.3**	76.0	80.7	**63.4**	69.2	81.3	74.9	83.2	54.3	74.8	46.4	80.3	80.2	76.6	80.8	53.3	77.2	50.1	75.9	66.8	71.4
	R-CASENet	82.8	**76.6**	**82.3**	62.5	**69.2**	**82.2**	**75.3**	**84.4**	**54.4**	**75.2**	**48.9**	**81.7**	**81.6**	**77.3**	**83.2**	**54.9**	**77.8**	**50.1**	76.4	**67.8**	**72.2**

category is not balanced and the number of samples in some categories is relatively small, we have expanded the dataset during the training process and set 5 different scales: 0.5, 0.75, 1.0, 1.25, 1.5.

4.2 Experiment Details

Our comparison algorithms include InvDet [10], HFL [3], DSN, CASENet. We also trained the VGG16 version of R-CASENet. Our pre-training model is ResNet 101 and VGG16 trained on the MS COCO dataset [15].

For experimental hyper-parameter settings, ResNet101's learning rate is $1e-7$, VGG16's learning rate is $1e-8$, fusion layer's learning rate is $1e-7$, fusion layer's channel number is 25, and training iterations are 22,000 times. We use the test code provided by the SBD dataset. The results is measured by maximum F-measure (MF).

4.3 Experimental Results

Results Analysis. Table 1 shows the optimal dataset scale (ODS) results on SBD dataset, our method performs better than previous methods. Through comparison, R-CASENet has a significant improvement over CASENet, which proves the correctness of our ideas. Binary edge detection algorithm such as HED and RCF uses VGG16, reaching excellent on public dataset. But as we can see from Table 1, VGG16 is not enough for multi-category edge detection. Our network performs much better than CASENet in category person (83.2 vs 80.8), which has the largest number of samples in test set (much more than other categories). Figure 3 is the precision-recall curves on class person, R-CASENet outperforms other methods.

Visualized Side Feature Map. Figure 4 shows the visualized feature map of R-CASENet. The side 4 and side 5 feature map of DSN look almost the same, which means side 4 is not necessary in this architecture. As we can see from side feature map of R-CASENet, side5-dog is much more brighter than side5-mbike because the target is a dog. Besides edges of the dog, some other edges in side 1, side 2 and side 3 of DSN are bright, which have similar type as side 4 and side 5. These bright pixels will cause more non-edge pixels classified as edge.

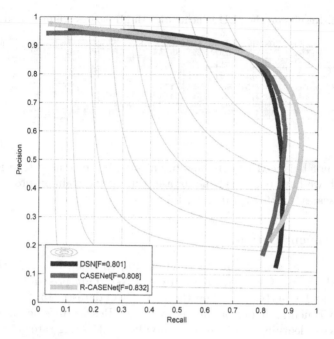

Fig. 3. Precision-recall curves of the proposed methods and baselines on the SBD dataset (class person).

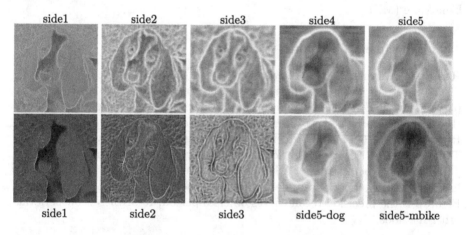

Fig. 4. Visualized side feature map, rows correspond to DSN and R-CASENet.

4.4   Exploration of Network Structure

We did some comparative experiments to observe the influence of factors on accuracy:

Table 2. Different fusion types

Fused block	Origin block	ODS	OIS
2, 3	5	83.1	87.7
2, 5	3	82.8	87.5
3, 5	2	83.0	87.7
3	2, 5	82.8	87.1
2	3, 5	82.7	87.3
-	2, 3, 5	80.8	86.1
2, 3, 5	-	83.2	88.0

Table 3. Different channel numbers

Channels number	ODS	OIS
1	83.1	88
5	83.1	87.9
15	83	87.9
20	83.1	88
25	83.2	88
35	83.1	88.1
50	83.2	87.9

- How different fusion types affects the results
- When building blocks in the composite block are fused, the number of down-sampled channels

Due to the large number of categories that need to be compared, for simplicity, we only conducted experimental comparisons in the category of people on the SBD dataset.

Tables 2 and 3 shows the optimal dataset scale (ODS) and optimal image scale (OIS) results on category person of SBD dataset. From Table 2 we can see, the more composite blocks fused, the higher the MF results. What's more, the 5th Composite block has the least impact on results, because it comes from high-level layer and mainly used for semantic classification. From Table 3, the number of down-sampled channels has no obvious effect on the experimental results, indicating that fusing building blocks is the key to improve accuracy.

5 Conclusion

In this paper, we propose a new network structure R-CASENet, which fuses the building blocks in each composite block to obtain feature maps at different scales and makes full use of the information of each block in the ResNet. Our network is an improved version of CASENet. It has achieved much better results on one large public dataset. Our future work is to combine the image segmentation network with the edge detection network to further improve the accuracy of the detection results (Fig. 5).

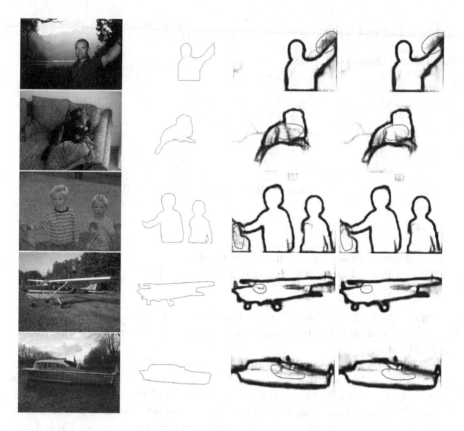

Fig. 5. Some results on SBD dataset, columns correspond to original image, ground truth, results of CASENet, and results of R-CASENet. We mark the differences in results with red lines. (Color figure online)

Acknowledgements. We gratefully acknowledge the NVIDIA Corporation with the donation of the Tesla K40 GPU for this research. The work is partially supported by the Natural Science Foundation for China (NSFC) (No. 61772296) and Shenzhen fundamental research fund (Grant Nos. JCYJ20160531194840025 and JCYJ20170412170438636).

References

1. Arbelaez, P., Maire, M., Fowlkes, C., Malik, J.: Contour detection and hierarchical image segmentation. IEEE Trans. Pattern Anal. Mach. Intell. **33**(5), 898–916 (2011)
2. Bertasius, G., Shi, J., Torresani, L.: DeepEdge: a multi-scale bifurcated deep network for top-down contour detection. In: 2015 IEEE Conference on Computer Vision and Pattern Recognition (CVPR), pp. 4380–4389. IEEE (2015)

3. Bertasius, G., Shi, J., Torresani, L.: High-for-low and low-for-high: efficient boundary detection from deep object features and its applications to high-level vision. In: Proceedings of the IEEE International Conference on Computer Vision, pp. 504–512 (2015)

4. Canny, J.: A computational approach to edge detection. In: Readings in Computer Vision, pp. 184–203. Elsevier (1987)

5. Chen, L.C., Papandreou, G., Kokkinos, I., Murphy, K., Yuille, A.L.: DeepLab: semantic image segmentation with deep convolutional nets, atrous convolution, and fully connected CRFs. IEEE Trans. Pattern Anal. Mach. Intell. **40**(4), 834–848 (2018)

6. Comaniciu, D., Meer, P.: Mean shift: a robust approach toward feature space analysis. IEEE Trans. Pattern Anal. Mach. Intell. **24**(5), 603–619 (2002)

7. Dollar, P., Tu, Z., Belongie, S.: Supervised learning of edges and object boundaries. In: 2006 IEEE Computer Society Conference on Computer Vision and Pattern Recognition, vol. 2, pp. 1964–1971. IEEE (2006)

8. Everingham, M., Van Gool, L., Williams, C., Winn, J., Zisserman, A.: The Pascal visual object classes challenge 2012 (voc2012) results (2012). http://www.pascal-network.org/challenges/VOC/voc2011/workshop/index.html (2011)

9. Ganin, Y., Lempitsky, V.: N^4-fields: neural network nearest neighbor fields for image transforms. In: Cremers, D., Reid, I., Saito, H., Yang, M.-H. (eds.) ACCV 2014. LNCS, vol. 9004, pp. 536–551. Springer, Cham (2015). https://doi.org/10.1007/978-3-319-16808-1_36

10. Hariharan, B., Arbeláez, P., Bourdev, L., Maji, S., Malik, J.: Semantic contours from inverse detectors. In: 2011 IEEE International Conference on Computer Vision (ICCV), pp. 991–998. IEEE (2011)

11. He, K., Zhang, X., Ren, S., Sun, J.: Deep residual learning for image recognition. In: Proceedings of the IEEE Conference on Computer Vision and Pattern Recognition, pp. 770–778 (2016)

12. Kittler, J.: On the accuracy of the sobel edge detector. Image Vis. Comput. **1**(1), 37–42 (1983)

13. Konishi, S., Yuille, A.L., Coughlan, J.M., Zhu, S.C.: Statistical edge detection: learning and evaluating edge cues. IEEE Trans. Pattern Anal. Mach. Intell. **25**(1), 57–74 (2003)

14. Lim, J.J., Zitnick, C.L., Dollár, P.: Sketch Tokens: a learned mid-level representation for contour and object detection. In: 2013 IEEE Conference on Computer Vision and Pattern Recognition (CVPR), pp. 3158–3165. IEEE (2013)

15. Lin, T.-Y., et al.: Microsoft COCO: common objects in context. In: Fleet, D., Pajdla, T., Schiele, B., Tuytelaars, T. (eds.) ECCV 2014. LNCS, vol. 8693, pp. 740–755. Springer, Cham (2014). https://doi.org/10.1007/978-3-319-10602-1_48

16. Liu, Y., Cheng, M.M., Hu, X., Wang, K., Bai, X.: Richer convolutional features for edge detection. In: 2017 IEEE Conference on Computer Vision and Pattern Recognition (CVPR), pp. 5872–5881. IEEE (2017)

17. Long, J., Shelhamer, E., Darrell, T.: Fully convolutional networks for semantic segmentation. In: Proceedings of the IEEE Conference on Computer Vision and Pattern Recognition, pp. 3431–3440 (2015)

18. Martin, D.R., Fowlkes, C.C., Malik, J.: Learning to detect natural image boundaries using local brightness, color, and texture cues. IEEE Trans. Pattern Anal. Mach. Intell. **26**(5), 530–549 (2004)

19. Shen, W., Wang, X., Wang, Y., Bai, X., Zhang, Z.: DeepContour: a deep convolutional feature learned by positive-sharing loss for contour detection. In: Proceedings of the IEEE Conference on Computer Vision and Pattern Recognition, pp. 3982–3991 (2015)
20. Simonyan, K., Zisserman, A.: Very deep convolutional networks for large-scale image recognition. arXiv preprint arXiv:1409.1556 (2014)
21. Xie, S., Tu, Z.: Holistically-nested edge detection. In: Proceedings of the IEEE International Conference on Computer Vision, pp. 1395–1403 (2015)
22. Yu, Z., Feng, C., Liu, M.Y., Ramalingam, S.: CASENet: deep category-aware semantic edge detection. ArXiv e-prints (2017)
23. Zhao, H., Shi, J., Qi, X., Wang, X., Jia, J.: Pyramid scene parsing network. In: IEEE Conference on Computer Vision and Pattern Recognition (CVPR), pp. 2881–2890 (2017)

Convolutional Neuronal Networks Based Monocular Object Detection and Depth Perception for Micro UAVs

Wilbert G. Aguilar[1,2,3]([✉]), Fernando J. Quisaguano[1],
Guillermo A. Rodríguez[1], Leandro G. Alvarez[1], Alex Limaico[1],
and David S. Sandoval[1]

[1] CICTE Research Center, Universidad de las Fuerzas Armadas ESPE,
Sangolquí, Ecuador
wgaguilar@espe.edu.ec
[2] FIS Faculty, Escuela Politécnica Nacional, Quito, Ecuador
[3] GREC Research Group, Universitat Politècnica de Catalunya,
Barcelona, Spain

Abstract. In this work, we present the development of a system for the detection and depth estimation of objects in real time using the on-board camera in a micro-UAV through convolutional neuronal networks. Traditionally for the detection of obstacles shows the use of SLAM visual systems. However, to solve this problem, this level of complexity is not necessary, saving resources and execution time. The training with convolutional neural networks using stereo images for the depth estimation and in the same way training the detection of common observable objects can obtain an accurate detection of obstacles in a real time.

Keywords: Monocular · Depth estimation · Micro-UAV · Object detection

1 Introduction

In recent years, the research approach to Unmanned Autonomous Vehicles has constantly increased [1, 2], the detection of obstacles to prevent collisions is very important for autonomous flights [3, 4]. The interest that has awakened the UAVs in recent years has shown various applications, such as inspection, supervision [5, 6] and mapping the latter with methods such as SLAM [7–9]. So also in autonomous land vehicles using embedded systems for processing. In terms of autonomy the detection of obstacles for robots has been made using active detection such as LIDAR, structured light, sonar or IR [3]. These sensors can be useful in terms of real cost, but in terms of weight and power requirements, there is a notable deficiency when it comes to small flying vehicles, so we have chosen to use passive detection as stereo vision [10].

In stereo vision to obtain depth images has shown disadvantages for having a small base line between cameras in a micro-UAV because of its limited range. Today approaches that address this problem more efficiently, a step towards the more robust obstacle detection has been made using monocular depth estimation methods based on

© Springer Nature Switzerland AG 2018
Y. Peng et al. (Eds.): IScIDE 2018, LNCS 11266, pp. 401–410, 2018.
https://doi.org/10.1007/978-3-030-02698-1_35

Convolutional Neuronal Networks (CNNs) [11–14]. The methods based on deep learning have shown very promising results for the task of depth estimation in individual images. However, most existing approaches treat depth prediction as a supervised regression problem and, as a result, require large amounts of depth data. A new study suggests a novel method that allows CNNs to perform a depth estimation of a single unsupervised image [15].

At the same time, research has progressed in recent years in the detection of objects through (CNNs), modern object detectors based on these networks have Faster R-CNN [16], R-FCN [17], SSD [18] and YOLO [19] that are good enough to be implemented in consumer products (e.g., Google Photos, Pin-terest, Visual Search) [20]. Therefore the objective of this paper is to obtain the error in the estimation of the obstacle distance with the unsupervised method [15] using a Parrot Bebop micro-UAV, since it will be based on the images captured in the built-in camera without the need to use extra sensors that increase the power consumption. At the same time get to identify the obstacles closest to the micro-UAV through the training of CNNs [20]. Then, verify the execution time to predict a dense depth map and the detection of obstacles in a 512 × 256 image on a modern GPU.

This paper is organized as follows: First we provide the related works in Sect. 2. The details of the proposed method including the creation of the data set and the CNN architecture fully implemented in the UAV are presented in Sect. 3. In Sect. 4, we present real flight experiments and the behavior of this method. Finally, the document concludes in Sect. 5.

2   Related Works

Automating the visual detection and tracking of moving objects through intelligent autonomous systems in unmanned aerial vehicles (UAVs), has been a recurrent research topic during the last decades in computer vision. This investigation has diverse applications that extend from the army, the surveillance, the systems of security, the aerial photography, the search and the rescue, the recognition of objects, the automatic navigation to the interactions man-machine [21].

The ability to detect and avoid obstacles like birds flying in a forest is fascinating and has been the subject of much research for its multiple applications [22–24]. To imitate human behavior, algorithms based on learning have been developed to predict paths directly from an RGB image, especially with the popularity of convolutional neural networks (CNN) [25]. With individual images they were studied for the first time for depth extraction, using "Supervised Learning" [26].

Some of these methods have been recently tested in obstacle detection and autonomous flight applications [27]. However, the aforementioned methods solve the depth estimation task and from it derive the obstacle map. Following these approaches, we propose a solution to the problem by training a model for depth estimation and obstacle detection. While they would appear to be two independent branches, the extraction of features from their RGB input is common and shared for both purposes [28].

3 Method

In our work, it is based on learning methods that can estimate the depth in the individual images and also in the detection of objects using CNNs. Our contribution consists of the combined use of these current methods in the micro-UAV.

3.1 Unsupervised Depth Estimation

Given the position of the eyes in humans and the way to move them, the images received in each eye are practically the same, with a difference in the relative position of the objects. These relative differences in the position in each image (the disparity), has a direct relationship with the distance (depth) [29].

Starting a simple image I, the objective is to obtain a function that proves to predict the depth of each pixel p (see Fig. 1) and can predict the depth of every scene per pixel $\hat{d} = f(I)$. For training we need two points of view O_L and O_R specifically, access to two images I^l and I^r, corresponding to the left and right color images from a calibrated stereo pair, captured at the same moment in time. The disparity is the difference in the horizontal coordinates of the points p_L and p_R that is $d = x_L - x_R$ that corresponds disparity—a scalar value per pixel. Instead of trying to directly predict the depth, we attempt to find the dense correspondence field d^r that, when applied to the left image, would enable us to reconstruct the right image.

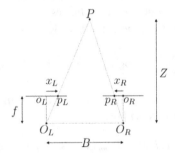

Fig. 1. Geometric relationship between the parameters of the stereo pair to obtain the depth Z from the disparity d.

We will refer to the reconstructed image $I^l(d^r)$ as $\tilde{I}^r$. Similarly, we can also estimate the left image given the right one, $\tilde{I}^l = I^r(d^l)$ [14]. At the pixel level by similarity between triangles PO_LO_R, $p_RO_RO_R$ and $p_LO_LO_L$ is obtained:

$$Zd = fB \tag{1}$$

To obtain the proposed distance function we have the baseline distance B between the cameras and the camera focal length f, we can then trivially recover the depth $\hat{d}$ from the predicted disparity.

$$\hat{d} = fB/d \tag{2}$$

3.2   Object Detection

The method is derived from the MultiBox fundamental idea is to train a convolutional network that outputs the coordinates of the object bounding boxes directly [30] but is extended to handle multiple object categories like a The SSD (The Single Shot Detector) training [18] therefore we use the VGG-16 network as a base, but other networks should also produce good results. For training, you must select the predetermined boxes correspond to a true detection on the ground and train the network.

Training Object
For each image we make an adjustment between the predictions and ground-truth boxes. First we show $x_{ij} = 1$ to indicate that the $i-th$ prediction is matched to the $j-th$ ground-truth and $x_{ij} = 0$ otherwise. Given a matching between predictions and ground-truth, the location loss term can be written as [30]

$$L_{loc}(x, l, g) = \frac{1}{2} \sum_{i,j} x_{ij} \|l_i - g_j\|_2^2 \tag{3}$$

Let $l_i \in \mathbb{R}^4$ be the $i-th$ set of predicted box coordinates for an image, and let $g_j \in \mathbb{R}^4$ be the $j-th$ ground-truth box coordinates.

Given the predicted scores c_i, the confidence loss term can be written as follows:

$$L_{conf}(x, c) = -\sum_{i,j} x_{ij} \log(c_i) - \sum_i \left(1 - \sum_j x_{ij}\right) \log(1 - c_i) \tag{4}$$

The overall objective loss function is weighted sum of the localization loss (loc) and the confidence loss (conf) [18]:

$$L(x, c, l, g) = L_{conf}(x, c) + \alpha L_{loc}(x, l, g) \tag{5}$$

3.3   Combination of Methods

Our system is composed of the bebop 2 mini drone that has a camera so it is controlled from a ground station from where the orders are sent, where the information provided during the flight is processed.

To obtain images we used the autonomy_bebop driver based on the official Parrot ARDrone_SDK3 SDK (see Fig. 2). This allows the use of several functions of the drone translating in topics, services, messages and parameters as well as access to the images of the drone. One of the most important points of Bebop Parrot ARDroneSDK3 is that the quality of the video is limited to 640 × 368 image at 30 Hz [31–33].

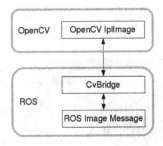

Fig. 2. Image conversion between ROS message type and OpenCV.

Once the image is obtained, the characteristics are extracted and sent to two branches depending on the task: a branch of depth prediction and an obstacle detector branch (see Fig. 3).

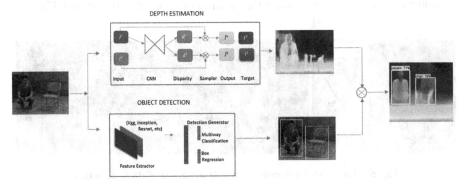

Fig. 3. Given a RGB input, it extracts the characteristics and then introduces them into the branches of the depth estimation and obstacle detection to produce the maps of the depth of the box with the limits of the obstacles.

The object detection branch is composed by 7 convolutional layer. The detection methodology is similar to the one present in [18]. And in the same way for depth estimation present in [14]. Therefore the architecture is composed of two CNNs that perform different tasks (see Fig. 3).

At runtime, estimations require about 0.4 s per frame on a NVIDIA GTX 1060 Q-MAX GPU.

4 Result and Discussion

4.1 Datasets

Imaging models are used for depth estimation. Results of the disparity images of the KITTI 2015 stereo training set 200 [34]. For object detection we use the SSD300 and

SSD512 architectures in the COCO dataset [35]. Allowing us to detect obstacles and the depth estimation of up to 15 m from the micro-UAV camera (see Fig. 4).

Fig. 4. The obstacle detection at 15 m (left) and depth estimation (right).

4.2 Metric of Evaluation

First we will obtain the $(x; y)$ coordinates of the bounding box center, the bounding box width w and height h(see Fig. 5). The average distance of the detected obstacle from the camera of the micro-UAV m and the variance of its depth distribution v [28].

Fig. 5. Taking of depth estimation data given the obstacle detection

To evaluate the depth estimator branch performance. Linear RMSE and Scale Invariant Log RMSE.

$$\frac{1}{n}\sum_i d_i^2 - \frac{1}{n^2}\left(\sum_i d_i\right)^2 \tag{6}$$

Detection RMSE on Obstacles (Mean/Variance): For each detected obstacle, we compare its estimated obstacle depth statistics (mean and variance) with the closest obstacle ones by using linear RMSE [28] (see Table 2).

We will test the difference in measurements in the error in the distance to a reference point measured by the system. We will also test the response of the system to different light intensity variations.

4.3   Distance Estimation Error

To estimate the error in the distance measured by the drone, we based on a known distance to a reference point, we calculate the error at different instances of the flying process. The results are shown in Table 1.

Table 1. Accuracy of depth estimation in detected objects

Obstacle	Real distance (m)	Depth monocular distance (m)	% error
Person 1	1.489	1.55	4.096
Person 2	1.124	1.22	8.540
Chair	1.245	1.29	6.451

We calculate the area defined by the compensated vertex of the rectangle w to h. To estimate the mass center Mc by the average value of the x and y coordinates. With the coordinates of that pixel an estimation of the depth of the detected object is taken (see Fig. 6).

Fig. 6. Test with obstacles at a short distance from the micro-UAV.

In terms of location of objects we observe good results but with relatively small objects it has problems mainly due to the quality of the post-processed image (see Fig. 7).

Table 2. Accuracy in the detection of trained obstacles

Category	Mean average precision	Precision
Person	0.74	71.2%
Chair	0.79	79.5%
Plant	0.61	56.7%

Fig. 7. Test with obstacles at a short distance from the micro-UAV. Although the person is squatting, the model recognizes him.

4.4   Refresh Time

The time obtained between each frame is 0.815 s, knowing that the transmission delay was 0.254 s. This satisfied us for the following investigation as it is the evasion of obstacles in drones.

5   Conclusions and Future Work

In this paper, we propose an obstacle detection system based on CNN applicable to various environments. But above all with methods superior to many previously shown and therefore the error of the distance obtained in a monocular image is small.

This joint detection of obstacles and depth estimation. We demonstrate its effectiveness in detecting obstacles in synthetic and world datasets real.

The time of transmission and processing of each frame so small would help in the development of an autonomous system to avoid obstacles designed for a drone.

In future research, the objective will be to obtain, from convolutional neural networks, trajectory planning for the evasion of the obstacles already recognized in this paper in order to model and control a totally autonomous micro-UAV.

Acknowledgement. This work is part of the project Perception and localization system for autonomous navigation of rotor micro aerial vehicle in gps-denied environments, VisualNav-Drone, 2016-PIC-024, from the Universidad de las Fuerzas Armadas ESPE, directed by Dr. Wilbert G. Aguilar.

References

1. Aguilar, W.G., Angulo, C.: Real-time model-based video stabilization for microaerial vehicles. Neural Process. Lett. **43**(2), 459–477 (2016)
2. Aguilar, W.G., Angulo, C.: Real-time video stabilization without phantom movements for micro aerial vehicles. EURASIP J. Image Video Process. **2014**(1), 46 (2014)
3. Gageik, N., Benz, P., Montenegro, S.: Obstacle detection and collision avoidance for a UAV with complementary low-cost sensors. IEEE Access **3**, 599–609 (2015)
4. Yang, S., Konam, S., Ma, C., Rosenthal, S., Veloso, M., Scherer, S.: Obstacle avoidance through deep networks based intermediate perception (2017)
5. Aguilar, W.G., et al.: Pedestrian detection for UAVs using cascade classifiers and saliency maps. In: Rojas, I., Joya, G., Catala, A. (eds.) IWANN 2017. LNCS, vol. 10306, pp. 563–574. Springer, Cham (2017). https://doi.org/10.1007/978-3-319-59147-6_48
6. Aguilar, W.G., et al.: Cascade classifiers and saliency maps based people detection. In: De Paolis, L.T., Bourdot, P., Mongelli, A. (eds.) AVR 2017. LNCS, vol. 10325, pp. 501–510. Springer, Cham (2017). https://doi.org/10.1007/978-3-319-60928-7_42
7. Aguilar, W.G., Rodríguez, G.A., Álvarez, L., Sandoval, S., Quisaguano, F., Limaico, A.: Visual SLAM with a RGB-D camera on a quadrotor UAV using on-board processing. In: Rojas, I., Joya, G., Catala, A. (eds.) IWANN 2017. LNCS, vol. 10306, pp. 596–606. Springer, Cham (2017). https://doi.org/10.1007/978-3-319-59147-6_51

8. Aguilar, W.G., Rodríguez, G.A., Álvarez, L., Sandoval, S., Quisaguano, F., Limaico, A.: Real-time 3D modeling with a RGB-D camera and on-board processing. In: De Paolis, L.T., Bourdot, P., Mongelli, A. (eds.) AVR 2017. LNCS, vol. 10325, pp. 410–419. Springer, Cham (2017). https://doi.org/10.1007/978-3-319-60928-7_35

9. Aguilar, W.G., Rodríguez, G.A., Álvarez, L., Sandoval, S., Quisaguano, F., Limaico, A.: On-board visual SLAM on a UGV using a RGB-D camera. In: Huang, Y.A., Wu, H., Liu, H., Yin, Z. (eds.) ICIRA 2017. LNCS (LNAI), vol. 10464, pp. 298–308. Springer, Cham (2017). https://doi.org/10.1007/978-3-319-65298-6_28

10. Oleynikova, H., Honegger, D., Pollefeys, M.: Reactive avoidance using embedded stereo vision for MAV flight. In: Proceedings of IEEE International Conference on Robotics and Automation, vol. 2015, pp. 50–56 (2015)

11. Eigen, D., Puhrsch, C., Fergus, R.: Depth map prediction from a single image using a multi-scale deep network. In: NIPS, pp. 1–9 (2014)

12. Liu, F., Shen, C., Lin, G., Reid, I.: Learning depth from single monocular images using deep convolutional neural fields. IEEE Trans. Pattern Anal. Mach. Intell. **38**(10), 2024–2039 (2016)

13. Chakravarty, P., Kelchtermans, K., Roussel, T., Wellens, S., Tuytelaars, T., Van Eycken, L.: CNN-based single image obstacle avoidance on a quadrotor. In: Proceedings of IEEE International Conference on Robotics and Automation (ICRA), pp. 6369–6374 (2017)

14. Aguilar, W.G., Quisaguano, F., Álvarez, L., Pardo, J., Proaño, Z.: Monocular depth perception on a micro-UAV using convolutional neuronal networks. Accepted

15. Godard, C., Mac Aodha, O., Brostow, G.J.: Unsupervised monocular depth estimation with left-right consistency (2016)

16. Ren, S., He, K., Girshick, R., Sun, J.: Faster R-CNN: towards real-time object detection with region proposal networks. In: NIPS, pp. 91–99 (2015)

17. Dai, J., Li, Y., He, K., Sun, J.: R-FCN: object detection via region-based fully convolutional networks. In: NIPS (2016)

18. Liu, W., et al.: SSD: single shot multibox detector, pp. 21–37 (2016)

19. Redmon, J., Divvala, S., Girshick, R., Farhadi, A.: You only look once: unified, real-time object detection (2015)

20. Cvpr, A., Id, P.: Speed/accuracy trade-offs for modern convolutional object detectors. In: CVPR, vol. 3562, pp. 7310–7319 (2017)

21. Yilmaz, A., Javed, O., Shah, M.: Object tracking. ACM Comput. Surv. **38**(4), 13 (2006)

22. Sebastian, S., Mori, T.: First results in detecting and avoiding frontal obstacle from monocular camera for micro unmanned aerial vehicles. In: 2013 IEEE International Conference on Robotics and automation (ICRA), vol. 53, no. 9, pp. 1689–1699 (2013)

23. Aguilar, W.G., Morales, S.G.: 3D environment mapping using the Kinect V2 and path planning based on RRT algorithms. Electronics **5**(4), 70 (2016)

24. Aguilar, W.G., Morales, S., Ruiz, H., Abad, V.: RRT* GL based optimal path planning for real-time navigation of UAVs. In: Rojas, I., Joya, G., Catala, A. (eds.) IWANN 2017. LNCS, vol. 10306, pp. 585–595. Springer, Cham (2017). https://doi.org/10.1007/978-3-319-59147-6_50

25. Tokui, S., Oono, K., Hido, S., Clayton, J.: Chainer: a next-generation open source framework for deep learning. In: Proceedings of Workshop on Machine Learning Systems, Twenty-Ninth Annual Conference on Neural Information Processing Systems, pp. 1–6 (2015)

26. Saxena, A., Chung, S.H., Ng, A.Y.: Learning depth from single monocular images. Adv. Neural. Inf. Process. Syst. **18**, 1161–1168 (2006)

27. Mancini, M., Costante, G., Valigi, P., Ciarfuglia, T.A., Delmerico, J., Scaramuzza, D.: Toward domain independence for learning-based monocular depth estimation. IEEE Robot. Autom. Lett. 2(3), 1778–1785 (2017)
28. Mancini, M., Costante, G., Valigi, P., Ciarfuglia, T.A.: J-MOD^2: joint monocular obstacle detection and depth estimation, vol. 3766, no. c, pp. 1–8 (2017)
29. Lecumberry, F.: Cálculo de disparidad en imágenes estéreo, una comparación. XI Congr. Argentino Ciencias la Comput (2005)
30. Szegedy, C., Reed, S., Erhan, D., Anguelov, D., Ioffe, S.: Scalable, high-quality object detection (2014)
31. Aguilar, W.G., Salcedo, V.S., Sandoval, D.S., Cobeña, B.: Developing of a video-based model for UAV autonomous navigation. In: Barone, D.A.C., Teles, E.O., Brackmann, C.P. (eds.) LAWCN 2017. CCIS, vol. 720, pp. 94–105. Springer, Cham (2017). https://doi.org/10.1007/978-3-319-71011-2_8
32. Aguilar, W.G., Casaliglla, V.P., Pólit, J.L.: Obstacle avoidance based-visual navigation for micro aerial vehicles. Electronics 6(1), 10 (2017)
33. Aguilar, W.G., Casaliglla, V.P., Pólit, J.L., Abad, V., Ruiz, H.: Obstacle avoidance for flight safety on unmanned aerial vehicles. In: Rojas, I., Joya, G., Catala, A. (eds.) IWANN 2017. LNCS, vol. 10306, pp. 575–584. Springer, Cham (2017). https://doi.org/10.1007/978-3-319-59147-6_49
34. Geiger, A., Lenz, P., Urtasun, R.: Are we ready for autonomous driving? The KITTI vision benchmark suite. In: Conference on Computer Vision and Pattern Recognition, pp. 3354–3361 (2012)
35. Lin, T.-Y., et al.: Microsoft COCO: common objects in context. In: Fleet, D., Pajdla, T., Schiele, B., Tuytelaars, T. (eds.) ECCV 2014. LNCS, vol. 8693, pp. 740–755. Springer, Cham (2014). https://doi.org/10.1007/978-3-319-10602-1_48

Improving Minority Language Speech Recognition Based on Distinctive Features

Tong Fu[⊠], Shaojun Gao, and Xihong Wu

Key Laboratory on Machine Perception (Ministry of Education),
Speech and Hearing Research Center, Peking University, Beijing, China
{fut,gaosj,wxh}@cis.pku.edu.cn

Abstract. With the development of deep learning technology, speech recognition based on deep neural networks has been continuously improved in recent years. However, the performance of minority language speech recognition still cannot compare with that on majority language whose data can be collected and transcribed easily relatively. Therefore, we attempt to work out an effective data sharing method cross different languages to improve the performance of minority language speech recognition. We proposed a speech attribute detector model under an end-to-end framework, and then we utilized the detector to extract features for minority language speech recognition. To the best of our knowledge, this is the first end-to-end model extracting distinctive features. We implemented our experiments on Tibetan and Mandarin. The results showed the significant improvements were achieved on Tibetan phoneme recognition via utilizing the Mandarin data.

Keywords: Speech recognition · Minority language
Distinctive features · End-to-end · Tibetan

1 Introduction

In recent years, with the development of deep learning technology, speech recognition performance based on DNN (Deep Neural Networks, DNN) has been continuously improved. At present, with a large amount of speech data and corresponding transcriptions, we have been able to obtain speech recognition systems whose performance has reached or even surpassed the human level. However, for most of minority languages, due to the influence of factors like the number of users and speech collection methods, it is often difficult to obtain large amounts of speech and transcriptions. Therefore, how to effectively improve the speech recognition performance on minority languages has been widely concerned by researchers.

Researchers have proposed some effective modeling methods, such as the traditional Hidden Markov model state tying method [1], the subspace Gaussian

© Springer Nature Switzerland AG 2018
Y. Peng et al. (Eds.): IScIDE 2018, LNCS 11266, pp. 411–420, 2018.
https://doi.org/10.1007/978-3-030-02698-1_36

mixture model parameter sharing method [2], and a series of models and methods based on neural networks [3–5].

The proposed model is based on the neural network model, we thus focus on DNN based methods in the followings. Under DNN framework, most of solutions are based on constructing multilingual or cross-lingual acoustic models, which are trained on multiple languages speech data. Therefore, it is possible to utilize other languages data to alleviate the problem of lacking data for minority languages.

Huang et al. [3] proposed a method of sharing parameters based on deep neural network model called Shared Hidden Layer Multilingual Deep Neural Networks (SHL-MDNN). It uses the multi-task learning framework where different languages have their own output layers to map the outputs from the shared hidden layer to their own phonemes respectively. The shared hidden layers are considered to extract shared representations across different languages. In this way, SHL-MDNN can use a large amount of other language corpus to help minority languages speech recognition.

Heigold et al. [4] proposed a adaptation method based on SHL-MDNN, which may be called a cross-lingual transfer learning method. The method has the same training procedure as the SHL-MDNN to obtain multilingual shared representations. Then it removes the original output layer and stack a new output layer on the shared hidden layers for another language. It then only trains the new added layer and keeps the original hidden layers unchanged to adapt the model for another language.

Grezl et al. [5] introduced the bottleneck layer [6] into the SHL-MDNN structure to achieve a low-dimensional representation of the shared representation. They combined these low-dimensional bottleneck features with traditional features to train the target language acoustic model and improve the performance.

All above methods are based on multi-task framework. Each language uses its own phonemes to construct the multilingual acoustic model. In fact, we can use language-independent phonetic units as the supervision symbols to enable representations to be shared further. As a result, researchers attempted to establish unified phonemes for different languages so that models can be with one shared output layer.

Schultz et al. [7] constructed a custom global symbol set, which can be shared among different languages to some extent. However, as phonemes are designed on researchers' own understanding, they may be altered by different researchers. And it introduces more confusion when sharing phonemes between different pronunciations in different languages, which will influence the performance of acoustic models. Other researchers attempted to utilize IPA (International Phoneme Alphabeta, IPA) [8] as the unified phoneme set [9,10]. However, the distribution of IPA phonemes for different languages are still different, which often results in certain phonemes in one language cannot be adequately modelled.

Lee et al. proposed the ASAT (Automatic Speech Attribute Transcription, ASAT) system [11,12] from the perspective of speech attributes. In the framework of the DNN-HMM hybrid system, it uses the GMM-HMM model to align the speech so that frames are labelled by corresponding phonemes. The phonemes are then mapped to the speech attributes (distinctive features)[1], which are used as supervision labels to train a series of DNNs for detecting speech attributes, and these DNNs are called attribute detectors. Afterwards, these detectors are used to extract the attribute features, which are then spliced with the original features to train the back-end acoustic models.

However, the ASAT system is based on hybrid DNN-HMM approach, which remains a complicated task where training DNNs relies on GMMs to obtain frame-level labels. In addition, building GMMs normally goes through multiple stages and is expertise-intensive.

Previous work has attempted to reduce the complexity of ASR. Research has focused on end-to-end ASR [13–15], i.e., modeling the mapping between speech and labels (words, phonemes, etc.). Therefore, we introduced distinctive features to build the attribute detector model under an end-to-end framework. We then applied this detector to extract attribute features for the minority language speech recognition task. The rest of the paper is arranged as follows: In Sect. 2, the baseline end-to-end model is introduced. In Sect. 3, the proposed attribute detector model is presented and the experiments are carried out to evaluate the performance of the proposed method for minority language speech recognition in Sect. 4. We conclude and discuss the method in Sect. 5.

2   The Baseline End-to-End Model

The baseline model in this paper is deep bidirectional RNNs trained with the CTC (Connectionist Temporal Classification, CTC) objective function [16].

Compared with the feedforward networks, RNNs have the advantage of modelling complex temporal dynamics of sequences. Given an input sequence $X = (x_1, ..., x_T)$, a recurrent layer computes the *forward* sequence of hidden states $\overrightarrow{H} = (\overrightarrow{h}_1, ..., \overrightarrow{h}_T)$ from $t = 1$ to T.

$$\overrightarrow{h}_t = \sigma(\overrightarrow{W}_{hx}x_t + \overrightarrow{W}_{hh}\overrightarrow{h}_{t-1} + \overrightarrow{b_h}) \tag{1}$$

where $\overrightarrow{W}_{hx}$ is the input-to-hidden parameters, $\overrightarrow{W}_{hh}$ is the hidden-to-hidden parameters. In addition to the inputs x_t, the hidden activation h_{t-1} from the previous time step is also fed. In a bidirectional RNN, an additional recurrent layer computes the *backward* sequence of hidden outputs $\overleftarrow{H}$ from $t = T$ to 1.

$$\overleftarrow{h}_t = \sigma(\overleftarrow{W}_{hx}x_t + \overleftarrow{W}_{hh}\overleftarrow{h}_{t-1} + \overleftarrow{b_h}) \tag{2}$$

When building a deep architecture, we stack multiple bidirectional recurrent layers. At each frame t, we concatenate the forward and backward hidden outputs $[\overrightarrow{h}_t, \overleftarrow{h}_t]$ from the current layer to input to the next recurrent layer. To overcome the vanishing gradients problem [17], we apply the LSTM (Long Short-Term

---

[1] The distinctive features are a set of distinguishing attributes that are summarized by linguists to differentiate phonemes, and reflect the different states of speech organs.

Memory, LSTM) units [18] as the building block to build the deep BLSTM (Bidirectional LSTM) neural networks.

Unlike the hybrid approach, the RNN model in baseline is not trained using frame-level labels with respect to the CE (Cross Entropy, CE) criterion. Instead, we adopt the CTC criterion to automatically learn the alignments between speech frames and their label sequences.

3   The Proposed Speech Attribute Detector

The proposed speech attribute detector is built under the end-to-end framework. We give a brief introduction to the concept of IPA phonemes and distinctive features in Sect. 3.1, and describe the structure of the proposed detector model in Sect. 3.2.

3.1   IPA Phonemes and Distinctive Features

The most commonly used phoneme set is IPA, which is designed and maintained by the International Phonetic Association. We can transcribe any speech segment into a sequence of phonemes, and the differences between phonemes can be analyzed via the distinctive features. These features reflect the state of the speech organs to some degree. We attempted to use 10 distinctive features to construct the proposed attribute detector model, and these 10 features are: consonantal, tense, continuant, nasal, back (tongue), front (tongue), high (tongue), labiodental, strident, round.

Table 1. 10 distinctive features of phoneme "b" and "t". "Cons", "Cont", "Lab" and "Str" represent consonantal, continuant, labiodental and strident respectively.

Phoneme	Cons	Tense	Cont	Nasal	Back	High	Lab	Str	Front	Round
b	+	?	−	−	?	?	−	?	?	−
t	+	?	−	−	?	?	?	−	?	?

Table 1 shows the example of the differences between phoneme "b" and "t" via these 10 distinctive features. "+" means the phoneme has this attribute, "−" means it does not have, and "?" represents this attribute is not discriminative in the phoneme. The mappings between IPA phonemes and distinctive features are collected from the website PHOIBLE Online [19].

3.2   The Speech Attribute Detector and Backend Acoustic Model

The structure of the proposed speech attribute detector is illustrated in Fig. 1. We utilize the multi-task learning framework, with an iterative training method, to incorporate phoneme recognition task and attribute detection tasks into a

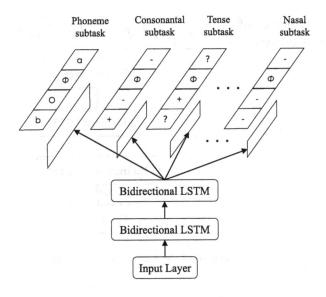

Fig. 1. The structure of the proposed speech attribute detector

unified model. In the subtask of phoneme recognition, we use phonemes as supervision symbols, and the objective function is CTC. When it comes to attribute detection subtasks, we construct different output layers to model different attributes, i.e., we build 10 subtasks for 10 distinctive features respectively with CE criterion. The supervision labels for each attribute detection subtask are alignments made by the phoneme recognition subtask.

In the iterative training, the CTC phoneme recognition subtask will output the corresponding phoneme for each frame, and then we map phonemes to distinctive features to produce labels for each attribute detection subtask. Since there are "blank" (Ø) labels in CTC alignments, we also add a "blank" (Ø) label for each attribute detection subtask(the output dimension of each attribute subtask is 4 now), so as to ensure the training continue.

After training, the attribute detector is used to extract the attribute features. We splice the outputs of the 10 detection subtasks to obtain the final "detected features", which is 40-dimensional. Then, we further use the detected features to train the back-end acoustic model for speech recognition. The structure of the acoustic model is shown in Fig. 2, which is consistent with the baseline model.

In practice, in order to ensure the accuracy of the alignments for distinctive features, we did not train the attribute detection subtasks in the early stage of training, but only after a certain number of epochs did we align the speech to obtain the alignments for these subtasks and start the iterative training.

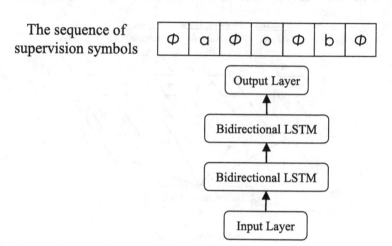

Fig. 2. The structure of the back-end acoustic model

4   Experiments

We implemented experiments on Tibetan to validate the proposed method for minority language speech recognition. In Sect. 4.1, the configuration of the baseline model is described. We experimented on the proposed speech attribute detector and utilized the extracted attribute features for Tibetan phoneme recognition in Sect. 4.2. In Sect. 4.3, we attempted to use Mandarin speech to enhance the detector for improving the performance of Tibetan phoneme recognition.

4.1   The Baseline Model for Tibetan Phoneme Recognition

The Tibetan corpus we used in the experiment contains about 10 h of speech, i.e., 9352 sentences of telephone conversational speech in Tibetan. We selected 200 sentences as the test set randomly. The speech was represented by 40-dimensional Mel-scale log-filterbank features (excluding the energy value), along with the first-order and second-order temporal derivatives. The frame length is 25 ms and the frame shift is 10 ms. The number of deep BLSTM layers is 3, and each layer has 256 forward and backward units, i.e., 512 units. The BLSTM does not use peephole connections. The initial training rate is 0.0001 and the Adam optimization algorithm is adopted. The result PER (Phoneme Error Rate, PER) is 34.9%.

4.2   Using Attribute Features for Phoneme Recognition

Firstly, we train a speech attribute detector model, using the Tibetan Mel-scale log-filterbank features in Sect. 4.1. The structure of the detector has 3 BLSTM hidden layers, i.e., the number of shared hidden layers is 3, and the number of forward and backward units in each hidden layer is 256, without peephole connections. The initial learning rate for the phoneme subtask network is set to 0.0001, while it is 0.00001 for attribute subtasks. The Adam optimization algorithm is adopted. In practice, in order to ensure the stability, only the phoneme subtask is trained during the first 5 epochs.

We uses the same structure as the baseline for the back-end acoustic model. The inputs are the 40-dimensional attribute features (splicing outputs of the 10 attribute subtasks) extracted by the detector. The initial learning rate is 0.0001, and the Adam optimization algorithm is also adopted. The result PER is 37.1%.

Table 2. Phoneme error rate (PER) of baseline and back-end acoustic model using attribute features.

Model	PER(%)
Baseline	34.9
Back-end	37.1

As shown in Table 2, the back-end model using the attribute features does not achieve better performance than the baseline. One of the possible reasons is that it is difficult for the attribute subtasks to be trained sufficiently with only 10-h speech, so that attribute features lack effectiveness. The most direct way is to augment the dataset, while which is not always accessible for most of minority languages. As the distinctive features are language-independent, we attempt to utilize other language speech data to help train the detector.

4.3   Using Mandarin Speech to Improve Performance

We experimented on HKUST, a Mandarin Chinese conversational telephone speech recognition corpus, as the augmentation dataset. Here the structure of the detector is slightly different from the one in Sect. 4.2. Based on the original detector structure, we added a Mandarin phoneme recognition subtask (CTC criterion) to enable Tibetan and Mandarin data to be trained at the same time. The remaining configurations are the same as Sect. 4.2. We also only trained the two phoneme subtasks during the first 5 epochs, i.e., the whole model is optimized together from the 6th epoch. We experimented on different data ratios of Tibetan and Mandarin to train detectors, and then used the detector to extract Tibetan attribute features for training the back-end Tibetan phoneme acoustic model. The configuration of the back-end acoustic model is consistent with that in Sect. 4.2.

In Table 3, the first column describes the detectors, where "MD(50 h)-TB(0 h)" means the detector trained on 50-h Mandarin data and 0-h Tibetan data. The second column represents the back-end acoustic model. Note that detectors of "MD(50 h)-TB(0 h)" and "MD(0 h)- TB(10 h)" have only one phoneme subtask as they do not use another language data.

Table 3. Phoneme error rate (PER) of back-end acoustic models using different attribute features.

Detector model	Back-end model	PER(%)
MD(50 h)-TB(0 h)	TB(10 h)	49.7
MD(0 h)-TB(10 h)	TB(10 h)	37.1
MD(10 h)-TB(10 h)	TB(10 h)	36.5
MD(30 h)-TB(10 h)	TB(10h)	35.1
MD(50 h)-TB(10 h)	TB(10h)	33.1

By comparing the results of "MD(50 h)-TB(0 h)" with "MD(0 h)-TB(10 h)", we can find that the performance of Tibetan phoneme recognition is not ideal using "MD(50 h)-TB(0 h)" detector attribute features, that is to say "MD(50 h)-TB(0 h)" trained only on non-target language (Mandarin) cannot extract effective attribute features from the target language (Tibetan).

We improved the performance by increasing the amount of non-target language (Mandarin) data, and attained the best PER of 33.1% using the attribute features extracted by the "MD(50 h)-TB(10 h)" detector. It means a 5.16% relative improvement compared with the baseline model (34.9%).

As there are two phoneme subtasks in the detector model when using non-target language data, this local network itself can be viewed as a minority language phoneme recognition model based on multi-task. Therefore, we tested this method for Tibetan phoneme recognition. The structure of the model is like the baseline in Sect. 4.1, in addition to having an extra subtask for Mandarin phoneme classification. This model was trained on 50-h Mandarin data and 10-h Tibetan data with the Adam optimization algorithm. The initial learning rates for the two subtasks are set to 0.0001. We directly used the Tibetan subtask for the final phoneme recognition. The result PER is 34.8%, which is slightly better than the baseline model and the proposed method based on attribute features achieved a relative improvement of 4.89% over this model.

5   Conclusion

This paper proposed and utilized a speech attribute detector model to improve the performance of speech recognition on Tibetan. To a certain extent, the proposed method alleviated the problem of lacking data in minority languages speech recognition. At present, this work is only a preliminary exploration. In order to verify this method quickly, we only utilize 10 distinctive features. We consider to use more distinctive features and propose method to select which distinctive features should be modelled in the future work. In addition, the ratios between the target and non-target language data will also be explored.

Actually, the extracted attribute features can be spliced with traditional features to train back-end acoustic models. This splicing method is adopted in most of the previous research. However, in this paper, we only utilize attribute features for speech recognition and achieve significant improvements, which demonstrates the feasibility of using the attribute feature alone for speech recognition to some degree.

Acknowledgments. The work is supported in part by the National Natural Science Foundation of China (No. 11590773, No. U1713217), the Key Program of National Social Science Foundation of China (No. 12 & ZD119).

References

1. Cohen, P., Dharanipragada, S., Gros, J., Monkowski, M.: Towards a universal speech recognizer for multiple languages. In: 1997 IEEE Workshop on Automatic Speech Recognition and Understanding Proceedings, pp. 591–598 (1997)
2. Burget, L., Schwarz, P., Agarwal, M., Akyazi, P., Feng, K., Ghoshal, A., et al.: Multilingual acoustic modeling for speech recognition based on subspace Gaussian mixture models. In: IEEE International Conference on Acoustics Speech and Signal Processing, vol. 130, pp. 4334–4337 (2010)
3. Huang, J.T., Li, J., Yu, D., Deng, L., Gong, Y.: Cross-language knowledge transfer using multilingual deep neural network with shared hidden layers. In: IEEE International Conference on Acoustics, Speech and Signal Processing, pp. 7304–7308 (2013)
4. Heigold, G., Vanhoucke, V., Senior, A., Nguyen, P., Ranzato, M., Devin, M., et al.: Multilingual acoustic models using distributed deep neural networks. In: IEEE International Conference on Acoustics, Speech and Signal Processing, pp. 8619–8623 (2013)
5. Grezl, F., Karafiat, M., Vesely, K.: Adaptation of multilingual stacked bottle-neck neural network structure for new language. In: IEEE International Conference on Acoustics, Speech and Signal Processing, pp. 7654–7658 (2014)
6. Yu, D., Seltzer, M.L.: Improved bottleneck features using pretrained deep neural networks. In: INTERSPEECH 2011, Conference of the International Speech Communication Association, Florence, Italy, August, pp. 237–240 (2011)
7. Schultz, T., Waibel, A.: Language-independent and language-adaptive acoustic modeling for speech recognition. Speech Commun. **35**(1–2), 31–51 (2001)
8. International Phonetic Association: Handbook of the International Phonetic Association: A Guide to the Use of the International Phonetic Alphabet. Cambridge University Press, Cambridge (1999)
9. Niesler, T.: Language-dependent state clustering for multilingual acoustic modelling. Speech Commun. **49**(6), 453–463 (2007)
10. Lin, H., Deng, L., Yu, D., Gong, Y., Acero, A., Lee, C.H.: A study on multilingual acoustic modeling for large vocabulary ASR. In: IEEE International Conference on Acoustics, Speech and Signal Processing, pp. 4333–4336 (2009)
11. Lee, C.H., Siniscalchi, S.M.: An information-extraction approach to speech processing: analysis, detection, verification, and recognition. Proc. IEEE **101**(5), 1089–1115 (2013)
12. Siniscalchi, S.M., Svendsen, T., Lee, C.H.: A bottom-up stepwise knowledge-integration approach to large vocabulary continuous speech recognition using weighted finite state machines, pp. 901–904 (2011)
13. Graves, A., Jaitly, N.: Towards end-to-end speech recognition with recurrent neural networks. In: International Conference on Machine Learning, pp. 1764–1772 (2014)

14. Maas, A., Xie, Z., Dan, J., Ng, A.: Lexicon-free conversational speech recognition with neural networks. In: Conference of the North American Chapter of the Association for Computational Linguistics: Human Language Technologies, pp. 345–354 (2015)
15. Chan, W., Jaitly, N., Le, Q., Vinyals, O.: Listen, attend and spell: a neural network for large vocabulary conversational speech recognition. In: IEEE International Conference on Acoustics, Speech and Signal Processing, pp. 4960–4964 (2016)
16. Graves, A., Gomez, F.: Connectionist temporal classification: labelling unsegmented sequence data with recurrent neural networks. In: International Conference on Machine Learning, vol. 2006, pp. 369–376 (2006)
17. Bengio, Y., Simard, P., Frasconi, P.: Learning long-term dependencies with gradient descent is difficult. IEEE Trans. Neural Netw. **5**(2), 157–166 (2002)
18. Hochreiter, S., Schmidhuber, J.: Long short-term memory. Neural Comput. **9**(8), 1735–1780 (1997)
19. PHOIBLE Online. http://phoible.org

Classification and Clustering

A Hybrid Approach Optimized by Tissue-Like P System for Clustering

Shaolin Wang, Laisheng Xiang, and Xiyu Liu[(✉)]

School of Management Science and Engineering, Shandong Normal University,
Jinan, Shandong, China
wangshaolin1116@163.com, xls3366@163.com,
sdxyliu@163.com

Abstract. K-medoids algorithm is a classical algorithm used for clustering, it is developed from K-means algorithm and it is more robust compared to K-means algorithm for noises and outliers. But it takes more time to achieve a better result. In this paper, we proposed a hybrid algorithm to overcome the drawbacks. We combined K-means algorithm and an improved K-medoids algorithm together. Firstly, to have an elementary clustering results, we run K-means algorithm for the data set. Then, the improved K-medoids algorithm are used to optimize the results to make it more robust. Furthermore, we designed a Tissue-like P system for the proposed approach, the Tissue-like P system operates in a parallel way thus can improve time efficiency greatly. We tested the efficiency and effectiveness of our approach on some data sets of the well-known UCI benchmark and compared the approach with the K-means and the K-medoids algorithm.

Keywords: K-means · K-medoids · Tissue-like P system · Clustering
Membrane computing

1 Introduction

In recent years, people pay more and more attentions to clustering because it has shown its usefulness in several fields like industry, business and processing. Many clustering algorithms have been proved effective but unfortunately, they also present some drawbacks. We devoted to explore some of these algorithms and try to improve them. We developed a new algorithm based on the most common clustering algorithms, that is, the K-means and K-medoids algorithms. To improve time efficiency, we modified the K-medoids algorithm in some aspects and designed a Tissue-like P system for it. The Tissue-like P system operates in a parallel way thus can improve time efficiency greatly. The idea is to run K-means algorithm first, then operate the K-medoids algorithm several times to make the results more robust when get the centroids. But as k-means can yield non object centroids, we integrate a process between them to get the closest objects from these centroids.

Experiments were performed on some data sets of the well-known UCI benchmark used by the data mining community, for K-means, K-medoids and proposed algorithm to test the efficiency and effectiveness of the latter.

© Springer Nature Switzerland AG 2018
Y. Peng et al. (Eds.): IScIDE 2018, LNCS 11266, pp. 423–432, 2018.
https://doi.org/10.1007/978-3-030-02698-1_37

The article is organized as follows. Section 2 presents the improved K-medoids algorithm and the proposed new clustering algorithm. Section 3 shows the Tissue-like P system we designed for the algorithm. Section 4 provides an experimental validation of the proposed algorithm. Section 5 draws the conclusions.

2 The Approach Based on an Improved K-Medoids Algorithm for Clustering

2.1 The Improved K-Medoids Algorithm

K-medoids algorithm is a classical algorithm used for clustering, it is developed from K-means algorithm, but it is more robust compared to K-means algorithm for noises and outliers.

Let $X = \{x_1, x_2, \ldots, x_n\}$ be a data set of n data, it can be separated into K clusters $C = \{C_1, C_2, \ldots, C_K\}$ by K-medoids algorithm. The K-medoids algorithm is performed based on the principle of minimizing the sum of the dissimilarities between each object x_i and its corresponding representative object. To measure the quality of the clustering, an objective function is used, defined as:

$$E = \sum_{j=1}^{K} \sum_{x_i \in C_j} dist(x_i, o_j) \tag{1}$$

Where, E is the sum of the absolute error for all objects in the data set, x_i is the object in the data set, o_j is the medoid of cluster C_j and $dist(x_i, o_j)$ represents the Euclidean distance between the objects x_i and o_j.

The steps of the K-medoids algorithm are as follows:

1. Initialization: Randomly choose K objects in the data set X as the initial medoids.
2. Assignment step: Assign each remaining object to the cluster with the nearest medoid evaluated by the Euclidean distance measures.
3. Update step: Randomly select a non-medoid object o'_j, calculate the objective function E' when o_j is instead of o'_j, swap o_j with o'_j if $E' < E$.
4. Stopping criterion: Repeat steps 2 and 3 until no change.

Although the traditional K-medoids algorithm is more robust compared to K-means algorithm, but for large values of n and k, it becomes very costly thus limits the application. To improve the time efficiency, we choose the candidate medoid o'_j from its q-nearest neighbors. Furthermore, we only compare the value of $\sum_{x_i \in C_j} dist(x_i, o_j)$ and $\sum_{x_i \in C_j} dist(x_i, o'_j)$ when the candidate medoid o'_j and the medoid o_j are compared in step 3.

The steps of the improved K-medoids are as follows:

1. Initialization: Randomly choose K objects in the data set X as the initial medoids.
2. Assignment step: Assign each remaining object to the cluster with the nearest medoid evaluated by the Euclidean distance measures.
3. Update step: Randomly select a non-medoid object o_j' from the q-nearest neighbors of the initial medoid o_j, calculate $\sum_{x_i \in C_j} dist(x_i, o_j)$ and $\sum_{x_i \in C_j} dist\left(x_i, o_j'\right)$ when o_j is instead of o_j', swap o_j with o_j' if $\sum_{x_i \in C_j} dist\left(x_i, o_j'\right) < \sum_{x_i \in C_j} dist(x_i, o_j)$.
4. Stopping criterion: Repeat steps 2 and 3 until no change.

2.2   The Hybrid Approach Proposed for Clustering

In this part, we propose a hybrid approach for clustering, which combines the K-means algorithm with the K-medoids algorithm.

As shown in Algorithm 1, the steps of the hybrid approach are as follows:

1. Call K-means on the initial data set X, finally we can get a set of clusters with means centers.
2. Calculating the distance between the means centers and each object of the cluster, select the closest object to the means centers as new centroid.
3. Call the improved K-medoids algorithm several times on the new centers representing initial medoids.

Algorithm 1

Input: Data set X; Number of clusters k; execution times S

Output: k clusters

1. K-means input(value k, Data set X)
2. Choose k data points $\{c_1,\dots,c_k\}$ as initial centers randomly
3. While (new center == original center)
4. FOR(each data point X_i){
5. Compute d_{ik} (data point X_i to center c_k)
6. IF(MinDis(k)= d_{ik}){
7. Center $c_k \leftarrow$ data point X_i
8. }
9. }
10. ENDFOR

11. Compute new center c_i=Mean(data points (Xi &&(Xi $\in$ cluster C_i)))

12.   }

13. ENDWHILE

14. FOR(each center c_i){

15. Compute d_{ij} (center c_i to data point X_j)

16. IF(Mindis(j)=d_{ij}){

17. New medoid= data point X_j

18. }

19.   }

20. ENDFOR

21. K-medoids input(value k, Data set X, k intial medoids, execution times S)

22. While (execution times!=S)

23. FOR(each data point X_i){

24. Compute d_{ik} (data point X_i to medoid c_k)

25. IF(MinDis(k)= d_{ik}){

26. Medoid c_k $\leftarrow$data point X_i

27. }

28.   }

29. ENDFOR

30. FOR(each medoid c_i){

31. Compute sum of d_{ij} (medoid c_i to data points (X_j &&(X_j $\in$ cluster C_j)))

32. FOR(each nearest non-medoid o of medoid c_i){

33. Compute sum of d_{oj} (medoid o to data points (X_j &&(X_j $\in$ cluster C_j)))

34. }

35. ENDFOR

36. IF(sum of d_{oj} less than sum of d_{ij}){

37. Swap c_i and o

38. }

39.   }

40. ENDFOR

41. ENDWHILE

The quality of clustering results is mainly depend on the value of execution times S, it cannot produce the most robust results if the value S is small and the computation time would be large if the value S is large. The k-medoids algorithm even does not need to execute in some extreme cases. In the following section, to handle this problem, a tissue-like P system is used.

3 A Tissue-Like P System for the Approach Proposed

3.1 Basic Concept of P System

Membrane computing is a new branch of natural computing, which derived from the construct and functions of cells or tissues. The membrane are arranged as a hierarchical structure and operates in a parallel way thus can reduce the time complexity greatly.

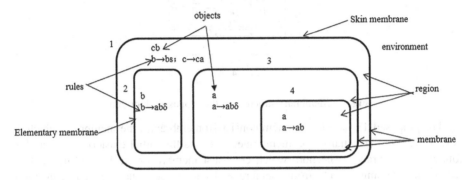

Fig. 1. The structure of a P system

As depicted in Fig. 1, a membrane m with no upper neighbor is called a skin membrane and a membrane m with no lower neighbor is called elementary membrane. The membranes always have objects and rules with them. The whole system is divided into different regions by the membranes. The space "outside" the skin membrane is called environment. A region is either a space delimited by an elementary membrane or a space delimited by non-elementary membranes. In the membrane, there are some rules and objects to execute the algorithm.

A system of degree $m \geq 1$ is a construst

$$\Pi = (O, H, \mu, \omega_1, \ldots, \omega_m, R_1, \ldots, R_m, i_0),$$

where:

1. O is the alphabet of objects;
2. H is the alphabet of membrane labels;
3. μ is a membrane structure of degree m;
4. $\omega_1, \ldots, \omega_m \in O^*$ are the multiset rewriting rules associated with the m regions of μ;
5. R_i, $1 \leq i \leq m$, are the multisets of objects associated with the m regions of μ;
6. $i_0 \in H \cup \{e\}$ specifies the input/output region of Π, where e is a reserved symbol not in H.

3.2   The Structure of the Used Tissue-Like P System

In this study, the Tissue-like P system is used to operate the proposed algorithm. Figure 2 depicts the membrane structure designed for the algorithm.

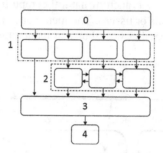

Fig. 2. Membrane structure of the proposed algorithm

The proposed Tissue-like P system consist 10 membranes, the membrane which is labeled by 0 is the initialization membrane, it selects the initial clustering objects for following membranes. The pool which consist 4 membranes are labeled by 1, they receive the passing objects from membrane 0 and execute the K-means algorithm independently. One of the membranes in pool 1 is delivered the final objects to membrane 3 directly and the remaining membranes deliveries the final objects to the pool 2 which consists 3 membranes. Each membrane in the pool 2 executes the K-medoids algorithm only one time when objects pass, then they pass the optimized objects to their neighbors and delivery to membrane 3. The membrane which labeled by 3 is the selection membrane. Selection membrane selects the most robust objects and delivery them to the output region which is labeled by 4. The overall execution process of the P system is shown in Fig. 3.

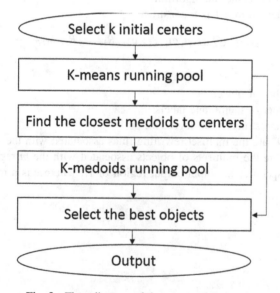

Fig. 3. Flow diagram of the proposed P system.

3.3 Rules of the Tissue-Like P System

In the designed system, the best objects, that is cluster centers, which minimize the objective function E in the jth membrane of pool 1 on the ith iteration, are denoted as O_{jlbest}^i. The membrane 0 are used to select the initial objects and membrane 4 stores the global best objects, which denoted as O_{gbest}.

(1) Evolution rules: The evolution rules are stored in the membranes of pool 1. The evolution strategy is

$$O_{jlbest}^{i+1} = \begin{cases} O_{jlbest}^{i+1}, & if\, f\left(O_{jlbest}^{i+1}\right) < f(O_{jlbest}^i) \\ O_{jlbest}^i, & otherwise \end{cases}, j = 1, 2, 3, 4.$$

Where $f\,(O)$ denotes the objective function.

(2) Selection rules: The selection rules are stored in the membrane 3. The selection strategy is

$$O_{gbest} = argmin f(O_{lbest}),$$

Where O_{lbest} denotes objects passed from the pool 1 and 2.

(3) Communication rules: For each iteration, membranes in pool 1 select best local objects O_{jlbest}^i and send it to pool 2 and membrane 4, then pool 2 executes the improved K-medoids algorithm and select local best objects and send it to membrane 4. The membranes in the pool 2 also commutate with each other to exchange objects.

4   Experiments Results

4.1   Experimental Data Sets

Three real-life data sets were obtained from the UCI Machine Learning Repository.

(a) Iris: This data set consists of 150 data points distributed over three clusters (Setosa, Versicolor, and Virginica), the dataset was in a four-dimensional space (sepal length, sepal width, petal length and petal width). Two classes (Versicolor and Virginica) showed a large amount of overlap while the class Setosa was linearly separable from the other two.

(b) Balance: The dataset was generated to model psychological experimental results. Each example was classified as having a balance scale tip to the right, to the left, or remain in balance. The dataset includes four features, three classes and has 625 examples.

(c) Escherichia Coli: The original dataset had 336 examples formed of 8 classes; however, three classes were represented with only 2, 2, 5 examples. Therefore, these nine examples were omitted and a total of 327 have been used. The dataset contains 327 examples with seven features and five classes.

4.2   Experimental Results

We tested K-means, K-medoids and the proposed algorithm on the Iris, Balance and Ecoli data set provided by the UCI Archive. We compared the performances of the three algorithms considering the criteria of execution time and correct points/rates of clustering. The clustering algorithms contain some stochastic and/or random factors. Thus, the three clustering algorithms were executed 100 times on each data set for the execution time and mean were calculated for each algorithm over the 100 runs for the correct points/rates. The proposed algorithm were executed by the Tissue-like P system we designed.

Table 1. The correct points of the three approaches in clustering Iris, Balance and Ecoli data sets.

Approach	Iris	Balance	Ecoli
K-means	132	327	244
K-medoids	133	333	231
Proposed approach	135	395	274

Table 1 depicts the average performance of the three approaches in the three clustering data sets. The value in the Table 1 represent the correct points and the corresponding correct rates are depicted in the Fig. 4.

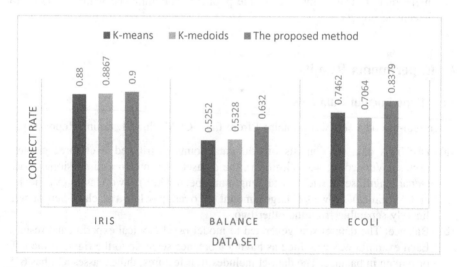

Fig. 4. The correct rates of the three approaches in clustering Iris, Balance and Ecoli data sets.

From the Table 1 and Fig. 4, we can see there is not much difference between the three algorithms when the Iris data set was executed, the proposed approach just a little better than the others. But for the Balance and Ecoli data sets, the proposed approach has the much better results. According to this set of experiments, we can conclude that for Iris, Balance and Ecoli data set, the proposed approach yields a better results than traditional K-means and K-medoids algorithm.

Figure 5 shows the execution time for the three approaches on three data sets. We observe that the proposed approach has the minimum value of execution time for Ecoli data set. However, for Iris and Balance data sets, the execution time is little bit than K-means algorithm, but less than K-medoids algorithm.

The proposed tissue-like P system, however, could not convergence as quick as K-means algorithm in some cases. But it can easily produce a more robust result compared to the K-means and K-medoids algorithm.

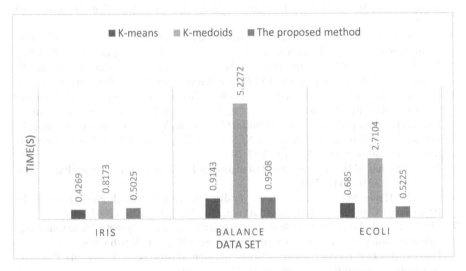

Fig. 5. The execution time of the three approaches in clustering Iris, Balance and Ecoli data sets.

5 Conclusion

The paper presents a hybrid algorithm operated by Tissue-like P system with special membrane structure in the framework of the cellular computing with membranes. Membrane computing is a new biological computing approach and Tissue-like P system has great parallelism which can immensely improve the time efficiency greatly.

The experimental validation of this new algorithm was tested on the Iris, Balance and Ecoli data set of the UCI archive. Under the control of the membrane rules, the objects are distributed into different membranes in the Tissue-like P system. Membranes update the local best objects and the global best object synchronously until system reach the halt conditions. The experimental results verify the advantages of the proposed Tissue-like P system with the hybrid algorithm.

Acknowledgments. This work is supported by the Natural Science Foundation of China (nos. 61472231, 61502283, 61170038), Ministry of Education of Humanities and Social Science Research Project, China (12YJA630152), Social Science Fund Project of Shandong Province, China (16BGLJ06, 11CGLJ22).

References

1. Paun, G., Rozenberg, G., Salomaa, A.: The Oxford Handbook of Membrane Computing. Oxford University Press Inc., Oxford (2010)
2. Freund, R., Păun, G., Pérez-Jiménez, M.J.: Tissue tissue-like P systems with channel states. Theor. Comput. Sci. **330**(1), 101–116 (2005)
3. Hartigan, J.A., Wong, M.A.: A K-means clustering algorithm. Appl. Stat. **28**(1), 100–108 (1979)
4. Kaufmann, L., Rousseeuw, P.J.: Clustering by means of medoids. In: Statistical Data Analysis Based on the L1-Norm & Related Methods, North-Holland, pp. 405–416 (1987)
5. Ng, R.T., Han J.: Efficient and effective clustering methods for spatial data mining. In: Proceedings of the 20th International VLDB Conference, vol. 88, no. 9, pp. 144–155 (1994)
6. Lichman, M.: UCI machine learning repository, School of Information and Computer Science, University of California, Irvine, CA, USA (2013). http://archive.ics.uci.edu/ml
7. Zhang, Q., Couloigner, I.: A new and efficient K-medoid algorithm for spatial clustering. In: Gervasi, O., et al. (eds.) ICCSA 2005. LNCS, vol. 3482, pp. 181–189. Springer, Heidelberg (2005). https://doi.org/10.1007/11424857_20
8. Park, H.S., Jun, C.H.: A simple and fast algorithm for K-medoids clustering. Expert Syst. Appl. **36**(2), 3336–3341 (2009)
9. Grira, N., Houle, M.E.: Best of both: a hybridized centroid-medoid clustering heuristic. In: International Conference on Machine Learning, pp. 313–320. ACM (2007)
10. Liu, X., Liu, H., Duan, H.: Particle swarm optimization based on dynamic Niche technology with applications to conceptual design. Comput. Sci. **38**(10), 668–676 (2006)
11. Liu, X., Xue, J.: A cluster splitting technique by hopfield networks and tissue-like P systems on simplices. Neural Process. Lett. **46**(1), 171–194 (2017)
12. Liu, X., Zhao, Y., Sun, M.: An improved apriori algorithm based on an evolution-communication tissue-like tissue-like P system with promoters and inhibitors. Discret. Dyn. Nat. Soc. **2017**(1), 1–11 (2017)
13. Zhao, Y., Liu, X., Wang, W.: Spiking neural tissue-like P systems with neuron division and dissolution. Sci. China **11**(9), e0162882 (2016)

An Improved Spectral Clustering Algorithm Based on Dynamic Tissue-Like Membrane System

Xuewei Hu and Xiyu Liu[✉]

College of Management Science and Engineering, Shandong Normal University,
Jinan, China
huxuewei486@163.com, sdxyliu@163.com

Abstract. With vast amount of data generated, it is becoming a main aspect to mine useful information from such data. Clustering research is an important task of data mining. Traditional clustering algorithms such as K-means algorithm are too old to propose high-dimensional data, so an efficient clustering algorithm, spectral clustering is generated. In recent years, more and more scholars has been firmly committing to studying spectral clustering algorithm for its solid theoretical foundation and excellent clustering results. In this paper we propose an improved spectral clustering algorithm based on Dynamic Tissue-like P System abbreviated as ISC-DTP. ISC-DTP algorithm takes use of the advantages of maximal parallelism in tissue-like membrane system. Experiment is conducted on an artificial data set and four UCI data sets. And we compare the ISC-DTP algorithm with original spectral clustering algorithm and K-means algorithm. The experiments demonstrate the effectiveness and robustness of the proposed algorithm.

Keywords: Spectral clustering algorithm · ISC-DTP algorithm
Tissue-like membrane system · Data mining

1 Introduction

With the rapid development of researches on Intelligent Computation, there is more and more data get generated. Clustering Analysis is an important segment of data mining which means dividing discrete data into responding clusters according to an efficient similarity cutting measure and makes the similarity between the data points in the same cluster is as large as possible [1]. Traditional clustering algorithms are based on different measures. Some of the existing clustering models are distance-based clustering analysis such as K-mean algorithm, there are also DBSCAN algorithm based on density. Although the calculation principle of above clustering algorithms is simple, they are not good at dealing with complex data sets.

The new developed spectral clustering method which is based on spectral graph theory has got more and more attention for its virtues such as overcoming above algorithms' shortcomings and ensuring that the global optimal solution can be obtained [2]. Spectral clustering method translates data clustering into graph cutting with graph theory.

© Springer Nature Switzerland AG 2018
Y. Peng et al. (Eds.): IScIDE 2018, LNCS 11266, pp. 433–442, 2018.
https://doi.org/10.1007/978-3-030-02698-1_38

P system means membrane computing, an efficient and novel distributed parallel computing model, which is based on the working mechanism of biological cells put forward by Gheorghe Paun in 1999. In a P system, there are three main ingredients (i) the hierarchical membrane structure separating compartments where (ii) multisets of data objects evolve according to prescribed (iii) rules of computing purposes [3]. There are three kinds of membrane computational models: cell-like membrane system, tissue-like membrane system and neural membrane system. In this paper, we choose dynamic tissue-like membrane system to combine with improved spectral clustering algorithm.

The remainder of this paper is arranged as follows. Section 2 describes graph partitioning and proposes a brief introduction of tissue-like P system solving clustering problem in a spectral clustering algorithm. The details of the improved spectral clustering algorithm based on dynamic tissue-like P System will be described in Sect. 3. Section 4 presents the results and analysis of experiments on an artificial data set and four UCI data sets. Finally, conclusions are drew in Sect. 5.

2   Preliminaries

2.1   Clustering

At present, it has been realized that only by converting data into information or digging out knowledge from data cannot find the value of data. Transforming data into corresponding graph and partitioning it is a data clustering method using graph theory. It has been proved to be a NP hard problem. Spectral clustering can loose data in a discrete optimization problem into responding real number field, and then above problem would be converted to a problem that can be solved in linear time [4]. The essence of graph partitioning is the maximum or minimization of matrix traces, and the spectral clustering algorithm can be used to solve the problem.

2.2   Spectral Clustering Algorithm

Spectral representation always uses Laplasse mapping to structure a laplacian matrix L corresponding to data set and calculates the eigenvalues and eigenvectors of L. In this thought, we reduce the dimension of the original data into a lower dimension using one or several eigenvalues. And then, K-means or other primary clustering algorithm is usually used for clustering for characteristic vectors of above laplacian matrix, so original data is partitioned into different classes.

In general, spectral clustering algorithm can be actualized in three steps which are data preprocessing, spectral representation and clustering [2]. In the first step of data preprocessing, the given data $X = \{x_1, x_2, \ldots\ldots, x_n\}$ is represented as vertices of the undirected graph G (V, E), and $E = \{w_{ij}\}$ is the aggregation of similarity between data points x_i and x_j. In actually $w_{ij} \geq 0$. When $w_{ij} = 0$, there is no connection between x_i and x_j. So the similarity matrix can be described as below.

$$W = (w_{ij})(i, j = 1, 2, \ldots, n) \tag{1}$$

Degree of vertex x_i is the sum of weights connecting to this x_i

$d_i = \sum_j w_{ij}, j \in \text{adjacent (i)}$

D is the degree matrix of G, and

$$D = \text{diag}(d_1, d_2, \ldots, d_n) \tag{2}$$

$L = D - A$ is unnormalized laplacian matrix used by most spectral clustering algorithms and Ng et al. have improved normalized laplacian matrix is more suitable for clustering. Normalized Laplacian matrix has two forms

$$L_{sym} = D^{-1/2}LD^{-1/2} = I - D^{-1/2}WD^{-1/2} \tag{3}$$

$$L_{rw} = D^{-1}L = I - D^{-1}L \tag{4}$$

where I is the identity matrix.

To divide data into K clusters, multi-way partitioning algorithm is a good choice. This algorithm takes use of the eigenvectors corresponding to the K maximum eigenvalues of the Laplacian matrix and reduce dimensions by mapping original data into these eigenvectors [5]. This method could use more eigenvectors in clustering than two-way partitioning algorithm.

2.3 Design of Membrane System

Membrane clustering is an efficient and novel computational model, which is based on the working mechanism of biological cells. It can be classified into three computational models: cell membrane system, tissue membrane system and neural membrane system. A common cell-like membrane system introduced in [6] is showed in Fig. 1.

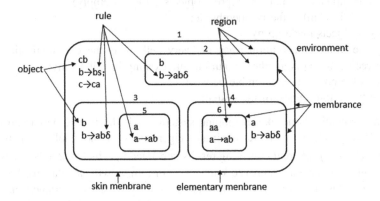

Fig. 1. A common cell-like membrane system

This cell-like membrane system named $\Pi 1$, the membrane can also be described in a function:

$$\Pi 1 = \{V, O, H, \mu, w_1, w_2, w_3, w_4, R_1, R_2, R_3, R_4, i_0\} \tag{5}$$

in which,

(1) $V = \{a, b, c, s\}$ is the alphabet;
(2) $O = \{s\}$ is the collection of output objects;
(3) $H = \{1, 2, 3, 4, 5, 6\}$ is membrane's number;
(4) $\mu = \left\{\left[[]_2 [[]_5]_3 [[]_6]_4 \right]_1 \right\}$ is the membrane structure;
(5) $w_1 = cb$, $w_2 = b$, $w_3 = b$, $w_4 = a$, $w_5 = a$, $w_6 = aa$ are objects in different membrane;
(6) $R_1 = \{b \rightarrow bs; c \rightarrow ca\}$, $R_2 = \{b \rightarrow ab\delta\}$, $R_3 = \{b \rightarrow ab\delta\}$, $R_4 = \{a \rightarrow ab\delta\}$, $R_5 = \{a \rightarrow ab\}$, $R_6 = \{a \rightarrow ab\delta\}$ are evolution rules;
(7) $i_0 = 1$ means the outputting membrane.

The tissue-like membrane system is an extension of cell-like membrane system inspired by behavior of multiple single-membrane cells which evolve in a public biological environment. In the environment, each cell processes objects in it according to given two kinds rules: one is evolution rules, the other is communication rules. Objects in different cells can move along pre-specified channels between cells [7]. Some basic concepts and mechanisms of how a tissue-like membrane system works are mentioned here briefly [6] and [8] give more information about this task.

As showed in Fig. 2, the construct of a tissue-like membrane system with m degrees is formally defined as follows:

$$\Pi = (O, \sigma_1, \ldots, \sigma_m, w_1, w_2, r_1, r_2, r_3, r_0, syn, i_0) \tag{6}$$

(1) O is an alphabet which can express objects (O is not empty);
(2) σ_i $(1 \leq i \leq m)$ is the ith membrane;
(3) $w_{1,2}$ is objects made up by O;
(4) $r_{1,2}$ are different evolution rules in every cell; r_0 is the communication rules between cells; r_3 is another evolution rule existing in cell 1 only;
(5) syn is the connection assemble between cells;
(6) i_0 is the output region of the system.

In this dynamic tissue-like membrane system, w_2 is the first K maximum eigenvectors corresponding to the K maximum eigenvalues of the Laplasse matrix, w_1 contains $\left(\sqrt{N} - 1\right)$ different clustering results generated in cells No. 2 to No. $\sqrt{N}$. r_1 is the rule which can operate improved k-means algorithm. r_2 will calculate the DB index to value the clustering result with different presetting clustering numbers namely i $(2 \leq i \leq m)$ in respect cell r_3 could compare the DB index and the smaller it is, the better clustering we will get. Then the best result is sent into the environment and i_0 is the final outcome of the whole clustering algorithm.

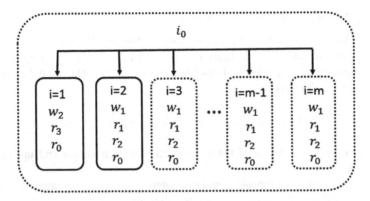

Fig. 2. A dynamic tissue-like membrane system with m cells

Because of creating or dissolving membranes, this system is a dynamic tissue-like membrane system. In the beginning there are only two cells σ_1 and σ_2 in the environment, once the first K maximum eigenvectors are transported into the environment, cell σ_2 could copy another $\left(\sqrt{N} - 2\right)$ cells. But the $\left(\sqrt{N} - 2\right)$ new cells' default clustering numbers are vary from 2 to $\sqrt{N}$. When rules r_1, r_2 perform already, the new created $\left(\sqrt{N} - 2\right)$ cells will dissolve according rule r_0, just cell σ_1 and σ_2 are left, they will be in a dormant state and wait for new missions.

3 Improved Spectral Clustering Algorithm with More Efficient K-Means Algorithm

3.1 More Efficient K-Means Algorithm

K-means algorithm is one of most popular unsupervised learning methods due to its simplicity and effectiveness, it was first introduced in the literature [9] and has been used in a wide variety of areas such as pattern recognition, data mining, and bioinformatics. However, k-means has several drawbacks: (1) the number of clusters must be given before clustering, then according to k, k initial points is chosen randomly as the initial clustering centers but these points are crucial to clustering result; (2) K-means algorithm is not so efficiently in dealing with non-convex data sets; (3) outliers may have a bad effect to the clustering result; (4) easy to trap into local optima [10]. Even though it is still widely used partitioning clustering algorithm. Because of its high efficiency, the method is relatively efficient and scalable in handling big datasets. The k-means algorithm takes the input parameter k, starts with k randomly selected objects as initial cluster centers, and partitions a data set into k clusters by iteratively updating the cluster centers.

The basic k-means algorithm can be described as follow:

(1) For a dataset $X = \{x_p\}$, $p \in [1, n]$ it gives the number of clusters to generate, it starts with selecting k initial cluster centers randomly, denoted by $\{m_i\}$, $i \in [1, k]$.

(2) Assign each data object in the dataset to cluster center to which the data object is most similar, denoted by

$$Ci = \{xp : \|x_p - m_i\|^2 \le \|x_p - m_j\|^2, 1 \le j \le k\} \tag{7}$$

(3) Calculate the mean value of the data objects for each cluster and updating the cluster centers.

(4) Step 2 and Step 3 are repeated until the criterion function converges. Usually, k-means algorithm adopts square error criterion, denoted by

$$E = \sum_{i=1}^{k} \sum_{x_p \in C_i} \|x_p - m_i\|^2 \tag{8}$$

The method of spectral cluster can give a better solution in clustering non-convex data sets and the results of conducting experiments on a double-loop artificial data set with improved spectral clustering algorithm and K-means algorithm respectively show in part 4.

In this paper, we employ a new way to select the k initial cluster centers. First, we chose a data point a randomly in data set. Then, calculate the distances between other data points and a, select the data point with the biggest distance marking it with b. What's more, calculate the distances between the rest data points and a and b respectively, according step two, the third point is confirmed. Finally repeat step 1 to 3 until the kth point is selected. In ISC-DTP algorithm K-means is used to cluster the first k eigenvectors corresponding to the largest eigenvalue comparing to the original data set, it handles much smaller number of data so we can ignore the shortcomings of outliers and local optima.

To decide how many clusters data will be divided into before operation is also a serious problem. Membrane computing can easily solve it. Because of the ability of maximum parallel, a tissue-like P system is used to determine the most appropriate number of clusters [7]. In above P system, there are $(\sqrt{N} - 2)$ cells operation K-means algorithm and calculate the DB index with k varying from 2 to $\sqrt{N}$ respectively. DB index is one of the most commonly used measurement indexes. It measures both similarity in a cluster and dissimilarity among clusters [17]. It can be described as:

$$DB = \frac{1}{K} \sum_{1}^{k} \max \left(\frac{\bar{C}_i + \bar{C}_j}{\|w_i - w_j\|^2} \right) (i \ne j) \tag{9}$$

In this function k is the number of clusters, $\bar{C}_i$ is the average sum of Euclidean distance in a cluster, $\|w_i - w_j\|^2$ is Euclidean distance between different clisters. The smaller it is, the better the clustering result is. Then $(\sqrt{N} - 2)$ DB index will be transported in to one another cell and in this cell DB index is compared and the kth cell's clustering outcome with the lest DB index is the final of whole ISC-DTP algorithm.

3.2   Description of ISC-DTP Algorithm

The ISC-DTP algorithm combines the advantages of spectral graph theory, tissue-like membrane system and K-means. The whole describe of ISC-DTP algorithm is showed as follows.

ISC-TP algorithm

Input : a data set with N data and each data has n attributes ;

Output : k clusters;

1   calculate Euclidean distance between data and structure similarity matrix W;

2   make the diagonal values of matrix W into 0, such as W(i,i)=0;

3   vol $\leftarrow \sum_{i=1}^{N} \sum_{j=1}^{N} W_{ij}$, D $\leftarrow$ diag$\sum_{j=1}^{N} W_{ij}$;

4   $\bar{L} \leftarrow I - D^{-1/2} W D^{-1/2}$;

6   $\lambda_{imax}(\bar{L}) \leftarrow$ the ith largest eigenvalue of $\bar{L}$ (1≤i≤k)

7   choose the first k maximum eigenvectors corresponding to the K maximum eigenvalues of $\bar{L}$ X $= [v_1, v_1, ..., v_k]$

8   normalize X→Y

9   do improved k-means algorithm in a tissue-like membrane system

---

Improved k-means algorithm in a tissue-like membrane system

Input : k maximum eigenvectors of $\bar{L}$;

Output : k clusters;

1   Choose the first data a randomly;

2   b $\leftarrow$ max$d_{a,b} = $ max$\sqrt{\sum_{i=1}^{n}(x_{ai} - x_{bi})^2}$,
    repeat until k initial clustering centers are got;

3   creat $\sqrt{N} - 3$ cells ,do clustering in $\sqrt{N} - 2$ cells parallelly and calculate DB index;

4   transport DB indexes into a cell , output the best number of clusters and k clusters;

5   dissolve new built $\sqrt{N} - 3$ cells.

---

4   Experiment Result

To investigate the performance of ISC-DTP algorithm, a number of experiments are conducted on an artificial data set and four UCI data sets. First, we compare the abilities of partitioning convex data sets using artificial between k-means algorithm and ISC-DTP algorithm. Then, we compare the effectiveness of the proposed ISC-DTP algorithm, the traditional k-means algorithm and traditional spectral clustering algorithm (short for SC) with k-means using four UCI data sets. To structure the similarity metric, Euclidean distance is used. What's more, in order to ensure the fairness of the comparison, all the experiments are performed on the same MATLAB platform. For all reported results, the test platform is a Pentium(R) Dual-Core CPU 3.2 GHz desktop with 64-bit 4 GB RAM. And the version of test platform is MatlabR2014b. Table 1. shows feature descriptions of above data sets.

Table 1. Some characteristics of data sets

Data sets	Source	Objects	Attributes	Classes
Allpts	Artificial	50	2	2
Balance	UCI	625	4	2
Iris	UCI	150	4	3
Ecoli	UCI	327	7	5
Wine	UCI	178	13	3

As for the artificial data set Allpts, it has 2 attributes and should be divided into two classes, these data points forming two circle distributions, one circle is a class and the other circle is another. Figure 3 describes Allpts, Fig. 4 shows the different clustering results between k-means algorithm and ISC-DTP algorithm.

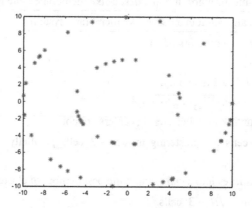

Fig. 3. Data in Allpts

To value the effectiveness of ISC-DTP algorithm, experiments are conducted on the five data sets over 100 times and the result of correct clustering is showed on Table 2. Shown as follows, compare the calculation results with the original data sets, ISC-DTP algorithm always performs better than original spectral clustering algorithm and basic K-means.

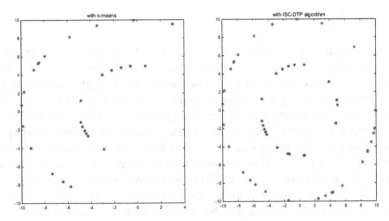

Fig. 4. Clustering results using k-means algorithm and ISC-DTP algorithm

Table 2. The average of correct clustering over 100 times

Data sets	ISC-TP	SC	K-means
Allpts	**50**	50	25
Balance	**381**	327	291
Iris	**134**	128	112
Ecoli	**200**	182	137
Wine	**145**	123	98

In this passage, we propose an active tissue-like membrane clustering algorithm to determine the number of clustering an unknown data set. Because of appointing clustering numbers before calculating, we should try different clustering numbers ranging from 2 to $\sqrt{N}$ and calculate the DB index respectively. But the actually clusters of the four UCI data sets are pretty small, so experiments are conducted with k ranging from 2 to 6 (bigger than all best clustering numbers). After running over 10 times, the running time of three algorithms begin to be pretty different, shown as Table 3. Although ISC-DTP algorithm takes more time than K-means, it always gets more accurate results than it and our algorithm runs faster than SC algorithm.

Table 3. The running time of three algorithms over 10 times (second)

Data sets	ISC-TP	SC	K-means
Allpts	11.467	12.998	5.755
Balance	10.235	210.974	11.003
Iris	8.527	9.11	8.823
Ecoli	14.427	21.018	10.604
Wine	13.168	22.638	9.019

5 Further Effort

Due to the extremely powerful computing ability and the clustering effect, spectral clustering algorithm has been widely used in many fields, especially in the field of computer graphics and complex network. Based on my own research direction, this paper puts forward the following ideas about spectral clustering algorithm. Most existing spectral clustering algorithms are hard partitioning methods, the application of fuzzy set theory in spectral clustering is of great significance to study the effective fuzzy spectral clustering algorithm. Clustering ensemble can make full use of the results of learning algorithm under different conditions, and find the clustering combination which cannot be obtained by single clustering algorithm. Therefore, spectral clustering ensemble algorithm can improve the quality and stability of clustering results, and has stronger robustness to noise, outliers and sample changes. In addition, we plan to implement more efficient membrane system in spectral clustering to further improve its efficiency.

Acknowledgment. This research project was partly supported by the National Natural Science Foundation of China (61472231, 61502283, 61640201), the Ministry of Education of Humanities and Social Science of China (12YJA630152), and the Shandong Social Science Fund of China (16BGLJ06, 11CGLJ22).

References

1. Liu, X., Xue, J.: Spatial cluster analysis by the bin-packing problem and DNA computing technique. Discret. Dyn. Nat. Soc. **5187**, 845–850 (2013)
2. Ng, A.Y., Jordan, M.I., Weiss, Y.: On spectral clustering: analysis and an algorithm. Adv. Neural. Inf. Process. Syst. **14**, 849–856 (2002)
3. Păun, G.: Computing with membranes. J. Comput. Syst. Sci. **61**, 108–143 (2000)
4. Wang, T.-S., Lin, H.-T.: Weighted-spectral clustering algorithm for detecting community structures in complex networks. Artif. Intell. Rev. **47**, 463–483 (2017)
5. Higham, D.J., Kibble, M.: A unified view of spectral clustering. University of Strathclyde Mathematics Research Report 02 (2004)
6. Păun, G.: From cells to computer: computing with membranes. BioSystems **59**, 139–158 (2001)
7. Liu, X., Xue, J.: A cluster splitting technique by Hopfield networks and tissue-like P systems on simplices. Neural Process. Lett. **46**, 171–194 (2017)
8. Liu, X., Zhao, Y., Sun, M.: An improved apriori algorithm based on an evolution-communication tissue-like tissue-like P system with promoters and inhibitors. Discrete Dyn. Nat. Soc. **1**, 1–11 (2017)
9. Le, J., Jiang, T.: Robust K-means algorithm with automatically splitting and merging clusters and its applications for surveillance data. Multimed. Tools Appl. **75**, 12043–12059 (2016)
10. Wang, B., Zhang, L.: Spectral clustering based on similarity and dissimilarity criterion. Pattern Anal. Appl. **20**, 495–506 (2017)
11. Lichman, M.: UCI machine learning repository. University California, School of Information and Computer Science, Irvine, CA, USA (2013). http://archive.ics.uci.edu/ml
12. Wang, J., Shi, P., Peng, H.: Weighted fuzzy spiking neural P systems. IEEE Trans. Fuzzy Syst. **21**(2), 209–220 (2013)
13. Han, R., Liu, J.: Spectral clustering and genetic algorithm for design of district metered areas in water distribution systems. Procedia Eng. **186**, 152–159 (2017)

Co-learning Binary Classifiers for LP-Based Multi-label Classification

Jincheng Shan, Chenping Hou$^{(\boxtimes)}$, Wenzhang Zhuge, and Dongyun Yi

College of Liberal Arts and Sciences, National University of Defense Technology,
Changsha 410073, China
njusjc@sina.cn, hcpnudt@hotmail.com, zgwznudt@yeah.net,
dongyun.yi@gmail.com

Abstract. A simple yet practical multi-label learning method, called label powerset (LP), considers each different combination of labels that appear in the training set as a different class value of a single-label classification task. However, because those classes source from multiple labels, there may be some inherent relationships among them. To tackle this challenge, we propose a novel model which aims to co-learn binary classifiers, by combining the training of binary classifiers and the characterizing the relationship among them into a unified objective function. In addition, we develop an alternating optimization algorithm to solve the proposed problem. Extensive experimental results on various kinds of datasets well demonstrate the effectiveness of the proposed model.

Keywords: Multi-label classification · Label powerset
Binary classifiers · Co-learning

1 Introduction

Multi-label learning is an extension of the standard supervised learning setting. In contrast to standard supervised learning where each instance is associated with one label that indicates its concept class belongingness, multi-label learning deals with data associated with multiple labels simultaneously. For example, in text categorization [1], a news document on presidential election can cover multiple topics such as politics, economics and culture simultaneously; In automatic image annotation [2] or video annotation [3], an image showing a house by the sea is associated with several annotated words such as house, beach and sea simultaneously; In music information retrieval [4], a piece of music may generate from different musical instruments such as piano, violin and guitar. Research into this important problem emerged in early 2000 and significant amount of progresses [5–8] have been made towards this emerging machine learning paradigm.

This paper focuses on the label powerset (LP) multi-label learning method [9,10], which considers each subset of a set of disjoint labels L, called labelset, that exists in the training set as a different class value of a single-label classification task. By LP, Multi-label classification are transformed into multi-class

© Springer Nature Switzerland AG 2018
Y. Peng et al. (Eds.): IScIDE 2018, LNCS 11266, pp. 443–453, 2018.
https://doi.org/10.1007/978-3-030-02698-1_39

classification, which can be solved by two common approaches One-Versus-All (OVA) [11], One-Versus-One (OVO) [12]. LP is an important approach to study, as it has the advantage of taking label correlations into consideration. This way it can, in some cases, achieve better performance compared to computationally simpler approaches like binary relevance (BR), which learns a binary model for each label independently of the rest [6].

Both in OVA or OVO, multiple binary classifiers are usually directly taken from many existing classifiers (e.g., SVM, C4.5) and trained separately [11–14]. In other words, these classifiers usually share the same pool of training data and the same learning algorithm. However, as labelsets sharing most same labels often dictates samples associated with them having similar structure in feature space (e.g., in multi-label scene classification, pictures annotated with beach, sea, blue sky and coconut tree are obviously similar to pictures annotated with beach, sea, blue sky, coconut tree and monkey), there may be inherent relationships among classifiers generated from new classes in LP methods. In this way, it is inappropriate to train multiple binary classifiers for new classes separately in the way described above. Intuitively, exploiting such relationships can improve the learning performances of LP methods.

In this work, we explore to utilize the inherent relationships among binary classifiers to improve the performances of LP methods. Specifically, we first propose a co-learning binary classifiers (CBC) method, where the training of binary classifiers and the learning of the relationships among these classifiers are formulated into an unified objective function. Then, an efficient alternating optimization algorithm is developed for the proposed CBC model. Experiments results validate that exploiting the relationships among binary classifiers can improve the performances of LP methods.

2   Related Work

Multi-label learning methods can be grouped into two categories [5]: problem transformation and algorithm adaptation. Methods of the first group transform the learning task into one or more single-label classification or ranking tasks, for which a large number of learning algorithms exists. The second group of methods adapt classical learning algorithms in order to fit multi-label learning.

2.1   Problem Transformation Methods

Binary Relevance (BR) [6], which trains one binary classifier separately for each label. For the classification of a new instance, BR outputs the union of the labels that are predicted by those classifiers. Goldbole et al. proposed Label Power-set (LP) [9], where each unique set of labels is considered as one of the classes of a new single-label classification task. Given a new instance, the learned single-label classifier predicts the most possible class, which actually represents a set of labels. RAndom k-labELsets (RAkEL) [15] is proposed to improve LP by breaking the initial set of labels into a number of small random subsets, called labelsets and employing LP to train a corresponding classifier. Another multi-label learning method [16] works by transforming the task into a ranking problem, trying to rank the proper labels before other labels for each instance.

2.2   Algorithm Adaptation Methods

AdaBoost.MH and AdaBoost.MR [17] are two extensions of AdaBoost for multi-label data. A combination of AdaBoost.MH with an algorithm for producing alternating decision trees was presented in [18]. The main motivation was to product multi-label models that can be understood by humans. An SVM algorithm that minimizes the ranking loss is introduced in [19]. [20] proposed a max-margin multi-label formulation to learn correlated predictors for labels. BP-MLL [21] is an adaptation of the popular backpropagation algorithm for multi-label learning. This algorithm mainly introduces a new error function that takes multiple labels into account. ML-RBF [22] adapts radial basis function networks to multi-label data. Several approaches [7,8] are based the popular k-Nearest Neighbors (kNN) lazy learning algorithms. An approach that combines lazy and associative learning is proposed in [23], where the inductive process is delayed until an instance is given for classification.

3   The Proposed Co-learning Binary Classifiers Model

In this section, we first introduce the notation in Sect. 3.1, and then present the proposed CBC model in Sect. 3.2. The optimization algorithm for the proposed method is described in Sect. 3.3.

3.1   Notation

Throughout this paper, vectors and matrices are written in boldface uppercase letters and boldface lowercase letters, respectively. Let $\mathcal{L} = \{\gamma_j : j = 1...m\}$ to denote the finite set of labels in a multi-label learning task and $(\mathbf{X}, \mathcal{Y}) = \{(\mathbf{x}_i, \mathbf{y}_i)\}_{i=1}^n$ denote a set of multi-label training examples, where $\mathbf{x}^i$ is a d-dimensional row vector and $\mathbf{y}_i \subseteq \mathcal{L}$ is the set of labels of the i-th example. As to LP methods, all different labelsets will be seen as c classes of a new single-label classification task. Then multi-class training examples are denoted as $(\mathbf{X}, \mathbf{z}) = \{(\mathbf{x}_i, z_i)\}_{i=1}^n$, where $z_i \in \{1, 2, \cdots, c\}$. To solve multi-class classification task, OVO and OVA are two common and effective approaches which aim to learn binary classifiers for each class. Denote $\mathbf{x}_i^l$ as the i-th instance in the l-th binary classification problem. For each of L binary classifiers, we want to learn a linear function $g(\mathbf{x}_i^l) = \mathbf{w}_l^T \mathbf{x}_i^l + b_l$, where $\mathbf{w}_l^T$ is the weight vector and b_l is the bias term, respectively.

3.2   Formulation

Follow the notation in the previous sub-section, and denote $\mathbf{W} = [\mathbf{w}_1^T, \cdots, \mathbf{w}_L^T] \in \mathbf{R}^{d \times L}$ where $\mathbf{w}_i \in \mathbf{R}^d (i = 1, \cdots, L)$. Let $\mathbf{b} = [b_1, \cdots, b_L]$ be the bias term, and N_l represents the number of instances in the l-th binary classifiers. Given the training data set $\mathbf{X}^l = [\mathbf{x}_1^l, \cdots, \mathbf{x}_{N_l}^l]$ and corresponding class label $\mathbf{z}^l = [z_1^l, \cdots, z_{z_l}^l]$ for the l-th binary classifier, we propose the following object function to learn these classifiers jointly:

$$\min_{\mathbf{W}} \sum_{l=1}^{L} \sum_{i=1}^{N_l} loss(z_i^l, f_l(\mathbf{x}_i^l)) + \lambda_1 \sum_{l=1}^{L} \Omega(\mathbf{w}_l), \tag{1}$$

where the term $loss(z_i^l, f_l(\mathbf{x}_i^l))$ is a loss function that measures the mismatch between z_i^l and the predicted value $f_l(\mathbf{x}_i^l)$, $\Omega(\mathbf{w}_l)$ is a regularizer that controls the complexity of the weight vector $\mathbf{w}_l$, and λ_1 is a regularization parameter to tune the tradeoff between the empirical loss and the regularization term. Actually, the model defined in Eq. (1) is a general framework for co-learning binary classifier, as one can use various loss functions and regularizers to adapt to the problems. For simplicity, we adopt the least squares loss function and the F-norm regularizer in this paper. Thus, the optimization problem defined in Eq. (1) can be rewritten as follows:

$$\min_{\mathbf{W}, \mathbf{b}, \{\rho_i^l\}} \sum_{l=1}^{L} \sum_{i=1}^{N_l} (\rho_i^l)^2 + \frac{\lambda_1}{2} \|\mathbf{W}\|_F^2$$

$$s.t. \ z_i^l(\mathbf{w}_l^T \mathbf{x}_i^l + b_l) = 1 - \rho_i^l, \forall i, l = 1, \ldots, L. \tag{2}$$

where ρ_i^l is a slack variable.

In order to exploit the relationships among binary classifiers corresponding to different labelsets, we adopt the column covariance matrix of the weight matrix $\mathbf{W}$. Thus, we further get:

$$\min_{\mathbf{W}, \mathbf{b}, \mathbf{M}\{\rho_i^l\}} \sum_{l=1}^{L} \sum_{i=1}^{N_l} (\rho_i^l)^2 + \frac{\lambda_1}{2} tr(\mathbf{W}\mathbf{W}^T) + \frac{\lambda_2}{2} tr(\mathbf{W}\mathbf{M}^{-1}\mathbf{W}^T)$$

$$s.t. \ z_i^l(\mathbf{w}_l^T \mathbf{x}_i^l + b_l) = 1 - \rho_i^l, \forall i, l = 1, \ldots, L \tag{3}$$

$$\mathbf{M} \geq 0, tr(\mathbf{M}) = 1.$$

where λ_1 and λ_2 are two regularization parameters, and $\mathbf{M}$ is the column covariance matrix. What needs to be stressed is that $tr(\mathbf{W}\mathbf{M}^{-1}\mathbf{W}^T)$ is used to describe the relevance among binary classifiers, where the inverse covariance matrix aims to couple pairs of weight vectors. The term $tr(\mathbf{M}) = 1$ is used for penalizing $\mathbf{W}$'s complexity, and the constraint $\mathbf{M} \geq 0$ restrict $\mathbf{M}$ to be positive semi-definite due to it denotes the covariance matrix. In addition, imbalance problem is common to see in multi-label task, which means that extremely few training examples appear in some labelsets, so it will happen to the classes in corresponding multi-class task. Then the result is that the empirical loss will be dominated by many instances in one binary classifier. In order to avoid this problem, we adopt the strategy introduced in [24] and define the importance of the example $(\mathbf{x}_i, z_i)$ in the l-th binary classifier:

$$c_i^l = \frac{max(N_+^t, N_-^t)}{n_i|\phi_l(i)|},$$

where N_+^t and N_-^t denote the positive and negative examples in the l-th binary classifier, and n_i denotes number of examples associated with z_i in the whole training data. What's more, $\phi_l(i)$ denote the number of sub-classes having same positive and negative as z_i. Then we get the final formulation:

$$\min_{\mathbf{W},\mathbf{b},\mathbf{M},\{\rho_i^l\}} \sum_{l=1}^{L}\sum_{i=1}^{N_l} c_i^l(\rho_i^l)^2 + \frac{\lambda_1}{2}tr(\mathbf{W}\mathbf{W}^T) + \frac{\lambda_2}{2}tr(\mathbf{W}\mathbf{M}^{-1}\mathbf{W}^T)$$

$$s.t. \ \ z_i^l(\mathbf{w}_l^T\mathbf{x}_i^l + b_l) = 1 - \rho_i^l, \forall i, l = 1, \dots, L \tag{4}$$

$$\mathbf{M} \geq 0, tr(\mathbf{M}) = 1,$$

In summary, in order to improve the capability of independent classifiers in LP methods, we learn all the binary classifiers jointly by exploiting relationships among them. Moreover, we takes the imbalance problem into consideration when learning classifiers.

3.3 The Optimization

As seen in Eq. (4), our proposed objective function is obviously jointly convex in all the variables. In the following, we describe an iterative updating algorithm to solve the problem. Firstly, the basis matrices are initialized by the initialization step and then the following steps are repeated until convergence:

(1) Optimize W and b: Fixing M. With $\mathbf{M}$ fixed, the objective function can be reduced to:

$$\min_{\mathbf{W},\mathbf{b},\{\rho_i^l\}} \sum_{l=1}^{L}\sum_{i=1}^{N_l} c_i^l(\rho_i^l)^2 + \frac{\lambda_1}{2}tr(\mathbf{W}\mathbf{W}^T) + \frac{\lambda_2}{2}tr(\mathbf{W}\mathbf{M}^{-1}\mathbf{W}^T)$$

$$s.t. \ \ z_i^l(\mathbf{w}_l^T\mathbf{x}_i^l + b_l) = 1 - \rho_i^l, \forall i, l = 1, \dots, L \tag{5}$$

The Lagrangian of above problem can be written as follows:

$$\mathcal{O} = \sum_{l=1}^{L}\sum_{i=1}^{N_l} c_i^l(\rho_i^l)^2 + \frac{\lambda_1}{2}tr(\mathbf{W}\mathbf{W}^T) + \frac{\lambda_2}{2}tr(\mathbf{W}\mathbf{M}^{-1}\mathbf{W}^T)$$

$$- \sum_{l=1}^{L}\sum_{i=1}^{N_l} \alpha_i^l[z_i^l(\mathbf{w}_l^T\mathbf{x}_i^l + b_l) - 1 + \rho_i^l] \tag{6}$$

After calculating the partial derivative of $\mathcal{O}$ with respect to $\mathbf{W}$, b_l and ρ_i^l, and setting them to 0, we get the following equations:

$$\frac{\partial \mathcal{O}}{\partial \mathbf{W}} = 0 \Rightarrow \mathbf{W} = \sum_{l=1}^{L} \sum_{i=1}^{N_l} \alpha_i^l z_i^l \mathbf{x}_i^l (\mathbf{e}_i^l)^T (\lambda_1 \mathbf{I}_L + \lambda_2 \mathbf{M}^{-1})^{-1} \tag{7}$$

$$\frac{\partial \mathcal{O}}{\partial b_l} = 0 \Rightarrow \sum_{i=1}^{N_l} \alpha_i^l z_i^l = 0 \tag{8}$$

$$\frac{\partial \mathcal{O}}{\partial \rho_i^l} = 0 \Rightarrow \rho_i^l = \frac{\alpha_i^l}{2c_i^l} \tag{9}$$

where $\mathbf{e}_i^l$ is the l-th column vector of identity matrix $\mathbf{I}_L$. Putting Eqs. (7)–(9) into Eq. (5), the following linear problem can be obtained:

$$\begin{pmatrix} K + \frac{1}{2\mathbf{C}} \mathbf{P}_{12} \\ \mathbf{P}_{21} \quad \mathbf{0} \end{pmatrix} \begin{pmatrix} \boldsymbol{\alpha} \\ \mathbf{b} \end{pmatrix} = \begin{pmatrix} \mathbf{1} \\ \mathbf{0} \end{pmatrix} \tag{10}$$

where $\mathbf{C}$ is a diagonal matrix and its elements value c_i^l if the corresponding instance belongs to the l-th binary classifier, and $\boldsymbol{\sigma} = [\alpha_1^1, \ldots, \alpha_{N_1}^1, \ldots, \alpha_1^L, \ldots, \alpha_{N_L}^L]$. K is a kernel matrix on all instances for all binary classifiers, with elements defined as:

$$\kappa(\mathbf{x}_{i1}^{l1}, \mathbf{x}_{i2}^{l2}) = [(\mathbf{e}_{l1}^T)(\lambda_1 \mathbf{I}_L + \lambda_2 \mathbf{M}^{-1})^{-1} \mathbf{e}_{l2}][z_{i1}^{l1}, z_{i2}^{l2}][(\mathbf{x}_{i1}^{l1})^T \mathbf{x}_{i2}^{l2}] \tag{11}$$

Given the label vector $\mathbf{z}^l$ of training data for the l-th classifier, $\mathbf{P}_{12}$ and $\mathbf{P}_{21}$ are defined as following:

$$\mathbf{P}_{12} = \mathbf{P}_{21}^T = \begin{bmatrix} \mathbf{z}^l & 0 & 0 \\ 0 & \ddots & 0 \\ 0 & 0 & \mathbf{z}^L \end{bmatrix} \tag{12}$$

(2) Optimize M: Fixing W and b. Fixing $\mathbf{W}$ and $\mathbf{b}$, the problem can be reduced to:

$$\min_{\mathbf{W}} tr(\mathbf{W}\mathbf{M}^{-1}\mathbf{W}^T) \tag{13}$$
$$s.t. \ \mathbf{M} \geq 0, tr(\mathbf{M}) = 1,$$

so we can easily obtain the solution:

$$\mathbf{M} = \frac{(\mathbf{W}^T\mathbf{W})^{\frac{1}{2}}}{tr((\mathbf{W}^T\mathbf{W})^{\frac{1}{2}})}. \tag{14}$$

To sum up, for the complete algorithm, two steps alternately update until converge or reach the maximum iterations. Although what we have discussed is the linear case in CBC model, it is easily extended to kernel extension with same method of solving steps.

4 Experiments

4.1 Experimental Settings

Datasets. We conducted experiments on four benchmark data sets with different scale and from different application domains[1]. Short descriptions of these data sets are given in the following paragraphs.

- **Scene** dataset [9] contains 2407 images annotated with up to 6 concepts such as beach, mountain and field. Each image is described with 294 visual features.
- **Emotions** dataset [25] is a music classification problem containing 593 songs. Songs are labeled with six emotions: amazed, happy, relaxed, quiet, sad and angry. Each song is represented by 72 rhythmic and timbre features.
- **Flags** dataset [26] contains details of 194 various nations and their flags which are annotated with 30 possible class labels.
- **Bibtex** dataset [27] contains metadata for the bibtex items like the title of the paper, the authors, etc. There are 7395 instances and 159 class labels. We randomly choose 10 class labels and corresponding instances in experiments.

Evaluation Measures. The F1 measure is the harmonic mean of precision and recall and is a popular evaluation measure in the research area of information retrieval. Micro F1 and macro F1 are the micro-averaged and macro-averaged versions of F1 respectively. Micro F1 aggregates true positives/negatives and false positives/negatives over labels, and then calculates an F1 from them. Macro F1 calculates F1 for each label and then takes the average over labels.

To evaluate the results, we used MacroF1 and MicroF1 as the evaluation measurements defined as:

$$\text{Macro F1:} = \frac{1}{q} \sum_{j=1}^{q} \frac{2 \sum_{i=1}^{n} \hat{\mathbf{Y}}_{ij} \cdot \mathbf{Y}_{ij}}{\sum_{i=1}^{n} \hat{\mathbf{Y}}_{ij} + \sum_{i=1}^{n} \mathbf{Y}_{ij}},$$

$$\text{Macro F1:} = \frac{2 \sum_{j=1}^{q} \sum_{i=1}^{n} \hat{\mathbf{Y}}_{ij} \cdot \mathbf{Y}_{ij}}{\sum_{j=1}^{q} \sum_{i=1}^{n} \hat{\mathbf{Y}}_{ij} + \sum_{j=1}^{q} \sum_{i=1}^{n} \mathbf{Y}_{ij}}.$$

Compared Methods. In this part, all algorithms are evaluated on the same five-fold partition of the same datasets. Benchmark approaches are evaluated and compared.

- **BR:** A simple baseline method under one-vs-all encoding which trains q classifiers for q label independently with all training data.
- **RANK-SVM:** It is an SVM style multi-label classification algorithm which minimizes ranking loss directly and has also exhibited excellent performance in previous studies.

---

[1] http://mulan.sourceforge.net/datasets-mlc.html.

- **MLKNN**[2]**:** The ML-KNN is a k-NN style multi-label classification algorithm which often outperforms other existing multi-label algorithms.
- **RAkEL:** This method learns an ensemble of LP classifiers, each one targeting a different small random subset of the set of labels.
- **FRAkEL**[3]**:** This methods proposes a fast algorithm for Random k-labelsets strategy by employing a two-stage classification to reduce the number of instances used in each local model.

For RANK-SVM the best parameters reported in the literature [19] are used; For MLkNN, the number of neighbors is set to 10 and the smoothing factor is set to 1 as recommended in [7]; For RAkEL and FRAkEL, a rule-of-thumb setting is: $k = 3$ and $n = 2q$. LIBSVM[4] software are used to learn base binary classifiers in this work. The parameters λ_1 and λ_2 for the proposed CBC model are chosen from $\{1e-4, 1e-3, 1e-2, 1e-1, 1e0, 1e1\}$ through 5-fold cross-validation on the training data.

4.2   Experimental Results

Tables 1 and 2 give the five-fold cross-validation performance of all compared algorithms on four datasets. On each data set, we perform 20 random runs and each we use a random 80% of each data set for training and the rest for evaluation. We report means and standard errors of each evaluation measure. The best results are in boldface.

From the experimental results in Tables 1 and 2, we can draw the following observations. (1) For the reason that FRAkEL is a fast and improved version

Table 1. MacroF1 of multi-label methods on four data sets. The highest mean MacroF1 is boldfaced.

Data sets	BR	ranksvm	MLKNN	RAkEL	FRAkEL	CBC
Emotions	.6071 ± .0214	.4807 ± .0322	.3787 ± .0275	.6375 ± .0291	.6554 ± .0318	**.6732 ± .0217**
Scene	.6951 ± .0143	.6378 ± .0173	.7118 ± .0155	.7202 ± .0264	.7182 ± .0170	**.7276 ± .0205**
Flags	.6080 ± .0384	.4666 ± .0521	.4465 ± .0166	.6105 ± .0488	.6229 ± .0346	**.6543 ± .0247**
bibtex	.6245 ± .0295	.4378 ± .0317	.4974 ± .0243	.6233 ± .0256	.6318 ± .0304	**.6436 ± .0231**

Table 2. MicroF1 of multi-label methods on four data sets. The highest mean MicroF1 is boldfaced.

Data sets	BR	ranksvm	MLKNN	RAkEL	FRAkEL	CBC
Emotions	.6540 ± .0139	.5108 ± .0321	.4856 ± .0302	.6556 ± .0234	.6708 ± .0219	**.6915 ± .0171**
Scene	.6946 ± .0132	.6234 ± .0167	.7112 ± .0128	.7231 ± .0241	.7203 ± .0143	**.7272 ± .0231**
Flags	**.7254 ± .0182**	.5133 ± .0546	.6541 ± .0201	.7033 ± .0216	.7061 ± .0208	.7125 ± .0129
bibtex	.6476 ± .0263	.5346 ± .0328	.5123 ± .0136	.6701 ± .0318	.6723 ± .0122	**.6857 ± .0231**

---

[2] http://cse.seu.edu.cn/PersonalPage/zhangml/files/ML-kNN.rar.
[3] https://github.com/KKimura360/fast_RAkEL_matlab.
[4] https://github.com/cjlin1/libsvm.

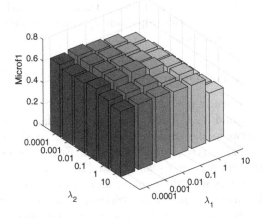

Fig. 1. Influence of CBC in MicroF1 with different selection of parameters on Emotions dataset

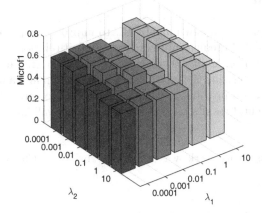

Fig. 2. Influence of CBC in MicroF1 with different selection of parameters on Flags dataset

of RAkEL, they always show the same performance on all datasets. In most cases, these two LP-based methods outperform BR, ranksvm and MLKNN, which shows that the ensemble of classifiers with different training data may boost the performance. FRAkEL has a little improvement in comparison with RAkEL because FRAkEL reduces the number of instances used in each base classifier. (2) It is interesting to notice that the baseline BR presents the second best average rank in both metrics on the whole, which suggests that BR is a simple but relatively effective methods. (3) We observe that CBC exhibits the best results on both four datasets except the second best result (which is just worse than BR) on Flags dataset as to MicroF1. On Emotions dataset, CBC has increased at least 2% in both measures than other methods. On Flags dataset,

CBC has increased at least 3% in MacroF1 than other methods. It suggests that utilizing the inherent relationships among binary classifiers can improve the performances of LP methods.

4.3 Parameter determination

In this subsection, we study how the performance of CBC will be affected by different values of λ_1 and λ_2. λ_1 and λ_2 are two regularization parameters to tune the tradeoff between the empirical loss and the regularization term. We conduct the parameter analytical experiments on Emotions dataset and Flags dataset. The parameters are set to vary from {1e−4, 1e−3, 1e−2, 1e−1, 1e0, 1e1}. The MicroF1 results with different λ_1 and λ_2 are shown in Figs. 1 and 2.

As seen from Figures, parameter determination takes influence on the performance of CBC. Different combinations of parameters may result in different models. Then, the results of classification change.

5   Conclusion

In this paper, we propose a new model CBC to improve the performances of LP methods for multi-label prediction. In order to exploit relationships between new classes which source from multiple labels, we combine the training of binary classifiers and the characterizing the relationship among them into a unified objective function. Moreover, we provide an efficient algorithm to solve the optimization problem. Experimental results on real-world data sets validate the effectiveness of CBC.

Acknowledgments. This work was supported by NSF China (No. 61473302, 61503396).

References

1. Ueda, N., Saito, K.: Parametric mixture models for multi-labeled text. In: International Conference on Neural Information Processing Systems, pp. 737–744 (2002)
2. Zhang, D., Islam, M.M., Lu, G.: A review on automatic image annotation techniques. Pattern Recogn. **45**(1), 346–362 (2012)
3. Qi, G.J., Hua, X.S., Rui, Y., Tang, J., Mei, T., Zhang, H.J.: Correlative multi-label video annotation. In: ACM International Conference on Multimedia, pp. 17–26 (2007)
4. Jiang, W., Cohen, A., Raś, Z.W.: Polyphonic music information retrieval based on multi-label cascade classification system. ProQuest LLC **17**(5–6), 452–470 (2009)
5. Zhang, M.L., Zhou, Z.H.: A review on multi-label learning algorithms. IEEE Trans. Knowl. Data Eng. **26**(8), 1819–1837 (2014)
6. Godbole, S., Sarawagi, S.: Discriminative methods for multi-labeled classification. In: Dai, H., Srikant, R., Zhang, C. (eds.) PAKDD 2004. LNCS (LNAI), vol. 3056, pp. 22–30. Springer, Heidelberg (2004). https://doi.org/10.1007/978-3-540-24775-3_5

7. Zhang, M.L., Zhou, Z.H.: ML-KNN: a lazy learning approach to multi-label learning. Pattern Recogn. **40**(7), 2038–2048 (2007)
8. Spyromitros, E., Tsoumakas, G., Vlahavas, I.: An empirical study of lazy multilabel classification algorithms. In: Darzentas, J., Vouros, G.A., Vosinakis, S., Arnellos, A. (eds.) SETN 2008. LNCS (LNAI), vol. 5138, pp. 401–406. Springer, Heidelberg (2008). https://doi.org/10.1007/978-3-540-87881-0_40
9. Boutell, M.R., Luo, J., Shen, X., Brown, C.M.: Learning multi-label scene classification. Pattern Recogn. **37**(9), 1757–1771 (2004)
10. Tsoumakas, G., Katakis, I., Taniar, D.: Multi-label classification: an overview. Int. J. Data Wareh. Min. **3**(3), 1–13 (2009)
11. Poladi, I., Ishwardas, H.: Review paper on error correcting output code based on multiclass classification. Int. J. Sci. Res. **2**(2), 134–136 (2012)
12. Hastie, T., Tibshirani, R.: Classification by pairwise coupling. In: Conference on Advances in Neural Information Processing Systems, pp. 507–513 (1998)
13. Dietterich, T.G., Bakiri, G.: Solving multiclass learning problems via errorcorrecting output codes. J. Artif. Intell. Res. **2**(1), 263–286 (1995)
14. Pujol, O., Radeva, P., Vitria, J.: Discriminant ECOC: a heuristic method for application dependent design of error correcting output codes. IEEE Trans. Pattern Anal. Mach. Intell. **28**(6), 1007 (2006)
15. Tsoumakas, G., Katakis, I., Vlahavas, I.: Random k-labelsets for multilabel classification. IEEE Trans. Knowl. Data Eng. **23**(7), 1079–1089 (2011)
16. Brinker, K.: Multilabel classification via calibrated label ranking. Mach. Learn. **73**(2), 133–153 (2008)
17. Singer, Y., Schapire, R.E.: BoosTexter: a boosting-based system for text categorization. In: Machine Learning, pp. 135–168 (2000)
18. Comité, F.D., Gilleron, R., Tommasi, M.: Learning multi-label alternating decision trees from texts and data. In: International Conference on Machine Learning and Data Mining in Pattern Recognition, pp. 35–49 (2003)
19. Elisseeff, A., Weston, J.: A kernel method for multi-labelled classification. In: International Conference on Neural Information Processing Systems: Natural and Synthetic, pp. 681–687 (2001)
20. Hariharan, B., Zelnik-Manor, L., Vishwanathan, S.V.N., Varma, M.: Large scale max-margin multi-label classification with priors. In: International Conference on Machine Learning, pp. 423–430 (2010)
21. Zhang, M.L., Zhou, Z.H.: Multilabel neural networks with applications to functional genomics and text categorization. IEEE Trans. Knowl. Data Eng. **18**(10), 1338–1351 (2006)
22. Zhang, M.L.: ML-RBF: RBF neural networks for multi-label learning. Neural Process. Lett. **29**, 61 (2009)
23. Veloso, A., Meira, W., Gonçalves, M., Zaki, M.: Multi-label lazy associative classification. Knowl. Discov. Databases PKDD **181**(13), 605–612 (2007)
24. Liu, X.Y., Li, Q.Q., Zhou, Z.H.: Learning imbalanced multi-class data with optimal dichotomy weights. In: IEEE International Conference on Data Mining, pp. 478–487 (2014)
25. Trohidis, K., Tsoumakas, G., Kalliris, G., Vlahavas, I.: Multilabel classification of music into emotions. Blood **90**(9), 3438–3443 (2008)
26. Plastino, A., Freitas, A.A.: A genetic algorithm for optimizing the label ordering in multi-label classifier chains. In: IEEE International Conference on TOOLS with Artificial Intelligence, pp. 469–476 (2013)
27. Katakis, G.T., Ioannis, V.I.: Multilabel text classification for automated tag suggestion. In: Proceedings of the ECML/PKDD Discovery Challenge, vol. 18 (2008)

Generalized Multiview Discriminative Projections with Spectral Reconstruction

Yun-Hao Yuan[1(✉)], Yun Li[1], Jipeng Qiang[1], Bin Li[1], Yuhui Zheng[2], Hongkun Ji[3], and Lixing Gou[4]

[1] School of Information Engineering, Yangzhou University, Yangzhou 225127, China
{yhyuan,liyun}@yzu.edu.cn
[2] School of Computer, Nanjing University of Information Science and Technology, Nanjing 210044, China
[3] Huatai Securities, Nanjing 210000, China
[4] School of Beijing New Oriental Foreign Language at Yangzhou, Yangzhou 225000, China

Abstract. In image recognition, there are a large number of small sample size problems in which the number of training samples is less than the dimension of feature vectors. For such problems, generalized multiview linear discriminant analysis (GMLDA) usually fails to achieve good learning performance for many classification tasks. With the idea of fractional order embedding, this paper proposes a new multiview feature learning method via fractional spectral modeling, namely, fractional-order generalized multiview discriminant analysis (FGMDA), which is able to subsume GMLDA as a special case. Experimental results on visual recognition have demonstrated the effectiveness of the proposed method and shown that FGMDA outperforms GMLDA.

Keywords: Image recognition · Multiview feature learning
Discriminant analysis · Feature representation

1 Introduction

In real world, multiple view data [1,2] with multiple representations from different high-dimensional feature spaces have widely arisen in numerous applications such as image recognition, information retrieval, and multimedia analysis. Since multi-view data usually have very different statistical properties and complementary nature, analyzing such data called multi-view data learning is a very meaningful direction, which has attracted more and more attention.

Supported by the National Natural Science Foundation of China under Grant Nos. 61402203, 61472344, 61611540347, and 61703362, Natural Science Fund of Jiangsu under Grant Nos. BK20161338 and BK20170513, and Yangzhou Science Fund under Grant Nos. YZ2017292 and YZ2016238. Moreover, it is also sponsored by the Excellent Young Backbone Teacher (Qing Lan) Fund and Scientific Innovation Research Fund of Yangzhou University under Grant No. 2017CXJ033.

© Springer Nature Switzerland AG 2018
Y. Peng et al. (Eds.): IScIDE 2018, LNCS 11266, pp. 454–462, 2018.
https://doi.org/10.1007/978-3-030-02698-1_40

With different feature extractors in image recognition, one image can be extracted multiple kinds of high-dimensional features which are taken as multiple views of the image. In high-dimensional space of each view, it is difficult to discriminate images from different classes, thus resulting in the curse of dimensionality problem [3]. A conventional method for multi-view feature learning is to simply concatenate feature vectors of different views as a new vector and then apply single view-based feature learning algorithms directly on the concatenated vectors. Despite of the simplicity of this concatenation, it may still not perform well for feature learning and recognition. This is because simple concatenation neglects the diversity and specific statistical properties of different representations.

To solve the aforementioned issue, a number of dedicated methods, see for example [4–11], have been developed to learn discriminative low-dimensional features for multi-view high-dimensional data. Thereinto, an attractive work is the Generalized Multiview Analysis (GMA) [10], which is a generic framework for multi-view feature learning and unifies many popular feature extraction approaches such as Canonical Correlation Analysis (CCA) [12], Bilinear Model (BLM) [13], and Partial Least Squares (PLS) [14] as specific instances. With the GMA framework, Sharma et al. [10] further proposed a specific multi-view feature extraction technique called Generalized Multiview Linear Discriminant Analysis (GMLDA), which has been proven to be more powerful than existing feature extraction techniques (such as CCA, BLM, and PLS) for face recognition [15] and text-image retrieval.

For image recognition tasks, it is well-known that there are many small sample size problems where the number of training samples is less than the dimension of feature vectors. In general, the limited samples are not able to accurately depict the geometric distribution of the data. In this situation, it is hard for the GMLDA method to obtain good performance for classification tasks. A recent technique called fractional order embedding [16] has been applied to canonical correlation-based feature learning algorithms. It has been shown that the learning performance can be significantly enhanced if the fractional-order idea is considered in small-sample-size cases.

Motivated by the idea of fractional order embedding, we in this paper improve the GMLDA method via fractional spectral modeling in small sample size problems, and thus present a new multi-view feature learning method called Fractional Generalized Multiview Discriminant Analysis or FGMDA. It can subsume the GMLDA method as a special case, which suggests that FGMDA is a general multi-view feature reduction approach. Experimental results on two benchmark databases have shown that our proposed FGMDA is more discriminative than the GMLDA method.

2   GMLDA

As our method is closely related to the GMLDA method, we in this section provide its detailed description. Assume that $\mathcal{X}^{(i)} = \{x_{lk}^{(i)} \in \mathcal{R}^{d_i} | l = 1, 2, \ldots, c, k = $

$1, 2, \ldots, n_l^{(i)}\}$ is the sample set from the i-th view of the same objects, $i = 1, 2, \ldots, m$, where $x_{lk}^{(i)}$ is the k-th sample of the l-th class in i-th view, d_i is the dimensionality of sample vectors, c is the number of classes, $n_l^{(i)}$ is the number of samples of the l-th class in i-th view, and m is the number of views. Let $X^{(i)} = [x_1^{(i)}, x_2^{(i)}, \ldots, x_{n^{(i)}}^{(i)}] \in \mathcal{R}^{d_i \times n_i}$ be the observation matrix in view i with $n^{(i)} = \sum_{l=1}^{c} n_l^{(i)}$. The goal of GMLDA is to find a set of projection directions $\{v^{(i)} \in \mathcal{R}^{d_i}\}_{i=1}^{m}$ that try to separate different subjects' class means and unite different views of the same class in the common subspace. More specifically, the projection directions $\{v^{(i)}\}_{i=1}^{m}$ can be obtained by the following optimization problem:

$$
\begin{aligned}
\max &\sum_{i=1}^{m} \mu_i v^{(i)T} X^{(i)} W^{(i)} X^{(i)T} v^{(i)} + \sum_{i=1}^{m} \sum_{j=1, j \neq i}^{m} \lambda_{ij} v^{(i)T} M^{(i)} M^{(j)T} v^{(j)} \\
s.t. &\sum_{i=1}^{m} \gamma_i v^{(i)T} X^{(i)} D^{(i)} X^{(i)T} v^{(i)} = 1,
\end{aligned}
\tag{1}
$$

where $\{\mu_i \in \mathcal{R}\}_{i=1}^{m}$, $\{\gamma_i \in \mathcal{R}\}_{i=1}^{m}$, and $\{\lambda_{ij} \in \mathcal{R}\}_{i,j=1, j \neq i}^{m}$ are separately model balance parameters with $\mu_1 = 1$ and $\gamma_1 = 1$, $M^{(i)} \in \mathcal{R}^{d_i \times c}$ is defined as the matrix with columns that are class means in view i, $W^{(i)} \in \mathcal{R}^{n^{(i)} \times n^{(i)}}$ is a weight matrix with (k, t)-th entry as

$$
W_{kt}^{(i)} = \begin{cases} 1/n_l^{(i)}, & \text{if } x_k^{(i)} \text{ and } x_t^{(i)} \text{ belong to class } l, \\ 0, & \text{otherwise}, \end{cases}
$$

and $D^{(i)} = I_i - W^{(i)}$ with I_i being the identity matrix in the i-th view. The optimization problem (1) can be solved by a generalized eigenvalue problem [10].

3   Proposed Approach

In this section, two fractional-order matrices are built via sample spectrum reconstruction. After that, a generalized multi-view feature learning approach that we call Fractional Generalized Multiview Discriminant Analysis (FGMDA) is proposed for recognition purpose.

3.1   An Equivalent Form of GMLDA

Before describing our FGMDA method, let us first give a compact form of GMLDA. Let us denote

$$
A = \begin{bmatrix} A_{11} & \lambda_{12} A_{12} & \cdots & \lambda_{1m} A_{1m} \\ \lambda_{21} A_{21} & \mu_2 A_{22} & \cdots & \lambda_{2m} A_{2m} \\ \vdots & \vdots & \ddots & \vdots \\ \lambda_{m1} A_{m1} & \lambda_{m2} A_{m2} & \cdots & \mu_m A_{mm} \end{bmatrix} \in \mathcal{R}^{d \times d}
\tag{2}
$$

and

$$B = \begin{bmatrix} B_{11} & & & \\ & \gamma_2 B_{22} & & \\ & & \ddots & \\ & & & \gamma_m B_{mm} \end{bmatrix} \in \mathcal{R}^{d \times d}, \tag{3}$$

where $A_{ij} = \begin{cases} X^{(i)} W^{(i)} X^{(i)T}, & i = j \\ M^{(i)} M^{(j)T}, & i \neq j \end{cases}$, $B_{ii} = X^{(i)} D^{(i)} X^{(i)T}$, and $d = \sum_{i=1}^{m} d_i$.

Using (2) and (3), we can reformulate the optimization problem (1) as the following concise form:

$$\begin{aligned} v^* &= \arg\max_v v^T A v \\ \text{s.t. } & v^T B v = 1, \end{aligned} \tag{4}$$

where $v^T = [v^{(1)T}, v^{(2)T}, \dots, v^{(m)T}] \in \mathcal{R}^d$. It is obvious that this maximization problem (4) is a quadratically constrained quadratic program (QCQP), which can be solved efficiently by the following:

$$Av = \lambda B v, \tag{5}$$

where λ is the eigenvalue corresponding to the eigenvector v. In GMLDA, we can get the following important theorem:

Theorem 1. *The rank of A is at most $m(c - 1)$.*

Proof. Let $X^{(i)} = [X_1^{(i)}, X_2^{(i)}, \dots, X_c^{(i)}]$ with $X_l^{(i)}$ as the sample matrix from l-th class of i-th view. Then, we have

$$\begin{aligned} A_{ii} &= X^{(i)} W^{(i)} X^{(i)T} \\ &= \sum_{l=1}^{c} \frac{1}{n_l^{(i)}} X_l^{(i)} \mathbf{1}_{n_l^{(i)}} \mathbf{1}_{n_l^{(i)}}^T X_l^{(i)T} \\ &= \sum_{l=1}^{c} n_l^{(i)} \left(\frac{1}{n_l^{(i)}} \sum_{k=1}^{n_l^{(i)}} x_{lk}^{(i)} \right) \left(\frac{1}{n_l^{(i)}} \sum_{k=1}^{n_l^{(i)}} x_{lk}^{(i)} \right)^T \\ &= \sum_{l=1}^{c} n_l^{(i)} m_l^{(i)} m_l^{(i)T} \\ &= M^{(i)} G^{(i)} M^{(i)T}, \end{aligned} \tag{6}$$

where $\mathbf{1}_{n_l^{(i)}} \in \mathcal{R}^{n_l^{(i)}}$ is the vector of all ones, $G^{(i)} \in \mathcal{R}^{c \times c}$ is a diagonal matrix with l-th diagonal entry as $n_l^{(i)}$, and $m_l^{(i)} = \left(1 / n_l^{(i)} \right) \sum_{k=1}^{n_l^{(i)}} x_{lk}^{(i)}$.

Let $M = diag(M^{(1)}, M^{(2)}, \dots, M^{(m)})$ and $K = [K_{ij}]_{mc \times mc}$ with K_{ij} as $\lambda_{ij} I_c$ if $i \neq j$ and $\mu_i G^{(i)}$ otherwise, where $I_c \in \mathcal{R}^{c \times c}$ is the identity matrix. Then, we get

$$A = MKM^T. \tag{7}$$

We thus have $rank(A) = rank(MKM^T) \leq \min(rank(M), rank(K))$. Notice that each $X^{(i)}$ has been centered, i.e., $\sum_{j=1}^{n^{(i)}} x_j^{(i)} = 0$, and $d_i \gg c$ in practice. Hence

$$rank(M^{(i)}) \leq c - 1.$$

It follows that

$$rank(M) \leq \sum_{i=1}^{m} rank(M^{(i)}) \leq m(c - 1).$$

Together with $rank(K) \leq mc$, we have

$$rank(A) \leq m(c - 1).$$

Theorem 1 indicates that GMLDA can extract at most $m(c - 1)$ dimensions in each view for recognition tasks. To the best of our knowledge, this conclusion is new.

3.2   Fractional Matrix Construction

It is clear both A and B in (2) and (3) are associated with multi-view high-dimensional training samples. According to [16], matrices A and B will deviate from real ones when training samples are limited or disturbed by noise. The fractional-order strategy [16] has been demonstrated to be effective for alleviating the problem of the sample matrix deviation. Consequently, we employ the fractional-order technique to correct sample matrices A and B.

Concretely, let the thin singular value decomposition (SVD) of matrix A be the following:

$$A = U\Lambda V^T, \Lambda = diag(\sigma_1, \sigma_2, \ldots, \sigma_r), \tag{8}$$

where U and V are left and right singular vector matrices of matrix A, respectively, thus satisfying $U^T U = V^T V = I_r$, and $\sigma_1 \geq \sigma_2 \geq \ldots \geq \sigma_r > 0$ are singular values of A, r is the rank of A, i.e., $r = rank(A)$. Similarly, let the thin SVD of matrix B be as follows:

$$B = \tilde{U} \tilde{\Lambda} \tilde{V}^T, \tilde{\Lambda} = diag(\tilde{\sigma}_1, \tilde{\sigma}_2, \ldots, \tilde{\sigma}_{\tilde{r}}), \tag{9}$$

where $\tilde{U}$ and $\tilde{V}$ are left and right singular vector matrices of B with $\tilde{U}^T \tilde{U} = \tilde{V}^T \tilde{V} = I_{\tilde{r}}$, and $\tilde{\sigma}_1 \geq \tilde{\sigma}_2 \geq \ldots \geq \tilde{\sigma}_{\tilde{r}} > 0$ are the sorted singular values of matrix B, and $\tilde{r} = rank(B)$.

In terms of (8) and the fractional-order idea in [16], fractional order A can now be defined as

$$A^\alpha = U\Lambda^\alpha V^T, \Lambda^\alpha = diag(\sigma_1^\alpha, \sigma_2^\alpha, \ldots, \sigma_r^\alpha), \tag{10}$$

where $\alpha \in \mathcal{R}$ is a fraction with $0 \leq \alpha \leq 1$, U, V, Λ, and r are defined in (8). Likewise, we are able to define the fractional order B using (9) as

$$B^\kappa = \tilde{U} \tilde{\Lambda}^\kappa \tilde{V}^T, \tilde{\Lambda}^\kappa = diag(\tilde{\sigma}_1^\kappa, \tilde{\sigma}_2^\kappa, \ldots, \tilde{\sigma}_{\tilde{r}}^\kappa), \tag{11}$$

where $\kappa \in \mathcal{R}$ is a fraction satisfying $0 \leq \kappa \leq 1$, $\tilde{U}$, $\tilde{V}$, $\tilde{\Lambda}$, and $\tilde{r}$ are defined in (9).

Apparently, the SVD in (8) or (9) can be viewed as a special case of fractional-order decomposition in (10) or (11). This implies that fractional-order decomposition is a more general way, in contrast with conventional SVD.

3.3   Formulation with Fractional Order

Replacing A and B with fractional-order matrices A^α and B^κ in (4) leads to the optimization formulation of our FGMDA below:

$$
\begin{aligned}
v^* &= \arg\max_v v^T A^\alpha v \\
s.t. \; & v^T B^\kappa v = 1,
\end{aligned}
\tag{12}
$$

which can be solved efficiently by the following generalized eigenvalue problem:

$$
A^\alpha v = \eta B^\kappa v. \tag{13}
$$

Note that when the factional-order matrix B^κ is singular in (13), the classical algorithms can not be directly used any more to solve the generalized eigenvalue problem. In this situation, we use the following technique to regularize the matrix B^κ:

$$
B^\kappa \leftarrow B^\kappa + \tau I, \tag{14}
$$

where I is the identity matrix and $\tau > 0$ is a regularization parameter and chosen as 0.001 in this paper.

3.4   Complexity Analysis

For given fractional-order parameters α and κ, if we use the big O notation to express the time complexity of one algorithm, our proposed FGMDA and GMLDA have the same computational complexity, i.e., $O(d^3)$. But, this is not precise enough to differentiate between the complexities of FGMDA and GMLDA. Actually, FGMDA needs more arithmetic operations than GMLDA because of extra matrix decompositions and multiplications for fractional matrix construction. Concretely, in contrast with GMLDA, our method needs to decompose extra two $d \times d$ matrices A and B according to (8) and (9). This scales as $O(d^2(r + \tilde{r}))$. Its subsequent part requires the computation of matrix vector product in (10) and (11), the time cost of which is $O((1 + d + d^2)(r + \tilde{r}))$. As a result, our proposed method consumes extra time in practical implementation scaled as $O((1 + d + 2d^2)(r + \tilde{r}))$.

4   Experiments

In this section, we carry out the recognition experiments using two benchmark classification datasets to validate the performance of our proposed FGMDA and

Table 1. Recognition accuracy (%) of GMLDA and our FGMDA with the cosine nearest neighbor classifier on multiple feature dataset.

Method	Pix-Fac-Fou	Pix-Fac-Kar	Pix-Fac-Zer	Pix-Fac-Mor	Pix-Fou-Kar
GMLDA	86.16	82.63	**84.00**	85.79	**84.11**
FGMDA	**87.05**	**82.68**	**84.00**	**87.37**	**84.11**
Method	Pix-Fou-Zer	Pix-Fou-Mor	Pix-Kar-Zer	Pix-Kar-Mor	Pix-Zer-Mor
GMLDA	87.16	91.05	85.05	84.00	85.58
FGMDA	**87.79**	**93.00**	**85.37**	**87.26**	**89.42**
Method	Fac-Fou-Kar	Fac-Fou-Zer	Fac-Fou-Mor	Fac-Kar-Zer	Fac-Kar-Mor
GMLDA	78.42	69.58	**84.63**	79.32	83.90
FGMDA	**91.63**	**91.79**	**84.63**	**88.37**	**86.47**
Method	Fac-Zer-Mor	Fou-Kar-Zer	Fou-Kar-Mor	Fou-Zer-Mor	Kar-Zer-Mor
GMLDA	72.32	83.53	86.47	77.32	83.84
FGMDA	**81.26**	**91.63**	**89.42**	**80.16**	**87.00**

compare it with the state-of-the-art algorithm GMLDA. In our FGMDA, there are two fractional-order parameters α and κ which play an important role for the recognition performance. In all the experiments, we empirically select them from $\{0.1, 0.2, \ldots, 1\}$. The parameters included in GMLDA are set to the same as those used in [10]. The classification is performed by the nearest neighbor (NN) classifier with cosine distance metric.

4.1   Multiple Feature Dataset

The Multiple Feature Dataset (MFD)[1] has been utilized extensively to evaluate the performance of multiple view learning approaches. This dataset consists of six-set features of handwritten numbers (i.e., '0'–'9') extracted from a collection of Dutch utility maps. Each digit (class) contains 200 samples and the total number of samples is 2000. The six sets of features are, respectively, Fourier coefficients (Fou, 76), profile correlations (Fac, 216), Karhunen-Loève coefficients (Kar, 64), pixel averages (Pix, 240), Zernike moments (Zer, 47), and morphological features (Mor, 6). Note that, in each bracket, the former is feature abbreviation and the latter denotes feature dimension.

On this dataset, we arbitrarily select three different feature sets to form three views. Thus, there are totally 20 feature combinations. For each combination, the first 10 samples per class are chosen to form the training set, while the remaining 190 samples are considered as the testing set. Thus, the number of training and testing samples is 100 and 1900, respectively. Table 1 summarizes the recognition accuracies of GMLDA and our FGMDA with different feature combinations under the cosine NN classifier.

---

[1] http://archive.ics.uci.edu/ml/datasets/Multiple+Features.

Table 2. Average recognition accuracy (%) of GMLDA and FGMDA across 10-run tests on the AT&T database and the corresponding standard deviations.

Method	GMLDA	FGMDA
Accuracy	94.2	**97.7**
Standard deviation	2.0	**1.0**

As can be seen, our FGMDA method is superior to the GMLDA method on most cases. FGMDA achieves better results than GMLDA on seventeen cases and same results as GMLDA on the remaining three cases. This implies that our proposed FGMDA can yield more discriminative low-dimensional feature representations than GMLDA for multi-view handwritten digit recognition tasks.

4.2   AT&T Face Database

The AT&T database[2] consists of 400 face images from 40 distinct individuals. Each individual has ten different grayscale images and the size of each image is 92×112 pixels. For some individuals, the images are taken at different times, varying the lighting, facial expressions, and facial details. In addition, the images are taken with a tolerance for some tilting and rotation of the face up to $20°$, and have some variation in the scale up to about 10%.

To evaluate the proposed FGMDA in comparison with GMLDA, we first carry out the Coiflets, Daubechies, and Symlets orthonormal wavelet transforms to get three-set low-frequency subimages for forming three views, as used in [17]. The K–L transform is then used to respectively reduce their dimensions to 150. The K–L-transformed three feature sets are adopted for the performance evaluation.

In this experiment, five images per class are randomly chosen for training, while the rest are used for testing. Thus, the number of training samples and testing samples is totally 200 and 200, respectively. Ten independent classification tests are implemented to verify the recognition performance of GMLDA and our FGMDA. Table 2 reports the average recognition results and corresponding standard deviations across 10-run tests under cosine NN classifier. As seen, FGMDA achieves again better recognition rate than state-of-the-art GMLDA. This reveals that the extracted features by FGMDA are powerful for recognition purpose.

5   Conclusion

This paper proposes a general multi-view discriminant analysis approach called FGMDA for multi-view feature learning and recognition purpose. The central idea of FGMDA is to use the fractional-order idea to reconstruct the spectra of sample matrices. The proposed FGMDA can subsume GMLDA as a special case. That is, when fractional-order parameters α and κ are one, FGMDA reduces to

---

[2] http://www.cl.cam.ac.uk/research/dtg/attarchive/facedatabase.html.

GMLDA. Based on benchmark datasets, two experiments have demonstrated the effectiveness of our proposed FGMDA method. A future interesting direction is how to theoretically choose the optimal fractional-order parameters for recognition tasks in practice.

References

1. Long, B., Yu, P.S., Zhang, Z.: A general model for multiple view unsupervised learning. In: Proceedings of SIAM International Conference on Data Mining, pp. 822–833. SIAM Press, Philadelphia (2008)
2. Chen, N., Zhu, J., Sun, F., Xing, E.P.: Large-margin predictive latent subspace learning for multi-view data analysis. IEEE TPAMI **34**(12), 2365–78 (2012)
3. Korn, F., Pagel, B., Faloutsos, C.: On the dimensionality curse and the self-similarity blessing. IEEE TKDE **13**(1), 96–111 (2001)
4. Xia, T., Tao, D., Mei, T., Zhang, Y.: Multiview spectral embedding. IEEE Trans. Syst. Man Cybern.–Part B: Cybern. **40**(6), 1438–1446 (2010)
5. Han, Y., Wu, F., Tao, D., Shao, J., Zhuang, Y., Jiang, J.: Sparse unsupervised dimensionality reduction for multiple view data. IEEE TCSVT **22**(10), 1485–1496 (2012)
6. Hou, C., Zhang, C., Wu, Y., Nie, F.: Multiple view semi-supervised dimensionality reduction. Pattern Recogn. **43**, 720–730 (2010)
7. Yuan, Y.-H., Sun, Q.-S., Zhou, Q., Xia, D.-S.: A novel multiset integrated canonical correlation analysis framework and its application in feature fusion. Pattern Recogn. **44**(5), 1031–1040 (2011)
8. Kan, M., Shan, S., Zhang, H., Lao, S., Chen, X.: Multi-view discriminant analysis. In: Fitzgibbon, A., Lazebnik, S., Perona, P., Sato, Y., Schmid, C. (eds.) ECCV 2012. LNCS, vol. 7572, pp. 808–821. Springer, Heidelberg (2012). https://doi.org/10.1007/978-3-642-33718-5_58
9. Yu, J., Wang, M., Tao, D.: Semisupervised multiview distance metric learning for cartoon synthesis. IEEE TIP **21**(11), 4636–4648 (2012)
10. Sharma, A., Kumar, A., Daume III, H., Jacobs, D.W.: Generalized multiview analysis: a discriminative latent space. In: Proceedings of CVPR, pp. 2160–2167. IEEE, New York (2012)
11. Gao, H., Nie, F., Li, X., Huang, H.: Multi-view subspace clustering. In: Proceedings of ICCV, pp. 4238–4246. IEEE, New York (2015)
12. Hardoon, D., Szedmak, S., Shawe-Taylor, J.: Canonical correlation analysis: an overview with application to learning methods. Neural Comput. **16**(12), 2639–2664 (2004)
13. Tenenbaum, J.B., Freeman, W.T.: Separating style and content with bilinear models. Neural Comput. **12**(6), 1247–1283 (2000)
14. Sharma, A., Jacobs, D.W.: Bypassing synthesis: PLS for face recognition with pose, low-resolution and sketch. In: Proceedings of CVPR, pp. 593–600. IEEE, New York (2011)
15. Zhu, W., Lu, J., Zhou, J.: Nonlinear subspace clustering for image clustering. Pattern Recogn. Lett. **107**(1), 131–136 (2018)
16. Yuan, Y.-H., Sun, Q.-S., Ge, H.-W.: Fractional-order embedding canonical correlation analysis and its applications to multi-view dimensionality reduction and recognition. Pattern Recogn. **47**, 1411–1424 (2014)
17. Yuan, Y.-H., Sun, Q.-S.: Multiset canonical correlations using globality preserving projections with applications to feature extraction and recognition. IEEE TNNLS **25**(6), 1131–1146 (2014)

Vector Similarity Measures About Hesitant Fuzzy Sets and Applications to Clustering Analysis

Min Wu[1(✉)], Hongwei Xu[2], Qiang Zhou[2], and Yubao Sun[2]

[1] Jinling Hospital, Medical School of Nanjing University, Nanjing 210002, China
njzywm@163.com
[2] Collaborative Innovation Center on Atmospheric Environment and Equipment Technology, B-DAT, Nanjing University of Information Science and Technology, Nanjing 210044, China
sunyb@nuist.edu.cn

Abstract. In this paper, some existing distance-induced similarity measures between hesitant fuzzy sets (HFSs) are firstly verified to have the drawback of low discrimination ability, and thus, we propose some new vector similarity measures between HFSs, which are proved to have stronger discrimination ability than some existing ones in pattern recognition problems by numerical examples validation. Then, a maximum spanning tree (MST) clustering method for HFSs is proposed, based on the vector similarity measures and the fuzzy graph theory. At last, the effectiveness of the proposed method is illustrated by numerical examples.

Keywords: Clustering analysis · Hesitant fuzzy sets
Maximum spanning tree (MST) · Similarity measure

1 Introduction

Uncertainty in real life is usually an unavoidable problem and fuzzy set proposed by Zadeh is a powerful tool to solve the uncertain problems [1,2]. A similarity measure is an important tool for determining the degree of similarity between two objects [3]. Similarity measures between fuzzy sets, as an important content in fuzzy mathematics, have gained attention from researchers for their wide applications in various fields, such as decision making, pattern recognition, clustering [4–7].

This work was supported in part by the projects of the former Nanjing Military Region under Grant Number 15MS129, in part by 333 high level talent training project of Jiangsu Province, China under Grant Number [BRA2015527,BRA2017594], in part by the Natural Science Foundation of China under Grant Number 61672292, in part by the Six Talent Peaks Project of Jiangsu Province, China under Grant Number DZXX-037.

© Springer Nature Switzerland AG 2018
Y. Peng et al. (Eds.): IScIDE 2018, LNCS 11266, pp. 463–475, 2018.
https://doi.org/10.1007/978-3-030-02698-1_41

Nevertheless, the modeling tools of the ordinary fuzzy sets are not capable of handling vague and imprecise information, where two or more sources of vagueness appear simultaneously. As a generalization of the fuzzy sets, HFS is a very useful tool to deal with uncertainty [8,9]. Moreover, when giving the membership degree of an element, it is not difficult to establish the membership degree because we have a margin of error, or some possibility distribution on the possibility values. However, in order cope with the issue of multiple degree values, the HFS method describes each degree value as a hesitant fuzzy in terms of the opinions of decision makers [10,11].

Recently, some effort has been devoted to the study of the similarity measures between HFSs. In [12], some distance measures between HFSs were proposed. The relation between distance measures and similarity measures about HFSs was discussed, based on which, the similarity measures between HFSs can be obtained by transforming the corresponding distance measures. In [4], Farhadinia studied the relationships among the entropy, the similarity measure and the distance measure for HFSs, and the systematic transformation of the distance into the similarity measure for HFSs was discussed. Similarity measures between HFSs defined in [4] and [12] were obtained by transforming the corresponding distance measures. With counterexamples, it is shown that the above mentioned similarity measures may not be suitable for discriminating some HFSs, even though they are apparently different.

Notice that, in vector space, especially the Jaccard, Dice, and Cosine similarity measures [13–15] are often used in information retrieval, citation analysis, and automatic classification, which do not mainly depend on the distance measures. However, these similarity measures are not capable of dealing with the similarity measures for hesitant information. Thus, in this paper, at first, some vector similarity measures between HFSs will be introduced. A numerical example will show that the new proposed vector similarity measures between HFSs have stronger discrimination ability than the existing ones.

As one of the widely-adopted key tools in handling data information, clustering analysis has been applied to solve the real world problems concerning social, medical, biological, climatic, financial systems [16–20]. While, there are only few papers related to cluster hesitant fuzzy information. In [12], Xia proposed a clustering technique based on the correlation coefficient between HFSs, by transforming a correlation matrix into an equivalent correlation matrix, and a *Lambda*-cutting matrix is used to cluster the given HFSs. The same idea was adopted by Farhadinia [4] to give a clustering technique based on the distance-induced similarity measures between HFSs.

However, the methods given in [4,12] took too much computational effort in the process of calculation. The graph theory-based clustering algorithm is an active research area [17,21,22]. Zahn proposed the clustering algorithm using the minimum spanning tree of the graph [23]. Chen et al. introduced the concept of maximum spanning tree of the fuzzy graph by constructing the fuzzy similarity relation matrix and used the threshold of fuzzy similarity relation matrix to cut maximum spanning tree, and then obtained the classification on level [24].

Zhao et al. investigated graph theory-based clustering techniques for intuitionistic fuzzy sets and interval-valued intuitionistic fuzzy sets [25]. With the help of the fuzzy graph, we will apply the new proposed vector similarity measures to give a new clustering algorithm for hesitant fuzzy information, and the algorithm is proved to outperform the existing one.

The rest of the paper is organized as follows. In Sect. 2, we review some basic notions for HFSs, which will be used in the analysis throughout this paper. Section 3 is devoted to the main results concerning vector similarity measures between HFSs. In Sect. 4, a fuzzy graph based clustering algorithm under hesitant fuzzy environment are developed in which the vector similarity measures of HFSs are the applied indices in data analysis and classification. Moreover, a practical example is provided to compare the proposed method with the existing one. This paper is concluded in Sect. 5.

2   Preliminary

In this section, we will review some basic notions related to our following discussions.

Throughout the paper, let $U = \{u_1, u_2, \cdots, u_n\}$ be a finite universe.

Definition 1 [8]. *Let U be a universe, a hesitant fuzzy set (HFS) A on U is in terms of a function that when applied to U returns a subset of $[0, 1]$. To be easily understood, we express the HFS by a mathematical symbol:*

$$A = \{<u, h_A(u)> \mid u \in U\}$$

where $h_A(u)$ is a set of some values in $[0, 1]$, denoting the possible membership degrees of the element $u \in U$ to the set A. For convenience, for any $u \in U$, we call $h_A(u)$ a hesitant fuzzy element (HFE).

For any HFE $h_A(u)$, we denote $l(h_A(u))$ as the number of elements in l $h_A(u)$.

Definition 2 [26]. *Let U be a universe, an intuitionistic fuzzy set A on U is defined as:*

$$H = \{<u, \mu_A(u), \nu_A(u)> \mid u \in U\}$$

where the functions $\mu_A(u)$ and $\nu_A(u)$ denote the degrees membership and non-membership of the element $u \in U$ to the set H, respectively, satisfying $0 \leq \mu_A(u) + \nu_A(u) \leq 1$ and $\mu_A(u), \nu_A(u) \in [0, 1]$. And $\alpha = (\mu_\alpha, \nu_\alpha)$ is named as an intuitionistic fuzzy value.

Torra [8] also introduced the concept of the envelop of a HFE as:

Given a HFE $h_A(u)$, its envelope is defined as $A_{env}(h) = (h_A^-(u), 1 - h_A^+(u))$, where, $h_A^-(u) = \min h_A(u)$ is the lower bound of $h_A(u)$ and $h_A^+(u) = \max h_A(u)$ is the upper bound of $h(u)$, respectively.

In [27], Chen et al. pointed out that envelope of a HFE is just an intuitionistic fuzzy value.

Xu and Xia gave the axioms for similarity measures and distance measures between HFSs as follows [12]:

Definition 3 [12]. *Let M and N be two HFSs on $U = \{u_1, u_2, \cdots, u_n\}$, the distance between M and N is defined as $d(M, N)$, which satisfies the following properties:*

(1) $0 \leq d(M, N) \leq 1$;
(2) $d(M, N) = 0$ *if and only if* $M = N$;
(3) $d(M, N) = d(N, M)$.

Definition 4 [12]. *Let M and N be two HFSs on $U = \{u_1, u_2, \cdots, u_n\}$, the similarity between M and N is defined as $s(M, N)$, which satisfies the following properties:*

(1) $0 \leq s(M, N) \leq 1$;
(2) $s(M, N) = 1$ *if and only if* $M = N$;
(3) $s(M, N) = s(N, M)$.

Remark 1. *It is noted that, in most cases, $l(h_M(u_i)) \neq l(h_N(u_i))$, and let $l_{u_i} = \max\{l(h_M(u_i)), l(h_N(u_i))\}$ for any $u_i \in U$ for convenience. When discuss the distance(similarity) measures about two HFSs, to operate correctly, the shorter one should be extended to the same length as the longer one [12].*

Hereafter, the following assumptions are made:

(1) *All the elements in each $h_M(u_i)$ are arranged in increasing order, and then $h_M^{\sigma(j)}(u_i)$ is referred to as the jth smallest value in $h_M(u_i)$.*
(2) *If there are fewer elements in $h_M(u_i)$ than in $h_N(u_i)$, an extension of $h_M(u_i)$ should be considered optimistically by repeating its maximum element until it has the same length with $h_N(u_i)$.*

In [12], Xu and Xia pointed out that $s(M, N) = 1 - d(M, N)$, accordingly. Thus, they only defined the following distance measures between HFSs M, N and the corresponding similarity measures can be obtained easily.

The normalized Minkowski distance $d_{nm}(M, N)$ and the normalized Hausdorff distance d_{nhd} between HFSs M and N are given as follows:

(1) $\quad d_{nm}(M, N) = \left[\frac{1}{n} \sum_{i=1}^{n} \left(\frac{1}{l_{u_i}} \sum_{j=1}^{l_{u_i}} |h_M^{\sigma(j)}(u_i) - h_N^{\sigma(j)}(u_i)|^\lambda \right) \right]^{1/\lambda}$;

(2) $\quad d_{nhd}(M, N) = \left[\frac{1}{n} \sum_{i=1}^{n} \left(\max_j |h_M^{\sigma(j)}(u_i) - h_N^{\sigma(j)}(u_i)|^\lambda \right) \right]^{1/\lambda}$,

where, $\lambda > 0$, $l_{u_i} = \max\{l(h_M(u_i)), l(h_N(u_i))\}$. $h_M^{\sigma(j)}(u_i)$ and $h_N^{\sigma(j)}(u_i)$ are the jth smallest values of $h_M(u_i)$ and $h_N(u_i)$, respectively.

Furthermore, in [4], Farhadinia investigated the generalized theorem on how distance measures between HFSs can be transformed to similarity measures as follows:

Theorem 1 [4]. *Let $Z : [0, 1] \rightarrow [0, 1]$ be a strictly monotone decreasing real function, and d be a distance between HFSs. Then, for any HFSs M and N on U*

$$s_d(M, N) = \frac{Z(d(M, N)) - Z(1)}{Z(0) - Z(1)}$$

is a similarity measure of HFSs based on the corresponding distance d.

By Theorem 1, different formulas can be developed to calculate the similarity measure between HFSs using different strictly monotone decreasing functions $Z : [0, 1] \rightarrow [0, 1]$ based on the corresponding distance measures, for instance,

(1) $Z(t) = 1 - t$; (2) $Z(t) = \frac{1-t}{1+t}$; (3) $Z(t) = 1 - te^{t-1}$; (4) $Z(t) = 1 - t^2$.

We uniformly call the similarity measures between HFSs obtained by Theorem 1 the distance-induced similarity measures between HFSs, and denoted the distance-induced similarity measures between HFSs M, N as $s_d(M, N)$.

3 Vector Similarity Measures Between the HFSs

In this section, the drawback of applying the distance-induced similarity measures between HFSs for pattern recognition problems will be pointed out. Then, some vector similarity measures between HFSs with stronger discrimination ability will be introduced.

Firstly, we analyse the principle of classifying HFSs based on the similarity measures.

Pattern Recognition Principle to HFSs Based on the Similarity Measures

Assume that there exist m patterns, which are represented by HFSs $M_k(k = 1, 2, \cdots, m)$ in the universe $U = \{u_1, u_2, \cdots, u_n\}$, and there is a sample to be recognized which is represented by a HFS N in U. Set

$$s(M_{k_0}, N) = \max_{1 \leq k \leq m} \{s(M_k, N)\},$$

where $s(M_k, N)$ expressed the similarity measure between M_k and N. Then we decide that the sample N should belong to the pattern M_{k_0}.

Example 1. *Consider two patterns represented by HFSs in $U = \{u_1, u_2, u_3\}$:*
$M_1 = \{\{0.1, 0.2\}, \{0.8, 0.9\}, \{0.5, 0.8\}\}$, $M_2 = \{\{0.1, 0.6\}, \{0.5, 0.9\}, \{0.5, 0.7\}\}$, and a sample $N = \{\{0.2, 0.4\}, \{0.7, 0.8\}, \{0.4, 0.6\}\}$ will be recognized.

Decide which pattern N should belong to, using the pattern recognition principle to HFSs based on the distance-induced similarity measures.

According to Theorem 1, choose the strictly monotone decreasing function $Z(t) = 1 - t$ to get the corresponding distance-induced similarity measures about the normalized Minkowski distance d_{nm} and the normalized Hausdorff distance d_{nhd} and set $\lambda = 1$, then we get

$$s_{nh}(M_1, N) = s_{nh}(M_2, N) = 0.8667,$$
$$s_{nhd}(M_1, N) = s_{nhd}(M_2, N) = 0.8333$$

The results show that, we can not tell which pattern N should belong to on the sense of the distance-induced similarity measures between HFSs. In fact, no matter which the strictly monotone decreasing function [4] and the distance with different parameter λ given in paper [12] chosen, it all follows that the distance-induced similarity measure between the HFSs M_1 and N is equal to that between the HFSs M_2 and N. That is to say that using the distance-induced similarity measures given in [4] and [12], in that case, we can not discriminate which pattern N should belong to.

Thus, some new similarity measures between HFSs with stronger discrimination ability should be introduced. As we mentioned above, in vector space, especially the Jaccard, Dice, and Cosine similarity measures are widely used, moreover, they do not depend on the distance measures at all. However, these similarity measures do not deal with the similarity measures for vague information. Therefore, Ye proposed the dice similarity between intuitionistic fuzzy sets and applied it to decision making [3] and the cosine similarity measure between intuitionistic fuzzy sets and applied it to pattern recognition and medical diagnosis [6], respectively.

Considering the relation between HFSs and intuitionistic fuzzy sets mentioned in Section 2, similar to the existing works [3,6], next, we will introduce the vector similarity measures, such as the Jaccard, Dice, and Cosine similarity measures between HFSs, respectively.

Definition 5. *Let M, N be two HFSs on the universe $U = \{u_1, u_2, \cdots, u_n\}$, the Jaccard similarity $s_J(M, N)$ between M and N is defined as follows:*

$$s_J(M, N) = \frac{1}{n} \sum_{i=1}^{n} \frac{\sum\limits_{j=1}^{l_{u_i}} h_M^{\sigma(j)}(u_i) h_N^{\sigma(j)}(u_i)}{\sum\limits_{j=1}^{l_{u_i}} \left(\begin{matrix} (h_M^{\sigma(j)}(u_i))^2 + (h_N^{\sigma(j)}(u_i))^2 \\ - h_M^{\sigma(j)}(u_i) h_N^{\sigma(j)}(u_i) \end{matrix} \right)}$$

where, $l_{u_i} = \max\{l(h_M(u_i)), l(h_N(u_i))\}$. $h_M^{\sigma(j)}(u_i)$ and $h_N^{\sigma(j)}(u_i)$ are the jth smallest values of $h_M(u_i)$ and $h_N(u_i)$, respectively.

Definition 6. *Let M, N be two HFSs on the universe $U = \{u_1, u_2, \cdots, u_n\}$, the Dice similarity $s_D(M, N)$ between M and N is defined as follows:*

$$s_D(M, N) = \frac{1}{n} \sum_{i=1}^{n} \frac{2 \sum\limits_{j=1}^{l_{u_i}} h_M^{\sigma(j)}(u_i) h_N^{\sigma(j)}(u_i)}{\sum\limits_{j=1}^{l_{u_i}} \left(\left(h_M^{\sigma(j)}(u_i) \right)^2 + \left(h_N^{\sigma(j)}(u_i) \right)^2 \right)}$$

where, $l_{u_i} = \max\{l(h_M(u_i)), l(h_N(u_i))\}$. $h_M^{\sigma(j)}(u_i)$ and $h_N^{\sigma(j)}(u_i)$ are the jth smallest values of $h_M(u_i)$ and $h_N(u_i)$, respectively.

Definition 7. *Let M, N be two HFSs on the universe $U = \{u_1, u_2, \cdots, u_n\}$, the Cosine similarity $s_C(M, N)$ between M and N is defined as follows:*

$$s_C(M, N) = \frac{1}{n} \sum_{i=1}^{n} \frac{\sum_{j=1}^{l_{u_i}} h_M^{\sigma(j)}(u_i) h_N^{\sigma(j)}(u_i)}{\sqrt{\sum_{j=1}^{l_{u_i}} \left(h_M^{\sigma(j)}(u_i)\right)^2} \sqrt{\sum_{j=1}^{l_{u_i}} \left(h_N^{\sigma(j)}(u_i)\right)^2}}$$

where, $l_{u_i} = \max\{l(h_M(u_i)), l(h_N(u_i))\}$. $h_M^{\sigma(j)}(u_i)$ and $h_N^{\sigma(j)}(u_i)$ are the jth smallest values of $h_M(u_i)$ and $h_N(u_i)$, respectively.

Definition 8. *Set A is a HFS on the universe $U = \{u_1, u_2, \cdots, u_n\}$, if for any $u_i \in U (i = 1, 2, \cdots, n)$, $h_A(u_i) = \{0\}$, then A is called the zero hesitant fuzzy set on U.*

Remark 2. *(1) We uniformly call the above three similarity measures between HFSs the vector similarity measures between HFSs, and denoted the vector similarity measures between HFSs M, N as $s_v(M, N)$.*

(2) The vector similarity measures between HFSs are similar in the sense that they take values in the interval $[0, 1]$. Jaccard and Dice formulas are undefined if M and N both are the zero hesitant fuzzy sets, and then we let the two measure values be zero when M and N both are the zero hesitant fuzzy sets. However, the cosine formula is undefined if any of M and N is the zero hesitant fuzzy set, and then we let the cosine measure value be zero when of M and N is the zero hesitant fuzzy set.

(3) It is easy to prove that all the three vector similarity measures between HFSs proposed above are satisfied with the similarity axioms about HFSs given in Definition 4.

Example 2 *(continued). Using the vector similarity measures between HFSs to decide which pattern N should belong to.*

According to the Jaccard, Dice, and Cosine similarity measures between HFSs given in Definitions 5–7, we get

$$s_C(M_1, N) = 0.9998, s_C(M_2, N) = 0.9776,$$
$$s_D(M_1, N) = 0.9189, s_D(M_2, N) = 0.9579,$$
$$s_J(M_1, N) = 0.8609, s_J(M_2, N) = 0.9209,$$

Remark 3. *(1) Comparing the results in Example 1 and Example 2, we see that the new proposed vector similarity measures between HFSs have stronger discrimination ability than the existing distance-induced ones.*

(2) According to the pattern recognition principle to HFSs based on the similarity measures, we see that in the sense of the Jaccard or Dice similarity measures between HFSs, the sample N should belong to pattern M_2, while in the sense of the Cosine similarity measures between HFSs, the sample N should belong

to pattern M_1. The reason is that the Cosine similarity measure is defined from different point with that of Jaccard or Dice similarity measures. In practice, researchers can choose the appropriate vector similarity between HFSs according to the concrete case.

4   Hesitant MST Clustering Method Based on the Vector Similarity Measures

Similarity measures are widely used in clustering problems to indicate the similarity degree between samples. While, there is little effort to study the clustering method to hesitant fuzzy information related to similarity measures. In fact, there is only one paper discuss the clustering method to hesitant fuzzy information related to similarity measures [4]. However, the method requires much computational effort, and more importantly, the drawbacks about the discrimination ability of distance-induced similarity measures between HFSs will affect the effectiveness and accuracy of the clustering results. Thus, in this section, a graph-based MST clustering method related to the vector similarity measures between HFSs will be proposed, a numerical case will show that the new algorithm not only requires less computational effort but also is more accurate than the existing one.

Firstly, we will review some related notions about graph and fuzzy graph [21, 22]:

A graph is an ordered pair $G = (V, E)$, where V is the set of vertices of G and E is the set of edges of G. Two vertices x and y in a graph G are said to be adjacent in G if $\{x, y\}$ is in an edge of G. If a crisp relation over $V \times V$ is defined, that is to say, if there exists an edge between x and y, then the membership degree equals 1, otherwise, the membership degree equals 0, and such a graph is called a normal graph. If a fuzzy relation over $V \times V$ is defined, that is to say, if there exists an edge between x and y, then the membership degree takes various values from 0 to 1, and such a graph is called a fuzzy graph.

Similar to the fuzzy graph-based clustering algorithm to fuzzy sets [24], next we will propose a fuzzy graph-based MST clustering method to hesitant fuzzy information related to the vector similarity measures between HFSs as follows:

Set $\{H_1, H_2, \cdots, H_m\}$ be a set of HFSs in $U = \{u_1, u_2, \cdots, u_n\}$:

Algorithm 1: Hesitant MST clustering method based on the vector similarity measures

Step 1. Construct the hesitant fuzzy vector similarity matrix.

Choose the vector similarity between the HFSs and calculate the similarity $s_v^{ij} = s_v(H_i, H_j)$. We get the hesitant fuzzy vector similarity matrix $S_v = (s_v^{ij})_{m \times m}$.

Step 2. Construct the fuzzy graph.

Construct the fuzzy graph $G = (V, E)$ with m nodes associated to the samples $H_i (i = 1, 2, \cdots, m)$ to be clustered which are expressed by HFSs and every edge between H_i and H_j having the weight s_v^{ij}, which is an element of the hesitant fuzzy vector similarity matrix S_v and denotes the similarity degree between the samples H_i and H_j.

Step 3. Compute the MST of the fuzzy graph $G = (V, E)$ by Kruskal method [29](or Prim method [30]):

(1) Arrange the edges of G in order from the biggest weight to the smallest one.
(2) Select the edge with the biggest weight.
(3) Select the edge with the biggest weight from the rest edges which do not form a circuit with those already chosen.
(4) Repeat the process (3) until $m - 1$ edges have been selected where m is the number of the nodes in $G = (V, E)$. Thus we get the MST of the fuzzy graph $G = (V, E)$.

Step 4. Group the nodes (sample points) into clusters.

By cutting down all the edges of the MST with weights smaller than a threshold λ, we can get a certain number of sub-trees (clusters) automatically.

The clustering results induced by the sub-trees do not depend on some particular MST [28].

Now, the same case in [4] is employed to demonstrate the effectiveness of the proposed algorithm.

Example 3 *(See [4]). Software evaluation and classification is an increasingly important problem in any sector of human activity. Industrial production, service provisioning and business administration heavily depend on software which is more and more complex and expensive. A CASE tool to support the production of software in a CIM environment has to be selected from the ones offered on the market. CIM software typically has responsibility for production planning, production control and monitoring. To better evaluate different types of CIM softwares $H_i(i = 1, 2, \cdots, 7)$ on the market, we perform clustering for them according to four attributes: u_1: functionality, u_2: usability, u_3: portability, and u_4: maturity. Given the experts who make such an evaluation have different backgrounds and levels of knowledge, skills, experience and personality, etc., this could lead to a difference in the evaluation information. To clearly reflect the differences of the opinions of different experts, the data of evaluation information are represented by the HFSs and listed in Table 1.*

Using Algorithm 1, the clustering problem is now solved as the following steps:

Step 1. Construct the hesitant fuzzy vector similarity matrix. We choose the Cosine similarity between HFSs to get the hesitant fuzzy Cosine similarity matrix $S_C = (s_C^{ij})_{7 \times 7}$ as follows,

$$S_C = \begin{pmatrix} 1.0000 & 0.9975 & 0.9915 & 0.9925 & 0.9935 & 0.9926 & 0.9825 \\ 0.9975 & 1.0000 & 0.9888 & 0.9901 & 0.9919 & 0.9834 & 0.9784 \\ 0.9915 & 0.9888 & 1.0000 & 0.9887 & 0.9960 & 0.9900 & 0.9747 \\ 0.9925 & 0.9901 & 0.9887 & 1.0000 & 0.9910 & 0.9894 & 0.9958 \\ 0.9935 & 0.9919 & 0.9960 & 0.9910 & 1.0000 & 0.9927 & 0.9770 \\ 0.9926 & 0.9834 & 0.9900 & 0.9894 & 0.9927 & 1.0000 & 0.9809 \\ 0.9825 & 0.9784 & 0.9747 & 0.9958 & 0.9770 & 0.9809 & 1.0000 \end{pmatrix}$$

Table 1. Hesitant fuzzy information

	u_1	u_2	u_3	u_4
H_1	$\{0.8, 0.85, 0.9\}$	$\{0.7, 0.75, 0.8\}$	$\{0.65, 0.8\}$	$\{0.3, 0.35\}$
H_2	$\{0.85, 0.9\}$	$\{0.6, 0.7, 0.8\}$	$\{0.2\}$	$\{0.15\}$
H_3	$\{0.2, 0.3, 0.4\}$	$\{0.4, 0.5\}$	$\{0.9, 1.0\}$	$\{0.45, 0.5, 0.65\}$
H_4	$\{0.8, 0.95, 1.0\}$	$\{0.1, 0.15, 0.2\}$	$\{0.2, 0.3\}$	$\{0.6, 0.7, 0.8\}$
H_5	$\{0.35, 0.4, 0.5\}$	$\{0.7, 0.9, 1.0\}$	$\{0.4\}$	$\{0.2, 0.3, 0.35\}$
H_6	$\{0.5, 0.6, 0.7\}$	$\{0.8, 0.9\}$	$\{0.4, 0.6\}$	$\{0.1, 0.2\}$
H_7	$\{0.8, 1.0\}$	$\{0.15, 0.2, 0.35\}$	$\{0.1, 0.2\}$	$\{0.7, 0.85\}$

Step 2. Construct the fuzzy graph. We get the fuzzy graph in Fig. 1.

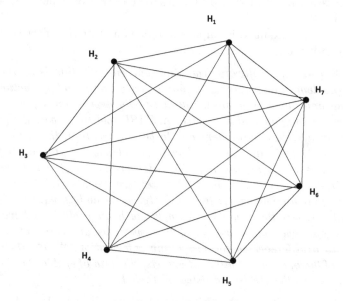

Fig. 1. Fuzzy graph

Step 3. Compute the MST of the fuzzy graph $G = (V, E)$:

(1) Arrange the edges of G in order from the biggest weight to the smallest one.

We get:

$$s_C^{12} > s_C^{35} > s_C^{47} > s_C^{15} > s_C^{56} > s_C^{16} > s_C^{14} > s_C^{25} > s_C^{13} > s_C^{45} > s_C^{24} >$$
$$s_C^{36} > s_C^{46} > s_C^{23} > s_C^{34} > s_C^{26} > s_C^{17} > s_C^{67} > s_C^{27} > s_C^{37} > s_C^{57}$$

(2)–(4) We get the MST of the fuzzy graph $G = (V, E)$ in Fig. 2.

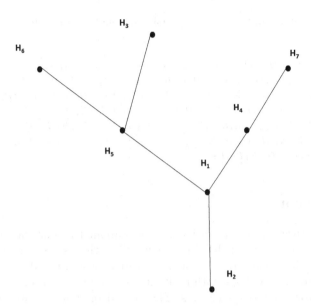

Fig. 2. MST of the fuzzy graph G

Step 4. Group the nodes (sample points) into clusters.
We get the clustering results as follows:

(1) If $\lambda = 1$, then we get $\{H_1\}, \{H_2\}, \{H_3\}, \{H_4\}, \{H_5\}, \{H_6\}, \{H_7\}$;
(2) If $\lambda = 0.9975$, then we get $\{H_1, H_2\}, \{H_3, H_4, H_5, H_6, H_7\}$;
(3) If $\lambda = 0.9960$, then we get $\{H_1, H_2\}, \{H_3, H_5\}, \{H_4, H_6, H_7\}$;
(4) If $\lambda = 0.9958$, then we get $\{H_1, H_2\}, \{H_3, H_5\}, \{H_4, H_7\}, \{H_6\}$;
(5) If $\lambda = 0.9935$, then we get $\{H_1, H_2, H_3, H_5\}, \{H_4, H_7\}, \{H_6\}$;
(6) If $\lambda = 0.9927$, then we get $\{H_1, H_2, H_3, H_5, H_6\}, \{H_4, H_7\}$;
(7) If $\lambda = 0.9925$, then we get $\{H_1, H_2, H_3, H_4, H_5, H_6, H_7\}$;

Comparing with the clustering algorithm given in [4], we see that the new algorithm requires less computational effort, by introducing the fuzzy graph theory.

On the other hand, it follows that $s_d(H_2, H_6) = s_d(H_2, H_7)$ using any distance-induced similarity measure between HFSs in this case, which shows the drawbacks about the discrimination ability of the distance-induced similarity measures between HFSs. While, the new proposed Cosine similarity measures between HFSs has avoided the drawback.

Using the Jaccard and Dice similarity measures between HFSs in Algorithm 1, we will get the clustering results in Table 2.

We see that the clustering results related to the Jaccard or Dice similarity measures are different to those related to the Cosine similarity measure between HFSs. As we mentioned above, the reason is that the Cosine similarity measures defined from the different point. In practice, researchers can choose the appropriate vector similarity between HFSs according to the concrete case.

Table 2. Clustering results related to Jaccard and Dice similarity measures

Classes	s_J	s_D
1	$\{H_1\},\{H_2\},\{H_3\},\{H_4\},\{H_5\},\{H_6\},\{H_7\}$	$\{H_1\},\{H_2\},\{H_3\},\{H_4\},\{H_5\},\{H_6\},\{H_7\}$
2	$\{H_4, H_7\},\{H_1\},\{H_2\},\{H_3\},\{H_5\},\{H_6\}$	$\{H_4, H_7\},\{H_1\},\{H_2\},\{H_3\},\{H_5\},\{H_6\}$
3	$\{H_4, H_7\},\{H_5, H_6\},\{H_1\},\{H_2\},\{H_3\}$	$\{H_4, H_7\},\{H_5, H_6\},\{H_1\},\{H_2\},\{H_3\}$
4	$\{H_4, H_7\},\{H_1, H_5, H_6\},\{H_2\},\{H_3\}$	$\{H_4, H_7\},\{H_1, H_5, H_6\},\{H_2\},\{H_3\}$
5	$\{H_4, H_7\},\{H_1, H_2, H_5, H_6\},\{H_3\}$	$\{H_4, H_7\},\{H_1, H_2, H_5, H_6\},\{H_3\}$
6	$\{H_4, H_7\},\{H_1, H_2, H_3, H_5, H_6\}$	$\{H_4, H_7\},\{H_1, H_2, H_3, H_5, H_6\}$
7	$\{H_1, H_2, H_3, H_4, H_5, H_6, H_7\}$	$\{H_1, H_2, H_3, H_4, H_5, H_6, H_7\}$

5   Conclusions

In this paper, three vector similarity measures about HFSs are proposed, which are proved to have stronger discrimination ability than some existing distance-induced similarity about HFSs. Moreover, a fuzzy graph based MST clustering method is introduced to cope with hesitant fuzzy information related to the new proposed vector similarity measures. The new algorithm is proved to be more effective and accurate than the existing one.

References

1. Zadeh, L.A.: Fuzzy sets. Inform. Control **8**, 338–353 (1965)
2. Zhang, F., Li, J., Chen, J., Sun, J., Attey, A.: Hesitant distance set on hesitant fuzzy sets and its application in urban road traffic state identification. Eng. Appl. Artif. Intell. **61**, 57–64 (2017)
3. Ye, J.: Multicriteria decision-making method using the Dice similarity measure based on the reduct intuitionistic fuzzy sets of interval-valued intuitionistic fuzzy sets. Appl. Math. Model. **36**, 4466–4472 (2012)
4. Farhadinia, B.: Information measures for hesitant fuzzy sets and interval-vlued hesitant fuzzy sets. Inf. Sci. **240**, 129–144 (2013)
5. Hwang, C.M., Yang, M.S., Hung, W.L.: A similarity measure of intuitionistic fuzzy sets based on the Sugeno integral with its application to pattern recognition. Inf. Sci. **189**, 9–109 (2012)
6. Ye, J.: Cosine similarity measures for intuitionistic fuzzy sets and their applications. Math. Comput. Model. **53**, 91–97 (2011)
7. Sun, G., Guan, X., Yi, X., Zhou, Z.: Grey relational analysis between hesitant fuzzy sets with applications to pattern recognition. Expert Syst. Appl. **92**, 521–532 (2018)
8. Torra, V.: Hesitant fuzzy sets. Int. J. Intell. Syst. **25**, 529–539 (2010)
9. Torra, V., Narukawa, Y.: On hesitant fuzzy sets and decision. In: The 18th IEEE International Conference on Fuzzy Systems, Jeju Island, Korea, pp. 1378–1382 (2009)
10. Zhang, N., Wei, G.W.: Extension of VIKOR hesitant fuzzy set. Appl. Math. Model. **37**, 4938–4947 (2013)
11. Arora, R., Garg, H.: A robust correlation coefficient measure of dual hesitant fuzzy soft sets and their application in decision making. Eng. Appl. Artif. Intell. **72**, 80–92 (2018)

12. Xia, M.M., Xu, Z.S.: Distance and similarity measures for hesitant fuzzy sets. Inf. Sci. **181**(11), 2128–2138 (2011)
13. Dice, L.R.: Measures of the amount of ecologic association between species. Ecology **26**(140), 297–302 (1945)
14. Jaccard, P.: Distribution de la flore alpine dans le Bassin des Drouces et dans quelques regions voisines. Bull de la Sociètè Vaudoise des Sciences Naturelles **37**(140), 241–272 (1901)
15. Salton, G., McGill, M.J.: Introduction to Modern Information Retrieval. McGraw-Hill, New York (1987)
16. Chaira, T.: A novel intuitionistic fuzzy C means clustering algorithm and its application to medical images. Appl. Soft Comput. **11**, 1711–1717 (2011)
17. Kumar, N., Nasser, M., Sarker, S.C.: A new singular value decomposition based robust graphical clustering technique and its application in climatic data. J. Geogr. Geol. **3**, 227–238 (2011)
18. Nikas, J.B., Low, W.C.: Application of clustering analyses to the diagnosis of Huntington disease in mice and other diseases with well-defined group boundaries. Comput. Methods Programs Biomed. **104**, 133–147 (2011)
19. Zhao, B., He, R.L., Yau, S.T.: A new distribution vector and its application in genome clustering. Mol. Phylogenet. Evol. **59**, 438–443 (2011)
20. Zhao, P., Zhang, C.Q.: A new clustering method and its application in social networks. Pattern Recogn. Lett. **32**, 2109–2118 (2011)
21. Bhutani, K.R.: On automorphism of fuzzy graphs. Pattern Recogn. Lett. **9**, 159–162 (1986)
22. Mordeson, J.N.: Fuzzy line graphs. Pattern Recogn. Lett. **14**, 381–384 (1993)
23. Zahn, C.T.: Graph-theoretical methods for detecting and describing gestalt clusters. IEEE Tran. Comput. **20**, 68–86 (1971)
24. Chen, D.S., Li, K.X., Zhao, L.B.: Fuzzy graph maximal tree clustering method and its application. Oper. Res. Manag. Sci. **16**, 69–73 (2007)
25. Zhao, H., Xu, Z.S., Liu, S.S.: Intuitionistic fuzzy MST clustering algorithms. Comput. Ind. Eng. **62**, 1130–1140 (2012)
26. Atanassov, K.T.: Intuitionistic fuzzy sets. Fuzzy Sets Syst. **20**, 87–96 (1986)
27. Chen, N., Xu, Z.S., Xia, M.M.: Correlation coefficients of hesitant fuzzy sets and their applications to clustering analysis. Appl. Math. Model. **37**, 2197–2211 (2013)
28. Gaertler, M.: Clustering with spectral methods. Master's thesis, Universitat Konstanz (2002)
29. Kruskal, J.B.: On the shortest spanning subtree of a graph and the traveling salesman problem. Proc. Am. Math. Soc. **7**, 48–50 (1956)
30. Prim, R.C.: Shortest connection networks and some generalizations. Bell Syst. Technol. J. **36**, 1389–1401 (1957)

A Fuzzy Density Peaks Clustering Algorithm Based on Improved DNA Genetic Algorithm and K-Nearest Neighbors

Wenqian Zhang and Wenke Zang[(✉)]

Shandong Normal University, Jinan, China
wink@sdnu.edu.cn

Abstract. In recent times, a density peaks based clustering algorithm (DPC) that published in Science was proposed in June 2014. By using a decision graph and finding out cluster centers from the graph can quickly get the clustering results, easy and efficient. While, in terms of local density measurement, DPC does not adopt uniform density metrics. Instead, uses different local density metrics according to the dataset size. In addition, when the size is small, the subjective choice of the cutoff distance dc has a greater impact on the clustering results. In order to make up for the defects of DPC and improve the performance of this algorithm, we propose a fuzzy density peaks clustering algorithm based on improved DNA genetic algorithm and K-nearest neighbors (named as FDPC+IDNA). On one hand, FDPC+IDNA uses fuzzy neighborhood relation to unify the local density metric which combines the high efficiency of DPC algorithm with the robustness of fuzzy theory. On the other hand, we introduce the idea of K-nearest neighbors and an improved DNA genetic algorithm to compute the global parameter dc that improves the shortcomings of empirical judgment. Experiments on synthetic and real-world datasets demonstrate that the proposed clustering algorithm outperforms DPC, DBSCAN and K-Means.

Keywords: Density peaks clustering · DNA genetic algorithm
K-nearest neighbors · Fuzzy neighborhood relation

1 Introduction

As a core topic in data mining, data clustering has been widely used in many fields, such as machine learning, pattern recognition, information extraction, image analysis and computer graphics [1–4]. The existence of massive and diverse data promotes the study of clustering algorithms with automatic understanding, processing and summarizing data imminent. Many kinds of clustering methods have been proposed in the past years and they can be classified into four types: partitioning, hierarchical, density-based and grid-based [5]. Density based clustering techniques like DBSCAN can find arbitrary shaped clusters along with noisy outliers [6–9]. However, unreasonable selection of input parameters will influence the clustering results severely; besides, this method cannot get satisfactory performance in overlapping densities [10]. Many kinds of

© Springer Nature Switzerland AG 2018
Y. Peng et al. (Eds.): IScIDE 2018, LNCS 11266, pp. 476–487, 2018.
https://doi.org/10.1007/978-3-030-02698-1_42

variant have been proposed to overcome these limitations, for example, ST-DBSCAN [6], P-DBSCAN [7], I-DBSCAN [8], MR-DBSCAN [9] and so on.

Rodriguez and Laio proposed a clustering by fast search and find of density peaks (DPC) algorithm in 2014 [11], which can quickly find the density peak point of any shape dataset, that is the clustering center point. The first step of the algorithm, which is the core of the DPC, is setting up a decision graph using two quantities of each point i: the local density ρ and the local distance δ from points of higher density. Then, the clustering center point will be selected by the decision graph. Next, the remaining points will be assigned to the cluster its nearest neighbor with higher density [12]. However, this algorithm still has some shortcomings: DPC does not adopt uniform density metrics. Instead, uses different local density metrics according to the dataset size. In addition, when the size is small, the subjective choice of the cutoff distance dc has a greater impact on the clustering results. In order to overcome these problems, Liang et al. proposed the 3DC clustering that selects the-highest confidence objects recursively [13]; Zang et al. proposed an automatic density peaks clustering approach using DNA genetic algorithm optimized data field and Gaussian process that can extract the optimal value of threshold with the potential entropy of data field [14]; Zhou et al. introduced the Fruit Fly Optimization Algorithm (FOA) to optimize the cut-off distance and clustering centers [15]. Du et al. proposed a new local density metric using fuzzy joint point, Nevertheless, it still select parameter dc by subjective judgment, so the drawback still exist.

Therefore, the introduction of heuristic algorithm is a trend to optimize the algorithm [16, 17]. In this paper, we propose the fuzzy density peaks clustering algorithm based on improved DNA genetic algorithm and K-nearest neighbors (named as FDPC +IDNA). The FDPC+IDNA also introduces the idea of fuzzy neighbors relation to compute the local density. Different from traditional crisp clustering methods, fuzzy partition can assign data points more rationally. What's more, we define the metric of cutoff distance using K-nearest neighbors, and we use the heuristic optimization algorithm-DNA genetic algorithm to obtain the optimal value of dc quickly.

The rest of this paper is organized as follows. In Sect. 2, the DPC algorithm is mentioned and some background knowledge about DNA genetic algorithm, K-nearest neighbors and fuzzy neighbors relation theory will be described briefly. In Sect. 3, a flowchart of FDPC + IDNA is given. In Sect. 4, experimental results are presented on synthetic data sets and real world data sets. Some conclusions and the future work are given in the last section.

2 Related Work

2.1 Density Peaks Clustering

The DPC algorithm bases on the assumption that a cluster center has higher local density than those of its neighbors and a relatively large distance from the other centers [11]. For the purpose of setting up a decision graph and then find the vintage clustering centers, DPC algorithm computes two quantities of each point i: local density ρi using (1) and local distance δi using (2). Where dij represents the Euclidean distance between points i and j that computed by (3). After ρ and δ values of all points are calculated, DPC will set up the decision graph. One can detect the clustering centers from the upper-right region of graph. With clustering centers, DPC can assign the remaining points nearest to the centers in a single step.

$$\rho_i = \sum_j \chi(d_{ij} - d_c), \chi(t) = \begin{cases} 1.if, t<0; \\ 0.otherwise \end{cases} \tag{1}$$

$$\delta_i = \begin{cases} \min_{\rho_i<\rho_j} d(x_i, x_j), if \ \rho_i < \rho_j \\ \max_j d(x_i, x_j), otherwise \end{cases} \tag{2}$$

$$d_{ij} = \left(\sum_{k=1}^{m} \left(x_{i,k} - x_{j,k}\right)^2 \right)^{1/2} \tag{3}$$

For "small" datasets, DPC adopts the Formula (4) to calculate the local density [11], but there is no objective metric to decide whether the dataset is small or large and clustering by using the two density metrics will produce very different results. In addition, for small datasets, the clustering results of DPC can be greatly affected by the cutoff distance dc even using (4) to calculate the local density [12, 18].

$$\rho_i = \sum_{j \neq i} \exp(-\frac{d_{ij}^2}{d_c^2}) \tag{4}$$

2.2   Neighborhood Relationship

To eliminate the influence from the cutoff distance dc and give a uniform density metric for datasets with any size, many researchers introduce the idea of the K-nearest neighbor into the local density calculation and defined different kinds of metrics [12, 18, 19]. In this paper, we adopt K-nearest neighbor too and use it to define the metric of cutoff distance; besides, we consider and introduce the fuzzy neighbors relation to define the metric of local density, which consider the membership of the member points to the clustering centers further.

We introduce the metric of cutoff distance based on K-nearest neighbors [12],

$$d_c = \mu^K + \sqrt{(\frac{1}{N-1} \sum_{i=1}^{N} (\varepsilon_i^K - \bar{\varepsilon})^2)} \tag{5}$$

$$\varepsilon_i^K = \max_{j \in KNN_i} d_{(x_i, x_j)} \tag{6}$$

Where $\bar{\varepsilon}$ is the mean value of ε_i^K of all member points.

Meanwhile, we use the fuzzy neighborhood relationship that can identify the different values of the neighborhood membership degrees of the points with respect to different distances from core point to replace the crisp neighborhood relation [20, 21]. According to fuzzy neighbors theory, the clustering center point of the right has a higher membership degree than the left one in the Fig. 1.

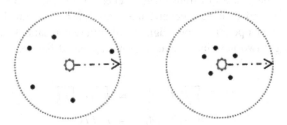

Fig. 1. Fuzzy neighborhood relationship

And such a neighborhood membership function is defined as:

$$\mu'_{x_i}(x_j) = \begin{cases} 1 - \frac{d(X_i, X_j)}{d_c}, & if, d(x_i, x_j) < d_c \\ 0 & otherwise \end{cases} \tag{7}$$

Where dc represents cutoff distance and the metric of the local density is turned into the following:

$$\rho_i = \sum_j \mu'_{x_i}(x_j) \tag{8}$$

2.3   The Improved DNA Genetic Algorithm

DNA Coding and Decoding
During the process of evolution, DNA genetic algorithm using base-based coding scheme [22]. Mathematically, a string composed of the four symbols A, G, C and T is used to encode a parameter of an optimization problem. This can be denoted by the four decimal codes: 0, 1, 2, and 3. There are two complementary pairs such as A–T and C–G, which corresponds to 0–3 and 1–2 after encoding. Through this correspondence, a sequence of DNA molecules that stores genetic information can be converted into a string of numbers that can be identified and computed by the computer.

The Selection Operation
Selection operation determines which individual can be kept to next generation and how many individuals are replicated to offspring [23], this paper uses a tournament

selection method to generate a new generation of population. The basic idea is to choose one of the two individuals with a better fitness to the next generation, thus ensuring the efficiency and convergence of the algorithm.

The Crossover Operation

The crossover operation is mainly responsible for the global searching ability of the algorithm [24]. In the process of transferring, genes far apart from one another can be combined together, and then new genetic materials can be generated. See as Fig. 2, through the crossover operation, the fragment of different positions can be replaced, and then we can get two new individuals that not appeared before.

$$C_1 = A\boxed{CT}GC\boxed{AT}, C_2 = G\boxed{CA}TC\boxed{GA}$$
$$C_1' = ACAGCGA, C_2' = GCTTCAT$$

Fig. 2. An example of crossover operation

The Mutation Operation

Mutation operation can generate new genetic material and maintain the population diversity. The deletion and insertion operators can be regarded as the most common mutation operation. The purpose of this operation is to replace those bad individuals with better ones at the early stage of evolutionary process.

The Splicing Operation

Inspired by the biologic DNA fragment splicing phenomenon [25] and the splicing of RNA in the research field of biomedical, we propose a novel DNA genetic operation in this section: splicing operation. We define two kinds of splicing: external splicing operator that base on the biomedical fact of variation of the left part of the chromosome is more likely to cause the change of fitness than the variation in the right part [26] and internal splicing-inversion operator, which can be demonstrated in Figs. 3 and 4 clearly. By introducing this new splicing operation that including two kinds of operator, the traditional genetic operation of standard DNA genetic algorithm has been improved further that the algorithm can converge faster.

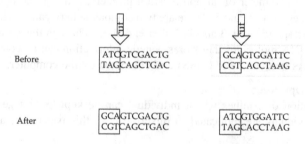

Fig. 3. An example of external splicing operator

Three bases of two different chromosomes head are swapped according to a certain probability, which can further increase population diversity. Two new different double-strands DNA molecules generation processes are shown in Fig. 3.

Fig. 4. An example of internal splicing-inversion operator

Besides, we set up a cell-like P system with dynamic membrane structure as a computational framework. This P system has a variety of membrane computing rules. By merging membrane computing and DNA genetic operations, not only the improved DNA genetic algorithm can avoid the disadvantage of easily getting to local optional solutions, but also converges rapidly. At the same time, compared with the traditional DNA algorithm, the ability of parallelism of this method can be improved further.

In the previous section, we have defined the metric of cutoff distance. In order to obtain the most reasonable and suitable value of dc, we make full use of the high efficiency of heuristic algorithm and we take Formula (5) as the fitness function.

3 The Proposed Algorithm: FDPC+IDNA

There are still existing shortcomings of DPC and those improved DPC algorithm. In order to solve these problems to a certain degree, a new idea is proposed. First, we define the metric of cutoff distance using the K-nearest neighbors and adopt an improved DNA genetic algorithm to find the most suitable value dc of every test dataset quickly. Second, we use the fuzzy neighbors relationship to uniform the local density metric ρ.

The process of the FDPC + IDNA algorithm can be described as follows in detail:

Step1. Preprocess data including normalizing and reducing dimensions;
Step2. Compute the Euclidean distance using (3) and calculate the cutoff distance dc using (5);
Step3. Calculate ρi and δi for point i using (8) and (2), respectively;
Step4. Plot decision graph and select clustering centers;
Step5. Assign each remaining point to the cluster;
Step6. Return the clustering C.

During the process of improved DNA algorithm, the termination condition is that the number of generations reaches the maximum. If this condition is met, the algorithm will stop and the current best result that is the suitable dc will be output to calculate the local metric next. And then we start to calculate the local metric. The next steps can be seen clearly from the following Fig. 5.

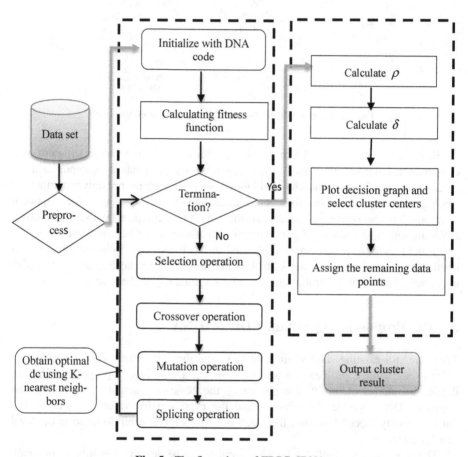

Fig. 5. The flow chart of FDPC+IDNA

4   Experiments and Analysis

In order to demonstrate the effectiveness of the proposed algorithm, 3 synthetic datasets with various shapes and densities are used. Moreover, by experiments on real-world datasets, the proposed method is compared with the comparison algorithms in terms of clustering accuracy (CA), normalized mutual information (NMI) and Rand index (RI), which are all classical clustering evaluation indexes. We conduct experiments in a desktop computer with MATLAB R2014b.

4.1 Experiments on Synthetic Datasets

As shown in Figs. 6, 7, 8, the proposed method does an excellent job in synthetic datasets with spherical or ellipsoidal shape. As these clustering result distribution graphs illustrate our algorithm is effective in finding clusters of arbitrary shape, density, distribution and number.

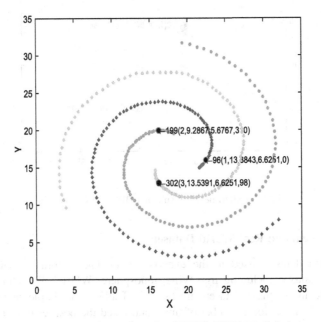

Fig. 6. Experimental result of FDPC+IDNA on Spiral

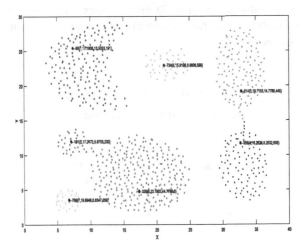

Fig. 7. Experimental result of FDPC+IDNA on Aggregation

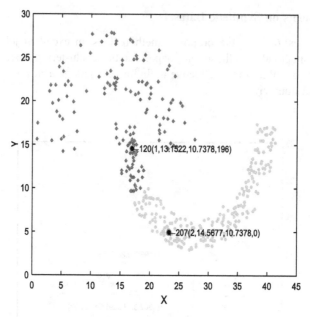

Fig. 8. Experimental result of FDPC+IDNA on Jain

4.2 Experiments on Real World Datasets

The real-world datasets used in the experiment are taken from the UCI Machine Learning Repository, including Iris, Wine, Ionosphere, WDBC, Waveform and Pen-based. The details of these datasets are listed in Table 1. We implemented the algorithms on each dataset for a number of times and listed the best result of each method out. The parameters of FDPC + IDNA, DPC, DBSCAN and K-means were carefully chosen for every implementation. Three popular criteria CA (Cluster Accuracy), NMI (Normalized Mutual Information) and RI (Rand index) were used to evaluate the performances of the above clustering algorithms. The larger of the value, the better of the clustering result is.

Table 1. The details of real-world datasets

Data sets	Cluster	Attribute	Number
Iris	3	4	150
Wine	3	13	178
Ionosphere	2	34	351
WDBC	2	30	569
Waveform	3	21	5000
Pen-based	10	16	10992

Table 2 lists parameters settings for the four clustering algorithms and their results in terms of the CA, NMI and RI on datasets. In the following tables, the numbers highlighted in bold indicate that the corresponding algorithm has the best performance in terms of its corresponding evaluation. The bar '−' represents there are no corresponding values.

Table 2. The comparison of clustering algorithms in terms of CA, NMI and RI on Real-world datasets

Algorithm	Iris				Wine			
	CA	NMI	RI	Par	CA	NMI	RI	Par
FDPC+IDNA	**0.975**	**0.922**	**0.914**	7	**0.947**	**0.828**	**0.833**	7
DPC	0.877	0.765	0.719	2	0.872	0.716	0.682	2
DBSCAN	0.892	0.776	0.742	0.14/9	0.847	0.689	0.662	−
K-means	0.824	0.693	0.658	3	0.923	0.804	0.832	−
Algorithm	Ionosphere				WDBC			
	CA	NMI	RI	Par	CA	NMI	RI	Par
FDPC+IDNA	**0.739**	**0.256**	**0.341**	8	**0.946**	**0.679**	**0.791**	7
DPC	0.679	0.247	0.165	0.65	0.615	0.008	0.011	9
DBSCAN	0.605	0.087	0.037	0.2/7	0.872	0.421	0.543	0.27/7
K-means	0.711	0.128	0.177	2	0.923	0.621	0.732	2
Algorithm	Waveform				Pen-based			
	CA	NMI	RI	Par	CA	NMI	RI	Par
FDPC+IDNA	**0.699**	**0.374**	**0.342**	5	**0.767**	**0.748**	**0.645**	8
DPC	0.579	0.318	0.269	0.5	0.718	0.593	0.589	0.16
DBSCAN	−	−	−	−	−	−	−	−
K-means	0.511	0.322	0.257	3	0.698	0.542	0.376	3

5   Conclusions

In this paper, we propose a fuzzy density peaks clustering algorithm based on K-nearest neighbors and an improved DNA genetic algorithm, which is a popular and effective heuristic intelligent algorithm. The uniform local density metric is defined by using fuzzy neighbor relationship theory. And we introduce a metric of cutoff distance using the K nearest neighbors and calculate by the improved DNA genetic algorithm automatically and effectively. By defining the cutoff distance as a function of the parameter K, FDPC+IDNA is more efficient than DPC, DBSCAN and K-means. Experiments on several synthetic datasets and real-world datasets show FDPC+IDNA outperforms other algorithms referenced in this paper.

However, we did not improve the aggregating strategy of traditional DPC algorithm, and this method can also be applied to many practical problems such as image processing and text clustering. These are the directions we will work hard in the future.

References

1. Menéndez, H.D., Barrero, D.F., Camacho, D.: A genetic graph-based approach to the partitional clustering. Int. J. Neural Syst. **24**(3), 1430008 (2014)
2. Peng, H., Wang, J., Shi, P., Pérez-Jiménez, M.J., Riscos-Núñez, A.: An extended membrane system with active membranes to solve automatic fuzzy clustering problems. Int. J. Neural Syst. **26**(03), 1650004 (2016)
3. Bajer, D., Martinović, G., Brest, J., Bajer, D., Martinović, G., Brest, J.: A Population initialization method for evolutionary algorithms based on clustering and cauchy deviates. Expert Syst. Appl. **60**(C), 294–310 (2016)
4. Aksehirli, E., Goethals, B., Müller, E.: Efficient cluster detection by ordered neighborhoods. In: Madria, S., Hara, T. (eds.) DaWaK 2015. LNCS, vol. 9263, pp. 15–27. Springer, Cham (2015). https://doi.org/10.1007/978-3-319-22729-0_2
5. Han, J., Kamber, M.: Data mining: concepts and techniques. Data Min. Concepts Models Methods Algorithms Second Ed. **5**(4), 1–18 (2011)
6. Birant, D., Kut, A.: ST-DBSCAN: an algorithm for clustering spatial-temporal data. Data Knowl. Eng. **60**(1), 208–221 (2007). (Elsevier Science Publishers B. V.)
7. Kisilevich, S., Mansmann, F., Keim, D.: P-DBSCAN: a density based clustering algorithm for exploration and analysis of attractive areas using collections of geo-tagged photos. In: International Conference and Exhibition on Computing for Geospatial Research and Application, p. 38 (2010)
8. Viswanath, P., Pinkesh, R.: l-DBSCAN: a fast hybrid density based clustering method. In: International Conference on Pattern Recognition, pp. 912–915 (2006)
9. He, Y., Tan, H., Luo, W., Feng, S., Fan, J.: MR-DBSCAN: a scalable MapReduce-based DBSCAN algorithm for heavily skewed data. Front. Comput. Sci. **8**(1), 83–99 (2014)
10. Bie, R., Mehmood, R., Ruan, S., Sun, Y., Dawood, H.: Adaptive fuzzy clustering by fast search and find of density peaks. Pers. Ubiquit. Comput. **20**(5), 785–793 (2016)
11. Rodriguez, A., Laio, A.: Clustering by fast search and find of density peaks. Science **344**(6191), 1492–1496 (2014)
12. Yaohui, L., Zhengming, M., Fang, Y.: Adaptive density peak clustering based on K-nearest neighbors with aggregating strategy. Knowl.-Based Syst. **133**, 208–220 (2017)
13. Liang, Z., Chen, P.: Delta-density based clustering with a divide-and-conquer strategy: 3DC clustering. Pattern Recogn. Lett. **73**(C), 52–59 (2016)
14. Zang, W., Ren, L., Zhang, W., Liu, X.: Automatic density peaks clustering using DNA genetic algorithm optimized data field and Gaussian process. Int. J. Pattern Recognit Artif Intell. **31**(08), 1750023 (2017)
15. Zhou, R., Liu, Q., Xu, Z., Wang, L., Han, X.: Improved fruit fly optimization algorithm-based density peak clustering and its applications. Tehnicki Vjesnik **24**(2), 473–480 (2017)
16. Zhang, W., Niu, Y., Zou, H., Luo, L., Liu, Q., Wu, W.: Accurate prediction of immunogenic T-cell epitopes from epitope sequences using the genetic algorithm-based ensemble learning. Plos One **10**(5), e0128194 (2015)
17. Li, D., Luo, L., Zhang, W., Liu, F., Luo, F.: A genetic algorithm-based weighted ensemble method for predicting transposon-derived piRNAs. BMC Bioinf. **17**(1), 329 (2016)
18. Xie, J., Gao, H., Xie, W., Liu, X., Grant, P.W.: Robust clustering by detecting density peaks and assigning points based on fuzzy weighted K-nearest neighbors. Inf. Sci. **354**(C), 19–40 (2016)
19. Du, M., Ding, S., Jia, H.: Study on density peaks clustering based on k-nearest neighbors and principal component analysis. Knowl.-Based Syst. **99**, 135–145 (2016)

20. Du, M., Ding, S., Xue, Y.: A robust density peaks clustering algorithm using fuzzy neighborhood. Int. J. Mach. Learn. Cybern. **9**(7), 1131–1140 (2017)
21. Nasibov, E.N.: Robustness of density-based clustering methods with various neighborhood relations. Elsevier North-Holland, Inc. (2009)
22. Adleman, L.M.: Molecular computation of solutions to combinatorial problems. Science **266** (5187), 1021–1024 (1994)
23. Dai, K., Wang, N.: A hybrid DNA based genetic algorithm for parameter estimation of dynamic systems. Chem. Eng. Res. Des. **90**(12), 2235–2246 (2012)
24. Li, Y., Lei, J.: A feasible solution to the beam-angle-optimization problem in radiotherapy planning with a DNA-based genetic algorithm. IEEE Trans. Bio-Med. Eng. **57**(3), 499–508 (2010)
25. Rogozhin, Y., Verlan, S.: Computational models based on splicing. In: Adamatzky, A. (ed.) Automata, Universality, Computation. ECC, vol. 12, pp. 237–257. Springer, Cham (2015). https://doi.org/10.1007/978-3-319-09039-9_11
26. Neuhauser, C., Krone, S.M.: The genealogy of samples in models with selection. Genetics **145**(2), 519–534 (1997)

Imaging

Multiple Classifier System for Remote Sensing Images Classification

Yunqi Miao[1], Hainan Wang[1,4], and Baochang Zhang[1,2,3(✉)]

[1] School of Automation Science and Electrical Engineering, Beihang University,
Beijing, China
bczhang@139.com
[2] State Key Laboratory of Satellite Navigation System and Equipment
Technology, Shijiazhuang, Hebei, China
[3] Shenzhen Academy of Aerospace Technology, Shenzhen, China
[4] School of Mechanical Engineering, Guizhou University,
Guiyang, Guizhou, China

Abstract. A new multiple classifier system (MCS) is proposed to address the land cover classification problem with remote sensing images. The system considers both effectiveness and efficiency by combining the classifiers pruning and ensemble at the same time, which is realized by transferring these tasks to an optimization problem. Experimental results show that proposed MCS can successfully classify the remote sensing images with high accuracy as well as reduce the computation cost. Besides, the given system generally outperforms the individual classifiers and majority vote scheme applied on all classifiers on different datasets. According to the experiment results, we conclude that our MCS is a promising method for remote sensing images classification problem, especially when the feature is not sufficient.

Keywords: Multiple classifiers system · Remote sensing · Ensemble learning
Classifiers pruning · Image classification

1 Introduction

Remote sensing images are widely used for land cover classification. However, with the characteristic of low illumination quality and low spatial resolution of remote sensing images themselves as well as the rapidly changing environment conditions, the classification of remote sensing images has become a challenge [1, 3]. One effective solution is to generate an ensemble classifier system to combine individual classifiers, termed multiple classifier system (MCS). MCS has been utilized in the remote sensing image classification recently, which is considered to be effective to improve classification performance of remotely sensed images due to its capability of improving both classification accuracy and efficiency. It is one of the fundamental issues in ensemble learning. A powerful system is expected to accurately recognize and classify the land cover from remote sensing images within short time [2].

The contribution of this paper is threefold. Firstly, a multiple classifier system (MCS) is proposed for remote sensing image classification, which combines the

© Springer Nature Switzerland AG 2018
Y. Peng et al. (Eds.): IScIDE 2018, LNCS 11266, pp. 491–501, 2018.
https://doi.org/10.1007/978-3-030-02698-1_43

classifiers pruning and ensemble, further reducing the computation cost as well as increasing the classification accuracy, whose satisfying performance has been given on two public datasets. Secondly, a classifier pruning method based on genetic algorithm is adapted and improved on the extreme learning machine (ELM). Finally, an optimization of the classifiers ensemble objective function is proposed to get the ensemble weight vector of the selected classifiers. Experiment results show that our framework, combining the advantages of classifiers pruning and ensemble, is highly efficient for remote sensing images classification, and achieves much better performance state-of-art approaches in terms of classification accuracy and computation cost.

The rest of this paper is organized as follows. Section 2 provides some related work on MCS in remote sensing image classification. Section 3 presents our proposed MCS system and improved algorithms in detail. Experiments to testify the validation of our approaches are introduced in Sect. 4 and experimental results as well as some discussion are presented in Sect. 5. Section 6 gives the conclusion of this paper.

2   Related Work

MCS was firstly introduced to tackle the classification problem of remote sensing images by Doan and Foody, who combined soft classification methods and found that multiple classifiers combination could improve the accuracy [5]. Existing multiple classifier systems improve the classification performance mainly from two aspects: (1) Increasing the diversity of single classifiers. (2) Improving combining scheme.

The diversity usually can be achieved by extending the source of data, the types of classifiers [12] and manipulating samples [6–8]. Briem et al. applied multiple classifiers system to multi-source remote sensing data [7]. Debeir et al. integrated multiple features and experimented Bagging and subset selection algorithms with high resolution remotely sensed images [6].

Also, the performance of MCS is highly dependent on the classifiers combination scheme [2]. Some early studies utilized the simple averaging or majority voting strategy to decide the result. As the development of ensemble learning, some widely used ensemble methods, such as Bagging, Boosting etc. and their improvement are introduced to this field [9, 10, 13]. Nowakowski [10] used Boosting scheme to combine the simple threshold classifiers based on both spectral and contextual information for land cover classification. Meanwhile, research on other ensemble methods encourages the development of MCS [11–15, 24, 25]. Ceamanos et al. designed a classifier ensemble for classifying hyperspectral data [15]. The output of a set of SVM classifiers are combined together with an additional SVM and achieved a better result than the single classifier.

So far, there are many existing diversity measures, but in the remote sensing literatures, they are rarely used and compared for remote sensing image classification by MCS [4].

3   Multiple Classifiers System

3.1   Architecture of Ensemble Classifiers System

The architecture of the proposed system can be seen in Fig. 1, which considers computation cost and classification accuracy at the same time by combining classifiers pruning and ensemble.

Firstly, features are extracted from the remote sensing images and then downsampled to a lower dimension with algorithm, such as PCA. After the feature reduction, the fixed, but less number of features are selected from the whole feature space to form the training samples of multi-classifiers system. To reduce the redundancy and save the computing time, classifiers are then pruned with a genetic algorithm. Thus, classifiers with large diversity will be selected to constitute the final classifiers subset. Finally, an improved ensemble scheme is proposed to further increase the accuracy of classification.

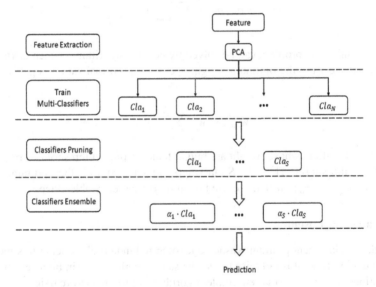

Fig. 1. Architecture of ensemble classifiers system.

3.2   Classifiers Pruning

As mentioned above, the classification results to be combined should be diverse. A successful MCS system depends to a large extent on the proper selection of diverse classifiers for incorporation [4]. Therefore, diversity is a vital requirement for the success of ensemble [21]. Classifiers with the same results are considered to be redundant, which need to be excluded in the ensemble to reduce the computation cost. GASEN [17] is a classifiers pruning method which is used to find out the individual neural networks that should be selected to constitute an ensemble.

In the classification problem, considering there are N base classifiers, each of which is expected to approximate a function $f : R^D \to y$, y is the set of class labels. Firstly, assigning each base classifier with a random weight $w_i (i = 1, 2, \ldots, N)$, presenting its importance in the ensemble, which satisfies Eqs. (1) and (2):

$$\sum_{i=1}^{N} w_i = 1. \tag{1}$$

$$0 < w_i < 1. \tag{2}$$

Then, selecting a subset of classifiers to minimize the generalization error. So the optimum weight vector $w_{opt} = (w_1, w_2, \ldots w_N)$ of base classifiers should satisfy Eq. (3), where C_{ij} presents the correlation of between the i-th and the j-th base classifier.

$$w_{opt} = arg\,min \left(\sum_{i=1}^{N} \sum_{J=1}^{N} w_i w_j C_{ij} \right). \tag{3}$$

This optimization problem can be solved by genetic algorithms, which is known as GASEN. After getting the optimum weight vector w_{opt}, normalization is performed as Eq. (4).

$$w_{opt,i} = w_i / \sum_{i=1}^{N} w_i. \tag{4}$$

Finally, classifiers whose weight are bigger than the pre-set threshold λ are selected to constitute the subset. Then the S classifiers are selected with the optimum weight $w_{opt} = \{w_1, w_2, \ldots, w_S\}$, which are used to initialize the ensemble stage.

3.3 Classifiers Ensemble

Although the classifiers pruning washes out some redundant classifiers, thus reducing the computation time, it is not wise to use the same weight vector in both selection and combination stage [18]. So an ensemble algorithm is proposed to reassign the weight vectors of selected classifiers to further improve the classification accuracy, which is also realized by considering the ensemble process as an optimization problem.

As is known to all, Adaboost is a widely used ensemble algorithm, whose objective function (Eq. (5)) is to minimize the exponential loss function:

$$\min_{w'} \sum_{i=1}^{S} \exp(-yF(w', x)), s.t.\, w' \geq 0, \|w'\|_1 = \delta. \tag{5}$$

$$F(w', x) = \sum_{i=1}^{S} w'_i h_i(x). \tag{6}$$

where y is the label of a sample and x is its feature vector. $h_i(x)$ is the output of the i-th classifier. w' is the weight vector of the selected classifiers subset. The constraint $\|w'\|_1 = \delta$ avoids the arbitrary enlargement of the vector. On the basis of Eq. (5), inspired by [20], linear objective function and multiple class constraint can be incorporated in the given objective function. Thus, an enhanced function is proposed as Eq. (7):

$$\min_{w'} \sum_{i=1}^{S} \exp(-yF(w', \mathbf{x})) + \frac{1}{N_q} \sum_{q=1}^{C} \left(\overline{w_m'^T} \cdot x_p^q - w_q'^T \cdot x_p^q \right) + \lambda \|w'\|_1. \qquad (7)$$

where x_p^q presents the p-th sample in the q-th class with N_q samples. $w_q'^T$ denotes the transpose of the weight vector on the q-th class, and $\overline{w_m'^T} (m \neq j)$ denotes the average of weight on other classes. Interior point method is used to solve the optimum w'.

4 Experiments

4.1 Experimental Datasets

Two remote sensing image benchmarks, UC Merced dataset and NWPU-RESISC45 respectively, were used to verify the performance of proposed algorithm. Figure 2 gives some example images of the two datasets.

Fig. 2. Class representatives of the UC Merced dataset (a, b, c) and NWPU-RESISC45 dataset (d, e, f). (a) airplane (b) intersection (c) forest (d) beach (e) bridge (f) church

The UC Merced dataset [22] consists of 21 land-use types with 100 images in each class. In the experiment, images in each class were further divided randomly into 3 subsets: 80 images for train set and 10 for two test sets respectively.

The NWPU-RESISC45 dataset [23] covers 45 scene classes with 700 images in each class, where we chose 10 classes in the experiment. For each class, 200 images are randomly selected for classifiers training and 50 images for test in both pruning and ensemble stage.

4.2 Experiment Variables and Parameter

The experiment was implemented with Matlab2017a installed in a computer which has i5 processor of 2.4 GHz CPU clock speed with 8 GB RAM. Firstly, multi-scale completed local binary patterns(MS-CLBP) [16] operator was used to extract the shallow feature of the remote sensing images, which was realized by applying CLBP operators with the number of neighbors $m = 8$ and radius $r = \{1, 2, 3, 4, 5, 6\}$ on images with 3 scales, i.e. $s = \{1/3, 1/2, 1\}$. Then, the Fisher vector was further extracted to describe the deeper feature, whose dimension was still not compacted enough for the computation. Therefore, the Fisher vector dimension was further reduced by PCA algorithm. Then, training samples are used for training the base classifiers and two test sets are used to evaluate the diversity of base classifiers and the learn ensemble weight respectively.

To increase the diversity of the base classifiers, the sub-trainset of each classifier was formed by randomly selecting the same but less number of feature from the whole feature space. In the experiments, 5 selected dimension numbers were tested, $F = \{200, 300, 400, 600, 800\}$ for the UC Merced dataset and $F = \{100, 300, 500, 700, 900\}$ for the NWPU-RESISC45 dataset respectively. At the same time, for each dataset, three different number of base classifiers M were chosen to prove the validation of the method in terms of accuracy as well as computation cost. Kernel ELM [19] was chose here as the base classifiers.

In the process of classifier pruning with GASEN algorithm, the threshold used to remove redundant classifiers is set as $\lambda = 1/M$. M is the total number of single classifiers before pruning. Furthermore, the improved ensemble scheme was applied to ensemble the selected classifiers.

5 Results and Analysis

5.1 UC Merced Dataset

Classification Accuracy. The same test set was applied on different stages and the corresponding classification accuracies are given in the Table 1. In the experiment, 30 base classifiers were firstly used. Here, D stands for the number of feature dimension and Sin is the average accuracy of all base classifiers without MCS. Similarly, Pru and Ens are the classification accuracy calculated after classifiers pruning and ensemble process separately. It can be seen that the proposed pruning and ensemble scheme can improve the classification accuracy effectively. To be more intuitive, Fig. 3 shows the rise in the accuracy after the pruning and ensemble process, which shows that ensemble scheme further improves the accuracy after applying the pruning all base classifiers.

From the result, a conclusion can be drawn that compared with simply combining all classifiers, the proposed multiple classifiers system can improve the classification accuracy significantly, especially for the lower feature dimension.

Table 1. Classification accuracy in different stages in MCS on UC Merced dataset

Stage \ Acc \ D	200	300	400	600	800
Sin	50.18%	66.45%	77.69%	85.58%	88.06%
Pru	83.81%	88.10%	87.86%	89.52%	90.71%
Ens	87.86%	90.00%	89.52%	90.71%	91.67%

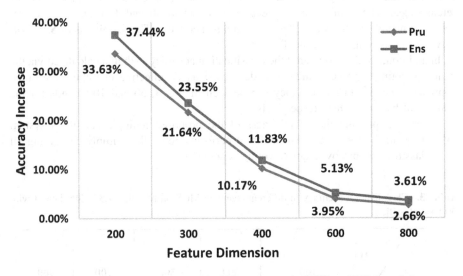

Fig. 3. Classification accuracy rise with different stages in MCS on UC Merced dataset

Computation Cost. Other than classification accuracy, the computation time is another essential measure of MCS. Table 2 gives the ratio of computation time of pruned classifiers T_{pru} to all base classifiers T_{all}. D stands for the feature dimension and N stands for the number of all base classifiers.

Combined with the result in the Fig. 3, it can be found that after the pruning, the accuracy increased remarkably and the computation time was greatly shortened, with the time not exceeding a half of the original time. Particularly, the reduction is evident when the feature dimension is high or the number of base classifiers is large.

Table 2. Ratio of the computation time of pruned classifiers to all base classifiers on the UC Merced dataset

Ratio D N	200	300	400	600	800
30	0.51/1	0.54/1	0.37/1	0.47/1	0.24/1
50	**0.44/1**	0.5/1	0.49/1	0.46/1	0.23/1
100	0.46/1	**0.49/1**	**0.35/1**	**0.16/1**	**0.035/1**

5.2   NWPU-RESISC45 Dataset

Classification Accuracy. Like the first dataset, the same test set was applied on different stages. The number of base classifiers is set to 50 and 100, and their corresponding results are shown in Tables 3 and 4 respectively. The definitions of *Sin, Pru, Ens* are the same as that in the Table 1.

In addition, we also compared the classification accuracy with that calculated via the majority voting scheme, which is widely used as an effective classifiers ensemble approach. The rise in the accuracy can be seen in Fig. 4, 50 and 100 stands for the number of base classifiers respectively.

Other than proving the effectiveness of the proposed multiple classifiers system (MCS) again, the results also show that it performs better than simply combining all base classifiers output by majority voting scheme.

Table 3. Classification accuracy in different stages in MCS on NWPU-RESISC45 dataset with 50 base classifiers

Acc D Stage	100	300	500	700	900
Sin	22.97%	40.22%	59.81%	70.65%	78.52%
Pru	65.80%	82.80%	90.80%	92.80%	91.00%
Ens	**82.40%**	**91.40%**	**91.20%**	**93.60%**	**93.40%**

Table 4. Classification accuracy in different stages in MCS on NWPU-RESISC45 dataset with 100 base classifiers

Stage \ Acc \ D	100	300	500	700	900
Sin	23.31%	41.93%	56.30%	71.78%	78.31%
Pru	75.20%	90.20%	91.60%	91.80%	92.80%
Ens	**87.60%**	**91.60%**	**93.60%**	**93.80%**	**93.60%**

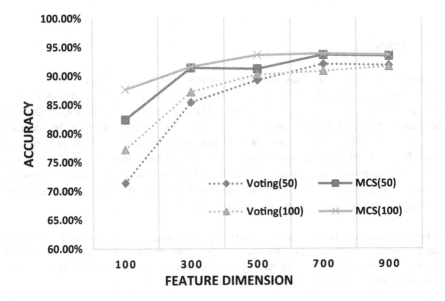

Fig. 4. Classification accuracy comparison between majority voting and MCS on NWPU-RESISC45 dataset with 50 and 100 base classifiers

Computation Cost. The computation cost of the second dataset was also considered and the ratio of computing time before and after pruning process, i.e. T_{pru}/T_{all}, can be found in Table 5. The definitions of D, N are the same as that in the Table 2.

It can be seen that, after the pruning process, the computation time was largely reduced, reaching to only 15% of the time spent without the proposed system.

Table 5. Ratio of the computation time of pruned classifiers to all base classifiers on the NWPU-RESISC45 dataset

N \ Ratio \ D	100	300	500	700	900
30	0.122/1	0.173/1	0.143/1	0.145/1	0.122/1
50	0.119/1	0.150/1	0.146/1	0.157/1	0.127/1
70	0.120/1	0.141/1	0.152/1	0.138/1	0.139/1
100	**0.117/1**	**0.137/1**	**0.145/1**	**0.092/1**	**0.120/1**

6   Conclusion

The paper proposed a multiple classifiers system (MCS) for remote sensing image classification task. The effectiveness and efficiency of the proposed system is verified by 5-fold cross validation on the UC Merced Dataset and NWPU-RESISC45 Dataset. The results indicate that the proposed system can improve the classification accuracy effectively as well as reduce the computation cost efficiently in both pruning and ensemble stage. Meanwhile, according to the observation that proposed MCS can significantly improve the accuracy especially when the feature dimension is low, an inspiration is given that it can be applied to the situation where image features' dimension is not sufficient to achieve a satisfactory result.

Acknowledgements. The work was supported by the Natural Science Foundation of China under Contract 61672079 and 61473086, and Shenzhen Peacock Plan KQTD2016112515134654. This work is supported by the Open Projects Program of National Laboratory of Pattern Recognition, and Shenzhen peacock plan.

References

1. Zhang, Y.: Ten years of technology advancement in remote sensing and the research in the CRC-AGIP lab in GCE. Geomatical **64**, 173–189 (2010)
2. Du, P., Xia, J., Zhang, W., et al.: Multiple classifier system for remote sensing image classification: a review. Sensors **12**(4), 4764–4792 (2012)
3. Ho, T.K.: A theory of multiple classifier systems and its application to visual word recognition (1992)
4. Ranawana, R., Palade, V.: Multi-Classifier Systems: Review and A Roadmap for Developers. IOS Press, Amsterdam (2006)
5. Doan, H.T.X., Foody, G.M.: Increasing soft classification accuracy through the use of an ensemble of classifiers. Int. J. Remote Sens. **28**(20), 4609–4623 (2007)

6. Debeir, O., Latinne, P., Van Den Steen, I.: Remote sensing classification of spectral, spatial and contextual data using multiple classifier systems (2001)
7. Briem, G.J., Benediktsson, J.A., Sveinsson, J.R.: Multiple classifiers applied to multisource remote sensing data. IEEE Trans. Geosci. Remote Sens. **40**(10), 2291–2299 (2002)
8. Tan, K., Jin, X., Plaza, A., et al.: Automatic change detection in high-resolution remote sensing images by using a multiple classifier system and spectral-spatial features. IEEE J. Sel. Top. Appl. Earth Obs. Remote Sens. **9**(8), 3439–3451 (2016)
9. Chen, Y., Dou, P., Yang, X.: Improving land use/cover classification with a multiple classifier system using AdaBoost integration technique. Remote Sens. **9**(10), 1055 (2017)
10. Nowakowski, A.: Remote sensing data binary classification using boosting with simple classifiers. Acta Geophys. **63**(5), 1447–1462 (2015)
11. Kumar, S., Ghosh, J., Crawford, M.M.: Hierarchical fusion of multiple classifiers for hyperspectral data analysis. Pattern Anal. Appl. **5**(2), 210–220 (2002)
12. Maulik, U., Chakraborty, D.: A Robust Multiple Classifier System for Pixel Classification of Remote Sensing Images. IOS Press, Amsterdam (2010)
13. Li, F., Xu, L., Siva, P., et al.: Hyperspectral image classification with limited labeled training samples using enhanced ensemble learning and conditional random fields. IEEE J. Sel. Top. Appl. Earth Obs. Remote Sens. **8**(6), 2427–2438 (2015)
14. Uma Shankar, B., Meher, S.K., Ghosh, A., Bruzzone, L.: Remote sensing image classification: a neuro-fuzzy MCS approach. In: Kalra, P.K., Peleg, S. (eds.) ICVGIP 2006. LNCS, vol. 4338, pp. 128–139. Springer, Heidelberg (2006). https://doi.org/10.1007/11949619_12
15. Ceamanos, X., Waske, B., Atlibenediktsson, J., et al.: A classifier ensemble based on fusion of support vector machines for classifying hyperspectral data. Int. J. Image Data Fusion **1**(4), 293–307 (2010)
16. Chen, C., Zhang, B., Su, H., et al.: Land-use scene classification using multi-scale completed local binary patterns. Sig. Image Video Process. **10**(4), 745–752 (2016)
17. Zhou, Z., Wu, J., Tang, W., et al.: Ensembling neural networks: many could be better than all. Artif. Intell. **137**(1–2), 239–263 (2002)
18. Zhou, Z.H., Wu, J.X., Jiang, Y., et al.: Genetic algorithm based selective neural network ensemble. In: International Joint Conference on Artificial Intelligence, pp. 797–802. Morgan Kaufmann Publishers Inc., Burlington (2001)
19. Huang, G.B., Zhou, H., Ding, X., et al.: Extreme learning machine for regression and multiclass classification. IEEE Trans. Syst. Man Cybern. Part B **42**(2), 513 (2012)
20. Zhang, B., Yang, Y., Chen, C., et al.: Action recognition using 3D histograms of texture and a multi-class boosting classifier. IEEE Trans. Image Process. **26**(10), 4648–4660 (2017)
21. Zhou, Z.H.: Ensemble Methods: Foundations and Algorithms. Taylor & Francis, Boca Raton (2012)
22. Yang, Y., Newsam, S.D.: Bag-of-visual-words and spatial extensions for land-use classification. In: Advances in Geographic Information Systems, pp. 270–279 (2010)
23. Cheng, G., Han, J., Lu, X.: Remote sensing image scene classification: benchmark and state of the art. Proc. IEEE **105**(10), 1865–1883 (2017)
24. Zhang, B., Luan, S., Chen, C., et al.: Latent constrained correlation filter. IEEE Trans. Image Process. **99**, 1 (2017)
25. Zhang, B., Perina, A., Li, Z., et al.: Bounding multiple Gaussians uncertainty with application to object tracking. Int. J. Comput. Vis. **118**(3), 364–379 (2016)

Joint Subspace and Dictionary Learning with Dynamic Training Set for Cross Domain Image Classification

Yufeng Qiu, Songsong Wu$^{(\boxtimes)}$, Kun Wang, Guangwei Gao, and Xiaoyuan Jing

College of Communication Information Enginering, NUPT, Nanjing, China
{b15011317,sswu}@njupt.edu.cn

Abstract. The assumption that training samples and test samples obey the same distribution is grossly violated when images are from distinct domains, which can lead to degradation in classification performance. In this paper, we propose a sparse representation based domain adaptation approach to tackle the cross-domain image classification problem. Specifically, our method aims to alleviate the distribution discrepancy and preserve the discriminative information from source domain, which is achieved by jointly learning a common subspace and a discriminative dictionary. We update the subspace and dictionary with dynamic training samples by selecting the target samples with high prediction confidence and adding them to the source sample set. We evaluate the proposed method on two cross-domain image datasets under both the unsupervised and semi-supervised domain adaptation scenarios. The experimental results demonstrate our method is more effective than competing domain adaptation methods for cross-domain image classification.

Keywords: Domain adaptation · Image classification
Dictionary learning · Jointly learning

1 Introduction

Traditional machine learning methods have good performance under a common assumption: the training data in source domain and testing data in target domain are drawn from the same distribution, which is seldom satisfied in real applications. For example, the distribution divergence of images in visual recognition applications is caused by different sensor types, camera viewpoints, illumination conditions, degrades the classification performance. Most models need to be retrained by target domain data to avoid degradation, while it' s expensive to collect sufficient labeled data in some applications. Domain adaptation is desired to address this problem by transferring knowledge between domains, which has proven to be promising in image classification [9], image denoising [7], object recognition [2,16], face recognition [31], WIFI localization [21] etc.

The domain discrepancy is the major obstacle to the degradation of classifier performance in target domain. A natural idea to address this issue is reducing

© Springer Nature Switzerland AG 2018
Y. Peng et al. (Eds.): IScIDE 2018, LNCS 11266, pp. 502–517, 2018.
https://doi.org/10.1007/978-3-030-02698-1_44

the difference between two domains. The methods of reducing domain discrepancy can be divided into three categories: the classifier adaptation approach, the instance reweighting approach, and the feature adaptation approach.

The classifier adaptation aims to learn an adaptive classifier on target domain leveraging labeled data on source domain [3,11,28]. The classifier adaptation method optimize the objective function directly but they cannot transfer the function to novel categories. Some methods have been developed to solve this problem by adapting classifier and transforming feature jointly. The instance reweighting approach reduce sample bias by reweighting source instances according to the 'mismatch level' between source and target domain instances [5,16,34].

The feature adaptation method is the most frequently used approach in domain adaptation. A feature argument approach replicates N dimensional source and target domain on to $3N$ dimension [6]. The feature transformation approaches transform original feature to alleviate mismatch of two domains [1,9]. The subspace learning-based approaches find a latent space and project data onto intermediate subspace to reduce the mismatch. [10,15,17,21].

Dictionary learning, which belongs to feature adaptation, has shown good performance in domain adaptation owes to the fact that high-dimensional features can be well-represented by an over-complete dictionary for further processing [29]. Most domain adaptive unsupervised and semi-supervised dictionary learning methods learn a shared dictionary to reconstruct both source and target domain data [23,24]. Also, inspired by FDDL [32], some methods [25,26] learn a semi-supervised or unsupervised dictionary consisting of class-specific sub-dictionaries which have good discriminative ability for samples. [19] is a unsupervised domain adaptation method which gives out a pseudo-label to each unlabeled data on target domain. And use the pseudo-label to update the dictionary.

Recently, deep neural network has shown significant promotion of performance on supervised learning tasks. [33] shows that DNN have more transferable features for domain adaptation, which investigated how features are transferable between DNNs. Also, [18,30] use MMD to reduce the discrepancy of the hierarchical nonlinear representations of inputs.

As mentioned above, several problems on feature adaptation motivated us to propose the method:

- Previous works seldom address the discrepancy issue by both subspace learning and feature representation method such as dictionary learning. It's still an open set for domain adaptation. But sometimes the difference of distribution is so huge that we can't handle the discrepancy with either of them.
- Few unsupervised/semi-supervised dictionary learning approaches exploit the latent discriminative information hidden in the unlabeled target domain data. While those approaches that use unlabeled target domain data usually train the dictionary with fixed training set (i.e. the pseudo-label only labeled to target domain data once). And it can't assure the credibility of the pseudo-label.

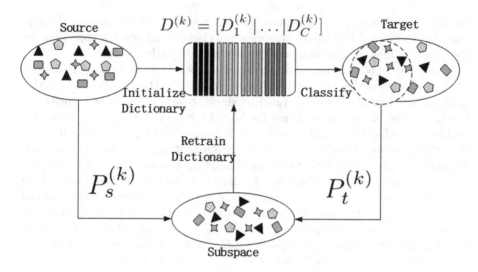

Fig. 1. The overview of our method

In this paper, we propose a dictionary learning method with a mechanism of dynamically constructing training set, which explicitly reduce the discrepancy between two domains. We jointly learn projections of data with the Maximum Mean Discrepancy constraining and a common dictionary to reduce the domain divergence. Then, we iteratively choose target samples with highly credible latent discriminative information and add them into training process to relearn the projection matrix and dictionary. In each iteration, we update the pseudo-label on target domain and dynamically construct training set by the credibility of pseudo-label. In summary, the main contribution of this work are twofold (Fig. 1).

1. Propose a dictionary learning-based method that can explicitly reduce the discrepancy between source and target domain by jointly learning a subspace with MMD constrained and a discriminative dictionary.
2. A dynamic training samples construction approach to make use of latent discriminative information on target domain samples with high credibility.

2   Related Work

In this section, we discuss prior works on transfer learning which are related to ours and analyze their differences.

According to the survey [22], domain adaptation is transfer learning which aims to improve the model performance on target domain which is different but related to source domain without leveraging the abundant labeled data. And as the categories divided above, our method falls down into the feature adaptation approach.

It is particularly relevant to our work that the idea of learning a feature representation which is domain-invariant. [21] use MMD to minimize the bias of data. [10] proposed a kernel-based technique and embed datasets into Grass-mann manifolds, which ingeniously import Grassmann manifolds into domain adaptation method. [8] aligns the source subspace with the target subspace to train a domain adaptive classifier. [17] use a nonparametric distance measure to minimize the distribution discrepancy.

Dictionary learning-based method linearly decompose images or signals as a combination of few atoms from an over-complete dictionary [29,35]. A typi-cal and significant algorithm is [32] which presents a novel method using Fisher discrimination criterion i.e. a class specified dictionary to classify data with the reconstruction error. [20] proposed a method to incrementally learn an inter-mediate domain dictionaries to handle the underlying domain shift in unsuper-vised scenario. [19] iteratively choose samples with its pseudo-label to reduce the domain mismatch smoothly. However, the approach would utilize the sam-ple with low credibility in early iteration which may mislead the classifier. [25] learn a discriminative dictionary in a common low-dimensional space, which the source and target data project on.

[4] introduce the concept of open set that source and target domain only share a subset of object classes whereas most samples of the target domain belong to classes not present in the source domain. Our work is in the frame of close set, we would like to expand our method onto open set scenario in our future work.

3   Proposed Method

In this section, we introduce a dictionary learning-based method which dynami-cally transfer samples for unsupervised learning scenario. We begin with describ-ing notations used in the paper.

We use $Y_s = [y_1, \ldots, y_{N_s}] \in \mathbb{R}^{n_1 \times N_s}$, $Y_t = [y_1, \ldots, y_{N_t}] \in \mathbb{R}^{n_2 \times N_t}$ to denote the data from source and target domain, where n_1, N_s denote the dimension and number of data in source domain respectively. n_2, N_t do so in target domain. Let $L = \{1 \cdots C\}$ represent the existing label set. Let $D^{(k)} = [D_1^{(k)} | \ldots | D_C^{(k)}] \in \mathbb{R}^{n \times K}$ denote the K-atoms dictionary on k^{th} iteration. $D_j^{(k)}$ represent the sub-dictionary that corresponds to class j. $X = [x_1, \ldots, x_N] \in \mathbb{R}^{K \times N}$ is the sparse representation of Y over D.

3.1   Joint Subspace and Dictionary Learning

Our method tries to exploit discriminative information on target domain with high credibility, which can reduce the discrepancy between source and target domain. However, the target domain data are unlabeled which means we can't use them to train the model directly. As the result, we use a discriminative dictionary $D^{(0)}$ trained by source domain data to associate pseudo-label for each target sample. Due to the bad performance of $D^{(0)}$, we choose samples with high credibility to project onto a sub-space. And we jointly learn the project matrix and dictionary $D^{(1)}$. In the same way, we classify the target domain data with $D^{(k-1)}$ and use those data with pseudo-label to relearn the $D^{(k)}$ iteratively, we named the samples whose pseudo-label associated by $D^{(k)}$ was different from label associated by $D^{(k-1)}$ as stable sample. The iteration stops only when the number of stable samples is more than Q. Then, We output the pseudo-label from k^{th} iteration as the result.

3.1.1   Initialize Dictionary and Pseudo-labels

In the first iteration, we train a discriminative dictionary $D^{(0)}$ on source domain data with FDDL [32] and associate pseudo-label on target domain data using reconstruction error defined below:

$$e_{ij}^{(0)} = \|y_i^t - D_j^{(0)} X_{ij}^{(0)}\|_2^2 \tag{1}$$

where $X_{ij}^{(0)}$ is the sparse coefficient vector of the ith sample associated with class j. The pseudo-label u_i of y_i classified by $D^{(0)}$ is made by:

$$u_i = \arg\min_j e_{ij} \tag{2}$$

We denoted the pseudo-labels as $U^{(0)} = [u_1, u_2, \cdots u_{N_t}]$. And we choose samples with credible pseudo-label as Y_t to retrain dictionary. The criterion of choosing named confidence are defined in following section.

3.1.2   Retrain Dictionary

The initialized approach is low-adaptive due to the distribution discrepancy between source and target domain. Our method iteratively learn a K-atoms discriminative dictionary $D^{(k)} = [D_1^{(k)}|\ldots|D_C^{(k)}] \in \mathbb{R}^{n \times K}$ on a common n-dimensional space, which is similar with [25]. In k^{th} iteration, We map source and whole target domain data onto the space by project matrix $P_s^{(k)} \in \mathbb{R}^{n \times n_s}$, $P_t^{(k)} \in \mathbb{R}^{n \times n_t}$ respectively. ($Y_s, Y_t, D, P_s P_t$ etc. was regard as $Y_s^{(k)}, Y_t^{(k)}, D^{(k)} P_s^{(k)} P_t^{(k)}$ in this section) We seek to minimize the following cost function for low representation error

$$\mathcal{C}_1 = \|P_s Y_s - DX_s\|_F^2 + \mu\|P_s Y_s - DX_{sin}\|_F^2 + v\|DX_{sout}\|_F^2$$
$$+ \|P_t Y_t - DX_t\|_F^2 + \mu\|P_t Y_t - DX_{tin}\|_F^2 + v\|DX_{tout}\|_F^2$$

where μ and v are the weights influence the dictionary. The update of target domain samples Y_t will be introduced in next section. And the matrices $\tilde{X}_{in}$ and $\tilde{X}_{out}$ are given as:

$$\tilde{X}_{in} = \begin{cases} \tilde{X}(i,j), & D_i, \tilde{Y}_j \in \text{same class} \\ 0, & \text{otherwise} \end{cases}$$

$$\tilde{X}_{out} = \begin{cases} \tilde{X}(i,j), & D_i, \tilde{Y}_j \in \text{different class} \\ 0, & \text{otherwise} \end{cases}$$

The formula (2) penalizes the reconstruction of the sub-dictionaries do not correspond to class of sample and encourages the corresponding one.

The data among the domains have large difference of distribution, even in the reduced space. We handle the domain shift by using Maximum Mean Discrepancy (MMD) to regularize project process.

$$C_2(P_s, P_t) = \|\frac{1}{N_s}\sum_{i=1}^{N_s} P_s^T x_{1i} - \frac{1}{N_t}\sum_{j=1}^{N_t} P_t^T x_{2j}\|_H^2$$

It's possible to lose information with the project process. To alleviate that, we add a regularization term of reversed projection which preserves energy in the original signal, given as:

$$C_3(P_s, P_t) = \|Y_s - P_s^T P_s Y_1\|_F^2 + \|Y_t - P_t^T P_t Y_t\|_F^2$$

The above cost $C_1 C_2 C_3$ can be rewritten after algebraic manipulations as:

$$C_1(D, \widetilde{P}, \widetilde{X}) = \|\widetilde{P}\widetilde{Y} - D\widetilde{X}\|_F^2 + \mu\|\widetilde{P}\widetilde{Y} - D\widetilde{X}_{in}\|_F^2 + v\|D\widetilde{X}_{out}\|_F^2 \quad (3)$$

$$C_2(\widetilde{P}) = tr((\widetilde{P}\widetilde{Y})\widetilde{M}(\widetilde{P}\widetilde{Y})^T) \quad (4)$$

$$C_3(\widetilde{P}) = -tr((\widetilde{P}\widetilde{Y})(\widetilde{P}\widetilde{Y})^T) \quad (5)$$

where $\widetilde{P} = [P_s P_t]$, $\widetilde{Y} = \begin{pmatrix} Y_s & 0 \\ 0 & Y_t \end{pmatrix}$, and $\widetilde{X} = [X_s X_t]$. The MMD matrix M is the form of:

$$M_{ij} = \begin{cases} \frac{1}{N_s^2}, & \widetilde{y}_i, \widetilde{y}_j \in Y_s \\ \frac{1}{N_t^2}, & \widetilde{y}_i, \widetilde{y}_j \in Y_t \\ -\frac{1}{N_s \times N_t}, & \widetilde{y}_i \in Y_s, \widetilde{y}_j \in Y_t \end{cases}$$

We enforce orthogonal constraint on columns of projection matrices P_s and P_t, in order to prevent the solution from becoming degenerate. Also enforce sparsity constraints on X_s and X_t.

The overall objective function is given as formula (6):

$$\{D^*, P^*, X^*\} = \arg\min_{D, \widetilde{P}, \widetilde{X}} C_1(D, \widetilde{P}, \widetilde{X}) + \lambda_1 C_2(\widetilde{P}) + \lambda_2 C_3(\widetilde{P})$$

$$\text{s.t.} P_i P_i^T = I (\text{i for s,t}) \text{ and } \|x_j\|_0 \leq T_0 \quad \forall j \quad (6)$$

where T_0 is the sparsity level.

3.1.3 Optimization

We solve the optimization problem (6) with proposition in [25] as follow:

$$P_i^* = (Y_i A_i)^T, \text{i for s,t} \quad (7)$$

$$D^* = \widetilde{P}\widetilde{Y}\widetilde{B} \quad (8)$$

where $\widetilde{P}^* = [P_s^*, P_t^*]$, for some $\widetilde{B} \in \mathbb{R}^{\sum N_i \times K}$ for some $A_i \in \mathbb{R}^{N_i \times n}$. The objective function can be rewritten with the proposition:

$$C_1(\widetilde{A}, \widetilde{B}, \widetilde{X}) = \|\widetilde{A}^T \mathcal{K}(I - \widetilde{B}\widetilde{X})\|_F^2 + \mu\|\widetilde{A}^T \mathcal{K}(I - \widetilde{B}\widetilde{X}_{in})\|_F^2 + \upsilon\|\widetilde{A}^T \mathcal{K}\widetilde{B}\widetilde{X}_{out}\|_F^2$$
$$C_2(\widetilde{A}) = tr((\widetilde{A}^T \mathcal{K})\widetilde{M}(\widetilde{A}^T \mathcal{K})^T)$$
$$C_3(\widetilde{A}) = -tr((\widetilde{A}^T \mathcal{K})(\widetilde{A}^T \mathcal{K})^T)$$

where, $\widetilde{K} = \widetilde{Y}^T \widetilde{Y}$ and $\widetilde{A} = [A_s^T, A_t^T]$, $\widetilde{M}$ is the MMD matrix mentioned before. The optimization problem can transform as:

$$\{A^*, B^*, X^*\} = \arg \min_{\widetilde{A}, \widetilde{B}, \widetilde{X}} C_1(\widetilde{A}, \widetilde{B}, \widetilde{X}) + \lambda_1 C_2(\widetilde{A}) + \lambda_2 C_3(\widetilde{A})$$
$$\text{s.t.} A_i^T \mathcal{K} A_i = I, \mathcal{K}_i = \widetilde{Y}_i^T \widetilde{Y}_i, \forall i = s, t \text{ and } \|x_j\|_1 \leq T \quad \forall j \tag{9}$$

As the [25] proposition, we update $\widetilde{A}$ as:

$$\widetilde{A}^* = V S^{-\frac{1}{2}} \widetilde{G} \tag{10}$$

where V and S are the production of eigendecomposition of $\widetilde{K} = VSV^T$, and $G^* \in \mathbb{R}^{n \times K \sum N_i \times n} = [G_s^*, G_t^*]$ is the optimal solution of the formula:

$$G^* = \arg \min_G \text{trace}[G^T H G]$$
$$\text{s.t.} G_i^T G_i = I \quad i = s, t \tag{11}$$

where,

$$H = S^{\frac{1}{2}} V^T (\lambda_1 \widetilde{M}I + (I - \widetilde{B}\widetilde{X})(I - \widetilde{B}\widetilde{X})^T + \mu(I - \widetilde{B}\widetilde{X}_{i}n)(I - \widetilde{B}\widetilde{X}_{i}n)^T$$
$$+ \upsilon(\widetilde{B}\widetilde{X}_{o}ut)\widetilde{B}\widetilde{X}_{o}ut)^T - \lambda_2 I) V S^{\frac{1}{2}} \tag{12}$$

The optimization problem involves the orthonormality condition on G_s and G_t. It would be solved on Stiefel manifold. We use the method in [25] to solve it.

Till now, we get the values of $\widetilde{A}$, then we fix $\widetilde{A}$ to update $\widetilde{B}$ and $\widetilde{X}$. The data can be rewrite as $Z = \widetilde{A}^T \widetilde{K}$, the dictionary $D = \widetilde{A}^T \widetilde{K}\widetilde{B}$. The framework in [32] can be used to solve D and X. Once the dictionary D is learned, we can update $\widetilde{B}$ as:

$$\widetilde{B} = (Z^T Z)^{-1} Z^T D \tag{13}$$

3.1.4   Kernellization

It is inefficient to project the original feature as the non-linear structure of data. We map the data onto a high-dimensional space before projecting them to solve the problem. Let $\Phi : \mathbb{R}^{n_i} \rightarrow \mathcal{H}$ be a mapping from source and target domain space to the reproducing kernel Hilbert space $\mathcal{H}$, the projection $\mathcal{P}_i : \mathcal{H} \rightarrow \mathbb{R}^n$ maps to the low dimensional space. Let $\mathcal{K} = \langle \Phi(\widetilde{Y}), \Phi(\widetilde{Y}) \rangle_{\mathcal{H}}$ be the kernel matrix. $\mathcal{P}_i$ can be represented as

$$\mathcal{P}_i = (\Phi(Y_i)A_i)^T$$

$$D^* = \widetilde{A}^T \mathcal{K} \widetilde{B}$$

for some matrix $A_i \in \mathbb{R}^{N_i \times n}$.

The cost function can be rewrite as:

$$\mathcal{C}_1(\widetilde{A}, \widetilde{B}, \widetilde{X}) = \|\widetilde{A}^T \mathcal{K}(I - \widetilde{B})\widetilde{X})\|_F^2 + \mu \|\widetilde{A}^T \mathcal{K}(I - \widetilde{B}\widetilde{X}_{in})\|_F^2 + v \|\widetilde{A}^T \mathcal{K} \widetilde{B} \widetilde{X}_{out}\|_F^2$$

$$\mathcal{C}_2(\widetilde{A}) = tr((\widetilde{A}^T \mathcal{K})\widetilde{M}(\widetilde{A}^T \mathcal{K})^T)$$

$$\mathcal{C}_3(\widetilde{A}) = -tr((\widetilde{A}^T \mathcal{K})(\widetilde{A}^T \mathcal{K})^T)$$

$$\{A^*, B^*, X^*\} = \arg \min_{\widetilde{A}, \widetilde{B}, \widetilde{X}} \mathcal{C}_1(D, \widetilde{A}, \widetilde{X}) + \lambda_1 \mathcal{C}_2(\widetilde{A}) + \lambda_2 \mathcal{C}_3(\widetilde{A}) \tag{14}$$

$$\text{s.t. } A_i^T \mathcal{K} A_i = I, \mathcal{K}_i = \langle \Phi(\widetilde{Y}), \Phi(\widetilde{Y}) \rangle_{\mathcal{H}}, \forall i = s, t$$

3.2 Dynamically Training Samples Constructing

For exploiting latent discriminative information in unlabeled target domain samples, we use the learned dictionary in $k^{(th)}$ to classify the whole target domain samples to update the pseudo-label $U_{(k+1)}$. Then we choose samples with highly credible pseudo-label to retrain the dictionary using following approach.

3.2.1 Pseudo-label Update

We update the pseudo-label U in k^{th} iteration by reclassifying target samples by $D^{(k)}$. For getting a better performance, we project the dictionary $D^{(k)}$ into the feature space with $P^{(k)}$ and assign the class j with minimum error as pseudo-label to target sample y_{ti}

$$u_i^{(k+1)} = \arg \min_j \|\Phi(y_{ti}) - \mathcal{P}_t^T D_j^{(k)} X_{ij}^{(k)}\|_F^2 \tag{15}$$

3.2.2 Target Samples Update

Previous works on unsupervised learning usually train classifier with fixed samples along with their pseudo-labels. It often come with the low credibility which would mislead the classifier. As a result, we determine the reconstruction error for each sample and each class. In $(k)^{th}$ iteration, we dynamically choose target domain samples with high credibility using the confidence matrix $F^{(k)}$ updated by current sub-dictionaries $D^{(k)} = [D_1^{(k)}| \dots |D_C^k]$. The confidence of pseudo-label for $i^{(th)}$ sample is defined as below:

$$f_{ij}^{(k)} = \begin{cases} \dfrac{\frac{1}{\sqrt{2}\sigma}exp(-\frac{e_{ij}^{(k)}}{2\sigma^2})}{\sum\limits_{l=1}^{C}\frac{1}{\sqrt{2}\sigma}exp(-\frac{e_{il}^{(k)}}{2\sigma^2})}, & \text{if } j = u_i^{(k)} \\ \\ 0 & \text{otherwise} \end{cases} \qquad (16)$$

where $u_i^{(k)}$ is the most likely class that sample i belong to i.e.pseudo-label and $e_{iu}^{(k)}$ denotes the reconstruction error of target samples over $D^{(k)_j}$ and σ^2 is a normalization parameter

$$e_{ij}^{(k)} = \begin{cases} \|y_i^t - D_j^{(0)}X_{ij}^{(0)}\|_2^2, & k = 0 \quad \text{i.e. initial iteration} \\ \|\Phi(y_{ti}) - \mathcal{P}_t^T D_j^{(k)}X_{ij}^{(k)}\|_2^2, & k = 1, 2\dots \end{cases} \qquad (17)$$

For each class that pseudo-labels are assigned to, we sort samples by their confidence and choose the Q% samples maximize the confidence using the approach like [19]. We select samples used to retrain dictionary by solving the following optimization problem:

$$W_j(k) = \arg\max_{W_j} tr(W_j F_J^{(k)})$$
$$\text{s.t.}\|W_j^{(k)}\|_0 = Q * N_j \qquad (18)$$

where $Wj \in \mathbb{R}^{N_t \times N_t}$ are diagonal matrices whose diagonal is the jth of W. And N_j denote the number of samples assign to $j^{(th)}$ class. The constraint ensures the most credible samples in each class are selected. As a result, the target domain samples selected to retrain the dictionary for $(k+1)^{th}$ iteration can be updated as follow:

$$Y_{jt}^{(k+1)} = [Y_t W_j^{(k)}] \quad j = 1, \cdots, C \qquad (19)$$

The selected samples have two properties. First, the selected samples are most credible in one iteration so that we use them to train a more reliable classifier. Second, the selected samples have less distribution discrepancy with source domain. We can train the a new classifier with low domain divergence.

3.2.3   Stop Criterion

The iteration keeps changing the pseudo-labels of samples with low credibility. With the iteration goes on, the classifier would more and more adaptive with target domain data. Therefore, the pseudo-label on target domain data will be changeless after several iterations. We named the sample whose pseudo-label in $k^{th}iteration$ is same as pseudo-label in $(k-1)^{th}$ as stable sample. The algorithm with stop when the number of stable samples is more than $Q * N_t$ in target domain.

The proposed approach is summarized in Algorithm 1.

Algorithm 1. Dynamic Samples Selection Domain Adaptation

Input:
    Data $\{Y_s\}$ with corresponding label C and $\{Y_t\}$, sparsity level T_0, parameter values
    $\lambda_1, \lambda_2, \mu, v$, dictionary size K, dimension n, and selection parameter Q
1: Initialize dictionary $D^{(0)}$ with FDDL trained on source domain data and initialize
    pseudo-labels $U^{(1)}$ of target domain samples with (2)
2: **repeat**
3:    Choose highly credible samples $Y_t^{(k)}$ with their pseudo-label by (16)(17)(18)(19)
4:    For k^{th} iteration.Retrain the dictionary $D^{(k)}$ using $Y_t^{(k)}$ and $U^{(k)}$ by solving (6)
    Procedure: 1. Initialize $\widetilde{A}$ with SVD to each kernel matrix $\widetilde{K}_i = V_i S_i V_i^T$, set $\widetilde{A}$
    as the matrix of eigen-vectors with top n eigen-values as columns
    2. Update $\widetilde{B}$ with (13)
    3. Update $\widetilde{A}$ with (10)
5:    Update pseudo-labels $U^{(k+1)}$ on whole target domain samples Y_t using (14)
6:    Compute the number of stable samples
7: **until** The number of stable samples is more than $Q*N_t$
Output:
    The pseudo-labels of whole target domain samples $U^{(k+1)}$ as the final classification
    result.

4 Experiments

We evaluate our method on both semi-supervised and unsupervised setting. Two
dataset are used for 2D object classification and face recognition respectively.
For object classification, we evaluate our method on standard benchmark dataset
Office+Caltech10. The CMU Multi-PIE dataset are used for face recognition.
We show the performance of our method comparing with the existing previous
domain adaptation algorithms.

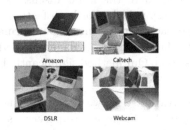
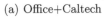
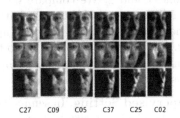

(a) Office+Caltech   (b) CMU-PIE

Fig. 2. (a) Example images from LAPTOP and COMPUTER-KEYBOARD in Amazon, Caltech, DSLR, Webcam. (b) Example images from CMU-PIE

4.1   Object Recognition

Office contains three domains of images: Amazon(A) which consists of images downloading from online merchants. High resolution images consist the DSLR(D). Webcam(W) consists of low resolution. Also tested on *Caltech* − 256(C) dataset as the fourth domain. We use 10 common classes in all domain: BACKPACK, TOURING-BIKE, CALCULATOR, HEADPHONES, COMPUTER-KEYBOARD, LAPTOP, COMPUTER-MONITOR, COMPUTER-MOUSE. For each class, the four database have about 100, 100, 15 and 30 images respectively. Figure 2(a) shows some samples from these datasets.

Unsupervised Setting. In unsupervised setting, following the original protocol in [10], we randomly select 20 samples per class for Amazon/Caltech, 8 samples per class for DSLR/Webcam as source domain. We use all data in A/W/C/D as target domain. The experiment runs 20 times for random splits and we use the average of them as result.

We use the precomputed 800-bin histograms with SURF feature provided in [14] for all the test. The feature obtained from datasets are grouped into 800 cluster using k-means. The centers of cluster to form 800-bin histograms. The final histograms are standardized to have zero mean and unit standard deviation in each dimension. Also, we use the deep-net extracted by VGG-net for deep-net feature experiments. The VGG-net feature provided by [12].

We compare the performance of the proposed method to previous unsupervised domain adaptation method as follow:

- FDDL: A class-specific dictionary with Fisher discrimination criterion [32]
- GFK: The Geodesic Flow Kernel algorithm, and use kernel-NNs to evaluate it [10]
- SA: The Subspace Alignment algorithm. Use 1-NN to evaluate it [8]
- CORAL: The Correlation Alignment algorithm [27]
- ILS: Invariant Latent Space algorithm. Results are evaluated using 1-NN [12].

In Tables 1 and 2, We compare our method to the works discussed above using SURF feature and VGG-FC6 feature respectively. For both feature types, our method outperforms the previous approaches in 6 domain transformations out of 12. It is noteworthy that, due to the lack of labels, our method performs well when the initialization process results good. Our method significantly boosts the performance over initialization.

Semi-supervised Setting. In semi-supervised setting, we also follow the original protocol on Office+Caltech10 provided in [14]. Same as unsupervised setting, we randomly select 20 samples per class for Amazon dataset, and 8 samples per class for other dataset as source domain. When it comes to target domain, we randomly select 3 labeled examples per class. We create 20 random splits and use the average result across them. We use the same feature extracted by SURF and VGG-FC6 as unsupervised setting.

We compare the performance of proposed method to some previous semi-supervised domain adaptation method as follow:

- 1-NN-t: 1 Nearest Neighbor (1-NN) classifier only test on target domain
- MMDT: Jointly learns a linear SVM along with a transformation between two domains [13]
- CDLS: The cross-domain landmark search algorithm [8]
- SDDL: The Shared Domain-adapted Dictionary Learning algorithm [25]
- ILS: Invariant Latent Space algorithm. Results are evaluated using 1-NN [12].

In Tables 3 and 4, we compare our method to the semi-supervised approaches discussed above using SURF feature and VGG-FC6 feature respectively. For both SURF feature and VGG-FC6 feature, our method outperforms the above approaches in 7 domain transformations out of 12, sometimes by a very large margin.

The better performance of our method compared to other methods is mainly attributed to the joint learning of dictionary and low-dimensional space. Good representations are beneficial for data classification. Moreover, the results show that the latent discriminative information in unlabeled target domain samples can improve the performance of the learned classifier.

Parameter Setting. In object recognition, The dimension of the common space is set to 70. Also, we set Q as 0.3 for Amazon and Caltech, and set Q as 0.5 for DSLR and Webcam for the little amount of two dataset samples. The number of atom for each class K is set to 4 to get the highest average classification accuracy. We set the normalization parameter σ as the average reconstruction error e_{ij}. The Fisher discriminative parameter μ and υ are set to 4 and 30 respectively. And $\lambda_1 = 100$ and $\lambda_2 = 2$ can get the highest average classification accuracy.

Table 1. Unsupervised domain adaptation performance comparison SURF on Office+Caltech10 dataset (D: DSLR, W: Webcam, A: Amazon, C: Caltech)

	A→W	A→D	A→C	W→A	W→D	W→C	D→A	D→W	D→C	C→A	C→W	C→D
FDDL	31.2	28.5	24.5	30.6	60.3	22.8	32.1	67.2	32.1	38.1	27	38.2
GFK	35.7	35.1	37.9	35.5	71.2	29.3	36.2	79.1	32.7	40.4	35.8	41.1
SA	38.6	37.6	35.3	37.4	80.3	32.3	38.0	83.6	32.4	39.0	36.8	39.6
CORAL	38.7	38.3	40.3	37.8	84.9	34.6	38.1	85.9	34.2	47.2	39.2	40.7
ILS	40.6	41	37.1	38.6	72.4	32.6	38.9	79.1	36.9	48.6	42	44.1
ours	41.8	36.8	33	42.86	79.6	35.7	39.1	78.4	33.1	49.2	36.25	50.2

Table 2. Unsupervised domain adaptation performance comparison with VGG-FC6 on Office+Caltech10 dataset (D: DSLR, W: Webcam, A: Amazon, C: Caltech)

	A→W	A→D	A→C	W→A	W→D	W→C	D→A	D→W	D→C	C→A	C→W	C→D
FDDL	72.6	64.2	63.9	75.7	87.9	58.9	54.8	81.9	51.2	77.5	65.7	57.5
GFK	74.1	63.5	77.7	77.9	92.9	71.3	69.9	92.4	64.0	86.2	76.5	66.5
SA	76.0	64.9	77.1	76.6	90.4	70.7	69.0	90.5	62.3	83.9	76.2	66.2
CORAL	74.8	67.1	79.0	81.2	92.6	75.2	75.8	94.6	64.7	89.4	77.6	67.6
ILS	82.4	72.5	78.9	85.9	87.4	77.0	79.2	94.2	66.5	87.6	84.4	73.0
ours	87.4	76.4	76.5	84	96.2	75.4	78.8	97.3	65.6	92	81.4	77.8

Table 3. Semi-supervised domain adaptation performance comparison with SURF on Office+Caltech10 dataset (D: DSLR, W: Webcam, A: Amazon, C: Caltech)

	A→W	A→D	A→C	W→A	W→D	W→C	D→A	D→W	D→C	C→A	C→W	C→D
1-NN-t	34.5	33.6	19.7	29.5	35.9	18.9	27.1	33.4	18.6	29.2	33.5	34.1
MMDT	64.6	56.7	36.4	47.7	67.0	32.2	46.9	74.1	34.1	49.4	63.8	56.5
CDLS	68.7	60.4	35.3	51.8	60.7	33.5	50.7	68.5	34.9	50.9	66.3	59.8
SDDL	72.3	60.7	29.5	51.2	62.4	31.8	52.4	73.4	32.2	49.74	60.5	75.9
ILS	59.7	49.8	43.6	54.3	70.8	38.6	55.0	80.1	41	55.1	62.9	56.2
ours	78.2	68.7	38.1	57.0	70.2	36.2	56.9	78.7	39.1	58.6	74.3	79.5

Table 4. Semi-supervised domain adaptation performance comparison VGG-FC6 on Office+Caltech10 dataset (D: DSLR, W: Webcam, A: Amazon, C: Caltech)

	A→W	A→D	A→C	W→A	W→D	W→C	D→A	D→W	D→C	C→A	C→W	C→D
1-NN-t	81.0	79.1	67.8	76.1	77.9	65.2	77.1	81.7	65.6	78.3	80.2	77.7
MMDT	82.5	77.1	78.7	84.7	85.1	73.6	83.6	86.1	71.8	85.9	82.8	77.9
CDLS	91.2	86.9	78.1	87.4	88.5	78.2	88.1	90.7	77.9	88.0	89.7	86.3
SDDL	85.3	66.7	69.7	80.2	89.7	72.9	79.6	90	73.8	82.6	75.6	86.7
ILS	90.7	87.7	83.3	88.8	94.5	82.8	88.7	95.5	81.4	89.7	91.4	86.9
ours	92.4	83.3	80.5	88.3	96.6	80.2	86.7	97.4	85.8	91.7	86.4	90.4

Parameters Sensitive Analysis. In this part, we will analyze the influence of some crucial parameter in our algorithm. Figure 3(a) shows the unsupervised classification accuracy with SURF changed by the atom of each class. The algorithm performs relatively well when the atom per class is between 4 to 6. Also, the ratio of transfer samples over all target samples Q is an important parameter in our algorithm. As the Fig. 3(b) shows, a relatively low Q (e.g.: $Q = 0.2$) will takes more iteration to converge. But high Q will make the algorithm converge so fast that the algorithm can't select the most credible target samples to retrain dictionary.

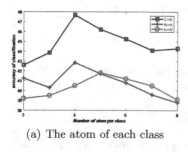

(a) The atom of each class

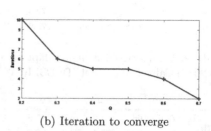

(b) Iteration to converge

Fig. 3. (a) Show the unsupervised classification accuracy with SURF affected by dictionary atoms of each classes. (b) Shows the average iterations needed for convergence as Q varies.

4.2   Face Recognition

In face recognition experiment, we use a benchmark database CMU-PIE to evaluate our method comparing with previous method mentioned above. The CMU Multi-PIE database is a comprehensive face dataset consisting of 337 subjects including four sessions, 15 poses, 20 illuminations and 6 expressions. Figure 2(b) shows the example picture from CMU-PIE.

Following the setting on [12], we use $C27$ (looking forward) as source domain. Also, set $C09$ (looking down) among with $C05, C37, C02, C25$ (looking toward left with increasing angle) as target domains. The increasing angle can be regard as the growing domain discrepancy between two domains. We use the feature provided by [12], which normalizes the images to 32×32 pixels and uses the vectoried gray-scale as feature. We set Q as 0.3 for the large amount of samples each domain. And the dimension of common subspace is set to 100 for the best classification accuracy.

Table 5 shows the result of our method comparing with previous work. The increasing angle leads to the decrease for performance of recognition accuracy. Our method achieves the top rank in the two domains transformation which has large domain discrepancy. Also, the PIE database is a large scale face database. The huge amount of target domain samples improve the accuracy of our method.

Table 5. The results of setting C27 (frontal face) as source domain

	C09	C05	C37	C25	C02
1-NN-s	92.5	55.7	28.5	14.8	11.0
GFK	92.5	74	32.1	14.1	12.3
SA	97.5	85.9	47.9	16.6	13.9
CORAL	91.4	74.8	35.3	13.4	13.2
ILS	96.6	88.3	72.9	28.4	34.8
ours	95.1	85.1	60.7	34.6	38.8

5   Conclusion

In this paper, we proposed a dictionary-based method which jointly learns a common subspace and discriminative dictionary D. And that explicitly reduce the discrepancy between source and target domain. Also, for exploiting the latent discriminative information in unlabeled target domain, we proposed a dynamically training set constructing approach. The extensive experiment on object recognition and face recognition showed the effectiveness of proposed methods.

References

1. Baktashmotlagh, M., Harandi, M.T., Lovell, B.C., Salzmann, M.: Unsupervised domain adaptation by domain invariant projection. In: 2013 IEEE International Conference on Computer Vision, pp. 769–776, December 2013
2. Bousmalis, K., Silberman, N., Dohan, D., Erhan, D., Krishnan, D.: Unsupervised pixel-level domain adaptation with generative adversarial networks. CoRR, abs/1612.05424 (2016)
3. Bruzzone, L., Marconcini, M.: Domain adaptation problems: a DASVM classification technique and a circular validation strategy. IEEE Trans. Pattern Anal. Mach. Intell. **32**(5), 770–787 (2010)
4. Busto, P.P., Gall, J.: Open set domain adaptation. In: 2017 IEEE International Conference on Computer Vision (ICCV), pp. 754–763, October 2017
5. Chu, W.S., Torre, F.D.L., Cohn, J.F.: Selective transfer machine for personalized facial action unit detection. In: 2013 IEEE Conference on Computer Vision and Pattern Recognition, pp. 3515–3522, June 2013
6. Daumé, H.: Frustratingly easy domain adaptation. CoRR, abs/0907.1815 (2007)
7. Elad, M., Aharon, M.: Image denoising via sparse and redundant representations over learned dictionaries. IEEE Trans. Image Process. **15**(12), 3736–3745 (2006)
8. Fernando, B., Habrard, A., Sebban, M., Tuytelaars, T.: Unsupervised visual domain adaptation using subspace alignment. In: 2013 IEEE International Conference on Computer Vision, pp. 2960–2967, December 2013
9. Ghifary, M., Balduzzi, D., Kleijn, W.B., Zhang, M.: Scatter component analysis: a unified framework for domain adaptation and domain generalization. IEEE Trans. Pattern Anal. Mach. Intell. **39**(7), 1414–1430 (2017)
10. Gong, B., Shi, Y., Sha, F., Grauman, K.: Geodesic flow kernel for unsupervised domain adaptation. In: 2012 IEEE Conference on Computer Vision and Pattern Recognition, pp. 2066–2073, June 2012
11. Guo, Z., Wang, Z.J.: Cross-domain object recognition via input-output kernel analysis. IEEE Trans. Image Process. **22**(8), 3108–3119 (2013)
12. Herath, S., Harandi, M., Porikli, F.: Learning an invariant Hilbert space for domain adaptation. In: 2017 IEEE Conference on Computer Vision and Pattern Recognition (CVPR), pp. 3956–3965, July 2017
13. Hoffman, J., Rodner, E., Donahue, J., Kulis, B., Saenko, K.: Asymmetric and category invariant feature transformations for domain adaptation. Int. J. Comput. Vis. **109**(1–2), 28–41 (2014)
14. Hoffman, J., Rodner, E., Donahue, J., Saenko, K., Darrell, T.: Efficient learning of domain-invariant image representations. CoRR, abs/1301.3224 (2013)
15. Long, M., Wang, J., Ding, G., Sun, J., Yu, P.S.: Transfer feature learning with joint distribution adaptation. In: 2013 IEEE International Conference on Computer Vision, pp. 2200–2207, December 2013
16. Long, M., Wang, J., Ding, G., Sun, J., Yu, P.S.: Transfer joint matching for unsupervised domain adaptation. In: 2014 IEEE Conference on Computer Vision and Pattern Recognition, pp. 1410–1417, June 2014
17. Long, M., Ding, G., Wang, J., Sun, J.-G., Guo, Y., Yu, P.S.: Transfer sparse coding for robust image representation. In: 2013 IEEE Conference on Computer Vision and Pattern Recognition, pp. 407–414 (2013)
18. Long, M., Wang, J.: Learning transferable features with deep adaptation networks. CoRR, abs/1502.02791 (2015)

19. Lu, B., Chellappa, R., Nasrabadi, N.M.: Incremental dictionary learning for unsupervised domain adaptation. In: BMVC (2015)
20. Ni, J., Qiu, Q., Chellappa, R.: Subspace interpolation via dictionary learning for unsupervised domain adaptation. In: 2013 IEEE Conference on Computer Vision and Pattern Recognition, pp. 692–699, June 2013
21. Pan, S.J., Tsang, I.W., Kwok, J.T., Yang, Q.: Domain adaptation via transfer component analysis. IEEE Trans. Neural Netw. **22**(2), 199–210 (2011)
22. Pan, S.J., Yang, Q.: A survey on transfer learning. IEEE Trans. Knowl. Data Eng. **22**(10), 1345–1359 (2010)
23. Qiu, Q., Chellappa, R.: Compositional dictionaries for domain adaptive face recognition. IEEE Trans. Image Process. **24**(12), 5152–5165 (2015)
24. Qiu, Q., Patel, V.M., Turaga, P., Chellappa, R.: Domain adaptive dictionary learning. In: Fitzgibbon, A., Lazebnik, S., Perona, P., Sato, Y., Schmid, C. (eds.) ECCV 2012. LNCS, vol. 7575, pp. 631–645. Springer, Heidelberg (2012). https://doi.org/10.1007/978-3-642-33765-9_45
25. Shekhar, S., Patel, V.M., Nguyen, H.V., Chellappa, R.: Generalized domain-adaptive dictionaries. In: 2013 IEEE Conference on Computer Vision and Pattern Recognition, pp. 361–368, June 2013
26. Shrivastava, A., Pillai, J.K., Patel, V.M., Chellappa, R.: Learning discriminative dictionaries with partially labeled data. In: 2012 19th IEEE International Conference on Image Processing, pp. 3113–3116, September 2012
27. Sun, B., Feng, J., Saenko, K.: Return of frustratingly easy domain adaptation. CoRR, abs/1511.05547 (2015)
28. Xinxiao, W., Dong, X., Duan, L., Luo, J., Jia, Y.: Action recognition using multilevel features and latent structural SVM. IEEE Trans. Circ. Syst. Video Technol. **23**(8), 1422–1431 (2013)
29. Xu, Y., Li, Z., Yang, J., Zhang, D.: A survey of dictionary learning algorithms for face recognition. IEEE Access **5**, 8502–8514 (2017)
30. Yan, H., Ding, Y., Li, P., Wang, Q., Xu, Y., Zuo, W.: Mind the class weight bias: weighted maximum mean discrepancy for unsupervised domain adaptation. In: 2017 IEEE Conference on Computer Vision and Pattern Recognition (CVPR), pp. 945–954, July 2017
31. Yang, J., Wright, J., Huang, T.S., Ma, Y.: Image super-resolution via sparse representation. IEEE Trans. Image Process. **19**(11), 2861–2873 (2010)
32. Yang, M., Zhang, L., Feng, X., Zhang, D.: Fisher discrimination dictionary learning for sparse representation. In: 2011 International Conference on Computer Vision, pp. 543–550, November 2011
33. Yosinski, J., Clune, J., Bengio, Y., Lipson, H.: How transferable are features in deep neural networks? CoRR, abs/1411.1792 (2014)
34. Yu, Y., Szepesvári, C.: Analysis of kernel mean matching under covariate shift. CoRR, abs/1206.4650 (2012)
35. Zhang, Z., Xu, Y., Yang, J., Li, X., Zhang, D.: A survey of sparse representation: algorithms and applications. IEEE Access **3**, 490–530 (2015)

A 3D Reconstruction System for Large Scene Based on RGB-D Image

Hongren Wang[1], Pengbo Wang[3], Xiaodi Wang[4], Tianchen Peng[4],
and Baochang Zhang[2,4,5(✉)]

[1] School of Electronics and Information Engineering,
Beihang University, Beijing, China
wanghongren616@hotmail.com
[2] State Key Lab Satellite Navigation System and Equipment Technology,
Shijiazhuang, Hebei, China
[3] School of Electronic Information and Electrical Engineering,
Shanghai Jiao Tong University, Shanghai, China
pengtianchen1031@gmail.com
[4] School of Automation Science and Electrical Engineering,
Beihang University, Beijing, China
bczhang@buaa.edu.cn
[5] Shenzhen Academy of Aerospace Technology, Shenzhen, China

Abstract. As an important research topic in the field of computer vision, 3D modeling of complex scene has an extensive application prospect. RGB-D sensor has been widely used to obtain the depth information in recent years. However, the existing processing system is simply suitable for small-scale scene modeling. In order to develop a better algorithm for large-scale complex scene modeling, this paper builds a 3D scene reconstruction system based on RGB-D images, achieving a better performance in accuracy and real time. SIFT algorithm is first to extract key points to match the descriptors between consecutive frames. By converting into three-dimensional space through the intrinsic matrix, the effective pixel points in the images are then reintegrated to establish the spatial point clouds model which is finally optimized by RANSAC algorithm. The experiments are based on the public database and propose the solution of the problems in the system, which provides a platform for basic research work.

Keywords: RGB-D · RANSAC · 3D reconstruction system

1 Introduction

The three-dimensional reconstruction of real scene has been widely applied in the aspect of human life, including game entertainment, virtual reality and smart home. Therefore, an accurate and rapid 3D modeling algorithm is not only of great significance but also full of challenge academically.

© Springer Nature Switzerland AG 2018
Y. Peng et al. (Eds.): IScIDE 2018, LNCS 11266, pp. 518–527, 2018.
https://doi.org/10.1007/978-3-030-02698-1_45

Multiple methods are proposed to solve various problems about 2D images, including OTO Network [14] for Remote Sense Image. Different from ordinary RGB image, we need the depth information to develop the spatial location and geometry information of the objects, which are the essential elements to convert 2D images into 3D space. Traditionally, there are two methods for obtaining depth information: one is the parallel binocular system, and the other is the laser ranging and scanning system. Both of these two either bring about various interference or ignore the color and texture information. With the development of 3D technology, RGB-D sensor equipment is becoming more and more popular because of its convenience and low price. Although RGB-D images have good performance for small scene modeling, the adaptability to complex large scenes still needs to be improved.

Currently, the research of 3D scene reconstruction based on RGB-D data is mainly based on two most widely used databases, New York University's NYU Depth data set [10] and Princeton University's SUN3D Database [13]. These two databases have gathered a large deal of RGB-D data taken from the perspective of human exploration scene, including classrooms, hotel rooms, libraries and laboratories, which provides a more realistic and effective data source for the subsequent research.

Considering the limitation of computation amount and speed, most researchers basically aim at small-scale scene reconstruction. For instance, Xiao et al. [12] applied the scene understanding and modeling to the place-centric scenarios without taking the high complexity into consideration. Likewise, the Kinect SDK which is exploited by Microsoft has developed the Kinect Fusion [8], a database which is not applicable to large complex scenes but only for the central scene modeling. Nevertheless, central scene modeling cannot solve the problems in the larger reconstruction tasks. More accurate and reliable algorithms are still in desperate need.

The research of 3D reconstruction of large-scale and complex scenes has also made some progress. Salas-Moreno et al. [9] proposed a SLAM based approach to achieve 3D reconstruction of large scenes. Bao and Savarese [1] proposed SSFM algorithm based on Semantics. Compared with traditional SfM method, the SSFM algorithm not only enables to recover 3D point clouds, but also can identify and estimate the high-level semantic model of the 3D model, such as the objects and blocks, at the same time. However, as these identification of objects happen simultaneously during the reconfiguration process, a sudden transformation of a scene will lead to the recognition distortion. In order to obtain a more reliable 3D reconstruction algorithm, Xiao et al. [13] put forward a approach combining SfM algorithm with real-time label for objects to achieve a more precise 3D reconstruction. More accurate though the manual calibration method is for 3D modeling, it is out of automatism. Besides, the more amounts of the scenes, the more complicated process will be produced, which is difficult to generalize this method.

In addition to the reliability of 3D scene reconfiguration, real time is also an important index for evaluating the system. In the small central scenes, many algorithms have achieved the real-time reconstruction. Microsoft's Kinect Fusion can not only achieve real-time modeling, but also can accomplish real-time recognition as well the replacement of objects [6] after modeling. However, for large scenes, because of a great deal of refactoring errors existing in the process of scanning scenes, we need to rescan and correct the deviation continuously in the process of reconstruction, which brings great challenges into realizing real time.

In order to get a better study on the method of 3D reconstruction based on RGB-D data in large and complex scenes, this paper builds a complete 3D scene reconstruction system, which will provide platform and reference for further research and exploration in this field. In this system, we firstly use Microsoft's Kinect equipment to acquire standard RGB-D data, and then SIFT algorithm is employed to realize the feature extraction and match the descriptors between the adjacent frames. The camera parameters can be described by the intrinsic matrix, through which it is very convenient to convert image pixels into the space coordinates to develop the point cloud, the camera taken as the origin point. Finally, the RANSAC algorithm is used to eliminate the reconstruction errors. In the process of system building, the design principles of interface and implementation are separated, where all parts are connected in series to make graphical interfaces to achieve the usability and integrity of the whole system. The core reconstruction algorithm is programmed based on MATLAB and the system integration is written in Python.

The rest of this paper are as follows: all of the modules of the system is listed in the second module while each of them is described based on the algorithm in the third section. In the fourth section, parts of the use of technology in this system are described. The results of experiments are also analyzed, based on the analysis of which we put forward the corresponding solutions. This paper finally summarizes the full text, analyzes the shortcomings of the current system, and proposes the following ways of improvement.

2   Reconstruction of 3D Scene Based on RGB-D

2.1   Matching the Selected Feature Points

Recently, there are more and more approaches for image feature extraction, not only hand-crafted methods but also using convolution, including the relatively new one: GoFs [7]. In this system, SIFT (Scale-invariant feature transform) algorithm is used to detect local feature points in the images which contain information of position, scale, direction and rotation invariants. Generally, there are four steps to acquire the key points of an image: (1) to compute potential extreme points in the scale space; (2) to locate key points; (3) to calibrate the key points; (4) to generate key points descriptors.

Accordingly, we initiate to match these creating key point descriptors between two consecutive frames. As is known that the SIFT descriptor is a 128 dimensional vector, it is unpractical to use the basic exhaustive nearest neighbor algorithm directly due to the computation complexity. Therefore, we use k-dimensional tree, that is, kd-tree [3] instead to perform the nearest neighbor search to reduce the computing time. However, the efficiency of kd-tree will greatly decrease if the data embraces a high dimensionality. To solve this problem, the Best bin first [2] algorithm was proposed by Beis and Lowe in 1997. By sorting the split search space based on kd-tree, BBF algorithm enable to return to the nearest neighbor point approximately and ensure the certain accuracy and efficiency in the high-dimensional space.

2.2   3D Point Cloud Reconstruction

Conversion from Plane Pixel to Camera Coordinate System. The effective key points which are selected by the SIFT algorithm need to be converted from the 2D plane to the three-dimensional camera coordinate space. It can be completed through the intrinsic matrix [5] of the camera. The intrinsic matrix consists of several basic parameters of the camera, including the focal length f_x, f_y, the offset of the camera hole x_0, y_0 and the distortion degree s, which is described as follows:

$$K = \begin{pmatrix} f_x & s & x_0 \\ 0 & f_y & y_0 \\ 0 & 0 & 1 \end{pmatrix} \tag{1}$$

where, $f_x = f_y$, $s = 0$.

The location of a pixel x on the image plane is (x', y'), and the depth value is d. From the theory of similar triangles, we can get the spatial location $X(x_c, y_c, z_c)$ of the point under the camera coordinates by Eqs. 2, 3, and 4 through the parameters in the intrinsic matrix.

$$x_c = \frac{(x' - x_0)d}{f_x} \tag{2}$$

$$y_c = \frac{(y' - y_0)d}{f_y} \tag{3}$$

$$z_c = d \tag{4}$$

By this transformation, we convert the key points in each frame into the camera coordinates to obtain the spatial information of the key points.

2.3   Conversion from Camera Coordinate System to Real Coordinate Space

To convert the camera coordinate space into the real coordinate space, the extrinsic matrix is necessary for the each corresponding image frame, which can display the position and the direction of the camera in the space. Typically, it is a rigid transformation matrix composed of the rotation matrix R and a transposed translation vector t, which is expressed as follows:

$$(R|t) = \begin{pmatrix} r_{1,1} & r_{1,2} & r_{1,3} & t_1 \\ r_{2,1} & r_{2,2} & r_{2,3} & t_2 \\ r_{3,1} & r_{3,2} & r_{3,3} & t_3 \end{pmatrix} \tag{5}$$

where R represents the rotation matrix of the real coordinate axis in the camera coordinate space, and t represents the origin point location of the real coordinate system in the camera coordinate space.

Provided that there is a point P in the space, its coordinate can be set as $P_w(x_w, y_w, z_w)$ relative to the origin point in the real coordinate system w, and $P_c(x_c, y_c, z_c)$ relative to the origin point in the camera coordinate system c. Thus, we can establish the mapping coordinate below based on the matrix $(R|t)$:

$$P_c = RP_w + t \tag{6}$$

Because the location of the real coordinate system on the space scale is arbitrary, we select the coordinates of the first frame in the camera coordinate system as the reference system and all the following frames are aligned to the first one. In other words, all the latter camera coordinates are regarded as the real coordinate system w, which are aligned to the first frame of the coordinate system c.

2.4   Optimization of Reconstruction Model

RANSAC (Random Sample Consensus) is a basic tool in the field of image processing, such as image splicing [5]. In this paper, the goal of RANSAC algorithm is to filter as many valid key points obtained by SIFT algorithm as possible to fit the rigid transformation matrix, so as to reduce the occurrence of mismatch.

RANSAC [4] is a model of iterative algorithm, of which the input consists of a set of data including inliers and outlier, a model exploited as the interpretation of the observational data and some confidence parameters. RANSAC algorithm keep carrying on the iteration of the following steps until there are enough inside points.

(1) A subset is randomly selected from the original data set, which is viewed as an assumed inliers set.
(2) The given data model is used to fit the assumed inliers set to calculate the specific model parameters.
(3) For every point in the original data set, we calculate the loss function $L(x)$ of the specific model. If its $L(x)$ satisfies certain condition, the point is identified as a real inlier.
(4) If there are enough assumed inliers identified as real ones, the model is therefore reasonable. Then return to this specific model.
(5) Otherwise, use the identified real inliers to resample, and then repeat step (1) to (5).

3 System Construction, Analysis and Improvement

3.1 System Construction

Data Acquisition. The RGB-D data used in this paper is collected from Microsoft's Kinect, the data-collecting interface showed as Fig. 1, of which the left side displays the obtained RGB image and the right side exhibits the depth image. Due to Microsoft's continuously updating the Kinect, the Kinect Sensor driver used in the open source field has been unable to meet the requirements of the latest version. Consequently, this paper employs Microsoft's official Kinect SDK for Windows to carry out the development of the RGB-D data collection. C# language is used for this development. At the same time, the public database, SUN3D Database, is also used in order to debug periodically and verify the applicability of the algorithm.

Color and Depth

Fig. 1. RGB-D Data acquisition interface.

The Kinect equipment is composed of three camera sensors, a color camera in the middle which is used to collect color images and a light depth sensor made from the cameras from two sides which is mainly applied to obtain depth data. The number of the frames collected by the equipment is 30 fps. For each frame, we will label the ID and time stamp. The resolution of both the RGB image and the depth image is 640 * 480, in which the distance unit of the pixel value in the depth image is millimeter, stored as the 16 bit gray PNG format. In order to facilitate direct checking, the value of the depth image has shifted 3 bit to the left in the storage process, otherwise the depth image will be too dark to recognize.

Due to the existence of parallax between the captured images, the problem of point coordinates mismatch will occur if they are used directly. After collecting RGB data and depth data which are read at the same time, we can acquire a mapping from RGB data matrix to depth data matrix by calling MapColorFrameToDepthFrame API interface [11]. By traversing each RGB data point, depth information can be found in the corresponding depth data. If there is no mapping in success, the depth value is set to 0, and the depth information of this pixel is invalid. Therefore, it is unavoidable to shape a black border in the aligned depth image.

3D Reconstruction. The 3D reconstruction algorithm is the research core of this system, in which part the programming language is exploited in MATLAB. In this paper, we call MATLAB in the way of the command line, requiring not to load the graphical interface at startup to save the system resources. In the specific design of the algorithm, we open multi-core computation based on MATLAB, so as to calculate the rigid transformation matrix between two adjacent frames in different process pools, accelerate parallel and shorten the reconstruction time.

System Integration. Three independent parts including data acquisition, 3D reconstruction and result display, should be systematically integrated, and Python language is used to build the system.

The system design should ensure simplicity and ease of use, and it should also be fault-tolerant for illegal input.

In the first step, enter the RGB-D data directory in the program and click the button of 'collect RGB-D data', and it will pop up the interface of the data acquisition in the partially program.

In the second step, after selecting the path of matlab.exe, clicking the button of 'start 3D reconstruction', the program will automatically call the matlab command line at the back-end and run the refiguration algorithm directly.

In the third step, after selecting the path of meshlab.exe, clicking the button of 'display reconstruction results', the program will automatically search for result.ply files in the RGB-D directory, and reproduce the 3D results by means of the meshlab program.

3.2 Systems Analysis

SIFT Descriptor Matching Speed Analysis. The main idea of SIFT descriptor matching algorithm is nearest neighbor method. We propose three specific methods: direct exhaustive traversal, traditional kd-tree and kd-tree with BBF. However, the time complexity of the first method is too high to be considered. The system compares the running time of the traditional kd-tree and the kd-tree with BBF method, the result of which is shown in Fig. 2.

As we can see in Fig. 2, the speed boost brought by the BBF algorithm is very stable and does not vary greatly with the number of descriptors. By calculating the average value of the percentage of 18 batches, the time used for BBF is 17.8% as much as the time used by traditional kd-tree.

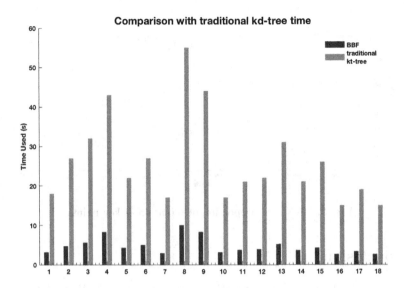

Fig. 2. Comparison with two kd-tree methods on running time.

RANSAC Optimization Analysis. In this paper, the method used to deal with the mismatch of descriptors is the RANSAC algorithm. The key points between the two frames may not be matched by one rigid transformation, or the totally different key points that are directly matched owing to the error problem. In the subsequent process of rigid transformation matrix estimation, if a mismatch happens, the error of the whole spatial coordinate will accumulate backwards, leading to the final distortion which will seriously destroy the 3D model. By using RANSAC, we remove those mismatched points and retain the effective points that can be explained by the model to reduce the error caused by the reconfiguration.

In the same way, we take the first 800 frames of the data to calculate the number of SIFT key points and the number of key points identified by RANSAC. The proportion of the number of the key points finally obtained through the RANSAC and the number of key points obtained by the SIFT algorithm is showed in Fig. 3.

As is seen from Fig. 3, RANSAC algorithm retains most of the SIFT key points, the general ratio within the 92%–98%. Through RANSAC, the key points of mismatch are eliminated effectively. In addition, the RANSAC estimation model is the transformation matrix between the views of two frames, in this way for those key points which do not fit the model, are also cut off, thus achieving the purpose of reducing the errors.

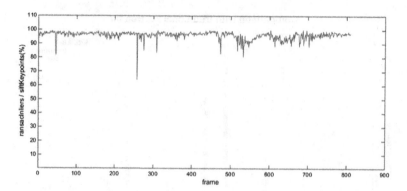

Fig. 3. RANSAC proportion of retention of key points.

4   Conclusion

In this paper, a 3D scene reconstruction system based on RGB-D data is built to study the reconstruction algorithm for large complex scenes. The system uses the SIFT algorithm to match the location of the key descriptors between the adjacent frames. The intrinsic matrix of the camera is used to transform the point coordinates from the image plane into the real coordinate system. Finally, the RANSAC algorithm is adopted to reduce the error and complete the establishment of the whole 3D reconstruction model. With the combination of data collection, 3D reconstruction and results display, a reconfiguration system is achieved. Moreover, we can simplify the operation process by adjusting and optimizing the system so that the entire system can be operated directly by clicking the button. By means of the separation design of the interface and the implementation, the system ensures extensibility and maintainability for the following steps.

The system built in this paper provides a basic platform for further research, but there are still shortcomings. For example, the 3D point cloud is generated, but the texture is lack of sense of reality, which leads to the lack of the true sense of the final model. At the same time, the speed of operation can not be done in real-time. It is necessary to further improve the optimization in the follow-up research.

Acknowledgments. The research is supported partially by the Natural Science Foundation of China under Contract 61672079, 61473086. The work of B. Zhang is supported partially by the Program for New Century Excellent Talents University within the Ministry of Education, China, as well by the Beijing Municipal Science and Technology Commission under Grant Z161100001616005.

References

1. Bao, S.Y.Z., Savarese, S.: Semantic structure from motion. In: CVPR, vol. 2011, pp. 2025–2032 (2011)
2. Beis, J.S., Lowe, D.G.: Shape indexing using approximate nearest-neighbour search in high-dimensional spaces. In: Proceedings of IEEE Computer Society Conference on Computer Vision and Pattern Recognition, pp. 1000–1006 (1997)
3. Bentley, J.L.: Multidimensional binary search trees used for associative searching. Commun. ACM **18**(9), 509–517 (1975)
4. Fischler, M.A., Bolles, R.C.: Random sample consensus: a paradigm for model fitting with applications to image analysis and automated cartography. In: Readings in Computer Vision: Issues, Problems, Principles, and Paradigms, pp. 726–740 (1987)
5. Hartley, R., Zisserman, A.: Multiple View Geometry in Computer Vision (2000)
6. Izadi, S., et al.: KinectFusion: real-time 3D reconstruction and interaction using a moving depth camera. In: Proceedings of the 24th Annual ACM Symposium on User Interface Software and Technology, pp. 559–568 (2011)
7. Luan, S., Chen, C., Zhang, B., Han, J., Liu, J.: Gabor convolutional networks. IEEE Trans. Image Process. 1 (2018)
8. Newcombe, R.A., et al.: KinectFusion: real-time dense surface mapping and tracking. In: 2011 10th IEEE International Symposium on Mixed and Augmented Reality, pp. 127–136 (2011)
9. Salas-Moreno, R.F., Newcombe, R.A., Strasdat, H., Kelly, P.H.J., Davison, A.J.: Slam++: simultaneous localisation and mapping at the level of objects. In: 2013 IEEE Conference on Computer Vision and Pattern Recognition, pp. 1352–1359 (2013)
10. Silberman, N., Hoiem, D., Kohli, P., Fergus, R.: Indoor segmentation and support inference from RGBD images. In: Fitzgibbon, A., Lazebnik, S., Perona, P., Sato, Y., Schmid, C. (eds.) ECCV 2012, Part V. LNCS, vol. 7576, pp. 746–760. Springer, Heidelberg (2012). https://doi.org/10.1007/978-3-642-33715-4_54
11. Smisek, J., Jancosek, M., Pajdla, T.: 3D with kinect. In: 2011 IEEE International Conference on Computer Vision Workshops (ICCV Workshops), pp. 1154–1160 (2011)
12. Xiao, J., Ehinger, K.A., Oliva, A., Torralba, A.: Recognizing scene viewpoint using panoramic place representation. In: 2012 IEEE Conference on Computer Vision and Pattern Recognition, pp. 2695–2702 (2012)
13. Xiao, J., Owens, A., Torralba, A.: SUN3D: a database of big spaces reconstructed using SFM and object labels. In: 2013 IEEE International Conference on Computer Vision, pp. 1625–1632 (2013)
14. Zhang, B., et al.: One-two-one networks for compression artifacts reduction in remote sensing. ISPRS J. Photogramm. Remote Sens. (2018)

Change Detection in Multispectral Remote Sensing Images Based on Optimized Fusion of Subspaces

Yuanyuan Chen[✉], Jianlong Zhang, and Xinbo Gao

School of Electronic Engineering, Xidian University, Xi'an, Shaanxi, China
chenyuanyuan_cyy@yeah.net

Abstract. In this paper, an effective approach is proposed for unsupervised change detection in multispectral remote sensing images. Firstly, the spectral-spatial information joint distribution of multispectral remote sensing images is achieved by multiscale morphological tools. Thus more geometrical details of images are extracted while exploiting the connections of a pixel and its adjacent regions. Subsequently, the difference images of change vector analysis and spectral angle mapper are generated according to the difference of spectral vectors magnitude and direction, respectively. Finally, the two difference images are combined by optimized fusion algorithm named affinity aggregation based on Nyträm spectral clustering to obtain the binary change mask. Experimental results show that the proposed method not only detects weak changes but also effectively maintains the integral geometry of objects.

Keywords: Change detection (CD) · Remote sensing (RS)
Change vector analysis (CVA) · Spectral angle mapper (SAM)

1 Introduction

Change detection (CD) utilizes multi-temporal remote sensing (RS) images in same geographical area at different time to identify changes of the state of ground objects or the difference of natural phenomena. In the past decades, CD has become an increasingly popular research topic due to its various practical applications, e.g., urban, forestry, agriculture, and disaster assessment.

In the previous literatures, numerous CD methods have been proposed, and they mainly could be divided into two categories, i.e., supervised and unsupervised. Supervised methods exploit the trained classifiers to identify changes. For instance, neural network (NN) and SVM are employed by CD methods in a supervised manner, and achieve good results in many cases [1, 2]. Although they are generally robust and effective to different datasets, the detection results seriously rely on the capacity of classifiers. Moreover, it requires the "ground truth", i.e., a large amount of labeled samples, for training classifier. However, it is a bit difficult to meet this requirement, which also impedes the extension of supervised approaches in many real practical applications. On the contrary, unsupervised CD methods do not require any ground truth, and make a direct comparison of multi-temporal images or analyze the different

Y. Peng et al. (Eds.): IScIDE 2018, LNCS 11266, pp. 528–538, 2018.
https://doi.org/10.1007/978-3-030-02698-1_46

image (DI) with pattern recognition techniques to detect the potential change. In general, the widely used unsupervised CD methods include image difference, image rationing, change vector analysis (CVA), image regression. CVA is one of classic methods for CD in multispectral RS images [3, 4]. However, the typical CVA takes the magnitude of spectral vector to extract change information for binary CD, while ignores the influence of the spectral angle. Spectral angle mapper (SAM) is employed for CD in Landsat-5 TM images that improves the accuracy [5]. A program is developed to obtain a better DI by combining SAM and CVA [6]. In the literature [7], an unsupervised distribution-free CD approach is presented based wavelet fusion strategy to generate a better DI. The rules based on an average operator and minimum local area energy are chosen to fuse the wavelet coefficients of low-frequency bands and high-frequency bands, respectively. This method has difficulty in confirming the optimum wavelet function and decomposition level. A novel DI is generated by using the automatic fusion strategy that defines the weight by employing the entropy of the difference images acquired by CVA and SAM in the literature [8]. Entropy fusion method only considers the average amount of information of DI, but ignores its spatial distribution.

On the one hand, traditional unsupervised CD methods neglect the spectral-spatial union-distribution. On the other hand, CVA has been widely used, whereas the usage of SAM is always ignored in binary CD, moreover, there is no an effective mechanism to combine SAM and CVA. For many computer vision applications, the datasets distribute on certain subspaces. A subspace clustering method is proposed to perform clustering on subspace representation of the datasets [9]. In this paper, an unsupervised method is proposed to tackle the aforementioned CD problems. Considering that geometrical structure is well extracted by morphological operators, we employ a series of morphological operators with different scales to construct multiscale subspaces that combines spectral and spatial information. Then, angle and magnitude in multiscale subspaces are extracted via CVA and SAM, furthermore, the two DIs are combined by AANSC. Experiments carried out on two real datasets confirm the effectiveness of the proposed method.

The rest of the paper is organized as follows. Section 2 describes the proposed method, Sect. 3 presents the experimental results, and Sect. 4 draws the conclusion.

2 The Proposed Method

The general scheme of the proposed CD approach is shown in Fig. 1. Let us consider two groups of multispectral images $\mathbf{X}_1$ and $\mathbf{X}_2$ of size $I \times J \times B$ acquired over the same geographical area at times t_1 and t_2, respectively, in which I, J and B are the number of rows, columns, and spectral bands of each image. Multiscale subspaces are built by applying morphological process with multiple structural elements (SEs) of different sizes to the bi-temporal images. Difference images are generated by CVA and SAM based on multiscale subspaces. Finally, two types of difference images are combined by AANSC algorithm to obtain the binary change mask.

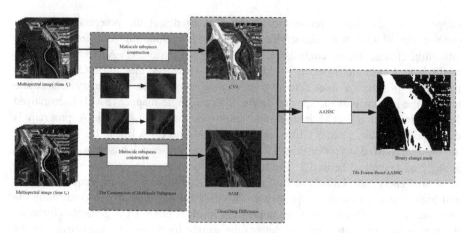

Fig. 1. The working diagram of the proposed CD approach

2.1   The Construction of Multiscale Subspaces

Due to the complexity of the scenes, RS images often contain various kinds of ground objects, each of which corresponds to an optimal expression scale [10]. So, it is more comprehensive that multiscale morphological process is introduced to extract geometrical structure of objects in multispectral image [11]. It can construct multiscale subspaces of original RS image to make spectral-spatial information jointed. Morphological process is defined as a sequence of opening (O) and closing (C) operations by utilizing different SE of different sizes on all bands of bi-temporal multispectral images $\mathbf{X}_1$ and $\mathbf{X}_2$. The operations are capable to suppress brighter and darker regions, respectively. Small isolated objects are fused into a surrounding local background and that main structure is kept while exploiting the interaction of a pixel with its adjacent regions. At a given scale r, i.e., SE having size r, for B-dimensional multispectral images $\mathbf{X}_k (k = 1, 2)$, its opening $O^r(\mathbf{X}_k)$ and closing $C^r(\mathbf{X}_k)$ are also B-dimensional:

$$O^r(\mathbf{X}_k) = \left\{ O^r(\mathbf{X}_k^1), \ldots, O^r(\mathbf{X}_k^b), \ldots, O^r(\mathbf{X}_k^B) \right\} \tag{1}$$

$$C^r(\mathbf{X}_k) = \left\{ C^r(\mathbf{X}_k^1), \ldots, C^r(\mathbf{X}_k^b), \ldots, C^r(\mathbf{X}_k^B) \right\} \tag{2}$$

where $b \in [1, B], k = \{1, 2\}$.

Let $OC^r(\mathbf{X}_k)$ be the stacking of $O^r(\mathbf{X}_k)$ and $C^r(\mathbf{X}_k)$ at a given size r, it has a dimensionality of $2 \times B$ and can be expressed as

$$OC^r(\mathbf{X}_k) = [O^r(\mathbf{X}_k), C^r(\mathbf{X}_k)]. \tag{3}$$

In addition, we define the multiscale space $\mathbf{PX}_k$ to express $\mathbf{X}_k$ that is processed by morphological tool, $\mathbf{PX}_k$ is shown as

$$\mathbf{PX}_k = [OC^\alpha(\mathbf{X}_k), \dots, OC^r(\mathbf{X}_k), \dots, OC^\beta(\mathbf{X}_k)] \tag{4}$$

where $r \in [\alpha, \beta]$.

Note that the size of SE r increases from α to β in order to implement a multiscale analysis and $\mathbf{PX}_k$ has a dimensionality of $L = 2 \times B \times (\beta - \alpha + 1)$.

Multiscale process builds a multiscale response filed via using morphological operators with different SEs, and it contributes to discover different changes with different scales.

2.2 Describing Difference of Multiscale Subspaces

In literature [12], a compressed change representation in a 2-D polar domain is defined to detect multiple changes by two change variables, i.e., the magnitude and the direction. CVA measures the total contribution of spectral change brightness, whereas it is not sensitive to the shape of spectral vectors. SAM measures the similarity between two given spectral vectors, particularly focuses on the shape of the spectrum. We take advantage of CVA and SAM techniques to generate two types of difference image for binary CD.

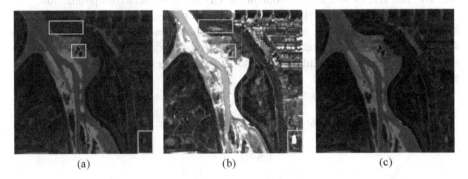

|(a)|(b)|(c)|

Fig. 2. (a) Difference image using SAM. (b) difference image using CVA. (c) the fused image.

Let $\mathbf{S}_1 = (x_1^1, x_1^2, \dots, x_1^L)$ and $\mathbf{S}_2 = (x_2^1, x_2^2, \dots, x_2^L)$ represent the spectral vectors associated to a specified position in considered images i.e., $\mathbf{PX}_1$ and $\mathbf{PX}_2$ of size $I \times J \times L$, respectively. Then, SAM can be calculated by (5)

$$\theta = \arccos \left[\frac{\sum_{b=1}^{L} x_1^b x_2^b}{\sqrt{\sum_{b=1}^{L} (x_1^b)^2} \sqrt{\sum_{b=1}^{L} (x_2^b)^2}} \right], \theta \in [0, 180°] \tag{5}$$

where x_1^b and x_2^b denote the spectral components in the band $b (= 1, 2, \dots, B)$ of multispectral images.

CVA can be calculated by the following Euclidean distance, i.e.,

$$\rho = \sqrt{\sum_{b=1}^{L} (\mathbf{X}_1^b - \mathbf{X}_2^b)^2} \tag{6}$$

Therefore, θ and ρ should be normalized in the range [0, 255] to obtain two grayscale difference images $\mathbf{DI}_1$ and $\mathbf{DI}_2$ individually. $\mathbf{DI}_1$ and $\mathbf{DI}_2$ are shown as Fig. 2 and the complementary information between the two images is mainly shown in the yellow framings.

2.3 DIs Fusion Based AANSC

Compared with other clustering algorithms, spectral clustering algorithm can converge to the global optimal solution in the sample space with arbitrary shape. An affinity aggregation spectral clustering (AASC) algorithm is proposed, and it extends spectral clustering to a setting with multiple feature spaces available in application of small samples clustering [13]. There could be several potentially serviceable difference images in RS image CD and thereby multiple affinity matrices. So we employ AASC to fuse $\mathbf{DI}_1$ and $\mathbf{DI}_2$ of size $I \times J$ to obtain the CD result by iterative optimization.

Due to the correlation between neighboring pixels, the 3×3 neighborhood information of $\mathbf{DI}_1$ and $\mathbf{DI}_2$ are extracted as a feature description, respectively. Each feature is arranged as a column vector $\mathbf{d}$ with size of 9×1, which is substituted into the Gaussian kernel to calculate pairwise affinity between $\mathbf{d}_i$ and $\mathbf{d}_j$, it's shown as formula (7).

$$\omega_{ij} = \exp(-\frac{(\mathbf{d}_i - \mathbf{d}_j)^T (\mathbf{d}_i - \mathbf{d}_j)}{2\sigma^2}) \tag{7}$$

$$1 \leq i, j \leq I \times J$$

Given $N (= I \times J)$ data points $\mathbf{d}_1, \mathbf{d}_2, \ldots, \mathbf{d}_N$ and some pairwise affinity ω_{ij}, spectral clustering divide these data into two clusters i.e., change and unchanged clusters, by finding N indicators $\mathbf{f}_1, \mathbf{f}_2, \ldots, \mathbf{f}_N$, whose objective function is expressed as formula (8). For satisfying the normalized spectral clustering, the constrain $\mathbf{f}^T \mathbf{D} \mathbf{f} = 1$ is required.

$$\min_{\mathbf{f}_1, \cdots, \mathbf{f}_N} \sum_{i,j} \omega_{ij} \| \mathbf{f}_i - \mathbf{f}_j \|^2 \tag{8}$$

In this paper, there are two types of difference image and thereby two symmetrical affinity matrices $\mathbf{W}_k (k = 1, 2)$ of size $N \times N$ are available, element $\omega_{ij;k}$ of $\mathbf{W}_k (k = 1, 2)$ represents the similarity between $\mathbf{d}_i$ and $\mathbf{d}_j$ in the feature space of $\mathbf{D}_k (k = 1, 2)$.

In order to find an appropriate weight distribution of these affinity matrices, we use $\mathbf{V} = [v_1, v_2]^T$ to represent the weight vector. AASC problem can be formulated as

$$\min_{\substack{\mathbf{f}_1,\cdots,\mathbf{f}_N \\ v_1,v_2}} \sum_k \sum_{i,j} v_k \omega_{i,j;k} \left\| \mathbf{f}_i - \mathbf{f}_j \right\|^2$$

$$= \min_{\substack{\mathbf{f}_1,\cdots,\mathbf{f}_N \\ v_1,v_2}} \sum_k v_k \mathbf{f}^T (\mathbf{D}_k - \mathbf{W}_k) \mathbf{f} \tag{9}$$

$$1 = \mathbf{f}^T \mathbf{D} \mathbf{f} = \mathbf{f}^T (v_1 \mathbf{D}_1 + \ldots + v_k \mathbf{D}_k) \mathbf{f} \equiv \sum_k \alpha_k v_k \tag{10}$$

where $\alpha_k = \mathbf{f}^T \mathbf{D}_k \mathbf{f}$, $\mathbf{D}_k$ is the k-th diagonal matrix, and the element of $\mathbf{D}_k$ is calculated as

$$D_{ii;k} = \omega_{i1;k} + \omega_{i2;k} + \ldots + \omega_{iN;k}. \tag{11}$$

To solve the above problem, there are two sets of variables, the indicator vector $\mathbf{f}$ and the weight vector $\mathbf{V}$. On the one hand, if the weight vector $\mathbf{f}$ is given, the problem becomes a conventional spectral clustering. On the other hand, if the indicator $\mathbf{f}$ is given and fixed, the problem can be reduced to a simple 1-D search problem by applying Lagrange multiplier method to the nonconvex quadratic constraint optimization problem.

However, due to the large memory usage of the similarity matrix and the high computational complexity of the Laplacian matrix decomposition process during each iteration, AASC algorithm cannot be directly applied in high resolution RS image analysis. The Nytröm method approximates the eigenvalues and eigenvectors of the matrix based on the integral equation theory [14]. In this paper, we introduce the Nytröm method into AASC model to construct AANSC algorithm to address CD problems for multispectral RS images.

Intuitively, with the assumption that the n samples are randomly chosen from N pixels $(n \ll N)$ in feature space of difference image. Consequently, partition the aggregated affinity matrix $\mathbf{W} = v_1 \mathbf{W}_1 + v_2 \mathbf{W}_2$ as formula (12).

$$\mathbf{W} = \begin{bmatrix} \mathbf{A} & \mathbf{B} \\ \mathbf{B}^T & \mathbf{C} \end{bmatrix} \tag{12}$$

where $\mathbf{A} \in IR^{n \times n}$, $\mathbf{B} \in IR^{(N-n) \times n}$, $\mathbf{C} \in IR^{(N-n) \times (N-n)}$, $\mathbf{A}$ represents the subblock of affinity among the random samples, $\mathbf{B}$ expresses the subblock of affinity from the random samples to the remaining data points, and $\mathbf{C}$ contains the affinity between all of the rest $N-n$ data points in feature space of the k-th difference image. Obviously, it is a challenge to calculate $\mathbf{C}$. $\mathbf{A}$ could be transformed into diagonal forms as $\mathbf{A} = \mathbf{U} \Lambda \mathbf{U}^T$. Letting $\bar{\mathbf{U}}$ denote the approximate eigenvectors of $\mathbf{W}$, the Nytröm extension defines as formula (13).

$$\bar{\mathbf{U}} = \begin{bmatrix} \mathbf{U} \\ \mathbf{B}^T \mathbf{U} \Lambda^{-1} \end{bmatrix} \tag{13}$$

and we use $\tilde{\mathbf{W}}$ to denote the associated approximation of $\mathbf{W_k}$. $\tilde{\mathbf{W}}$ is given as

$$\tilde{\mathbf{W}} = \bar{\mathbf{U}}\mathbf{\Lambda}\bar{\mathbf{U}}^{\mathrm{T}} = \begin{bmatrix} \mathbf{U} \\ \mathbf{B}^{\mathrm{T}}\mathbf{U}\mathbf{\Lambda}^{-1} \end{bmatrix} \mathbf{\Lambda}[\mathbf{U}^{\mathrm{T}} \quad \mathbf{\Lambda}^{-1}\mathbf{U}\mathbf{B}] = \begin{bmatrix} \mathbf{A} \\ \mathbf{B}^{\mathrm{T}} \end{bmatrix} \mathbf{A}^{-1}[\mathbf{A} \quad \mathbf{B}]. \tag{14}$$

AASC problem is solved using a two-step iterative algorithm which alternatively finds the optimal weight vector $\mathbf{V}$ and the optimal indicator $\mathbf{f}$. In the process of affinity matrix eigenvalue decomposition for each iteration, the Nytröm method is used to obtain the approximate eigenvectors of Laplacian matrix by normalized $\mathbf{A}$ and $\mathbf{B}$. Once optimization converge, k-means algorithm is applied to cluster optimal indicator $\mathbf{f}$ to obtain final CD result. Algorithm 1 summarizes the proposed AANSC algorithm.

Algorithm 1 Affinity Aggregation based on Nytröm Spectral Clustering (AANSC)
Input: affinity matrices $\mathbf{W}_k (k = 1, 2)$

      the number of clusters $K(= 2)$

Output: change mask

Initialize $v_k = 1 / 2$

Do until convergence
  1) fix weight vector $\mathbf{V}$ and find indicator $\mathbf{f}$
    form the aggregated affinity matrix $\mathbf{W}$ with $\mathbf{W} = \sum_k v_k \mathbf{W}_k$

    find indicator $\mathbf{f}$ by Nytröm method
  2) fix $\mathbf{f}$ and find weight vector $\mathbf{V}$
    find weight vector by Lagrange multiplier method
End while

run k-means on indicator $\mathbf{f}$ to cluster data into K clusters

---

3 Simulation Experiment

In order to validate the effectiveness of the proposed algorithm, two multispectral datasets are chosen as shown in Fig. 3(a) and (b), the size of each image is 2547×2548 pixels. They are Xi'an Chanba Ecological District multispectral images acquired in September 2011 and June 2012, respectively. They are too large to display the information in details. Therefore, we select two typical subsets with different kinds of change to verify the effectiveness of the proposed method.

The first dataset, as shown in Fig. 4(a) and (b), concludes two multispectral images of the same size of 301×301 pixels with four bands (R, G, B, and NIR) framing in red, denoting the riverway change of Chanhe River. The reference image, as shown in Fig. 4(c), is produced via manual marking by the combination of professional knowledge and surface prior information. It is used for quantitative evaluation purposes. The second dataset, as shown in Fig. 5, records the change of buildings and grasslands with the size of 189×192 pixels framing in yellow.

The proposed method is verified in the conditions that PC specifications are CPU pentium-i7 3.6 GHz and RAM 16 G. Preprocessing have been done including radiometric and atmospheric corrections and image-to-image co-registration (co-registration residual error within 0.5 pixel). Performance of the proposed method is assessed both qualitatively and quantitatively by a comparison with two reference multispectral images CD methods. They are AFS method that fuses two types of IDs by using the auto-adapted fusion based on entropy [8], the binary CD part of the improved C^2VA (denoted as M^2C^2VA) that preserves the geometrical information by multiscale morphological tools [11]. The CD results are displayed in the form of binary maps, in which the white and black pixels indicate the changed and unchanged pixels respectively.

We use four objective measures to assess the performance of different methods, i.e., false positive (FP), false negative (FN), percentage correct classification (PCC) and Kappa Coefficient (Kappa).

In the present experiments, the assignment of SE of disk shape has been verified robust in early scenarios [15], so we choose. The round SE sizes of different sizes, e.g., $SE_1 = [2,3], SE_2 = [1,2]$ to optimize the two original datasets and σ is all set to 130. A careful qualitative comparison is carried out based on the CD map obtained by the reference and the proposed approaches in our datasets, as shown in Figs. 6 and 7. One could clearly see that in the CM of the proposed approach, not only the identified change objects are well depicted according to their geometrical structures and shapes shown in Fig. 6 framing in red ellipse, but also the false alarms are effectively suppressed. The proposed method is proven the effective performance in identifying weak change regions [see Fig. 7 framing in red rectangle]. Meanwhile, according to the numeric results in Tables 1 and 2, we can see the phenomena that the proposed approach decreases FP and FN meantime results in greater PCC and Kappa values than two reference methods in all cases.

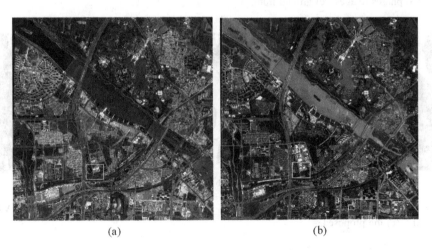

(a)   (b)

Fig. 3. The multispectral pseudo-color images of Xi'an Chanba Ecological District. (a) the image acquired on June, 2011. (b) the image acquired on September, 2012.

Fig. 4. The multispectral pseudo-color images and reference image of the first dataset. (a)–(b) two phases images. (c) ground truth.

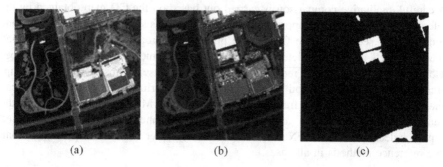

Fig. 5. The multispectral pseudo-color images and reference image of the second dataset. (a)–(b) two phases images. (c) ground truth.

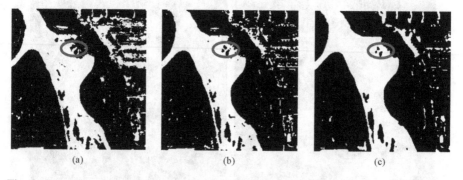

Fig. 6. Detection results for the first dataset produced by different methods. (a) M^2C^2VA (b) AFS (c) the proposed.

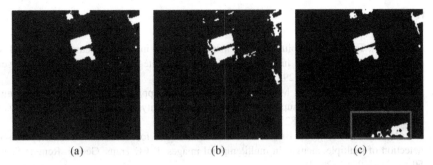

Fig. 7. Detection results for the second dataset produced by different methods. (a) M^2C^2VA (b) AFS (c) the proposed.

Table 1. Comparison of detection results for the first datasets in Fig. 6

Method	FP	FN	PCC (%)	Kappa
M^2C^2VA	4859	2916	91.42	0.7974
AFS	1743	1514	96.41	0.9136
Proposed	**1078**	**706**	**98.03**	**0.9527**

Table 2. Comparison of detection results for the second datasets in Fig. 7

Method	FP	FN	PCC (%)	Kappa
M^2C^2VA	185	1121	96.40	0.5882
AFS	587	853	96.03	0.5929
Proposed	**148**	**292**	**98.79**	**0.8863**

4 Conclusion

In this paper, we present a novel CD approach based on multiscale subspaces fusion for multispectral RS images. Subspace construction by multiscale morphological tool can capture the geographical information of RS images and optimized fusion process can combine the two DIs of CVA and SAM to obtain a better CD result by AANSC algorithm. The experimental results demonstrate that the proposed method not only accurately identify weak changes, but also preserves the multiscale geometric structure of the objects. Our future work will focus on investigating a multi-attribute morphological tool to flexibly capture spectral-spatial joint distribution for higher resolution RS images CD.

Acknowledgments. The author would like to thank supports from the National Natural Science Foundation of China under Projects 61571347 and 61471161.

References

1. Pacifici, F., Frate, F.D., Solimini, C., Emery, W.J.: An innovative neural-net method to detect temporal changes in high-resolution optical satellite imagery. IEEE Trans. Geosci. Remote Sens. **45**(9), 2940–2952 (2007)
2. Bovolo, F., Bruzzone, L., Marconcini, M.: A novel approach to unsupervised change detection based on a semi-supervised SVM and a similarity measure. IEEE Trans. Geosci. Remote Sens. **46**, 2070–2082 (2008)
3. Bovolo, F., Marchesi, S., Bruzzone, L.: A framework for automatic and unsupervised detection of multiple changes in multitemporal images. IEEE Trans. Geosci. Remote Sens. **50**(6), 2196–2212 (2012)
4. Bovolo, F., Bruzzone, L.: A theoretical framework for unsupervised change detection based on change vector analysis in the polar domain. IEEE Trans. Geosci. Remote Sens. **45**(1), 218–236 (2006)
5. Moughal, T.A., Yu, F.: An automatic unsupervised method based on context-sensitive spectral angle mapper for change detection of remote sensing images. In: ADMA, pp. 151–162 (2014)
6. Chen, J., Chen, X., Cui, X., Chen, J.: Change vector analysis in posterior probability space: a new method for land cover change detection. IEEE Geosci. Remote Sens. Lett. **8**(2), 317–321 (2011)
7. Gong, M., Zhou, Z., Ma, J.: Change detection in synthetic aperture radar images based on image fusion and fuzzy clustering. IEEE Trans. Image Process. **21**(4), 2141–2151 (2012)
8. Zhuang, H., Deng, K., Fan, H., Yu, M.: Strategies combining spectral angle mapper and change vector analysis to unsupervised change detection in multispectral images. IEEE Geosci. Remote Sens. Lett. **13**(5), 681–685 (2016)
9. Gao, H., Nie, F., Li, X., Huang, H.: Multi-view subspace clustering. In: CVPR, pp. 4238–4246 (2015)
10. Zhang, Y., Peng, D., Huang, X.: Object-based change detection for vhr images based on multiscale uncertainty analysis. IEEE Geosci. Remote Sens. Lett. **PP**(99), 1–5 (2017)
11. Liu, S., et al.: Multiscale morphological compressed change vector analysis for unsupervised multiple change detection. IEEE J. Sel. Top. Appl. Earth Observ. Remote Sens. **10**(9), 4124–4137 (2017)
12. Liu, S., Bruzzone, L., Bovolo, F., Zanetti, M., Du, P.: Sequential spectral change vector analysis for iteratively discovering and detecting multiple changes in hyperspectral images. IEEE Trans. Geosci. Remote Sens. **53**(8), 4363–4378 (2015)
13. Huang, H.C., Chuang, Y. Y., Chen, C.S.: Affinity aggregation for spectral clustering. In: CVPR, pp. 773–780 (2012)
14. Fowlkes, C., Belongie, S., Chung, F., Malik, J.: Spectral grouping using the Nyström method. IEEE Trans. Pattern Anal. Mach. Intell. **26**(2), 214–225 (2004)
15. Mura, M.D., Benediktsson, J.A., Bovolo, F., Bruzzone, L.: An unsupervised technique based on morphological filters for change detection in very high resolution images. IEEE Geosci. Remote Sens. Lett. **5**(3), 433–437 (2008)

Apple Surface Pesticide Residue Detection Method Based on Hyperspectral Imaging

Yaguang Jia[1,3], Jinrong He[1,3(✉)], Hongfei Fu[2,3], Xiatian Shao[1,3], and Zhaokui Li[4]

[1] College of Information Engineering, Northwest A&F University, Yangling 712100, Shaanxi, China
{jyg,hejinrong}@nwafu.edu.cn, sxt.25@qq.com
[2] College of Food Science and Engineering, Northwest A&F University, Yangling 712100, Shaanxi, China
fuhongfei@nwafu.edu.cn
[3] Sino-US Joint Research Center for Food Safety, Northwest A&F University, Yangling 712100, Shaanxi, China
[4] School of Computer, Shenyang Aerospace University, Shenyang 110136, Liaoning, China
lmy52wy@163.com

Abstract. In order to study the rapid and effective non-destructive detection method of pesticide residues on apple surface, this paper uses hyperspectral imaging technology to verify the feasibility of pesticide residue detection on apple surface. 225 apple samples from two groups were collected to construct the discriminant models of two pesticide residues, i.e., chlorpyrifos and carbendazim. The Hough circle transformation technique was used to determine the Region of Interest (ROI) automatically, and the averaged spectral value of the ROI is calculated as the representative spectrum of the sample. Then the Savitzky-Golay smoothing method was used for spectral denoising. Finally, the discriminant modeling is performed on the whole band with five methods: linear discriminant analysis, linear support vector machine, K nearest neighbor, decision tree and subspace discriminant ensemble. Furthermore, feature band selection was carried out by the successive projection algorithm and subspace discriminant ensemble method, then discriminant models were constructed on the feature band using linear discriminant analysis, linear support vector machine and K nearest neighbor. The experimental results show that the classification accuracy in both the whole band and the selected feature band for the detection of pesticide residues can be up to 95%. For the prediction of pesticide residue concentration, the subspace discriminant ensemble method based on the full band performs better, in which chlorpyrifos pesticide concentration prediction accuracy of up to 95%. The results confirmed the feasibility and effectiveness of hyperspectral imaging to detect pesticide residues on apple surface.

Keywords: Non-destructive detection · Pesticide residues
Hyperspectral imaging · Subspace discriminant ensemble

© Springer Nature Switzerland AG 2018
Y. Peng et al. (Eds.): IScIDE 2018, LNCS 11266, pp. 539–556, 2018.
https://doi.org/10.1007/978-3-030-02698-1_47

1 Introduction

Application of pesticides is increasing rapidly all over the world. Various pesticides are applied worldwide for crop protection to increase their quality, increase the yield as well as extend the storage time [1]. As people's living standards improve, consumers raise awareness of food safety issues. China is the country that uses the largest amount of pesticides in the world. Pesticides effectively reduce the hazards of pests, weeds and other diseases and improve the quality and yield of agricultural products [2]. Pesticide residues are unavoidable in the process of using pesticides. Pesticides directly attached to the surface of agricultural products can cause serious damage to human health. Control of pesticide residues has become an important part of food safety issues, and related issues of pesticide detection is also more and more important.

The application of hyperspectral imaging and computer vision technology in food processing and testing has become a hot research topic in the field of agriculture [3]. It has received widespread attention in food safety and quality testing and is widely used in the nondestructive testing of agricultural products.

Hyperspectral imaging technology is widely used in the detection of agricultural products diseases and quality [4, 5] (such as surface defects, sugar content, moisture content, solidity, internal defects and surface contamination) In 2006, Nicolaï et al. [6] proposed Non-destructive measurement of bitter pit in apple fruit using NIR hyperspectral imaging. In 2008, Zhao [7] detect subtle bruises on fruits with hyperspectral imaging. In 2011, Wang et al. [8] use hyperspectral imaging to detect external insect infestations in jujube fruit. In 2012, Rajkumar et al. [9] use hyperspectral imaging to study the banana fruit quality and maturity stages. Nanyam et al. [9] propose a decision-fusion strategy for fruit quality inspection using hyperspectral imaging. In 2013, Lorente et al. [11] use hyperspectral imaging to compare the ROC feature selection method for the detection of decay in citrus fruit. Haff et al. [12] detect fruit fly infestation with hyperspectral images of mangoes.

In contrast, Hyperspectral imaging technology is still in the experimental stage in pesticide residue detection. In 2008, Xue et al. [13] use Hyperspectral Imaging to detect pesticide residue on Navel Orange Surface. In 2010, Nansen et al. [14] use spatial structure analysis of hyperspectral imaging data to determine bioactivity of surface pesticide. Dai et al. [15] propose a nondestructive detection method for pesticide residue on Longan surface. In 2014, Liu et al. [16] use hyperspectral imaging to detect pesticide residues on Lingwu long jujubes' surface. In 2015, Wu et al. [17] use hyperspectral remote sensing to detect pesticide residues. Chen et al. [18] propose a novel technique of an unsupervised subpixel detection algorithm based on hyperspectral imaging for pesticide residue detection. In 2016, Nansen et al. [19] use hyperspectral imaging to characterize consistency of coffee brands. In 2017, Sun et al. [20] propose a quantitative detection method for lettuce leaves based on hyperspectral technique. Mohite et al. [21] apply hyperspectral techniques to the detection of grape pesticide (cyanomethacin) residues

At present, there is no report on the method of using hyperspectral technology to detect pesticide residues in apples. At the same time, hyperspectral image technology is still in the experimental stage for pesticide residue detection. Further research is needed to improve model accuracy, reduce data redundancy, select feature bands, etc. In this paper, we conducted an experimental study on the detection of apple pesticide residues based on near-infrared hyperspectral, using a variety of pesticide residue discriminant models to compare experimental results, and applied ensemble learning in the field of machine learning to experiments. The results verify that the method of using the hyperspectral imaging technology to detect pesticide residues on apple surfaces is feasible

2 Methods

2.1 Feature Band Selection Algorithm

2.1.1 Successive Projections Algorithm

The Successive Projections Algorithm (SPA) [22] is a novel variable selection strategy for multivariate calibration which uses simple operations in a vector space to minimize variable collinearity. SPA method starts from one feature of the data, then select a new feature at each iteration, until the required number of features is reached.

Let $X_{k(0)}$ be the initial iteration vector and N be the number of variables to be extracted. The spectral matrix has a total of J columns.

Step 1: Before iteration begins, randomly select one column j of the spectral matrix and assign the jth column of the data set to x_j, denoted $X_{k(0)}$

Step 2: The unselected column vector set denoted as S,

$$S = \{x_j, 1 \leq j \leq J, j \notin \{k(0), k(1), \cdots, k(n-1)\}\} \tag{1}$$

Step 3: Calculate the projection of x_j on the remaining column vectors respectively

$$Px_j = x_j - (x_j^T x_{k(n-1)}) x_{k(n-1)} (x_{k(n-1)}^T x_{k(n-1)})^{-1}, j \in S \tag{2}$$

Step 4: Calculate k(n)

$$k(n) = \arg(\max(\|Px_j\|)), j \in S \tag{3}$$

Step 5: Calculate x_j

$$x_j = Px_j, j \in S \tag{4}$$

Step 6: Let n increase by one, if n < N go back to Step 1 for next iteration
Step 7: The final extracted variables are F:

$$F = \{k(n); n = 0, \cdots N - 1\} \tag{5}$$

You can also calculate the root mean square error (RMSE) by Multiple linear regression (MLS) method to find the optimal initial vector $X_{k(0)}$ and number of features N by setting a range to N.

2.1.2   Subspace Discriminant Ensemble

Ensemble learning is a machine learning method that accomplishes learning tasks by combining multiple learners [23]. Usually ensemble learning is used to achieve better results than a single learner, and ensemble learning is especially effective for "weak learners".

In the Subspace Discriminant Ensemble (SDE) method, the selected weak learner is discriminant learner, and combined with the random subspace method [24–30] to extract features to classify the samples. This method is more suitable for the processing of high dimensional data.

The algorithm details are as follows:

Input:

Training data $D = \{(X_1, y_1), (X_2, y_2), \ldots, (X_m, y_m)\}$
Base learning algorithm: δ
The number of weak learners: N
Subspace dimension: d

Algorithm steps:

For $t = 1, 2, \ldots, T$ do
    $t = 1, 2, \ldots, T$
    $F_t = RS(D, d)$
    $D_t = Map_{F_t}(D)$
    $h_t = \delta(D_t)$
End For

Output:

$$H(x) = \arg\max \sum_{t=1}^{T} S(h_t(Map_{Ft}(x)) = y), y \in Y$$

In the input of the algorithm, D is the training data, X are the features of the sample, y is the label of the sample.

In the steps of the algorithm.

$F_t = RS(D, d)$ can randomly pick d-dimensional features from D.
$D_t = Map_{F_t}(D)$ can map selected subspace features from D.
$h_t = \delta(D_t)$ use training data to train weak learners.

In the output of the algorithm, $H(x)$ is the final classification result given by the Subspace Discriminant Ensemble method, where $S(x)$ is indicator function. If x is true, $S(x) = 1$, else $S(x) = 0$.

Considering that SDE selects the band for discriminative modeling every time, we can also apply this method to the selection of feature bands. Thus, by recording which bands are used by a well-performing classifier, these bands can be selected as feature bands.

2.2 Discriminant Modeling Algorithm

2.2.1 Linear Discriminant Analysis

The idea of linear discriminant analysis (LDA) is relatively simple. This method projects the sample to an optimal discriminant space. This discriminant space makes the projections of similar samples as close as possible, and the sample projections between classes and classes are as far away as possible, that is, there is a minimum intraclass distance and a maximum class spacing after projection to obtain the best separation.

2.2.2 Support Vector Machine

Support Vector Machine (SVM) classifies data by finding the best hyperplane that separates data points of one class from data points of another. The best hyperplane is the hyperplane with the largest margin between the two classes. The margin is the maximum width of the plate parallel to the hyperplane without internal data points. Linear SVM refers to a support vector machine method in which the kernel function is a linear function, and the method performs simple linear separation between classes.

2.2.3 K-Nearest Neighbor

K-Nearest Neighbor (KNN) classification algorithm is one of the simplest methods in data mining classification technology. The so-called K nearest neighbor is the meaning of the k nearest neighbors. It is said that each sample can be represented by its closest k neighbors.

2.2.4 Decision Tree

A decision tree is a decision analysis method that determines the feasibility of a project by determining the probability that the expected value of the net present value is greater than or equal to zero by constructing a decision tree based on the probability of occurrence of various conditions. It is a graphical method of using probability analysis intuitively. The depth of the control tree in this experiment is 4.

2.2.5 Subspace Discriminant Ensemble

In this experiment, we adopt a random subspace method. When specific to the experimental hyperspectral data, we randomly select n features as their characteristic subspaces in all features of the sample. For example, there are a total of 256 bands of information in the acquired hyperspectral data sample. For subspace discrimination integration methods, the specified subspace dimension is 128 and the number of weak learners is 30. When training each weak classifier, this method randomly selects 128 features from 256 features for classifying modeling, and trained a total of 30 weak classifiers. For test data, this method combines the classification results of all classifiers to give the final classification result.

3   Data Collection and Preprocessing

3.1   Apple Sample Collection

The test sample was collected from Yangling apple farming demonstration station. The apple tree was a biennial dwarfing rootstock densely-woven spindle-shaped Fuji. 36 trees were selected as the test object after a thorough spray cleaning. The samples were divided into four groups: control group, chlorpyrifos treatment group, carbendazim treatment group, chlorpyrifos and carbendazim combination group, with 9 trees in each group, of which 3 was a treatment concentration, divided into low, medium, High three doses, the control group sprayed with clean water. The dilution ratios of chlorpyrifos were 1: 500, 1: 1000 and 1: 2000, and the dilution ratios of carbendazim were 1: 600, 1: 1200 and 1: 2400 in turn. The dose of mixed group corresponded to the other two doses (1: 500 + 1: 600, 1: 1000 + 1: 1200, 1: 2000 + 1: 2400).

The experimental data is divided into two batches of acquisition, the first batch was taken 12 h after spraying, the second batch was collected 96 h after spraying. Two batches of different treatment methods to obtain the number of test samples shown in Table 1. And data processing flow chart shown in Fig. 1

Table 1. Details of different treatment samples

	Chlorpyrifos	Carbendazim	Mixed	Control
12 h	27	27	27	54
96 h	21	25	27	26

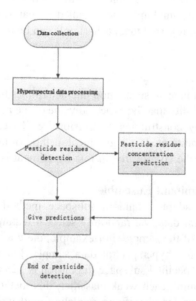

Fig. 1. The overall flow of data processing

3.2 Hyperspectral Data Acquisition

After sample processing was completed, the hyperspectral data of the sample was acquired using ZOLIX HyperSIS-VNIR-PFH hyperspectral equipment. Data acquisition equipment shown in Fig. 2. The device is capable of capturing spectra in the 865.11 (nm) −1711.71 (nm) range, with a single hyperspectral image at 3.32 nm and a total of 256 imaging times in different bands. In order to obtain more complete information, the apple was on both sides of the imaging. Six apple samples were imaged each time.

Fig. 2. Hyperspectral imaging equipment

The imaging equipment corrects the reflectivity of hyperspectral images using white and black correction images, white calibration image using a polytetrafluoroethylene-based calibration plate (with reflectivity close to 1). While black correction was achieved by blocking the imaging lens (its reflectivity close to 0) [30]. Equation (6) is the reflectivity correction formula. Where $P_{(i,j)}$ is the value of the jth column of the ith row before the image is corrected and $P'_{(i,j)}$ is the corresponding value of the jth column of the ith row of the after the image is corrected. $D_{(i,j)}$ and $W_{(i,j)}$ represent the corresponding values of the black correction image and the white correction image in the ith column and the jth column, respectively.

$$P'_{ij} = \frac{P_{ij} - D_{ij}}{W_{ij} - D_{ij}} \tag{6}$$

The resulting imagery of the device includes hyperspectral source files (raw format) and their corresponding header files (hdr format). Hyperspectral source files are stored in Band Interleaved by Line format (BIL). Hyperspectral header files contain information about the source band's imaging band, storage format, and more. This experiment uses matlab 2015b software and Python language for experimental analysis.

3.3 Hyperspectral Data Processing

After obtaining the hyperspectral image, in order to obtain the spectral information of the apple surface, a Region of Interest (ROI) is performed to obtain an image of the apple.

Detection of circles in images is an important task in computer vision and pattern recognition. Hough Transform (Hough Transform) is a method of boundary shape detection in image processing [31, 32]. It detects objects with specific shapes by a voting algorithm and is used to detect straight line segments, circles and ellipses. The basic idea of the Hough transform is to map the spatial shape information of the image into the parameter space. The constraint function of the image boundary point determines the value of the accumulator in the parameter space. The higher the value is, the bigger the probability that the boundary point forms the target shape. As a result, the problem of shape detection successfully transformed into the peak for the problem. Hough transform technique calculates boundary point information according to local measure, therefore, it has good fault tolerance and robustness to the circumstance that the boundaries are interrupted due to the noise of regional boundary or covered by other targets.

For circular detection, the three parameters of the circle, the coordinates of the circle center (x_0, y_0), and the length r_0 of the radius need to be determined in the image. For the circle in the image, the curve parameter can be written as (7). Where (x, y) is the point on the circumference.

$$(x - x_0)^2 + (y - y_0)^2 = r_0^2 \tag{7}$$

The first thing to do is to detect the edge of the image. Sobel operator can be used. Formula (8) is its matrix form, where Sx and Sy are the convolution templates in x direction and y direction respectively, K is the matrix of the target x and its neighborhood. The gradient of each point can be expressed as Gx in (9).

$$S_x = \begin{pmatrix} -1 & 0 & 1 \\ -2 & 0 & 2 \\ -1 & 0 & 1 \end{pmatrix}, S_y = \begin{pmatrix} 1 & 2 & 1 \\ 0 & 0 & 0 \\ -1 & -2 & -1 \end{pmatrix}, K = \begin{pmatrix} n_1 & n_2 & n_3 \\ n_4 & x & n_5 \\ n_6 & n_7 & n_8 \end{pmatrix} \tag{8}$$

$$\begin{cases} G_x = \sqrt{S_x^2 + S_y^2} \\ S_x = (n_3 + 2n_5 + n_8) - (n_1 + 2n_4 + n_6) \\ S_y = (n_1 + 2n_2 + n_3) - (n_6 + 2n_7 + n_8) \end{cases} \tag{9}$$

After determining the edge point and its gradient, you can calculate the direction of the gradient at that point. As shown in Eq. (5), for a point on the circumference, a straight line passing the gradient direction must pass the center of the circle. Calculate the number of each point through the gradient straight line, Every time a gradient passes through a coordinate point, the corresponding accumulator of the coordinate increases by one. After setting a threshold, coordinate points larger than the threshold are saved as the coordinates of the center point.

$$\theta = \arctan\left(\frac{G_y}{G_x}\right) \qquad (9)$$

After determining the coordinates of the center, the distance from the center to each edge point is calculated. The distance from the center to the edge point on the corresponding circle is equal to the radius r_0. Let $F(d)$ be the number of edge points at a distance d from the center of the circle. $F(d)$ satisfies Eq. (6) because the point at which the distance from the center of the circle is equal to its corresponding radius must be the largest. By constraining the radius, the number of circles in the image, the region of interest can be identified more accurately and quickly. By constraining the range of radius and the number of circles in the image, the region of interest can be identified more accurately and quickly.

$$\arg\max(F(d)) = r_0 \qquad (10)$$

For the acquired hyperspectral image, a band image that is more clearly and clearly detected by the apple can be selected. In this experiment, an image with a wavelength of 1515.83 nm (the 197th band) is selected as a standard image of the region of interest, and then the band Under the image is converted to a grayscale image. For the acquired hyperspectral image, a band image that is more clearly and easily detected by the apple can be selected. In this experiment, an image with a wavelength of 1515.83 nm (the 197th band) is selected as a standard image of the region of interest, and then the image of this band is transformed into a grayscale image. After the grayscale image is obtained, the ROI area is automatically selected according to the Hough circle detection process. In this experiment, the radius of the constrained target circle ranges from 30 to 80 pixels, and the number of target circles is six. ROI regional selection results shown in Fig. 3. Each image containing an apple, after the selection of the region of interest will get two sets of parameters, namely the coordinates of the center of the circle and the length of the radius. The blue circle in the image is the detected area of interest, which is in good agreement with the apple target area.

Considering that apple is not a standard circle, the circle detected by the region of interest cannot completely correspond to the target apple, so when calculating the spectral information, the radius is reduced by half and then the average spectrum is calculated. The circle to be used for calculation is shown in Fig. 4a, and the spectrum curve extracted in Fig. 4a is shown in Fig. 4b, where the abscissa is the wavelength and the ordinate is the reflectivity. Use different colors to draw the apple spectral information in different positions. The average spectral information for the two datasets for the 12-h and 96-h sampling is shown in Fig. 5. From the graph, the reflectance of samples taken after 96 h in different wavelengths is lower than the samples sampled after 12 h.

After obtaining the average spectrum of the region of interest, the hyperspectral data is likely to receive external interference during the acquisition, including the noise and environmental factors of the acquisition device itself. This will cause some unrelated interference information to be collected. In order to eliminate the interference noise information while ensuring that the spectral information is not lost as much as possible, data preprocessing is required for obtaining the average spectrum. This experiment uses Savitzky-Golay smoothing to smooth the data. Savitzky-Golay smoothing is a common algorithm for smoothing hyperspectral data.

The Savitzky-Golay smoothing method uses a polynomial to perform polynomial least-squares fitting on the data. In this experiment, a quadratic polynomial is selected for fitting, and the size of the moving window is selected as 15. The smoothing effect of Savitzky-Golay is shown in Fig. 6. Compared with the smoothing result, the smoothing effect is apparent near the bands of 800–900 nm and 1200–1300 nm. Different samples of information are drawn with different colored curves.

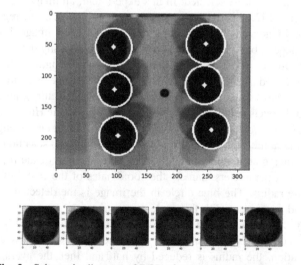

Fig. 3. Schematic diagram of ROI selection (Color figure online)

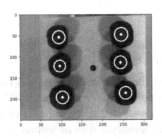

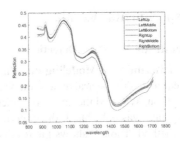

a. ROI for calculation

b. calculation results

Fig. 4. Average spectrum acquisition

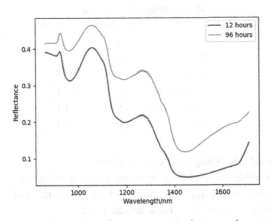

Fig. 5. Comparison of average spectra between data sets

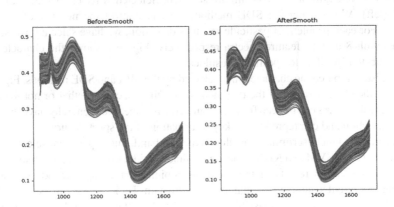

Fig. 6. Savitzky-Golay smoothing diagram

4 Experimental Results

4.1 Pesticide Residues Detection

4.1.1 Discriminant Modeling on Whole Band

After acquiring the spectral data of the valid region, the experimental data was modeled using Python. We used the above five methods for comparative experiments. When the accuracy rate was calculated, the experiment was conducted using a 5-fold cross validation method. Table 2 shows the results of the experiments with or without pesticide residues.

Table 2. The classification results of pesticide residue in different methods

Method name	Five-fold cross validation accuracy/%	
	96 h	12 h
LDA	98.5	83.5
SVM	99.0	96.5
Fine KNN	94.43	61.1
Decision tree	85.4	64.1
SDE	100.0	95.4

4.1.2 Discriminant Modeling on Feature Band

When using all the bands for modeling, there are some disadvantages such as large data volume and complex modeling. This paper adopts SDE method and SPA method mentioned previously to select the characteristic bands.

When using SPA, set the number of selected bands from 5 to 20. For the discrimination of whether or not there is a pesticide residue, 14 characteristic wavelengths such as a wavelength of 1568.94 nm are selected. Selected wavelengths are shown in Fig. 7 (left). When using the SDE method, we can control the number of selected features. For each problem of pesticide residues detection, we have selected two sets of features with 8 and 18 feature spectra respectively. Figure 8 shows the characteristic bands selected by SDE for pesticide residues.

We also compared the characteristics selected by the SPA and SDE methods. Figure 9 compares the characteristics of the two methods on the issue of whether or not there are pesticide residues. As can be seen from the figure, the two methods generally have the same feature selection and can represent the key information of the spectral curve.

According to the discriminant results on the full band, this experiment selected three methods of LDA, SVM and KNN to detect pesticide residues on the characteristic bands.

The classification results of the three discriminant modeling methods in the full band and different characteristic bands are shown in Table 3. As can be seen from Table 3, the LDA and SVM classifiers use the 18 characteristic bands selected by the SDE to determine whether or not there are pesticide residues give the best performance.

Table 3. Classification results for pesticide residues based on characteristic wavelength

Method name	Five-fold cross validation accuracy/%	
	96 h	12 h
Full Band + LDA	98.5	83.5
SDE 8 features + LDA	98.98	81.66
SDE 18 features + LDA	100	90.46
SPA 14 features + LDA	100	87.0
Full Band + SVM	99.0	96.5
SDE 8 features + SVM	98.5	64.5
SDE 18 features + SVM	100	95.0
SPA 14 features + SVM	99.5	90.45
Full Band + KNN	94.43	61.1
SDE 8 features + KNN	94.93	62.3
SDE 18 features + KNN	96.43	63.33
SPA 14 features + KNN	93.44	68.72

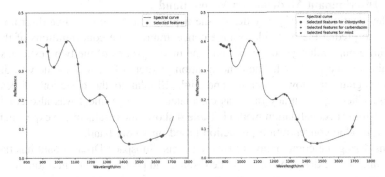

Fig. 7. The feature band selected by the SPA algorithm for pesticide residues (left) and different pesticide residue concentrations (right)

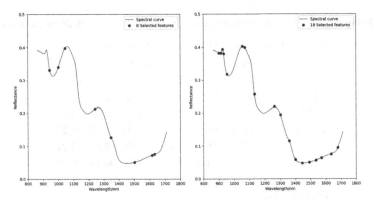

Fig. 8. The feature band selected by the SDE algorithm for pesticide residues (left: 8 features, right: 18 features)

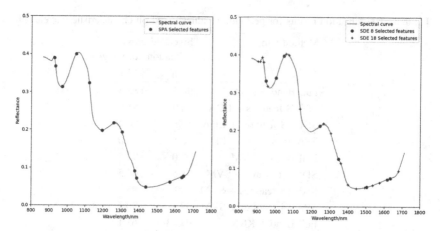

Fig. 9. SPA and SDE feature comparisons on whether or not there are pesticide residues

4.2 Pesticide Residue Concentration Prediction

4.2.1 Discriminant Modeling on Whole Band

After the identification of pesticides on the apple surface, we used five discriminating methods to predict the pesticide residue concentration. The concentrations of the carbendazim group, chlorpyrifos group, and the mixed groups of the two pesticides were discriminated and predicted. The concentration gradient of each pesticide was divided into three gradients: low, medium, and high. Similar to the detection of pesticide residues. When the accuracy rate was calculated, the experiment was also conducted using a 5-fold cross validation method. Table 4 shows the results of the experiments on pesticide residue concentration prediction based on whole band.

Considering the combination of two data sets, Subspace Discriminant has the best performance in the detection of pesticide residue and pesticide concentration. Some test results can reach 95% or more, and the accuracy of some detection items can reach 100%.

Table 4. Classification results of pesticide concentrations in different methods

Method name	Five-fold cross validation accuracy/%					
	96 h			12 h		
	A	B	C	A	B	C
LDA	84.63	69.32	72.35	84.56	82.51	75.47
SVM	71.24	74.1	57.44	76.99	82.66	64.60
Fine KNN	60.90	61.6	51.30	54.30	66.30	38.73
Decision tree	57.42	55.82	46.76	54.48	55.52	37.21
SDE	95.27	84.0	88.90	98.18	96.18	89.09

(A represents Chlorpyrifos, B represents Carbendazim, C represents Mixd)

4.2.2   Discriminant Modeling on Feature Band

When discriminating the concentration of pesticide residues, SPA selected eight characteristic spectral bands for chlorpyrifos pesticides, five characteristic bands for carbendazim, and eight characteristic bands for the mixed group. SPA selected characteristic bands for different discriminative purposes are shown in Fig. 7 (right).

Figure 10 shows the characteristic bands selected by SDE for different pesticide residue concentrations. The results of pesticide concentration prediction based on characteristic bands are shown in Table 5. In general, the prediction of pesticide concentration based on characteristic bands can achieve relatively high accuracy with a small number of bands. However, the full-band SDE method performs best in pesticide concentration prediction.

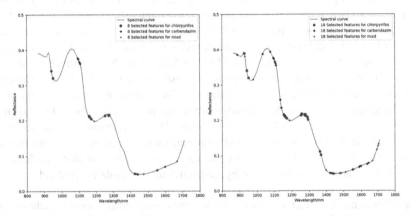

Fig. 10. The feature band selected by the SDE algorithm for different pesticide residue concentrations (left: 8 features, right: 18 features)

Table 5. Classification results for pesticide concentrations based on characteristic wavelength

Method name	Five-fold cross validation accuracy/%					
	96 h			12 h		
	A	B	C	A	B	C
SDE 8 features + LDA	85.55	62.19	65.55	75.50	73.33	73.33
SDE 18 features +LDA	83.33	72.65	65.0	87.47	74.44	77.77
SPA +LDA	81.11	66.34	60.08	79.13	86.66	78.88
SDE 8 features +SVM	63.33	61.11	52.22	55.0	60.0	45.55
SDE 18 features +SVM	71.11	69.06	61.11	56.66	61.66	53.33
SPA +SVM	69.07	68.3	53.26	48.93	63.40	43.42
SDE 8 features +KNN	52.22	58.20	44.31	73.08	66.11	38.33
SDE 18 features +KNN	67.77	60.42	51.66	69.19	70.55	38.88
SPA +KNN	65.21	55.72	56.22	59.19	61.32	39.19

(A represents Chlorpyrifos, B represents Carbendazim, C represents Mixd)

5 Conclusion

This paper aims at the non-destructive testing of apple pesticide residues, using hyperspectral images to find a suitable modeling method for pesticide residue detection. The subspace discriminant ensemble method has the best effect on the whole band, and the discrimination accuracy of the target apple with respect to pesticides can reach more than 95%. There is also a relatively high accuracy in the identification of pesticide concentration. In feature selection, the SDE algorithm is used to select the feature bands and the traditional SPA algorithm is compared. The experimental results show that in the discriminant modeling of pesticide residues, both methods perform well, in which the use of LDA + SDE18 features, LDA + SPA and SVM + SDE18 features on the data after 96 h of sampling, use 5-fold cross validation can achieve 100% accuracy.

From the comparison of sampling time, apples sampled 96 h after spraying were found to be higher in discrimination accuracy than apples sampled 12 h later. Compared with the prediction of pesticide residue concentration, the accuracy of prediction of pesticide residue concentration of chlorpyrifos is higher than that of carbendazim pesticide. Using the features selected by LDA + SDE to predict chlorpyrifos can reach 85.55%. The predictive effect of the mixture of pesticides has declined. Taken together, the discriminative modeling using the subspace discriminant ensemble method on the whole band can achieve the best results both in the presence of pesticides and in the prediction of pesticide residue concentrations. However, there is still room for improvement in the accuracy of pesticide concentration prediction. This work provides a new idea for the rapid non-destructive testing of apple pesticide residues, and demonstrates that it is feasible to use hyperspectral images to detect pesticide residues.

Acknowledgment. This work is supported by the China Postdoctoral Science Foundation under Grant No. 2018M633585, Natural Science Basic Research Plan in Shaanxi Province of China under Grant No. 2018JQ6060, Yangling Demonstration Zone Science and Technology Plan Project under Grant No. 2016NY-31, Shaanxi University Science and Technology Innovation Project under Grant No. S201710712127, Shaanxi Agricultural Science and Technology Innovation and Research Project under Grant No. 2015NY023.

References

1. Dhakal, S., Li, Y., Peng, Y., et al.: Prototype instrument development for non-destructive detection of pesticide residue in apple surface using Raman technology. J. Food Eng. **123**(2), 94–103 (2014)
2. Huang, H., Liu, L., Ngadi, M.O.: Recent developments in hyperspectral imaging for assessment of food quality and safety. Sensors **14**(4), 7248–7276 (2014)
3. Chen, Q., Zhang, C., Zhao, J., et al.: Recent advances in emerging imaging techniques for non-destructive detection of food quality and safety. Trends Anal. Chem. **52**(52), 261–274 (2013)
4. Pu, Y.-Y., Feng, Y.-Z., Sun, D.-W.: Recent progress of hyperspectral imaging on quality and safety inspection of fruits and vegetables: a review. Compr. Rev. Food Sci. Food Saf. **14**(2), 176–188 (2015)

5. Lorente, D., Aleixos, N., Gómez-Sanchis, J., et al.: Food Bioprocess Technol. **5**, 1121 (2012). https://doi.org/10.1007/s11947-011-0725-1
6. Nicolaï, B.M., Lötze, E., Peirs, A., et al.: Non-destructive measurement of bitter pit in apple fruit using NIR hyperspectral imaging. Postharvest Biol. Technol. **40**(1), 1–6 (2006)
7. Zhao, J.: Detecting subtle bruises on fruits with hyperspectral imaging. Trans. Chin. Soc. Agric. Mach. **39**(1), 106–109 (2008)
8. Wang, J., Nakano, K., Ohashi, S., et al.: Detection of external insect infestations in jujube fruit using hyperspectral reflectance imaging. Biosys. Eng. **108**(4), 345–351 (2011)
9. Rajkumar, P., Wang, N., Eimasry, G., et al.: Studies on banana fruit quality and maturity stages using hyperspectral imaging. J. Food Eng. **108**(1), 194–200 (2012)
10. Nanyam, Y., Choudhary, R., Gupta, L., et al.: A decision-fusion strategy for fruit quality inspection using hyperspectral imaging. Biosys. Eng. **111**(1), 118–125 (2012)
11. Lorente, D., Blasco, J., Serrano, A.J., et al.: Comparison of ROC feature selection method for the detection of decay in citrus fruit using hyperspectral images. Food Bioprocess Technol. **6**(12), 3613–3619 (2013)
12. Haff, R.P., Saranwong, S., Thanapase, W., et al.: Automatic image analysis and spot classification for detection of fruit fly infestation in hyperspectral images of mangoes. Postharvest Biol. Technol. **86**(8), 23–28 (2013)
13. Xue, L., Li, J., Liu, M.: Detecting pesticide residue on navel orange surface by using hyperspectral imaging. Acta Optica Sinica **28**(12), 2277–2280 (2008)
14. Nansen, C., Abidi, N., Sidumo, A.J., et al.: Using spatial structure analysis of hyperspectral imaging data and fourier transformed infrared analysis to determine bioactivity of surface pesticide treatment. Remote Sens. **2**(4), 908–925 (2010)
15. Dai, F., Hong, T., Zhang, K., et al.: Nondestructive detection of pesticide residue on longan surface based on near infrared spectroscopy. In: International Conference on Intelligent Computation Technology and Automation, pp. 781–783. IEEE (2010)
16. Liu, M.F., Zhang, L.B., Jian-Guo, H.E., et al.: Study on non-destructive detection of pesticide residues on Lingwu long jujubes' surface using hyperspectral imaging. Food Mach. **5**, 87–92 (2014)
17. Wu, C.C., Liao, Y.H., Lo, W.S., et al.: Band weighting spectral measurement for detection of pesticide residues using hyperspectral remote sensing. In: Geoscience and Remote Sensing Symposium, pp. 457–460. IEEE (2015)
18. Chen, S.Y., Liao, Y.H., Lo, W.S., et al.: Pesticide residue detection by hyperspectral imaging sensors. In: The Workshop on Hyperspectral Image & Signal Processing: Evolution in Remote Sensing, pp. 1–4 (2015)
19. Nansen, C., Singh, K., Mian, A., et al.: Using hyperspectral imaging to characterize consistency of coffee brands and their respective roasting classes. J. Food Eng. **190**, 34–39 (2016)
20. Sun, J., Cong, S., Mao, H., et al.: Quantitative detection of mixed pesticide residue of lettuce leaves based on hyperspectral technique. J. Food Process Eng. **41**(2), e12654 (2017)
21. Mohite, J., Karale, Y., Pappula, S., et al.: Detection of pesticide (cyantraniliprole) residue on grapes using hyperspectral sensing. In: SPIE Commercial + Scientific Sensing and Imaging, p. 102170P (2017)
22. Araújo, M.C.U., Saldanha, T.C.B., Galvão, R.K.H., et al.: The successive projections algorithm for variable selection in spectroscopic multicomponent analysis. Chemometr. Intell. Lab. Syst. **57**(2), 65–73 (2001)
23. Krawczyk, B., Minku, L.L., Woniak, M., et al.: Ensemble learning for data stream analysis. Inf. Fusion **37**(C), 132–156 (2017)
24. Ho, T.K.: Random subspace method for constructing decision trees. IEEE Trans. Pattern Anal. Mach. Intell. **20**(8), 832–844 (1998)

25. Zhou, Z.H., Yang, Q.: Machine Learning and Its Applications. Tsinghua University Press, Beijing (2011)
26. Boot, T., Nibbering, D.: Forecasting using random subspace methods. Tinbergen Institute Discussion Paper (2016)
27. Hang, R., Liu, Q., Song, H., et al.: Matrix-based discriminant subspace ensemble for hyperspectral image spatial-spectral feature fusion. IEEE Trans. Geosci. Remote Sens. **54**(2), 783–794 (2016)
28. Yang, H., Jia, X., Patras, I., et al.: Random subspace supervised descent method for regression problems in computer vision. IEEE Signal Process. Lett. **22**(10), 1816–1820 (2015)
29. Gu, J., Jiao, L., Liu, F., et al.: Random subspace based ensemble sparse representation. Pattern Recognit. **74**, 544–555 (2017)
30. Vásquez, N., Magan, C., Oblitas, J., et al.: Comparison between artificial neural network and partial least squares regression models for hardness modeling during the ripening process of Swiss-type cheese using spectral profiles. J. Food Eng. **219**, 8–15 (2017)
31. Panwar, S., Raut, S.: Survey on lane detection using Hough transform technique. Int. J. Adv. Res. Electr. Electron. Instrum. Eng. **4**(1), 401–405 (2015)
32. Vieira, L.H.P., Pagnoca, E.A., Milioni, F., et al.: Tracking futsal players with a wide-angle lens camera: accuracy analysis of the radial distortion correction based on an improved Hough transform algorithm. Comput. Methods Biomech. Biomed. Eng. Imaging Vis. **5**(3), 221–231 (2015)

Regression Analysis for Dairy Cattle Body Condition Scoring Based on Dorsal Images

Heng Pan[1], Jinrong He[1,2(✉)], Yu Ling[1], and Guoliang He[3]

[1] College of Information Engineering, Northwest A&F University,
Yangling 712100, China
hejinrong@nwafu.edu.cn
[2] College of Mechanical and Electronic Engineering,
Northwest A&F University, Yangling 712100, China
[3] School of Computer, Wuhan University, Wuhan 430070, China

Abstract. Body condition scoring (BCS) provides an objective assessment of the amount of subcutaneous fat appositions and energy storage of the dairy cattle, which has become a powerful tool for dairy industry management. Traditional BCS is estimated by technicians manually method involving visual and tactile aspects, which is high-cost and subjective. Hence, the, accurate and efficient BCS automatic evaluation technology is studied. In this paper, we proposed an effective dairy cattle body condition scoring method based on cow's dorsal 2D digital images. A bounding rectangle normalization method is used to extract cow's contour information and distance vectors are constructed to describe the shape. Then, six regression methods are discussed for BCS regression modeling. Experiment on a benchmark dataset demonstrated that elastic net obtained the best accuracy on the BCS task.

Keywords: Body condition scoring · Regression analysis · Dorsal images
Contour extraction

1 Introduction

The metabolic energy stored in fat and muscle of dairy cattle is essential to maintain its basic activity and milk production. Body condition score (BCS) is put forward as an important indicator to measure the energy status and changes of dairy cattle. There exist different dairy BCS evaluation systems in different countries and regions. Most of dairy farms in America adopt the five-grade marking system described by Wildman et al. [1], which is later modified by Edmonson et al. [2] and Ferguson et al. [3]. This system is based on a numerical range from 1.0 to 5.0, in increments of 0.25, with 1.0 denoting scraggy cattle and 5.0 denoting corpulent cattle. One point of BCS equals 100 to 140 lb gain in body weight [4].

Traditional BCS estimation method involves visual and tactile aspects, who estimates the scores by technicians manually. The drawback is obvious that the subjectivity in the judgement may lead to different scores for the same cow under the same conditions by individual valuators. In addition, it is high-cost to train a professional on-farm technician and time-consuming to grade to plentiful cows and record the BCS

© Springer Nature Switzerland AG 2018
Y. Peng et al. (Eds.): IScIDE 2018, LNCS 11266, pp. 557–566, 2018.
https://doi.org/10.1007/978-3-030-02698-1_48

data. Hence, an automatic, accurate and efficient BCS evaluation technology is desperately in need for dairy industry. In [5], Bewley et al. labeled altogether 23 anatomical feature points of cows from each 2-dimension digital image by hand to analysis the shape and contour of cows. Following the Bewley's work, Azzaro et al. developed a technique that was able to describe the body shape of cows in a reconstructive way in [6]. In the research, the 23 feature points were extracted automatically, and the shapes were reconstructed using kernel principal component analysis (KPCA) [7]. The average error of proposed model is 0.31. In [8], image recognition was applied on scoring the cow body condition. they locate the cow in the image, then preprocess the black and white speckle, and interrelated experiments show the feasibility.

Recently, 3-dimension computer version technique is introduced as an advanced tool into BCS evaluation system for the sack of improving accuracy. In [9], Weber et al. proposed a backfat thickness (BFT) automatic evaluation method which based on 3-dimension optical imaging system. In [10], Fischer et al. captured the cow's back surface depth image by using 3-dimension camera and labeled 4 key points manually from the depth image, which were used to train the multiple linear regression model. In [11], Wang et al. collected the 3-dimension information of dairy cattle in virtue of the Microsoft Kinect Sensor device, then carried through feature extraction for the 3-dimension information. In [12], Zhao et al. extracted cow's back area from depth image and rotated the image so that the cow's spine was parallel to horizontal axes, then located the iliac bone, hook bone, pin bone, sacral bone and obtained 4 feature values for prediction model. Compared to 2-dimension digital image, 3-dimension depth image has more information of the observed, but it has a higher hardware requirement at the mean time.

Inspired by Azzaro's work, we propose an effective cattle body condition scoring scheme based on cow's dorsal images in this paper. We do a bounding rectangle normalization for cow's contour and use distance vector to describe the shape. Experiment on contrasting 6 regression methods demonstrates that elastic net gained the best accuracy on the BCS database. The overview of regression analysis on the BCS database is shown as Fig. 1.

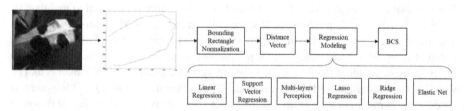

Fig. 1. Overview of regression analysis on the BCS database

2 Body Condition Scoring Estimation System

Body condition scoring is a widely used tool for dairy producers detecting metabolic status and health risks like subacute ketosis or fertility problems and managing the nutrition of their herd [13]. Arm and Hammer provides an excellent booklet on BCS as Table 1.

Table 1. Criterion for BCS

Score	Description
1.5	"The cow is ideal for demonstrating the key indicators. Each vertebra is sharp and distinct along the backbone."
2.0	"The cow is too thin. Its reproduction and milk production may suffer from a lack of body condition. Her backbones are easily seen, but they do not stand out as individual vertebra."
3.0	"This cow is in ideal condition for most stages of lactation. The vertebra is rounded, but the backbone can still be seen. Hook and pin bones are easily seen, but are round instead of angular."
4.0	"A BCS 4 cow looks fleshy. Her back appears almost solid, like a table top. The short ribs still form a shelf, but they cannot be seen as individual bones and only felt with deep palpation."
5.0	"Her backbone and short ribs cannot be seen and only felt with difficulty. The shelf formed by the short ribs is well-rounded. Her thurl is filled in."

3 Material and Modeling

3.1 Data Acquisition

We use The BCS Database for our experiment which is publicly available and can be obtained from http://www.corfilac/bcs/dataset.html. The data set contains 286 color images corresponding to 29 cow's body shapes, of which resolution is 704 * 480 pixels. A data sample is shown in Fig. 2a. The images were captured by the camera positioned 3 m above cows when the cows passed through the exit gate from the pair of milking robots. There were two technicians evaluating the body condition of cows at the exit alley of the milking robots simultaneously. The details of selecting appropriate images from video can be seen in [6]. Because different technicians could give different scores of the same cow, we take the average scores as the final score of the object.

Go a step further, the 23 anatomical points useful for BCS estimation according to Bewley et al. [5] were automatically labeled by an ad hoc JAVA application, which is available for download at http://www.corfilac/bcs/software.html. A shape sample is shown in Fig. 2b. There is a legend describing the meaning of all 23 anatomical landmarks in Table 2.

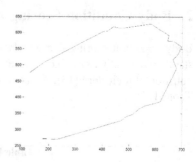

a)   original image   b)   shape from anatomical points

Fig. 2. Overview of the BCS database

Table 2. Meaning of 23 anatomical landmarks

Points	Description	Points	Description	Points	Description
1	Left forerib	2	Left short rib start	3	Left hook start
4	Left hook anterior midpoint	5	Left hook	6	Left hook posterior midpoint
7	Left hook end	8	Left thurl	9	Left pin
10	Left tailhead nadir	11	Left tailhead junction	12	Tail
13	Right tailhead junction	14	Right tailhead nadir	15	Right pin
16	Right thurl	17	Right hook end	18	Right hook posterior midpoint
19	Right hook	20	Right hook anterior midpoint	21	Right hook start
22	Right short rib start	23	Right forerib		

3.2   Bounding Rectangle Normalization

The shapes preliminarily extracted from original images exist difference in location, scale and rotation, which will become the interference factor for scoring the cow's body condition. Therefore, all shapes referred ought to be adjusted into a consistent representation. The alignment of shapes is handled by reestablishing and unifying the coordinate system of each shape.

Because of the anatomical landmarks describing the outline of the cow, we can calculate the minimum bounding rectangle of these points and regard it as the first quadrant of the coordinate system. Point 5 and point 19 indicate the left hook and right hook respectively, their distance can represent the body width of the cow (Fig. 3a). Hence, we use point 19 as the center of rotation to rotate the shape counterclockwise so that l_{p5p19} is parallel to the y axis of the coordinate (Fig. 3b), where l_{p5p19} is the line between point 5 and point 19, denoting the hook bone. The bounding rectangle is determined of which lower left point is (x_{min}, y_{min}) and upper right point is (x_{max}, y_{max}), where x_{min}, x_{max} are minimum and maximum values of x coordinates and y_{min}, y_{max} are minimum and maximum values of y coordinates. Then, the shape is translated with the lower left point of bounding rectangle translating to the origin (Fig. 3c). Finally, we normalize the rectangle with all of shapes showing an equal length base (Fig. 3d).

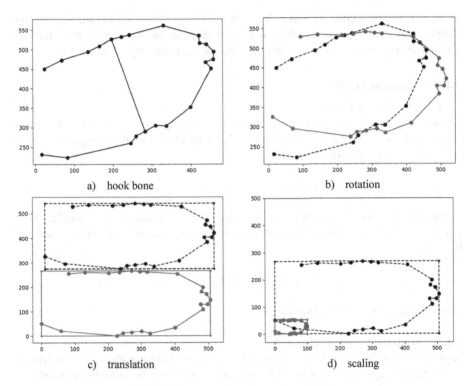

a) hook bone   b) rotation

c) translation   d) scaling

Fig. 3. The procedure shape normalization

3.3 Distance Vector and Kernel PCA Descriptor

After filtering out location, scale and rotational influence from the original shape data, each sample can be reformulated mathematically as a feature vector for describing its geometrical information and further model training. As for these anatomical points, we introduce distance vector which defined as below:

$$x = [dist_{uv}]^T,$$
$$dist_{uv} = \|p_u - p_v\|^2 e^{-\frac{1}{\|p_u - p_v\|^2}}, u, v \leq n \ and \ u \neq v \tag{1}$$

where n is the number of points and p_u denotes the uth point, $x \in R^{n*(n-1)/2}$. In this case, $n = 23$ and x is a 253-dimensional column vector. Then, for m samples, we can construct the data matrix $X = \{x_1, x_2, \ldots, x_m\}$.

The dimensions of distance vector are high and may contain redundant information. A widely used unsupervised feature reduction method is principal components analysis (PCA). However, PCA has a limitation in dealing with nonlinear cases, and later be extended to Kernel PCA (KPCA). The objective function of KPCA is:

$$\max_{V^T V=1} V^T X^T X V \tag{2}$$

where V is the projection direction, $X = \{\Phi(x_1), \Phi(x_2), \ldots, \Phi(x_m)\}$ is the data matrix mapped into kernel space and $\Phi(x)$ is the kernel function. Equation 2 is a standard generalized eigenvalue problem which can be solved using any eigen-solver.

3.4   Regression Models

3.4.1   Linear Regression

Linear regression [14] is one of the best-known modeling techniques of which form is simple and comprehensibility is favorable. The formulation of linear regression forecasting function is shown as below:

$$f(x_i) = w^T x_i + b, f(x_i) \approx y_i \tag{3}$$

where $x_i \in R^d$ is a d-dimension column vector denoting i^{th} sample, $w = (w_1, w_2, \ldots, w_d)$ and b are the regression parameters for training. The final solution of Eq. 3 is:

$$f(\hat{x}_i) = \hat{x}_i^T (X^T X)^{-1} X^T y \tag{4}$$

$X = \{x_1, x_2, \ldots, x_m\}$ is the data matrix.

3.4.2   Support Vector Regression

Support Vector Machine (SVR) [15] is a supervised learning model which has an application in regression analysis. The solution of SVR can be formulated as below:

$$f(x) = \sum_{i=1}^{m} (\hat{a}_i - a_i) k(x, x_i) + b \tag{5}$$

$k(x_i, x_j)$ is the kernel function, $a_i, \hat{a}_i$ are Lagrange multiplier and b is a parameter to be solved.

3.4.3   Multi-Layer Perceptron

Multi-layer perceptron (MLP) [16] is a kind of feedforward artificial neural network which consists of at least three layers of nodes. Each node is a neuron using a nonlinear activation function except input nodes. MLP is a supervised method.

3.4.4   Lasso Regression and Ridge Regression

Least absolute shrinkage and selection operator (Lasso) [17] is a regression analysis method which was proposed in 1996 based on Leo Breiman's nonnegative garrote.

Regularization term was applied in Lasso to reduce overfitting. The objective function of Lasso is

$$J_L(w) = \frac{1}{n}\|y - Xw\|^2 + \lambda\|w\|_1 \tag{6}$$

here λ is the parameter controlling the smoothness of regularizer.

Ridge regression [18] is an improvement of least squares by adding Tikhonov regularization. It is a biased estimate regression method specialized in collinear data analysis of which loss function is

$$J_L(w) = \|y - Xw\|^2 + \lambda\|w\|^2 \tag{7}$$

that the regularization term of ridge regression is L_2 norm while Lass is L_1 norm.

3.4.5 Elastic Net Regression

Elastic Net [19] is a linear regularized method that linearly combines the L_1 and L_2 penalties of Lasso regression and ridge regression methods. The estimates from elastic net method are defined as

$$\hat{w} = \arg\min_{w}(\|y - Xw\|^2 + \lambda_2\|w\|^2 + \lambda_1\|w\|_1) \tag{8}$$

The quadratic penalty term makes the loss function strictly convex, and it therefore has a unique minimum. This kind of estimation incurs a double amount of shrinkage, which leads to increased bias and poor predictions.

4   Experiment

4.1   Experimental Settings

In this section, six regression methods were evaluated on The BCS Database described in Sect. 2, which were linear regression (LR), support vector regression (SVR), multi-layer perceptron (MLP), Lasso regression, ridge regression, elastic net (EN) respectively. The leave one out cross validation (LOOCV) was chosen as the data splitting method. For different regression models, correlation coefficient and mean error were selected as the performance measure. Let $f(x)$ be the score predicted an y be the score estimated manually, the mean error is defined as:

$$mean\ error = \frac{1}{m}\sum_{i=1}^{m}|f(x_i) - y_i| \tag{9}$$

m is the number of samples. Form of correlation coefficient is

$$corr(X, Y) = \frac{Cov(X, Y)}{\sqrt{Var(X)Var(Y)}} \tag{10}$$

here $Cov(X, Y)$ is the covariation of X and Y, $Var(X)$ is the variance of X.

4.2 Result and Discussion

The result of experiment is shown in Fig. 4 and Table 3, with abscissa values representing the scores evaluated by technicians and ordinate values representing the scores predicted by models. Linear regression employed by Azzaro in [6] is the simplest method for regression. As we have seen, it obtained a biggest error (0.42) and a smallest correlation coefficient (0.58). In Fig. 4a, it leads to some extreme scores which go out of range by maximum value 5. Although this problem can be solved by truncation operation, linear regression model is still not accurate enough. Support vector regression outperformed linear regression with error 0.31 and correlation coefficient 0.59. Multilayer perceptron is a nonlinear method and reaches a more aggregated distribution, as shown in Fig. 4c. Lasso regression and ridge regression get 0.29, 0.27 error and 0.70, 0.71 correlation coefficient respectively. Elastic net considers the advantages of both lasso and ridge, thus get a smallest error of 0.27 and biggest correlation coefficient of 0.72. Experiment shows that elastic net performs the best on The BCS Database.

Table 3. Mean error and correlation coefficient with different regression methods

	LR	SVR	MLP	Lasso	Ridge	EN
Mean error	0.4208	0.3079	0.2887	0.2914	0.2733	0.2705
Corr	0.5844	0.5939	0.6897	0.7030	0.7100	0.7210

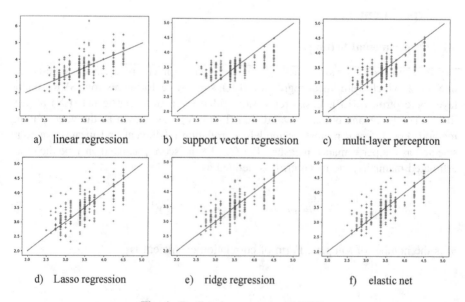

a) linear regression   b) support vector regression   c) multi-layer perceptron

d) Lasso regression   e) ridge regression   f) elastic net

Fig. 4. Predicted versus manual BCS

5   Conclusions

Body condition scoring provides an objective assessment of the amount of subcutaneous fat appositions and energy storage of the dairy cattle, which has become a powerful tool for dairy industry management. In this paper, we extend the work of [6] and propose a new shape normalization method. Distance vector is introduced to describe the shapes and kernel PCA is used for feature reduction. After getting the consistent representation of all shapes, totally 6 regression models are evaluated on BCS. Experiment shows that elastic net performs the best on The BCS Database. However, The BCS database only records the 2D-information of dairy cattle, ignoring the body surface information. Future work will focus on cow's depth images processing and improve the accuracy of estimating body condition of the dairy cattle.

Acknowledgement. This work was partially supported by the China Postdoctoral Science Foundation (2018M633585), Natural Science Basic Research Plan in Shaanxi Province of China (No. 2018JQ6060), the Doctoral Starting up Foundation of Northwest A&F University (No. 2452015302), and Students Innovation Training Project of China (201710712064).

References

1. Wildman, E.E., Jones, G.M., Wagner, P.E., et al.: A dairy cow body condition scoring system and its relationship to selected production characteristics. J. Dairy Sci. **65**(3), 495–501 (1982)
2. Edmonson, A.J., Lean, I.J., Weaver, L.D., et al.: A body condition scoring chart for Holstein dairy cows. J. Dairy Sci. **72**(1), 68–78 (1989)
3. Ferguson, J.D., Galligan, D.T., Thomsen, N.: Principal descriptors of body condition score in Holstein cows. J. Dairy Sci. **77**(9), 2695–2703 (1994)
4. Elanco Animal Health: Body Condition Scoring in Dairy Cattle. In: Body Condition Scoring in Dairy Cattle, 1st ed., p. 8 (2009)
5. Bewley, J.M., Peacock, A.M., Lewis, O., et al.: Potential for estimation of body condition scores in dairy cattle from digital images. J. Dairy Sci. **91**(9), 3439 (2008)
6. Azzaro, G., Caccamo, M., Ferguson, J.D., et al.: Objective estimation of body condition score by modeling cow body shape from digital images. J. Dairy Sci. **94**(4), 2126 (2011)
7. Sahbi, H.: Kernel PCA for similarity invariant shape recognition. Neurocomputing **70**(16–18), 3034–3045 (2007)
8. Liu, J., Wei, J., Guo, Y.: Study on application of image recognition technique in the cow body condition score. In: International Conference on Communication Technology, pp. 373–376. IEEE (2011)
9. Weber, A., Salau, J., Haas, J.H., et al.: Estimation of backfat thickness using extracted traits from an automatic 3D optical system in lactating Holstein-Friesian cows. Livestock Sci. **165**(1), 129–137 (2014)
10. Fischer, A., Luginbühl, T., Delattre, L., et al.: Rear shape in 3 dimensions summarized by principal component analysis is a good predictor of body condition score in Holstein dairy cows. J. Dairy Sci. **98**(7), 4465–4476 (2015)
11. Wang, L.: Application of dairy cow body condition scoring based on machine vision. Donghua University (2014)

12. Zhao, K.X.: Dairy cattle's information perception and behavior analysis based on machine vision. Northwest A&F University (2017)
13. Isensee, A., Leiber, F., Bieber, A., et al.: Comparison of a classical with a highly formularized body condition scoring system for dairy cattle. Anim. Int. J. Anim. Biosci. **8** (12), 1971–1977 (2014)
14. Kianifard, F.: Applied linear regression models. Technometrics **32**(3), 352–353 (1996)
15. Drucker, H., Burges, C.J.C., Kaufman, L., et al.: Support vector regression machines. Adv. Neural Inf. Process. Syst. **28**(7), 779–784 (1997)
16. Zhang, Z., Lyons, M., Schuster, M., et al.: Comparison between geometry-based and Gabor-wavelets-based facial expression recognition using multi-layer perceptron. In: Proceedings of IEEE International Conference on Automatic Face and Gesture Recognition, pp. 454–459. IEEE (1998)
17. Tibshirani, R.: Regression shrinkage and selection via the lasso: a retrospective. J. Roy. Stat. Soc. **73**(3), 273–282 (2011)
18. Marquardt, D.W., Snee, R.D.: Ridge regression in practice. Am. Stat. **29**(1), 3–20 (1975)
19. Zou, H., Hastie, T.: Regularization and variable selection via the elastic net. J. Roy. Stat. Soc. **67**(5), 301–320 (2005)

Intensive Positioning Network for Remote Sensing Image Captioning

Shengsheng Wang, Jiawei Chen[✉], and Guangyao Wang

College of Computer Science and Technology, Jilin University,
Changchun 130012, China
Jwchen16@mails.jlu.edu.cn

Abstract. This paper focuses on solving the problem of information loss during the generation of remote sensing image captions. In the field of artificial intelligence, the automatic description of remote sensing images is an important but rarely studied task. In the traditional framework, due to the higher pixels of the remote sensing image and the smaller target, when the image is processed and classified, the information is largely lost. In this case, we propose a new remote sensing image captioning framework using deep learning technology and attention mechanism. The experimental results show that the model can generate a full sentence description for remote sensing images.

Keywords: Intensive positioning network (IPN) · Attention mechanism
Deep learning · Image processing · Remote sensing image captioning

1 Introduction

Automatic remote sensing image captioning means that the machine processes remote sensing images and presents the user with as much useful information as possible. Although there have been many studies in the field of remote sensing in the past, most of them have been studied on scene classification and remote sensing target detection. Remote sensing image captioning differs from these tasks in that its goal is to produce sentences that are reliable and conform to human language logic rather than simply categorizing or predicting a label. To generate reliable and logical sentence descriptions, you must accurately identify every small target in the remote sensing image and use the spatial relationship between them.

Automatically generating image captioning has been a difficult AI problem for a long time in the past.

Benefit from the rapid development of image processing and natural language processing technology in these two years, many image captioning systems have been able to generate accurate and valuable sentences through information conveyed by images.

However, the processing of remote sensing image captioning is somewhat embarrassing. Because the pixels of remote sensing images are larger and the targets are smaller and more, multi-target recognition is more difficult. When using the same CNN framework as traditional image captioning processing, many small objects and background information are ignored, causing serious information loss.

© Springer Nature Switzerland AG 2018
Y. Peng et al. (Eds.): IScIDE 2018, LNCS 11266, pp. 567–576, 2018.
https://doi.org/10.1007/978-3-030-02698-1_49

In this case, we propose a new remote sensing image captioning framework.

In recent years, with the rise of deep learning, neural networks based on attention mechanism have become a hot topic in neural network research. The attention mechanism is a strategy that was first proposed in the visual image field. The idea of the attention mechanism is to increase the weight of useful information, so that the task processing system is more focused on finding useful information in the input data that is relevant to the current output, thereby improving the quality of the output.

By leveraging the recent popular attention mechanism and neural network, we propose a new network: intensive positioning network (IPN). This intensive positioning network can predict regions containing important information in the picture and output multiple region description blocks around these regions. Our remote sensing subtitle framework consists of two parts: (1) Multi-target recognition phase, remote sensing images are detected and identified by the IPN model, and output area characteristics and regional locations. (2) At the language generation stage, regional features and regional locations were integrated by the language model and output sentence descriptions. Figure 1 shows the specific flow of the framework.

The rest of this article is organized as follows. In Sect. 2, Introduced the target recognition model IPN; in Sect. 3, introduced the specific design of the loss function; in Sect. 4, Introduce the language model we use; in Sect. 5, experiment to evaluate the performance of the proposed model. Finally, in Sect. 6, summarized the conclusion of the experiment.

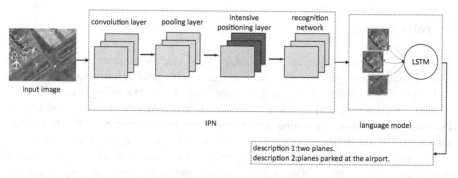

Fig. 1. Model overview.

2   IPN Model

In this section, we will detail our IPN model. Figure 2 shows the specific algorithm flow of the model.

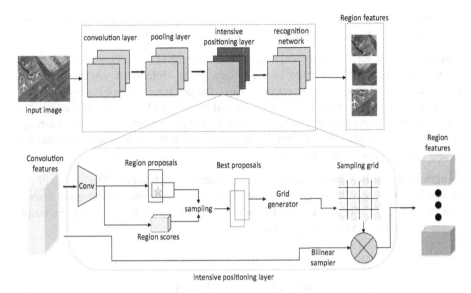

Fig. 2. Algorithmic flow of the proposed model.

2.1 Convolutional Layer and Pooling Layer

Our model body uses the VGG-16 architecture [1]. It consists of 13 convolutional layers and 5 pooling layers, with a convolution kernel size of 3 × 3 and a pool size of 2 × 2. We delete the last pooling layer, so after inputting an image with a shape of $3 \times W \times H$, We can get a set of features. These features have the size of $C \times W' \times H'$ Where $C = 512$, $W' = \lfloor \frac{W}{16} \rfloor$, and $H' = \lfloor \frac{H}{16} \rfloor$. Uniformly sampled images will be input into intensive positioning layer after passing through the convolution and pooling layers.

2.2 Intensive Positioning Layer

The intensive positioning layer receives input tensors from the convolution layer, extracts regions containing important information from it, and extracts representations from each region. We replaced the original pooling mechanism with bilinear interpolation [2], so that in our model, the gradient can be propagated backwards through the coordinates of the prediction region.

Inputs/Outputs. A tensor of size $C \times W' \times H'$ is input into the intensive positioning layer. Then select the B regions that contain the important information in the tensor and return the three output tensors about the region information:

1. **Coordinates:** The coordinates of the bounding box of each output area, the size of the matrix of B × 4.
2. **Scores:** A confidence score is given to each output area and a total of B scores are output.

3. **Features:** The characteristics of each output area, the size of the tensor $B \times C \times X \times Y$, expressed as the $X \times Y$ grid of C-dimensional features.

Anchor Return. First, we project points in the feature grid to the $W \times H$ image plane. Then center the point projected to the image plane, different k anchor boxes are obtained by transforming the aspect ratio. Next, the intensive positioning layer predicts the confidence scores of the k anchor boxes and records back from the anchor point to the predicted box coordinates. In the calculation, the input features are passed through 256 3 × 3 filters and 5k 1 × 1 filters. The specific result includes all anchor scores and offsets and the size is $5k \times W' \times H'$.

Length and Width Transformation. First, given an anchor box. The center of this anchor box is (x_a, y_a), the height is h_a, the width is w_a. Then, our model gives a set of scalars (t_x, t_y, t_w, h_h) for position and length-width transformations. so that the output region has center (x, y) and shape (w, h) given by

$$x = x_a + t_x w_a \quad y = y_a + t_y h_a \tag{1}$$

$$w = w_a \exp(t_w) \quad h = h_a \exp(h_w) \tag{2}$$

Sampling. For example, to process an image with W = 810, H = 620, when there are 14 anchor boxes, there are 18,280 region proposals. If the network is running for all region proposals, the resulting costs are too high, so it is necessary to sub-sample them. During training, we followed the method of [3]. We sample a small sample of B = 256 boxes. If an area coincides with another area with at least 0.7 and has some true value area, the area is positive, otherwise it is negative. Our small sample contains positive and negative regions, which are uniformly sampled from the collection of all areas at the time of sampling. In the test, we use the non maximum suppression (NMS) algorithm to select the size of B. When B = 300, the effect is optimal. After sampling, the proposed coordinate and confidence tensor will be output from the intensive positioning layer. The size of the coordinate tensor is B × 4, and the size of the confidence tensor is B.

Bilinear Interpolation. Since the image passes through the convolutional layer, the pooled layer, and the intensive positioning layer, it outputs different size region proposals, and the recognition network and language model can only accept fixed-size feature input. To extract fixed-size features, we replace the last pooling layer with bilinear interpolation. Firstly, given a $C \times W' \times H'$ shape feature map U and we insert the information of U to generate a $C \times X \times Y$ shape feature map V. Secondly, we compute a sampling grid G of shape $X \times Y \times 2$. If $G_{i,j} = (x_{i,j}, y_{i,j})$ then $V_{c,i,j} = U$; however since $(x_{i,j}, y_{i,j})$ are real-valued, we use the sampling kernel k to convolve and set

$$V_{c,i,j} = \sum_{i'=1}^{W} \sum_{j'=1}^{H} U_{c,i',j'} k(i' - x_{i,j}) k(j' - y_{i,j}) \tag{3}$$

$$k(d) = max(0, 1 - |d|) \tag{4}$$

The sampling grid uses a linear function of the proposal coordinates, so the gradient in the region coordinates can be propagated backwards. A $B \times C \times X \times Y$ tensor can be obtained by extracting all the region features using bilinear interpolation. This tensor is also the final output of the intensive positioning layer.

2.3 Recognition Network

The recognition network is a fully-connected neural network. It is used to process region features from the intensive positioning layer. The recognition network flattens the input area feature into a vector through two fully connected layers. In order to reduce the amount of computation, we use Dropout to identify adjustments in the recognition network. Thus each area will produce a code of size D = 4096. The code includes its visual appearance. The network collects all positive area codes into a $B \times D$ matrix and passes them to the language model. In addition, we also allow the recognition network to adjust the confidence and position of each proposal region again. Finally, the recognition network will output the final scalar confidence of each proposed region and four scalars encoding a final spatial offset.

3 Loss Function

During training, our input and output consist of boxes and descriptions. It should be noted that in the intensive positioning layer and recognition network, the positions and confidences of our sampling regions were adjusted twice. We use binary logistic losses to train positive and negative confidence in the sample. For the box regression, we use the L1 loss in [3]. We use the cross-entropy term for each time step of the language model as the fifth item of the loss function. During initialization, we tested the range of values of these weights and found that the reasonable setting is that the weights of the first four criteria are 0.1 and the weight of the captioning is 1.0.

4 Language Model

The previous work was given in [4–8]. In order to make the RNN language model [9–11] more in line with our framework, we use the region codes to adjust it. Specifically, we give a training sequence of tokens $s_1, \ldots, s_t$ and provide the RNN $T + 2$ word vectors. The number of the vector is $x_{-1}, x_0, x_1, \ldots, x_T$. Except x_{-1}, the rest is encoded using the ReLU function. x_0 is marked as start, and x_t encode each of the tokens s_t, $t = 1, \ldots, T$. Here we use the recursion formula in LSTM [12] to calculate the hidden state h_t and output vector y_t in the network. The size of the output vector y_t is $|V| + 1$, where V is the token vocabulary. Here we attach an END token. We use the average cross-entropy as a loss function. The size of our tokens is 512. At each step, we sample important information in the next step and provide it to the RNN in the next step, repeating the process until the END mark is sampled.

5 Experiments

Due to the lack of research on the captioning task of remote sensing images, there is no unified data set and evaluation standard in the field of remote sensing. Therefore, we use the data set and evaluation criteria in [13]. In this section, the effectiveness of the proposed model will be verified using the Google Earth image and the GaoFen-2 (GF-2) image from [13].

5.1 Experiment Settings

Our experimental data set uses the image in [13]. The Google Earth image has a resolution of 0.5 m/pixel and the GF-2 image has a resolution of 0.8 m/pixel. To test the performance of the algorithm in different size images. We divide the picture into different sizes and fixed resolutions. The size includes two kinds, one is 1000–8000 pixels, and the other is 480 × 640 pixels. Table 1 lists detailed information about these images.

Table 1. Experimental image details.

Source	Size	Training	Testing	Resolution
Google Earth	1000–8000pxl	310	10	0.5 m/pxl
	480 × 640pxl	0	100	
GF-2 (fused)	1000–3000pxl	0	10	0.8 m/pxl
	480 × 640pxl	0	100	

Data Preprocessing: Since GF-2 images are fused, we extract the three visible bands and superimpose them onto the pseudo RGB image. Here in order to facilitate our experimental calculations, we use 3-channel RGB images. But it needs to be emphasized that our model itself does not limit the data types. We only need to make some minor changes to the first-level convolution filter to handle different image types.

In our data set, the distribution of training sets and test sets is detailed in Table 1. During training, we use large images to train key cases. When training environmental factors, we trained the image further by cutting it into a set of smaller images. Before training the images, the category labels and bounding boxes of all key instances and elements (such as airports, ports, oceans, and cities) have been manually tagged. Some ambiguous examples are also directly deleted from our data set. Examples of these exclusions are boats and aircraft less than 20 pixels in length or less than 10 pixels in radius.

To detect the direction of the instance, we rotate each positive training window several times. We randomly sample a set of non-instance windows from the data set as our negative training sample. By the above method, we have obtained a positive window of more than 40 k and a negative window of 400 k. In order to reduce over-fitting of the network to instances of the same class, each image is down-sampled at the beginning of the detection process at four scaling ratios, with a uniform down-sampling rate of 1.2. For environmental analysis, each image is down sampled at a rate of 3.0 to obtain a larger perceived field. Finally, use a 190 k patch to train its softmax weights. Here, we set the learning rate $\mu = 0.001$. With this setting, the objective function value can converge to a constant within a few minutes.

5.2   Experimental Results and Evaluation

Figure 3 shows some of the images of the test set and their description. In order to evaluate the generated description, we have introduced the evaluation criteria in [6]. The advantages and disadvantages of the description are divided into four levels. We give an example of this standard in Fig. 4. For each image description, it is evaluated by 10 different people [13]. The results of the evaluation will be given in the Comparative Experiment section.

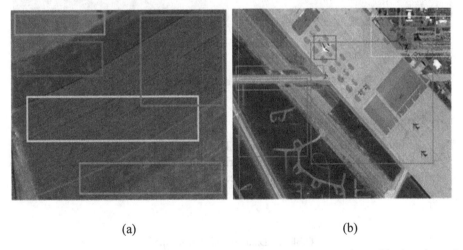

(a)   (b)

Fig. 3. Example images and descriptions. (a) description: a meadow; a piece of land; a piece of farmland; a piece of farmland; a meadow. (b) description: an airplane; a long runway; This area consists of some buildings and structures; a lawn adjacent to an airstrip; several planes parked on the airfield.

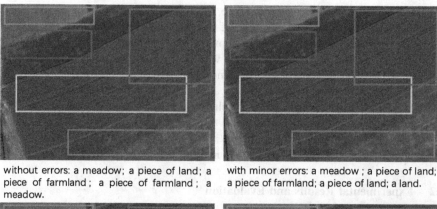

without errors: a meadow; a piece of land; a
piece of farmland ; a piece of farmland ; a
meadow.

with minor errors: a meadow ; a piece of land;
a piece of farmland; a piece of land; a land.

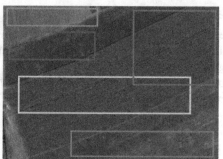

Related to the image: a meadow; a meadow;
a meadow; a piece of farmland; a meadow.

unrelated to the image: an airplane; an
airplane; an airplane; an airplane; an airplane.

Fig. 4. An example of an evaluation criterion.

5.3   Experimental Comparison

The experimental code in this article is A. The literature [13] experiment code is B.

We follow the subjective assessment criteria described in the previous section to
score the descriptions obtained in this article. The detailed data comparison is shown in
Table 2.

Table 2. Data comparison results.

Experiment name	A		B	
Source	Google earth	GF-2	Google earth	GF-2
Without errors	80%	73%	63%	48%
With minor errors	12%	18%	22%	23%
Related to the image	6%	7%	10%	19%
Unrelated to the image	2%	2%	5%	10%

We run our model on a PC with 32G RAM, Nvidia Taitan X graphics, and an Intel Xeon CPU. The programming platform is TORCH + CUDA.

We calculate the average calculation time for each phase. When entering a 480 × 640 image, our model takes only about 1 s to complete the entire subtitle process. When you enter a 1600 × 2400 image, the runtime is less than 5 s. Obviously, our model is highly computationally efficient thanks to our ICN network. The ICN can calculate the convolution features of the entire input image and then reduce the computational redundancy by sharing the extracted vectors of the feature map. If the remote sensing image is large, such as 10000 × 10000 pixels, we recommend processing the image after slicing. Here, we use time1 to represent the time passed by the model, time2 indicates the time through the language model, and time3 indicates the total time to pass the model. Table 3 shows a comparison of the detailed time at different stages in this paper and in [13].

Table 3. Detailed execution time comparison.

Experiment name	A		B	
Image size(pxl)	GPU time(s)	CPU time(s)	GPU time(s)	CPU time(s)
600 × 480	time1 = 1.10	time1 = 7.30	time1 = 1.25	time1 = 8.12
	time2 = 0.01	time2 = 0.01	time2 = 0.01	time2 = 0.01
	time3 = 1.11	time3 = 7.31	time3 = 1.26	time3 = 8.13
2400 × 1600	time1 = 3.61	time1 = 112.10	time1 = 5.71	time1 = 139.10
	time2 = 0.09	time2 = 0.09	time2 = 0.13	time2 = 0.13
	time3 = 3.70	time3 = 112.19	time3 = 5.84	time3 = 139.23

6 Conclusion

In this paper, we study the problem of information loss during the generation of remote sensing image captions. To solve this problem, we propose a remote sensing image captioning framework. The experimental results based on a series of remote sensing images prove the superiority and transferability of the model. Although there is still room for improvement in our model, we think this is a fast, powerful and promising framework. Our future work will focus on identifying blurry super-resolution images and generating richer language descriptions.

Acknowledgments. This work is supported by the National Natural Science Foundation of China (61472161), Science & Technology Development Project of Jilin Province (20180101334JC).

References

1. Simonyan, K., Zisserman, A.: Very deep convolutional networks for large-scale image recognition. arXiv preprint arXiv:1409.1556 (2014)
2. Jaderberg, M., Simonyan, K., Zisserman, A., Kavukcuoglu, K.: Spatial transformer networks. In: 28th Annual Conference on Neural Information Processing Systems, pp. 2017–2015. MIT Press, Montreal (2015)
3. Ren, S., He, K., Girshick, R., Sun, J.: Faster R-CNN: towards real-time object detection with region proposal networks. In: 28th Annual Conference on Neural Information Processing Systems, pp. 91–99. MIT Press, Montreal (2015)
4. Mao, J., Xu, W., Yang, Y., Wang, J., Yuille, A.L.: Explain images with multimodal recurrent neural networks. arXiv preprint arXiv:1410.1090 (2014)
5. Karpathy, A., Fei-Fei, L.: Deep visual-semantic alignments for generating image descriptions. In: 25th IEEE Conference on Computer Vision and Pattern Recognition, pp. 3128–3137. IEEE, Boston (2015)
6. Vinyals, O., Toshev, A., Bengio, S., Erhan, D.: Show and tell: a neural image caption generator. In: 25th IEEE Conference on Computer Vision and Pattern Recognition, pp. 3156–3164. IEEE, Boston (2015)
7. Donahue, J., et al.: Long-term recurrent convolutional networks for visual recognition and description. In: 25th IEEE Conference on Computer Vision and Pattern Recognition, pp. 2625–2634. IEEE, Boston (2015)
8. Chen, X., Zitnick, C.L.: Mind's eye: a recurrent visual representation for image caption generation. In: 25th IEEE Conference on Computer Vision and Pattern Recognition, pp. 2422–2431. IEEE, Boston (2015)
9. Graves, A.: Generating sequences with recurrent neural networks. arXiv preprint arXiv: 1308.0850 (2013)
10. Mikolov, T., Karafiát, M., Burget, L., Cernocký, J., Khudanpur, S.: Recurrent neural network based language model. In: 11th Annual Conference of the International Speech Communication Association. ISCA, Makuhari (2010)
11. Sutskever, I., Martens, J., Hinton, G.E.: Generating text with recurrent neural networks. In: 28th International Conference on Machine Learning, pp. 1017–1024. ACM, Bellevue (2011)
12. Hochreiter, S., Schmidhuber, J.: Long short-term memory. Neural Comput. 9(8), 1735–1780 (1997)
13. Shi, Z., Zou, Z.: Can a machine generate humanlike language descriptions for a remote sensing image? IEEE Trans. Geosci. Remote Sens. 55(6), 3623–3634 (2017)

A Level Set Algorithm Based on Probabilistic Statistics for MR Image Segmentation

Jin Liu$^{(\boxtimes)}$, Xue Wei$^{(\boxtimes)}$, Qi Li, and Langlang Li

School of Electronic Engineering, Xidian University, Xi'an 710071, China
{jinliu,liqi}@xidian.edu.cn, wxyyxz@163.com,
langzgy@163.com

Abstract. MR image segmentation is of great importance in medical image application. MR images have the characteristics of intensity inhomogeneities, strong background interference and blurred target area. These characteristics will greatly increase the difficulty of segmentation and affect image segmentation results. To obtain the satisfied performance of MR image segmentation, a level set algorithm based on probabilistic statistics for MR image segmentation is proposed. Because of the intensity inhomogeneity of the image, a bias field is used to describe the image in the proposed model. But the addition of a bias field will increase the amount of computation. Therefore, combining with the probabilistic statistical theory, the energy function is defined by the pixel distribution probability to improve operational efficiency. In addition, a new rule item is added to enhance the edge information of the image to highlight the edge segmentation curve. Experimental results show that the proposed model behaves well in segmenting MR images.

Keywords: Probability statistics · Level set · Intensity inhomogeneity
MR image segmentation

1 Introduction

Image segmentation is a key step from image processing to image analysis. MR image segmentation is of great significance in clinical diagnosis, pathological analysis and treatment. It can be used not only for the analysis of biomedical images, but also for compressing and transmitting data without loss of useful information. The commonly used MR image segmentation methods are the following: (1) Manual segmentation (2) Intensity-based methods, including clustering, region growth, classification, threshold, etc. [1–4] (3) Surface-based methods, including active contour models and Multiphase active contour model [5, 6] (4) Hybrid segmentation methods [7–9].

In the active contour model, applying the level set model for segmentation achieves good results. In terms of improving the level set model, Li et al. introduced the window function and bias field into the level set model. He used the local energy term to solve the image in a small range, which was extended to the whole image [10–12]. By solving the variables within a small range of the image, the deviation of the mean intensity of the pixel points in the corresponding region is corrected. The correctness of the evolution of the level set algorithm with the regional mean energy is guaranteed,

© Springer Nature Switzerland AG 2018
Y. Peng et al. (Eds.): IScIDE 2018, LNCS 11266, pp. 577–586, 2018.
https://doi.org/10.1007/978-3-030-02698-1_50

and the accurate segmentation result is obtained. However, the algorithm uses the step function and the impulse function as the function support of the level set evolution. This makes it easier for the algorithm produces deviation in the region of the contour jump, and therefore it cannot outline the sharp contour well. Zhang et al. preprocessed the image firstly, using the window function to map the original image to another domain, and then defined a maximum likelihood function combining the bias field and the level set function. At the same time as the pixel mean deviation in the area is corrected, the energy function is solved [13–15]. However, the calculation model of this method is complicated and takes a long time.

Inspired by these studies and the characteristics of MR images, a level set algorithm based on probabilistic statistics for MR image segmentation is proposed. The idea of the proposed algorithm is stated as follows. First, the relational term between pixels in the probabilistic model is added to the level set model as an external thrust. This can increase the iterative step size of the level set algorithm and make it converge faster to the target area. The probability of pixels formula comes from [16–18]. Second, since the bias field can effectively correct the mean deviation caused by the intensity inhomogeneity of MR image, the bias field is still applied as an auxiliary function of evolution to join in the energy function. Finally, while the algorithm reduces image noise interference, it also weakens the edge information of the image at the same time. Therefore, it is necessary to add a rule item to highlight the contour information of the image and constrain the evolution process of the algorithm.

The rest of the paper is organized as follows. In Sect. 2, a brief review of level set segmentation algorithm combining with local information and the design of the new energy function term are given. Then we describe the new algorithm in Sect. 3. The experimental results are presented in Sect. 4. Finally, Sect. 5 concludes the paper.

2   Related Works

2.1   Level Set Segmentation Method Combining Local Information

Li et al. introduced the bias field into the level set model, and proposed a level set segmentation method that combines local information [10]. It is assumed that the image is composed of a real image and the variation of the intensity inhomogeneity. The real image is an evenly distributed image with equal pixels in each region. Then the image I can be described as:

$$I = b \cdot J + n, \tag{1}$$

where J denotes the real image, which can be approximated as a piecewise constant; b denotes the component of intensity inhomogeneity, which is called bias field; and n denotes an additional noise, which is regarded as Gaussian white noise with a mean of 0.

Because of the intensity inhomogeneities of MR images, it is difficult to describe the average value of intensity in each region. But the property of the local intensity is very simple, and it can be used to segment images while estimating the bias field. In a circular neighborhood with a radius ρ centered at each point, defined by $O_y \triangleq \{x : |x - y| \le \rho\}$. Slowly changing bias field b can be considered as a constant value, $b(x) \approx b(y), x \in O_y$. Therefore, the image model can be described as:

$$I(x) \approx b(y)c_i + n(x), x \in O_y \cap \Omega_i. \tag{2}$$

Then the standard k-means algorithm is used to represent energy term. The neighborhood O can be divided into N classes, and the central values are $m_i \approx b(y)c_i, i = 1, \ldots, N$. It can be expressed as specifically:

$$E_{k\text{-means}} = \sum_{i=1}^{N} \int_{\Omega_i \cap O_y} |I(x) - b(y)c_i|^2 dx. \tag{3}$$

By introducing window functions $K(y - x) : K(y - x) = 0, x \notin O_y$, define a local clustering standard function for classifying pixels in the neighborhood of O_y:

$$E_{k\text{-means}} = \sum_{i=1}^{N} \int_{\Omega_i} K(y - x)|I(x) - b(y)c_i|^2 dx. \tag{4}$$

Then it is extended from the local to the global and define an energy item:

$$E = \int \sum_{i=1}^{N} \int_{\Omega_i} K(y - x)|I(x) - b(y)c_i|^2 dxdy. \tag{5}$$

Combining the level set theory, the variational method is used to solve the problem of energy minimization, and the expression of the level set function is obtained:

$$\frac{\partial \emptyset}{\partial t} = -\delta(\emptyset)(E_1 - E_2) + \mu \left[\nabla^2 \emptyset - div\left(\frac{\nabla \emptyset}{|\nabla \emptyset|}\right) \right] - \beta \delta_\epsilon(\emptyset)div\left(g\frac{\nabla \emptyset}{|\nabla \emptyset|}\right). \tag{6}$$

where the average of target area and background area are as follows respectively:

$$c_1 = \frac{\int_\Omega (b * K)I(x) \cdot H(\emptyset)dx}{\int_\Omega (b^2 * K)H(\emptyset)dx}, c_2 = \frac{\int_\Omega (b * K)I(x) \cdot (1 - H(\emptyset))dx}{\int_\Omega (b^2 * K)(1 - H(\emptyset))dx}, \tag{7}$$

the solution formula for the bias field is expressed as follows:

$$b = \frac{(I(x) \cdot (H(\phi) \cdot c_1 + (1 - H(\emptyset)) \cdot c_2)) * K}{\left(H(\emptyset) \cdot c_1^2 + (1 - H(\emptyset)) \cdot c_2^2\right) * K}, \tag{8}$$

where * denotes the convolution operation; K denotes the window function. In general, the window function is selected as the Gaussian window function, ensuring the weight of the point itself is significant, and it can better ensure that the original information of the image is not lost.

2.2   Energy Function Term Based on Probability Theory

In order to speed up the algorithm's speed in the case of using a bias field, we use probabilistic statistical theory when designing the energy function. In the image, the existence of each pixel is not independent, and the relationship between the pixels and the entire image can be described by a probabilistic model.

The probability of pixels in each region is:

$$P(\Omega_i) = \prod_{s \in S} \frac{1}{\sigma} K(x-y) \exp\left(\frac{\nabla I}{\max(\nabla I)} - 1\right), i = 1, 2, \cdots \cdots, n, \qquad (9)$$

where σ is the normalization factor and $K(x-y)$ is the window function. Here we use the Gaussian window function, which assigns a different value to each pixel in the neighborhood to prevent the pixel point information from being lost. ∇ denotes gradient operation. If the pixels in the area belong to the background or the target area, the probability value $P(\Omega_i)$ is small, whereas if the pixel is located at the edge, the probability value $P(\Omega_i)$ is large. The more points in the neighborhood of the pixel are at the edge, the greater the probability value, and the maximum value is 1.

The conditional probability can be described as:

$$P(I|\Omega_i) = 0.5 + 0.5 \cdot \frac{I - \frac{c_1 + c_2}{2}}{\max\left(\left|I - \frac{c_1 + c_2}{2}\right|\right)}, x \in \Omega, \qquad (10)$$

where c_1 and c_2 present the mean values of the target area and the background area respectively.

Assuming that the domain of image I is Ω, and the image is divided into the target region O and the background region B by a simple closed curve C, i.e., $O \cap B = 0$ and $O \cup B = \Omega$. The probability of any pixel can be described as:

$$P(I) = P(I|O)P(O) + P(I|B)P(B). \qquad (11)$$

According to the previous formula, the probability of the target and background in the image can be obtained as:

$$P(O) = P(B) = \prod_{s \in S} \frac{1}{\sigma} K(x-y) \exp\left(\frac{\nabla I}{\max(\nabla I)} - 1\right), \qquad (12)$$

$$P(I|O) = \frac{I - C_o}{\max(|I - C_o|)}, \quad P(I|B) = \frac{I - C_B}{\max(|I - C_B|)}. \qquad (13)$$

Due to the characteristics of intensity inhomogeneity, the bias field k still as an auxiliary function of evolution in the proposed model.

Traversing all the pixels of the input image matrix, combining the energy center value of the target area and the energy center value of the background area, an iterative formula of the bias field matrix is obtained:

$$k_i = \frac{\left(I\left(c_{1i}u_{i-1}^m + c_{2i}(1 - u_{i-1})^m\right)\right) * G(x)}{\left(c_{1i}^2 u_{i-1}^m + c_{2i}^2 (1 - u_{i-1})^m\right) * G(x)},$$ (14)

where u denotes the membership function. The concept of membership function of fuzzy clustering algorithm is introduced into the algorithm. In this model it is expressed as the membership matrix equal to the size of the image. The membership matrix assigns a degree of membership value u to each pixel of the image matrix to determine whether the corresponding pixel belongs to the target area. When $u > 0.5$, the pixel belongs to the background area. When $u < 0.5$, the pixel belongs to the target area. And When $u = 0.5$, the pixel is located on the target contour. So the new energy function we get is expressed as follows:

$$E = P(O) \cdot P(I|O) + P(B) \cdot P(I|B)$$
$$= \prod_{s \in S} \tfrac{1}{\sigma} K(x - y) \exp\left(\frac{\nabla I}{\max(\nabla I)} - 1\right) \cdot \left(\frac{u^m(I - k \cdot C_O)}{\max(I - k \cdot C_O)} + \frac{(1-u)^m(I - k \cdot C_B)}{\max(I - k \cdot C_B)}\right).$$ (15)

3 A Level Set Algorithm Based on Probabilistic Statistics for MR Image Segmentation

We add the probability statistic model to the design of the energy function in order to increase the running speed of the algorithm. And as mentioned before, the bias field will also weaken the edge information of the image. Here we add a rule item to highlight the outline information of the image and also restrict the evolution process of the algorithm. The rule term is expressed as follows:

$$E_{reg} = \frac{g' \cdot (1 - 2u(x))}{1 + e^{1-g'}} \int |H(0.5 - u(x))| dx.$$ (16)

Since the pixel point is judged to be in the background or the target is differentiated by the membership function, the new rule must also be strictly controlled between [0, 1]. We redefine the cut-off function and normalize it:

$$g = \frac{1}{1 + |\nabla(G_\delta * I)|}, g' = \frac{g}{g_{max}},$$ (17)

where G_δ is a Gaussian function with a variance of δ. The same gaussian function in the energy term can be selected to guarantee the consistency of the calculation. ∇ represents the gradient operation, * represents the convolution operation, and $|*|$ is an absolute value operation. The cut-off function is inversely proportional to the image gradient. When the image gradient becomes smaller, the value of the cut-off function increases to accelerate the evolution of the algorithm. When the gradient of the image becomes larger, the value of the cut-off function decreases so that the outline of the algorithm accurately converges near the target area. Let $\Psi = \prod_{s \in S} \tfrac{1}{\sigma} K(x - y)$ $\exp\left(\frac{\nabla I}{\max(\nabla I)} - 1\right)$, then the energy term is expressed as:

$$E_p = \Psi \cdot \left(\lambda_1 \int \frac{u^m (I - k \cdot C_o)^2}{max \left(|I - k \cdot C_O|^2 \right)} dxdy + \lambda_2 \int \frac{(1-u)^m (I - k \cdot C_B)^2}{max \left(|I - k \cdot C_B|^2 \right)} dxdy \right), \quad (18)$$

$$E = \xi E_p + \eta E_{reg}. \quad (19)$$

where ξ and η are weight terms, which determine the proportion of probability energy and the rule item in the segmentation. Minimize the energy function to get the iterative formula of u.

The main procedure of our proposed algorithm is shown in Table 1.

Table 1. The main procedure of our proposed algorithm

Input:	The original image
Output:	The segmented image
Step 1	Initialize the membership function u:
	$$u(x,\ t=0) = \begin{cases} d & x \in O \\ 1-d & x \in B \end{cases}$$
	where $d \in [0.5,1]$, O is the target area and B is the background area.
Step 2	Calculate the mean c_1 of the target area and the mean c_2 of the background area:
	$$c_1 = \frac{K_\sigma(x) * (H(x)I(x))}{K_\sigma(x) * H(x)}, \quad c_1 = \frac{K_\sigma(x) * [(1-H(x))I(x)]}{K_\sigma(x) * (1-H(x))}$$
Step 3	Update the bias field according to equation (14).
Step 4	Minimize the energy function and update the membership value corresponding to the pixel.
Step 5	Stop the iteration when N=80 and take the points of membership function u = 0.5 as the contour line.

4 Experiments

In this section, the performance of the proposed method is compared with LBF [10], Locally Statistical Level Set Method(LS-LSM) [13], RD [19] and GDRLSE3 [20] under the same experiment condition. All the experiments are conducted on MATLAB R2015b installed in a computer with a 3.40 GHz Intel Core i3 CPU and 4 GB of RAM. For all experiments, the proposed method sets the fuzzy degree of m = 2, the initial value of the membership function d is 2, the initial value of the bias field k is 1, the constant term $\beta = 0.001 \times 255$, the radius of the Gaussian window function $\rho = 3$, and the proportion of the external energy term $\lambda_1 = \lambda_2 = 1$.

Experiment 1: Brain model 90-degree slice image is selected for simulation experiment, which is from the website http://brainweb.bic.mni.mcgill.ca. The experimental results are shown in the figure below.

From the segmentation results shown in Fig. 1 the LBF algorithm and the GDRLSE3 algorithm can roughly segment the image, but they lack the outline at the edges. The LS-LSM algorithm cannot complete the segmentation of the image and stop the evolution prematurely. The RD algorithm can divide the middle part, but it ignores the edge completely. And the proposed algorithm can outline the gray matter, white matter and other parts.

Experiment 2: Brain model 140-degree slice image is selected for simulation experiment, which is from the website http://brainweb.bic.mni.mcgill.ca. The experimental results are shown in the figure below.

From the segmentation results shown in Fig. 2 and the time comparison in Table 2, the LS-LSM algorithm can depict the internal contours of the brain, but because of its computational model is too complex and the running time is too long. The RD algorithm fails to outline the internal curve of the brain and only draws the external curve. The LBF algorithm, GDRLSE3 algorithm and our proposed algorithm all correctly segment the image and the segmentation results are not much different. However, comparing with the running time, the proposed algorithm needs the least time, followed by the LBF algorithm, and the GDRLSE3 algorithm takes the longest time.

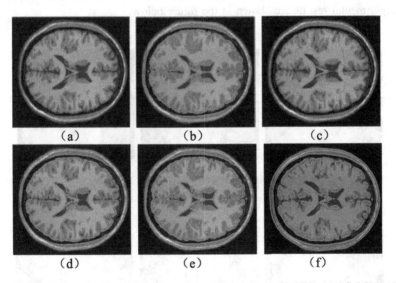

Fig. 1. Segmentation results on MR image: (a) original image, (b) LBF, (c) LS-LSM, (d) RD, (e) GDRLSE3, (f) the proposed algorithm.

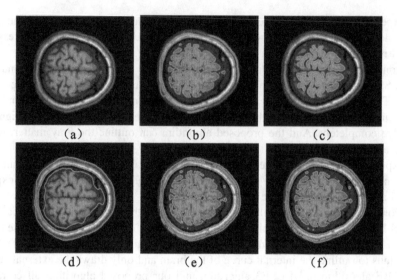

Fig. 2. Segmentation results on MR image: (a) original image, (b) LBF, (c) LS-LSM, (d) RD, (e) GDRLSE3, (f) the proposed algorithm.

Experiment 3: the MR image of human brain is selected for simulation experiment. The experimental results are shown in the figure below.

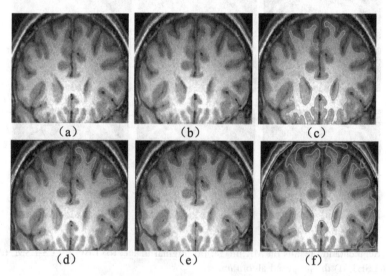

Fig. 3. Segmentation results on MR image: (a) original image, (b) LBF, (c) LS-LSM, (d) RD, (e) GDRLSE3, (f) the proposed algorithm.

From the segmentation results shown in Fig. 3 and the time comparison in Table 2, the LBF algorithm, RD algorithm, and GDRLSE3 algorithm take about the same time as our proposed algorithm. But comparing with the segmentation results, the LBF algorithm, RD algorithm, and GDRLSE3 algorithm cannot accurately outline the target contour at the bottom of the image, missing part of the target area. The LS-LSM algorithm can accurately outline the internal target area. However, the running time is too long. The proposed algorithm can not only obtain good segmentation results but also take a short time.

In addition, comparisons in terms of running time of the five algorithms are shown in Table 2. It can be seen that the algorithm we propose is faster than other algorithms. All the experiments show that the proposed algorithm is a real-time and effective segmentation algorithm.

Table 2. Running time of five algorithms (Seconds)

MR image	LBF	LS-LSM	RD	GDRLSE3	The proposed model
1	24.5	574.1	10.5	14.0	5.6
2	25.0	592.0	8.6	30.4	17.1
3	2.9	10.5	3.8	3.2	2.6

5 Conclusion

In this article, a level set algorithm based on probabilistic statistics for MR image segmentation is proposed. Because the level set algorithm has a free transformation topological structure, the membership matrix of fuzzy clustering algorithm is used during building model. The convergence point of the ordinary level set function is changed from 0 to 0.5, and the stability of the algorithm is also ensured when the target contour is segmented. Aiming at the intensity inhomogeneities of MR images, the bias field is introduced to adjust the mean value of the image area. At the same time, the Bayesian formula and other probabilistic theories are used to describe the regional distribution of image pixels. And the total probability formula is used to combine it with the level set algorithm to improve the algorithm's operation speed. Through segmentation experiments of brain MR images, it is shown that this algorithm can obtain a good segmentation effect, effectively highlighting the outline position of the target area in the image.

Acknowledgments. We would like to thank the associate editor and all anonymous reviewers for their constructive comments and suggestions. This research was partially supported by the National Science Foundation of China (Grant No. 61101246) and the Fundamental Research Funds for the Central Universities (Grant No. JB180208).

References

1. Despotovic, I., Goossens, B., Philips, W.: MRI segmentation of the human brain: challenges, methods, and applications. Comput. Math. Methods Med. **2015**(6), 1–23 (2015)
2. Banerjee, S., Mitra, S., Shankar, B.U.: Single seed delineation of brain tumor using multi-thresholding. Inf. Sci. **330**(C), 88–103 (2016)
3. Caldairou, B., Passat, N., Habas, P.A., Studholme, C., Rousseau, F.: A non-local fuzzy segmentation method. Pattern Recogn. **44**(9), 1916–1927 (2016)
4. Anitha, V., Murugavalli, S.: Brain tumour classification using two-tier classifier with adaptive segmentation technique. IET Comput. Vision. **10**(1), 9–17 (2016)
5. Kamaruddin, N.: Active contour model using fractional sync wave function for medical image segmentation. Surf. Sci. **363**(1–3), 321–325 (2017)
6. Khadidos, A., Sanchez, V., Li, C.T.: Active contours based on weighted gradient vector flow and balloon forces for medical image segmentation. In: IEEE International Conference on Image Processing, pp. 902–906 (2015)
7. Agrawal, R., Sharma, M., Singh, B.K.: Segmentation of brain lesions in MRI and CT scan images: a hybrid approach using k-means clustering and image morphology. J. Inst. Eng. **99** (2), 1–8 (2018)
8. Abdel-Maksoud, E., Elmogy, M., Al-Awadi, R.: Brain tumor segmentation based on a hybrid clustering technique. Egypt. Inf. J. **16**(1), 71–81(2015)
9. Lu, S., Lei, L., Huang, H., Xiao, L.: A hybrid extraction-classification method for brain segmentation in MR image. Int. Congr. Image Sig. Proc. 1381–1385 (2017)
10. Li, C., Huang, R., et al.: A level set method for image segmentation in the presence of intensity inhomogeneities with application to MRI. IEEE Trans. Image Process. **20**(7), 2007–2016 (2011)
11. Li, C., Xu, C., Anderson, Adam W., Gore, John C.: MRI tissue classification and bias field estimation based on coherent local intensity clustering: a unified energy minimization framework. In: Prince, J.L., Pham, D.L., Myers, K.J. (eds.) IPMI 2009. LNCS, vol. 5636, pp. 288–299. Springer, Heidelberg (2009). https://doi.org/10.1007/978-3-642-02498-6_24
12. Li, C., Gatenby, C., Wang, L., et al.: A robust parametric method for bias field estimation and segmentation of MR images. In: IEEE Conference on Computer Vision and Pattern Recognition, pp. 218–223 (2009)
13. Zhang, K., Zhang, L., Lam, K.M., Zhang, D.: A level set approach to image segmentation with intensity inhomogeneity. IEEE Trans. Cybern. **46**(2), 546–557 (2016)
14. Zhang, K., Liu, Q., Song, H., Li, X.: A variational approach to simultaneous image segmentation and bias correction. IEEE Trans. Cybern. **45**(5), 1426–1437 (2015)
15. Zhang, K., Zhang, L., Song, H., Zhou, W.: Active contours with selective local or global segmentation: a new formulation and level set method. Image Vis. Comput. **28**(4), 668–676 (2010)
16. Rajapakse, J., Giedd, J., Rapoport, J.: Statistical approach to segmentation of single-channel cerebral MR images. IEEE Trans. Med. Imaging **16**(2), 176–186 (1997)
17. Rajapakse, J., Kruggel, F.: Segmentation of MR images with intensity inhomogeneities. Image Vis. Comput. **16**(3), 165–180 (1998)
18. Yang, X., Gao, X., Li, X., et al.: An efficient MRF embedded level set method for image segmentation. IEEE Trans. Image Proc. **24**(1), 9 (2015)
19. Zhang, K., Zhang, L., Song, H., Zhang, D.: Re-initialization free level set evolution via reaction diffusion. IEEE Trans. Image Process. **22**(1), 258–271 (2012)
20. Xie, X.: Active contouring based on gradient vector interaction and constrained level set diffusion. IEEE Trans. Image Process. **19**(1), 154–164 (2010)

Biomedical Signal Processing

Microarray-Based Cancer Prediction Using Single-Gene Ensemble Classifier

Ziyi Yang⑩, Yanqiong Ren⑩, Hui Zhang⑩, and Yong Liang$^{(\boxtimes)}$⑩

Macau University of Science and Technology,
Building C407, Avenida Wai Long, Taipa, Macau
yliang@must.edu.mo

Abstract. With the microarray technology widely used in cancer diagnosis, various effective classification approaches have been proposed for gene selection and cancer classification. Generally, single classifiers with numerous genes have been widely used in cancer classification. In biology, many different genotypes can produce the same phenotype. In other words, there might be different pathogenic genes among individuals in various patients. Hence, single classifiers with a plenty of genes suffer from the disadvantage that it is not easy to identify the significant biomarkers for each patient. In this paper, we present a novel approach for cancer classification using ensemble classifier with the single gene to effectively identify candidate pathogenic biomarkers for every cancer sample. We applied our approach to three publicly available cancer datasets and compared classification accuracy to that for six standard methods. The single-gene ensemble classifier performs well in microarray-based cancer prediction.

Keywords: Cancer classification · Gene expression · Ensemble classifier Biomarker

1 Introduction

With the rapid development of microarray technology, this technology has been applied to prediction and diagnosis of cancer [1–5]. Accurate classification of cancer is a very important issue for cancer treatment. Various machine learning approaches have been developed for cancer classification, including Logistic Regression [6–8], Support Vector Machine (SVM) [9], K-Nearest Neighbor (K-NN) [10, 11], Artificial Neural Network (ANN) [12], Decision Tree (DT) [13] and so on. These machine-learning classification approaches were constructed by a single model with numerous genes. In the view of biology, there are different genotypes behind the same phenotype of human cancer disease. Therefore, various cancer patients are caused by different pathogenic genes due to different genotypes. It is difficult to identify significant biomarkers of related cancers by utilizing only one model to predict the classification of cancer for every sample. This limit the interpretability of classifiers and the application of cancer classification in identifying biomarkers for drug design.

To deal with the disadvantage of single classifiers with numerous genes, we proposed a novel ensemble approach for classification, which based on multi-independent classifiers. We construct independent classifiers according to different genotypes, and

© Springer Nature Switzerland AG 2018
Y. Peng et al. (Eds.): IScIDE 2018, LNCS 11266, pp. 589–600, 2018.
https://doi.org/10.1007/978-3-030-02698-1_51

each independent classifier can identify fewer genes to effectively classify the samples of cancer. Some researchers suggested that single-gene classifiers could perform well for cancer prediction in some situations [2, 7–9, 11]. Therefore, we utilize single gene to construct each independent classifier to improve the interpretability of the classier.

Our single-gene ensemble classifier is suitable for clinical application, which can identify the candidate biomarkers for every cancer sample. It is worthwhile because of the advantage for interpretability and applicability for biological study and medical use. We compared the performance of single-gene ensemble classifier to that of variety of single classifiers using three publicly available gene expression data. We validated that our single-gene ensemble classifier has superior performance to single classifiers in most cases examined. In the following section, we provide detailed illustration and comparison of our proposed method.

2   The General Framework and Implementation of Our Method

Our single-gene ensemble classifier includes two key procedures: feature selection and ensemble classifier construction. The pipeline of our method can be seen in Fig. 1.

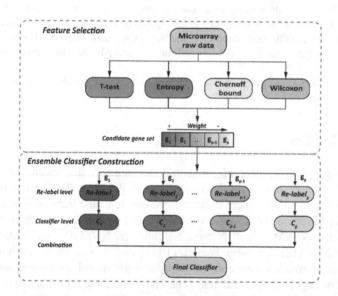

Fig. 1. The pipeline of single-gene ensemble classifier.

2.1   Feature Selection

Feature selection is particularly important issues in classification, which can efficiently reduce the complexity of the model learning process and make it more comprehensible. From the Fig. 1, it clearly demonstrates the process of feature selection. We selected the top-ranked 20 significant genes from each of the four filter methods, T-test, entropy,

Chernoff bound and Wilcoxon test, respectively. Hence, there were produced four candidate gene subsets. After that, we merged all the significant genes and generated a weighting candidate gene set. Candidate gene weights are calculated as follows:

Step 1: Calculating the weight of each gene in each candidate gene subset;

$$w_{ji} = \begin{cases} L_j - I_{ji} + 1, & \text{If gene } i \text{ exists in candidate gene subset } j^{th}; \\ 0, & \text{If gene } i \text{ not exists in candidate gene subset } j^{th}. \end{cases} \tag{1}$$

Step 2: Calculating the weight of each candidate gene.

$$W_i = \frac{\sum_{j=1}^{4} w_{ji}}{\sum_{j=1}^{4} L_j}, i = 1, 2, \ldots, p; j = 1, 2, \ldots, 4 \tag{2}$$

where I_{ji} is the index of gene i in the candidate gene subject j^{th}, L_j is the length of candidate gene subset j^{th}, and w_{ji} is the weight of gene i in candidate gene subset j^{th}. In this way, the selected genes can correspond with some biological meaning and might give out more accurate prediction about disease.

2.2 Ensemble Classifier Construction

The pipeline of single-gene ensemble classifier is clearly shown in the Fig. 1. Our ensemble classifier is composed of multiple independent single-gene classifiers, each of which is constructed by support vector machine classifier.

Each single-gene classifier is composed of two levels. The first is re-label level where the training data is partitioned into two classes by k-means clustering with squared Euclidean distance measure. Due to each independent classifier is constructed based on different genotypes, it is necessary to re-label the class of samples according to the different express of one gene between the two types of cancer. In other words, there are different genotypes behind the same phenotype of human cancer disease, and for a specific pathogenic gene, some patients may not be caused by this gene. Based on this reason, we designed re-label level to re-label the class of samples under a specific pathogenic gene. The second is the classifier level where independent SVM classifier is constructed based on re-labelled training data. Finally, combination strategy is used to predict the cancer type to which a cancer sample belonged.

Single-gene ensemble classifier predicts each cancer test sample by trained independent SVM models. Each SVM model is used to classify the test samples sequentially according to the gene weight of the constructed model. To improve the classification accuracy and remove the noise, each classifier deletes test samples which are classified to class 1. The classifier predicts the classification of test samples, until there is no sample belongs to patient class in the testing dataset. For example, test samples are firstly predicted using the SVM classifier that constructed by the highest weight gene from candidate gene set. The test samples are removed when their predictions belong to patient class. Using the classifier with second highest weight gene to predict the residual samples, and so on. Finally, our model combines all the prediction results of each independent classifier. Our single-gene ensemble classifier can effectively identify the candidate biomarkers for every cancer sample.

3   Analysis on Microarray Data

In this section, we compared our single-gene ensemble classifier (SGEC) with other six methods, single gene classifier with the t-test (SGC-t), single gene classifier with the WMW (SGC-W), SVM, sparse logistic regression with L1 penalty, K-NN and Random Forest (RF), on 3 publicly available lung cancer gene expression datasets (https://www. ncbi.nlm.nih.gov/gds/). A brief description of these lung cancer datasets is shown below.

GSE10072 Dataset. The original gene expression data is provided by Landi et al. [14] has 22283 genes per sample and contains 107 final expression samples from 58 tumors and 49 non-tumor tissues.

GSE19804 Dataset. The dataset is provided by Lu [15] and contains the expression profiles of 54676 genes for 60 tumors and 60 normal tumor tissues.

GSE19188 Dataset. The dataset contains 156 samples from 65 normal tumors and 91 tumor tissues and each sample has 54675 genes. The more information can be found in Hou et al. [12].

We randomly divided the datasets, about 70% of the datasets become training samples and the other 30% turn into testing samples. The detail information of these data is clearly listed in Table 1.

Table 1. The description of three publicly available lung cancer gene expression datasets.

Dataset	No. of genes	Classes (Class 0/Class 1)	No. of training samples (Class 0/Class 1)	No. of testing samples (Class 0/Class 1)
GSE10072	22284	Normal/lung cancer	75 (34/41)	32 (15/17)
GSE19804	54676	Normal/lung cancer	84 (42/42)	36 (18/18)
GSE19188	54675	Normal/lung cancer	110 (46/64)	46 (19/27)

Table 2 explores the classification performance of our approach and other six classifiers in three publicly lung cancer datasets. The experiments were repeated 30 times and we report the average value in Table 2. For GSE10072 dataset, SVM and Logistic with L_1 penalty give the highest classification accuracy compared with other methods. The classification accuracy of SGEC has reached 97.08%, which is inferior to SVM, L_1 regularized logistic regression and random forest but superior to SGC-t, SGC-W and K-NN. The sensitivity of all the methods up to 94%, and the specificity achieved at least 88% except for SGC-W. For GSE19804 dataset, SGEC achieves the best classification performance with the highest accuracy. SVM, L_1 regularized logistic regression and RF have the similar accuracy up to 90%. K-NN has the worst classification performance compared with all methods. The accuracy of SGC-t and SGC-W are better than K-NN but worse than the other four methods. SGC-W achieves the best sensitivity and other four approaches have similar sensitivity reached 90% except for K-NN. For the specificity, our approach SGEC achieves the best performance, L_1 regularized logistic regression and RF are inferior to SGEC but superior to SGC-t, SGC-W and K-NN. For GSE19188 dataset, SGEC with the highest accuracy, but SGC-W with the worst accuracy performance. The other four methods with the similar accuracy reached 85%. Except for L_1 regularized logistic regression and RF, other four methods with high specificity up to 95%.

Table 2. The classification performance of seven different methods for three lung cancer datasets.

Dataset	SGEC	SGC-t	SGC-W	SVM	Logistic + L_1	K-NN	RF
Accuracy (%)							
GSE10072	97.08	93.75	78.13	**100.00**	**100.00**	93.75	98.85
GSE19804	**97.87**	88.89	88.89	94.44	94.44	66.67	92.78
GSE19188	**96.88**	86.96	52.17	95.65	87.17	84.78	87.68
Sensitivity (%)							
GSE10072	95.02	94.12	**100.00**	**100.00**	**100.00**	**100.00**	97.96
GSE19804	95.89	93.75	**100.00**	94.44	94.44	87.50	91.46
GSE19188	95.83	95.65	**100.00**	**100.00**	86.52	**100.00**	84.95
Specificity (%)							
GSE10072	**100.00**	93.33	68.18	**100.00**	**100.00**	88.24	**100.00**
GSE19804	**100.00**	85.00	81.82	94.44	94.44	60.71	94.25
GSE19188	**97.98**	78.26	46.34	90.48	88.30	73.08	93.27

Table 3. The number of selected features by four methods from three lung cancer datasets.

Dataset	Method	No. of selected features
GSE10072	SGEC	5
	SGC-t	1
	SGC-W	1
	Logistic + L_1	22
GSE19804	SGEC	4
	SGC-t	1
	SGC-W	1
	Logistic + L_1	16
GSE19188	SGEC	18
	SGC-t	1
	SGC-W	1
	Logistic + L_1	20

Four of seven classification models possess feature selection capability. The number of genes selected by SGEC, SGC-t, SGC-W and Logistic with L_1 penalty are clearly shown in Table 3. Obviously, SGC-t and SGC-W belong to single-gene classifier that only identify one significant gene to predict the classification of cancer samples. And the number of significant genes selected by SGEC is fewer than Logistic with L_1 penalty in three cancer datasets.

Combining with Tables 2 and 3, we can conclude that single-gene classifier can perform well in some situations, but it might be influenced by noise gene and result in this classification accuracy is not ideal. For example, in GSE19188 dataset, SGC-W achieves the worst classification accuracy. Moreover, Logistic with L_1 penalty and our approach SGEC can be successfully applied to classification of cancer samples and with ideal accuracies. Note that, SGEC is constructed according to the different genotypes and can identify fewer genes to effectively classify the cancer samples than Logistic with L_1 penalty.

Figures 2, 3 and 4 directly show the significant genes that were selected by four different methods have large difference in expression between the two classes for GSE10072, GSE19804 and GSE19188 datasets respectively. From a statistical viewpoint, each selected feature has significant difference in the expression between two classes. Each boxplot shows the 25th quantile, median, 75th quantile in the distribution of expression value.

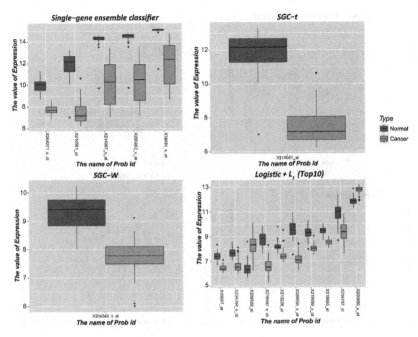

Fig. 2. Expression of significant features in tumors and normal in GSE10072 dataset.

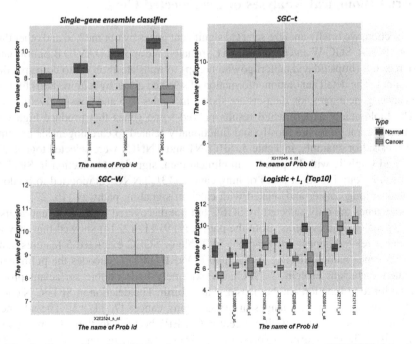

Fig. 3. Expression of significant features in tumors and normal in GSE19804 dataset.

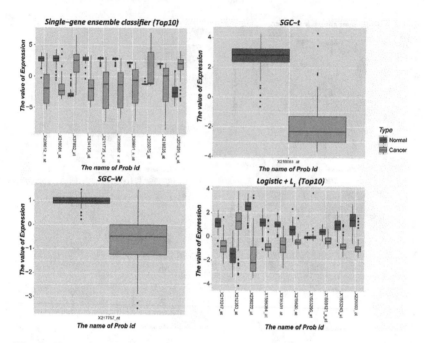

Fig. 4. Expression of significant features in tumors and normal in GSE19188 dataset.

4 Brief Biological Analyses of the Selected Genes

In this section, we briefly analyze selected significant genes by our method and other three methods SGC-t, SGC-W and Logistic with L_1 penalty. We applied DAVID Bioinformatics Resources 6.8 (https://david.ncifcrf.gov/home.jsp) to translate Prob ID to corresponding gene names. The detail annotation information of selected genes by four methods for three lung cancer gene expression datasets are shown in Tables 4, 5 and 6 respectively.

The biologically experimental results proved some genes that produce high classification accuracy rate are mostly and functionally related to carcinogenesis or tumor histogenesis. For example, in Table 5, SPTBN1 and TNNC1 were selected together by SGEC and Logistic with L_1 penalty. In clinical data, significant reduction in SPTBN1 expression is found in most cases of lung cancer [13]. TNNC1 is reported to be down regulated in lung cancer and related with calcium signaling pathway [16, 22].

Some genes are only selected by SGEC but not discovered by SGC-W and logistic with L_1 penalty. The evidence from the papers [7, 9, 11, 12, 14] showed that they are lung cancer related genes. For example, in Table 4, SGEC extracted 5 features annotated to 3 genes named SFTPC, AGER and EDNRB. SFTPC encodes the pulmonary-associated surfactant protein C (SPC), an extremely hydrophobic surfactant protein essential for lung function and homeostasis after birth. The deletion in SFTPC is one of the most common genetic changes in primary lung cancer [7, 11, 12]. The advanced glycosylation end-product specific receptor (AGER) belongs to the immunoglobulin superfamily, whose abnormal expression has been detected in lung cancer [21]. The Reduction of EDNRB expression may drive tumorigenesis [22].

Table 4. The significant genes found by the four methods from the GSE10072 dataset.

Method	Prob ID	Gene name
SGEC	214387_x_at	surfactant protein C(SFTPC)
	205982_x_at	surfactant protein C(SFTPC)
	210081_at	advanced glycosylation end-product specific receptor(AGER)
	38691_s_at	surfactant protein C(SFTPC)
	204271_s_at	endothelin receptor type B(EDNRB)
SGC-t	210081_at	advanced glycosylation end-product specific receptor(AGER)
SGC-W	204249_s_at	LIM domain only 2(LMO2)
Logistic + L_1 (Top 10)	200650_s_at	lactate dehydrogenase A(LDHA)
	40687_at	gap junction protein alpha 4(GJA4)
	204364_s_at	receptor accessory protein 1(REEP1)
	219597_s_at	dual oxidase 1(DUOX1)
	204787_at	V-set and immunoglobulin domain containing 4(VSIG4)
	209529_at	phospholipid phosphatase 2(PLPP2)
	219059_s_at	lymphatic vessel endothelial hyaluronan receptor 1(LYVE1)
	213228_at	phosphodiesterase 8B(PDE8B)
	218665_at	frizzled class receptor 4(FZD4)
	209555_s_at	CD36 molecule(CD36)

Table 5. The significant genes found by the four methods from the GSE19804 dataset.

Method	Prob ID	Gene name
SGEC	217046_s_at	advanced glycosylation end-product specific receptor(AGER)
	215918_s_at	spectrin beta, non-erythrocytic 1(SPTBN1)
	209904_at	troponin C1, slow skeletal and cardiac type(TNNC1)
	1557729_at	G protein-coupled receptor kinase 5(GRK5)
SGC-t	217046_s_at	advanced glycosylation end-product specific receptor(AGER)
SGC-W	202524_s_at	SPARC/osteonectin, cwcv and kazal like domains proteoglycan 2 (SPOCK2)
Logistic + L_1 (top 10)	212115_at	hematological and neurological expressed 1 like(HN1L)
	215918_s_at	spectrin beta, non-erythrocytic 1(SPTBN1)
	207302_at	sarcoglycan gamma(SGCG)
	1560879_a_at	synaptotagmin 15(SYT15)
	228540_at	QKI, KH domain containing RNA binding(QKI)
	210608_s_at	fucosyltransferase 2(FUT2)
	205941_s_at	collagen type X alpha 1 chain(COL10A1)
	209904_at	troponin C1, slow skeletal and cardiac type(TNNC1)
	217771_at	golgi membrane protein 1(GOLM1)
	223816_at	solute carrier family 46 member 2(SLC46A2)

Table 6. The significant genes found by the four methods from the GSE19188 dataset.

Method	Prob Id	Gene name
SGEC (Top 10)	210081_at	advanced glycosylation end-product specific receptor(AGER)
	38691_s_at	surfactant protein C(SFTPC)
	205982_x_at	surfactant protein C(SFTPC)
	211735_x_at	surfactant protein C(SFTPC)
	235075_at	desmoglein 3(DSG3)
	218835_at	surfactant protein a1(SFTPA1)
	37892_at	collagen type xi alpha 1 chain(COL11A1)
	214135_at	claudin 18(CLDN18)
	201291_s_at	topoisomerase (DNA) ii alpha(TOP2A)
	209612_s_at	alcohol dehydrogenase 1b (CLASS I), beta polypeptide (ADH1B)
SGC-t	210081_at	advanced glycosylation end-product specific receptor(ager)
SGC-W	217757_at	alpha-2-macroglobulin(A2 M)
Logistic + L_1 (Top 10)	213247_at	sushi, von willebrand factor type a, egf and pentraxin domain containing 1(SVEP1)
	1556364_at	adamts9 antisense rna 2(ADAMTS9-AS2)
	220003_at	leucine rich repeat containing 36(LRRC36)
	219820_at	solute carrier family 6 member 16(SLC6A16)
	1559121_s_at	Unknown
	1553296_at	adhesion g protein-coupled receptor g7(ADGRG7)
	238222_at	gastrokine 2(GKN2)
	1553243_at	inter-alpha-trypsin inhibitor heavy chain family member 5 (ITIH5)
	212353_at	sulfatase 1(SULF1)
	235301_at	kiaa1324 like(KIAA1324L)

5   Conclusions

To deal with the high-dimension DNA microarray data, a variety of cancer predictive classifiers have been proposed. Single models with numerous genes have been widely applied in classification problem. However, single models limit the interpretability of classifiers and the applicability for biological study and target-based drug design. Therefore, we proposed a useful and robust single-gene ensemble classifier. The most prominent contribution is our classification model can easily identify the candidate pathogenic biomarkers for every cancer sample. We examined three publicly available cancer datasets and six previously published classifiers and found that our single-gene ensemble classifier has superior performance to standard methods in most cases examined.

Someone might doubt the utility of our model because the pathogenesis of cancer is complex that it must be interact with multiple genes, nevertheless, our single-gene ensemble classifiers that select just one significant gene for every one tumor sample. We do not refute this argument. Clearly, our method might identify the candidate biomarker of every cancer sample easily with good classification performance. It is worthwhile because of the advantage for interpretability and applicability for biological study and medical use.

Acknowledgments. This work was supported by FDCT Grant No.003/2016/AFJ from the Macau Special Administrative Region of the People's Republic of China, the National Grand Fundamental Research 973 Program of China under Grant No. 2013CB329404 and the China NSFC projects under Contracts 61373114, 61661166011, 11690011, 61721002.

References

1. Golub, T.R., et al.: Molecular classification of cancer: class discovery and class prediction by gene expression monitoring. Science **286**(80), 531–537 (1999)
2. Gordon, G.J., et al.: Translation of microarray data into clinically relevant cancer diagnostic tests using gene expression ratios in lung cancer and mesothelioma. Cancer Res. **62**, 4963–4967 (2002)
3. Gordon, G.J., et al.: Using gene expression ratios to predict outcome among patients with mesothelioma. J. Natl. Cancer Inst. **95**, 598–605 (2003)
4. Schena, M., Shalon, D., Davis, R.W., Brown, P.O., et al.: Quantitative monitoring of gene expression patterns with a complementary DNA microarray. Science **270**, 467 (1995). York Then Washington
5. Van't Veer, L.J., et al.: Gene expression profiling predicts clinical outcome of breast cancer. Nature **415**, 530–536 (2002)
6. Wang, X., Gotoh, O.: Accurate molecular classification of cancer using simple rules. BMC Med. Genomics **2**, 64 (2009)
7. Wang, X., Gotoh, O.: Cancer classification using single genes. Genome Inform. **23**, 179–188 (2009)
8. Wang, X., Simon, R.: Microarray-based cancer prediction using single genes. BMC Bioinform. **12**, 391 (2011)
9. Geman, D., d'Avignon, C., Naiman, D.Q., Winslow, R.L.: Classifying gene expression profiles from pairwise mRNA comparisons. Stat. Appl. Genet. Mol. Biol. **3**, 1–19 (2004)
10. Jiang, K.: The application of multiple classifier systems in the analysis of gene microarray datasets (2008)
11. Wang, X., Gotoh, O.: Microarray-based cancer prediction using soft computing approach. Cancer Inform. **7**, 123 (2009)
12. Hou, J., et al.: Gene expression-based classification of non-small cell lung carcinomas and survival prediction. PLoS One **5**, e10312 (2010)
13. Baek, H.J., et al.: Inactivation of TGF-β signaling in lung cancer results in increased CDK4 activity that can be rescued by ELF. Biochem. Biophys. Res. Commun. **346**, 1150–1157 (2006)
14. Landi, M.T., et al.: Gene expression signature of cigarette smoking and its role in lung adenocarcinoma development and survival. PLoS One **3**, e1651 (2008)
15. Lu, T.-P., et al.: Identification of a novel biomarker SEMA5A for non-small cell lung carcinoma in non-smoking women. Cancer Epidemiol. Prev. Biomark. **19**, 2590–2597 (2010). cebp–0332
16. Xu, Q., Gao, Y., Liu, Y., Yang, W., Xu, X.: Identification of differential gene expression profiles of radioresistant lung cancer cell line established by fractionated ionizing radiation in vitro. Chin. Med. J. (Engl. Ed.) **121**, 1830 (2008)

17. Urgard, E., et al.: Metagenes associated with survival in non-small cell lung cancer. Cancer Inform. **10**, 175 (2011)
18. Li, R., et al.: Genetic deletions in sputum as diagnostic markers for early detection of stage I non-small cell lung cancer. Clin. Cancer Res. **13**, 482–487 (2007)
19. Jiang, F., Yin, Z., Caraway, N.P., Li, R., Katz, R.L.: Genomic profiles in stage I primary non small cell lung cancer using comparative genomic hybridization analysis of cDNA microarrays. Neoplasia **6**, 623–635 (2004)
20. Li, R., et al.: Identification of putative oncogenes in lung adenocarcinoma by a comprehensive functional genomic approach. Oncogene **25**, 2628–2635 (2006)
21. Pan, Z., et al.: Long non-coding RNA AGER-1 functionally upregulates the innate immunity gene AGER and approximates its anti-tumor effect in lung cancer. Mol. Carcinog. **57**, 305–318 (2017)
22. Knight, L., et al.: Hypermethylation of endothelin receptor type B (EDNRB) is a frequent event in non-small cell lung cancer (2007)

A Novel EEG Signal Recognition Method Using Modified Optimal Electrodes Recombination Strategy

Lijuan Duan[1,2,3], Song Cui[1], Lili Liu[1], and Yuanhua Qiao[4(✉)]

[1] Faculty of Information Technology, Beijing University of Technology,
Beijing 100124, China
[2] Beijing Key Laboratory on Integration and Analysis
of Large-Scale Stream Data, Beijing 100124, China
[3] National Engineering Laboratory for Critical Technologies
of Information Security Classified Protection, Beijing 100124, China
[4] College of Applied Science, Beijing University of Technology,
Beijing 100124, China
qiaoyuanhua@bjut.edu.cn

Abstract. Electroencephalogram (EEG) is a comprehensive indicator of human physiological activities. Because of its comprehensiveness and complexities, an electrode-covered collection device on the scalp cannot collect the discharge phenomenon of the activated brain area exhaustively. A single electrode analysis will miss a lot of important association information between different brain regions. This paper presents a novel strategy to solve this problem by combining the optimal electrodes. The whole method is divided into five steps: (1) input multi-electrodes EEG data, (2) utilize the Principal Component Analysis (PCA) method to obtain the optimal electrodes, (3) adopt the modified optimal electrodes recombination strategy and obtain optimal electrode combination strategy, (4) extract features by using Empirical Mode Decomposition (EMD), and (5) use Naive Bayes classifier to do the classification tasks. In order to evaluate the validity of the proposed method, we apply the proposed method in BCI Competition II datasets Ia. Experimental results show that our method improves recognition performance for BCI Competition II datasets Ia and the modified optimal electrodes recombination strategy is reasonable. This provides a new idea for analyzing the EEG features of other tasks.

Keywords: Electroencephalogram · Optimal electrodes selection
Optimal electrodes recombination · Principal component analysis
Empirical mode decomposition

1 Introduction

The brain is the controller of various human activities. When it produces mental activity, the cerebral cortex can produce a variety of waveforms; these EEG signals can be monitored externally. The study on brain development has been a continuous research work in recent decades. Brain Computer Interface (BCI) technologies make it possible to

© Springer Nature Switzerland AG 2018
Y. Peng et al. (Eds.): IScIDE 2018, LNCS 11266, pp. 601–613, 2018.
https://doi.org/10.1007/978-3-030-02698-1_52

control the outside world through the advent of thought. It is a hot spot and difficult point in the field of EEG study to improve classification precision of motor imagery BCI data.

An EEG signal is composed of a large number of brain cells signals superposition results. And thus, it is very complex because of its nonlinearity and overlap. This makes it difficult to obtain high signal accuracy in relation to the task being monitored since some signals are not of relevance.

The current EEG based BCI applications seek to recognize mental tasks by selecting electrodes related to these tasks, extracting relevant features from signals recorded through these electrodes, developing an algorithm that yields the highest classification rate, and providing this information to communication and control units through translation algorithms [1–3]. Most studies lay stress upon the feature extraction and classification algorithm. However, research shows that different mental tasks will activate special brain areas using functional magnetic resonance imaging (fMRI) [4] and Positron Emission Tomography (PET) technology [5]. What is more, the same mental task will activate different areas if experimental designs are different. Some experimental research demonstrates that the interaction effect exists between multi-areas of brain [6]. Therefore, selecting optimal electrodes combinations is of great importance for improving classification accuracy of BCI system

In previous study, Kayikcioglu et al. [3] only uses data in electrode A1 of BCI Competition II datasets Ia to analyze and recognize EEG data, and coefficients of the second order polynomial are used for feature extraction. Then, the features are classified using k-nearest neighbor (k-NN) algorithm and obtained 92.15% classification performance. Though classification accuracy shows encouraging result, there is a huge loss of information from other electrodes for EEG signal recognition using only one electrode. Duan et al. [7] proposed two feature selection strategies. One is based on combination of optimal single electrodes. This strategy is simple and effective, but classification performance is still weak and optimal electrode is not defined explicitly. The other is based on combination of optimal time spans. This strategy obtained good classification performance, but more tedious. Previously, we also proposed optimal electrodes recombination strategy [8], but did not obtain high classification accuracy.

In this study, we propose a novel EEG signal recognition method. This method uses modified optimal electrodes recombination strategy, and redefines optimal electrodes and optimal electrodes combination explicitly. PCA is used to select the optimal electrodes and EMD is adopted to extract features. Then these features are used as an input to a Naive Bayes classifier, which has two outputs: negative or positive signal. We apply the proposed method to BCI Competition II data sets Ia. The experiment results show that the classification performance of our method outperform the other methods for data set Ia in the BCI competition.

2   Related Works

2.1   Principal Methods

Different motor imagery tasks may activate specific human brain functional areas. Hence, some scalp electrodes are full of effective information and some are not. The so-called effective information is the frequency or energy imbedded in the signals that changes over time; which is helpful for signal recognition as soon as the task is performed. It is beneficial to do classification from only the effective information and reject inefficient information [9, 10]. Redundant information occupies a big part of the raw data, which makes features extraction difficult. Meanwhile the dimension of EEG data is a little bit high. PCA is a common way to analyze high-dimensional data. We consider to use the PCA to select the optimal electrodes and the classification accuracy based on a Naive Bayes classifier is deemed to be the judge standard of optimal electrodes selection.

Empirical mode decomposition (EMD) method is a kind of adaptive signal decomposition method. It can decompose any section of complex time series into a range of different, simple, non-sinusoidal intrinsic mode functions (IMF). The original EEG signal is decomposed by EMD method into a series of IMFs. Each IMF contains the time domain information and instantaneous frequency information of original EEG signal. Because the decomposition method of the EMD method is completely determined by the local frequency information of the original signal, it has strong adaptability and is suitable for processing nonlinear and non-stationary signals. Therefore, it is a good choice to use the EMD method to extract the signal features.

Naive Bayes classifier is a simple probabilistic classifier based on Bayesian theorem (from Bayesian statistics) with strong (naive) independence assumptions between the features [12]. Naive Bayes is not a single algorithm for training such classifiers, but a family of algorithms. Those algorithms base on a common principle: all naive Bayes classifiers assume that the value of a particular feature is independent of the value of any other feature, given the class variable.

2.2 Principal Component Analysis

Principal component analysis (PCA) is a well-established method for feature extraction and dimensionality reduction. Firstly, calculate the covariance matrix of the original sample, then calculate the eigenvalues and eigenvectors of the covariance matrix, and the corresponding eigenvectors are rearranged according to the order of the eigenvalues. Finally, the target dimension is obtained by setting the cumulative contribution rate threshold. The reduced dimension data is obtained by multiplying the original data by the characteristic vector. Therefore, PCA became a common method for removing the eye movement and blink signal interference on the EEG signal. However, PCA cannot guarantee complete deletion of potential noise from the EEG signal which is similar to its own waveform. So PCA is used as an unsupervised method in the process of feature mapping instead of using the information for the classification of internal data. Looking for features extracted by PCA is data effectively represent the direction rather than the data classification.

2.3 Empirical Mode Decomposition

The empirical mode decomposition algorithm is defined during the algorithm process instead of theory formula. The method is essentially a signal smoothing operation. Its purpose is to decompose the fluctuation or trend of different scales step by step and then produce a series of data sequences with different characteristic scales. Each sequence is called an intrinsic mode function component. Normally, the lowest frequency IMF component is the trend of the original signal or the raw data removes the mean of the data sequence. The test results show that the EMD method is the best way to extract the trend or mean of the data sequence. The basic process of EMD is: firstly, identify all the local extrema of the original data sequence and then using the cubic spline function to fit the original sequence of upper and lower envelope. Using the upper and lower envelope to calculate the mean envelope line; Let the original data sequence subtract the mean envelope data and a new sequence can be obtained. Finally keep modifying the new sequences until it satisfies the two conditions of IMF judgment mentioned above and we can get the data component. The remainder of the new data sequence of the high frequency component is obtained by subtracting the component from the original signal. It is decomposed into second intrinsic mode functions. Repeat the loop until the data sequence cannot be decomposed. Then we get the residual quantity which represents the trend or the mean of the original sequence of the data.

2.4 Naive Bayes Classifier

In machine learning, Naive Bayes classifiers are a family of simple probabilistic classifiers based on applying Bayesian theorem with strong (naive) independence assumptions between the features. This algorithm is a classical algorithm, which is widely used in many fields. In this algorithm, the probability distribution function and the overall probability of the sample in different categories is unknown. In order to obtain these two parameters, we always need a large enough sample set but the increased computation will lead to a long training time. In BCI real-time system, EEG signals have small amount of data and request for a short training time which cause many algorithms in BCI unable to play its role effectively in real-time system.

In this article, we firstly let the data dealt with PCA method to select the optimal electrodes and get the combination strategy. Then, using the EMD to modified the complicated EEG data into a simple and Non-sinusoidal IMF. Finally, the Naïve Bayes classifier charges the calculate work and provides the classification accuracy.

3   Framework of Proposed Method

In this paper, a novel EEG signal recognition method using modified optimal electrodes recombination strategy is proposed, and involves five processes: (1) input multi-electrode EEG data, (2) utilize the optimal electrodes selection method using PCA and obtain optimal electrodes, (3) adopt the modified optimal electrodes recombination strategy and obtain optimal electrode combinations, (4) extract features by using EMD, (5) use Naive Bayes classifier and obtain classification results. The framework of the proposed is shown as Fig. 1.

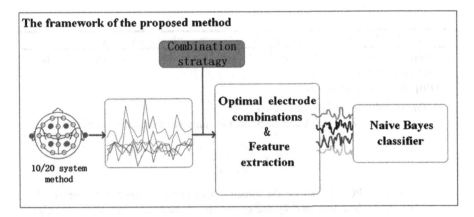

Fig. 1. The framework of the proposed method

Firstly, it is shown that different stimulus will activate special brain function areas. That is to say, only part electrodes of the EEG acquisition device (multi-electrodes cap) are related to the task. Therefore, it is necessary to confirm that which electrodes record the real EEG signals evoked by stimulus. And we called these electrodes as optimal ones. A significant method PCA is used to solve the selection problem.

Secondly, start with an idea that there are neural connections between different brain areas; we discuss the influence on the classification performance by analyzing the multi electrodes, simultaneously based on the optimal electrodes. It can transform the time-frequency information of EEG signals of multi electrodes to the same feature space by joining the properties of multiple electrodes and then carried out the feature extraction algorithm. In the final, it can speculate that whether there is relation between electrodes according to the final classification accuracy.

Thirdly, EEG signal is strong randomness, non-stationary, non-linear and high dimension, and it has high time resolution and low space resolution. Through specific arithmetic operation, EMD method decomposes the signal into a series of Intrinsic Mode Function (IMF) by identifying the oscillation mode contained by the signal itself [10]. Each IMF describes the characteristic of the original signal in time-frequency domain. Moreover, the decomposition of the EMD method completely decided by the local frequency information of the signal. The whole process has strong adaptability and it is very suitable for processing non-stationary and non-linear signals.

At last, the Naive Bayes classifier is used in the paper for classification.

3.1   The Optimal Electrodes Selection Method Using PCA

The PCA method successively reduce the dimension of the electrodes. Then calculate the cumulative contribution to find out the appropriate parameter. Using the threshold to choose the optimal electrodes. The details of the algorithm are shown in Algorithm 3-1.

Algorithm 3-1. The optimal electrodes selection method using PCA

Input: Raw EEG data, Threshold t1 = (1/c + 0.1), where c is the number of label categories in training set.

Output: Optimal electrodes

1. **For** signal E **in** all EEG electrodes:

  E is processed by PCA and calculate the cumulative contribution rate as:

$$r = \frac{\sum_{i=1}^{d} \rho_i}{\sum_{i=1}^{n} \rho_i}$$

  Using the PCA to reduce the dimension of single original EEG signal. The dimension is determined by the cumulative contribution rate of the principal component.

  While d increases:

    if (r >= 85% && r <= 95%)

      the corresponding dimension d is picked up;

    else:

      continue;

2. Using the Bayesian classifier to classify signals in different dimensions;

3. **If** Ac > t1:

    E is optimal electrode.

  Else:

    E is non-optimal electrodes.

4. According to the definition of the advantage electrode, the electrodes are divided into two parts, the dominant electrodes and the non-dominant electrodes.

End

3.2   The Modified Optimal Electrodes Recombination Strategy

Considering the interaction effect between brain regions, we deal with the multi optimal electrodes simultaneously. Usually, method of exhaustion is used to analyze all the possible combinations. However, it is a big workload and it is not suitable for applications of BCI. According to the physiology research conclusion, different motor imagery tasks have their intrinsic brain areas. Thus the modified optimal electrodes recombination strategy is based on the conclusion of intrinsic brain areas, including four steps as follows.

Algorithm 3-2. The modified optimal electrodes recombination strategy

---

Input: Raw EEG data, Threshold t2=80%

Output: Optimal combinations and non-optimal combinations

1. Select Top2 electrodes as C_o; Others as C_s;
    C_g = Combination C_o with $\{C \in C_s\}$
2. Using the Bayesian classifier to classify C_g;
    Aa = Bayesian classifier (C_g);

3. **If** Aa > t2:
        the combination C_g is optimal combination.
    Else:
        the combination C_g is non-optimal combination.
4. Obtain all the optimal combinations.
End

3.3 Feature Extraction Using EMD

Any complex time series is composed of several intrinsic mode functions (IMFs), which are different, simple and non-sinusoidal components. So a finite set of IMFs can be separated from the complex time series. Each IMF should satisfy two conditions [11]:

(1) in the whole dataset, the number of extrema and the number of zero-crossings must either be equal or differ at most by one;
(2) at any point, the mean value of the envelope defined by the local maxima and the envelope defined by the local minima is zero.

The process of generating these IMFs is referred to as the EMD decomposition. For the given EEG signal, the detail process of IMF extraction is shown in Algorithm 3-3.

Algorithm 3-3. Feature extraction using EMD

Input: Electrode combinations

Output: Features

1. Using $x'(t) = 0$ to identify all the local extrema of $x(t)$. Finding out all of the local maxima point with the cubic spline curve and then connecting them to generate the upper envelope. Lower envelope is generated in the same way. Then, calculate the mean envelope $m_1(t)$.
    2. The component $h_1(t)$ is produced by subtracting the $m_1(t)$ from the original data.
    3. **If** $h_1(t)$ not satisfies IMF (1)(2) conditions:
    Process $h_1(t)$ by step 1,2
    Else:
        Get the IMF $c_1(t) = h_k(t)$
4. Calculate $r_1(t) = x(t) - c_1(t)$.
5. Let the $r_1(t)$ as a new signal to process (1) - (4) steps.
End

3.4   Classification Using Naive Bayes Classifier

After extracting the features by EMD we can use the Naive Bayes classifier to classify the EEG data. The input should include the training data, training label, testing data and testing label. The algorithm details are shown as the Algorithm 3-4.

Algorithm 3-4. Classification using Naive Bayes classifier

Input: Sample set $X = \{x_1, x_2, \dots, x_N\}$;
The extracted EMD features for x_i are $a_i \in \mathbb{R}^n, (i = 1, 2, \dots, N)$;
Training set labels $T = \{T_1, T_2, \dots, T_c\}$;
Output: Prediction on testing set $L = \{L_1, L_2, \dots, L_u\}$;

1. Compute prior probabilities $P(Y = T_k) = \frac{N_k}{N}$, where N_k is the number of samples in class k;

And compute condition probabilities $P(A = a_{ij}| Y = T_k), (j = 1, 2, \dots, n)$ where a_{ij} is the j-th component of feature a_i;

2. Naive Bayes classifier judge X

For m **in** 1:c

   Compute

$$P(L_{test}|T_k) = P(T_k) \prod_{j=1}^{n} P(a_j^{test}|T_k)$$

Output label of naive Bayes classifier is:

$$T_{test} = argmax_{T_k} P(T_k) \prod_{j=1}^{n} P(a_j^{test}|T_k)$$

End

---

4   Experimental Results

4.1   Dataset Description

In this study, we used the publicly available BCI Competition II data sets Ia [13]. The datasets are acquired from a healthy subject. The task of the subject is to move a cursor up and down on a computer screen by imagination. Simultaneously, the cortical potentials of the subject are acquired. When the cortical potentials of the subject are recording, the subject receives visual feedback of their slow cortical potentials. Central parietal region electrode (Cz-Mastoids) is regarded as the reference electrode, and also records their slow cortical potentials. If the slow cortical potentials of the subject are positive, the cursor on the screen is moved up. And if the slow cortical potentials of the subject are negative, the cursor is moved down. EEG signals are collected with 6-channel electrodes. They are, respectively, A1-Cz (A1 = left mastoid), A2-Cz, two cm frontal of C3 (F3), two cm parietal of C3 (P3), two cm frontal of C4 (F4), and two cm parietal of C4 (P4). The sampling rate is 256 Hz. The experiment included 268 trials and each trial lasted six seconds without inter-trial intervals. In addition, there are 561 samples for EEG signal analysis. From 0.5 s to 6 s, there is a highlighted goal appeared at the top or the bottom of the screen to indicate negativity or positivity. From 2 s to 5.5 s, EEG signals are recorded for training and testing.

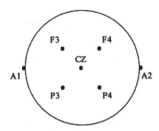

Fig. 2. The distribution diagram of the six electrodes in scalp potentials figure.

4.2   Parameter Selection of Optimal Electrodes Selection Using PCA

According to step 1 of the optimal electrodes selection method using PCA, 28 different dimensions are obtained, ranging from dimension 3 to dimension 30. From the 28 dimensions, 7 different dimensions is equidistantly selected, namely dimension 3, 5, 10, 15, 20, 25, 30. The raw EEG data of each electrode is reduced to those 7 different dimensions and we obtain 7 EEG data in each electrode. The 7 EEG data of each electrode are classified by using Naive Bayes classifier and we get the 7 accuracies in each electrode, shown as Fig. 2. The average accuracy of 7 accuracies of each electrode is calculated, seen as Table 1. There are two classes in EEG data used, so threshold t1 is set to 60%. The electrode is selected as optimal electrode if its average accuracy is above 60%. In this way, A1, A2, F3 and P3 are optimal electrodes.

Table 1. The average accuracy of each electrode

Electrode	A1	A2	F3	F4	P3	P4
Average accuracy (%)	78.89	82.10	65.43	53.63	65.24	54.12

4.3   Parameter Selection of the Modified Strategy

All the optimal electrodes are sorted by the average accuracy and we select the top two electrodes A1 and A2 as intrinsic brain areas electrodes. Then four different combinations that contain intrinsic brain areas electrodes are obtained, which are A1A2, A1A2P3, A1A2F3, and A1A2P3F3, respectively. Similarly, we utilize PCA to reduce the dimension of each combination and Naive Bayes classifier to obtain accuracies. The accuracies of each combination in different dimensions are plotted in Fig. 3.

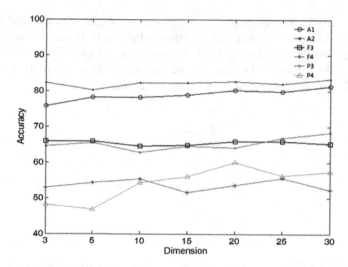

Fig. 3. The classification accuracies of each electrode in 7 different dimensions.

Then the average accuracies of four electrode combinations are obtained, shown in Table 2. The electrode combination is selected as optimal combination if its average accuracy is above 80%. In this way, A1A2, A1A2P3, A1A2F3, and A1A2P3F3 are optimal combinations.

Table 2. The average accuracies of four electrode combinations

Electrode combination	A1A2	A1A2P3	A1A2F3	A1A2P3F3
Average accuracy (%)	84.64	85.76	87.81	84.69

4.4   Result of the Proposed Method

For each optimal combination, the corresponding EEG data is extracted features by using EMD method. After that, the data that extracted features are classified by using Naive Bayes classifier. In this way, we get the final accuracies of all the optimal combinations, shown as Table 3.

Table 3. The final classification results of all the optimal combinations

Optimal combination	Classification accuracy(%)
A1A2	85.32
A1A2F3	90.44
A1A2P3	**93.86**
A1A2F3P3	87.03

There is obvious improvement of the classification performance of combination A1A2F3 and A1A2P3. So it is speculated that interaction effect exists between central region and frontal area, while also exists between central region and parietal region. Obviously, the front one is thinner than the last one. The classification performance of combination A1A2F3P3 is relatively weaker. So there is no significant interaction effect between central region, parietal and frontal area. A1 and A2 represent the information of the central electrode, so the performance of the combination A1A2 is close to the performance of the single electrode.

The experimental results obtained by this method are compared with the results of other researchers, and the results are compared with the results shown in the Table 4.

Table 4. Comparison with the results of other researchers

Methods	Classification accuracy(%)
Gamma band power spectrum and SCP + linear classifier [14]	88.70
Alpha band power spectrum and SCP + neural network [15]	91.47
Spectral centroid feature + PCA dimension reduction and Bayesian [16]	90.44
Wavelet transform + probabilistic neural network [17]	90.80
Polynomial fitting + KNN [3]	92.15
LDA-after-PCA+V-ELM [18]	93.52
Advantage combination + Bayesian	**93.86**

The first five lines shows the BCI 2003 Data set Ia data set of previous studies, where the first row shows the first prize results of competition, the EEG data in channel 4 is used for the extraction of gamma band power spectrum information and EEG data in channel 1 is used for the construction of features. The signal is classified by linear classifier method with recognition rate at 88.7%. On the second line, EEG data in channel 4 and channel 6 are used for extraction of α band spectrum information. EEG data in channel 1 is used for construction of features. The signal is classified by neural network method with recognition rate at 91.47%. The third line showed that spectral centroid feature of EEG data in total six channels is extracted to be dimensionally reduced firstly and then classified by Bayesian method with recognition rate at 90.44%. The fourth line illustrates results of recognition with probabilistic neural network method classifying signal after features extraction with wavelet transformation method. On the fifth line, data is analyzed with a combination method of Polynomial fitting and K-nearest neighbor. The algorithm structure is too complicated though the classification performance is improved.

5 Discussion and Conclusion

In this paper, we proposed a novel EEG signal recognition method using modified optimal electrodes recombination strategy. The main conclusions are summarized as follows:

(1) The method is applied to the data set Ia of BCI competition, which is also analyzed by other researchers. The winner acquired 88.7% using gamma band power combined with SCPs [14]. The best result (93.52%) is obtained by Duan et al. [15] based on LDA-after-PCA feature extraction method and V-ELM classifier. Our method makes a growth of 0.34% than the best one, which is simple and efficient.

(2) It is significant to select optimal electrodes. The results show that optimal electrodes selected by the proposed method are consistent with electrodes selected by psychology knowledge, furthermore, it is subtle.

In a word, the proposed method offers great potential for improving the classification accuracy of BCI datasets. What is more, there are some shortcomings should be overcome in the future. The data set Ia of BCI competition includes 6 electrodes only. If much more electrodes are included, the amount of calculation and complexity will increase. The concurrent computation and high-performance equipment may be expected to be an assistant.

The idea of this method is based on the brain processing towards tasks, considering the phenomenon that multiple regions of brain could be activated at the same time. Therefore, the relationship of different brain regions is deduced from analyzing the relationship of EEG data collected on corresponding regions by recording electrodes. This method combines the knowledge of psychology and information technology and it provided a new way of studying interactions between stimulating brain regions with certain significance.

Acknowledgements. This research is partially sponsored by Natural Science Foundation of China (Nos. 61672070, 81471770, 61572004), the Beijing Municipal Natural Science Foundation (grant number 4182005).

References

1. Pfurtscheller, G., Neuper, C.: Motor imagery and direct brain-computer communication. Proc. IEEE **89**(7), 1123–1134 (2001)
2. Kalicinski, M., Kempe, M., Bock, O.: Motor imagery: effects of age, task complexity, and task setting. Exp. Aging Res. **41**(1), 25–38 (2015)
3. Kayikcioglu, T., Aydemir, O.: A polynomial fitting and k-NN based approach for improving classification of motor imagery BCI data. Pattern Recogn. Lett. **31**(11), 1207–1215 (2010)
4. Cox, D.D., Savoy, R.L.: Functional magnetic resonance imaging (fMRI) "brain reading": detecting and classifying distributed patterns of fMRI activity in human visual cortex. Neuroimage **19**(2), 261–270 (2003)
5. Blanco, A., et al.: RPC-PET: a new very high resolution PET technology. In: 2004 IEEE Nuclear Science Symposium Conference Record, vol. 4, pp. 2356–2360. IEEE, October 2004
6. Cicinelli, P., Marconi, B., Zaccagnini, M., Pasqualetti, P., Filippi, M.M., Rossini, P.M.: Imagery-induced cortical excitability changes in stroke: a transcranial magnetic stimulation study. Cereb. Cortex **16**(2), 247–253 (2006)

7. Duan, L., et al.: An emotional face evoked EEG signal recognition method based on optimal EEG feature and electrodes selection. In: Lu, B., Zhang, L., Kwok, J. (eds.) ICONIP 2011. LNCS, vol. 7062, pp. 296–305. Springer, Heidelberg (2011). https://doi.org/10.1007/978-3-642-24955-6_36

8. Duan, L., Zhang, Q., Yang, Z., Miao, J.: Research on heuristic feature extraction and classification of EEG signal based on BCI Data Set. Res J Appl Sci Eng Technol 5(3), 1008–1014 (2013)

9. Li, S., Zhou, W., Cai, D., Liu, K., Zhao, J.: EEG signal classification based on EMD and SVM. Sheng wu yi xue gong cheng xue za zhi = Journal of biomedical engineering = Shengwu yixue gongchengxue zazhi. 28(5), 891–894 (2011)

10. Lal, T.N., et al.: Support vector channel selection in BCI. IEEE Trans. Biomed. Eng. 51(6), 1003–1010 (2004)

11. Rutkowski, T.M., Mandic, D.P., Cichocki, A., Przybyszewski, A.W.: EMD approach to multichannel EEG data—The amplitude and phase synchrony analysis technique. In: Huang, D.-S., Wunsch, D.C., Levine, D.S., Jo, K.-H. (eds.) ICIC 2008. LNCS, vol. 5226, pp. 122–129. Springer, Heidelberg (2008). https://doi.org/10.1007/978-3-540-87442-3_17

12. Murphy, K.P.: Naive Bayes classifiers. University of British Columbia 18 (2006)

13. Birbaumer, N.: Data Sets Ia for the BCI Competition II (2015). http://www.bbci.de/competition/ii/#datasets

14. Mensh, B.D., Werfer, J., Seung, H.S.: Combining gamma-band power with slow cortical potentials to improve single-trial classification of electroencephalographic signals. IEEE Trans. Biomed. Eng. 51(6), 1052–1056 (2004)

15. Wang, B-J., Jun, L., Bai, J., Peng, L., Li, Y., Li, G.: EEG recognition based on multiple types of information by using wavelet packet transform and neural networks. In: 2005 Engineering in Medicine and Biology Society. 27th Annual International Conference of the IEEE. IEEE-EMBS 2005, pp. 5377–5380 (2006)

16. Wu, T., Yan, G.-Z., Yang, B.-H., Sun, H.: EEG feature extraction based on wavelet packet decomposition for brain computer interface. Measurement 41(6), 618–625 (2008)

17. Sun, S., Zhang, C.: Assessing features for electroencephalographic signal categorization, acoustics, speech, and signal processing, 2005. Proceedings (ICASSP'05) IEEE international conference on IEEE, Vol. 5, pp. 417–420 (2005)

18. Duan, L., Zhong, H., Miao, J., Yang, Z., Ma, W., Zhang, X.: A voting optimized strategy based on ELM for improving classification of motor imagery BCI data. Cognitive Computation 6(3), 477–483 (2014)

Early Diagnosis of Alzheimer's Disease by Ensemble Deep Learning Using FDG-PET

Chuanchuan Zheng[1,2], Yong Xia[1,3(✉)], Yuanyuan Chen[1],
Xiaoxia Yin[2], and Yanchun Zhang[2]

[1] National Engineering Laboratory for Integrated Aero-Space-Ground-Ocean
Big Data Application Technology, School of Computer Science and Engineering,
Northwestern Polytechnical University, Xi'an 710072, China
yxia@nwpu.edu.cn
[2] Centre of Applied Informatics, Victoria University,
Melbourne, VIC 8001, Australia
[3] Centre for Multidisciplinary Convergence Computing (CMCC),
School of Computer Science and Engineering,
Northwestern Polytechnical University, Xi'an 710072, China

Abstract. Early diagnosis of Alzheimer's disease (AD) is critical in preventing from irreversible damages to brain cognitive functions. Most computer-aided approaches consist of extraction of image features to describe the pathological changes and construction of a classifier for dementia identification. Deep learning technique provides a unified framework for simultaneous representation learning and feature classification, and thus avoids the troublesome hand-crafted feature extraction and feature engineering. In this paper, we propose an ensemble of AlexNets (EnAlexNets) algorithm for early diagnosis of AD using positron emission tomography (PET). We first use the automated anatomical labeling (AAL) cortical parcellation map to detect 62 brain anatomical volumes, then extract image patches in each kind of volumes to fine-tune a pre-trained AlexNet, and finally use the ensemble of those well-performed AlexNets as the classifier. We have evaluated this algorithm against seven existing algorithms on an ADNI dataset. Our results indicate that the proposed EnAlexNets algorithm outperforms those seven algorithms in differentiating AD cases from normal controls.

Keywords: Alzheimer's disease · Deep learning · Computer-aided diagnosis
AlexNet

1 Introduction

Alzheimer's disease (AD) is a chronic and progressive decline in cognitive function due to the damage to brain cells. The number of patients suffering from AD worldwide is more than 30 million, and this number is expected to triple by 2050 due to the increasing life expectancy [1]. The onset of AD is typically insidious-cognitive decline generally begins years before reaching the threshold of clinical significance and functional impairment [2]. The prodromal stage of AD called mild cognitive impairment (MCI) is not bound to fall in the deterioration with treatments performed in time.

© Springer Nature Switzerland AG 2018
Y. Peng et al. (Eds.): IScIDE 2018, LNCS 11266, pp. 614–622, 2018.
https://doi.org/10.1007/978-3-030-02698-1_53

The conversion rate from MCI to AD is estimated to be between 10 and 25% per year [3]. Although there are no current disease modifying agents to halt the progression of AD, there are a number of clinical trials underway in patients with pre-symptomatic disease. Thus, as effective therapies become available, the early identification of patients with MCI will be of tremendous benefit to patients and their families.

The pathology of AD includes cortical and subcortical atrophy together with the deposition of β-amyloid. Molecular medical imaging, such as the positron emission tomography (PET), single-photon emission computed tomography (SPECT), and magnetic resonance imaging (MRI), offers the ability to visualize hypometaboism or atrophy introduced by AD, and hence has led to a revolution in early diagnosis of AD [4]. In particular, functional PET and SPECT can detect subtle changes in cerebral metabolism prior to anatomical changes are evident or a symptomatological diagnosis of probable AD can be made with MR imaging [5]. Comparing to SPECT, PET is able to provide higher resolution information than SPECT in evaluating patients with suspected AD. 18-Fluorodeoxyglucose (FDG) is a widely used tracer in PET imaging for studying glucose metabolism. The comparative study made by Silverman [6] demonstrated that FDG-PET is superior to perfusion SPECT in identifying early changes associated with AD and other neurodegenerative dementias.

However, early diagnosis of AD remains a challenging task, since the pathological changes can be subtle in the early course of the disease and there can be some overlap with other neurodegenerative disorders [7]. Most computer-aided approaches consist of extraction of image features to describe the pathological changes and construction of a statistical model of the disease from a set of training examples using supervised methods [8]. The features that have been considered include global features, which are computed by applying the entire brain volume to a linear transform, and local features, such as the statistics, histograms, and gradients calculated from volumes of interest (VOIs). Most pattern classification techniques, including the k-mean clustering [9], artificial neural network (ANN) [10] and support vector machine (SVM) [5], have been successfully applied to this task.

Recently, the deep learning technique has shown proven ability to analysis medical images. It provides a unified framework for simultaneous learning image representation and feature classification, and thus avoids the troublesome hand-crafted feature extraction and feature engineering. Suk et al. [11] used the deep model learned image representation to differentiate AD subjects from MCI ones based on the assumption that those deep features inherent the latent high-level information. They also fused multiple sparse regression networks as the target-level representation [12]. Ortiz et al. [13] applied the deep belief network to the early diagnosis of the Alzheimer's disease using both MRI and PET scans. Valliani et al. [14] employed the deep residual network (ResNet) that has been pre-trained on the ImageNet natural image dataset to identify dementia types.

In this paper, we propose an ensemble of AlexNets (EnAlexNets) model to diagnosis between AD and NC and differentiate between stages of MCI, mild MCI (mMCI) vs. severe MCI (sMCI). The uniqueness of our approach include: (1) using the automated anatomical labeling (AAL) cortical parcellation map to obtain 62 brain anatomical volumes; (2) extracting image patches from each cortical volume to train a candidate AlexNet, the champion model for the ILSVRC2012 image classification task; (3) selecting effective AlexNet models and ensemble them to generate more accurate diagnosis. The proposed EnAlexNets model was evaluated against seven methods on the ADNI dataset and achieved the state-of-the-art performance.

2   Dataset

Data used in preparation of this article were obtained from the Alzheimer's Disease Neuroimaging Initiative (ADNI) database (adni.loni.usc.edu). The ADNI was launched in 2003 as a public-private partnership, led by Principal Investigator Michael W. Weiner, MD. The primary goal of ADNI has been to test whether serial MRI, PET, other biological markers, and clinical and neuropsychological assessment can be combined to measure the progression of MCI and early AD.

We selected 962 FDG-PET studies from the ADNI database, including 241 AD cases, 306 mild MCI (mMCI) cases, 127 severe MCI (sMCI) cases and 288 normal controls (NCs). 185 MBq of 18F-Fludeoxyglucose (FDG) was injected to the subjects and PET scanning was commenced approximately 30 min after tracer injection that produced 3D scan consists of six 5 min frames. The scans used for this study are baseline or screening in the ADNI database, and had been through a pre-processing pipeline that includes co-registration, averaging, voxel normalization, and isotropic Gaussian smoothing [15]. This pre-processing work was done by the ADNI participants and it makes any subsequent analysis simpler as the data from different PET scanners are then uniform. The demographic information of the selected dataset is shown in Table 1.

Table 1. Demographics data of patients in our dataset

Diagnosis	Number	Gender (F/M)	Age
AD	241	99/142	75.1 ± 7.8
NC	288	137/151	74.4 ± 5.9
mMCI	306	135/171	71.4 ± 7.5
sMCI	127	42/85	75.6 ± 7.4

3   Method

The proposed algorithm consists of four major steps: anatomical volume detection, data augmentation, patch classification, and dementia differentiation. A diagram that summarizes this algorithm was shown in Fig. 1.

3.1   Anatomical Volume Detection

The AAL cortical parcellation map [16], which provides 62 brain anatomical volumes including 54 symmetrical and 8 asymmetrical volumes, is well aligned with the template brain PET image supplied with the statistical parametric mapping (SPM, Version 2012) package [17].

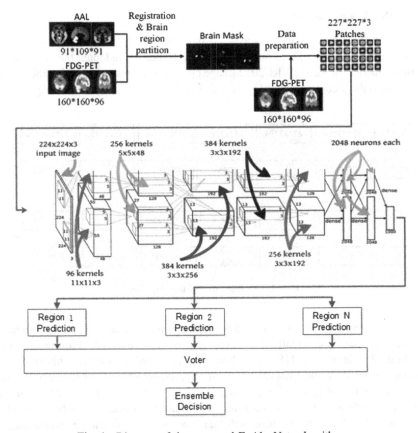

Fig. 1. Diagram of the proposed EnAlexNets algorithm.

To map the anatomical labels from the atlas onto each study and maintain the original brain image information, we spatially normalized the SPM brain PET template to FDG-PET images, which are self-aligned, using the spatial normalization procedure supplied with the SPM package. Thus each anatomical volume in the AAL atlas can be transferred to our FDG-PET images. For each volume, we located its geometry center on FDG-PET images and put a 67×67 window on that center for image patch extraction.

3.2 Data Augmentation

Since number of extracted image patches is limited, we employed three data augmentation techniques to enlarge our image patch dataset. First, we shifted the PET image horizontally and vertically with a step of 5 voxels before patch extraction. Second, we move the patch extraction window to its adjacent transverse slices forwardly and backwardly. Third, rotate the PET image clockwise and counter-clockwise with 5 and 10°. With these operations, each image patch has nine augmented copies.

3.3 Patch Classification

Considering the limited image patches we have, too complex deep neural networks may suffer from over-fitting to our training data. Hence, we adopted the AlexNet, the champion model for the ILSVRC2012 image classification task, as our classifier, which consists of only 5 convolutional layers and 3 fully-connected layers (see Fig. 2).

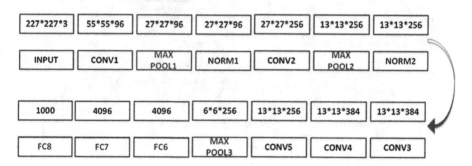

Fig. 2. Architecture of the AlexNet.

It has been widely acknowledged that the image representation learned from large-scale datasets can be efficiently transferred to generic visual recognition tasks, which have limited training data [18, 19]. Therefore, the AlexNet used for this study was previously trained on the ImageNet training dataset, which is a 1000-category large-scale natural image database. To adapt this model to our problem, we replaced its last three layers with a fully connected layer, a softmax layer and a classification output layer, and adjust the learning rates such that those new layers can be trained quickly and other layers are trained slowly.

Since the AlexNet takes an input image of size $227 \times 227 \times 3$, we duplicated each image patch three times and resize each copy to 227×227 by the bilinear interpolation algorithm.

3.4 Dementia Differentiation

Since we obtained 62 cortical volumes on each PET scan, we can use the image patches extracted in each volume to fine-tune a pre-trained AlexNet. As we know, not all cortical volumes play equally critical roles in dementia diagnosis. Therefore, the diagnosis capacity of each fine-tuned AlexNet varies a lot. We chose 30% of the fine-tuned AlexNets, which performed best on the validation dataset, and adopted the majority voting scheme to combine their decisions for dementia differentiation. Such ensemble learning acts as the role of expert consultation aiming to get a more accurate diagnosis.

3.5   Evaluation

We adopted the 5-folder cross-validation scheme to evaluate the performance of the proposed algorithm. We randomly sampled the instances from each class to form five folds, aiming to ensure the distribution of data in each fold is as similar as that of the whole dataset. In each run, we used 80% of patches for training, 20% for validation and 20% for testing. The data augmentation procedure was not applied to test patches.

4   Results

Table 2 gives the accuracy of the proposed algorithm and other seven state-of-the-art algorithms when applying them to differentiate AD cases from NCs. It shows that our algorithm and the method reported in [13] achieved the highest classification accuracy. However, the other algorithm used both PET and MRI data. Meanwhile, it reveals that our algorithm achieved the highest specificity and the second best sensitivity among those eight algorithms.

Table 2.   Accuracy of eight methods (AD vs. NC)

Methods	Accuracy	Sensitivity	Specificity
Karwath et al. [20]	0.89	0.85	0.91
Valliani et al. [21]	0.81	–	–
Vu et al. [22] (PET)	0.85	–	–
Vu et al. [22] (MRI + PET)	**0.91**	–	–
Ortiz et al. [13] (PET + MRI)	0.90	0.86	0.94
Suk et al. [11] (SAE-classifier)	0.89	–	–
Liu et al. [23]	0.87	**0.88**	0.87
Proposed	**0.91**	0.86	**0.95**

Figure 3 depicts the accuracy of the proposed algorithm when applying it to the mMCI-sMCI classification. It shows that our algorithm achieved a classification accuracy of 85% in this problem.

5   Discussion

The proposed EnAlexNets algorithm is based on the pre-trained and fine-tuned AlexNets and ensemble of these models. Our experiments indicate that this algorithm is robust in securing more distinguish features and achieving better performance than that of using multi-modalities images. The EnAlexNets algorithm enables to effectively compensate for the less amount of information afforded by single image mode and reduce the complexity of diagnosis process requirements for multi-mode images. These investigations can be further improved by improving the CNN structure and enlarging the dataset. As for the time-complexity, the training time of multiple AlexNets summed up to 111,600 s same level as other state-of-art deep learning based methods. The diagnosis time of one patient case in our method was around 10 s. With better computation resources, this time consumed can be further diminished.

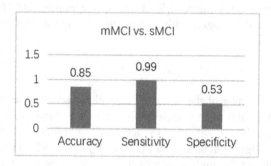

Fig. 3. The performance of proposed algorithm (mMCI vs. sMCI).

6   Conclusion

In this study, we proposed a multi-AlexNets based method for computer-aided diagnosis of AD and MCI. This method not only utilized the combination of different brain regions but also incorporated AlexNet with its strong feature representation ability. Better performance was observed on two classification tasks. Apart from the traditional diagnosis between AD and NC, we initially conducted the differentiation between different stages of MCI, mMCI vs. sMCI. Due to the latent progression from MCI to further stage, we intend to find the risk as early as possible, which is of great significance to both clinicians and patients. This study may be of great assistance to the computer-aided diagnosis of dementia and other biomedical fields.

Acknowledgements. This work was supported in part by the National Natural Science Foundation of China under Grants 61471297 and 61771397, in part by the China Postdoctoral Science Foundation under Grant 2017M623245, in part by the Fundamental Research Funds for the Central Universities under Grant 3102018zy031, and in part by the Australian Research Council (ARC) Grants.

Data collection and sharing for this project was funded by the Alzheimer's Disease Neuroimaging Initiative (ADNI) (National Institutes of Health Grant U01 AG024904) and DOD ADNI (Department of Defense award number W81XWH-12-2-0012). ADNI is funded by the National Institute on Aging, the National Institute of Biomedical Imaging and Bioengineering, and through generous contributions from the following: AbbVie, Alzheimer's Association; Alzheimer's Drug Discovery Foundation; Araclon Biotech; BioClinica, Inc.; Biogen; Bristol-Myers Squibb Company; CereSpir, Inc.; Cogstate; Eisai Inc.; Elan Pharmaceuticals, Inc.; Eli Lilly and Company; EuroImmun; F. Hoffmann-La Roche Ltd and its affiliated company Genentech, Inc.; Fujirebio; GE Healthcare; IXICO Ltd.; Janssen Alzheimer Immunotherapy Research & Development, LLC.; Johnson & Johnson Pharmaceutical Research & Development LLC.; Lumosity; Lundbeck; Merck & Co., Inc.; Meso Scale Diagnostics, LLC.; NeuroRx Research; Neurotrack Technologies; Novartis Pharmaceuticals Corporation; Pfizer Inc.; Piramal

Imaging; Servier; Takeda Pharmaceutical Company; and Transition Therapeutics. The Canadian Institutes of Health Research is providing funds to support ADNI clinical sites in Canada. Private sector contributions are facilitated by the Foundation for the National Institutes of Health (www. fnih.org). The grantee organization is the Northern California Institute for Research and Education, and the study is coordinated by the Alzheimer's Therapeutic Research Institute at the University of Southern California. ADNI data are disseminated by the Laboratory for Neuro Imaging at the University of Southern California.

References

1. Barnes, D.E., Yaffe, K.: The projected effect of risk factor reduction on Alzheimer's disease prevalence. Lancet Neurol. **10**(9), 819–828 (2011)
2. Ferreira, L.K., Rondina, J.M., Kubo, R., et al.: Support vector machine-based classification of neuroimages in Alzheimer s disease: direct comparison of FDG-PET, rCBF-SPECT and MRI data acquired from the same individuals. Revista Brasileira De Psiquiatria (2017)
3. Grand, J.H., Caspar, S., Macdonald, S.W.: Clinical features and multidisciplinary approaches to dementia care. Multidiscip. Healthc. **4**, 125–147 (2011)
4. Gomez-Isla, T., Price, J.L., McKeel Jr., D., et al.: Profound loss of layer II entorhinal cortex neurons occurs in very mild Alzheimer's disease. J. Neurosci. **16**(14), 4491–4500 (1996)
5. Fung, G., Stoeckel, J.: SVM feature selection for classification of SPECT images of alzheimer's disease using spatial information. Knowl. Inf. Syst. **11**(2), 243–258 (2007)
6. Silverman, D.H.S.: Brain 18F-FDG PET in the diagnosis of neurodegenerative dementias: comparison with perfusion SPECT and with clinical evaluations lacking nuclear imaging. J. Nucl. Med. **45**(4), 594–607 (2004)
7. Adeli, H., Ghosh-Dastidar, S., Dadmehr, N.: Alzheimer's disease and models of computation: imaging, classification, and neural models. J. Alzheimers Dis. **7**(3), 187–199 (2005)
8. Zheng, C., Xia, Y., Pan, Y., et al.: Automated identification of dementia using medical imaging: a survey from a pattern classification perspective. Brain Inform. **3**(1), 17–27 (2016)
9. Pagani, M., Kovalev, V.A., Lundqvist, R., et al.: A new approach for improving diagnostic accuracy in Alzheimer's disease and frontal lobe dementia utilising the intrinsic properties of the SPET dataset. Eur. J. Nucl. Med. Mol. Imaging **30**(11), 1481–1488 (2003)
10. Nagao, M., Sugawara, Y., Ikeda, M., et al.: Heterogeneity of cerebral blood flow in frontotemporal lobar degeneration and Alzheimer's disease. Eur. J. Nucl. Med. Mol. Imaging **31**(2), 162–168 (2004)
11. Suk, H., Shen, D.: Deep learning-based feature representation for AD/MCI classification. In: Mori, K., Sakuma, I., Sato, Y., Barillot, C., Navab, N. (eds.) MICCAI 2013. LNCS, vol. 8150, pp. 583–590. Springer, Heidelberg (2013). https://doi.org/10.1007/978-3-642-40763-5_72
12. Suk, H., Shen, D.: Deep ensemble sparse regression network for Alzheimer's disease diagnosis. In: Wang, L., Adeli, E., Wang, Q., Shi, Y., Suk, H. (eds.) MLMI 2016. LNCS, vol. 10019, pp. 113–121. Springer, Cham (2016). https://doi.org/10.1007/978-3-319-47157-0_14
13. Ortiz, A., Munilla, J., Jorge, et al.: Ensembles of deep learning architectures for the early diagnosis of the Alzheimer's disease. Int. J. Neural Syst. **26**(7), 1650025 (2016)
14. Valliani, A., Soni, A.: Deep residual nets for improved Alzheimer's diagnosis. In: Proceedings of the 8th ACM International Conference on Bioinformatics, Computational Biology and Health Informatics 2017, pp. 615–615 (2017)

15. ADNI/PET   Pre-processing.   http://adni.loni.usc.edu/methods/pet-analysis/pre-processing. Accessed 05 Feb 2016
16. Tzouriomazoyer, N., Landeau, B., Papathanassiou, D., et al.: Automated anatomical labeling of activations in SPM using a macroscopic anatomical parcellation of the MNI MRI Single-Subject Brain. NeuroImage **15**(1), 273–289 (2002)
17. Frackowiak, R.S.J.: Human brain function (2004)
18. Hu, J., Lu, J., Tan, Y.P., et al.: Deep transfer metric learning. IEEE Trans. Image Process. **25** (12), 5576–5588 (2016)
19. Oquab, M., Bottou, L., Laptev, I., et al.: Learning and transferring mid-level image representations using convolutional neural networks. In: IEEE Conference on Computer Vision and Pattern Recognition (CVPR) 2014, pp. 1717–1724. IEEE (2014)
20. Karwath, A., Hubrich, M., Kramer, S.: Convolutional neural networks for the identification of regions of interest in PET scans: a study of representation learning for diagnosing Alzheimer's disease. In: ten Teije, A., Popow, C., Holmes, J.H., Sacchi, L. (eds.) AIME 2017. LNCS (LNAI), vol. 10259, pp. 316–321. Springer, Cham (2017). https://doi.org/10. 1007/978-3-319-59758-4_36
21. Valliani, A., Soni, A.: Deep residual nets for improved Alzheimer's diagnosis. In: ACM International Conference on Bioinformatics, Computational Biology and Health Informatics 2017, pp. 615–615 (2017)
22. Vu, T.D., Yang, H.J., Nguyen, V.Q., et al.: Multimodal learning using convolution neural network and Sparse Autoencoder. In: IEEE International Conference on Big Data and Smart Computing (BigComp) 2017, pp. 309–312. IEEE (2017)
23. Liu, S., Liu, S., Cai, W., et al.: Early diagnosis of Alzheimer's disease with deep learning. In: 11th International Symposium on Biomedical Imaging (ISBI) 2014. IEEE (2014)

Study on Matthew Effect Based Feature Extraction for ECG Biometric

Gang Zheng$^{(\boxtimes)}$, Yu Wang, XiaoXia Sun, Ying Sun,
and ShengZhen Ji

School of Computer Science and Engineering, Tianjin University of Technology,
Tianjin, China
kenneth_zheng@vip.163.com,
jiayidoubao@163.com, sxxl5757118176@163.com,
sunying@tjut.edu.cn, shulinji@163.com

Abstract. Electrocardiogram (ECG) is a "live" signal and is very difficult to be copied or forged, which make it becoming a competitive biology material for biometric. But, due to ECG signals are easily be affected by the external environment and human states (physiological or psychological), therefore, finding stable features becomes one of the key issues in the research. The paper proposed a feature extraction method based on ECG superposition matrix of single heartbeat ECG. By matrix segmentation and similarity comparison, ECG stable feature distribution area can be selected, and stable feature sets are constructed. And through further study, it is found that a large number of heartbeats are need to build superposition matrix. For solving the problem, the paper proposed a Matthew effect based method, by which superposition matrix can be constructed by only 10 heartbeats. Experiments results showed that average TPR of 100 heartbeats was reaching 83.21, 83.93 and 80% respectively. And that of 10 heartbeats with Matthew Effect reach 80.36, 82.68 and 80.77% respectively, which is competitive compare to 100 heartbeats superposition matrix.

Keywords: ECG biometric · Matthew effect · Superposition matrix
Identity recognition

1 Introduction

The Electrocardiogram (ECG) had been used in heart disease diagnosis for one hundred years. Its waveform contains abundant information of heart structures and electronic conductivity of cardiac muscle cells. And this information just shows that ECG is highly personalized and has the basic conditions for identity recognition. And furthermore, ECG has the characteristic of "liveness". This make ECG signal was difficult to be forged in real scenery, and it can be expected to be competitive biometric compared with existing biometric methods, such as fingerprint, face, iris, palm print, vein, gait, speech, handwriting and so on. In the last twenty years, several intelligent techniques were developed for ECG based identification [1–4]. Usually, the processing procedures contained, Firstly, ECG signals collecting or public ECG data [5]. Traditionally, ECG collecting needed AgCl chest electrode, which is not convenient for user,

© Springer Nature Switzerland AG 2018
Y. Peng et al. (Eds.): IScIDE 2018, LNCS 11266, pp. 623–634, 2018.
https://doi.org/10.1007/978-3-030-02698-1_54

and it has becoming an obstruction in ECG based identification; Secondly, the ECG signal was normally enlarged by differential amplifier for 1000 times, by which noise and artifacts were introduced; Thirdly, ECG feature were extracted in two directions, one is depended on ECG waveform point, shown as PQRST, like the segment period, amplitude, and slope of these points, like QRS, T, etc. [6–9], the other direction is to transform ECG into single heartbeat waveform and extract morphological features after recognizing R wave in ECG [10], feature extraction methods contained, Empirical Mode Decomposition (EMD) [11], Fast Fourier Transform (FFT) [12], Wavelet Transform (WT) [13], reduced binary pattern (RBP) [14], Pulse Active Mean (PAM) [16], quantization, [17], Frequency and Rank Order Statistics (FROS) [18], Discrete Cosine Transform (DCT) [19], Vector Cardiograph [20], Sparse Representation (SR) [21], Chaos Theory [22]. In dimension reduction, LDA [23], Principal component analysis (PCA) [15] were used. Finally, all acquired features are used to learn or train classifiers to achieve recognition, such as LDA [23], Neural Network (NN) [6], Support Vector Machine (SVM) [7, 13], Dynamic Time Warping (DTW) [24], K Nearest Neighbors (KNN) [11], Hidden Markov Model (HMM) [25], Logistic Regression (LR) [26], Probabilistic nonlinear kernel classifier [27], Optimum Path Forest (OPF) [4], MultiLayer Perceptron (MLP) [28], Deep Neural Network (DNN) [29] and linear distance, like, Euclidean Metric [9], CC, Log-likelihood Ratio [12], non-linear distance, like Wavelet [22], Hamming Distance (HD) [30], and so on.

Despite the above big progress, it has been shown recently that identification methods were still facing with at least two challenges. One, compared to other biometric methods, the information contained in ECG waveform is relatively less. In one dimension, it is a time series data, which can be expressed by vector. In two dimensions, it is a pseudo periodic curve, which can be expressed by matrix, and the matrix is sparse.

Secondly, the more ECG waveforms of one individual were used to establish one's ECG pattern, the higher recognition rate will be gotten. Therefore, the number of ECG waveforms to build one's ECG pattern normally reached several hundreds, especially when the scale of individual amount is up to several thousands. This will hold back ECG identification in real application, since it is difficult to get such number of individual ECG waveform in short period of time.

2 Strategy Description

2.1 Man Idea

The paper proposes an approach of ECG based person identification by Matthew effect pattern superposition strategy. The target of the approach is to add more information on ECG data, especially add third dimension "deep" description; this will be enlarging the discrimination of ECG for identification. The approach only needs 10 single beats to establish an individual pattern matrix. In the paper, $T = \{(X_i, Y_i)\}_{i=1}^{n}$ was assumed as training set, it was composed of n single periodic ECG signals as input data. $X_i \in R^T$

stands for the ECG beat vector of dimension T, and $Y_i \in [1, K]$ is the label of individuals. Our target is to recognize the individual in the data set which is not belong to both training and testing set.

2.2 Feature Enhancement Method by Matthew Effect

The strategy follows Matthew effect theory. From the observation, one's ECG waveform of single heartbeat is nearly the same in short period of time (such as in one minute). The probability of generating waveform in other position other than the previous tracks is very low. This phenomenon follows Matthew effect, and it is used to establish individual data by our proposal. That is to say, one's ECG waveform had characteristic of pseudo periodicity. By comparing with everyone's ECG waveform, it can be seen that everyone's ECG waveform has its own trace, as shown in Fig. 1. ECG data were selected from MIT-BIH Database Distribution [5].

1(a) No.111   1(b) No.113   1(c) No.210   1(d) No.217

Fig. 1. ECG superposition channel of MIT-BIH database 1(a) patient No. 111; 1(b) patient No. 113; 1(c) patient No. 210; 1(d) patient No. 217.

The color changing was used to express the superposition times of the superposition points. The lighter the color, the more superposition times of the points drawing. It could be clearly seen that everyone's waveforms form a similar channel effect, and in this channel, the number of outside channel points were significantly lower than that of within the channel. This followed Matthew effect (A phenomenon in specific circles whereby one's accomplishments and reputation tend to snowball, and those with meager accomplishments have greater difficulty achieving accomplishments.)

2.3 Definition

Therefore, we proposed a method for describing the track of ECG waveform based on the Matthew effect, which does not require the superposition of the ECG waveform data generated by hundreds of heartbeats per person. Usually within 10–15 ECG waveforms, the expression of similar effects can be formed, which is possible for the practical application of the identification system with ECG waveform.

Firstly, we defined ECG data as set X*.

$$X^* = \left(X^1, X^2, \ldots, X^p\right)^T P \text{ is the number of heartbeat in } X^*, X^t \text{ is the } i^{th} \text{ heartbeat in } X^*$$
$$X^i = \left(a_1^i, a_2^i, \ldots a_t^i\right), \quad 1 \leq i \leq p, \, t \text{ is the dimension of heartbeat} \tag{1}$$

Secondly, to construct ECG superposition matrix, randomly selected 10 ECG data in X^* to map into a matrix $E(i, j)$. The horizontal axis of the matrix is the order of sampling points of the single heartbeat ECG data, sign as i, the ordinate is the voltage value of the ECG signal, sign as a_j^i. If a_j^i is greater than "0", the position of matrix E is assigned to "1", the other positions are assigned to "0", shown in Fig. 2.

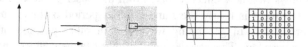

Fig. 2. Schematic diagram of ECG matrix projection

In the procedure of superposition, f is signed as the number of superposition, 10 ECG heartbeats are superimposed sequentially, if the original value of E is "0", then the value of that position is assigned to the value of same position of next heartbeat, f is assigned as "0"; If the original value is not 0, then $f = f + 1$, the position value is converted to the value at the original position plus twice times f. At the same time, the 8-adjacent position of this position plus f, as shown in (2).

$$\begin{cases} E\left[j, a_j^i\right] = E\left[j, a_{j+1}^i\right], f = 0 & a_j^i = 0 \\ f = f + 1 \\ E\left[j, a_j^i\right] = E\left[j, a_{j+1}^i\right] + f * 2 \\ E\left[j+1, a_j^i\right] = E\left[j+1, a_{j+1}^i\right] + f \\ E\left[j-1, a_j^i\right] = E\left[j-1, a_{j+1}^i\right] + f \\ E\left[j, a_j^i + 1\right] = E\left[j, a_{j+1}^i + 1\right] + f \\ E\left[j, a_j^i - 1\right] = E\left[j, a_{j+1}^i - 1\right] + f & a_j^i \neq 0 \\ E\left[j-1, a_j^i + 1\right] = E\left[j-1, a_{j+1}^i + 1\right] + f \\ E\left[j-1, a_j^i - 1\right] = E\left[j-1, a_{j+1}^i - 1\right] + f \\ E\left[j+1, a_j^i + 1\right] = E\left[j+1, a_{j+1}^i + 1\right] + f \\ E\left[j+1, a_j^i - 1\right] = E\left[j+1, a_{j+1}^i - 1\right] + f \end{cases} \quad (2)$$

For example, $E(2, 482)$ is greater than "0", after superposition, its value is increased by "2", and its 8-adjacent coordinates are increased by 1, so did $E(3, 483)$, shown in Fig. 3.

This will make the distribution of ECG waveform more significant. Through the feature enhancement method based on the Matthew effect, ECG stable region can quickly be formed, which makes the distribution frequency of the stable region and the unstable region of ECG waveform distribution more obvious and easier to distinguish. By this strategy, a small amount of ECG data can form the stable distribution of ECG signal, and the effect is better than the original superposition mechanism of the same

ECG heartbeats, and is like that of more than hundreds of ECG heartbeats. The acquisition length of 10 ECG signals varies about 6–10 s according to the heart rate, and the application of ECG signals was greatly improved.

484	0	0	0	0
483	1	1	1	0
482	1	2	1	0
481	1	1	1	0
	1	2	3	4

3(a) Value change in (482,2)

484	0	1	1	1
483	1	2	3	1
482	1	3	2	1
481	1	1	1	0
	1	2	3	4

3(b) Value change in (483,3)

Fig. 3. Schematic diagram of the formation mechanism of the ECG heartbeat superposition

3 ECG Feature Extraction on Matrix Segmentation

The ECG superposition matrix is confronted with the problem of large amount of data in ECG biometric. Since the sampling rate of ECG signal varies from 200 to 1000 Hz, the transverse coordinate of matrix is usually between 200 and 1000, which make superposition matrix in a huge size. The general ordinate is from 4096 to 65535. Such a calculation amount will delay the identification time. Therefore, this paper proposes a feature extraction method based on matrix segmentation to extract the local stable distribution of ECG signals. The main idea of this feature extraction method is to divide the ECG superposition matrix by 3 * 3 matrix, select all the sub matrices of the first ECG superposition matrix as the training template, the others as the matching data. Compare the sub matrices in the matching data with the matrix of the same position in the training template, if the similarity degree is greater than the threshold, the match is successful, and the number of successful sub matrices matching in the same position is higher, which represents the corresponding position of the sub matrix in the template matrix is stable. Then the content of corresponding position in matrix and its ECG superposition matrices is the stable feature we want to extract. The procedure is shown below:

3.1 Matrix Segmentation

E^* is defined as the ECG Template Library, which is used to preserve ECG identity template of individual. E_j is a matrix in $E^*(c, m)$, it is divided into e in size 3 * 3, the number of that is d, $d = (c * m)/9$. As shown in (3).

$$E_j = \begin{bmatrix} e^j_{\frac{c*m}{9} - \frac{c}{3}} & \cdots & e^j_d \\ \vdots & \ddots & \vdots \\ e^j_1 & \cdots & e^j_{\frac{c}{3}} \end{bmatrix},$$

$$e^*_j = \left(e^j_1, e^j_2, \ldots, e^j_d\right), e^j_d \begin{bmatrix} E_j(m-2, c-2) & \cdots & E_j(m-2, c) \\ \vdots & \ddots & \vdots \\ E_j(m, c-2) & \cdots & E_j(m, c) \end{bmatrix} \quad (3)$$

As shown in Fig. 4, the red area in the figure is the region in the matrix E_j that does not overlap by 0. That is, the corresponding distribution area of ECG in the matrix, the black border represents the sub matrix e_j^* in E_j, and the 3 * 3 matrix represents the sub matrix e_j^*, e_d^j represents dth sub matrix. The set of all templates in the E^* formed U^*, shown in (4)

$$U^* = \left(e_1^*, e_2^*, \ldots, e_j^*, \ldots e_f^*\right) \tag{4}$$

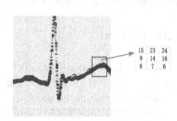

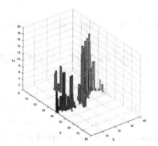

Fig. 4. Schematic of matrix segment (Color figure online)

Fig. 5. Three-dimensional schematic diagram of matrix

3.2   Matrix Similarity Comparison

All sub matrices of e_j^* in U^* are used as templates. The corresponding matrices in the other matrices set e_1^* are judged by similarity, result is saved in S, and S_r is given threshold. Matrix $T(m/3, c/3)$ is used to store the matrix comparison result, If $S > S_r$, the value of the corresponding coordinate in T is added by 1, as in (5).

$$S > S_r \ T\left(\frac{m}{3}, \frac{c}{3}\right) = T\left(\frac{m}{3}, \frac{c}{3}\right) + 1, \quad S = \text{Math}\left(e_d^j, e_d^1\right) \tag{5}$$

As shown in Fig. 5, the x and y axes represent the coordinate information of the sub matrices, y-axis intercepts the position with value of ECG signal. The z axis represents the value at the corresponding coordinates, the larger the value of on the z-axis, the higher the stability of the region corresponding to the sub matrix. For all value greater than $d/2$ in T, select the corresponding sub matrix to form feature set M, shown in (6).

$$M = \left(\ldots, e_n^1, \ldots\right) 1 \leq n \leq d \tag{6}$$

4 ECG Based Identity System

4.1 Extraction of Stable Area Identity Processing

Construction Superposition Matrix

According to the proposed strategy, ECG data is used to construct superposition *matrix*. Assign $D^* = (D_1, D_2, \ldots, D_h, \ldots, D_k)$ as test data set, D_h represents the hth test matrix, $1 \leq h \leq k$. Based on the location information of the feature set M, the sub matrices at the same position of the test matrix D_h used to form test feature set $M_h = (\ldots, d_n^h, \ldots)$, d_n^h is the sub matrix that is extracted from the same position of test matrix D_h, according to the position of e_n^1 in E_1

Similarity Computing

Compute the similarity of corresponding sub matrix in M and M_h, see (7)

$$S = Math\left(d_n^h, e_n^1\right) \tag{7}$$

S is similarity, S_r is predefined threshold, if $S > S_r$, then d_n^h is matched successfully, otherwise, match is fail.

Sub Matrix Counting

Calculate the number of successfully matched sub matrix in M_h, sign as t. If the number of sub matrices matched successfully is greater than 1/2, that is, $t > 1/2$, then the hth sub matrix is authenticated, whereas the identity authentication is fail.

4.2 S_r Selecting

In this paper's experiment, the self-collected ECG signal data and the MIT public dataset are randomly selected for 10 persons, each of which has 1000 heartbeats, each 100 of that constructs the ECG superposition matrix, one matrix of each person is chose as the template matrix, the remaining matrix are used as the test matrix. The matrix segmentation and similarity calculation function are identical with the authentication method based on the number of ECG superposition matrix. The similarity between the test matrix and the template sub matrix is calculated by using the angle cosine, and the K-nearest neighbor algorithm is used to determine the label of the sub Matrix and calculate the similarity coefficients of right labeled sub matrix. Through observation, the range of the similarity coefficients of the correct matrices is in [0.91, 0.95] on self-collected ECG data, while the range of that is in [0.93, 0.96] on MIT data. After calculating their mathematical expectations separately, S_r is assigned 0.93 and 0.95 on self-collected data and MIT data respectively.

5 Experiment and Analysis

In this section randomly selected 14 patient's ECG data from MIT ECG database. And ECG by self-collected of 28 volunteers in calm state, 13 in unlimited state (high pressed, satiation, lack of enough sleeping, and so on), six trails were carried out. In

Table 1, HS means Heartbeats amount to superposition; HSM means Heartbeats amount to superposition by Matthew Effect; HMT means Heartbeat amount of one subject to form matrix in training set; HMTE means Heartbeat amount of one subject to form matrix in test set; SPT means Subject in Positive test set; SNT means Subject in Negative test set.

Table 1. Experiment design

Experiment no.	HS	HSM	HMT	HMTE	MIT		Self-collected data set		Calm state data set		Unlimited state data set	
					SPT	SNT	SPT	SNT	SPT	SNT		
Exp1	10	10	10	20	1	13	1	13				
Exp2	100	10	10	20	1	13	1	13				
Exp3	10	10	10	20					1	27		
Exp4	100	10	10	20					1	27		
Exp5	10	10	10	20							1	13
Exp6	100	10	10	20							1	13

5.1 Result Analysis

Result Evaluation

TP (True Positive): when test sample is positive, the test result is positive. FN (False Negative): when test sample is positive, the test result is negative. FP (False Positive): when test sample is negative, the test result is positive. TN (True Negative): when test sample is negative, the test result is negative. TPR (True Positive Rate) and FPR (False Positive Rate) are used to evaluate authentication in the paper.

$$TPR = TP/(TP + FN) \tag{8}$$

$$FPR = FP/(FP + TN) \tag{9}$$

Experiment Results

The results of identity authentication using different number of heartbeats on MIT dataset are shown in Table 1 (Result of Exp. 1 and Exp. 2) (Table 4).

The results of identity authentication using different number of heartbeats on self-collect dataset in calm state are shown in Table 2 (Result of Exp. 3 and Exp. 4).

The results of identity authentication using different number of heartbeats on self-collect dataset in unlimited state are shown in Table 3 (Result of Exp. 5 and Exp. 6).

From Tables 1, 2 and 3, experiments on three different ECG data set including MIT-BIH, self-collected ECG data in calm state and unlimited state, 10 heartbeats superposition matrix is failing to get stable feature distribution, and in the authentication, the average TPR is below 50%. Therefore, 100 heartbeats superposition matrix can get stable feature distribution by superposition matrix. When it is used on authentication, average TPR is reaching 83.21, 83.93 and 80% respectively. And the similar result was gotten on 10 heartbeats superposition matrix constructed by Matthew Effect, the average TPR is reaching 80.36, 82.68 and 80.77% respectively.

Table 2. TPR and FPR on MIT data set (%)

Patient no.	10 heartbeats		100 heartbeats		10 heartbeats by Matthew effect	
	TPR	FPR	TPR	FPR	TPR	FPR
m1	25.00	0.00	90.00	0.00	85.00	0.00
m2	20.00	0.00	80.00	0.00	75.00	0.00
m3	20.00	5.00	75.00	0.00	70.00	0.00
m4	25.00	0.00	90.00	0.00	85.00	0.00
m5	30.00	10.00	85.00	0.00	85.00	0.00
m6	30.00	0.00	75.00	0.00	80.00	0.00
m7	40.00	5.00	85.00	0.00	85.00	0.00
m8	20.00	5.00	90.00	0.00	80.00	0.00
m9	35.00	0.00	85.00	0.00	75.00	0.00
m10	45.00	10.00	85.00	0.00	80.00	0.00
m11	15.00	0.00	70.00	0.00	75.00	0.00
m12	30.00	0.00	80.00	0.00	80.00	0.00
m13	25.00	0.00	85.00	0.00	85.00	0.00
m14	20.00	0.00	90.00	0.00	85.00	0.00

Table 3. TPR and FPR on self-collect data set in calm state (%)

Volunteer no.	10 heartbeats		100 heartbeats		10 heartbeats by Matthew effect	
	TPR	FPR	TPR	FPR	TPR	FPR
1	10.00	0.00	85.00	0.00	80.00	0.00
2	25.00	0.00	70.00	0.00	70.00	0.00
3	50.00	10.00	90.00	0.00	85.00	0.00
4	20.00	0.00	85.00	0.00	75.00	0.00
5	15.00	0.00	75.00	0.00	80.00	0.00
6	30.00	0.00	75.00	0.00	85.00	0.00
7	25.00	0.00	80.00	0.00	65.00	0.00
8	20.00	0.00	75.00	0.00	80.00	0.00
9	15.00	5.00	80.00	0.00	75.00	0.00
10	15.00	0.00	75.00	0.00	85.00	0.00
11	15.00	0.00	80.00	0.00	90.00	0.00
12	25.00	5.00	85.00	0.00	90.00	0.00
13	20.00	0.00	90.00	0.00	80.00	0.00
14	15.00	0.00	80.00	0.00	85.00	0.00
15	25.00	0.00	85.00	0.00	90.00	0.00
16	0.00	0.00	75.00	0.00	65.00	0.00
17	10.00	5.00	85.00	0.00	80.00	0.00

(*continued*)

Table 3. (*continued*)

Volunteer no.	10 heartbeats		100 heartbeats		10 heartbeats by Matthew effect	
	TPR	FPR	TPR	FPR	TPR	FPR
18	10.00	0.00	90.00	0.00	90.00	0.00
19	25.00	5.00	90.00	0.00	90.00	5.00
20	10.00	0.00	85.00	0.00	75.00	0.00
21	20.00	0.00	90.00	0.00	85.00	0.00
22	25.00	0.00	85.00	0.00	85.00	0.00
23	25.00	0.00	95.00	0.00	90.00	0.00
24	15.00	10.00	90.00	0.00	90.00	0.00
25	25.00	0.00	85.00	0.00	85.00	0.00
26	20.00	0.00	90.00	0.00	90.00	0.00
27	30.00	5.00	95.00	0.00	85.00	0.00
28	30.00	0.00	85.00	0.00	90.00	0.00

Table 4. TPR and FPR on self-collect data set in unlimited state (%)

Volunteer no.	10 heartbeats		100 heartbeats		10 heartbeats by Matthew effect	
	TPR	FPR	TPR	FPR	TPR	FPR
u1	10.00	0.00	85.00	0.00	80.00	0.00
u2	25.00	5.00	75.00	0.00	75.00	0.00
u3	35.00	5.00	70.00	0.00	75.00	0.00
u4	20.00	0.00	80.00	0.00	80.00	0.00
u5	15.00	0.00	85.00	0.00	90.00	0.00
u6	30.00	10.00	85.00	0.00	80.00	0.00
u7	25.00	0.00	85.00	0.00	85.00	0.00
u8	20.00	0.00	75.00	0.00	80.00	0.00
u9	15.00	5.00	70.00	0.00	80.00	0.00
u10	15.00	0.00	85.00	0.00	85.00	0.00
u11	15.00	0.00	80.00	0.00	75.00	0.00
u12	25.00	5.00	85.00	0.00	85.00	0.00
u13	30.00	5.00	80.00	0.00	80.00	0.00

6 Conclusion and Future Study

The paper proposed an ECG data superposition matrix strategy for ECG based bio-metric. And on the help of Matthew Effect, 10 heartbeats can construct a useable authentication template, by which ECG based biometric can go forward to real application.

In the future, the further reason of stable feature distribution forming will be studied more clearly, the authentication strategy can be improved by introducing deep learning algorithm, and the unlimited state can also be enlarged broadly, furthermore, ECG data amount will be expected to across ten thousand scales.

Acknowledgement. The paper was supported by TianJin Nature Science Foundation 16JCYBJC15300.

References

1. Ogawa, M., et al.: Fully automated biosignal acquisition system for home health monitoring. In: 1997 Proceedings of the 19th Annual International Conference of the IEEE Engineering in Medicine and Biology Society, vol. 6, pp. 2403–2405. IEEE, USA (1997)
2. Sidek, K.A., Khalil, I., Jelinek, H.F.: ECG biometric with abnormal cardiac conditions in remote monitoring system. IEEE Trans. Syst. Man Cybern. Syst. **44**(11), 1498–1509 (2014)
3. Zhao, Z., Yang, L., Chen, D., et al.: A human ECG identification system based on ensemble empirical mode decomposition. Sensors **13**(5), 6832–6864 (2013)
4. Belgacem, N., Fournier, R., et al.: A novel biometric authentication approach using ECG and EMG signals. J. Med. Eng. Technol. **39**(4), 226–238 (2015)
5. MIT-BIH Arrhythmia Database. http://physionet.org/physiobank/database/mitdb/
6. Sansone, M., Fratini, A., Cesarelli, M., et al.: Influence of QT correction on temporal and amplitude features for human identification via ECG. In: 2013 Biometric Measurements and Systems for Security and Medical Applications, pp. 22–27. IEEE, Naples (2013)
7. Choi, H.S., Lee, B., Yoon, S.: Biometric authentication using noisy electrocardiograms acquired by mobile sensors. IEEE Access **4**, 1266–1273 (2016)
8. Arteaga-Falconi, J.S., Osman, H.A., Saddik, A.E.: ECG authentication for mobile devices. IEEE Trans. Instrum. Meas. **65**(3), 591–600 (2016)
9. Singh, Y.N.: Human recognition using Fisher's discriminant analysis of heartbeat interval features and ECG morphology. Neurocomputing **167**, 322–335 (2015)
10. Islam, M.S., Alajlan, N., Bazi, Y., et al.: HBS: a novel biometric feature based on heartbeat morphology. IEEE Trans. Inf. Technol. Biomed. **16**(3), 445–453 (2012)
11. Kouchaki, S., Dehghani, A., Omranian, S., et al.: ECG-based personal identification using empirical mode decomposition and Hilbert transform. In: 2012 CSI International Symposium on Artificial Intelligence and Signal Processing, pp. 569–573. IEEE, Shiraz (2012)
12. Matos, A.C., et al.: Biometric recognition system using low bandwidth ECG signals. In: 2013 International Conference on E-Health Networking, Applications & Services, pp. 518–522. IEEE, Lisbon (2013)
13. Venugopalan, S., Savvides, M., Griofa, M.O., et al.: Analysis of low-dimensional radio-frequency impedance-based cardio-synchronous waveforms for biometric authentication. IEEE Trans. Biomed. Eng. **61**(8), 2324–2335 (2014)
14. Zeng, F., Tseng, K.K., Huang, H.N., et al.: A new statistical-based algorithm for ECG identification. In: 2012 Eighth International Conference on Intelligent Information Hiding and Multimedia Signal Processing, pp. 301–304. IEEE Computer Society, Piraeus (2012)
15. Hejazi, M., Al-Haddad, S.A.R., et al.: ECG biometric authentication based on non-fiducial approach using kernel methods. Digit. Signal Process. **52**(C), 72–86 (2016)
16. Safie, S.I., Soraghan, J.J., Petropoulakis, L.: Pulse active mean (PAM): a PIN supporting feature extraction algorithm for doubly secure authentication. In: 2011 International Conference on Information Assurance and Security, pp. 210–214. IEEE, Melaka (2011)

17. Coutinho, D.P., Silva, H., Gamboa, H., et al.: Novel fiducial and non-fiducial approaches to electrocardiogram-based biometric systems. IET Biom. **2**(2), 64–75 (2013)
18. Chen, H., Zeng, F., Tseng, K.K., et al.: ECG human identification with statistical support vector machines. In: 2012 International Conference on Computing, Measurement, Control and Sensor Network, pp. 237–240. IEEE Computer Society, Taiyuan (2012)
19. Merone, M., Soda, P., Sansone, M., et al.: ECG databases for biometric systems: a systematic review. Expert Syst. Appl. **67**, 189–202 (2017)
20. Sidek, K., Khali, I.: Biometric sample extraction using Mahalanobis distance in Cardioid based graph using electrocardiogram signals. In: 2012 Engineering in Medicine and Biology Society, pp. 3396–3399. IEEE, San Diego (2012)
21. Wang, J., She, M., Nahavandi, S., et al.: Human identification from ECG signals via sparse representation of local segments. IEEE Signal Process. Lett. **20**(10), 937–940 (2013)
22. Lin, S.L., Chen, C.K., Lin, C.L., et al.: Individual identification based on chaotic electrocardiogram signals during muscular exercise. Biom. IET **3**(4), 257–266 (2014)
23. Sidek, K.A., Mai, V., Khalil, I.: Data mining in mobile ECG based biometric identification. J. Netw. Comput. Appl. **44**(2), 83–91 (2014)
24. Page, A., et al.: Utilizing deep neural nets for an embedded ECG-based biometric authentication system. In: 2015 Biomedical Circuits and Systems Conference, pp. 1–4. IEEE, Atlanta (2015)
25. Venkatesh, N., et al.: Human electrocardiogram for biometrics using DTW and FLDA. In: 2010 International Conference on Pattern Recognition, pp. 3838–3841. IEEE, Istanbul (2010)
26. Yang, L.U., Bao, S., Zhou, X., et al.: Real-time human identification algorithm based on dynamic electrocardiogram signal. J. Comput. Appl. **35**(1), 262–264 (2015)
27. Rabhi, E., Lachiri, Z.: Biometric personal identification system using the ECG signal. In: 2014 Computing in Cardiology Conference, pp. 507–510. IEEE, Zaragoza (2014)
28. Sahadat, M.N., Jacobs, E.L., Morshed, B.I.: Hardware-efficient robust biometric identification from 0.58 second template and 12 features of limb (Lead I) ECG signal using logistic regression classifier. In: Conference Proceedings IEEE Engineering in Medicine and Biology Society, pp. 1440–1443 (2014)
29. Gutta, S., et al.: Joint feature extraction and classifier design for ECG based biometric recognition. IEEE J. Biomed. Health Inf. **20**(2), 460–468 (2016)
30. Hari, S., et al.: Design of a Hamming-distance classifier for ECG biometrics. In: 2013 IEEE International Conference on Acoustics, Speech and Signal Processing, pp. 3009–3012. IEEE, Vancouver (2013)

Medical Image Fusion Using Non-subsampled Shearlet Transform and Improved PCNN

Weiwei Kong[1(✉)] and Jing Ma[2]

[1] Xijing University, Xi'an 710123, China
kongweiwei@xijing.edu.cn
[2] Key Laboratory of Information Assurance Technology, Beijing 100072, China

Abstract. Image fusion is an effective method to increase the accuracy of clinical diagnosis, since it can combine the advantages of a series of diverse medical images. In this paper, a novel image fusion method based on non-subsampled shearlet transform (NSST) and improved pulse coupled neural network (PCNN) is proposed. As an efficient multi-resolution analysis tool, NSST is used to obtain a series of sub-bands with different scales and directions. Then, the traditional PCNN is improved to be a novel model with much less parameters. Certain fusion rules are utilized to complete the fusion process of sub-bands. Finally, the inverse NSST is conducted to obtain the final fused image. Experimental results demonstrate that the proposed method has much better performance than those typical ones.

Keywords: Image fusion · Non-subsampled shearlet transform
Pulse coupled neural network · Medical image

1 Introduction

Human beings are inevitably beset by the disease. For sake of examining and evaluating the patients' conditions as accurately and quickly as possible, a series of medical imaging technologies appears, such as computed tomography (CT), magnetic resonance imaging (MRI), single photon emission computed tomography (SPECT), and so on. However, each imaging technology has its own advantage, but also has inherent shortcomings. For instance, since the X-ray tube and the detector revolve around a certain part of the body, CT is only sensitive to the bones, while MRI is adept to tracing the water molecules of the body. Consequently, in order to make an accurate diagnosis on the lesion, the combination of several imaging technologies may be a good idea. It is noteworthy that higher requirements to the quality of fused image in the area of the clinical diagnosis is proposed compared with other fields.

So far, a variety of multi-modal medical image fusion methods [1–11, 13] have been proposed. On the whole, the mainstreamed methods can be classified into two chief branches, including spatial domain based ones and transform domain based one. From the former point of view, Zhu et al. used the dictionary learning model to complete the fusion process of medical images [1]. Manchanda et al. introduced the fuzzy transform into the area of multimodal medical image fusion [2]. Du et al. proposed a new image fusion rule that uses parallel saliency features and considers

© Springer Nature Switzerland AG 2018
Y. Peng et al. (Eds.): IScIDE 2018, LNCS 11266, pp. 635–645, 2018.
https://doi.org/10.1007/978-3-030-02698-1_55

extracting specific features from different medical images [3]. Reference [4] took the noisy source images into account, and developed a novel medical image fusion, denoising, and enhancement method based on low-rank sparse component composition and dictionary learning. Kavitha et al. [5] published an article called "Medical image fusion based on hybrid intelligence" where the swarm intelligence and neural network are combined to achieve a better fused output. Daniel et al. [6] used gray wolf optimization to deal with the issue of medical image fusion.

On the other hand, transform domain based methods have aroused widespread concern. Bhatnagar et al. proposed a novel framework for spatially registered multi-modal medical image fusion [7]. Xu et al. [8] proposed a new fusion algorithm for multi-modal medical images based on multi-level local extrema representation. Shah-doosti et al. [9] completed the fusion process of the medical images in the tetrolet domain. Du et al. [10] constructed a novel model called union Laplacian pyramid with multiple features to complete the medical image fusion. Liu et al. [11] utilized the non-subsampled shearlet transform (NSST) [12] to fuse the multimodal medical images. James et al. [13] summarized the existing fusion methods of medical images and concluded the development trends.

Compared with the past transform domain models especially the non-subsampled contourlet transform (NSCT) [14] and shearlet transform (ST) [15–17], NSST has not only much better feature capturing performance, but also less computational costs. In this paper, NSST is combined with pulse coupled neural network (PCNN) to fuse the medical images.

The rest of this paper is organized as follows. The related theories are reviewed in Sect. 2 followed by the improved PCNN model in Sect. 3. Experimental results are reported in Sect. 4. In Sect. 5, the concluding remarks are given in the end.

2   Related Work

In this section, we briefly review the important work on NSST and PCNN.

2.1   The Structure of NSST

Let dimension $n = 2$, the affine systems of ST can be expressed as follows.

$$\{\psi_{j,l,k}(x) = |\det A|^{j/2}\psi(S^l A^j x - k) : l, j \in Z, k \in Z^2\} \tag{1}$$

Where ψ is a collection of basis function and satisfies $\psi \in L^2(R^2)$; A denotes the anisotropy matrix for multi-scale partitions, S is a shear matrix for directional analysis. j, l, k are scale, direction and shift parameter respectively. A, S are both 2×2 invertible matrices and $|\det S| = 1$. For each $a > 0$ and $s \in \mathbf{R}$, the matrices of A and S are given as follows.

$$A = \begin{pmatrix} a & 0 \\ 0 & \sqrt{a} \end{pmatrix}, \quad S = \begin{pmatrix} 1 & s \\ 0 & 1 \end{pmatrix} \tag{2}$$

Assume $a = 4$, $s = 1$, Eq. (2) can be modified further.

$$A = \begin{pmatrix} 4 & 0 \\ 0 & 2 \end{pmatrix}, \quad S = \begin{pmatrix} 1 & 1 \\ 0 & 1 \end{pmatrix} \tag{3}$$

For any $\xi = (\xi_1, \xi_2) \in R^2$, $\xi_1 \neq 0$, the mathematical expression of basic function $\hat{\psi}^{(0)}$ for ST can be given

$$\hat{\psi}^{(0)}(\xi) = \hat{\psi}^{(0)}(\xi_1, \xi_2) = \hat{\psi}_1(\xi_1)\hat{\psi}_2(\xi_2/\xi_1) \tag{4}$$

Where ψ is the Fourier transform of ψ. $\psi_1 \in C^\infty(R)$, $\psi_2 \in C^\infty(R)$ are both wavelet, and supp $\psi_1 \subset [-1/2, -1/16] \cup [1/16, 1/2]$, supp $\psi_2 \subset [-1, 1]$. It implies that $\psi_0 \in C^\infty(R)$ and compactly supported with supp $\psi_0 \subset [-1/2, 1/2]^2$. In addition, we assume that

$$\sum_{j \geq 0} |\hat{\psi}_1(2^{-2j}\omega)|^2 = 1, \ |\omega| \geq 1/8 \tag{5}$$

And for each $j \geq 0$, ψ_2 satisfies that

$$\sum_{l=-2^j}^{2^j-1} |\hat{\psi}_2(2^j\omega - l)| = 1, \ |\omega| \leq 1 \tag{6}$$

From the conditions on the support of ψ_1, ψ_2 one can obtain that the function $\psi_{j,l,k}$ has the frequency support listed below:

$$\text{supp } \hat{\psi}^0_{j,l,k} \subset \{(\xi_1,\xi_2)|\xi_1 \in [-2^{2j-1}, -2^{2j-4}] \cup [2^{2j-4}, 2^{2j-1}], |\xi_2/\xi_1 + l2^{-j}| \leq 2^{-j}\} \tag{7}$$

That is, each element $\psi_{j,l,k}$ is supported on a pair of trapeziform zones, whose sizes all approximate to $2^{2j} \times 2^j$. The tiling of the frequency by shearlet and the size of the frequency support of $\psi_{j,l,k}$ are illustrated in Fig. 1. Note that Fig. 1b only shows the frequency support for $\xi_1 > 0$; the figure of the other support for $\xi_1 < 0$ is symmetrical.

NSST combines the non-subsampled Laplacian pyramid (NSLP) transform with several different combinations of the shearing filters (SF). It is commonly acknowledged that NSST is the shift-invariant version of ST essentially. In order to eliminate the courses of up-sampling and sub-sampling, NSST utilizes NSLP filters as a substitute for the Laplacian pyramid filters used in the ST mechanism, so that it has superior performance in terms of shift-invariance, multi-scale and multi-directional properties. The discretization process of NSST is composed of two phases including multi-scale factorization and multi-directional factorization. NSLP is utilized to complete multi-scale factorization. The first phase ensures the multi-scale property by using two-channel non-subsampled filter bank, and one low frequency image and one high frequency image can be produced at each NSLP decomposition level. The subsequent

NSLP decompositions are implemented to decompose the low frequency component available iteratively to capture the singularities in the image. The multi-directional factorization in NSST is realized via improved SF. These filters are formed by avoiding the sub-sampling to satisfy the property of shift-invariance. SF allows the direction decomposition with l stages in high frequency images from NSLP at each level and produces 2^l directional sub-images with the same size as the source image. Figure 2

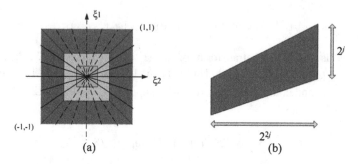

(a)   (b)

Fig. 1. The structure of the frequency tiling by the shearlet: (a) The tiling of the frequency plane R^2 induced by the shearlet. (b) The size of the frequency support of a shearlet $\psi_{j,l,k}$

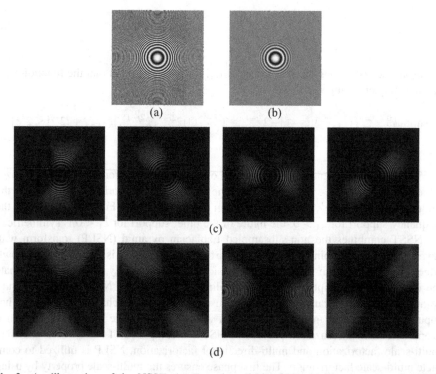

Fig. 2. An illustration of the NSST: (a) Zoneplate (256 * 256). (b) The approximate NSST coefficients. (c) Images of the detailed coefficients at level 1. (d) Images of the detailed coefficients at level 2.

illustrates the two-level NSST decomposition of an image. The number of shearing directions is chosen to be 4 and 4 from finer to coarser scale.

2.2 Traditional PCNN

As the famous third generation of artificial neural network, PCNN is a model based on the cat's primary visual cortex which is formed by the connection of lots of neurons. A pulse coupled neuron commonly denoted by N_{ij} is composed of three units: the receptive field, the modulation field and the pulse generator. Figure 3 shows the structure of a basic pulse coupled neuron, whose corresponding discrete mathematical expressions can be described as follows:

$$F_{ij}[n] = \exp(-\alpha_F)F_{ij}[n-1] + V_F \sum_{kl} M_{ijkl}Y_{kl}[n-1] + I_{ij} \tag{8}$$

$$L_{ij}[n] = \exp(-\alpha_L)L_{ij}[n-1] + V_L \sum_{kl} W_{ijkl}Y_{ij}[n-1] \tag{9}$$

$$U_{ij}[n] = F_{ij}[n](1 + \beta L_{ij}[n]) \tag{10}$$

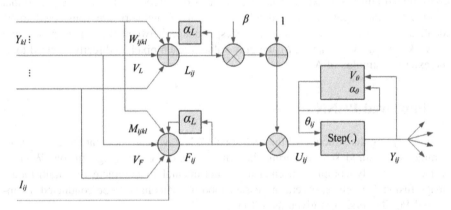

Fig. 3. The basic model of PCNN neuron

$$Y_{ij}[n] = \begin{cases} 1, & if \ U_{ij}[n] \geq \theta_{ij}[n] \\ 0, & else \end{cases} \tag{11}$$

$$\theta_{ij}[n] = \exp(-\alpha_\theta)\theta_{ij}[n-1] + V_\theta Y_{ij}[n] \tag{12}$$

As shown in Fig. 3, N_{ij} receives input signals via other neurons and external sources by two channels in the receptive field. One channel is the feeding input F_{ij} and the other one is the linking input L_{ij}, both of which correspond to Eqs. (8) and (9), respectively. U_{ij} combines the information of the above two channels in a second order mode to form the total internal activity, as shown in Eq. (10). Equation (11) indicates that U_{ij} is then compared with the dynamic threshold θ_{ij} to decide the value of the output Y_{ij}. If U_{ij} is larger than θ_{ij}, then the neuron N_{ij} will be activated and generate a pulse, which is characterized by $Y_{ij} = 1$, else $Y_{ij} = 0$. According to Eq. (12), the dynamic threshold θ_{ij} will decline with the iterative number n increasing. However, if $Y_{ij} = 1$, θ_{ij} will immediately raise with the function of V_θ whose value is relatively large, so that the behavior of firing of N_{ij} will stop at once and $Y_{ij} = 0$. Later, if θ_{ij} reduces to be equal to or less than U_{ij}, N_{ij} will fire again and make an impulse sequence. On the other hand, the relations between N_{ij} and its surrounding neurons exist, therefore, if N_{ij} is activated, those neurons having similar gray values around it may also be activated at the next iteration. The result is an auto wave expanding from an active neuron to the whole region.

Apart from the parameters mentioned above, there are still nine ones required explaining. I_{ij} is the input signal; β is the linking strength; V_F and V_L are the magnitude scaling terms. α_F, α_L and α_θ are the time decayed constants associated with F_{ij}, L_{ij} and θ_{ij}, respectively. M_{ijkl} and W_{ijkl} are both the linking matrices, which respectively correspond to F channel and L channel.

The PCNN used for image fusion is a single layer two-dimensional array of laterally linked pulse coupled neurons. The number of neurons in PCNN is equal to that of pixels in the input image, and all neurons in the network are considered to be identical. There exists a one-to-one correspondence between the image pixels and network neurons. Commonly, I_{ij}, the gray value of each pixel, is directly referred to as the external stimulus of N_{ij}.

3   Improved PCNN

After analyzing the structure of the traditional PCNN, we can find that there are lots of parameters required setting during the whole course of the image fusion. What is worse, the overly complex mechanism is detrimental to complete the multi-modal image fusion. Therefore, several necessary modifications have to be conducted. A improved PCNN version is given as follows.

(a) Since the previous feeding input has an weak effect on the feeding input this time, it is desirable to ignore $F_{ij}[n-1]$. Similarly, $L_{ij}[n-1]$ should also be omitted. For simplicity, the magnitude scaling terms including V_F and V_L are both set to be 1. As a result, Eqs. (8) and (9) can be written as:

$$F_{ij}[n] = \sum_{kl} M_{ijkl} Y_{kl}[n-1] + I_{ij} \tag{13}$$

$$L_{ij}[n] = \sum_{kl} W_{ijkl} Y_{ij}[n-1] \tag{14}$$

(b) Firstly, W_{ijkl} is the linking matrix which directly decides the participation extent of the surrounding neurons in a square area centered in another certain neuron. Then, it brings the output information of the surrounding neurons into the central one to decide the value of the internal activity for further comparison between U_{ij} and the dynamic threshold E_{ij}. According to the difference of the distance between each surrounding pixel and the central pixel, suppose that the size of the square area is 3×3 and the radius is 1, the distance between the central pixel with the corner one is $2^{1/2}$, for this reason, W_{ijkl} can be set as Eq. (15) after normalization. Of course, our matrix setting does not invalidate other setting measures absolutely. But since the influence of the closer pixel is larger to some extent than the distant pixel, absorbing the distance information into the matrix setting is more direct and more convenient.

$$W_{ijkl} = \begin{bmatrix} 0.1035 & 0.1465 & 0.1035 \\ 0.1465 & 0 & 0.1465 \\ 0.1035 & 0.1465 & 0.1035 \end{bmatrix} \tag{15}$$

Since the linking matrices W_{ijkl} and M_{ijkl} have the same function, W_{ijkl} is also set as Eq. (15).

(c) U_{ij} combines the information of the both feeding input and linking input in a second order mode to form the total internal activity. β is the linking strength and reflects the influential extent between the surrounding neurons and the central one. The larger the value of β is, the more obvious the extent of the contribution to the central neuron from the surrounding ones is. However, in most applications of the PCNN in the field of image processing, β is only set to be a constant. In other words, we consider the influential extent of the linking input to be the same regardless of the source images for the most part. In fact, in human vision, the responses to a region with notable features are stronger than to a region with non-notable ones. Therefore, the β of each neuron in the PCNN should not be the same but be related to the features of the corresponding pixels of the images. As a result, it is unreasonable to set the values of all neurons to be a constant.

After considering many evaluation metrics concerning the final fusion result, we decide to substitute the index of local visual contrast [18] for β, whose mathematical expressions are given as follows [18].

$$LVC(x, y) = \begin{cases} \left(\frac{1}{I'(x,y)}\right) \times \frac{SML(x,y)}{I'(x,y)}, & \text{if } I'(x, y) \neq 0 \\ SML(x, y), & \text{else} \end{cases} \tag{16}$$

where α is a constant ranging from 0.6 to 0.7, $I'(x, y)$ is the mean intensity value of the pixel (x, y) centered in the neighborhood window. $SML(x, y)$ denotes the Sum Modified Laplacian (SML) located at (x, y). The definition of SML is as follows:

$$SML(i, j) = \sum_{p=-P}^{P} \sum_{q=-Q}^{Q} [ML(i+p, j+q)] \tag{17}$$

where

$$ML(i, j) = |2I(i, j) - I(i - step, j) - I(i + step, j)|$$
$$+ |2I(i, j) - I(i, j - step) - I(i, j + step)| \quad (18)$$

where P and Q denote the size of the window which is $(2P + 1) \times (2Q + 1)$, step is the variable spacing between the coefficients and always equals 1. $I(x, y)$ is the pixel value of one coefficient located at (x, y).

4   Experimental Results

In order to verify the effectiveness of the proposed method, two pairs of brain source images are used which are shown in Fig. 4. It is noteworthy that all source images share the same size of 256×256 pixels and cover 256-level gray scale. All source images can be downloaded from the Harvard university site. Three recently published methods are compared with the proposed one to verify the effectiveness, which are based on adaptive sparse representation (ASR) [19], quantum-behaved particle swarm optimization PCNN (QPSO-PCNN) [20] and non-subsampled rotated complex wavelet transform (NSRCxWT) [21]. The decomposition stages of NSST is set as 4, and the direction number from coarser to finer scales is (4, 8, 8, 16). Apart from the subjective visual performance, objective evaluation is also necessary for providing a much more intuitive and convincing statistics. Liu et al. [22] did in-depth researches on a series of current efficient quality metrics, and proposed a detailed comparison among them.

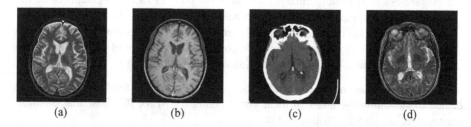

Fig. 4. Eight pairs of human brain images. Pair I: (a)(b). Pair II: (c)(d).

Here, considering the characteristics of multimodal medical images, three metrics mentioned in reference [22] are chosen to conduct the objective evaluation on the final fused results, including spatial frequency (Q_{SF}) [23], mutual information (Q_{MI}) [24], and phase congruency (Q_P) [25].

The fused images based on four methods including the proposed one are shown in Fig. 5. As can be observed from the final fused results, the fused image based on the proposed method has much better visual performance. In addition, the quantitative comparison is also conducted in this section in Table 1. The values in bold denote the best results among the methods. Overall, the proposed method outperform other three ones.

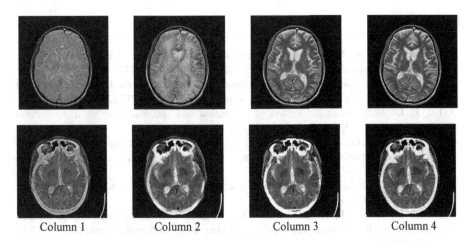

Column 1   Column 2   Column 3   Column 4

Fig. 5. Fused images based on four different methods. Column 1 [19], Column 2 [20], Column 3 [21] and Column 4 (Proposed method)

Table 1. Objective performance of the four methods

		ASR	QPSO-PCNN	NSRCxWT	Proposed
Pair I	Q_{SF}	34.8118	36.0681	42.4388	**43.7656**
	Q_{MI}	0.7083	0.7799	**1.1378**	1.1187
	Q_P	0.2387	0.3911	0.6577	**0.7012**
Pair II	Q_{SF}	40.8550	44.3366	49.7400	**50.5428**
	Q_{MI}	0.6974	0.9054	1.0025	**1.0315**
	Q_P	0.4107	0.5017	0.4138	**0.5642**

5   Conclusion

In this paper, a novel image fusion method based on NSST and improved PCNN is proposed. A series of simulation experiments are conducted to indicate that the proposed method has remarked superiorities over other current typical ones in terms of both visual performance and objective evaluation results. How to further enhance and optimize the fusion performance are the emphasis during our other work.

Acknowledgments. The authors thank all the reviewers and editors for their valuable comments and works. The work was supported in part by Foundation of Science and Technology on Information Assurance Laboratory under Grant KJ-17-105, in part by the Natural Science Foundation of Shannxi Provincial Department of Education under Grant 16JK2246, and the Foundation of Xijing University under Grant XJ16T03. I declare that the author has no conflicts of interest to this work.

References

1. Zhu, Z.Q., Chai, Y., Yin, H.P., Li, Y.X., Liu, Z.D.: A novel dictionary learning approach for multi-modality medical image fusion. Neurocomputing **214**(11), 471–482 (2016)
2. Manchanda, M., Sharma, R.: A novel method of multimodal medical image fusion using fuzzy transform. J. Vis. Commun. Image Represent. **40**(1), 197–217 (2016)
3. Du, J., Li, W.S., Xiao, B.: Fusion of anatomical and functional images using parallel saliency features. Inf. Sci. **S**(1), 567–576 (2018)
4. Li, H.F., He, X.G., Tao, D.P., Tang, Y.Y., Wang, R.X.: Joint medical image fusion, denoising and enhancement via discriminative low-rank sparse dictionaries learning. Pattern Recogn. **79**(1), 130–146 (2018)
5. Kavitha, C.T., Chellamuthu, C.: Medical image fusion based on hybrid intelligence. Appl. Soft Comput. **20**(7), 83–94 (2014)
6. Daniel, E., Anitha, J., Kamaleshwaran, K.K., Rani, I.: Optimum spectrum mask based medical image fusion using gray wolf optimization. Biomed. Signal Process. Control **34**(1), 36–43 (2017)
7. Bhatnagar, G., Wu, Q.M.J., Liu, Z.: A new contrast based multimodal medical image fusion framework. Neurocomputing **157**(1), 143–152 (2015)
8. Xu, Z.P.: Medical image fusion using multi-level local extrema. Inf. Fusion **19**(1), 38–48 (2014)
9. Shahdoosti, H.R., Mehrabi, A.: Multimodal image fusion using sparse representation classification in tetrolet domain. Digit. Signal Process. **S**(1), 1–14 (2018)
10. Du, J., Li, W.S., Xiao, B., Nawaz, Q.: Union Laplacian pyramid with multiple features for medical image fusion. Neurocomputing **194**(6), 326–339 (2016)
11. Liu, X.B., Mei, W.B., Du, H.Q.: Multi-modality medical image fusion based on image decomposition framework and nonsubsampled shearlet transform. Biomed. Signal Process. Control **40**(1), 343–350 (2018)
12. Easley, G., Labate, D., Lim, W.Q.: Sparse directional image representation using the discrete shearlet transforms. Appl. Comput. Harmonic Anal. **25**(1), 25–46 (2008)
13. James, A.P., Dasarathy, B.V.: Medical image fusion: a survey of the state of the art. Inf. Fusion **19**(1), 4–19 (2014)
14. da Cunha, A.L., Zhou, J., Do, M.N.: The nonsubsampled contourlet transform: theory, design, and applications. IEEE Trans. Image Process. **15**(10), 3089–3101 (2006)
15. James, M.M., Jacqueline, L.M., David, J.H.: Automatic image registration of multimodal remotely sensed data with global shearlet features. IEEE Trans. Geosci. Remote Sens. **54**(3), 1685–1704 (2016)
16. Azam, K., Rob, H., Paul, S.: Band-specific shearlet-based hyperspectral image noise reduction. IEEE Trans. Geosci. Remote Sens. **53**(9), 5054–5066 (2015)
17. Sun, G.M., Leng, J.S., Huang, T.Z.: An efficient sparse optimization algorithm for weighted shearlet-based method for image deblurring. IEEE Access **5**, 3085–3094 (2017)
18. Yang, Y., Yang, M., Huang, S., Que, Y., Ding, M., Sun, J.: Multifocus image fusion based on extreme learning machine and human visual system. IEEE Access **5**, 6989–7000 (2017)
19. Liu, Y., Wang, Z.F.: Simultaneous image fusion and denoising with adaptive sparse representation. IET Image Process. **9**(5), 347–357 (2015)
20. Xu, X.Z., Shan, D., Wang, G.Y., Jiang, X.Y.: Multimodal medical image fusion using PCNN optimized by the QPSO algorithm. Appl. Soft Comput. **46**(1), 588–595 (2016)
21. Chavana, S.S., Mahajan, A., Talbar, S.N., Desai, S., Thakur, M., Cruz, A.D.: Nonsubsampled rotated complex wavelet transform (NSRCxWT) for medical image fusion related to clinical aspects in neurocysticercosis. Comput. Biol. Med. **81**(2), 64–78 (2017)

22. Liu, Z., Blasch, E., Xue, Z.Y., Zhao, J.Y., Laganiere, R., Wu, W.: Fusion algorithms for context enhancement in night vision: a comparative study. IEEE Trans. Pattern Anal. Mach. Intell. **34**(1), 94–109 (2012)

23. Zheng, Y., Essock, E.A., Hansen, B.C., Haun, A.M.: A new metric based on extended spatial frequency and its application to DWT based fusion algorithms. Inf. Fusion **8**(2), 177–192 (2007)

24. Hossny, M., Nahavandi, S., Creighton, D.: Comments on 'information measure for performance of image fusion. IET Electron. Lett. **44**(18), 1066–1067 (2008)

25. Zhao, J., Laganiere, R., Liu, Z.: Performance assessment of combinative pixel-level image fusion based on an absolute feature measurement. Int. J. Innov. Comput. Inf. Control **3**(6), 1433–1447 (2007)

Dissociating Group and Individual Profile of Functional Connectivity Using Low Rank Matrix Recovery

Jian Qin, Hui Shen, LingLi Zeng, Kai Gao, and Dewen Hu[✉]

College of Artificial Intelligence, National University of Defense Technology,
410073 Changsha, Hunan, China
dwhu@nudt.edu.cn

Abstract. Brain connectivity network consists of general substrate and specific traits, yet their characteristic and relationships were still unknown. Here, we systematically investigate the substrate and traits of functional connectivity (FC) network. We calculated the resting-state functional magnetic resonance imaging-based FC using data from the Human Connectome Project. Subjects' FC was decomposed into general substrate and specific traits via a novel low rank matrix recovery method. Then we investigated the relationships between FC traits and the cognitive behaviors. We found that FC traits were significantly associated with the cognitive behaviors. Our findings suggest that individual differences in FC traits could mainly account for inter-subject variability of the cognition and behaviors. This could advance our understanding of substrate and traits of brain function.

Keywords: Connectome · Individual differences · Low-rank matrix recovery
Resting-state fMRI · Sparse representation

1 Introduction

Understanding brain functions is an important goal for studies of functional magnetic resonance imaging (fMRI) [1, 2]. Resting-state functional connectivity (RSFC) was a powerful tool used to evaluate the functional brain organization [3], such as functional connectivity (FC) network. Previous RSFC studies mainly focus on group general features by averaging individual specific features across many subjects, such as group-level RSFC network. Yet many individual specific features that may reflect the detailed aspect of brain system do not present in group general features [4, 5]. Recent studies suggested that FC networks exhibit striking inter-subject variability which may account for great individual differences in human cognition and behaviors [6–8]. More importantly, FC traits were demonstrated to be unique and reliable for a single individual, serving similarly as a fingerprint [9, 10].

Combining the studies about group general features and individual differences, it is reasonable to recognize that individual FC networks contain both general substrate and specific traits, which represent the fundamental and varying aspects of functional networks, respectively [4]. In contrast to the large amount of works only assessing general FC substrate or specific FC traits, few works have investigated their relationship and what

© Springer Nature Switzerland AG 2018
Y. Peng et al. (Eds.): IScIDE 2018, LNCS 11266, pp. 646–654, 2018.
https://doi.org/10.1007/978-3-030-02698-1_56

the roles they play in functional networks. Understanding how FC substrate and traits shape functional networks may add our knowledge of brain generality and characteristic.

Individual FC serves as subject's inherent brain features. Its relationship to the cognition and behaviors were extensively investigated [9]. Previous studies suggested that individual FC was associated with many cognitive behaviors, such as learning, emotion, sensor and motor. Yet previous works were based on the FC networks containing both general substrate and specific traits. We would like to investigate whether individual FC traits could mainly represent the cognitive behaviors when excluding FC substrate.

Here, we hypothesis that specific FC traits mainly represent the characteristic of the cognitive behaviors. We calculated the FC based on the fMRI data of 477 subjects from Human Connectome Project (HCP) [11–14]. Then, we decomposed subjects' FC networks into general substrate and specific traits via a low rank matrix recovery (LRMR) method [15, 16]. Further, we investigated whether FC traits could represent fluid intelligence when excluded FC substrate via the sparse representation [17] of FC traits and support vector regression (SVR) [18]. Notably, the sparse representation of FC traits was performed with two hypotheses that individual differences are multiple dimensions (or multiple kinds) and that one dimension of individual differences relates to sparse connections and nodes of brain.

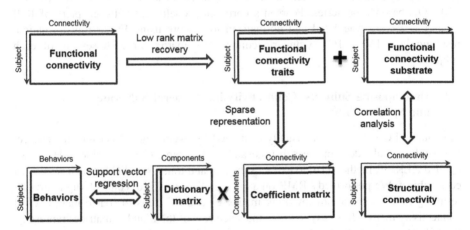

Fig. 1. Identification of analysis procedure. Given the subjects' FC matrix, we firstly decomposed it into the general substrate and specific traits via a low rank matrix recovery method. For each subject, we then calculated the correlations of SC to FC substrate and traits, and original FC. At the meanwhile, we calculated the sparse representation of FC traits X, resulting in a dictionary matrix D (individual differences patterns) and a sparse coefficient weight matrix α, i. e, X_i approximately modeled d_i x α_i ($i = 1, 2, \ldots k$). The d_i in ith column of D represents a dimension of individual differences patterns, while the α_i in ith row of α corresponding to d_i could be mapping back to whole brain connectivity networks. Then we investigated the relationships between individual differences patterns (dictionary matrix D) and the cognitive behaviors. For each cognitive behavior, the correlation coefficient between it and individual difference patterns in each dimension (each d_i) were calculated. The significant individual difference patterns ($P < 0.05$, FDR corrected) were selected as the features and regressed to the corresponding behavior via Support Vector Regression.

2 Materials and Methods

2.1 Data Acquisition and Preprocessing

We adopt the high-quality resting-state fMRI data from the HCP (Q1, Q2, Q3 and S500 releases) in this study [12, 13]. The full Q1, Q2, Q3 and S500 releases totally contain data on 520 subjects; we restricted our analysis to subjects for whom two sessions of fMRI data are available (n = 472, 195 males, age 22–35). The acquisition parameters of fMRI data are as follows: 90 × 104 matrix, 220 mm FOV, 72 slices, TR = 0.72 s, TE = 33.1 ms, flip angle = 52°, BW = 2290 Hz/Px, in-plane FOV = 208 × 180 mm, 2.0 mm isotropic voxels [11].

Preprocessing fMRI data. We adopt the publicly released resting-state fMRI data after the minimal preprocessing pipelines which are especially defined for high spatial and temporal resolution of HCP datasets [11]. Based on minimal preprocessing data, we then performed spatial smoothing using Gaussian kernel with 6 mm full width at half maximum and temporal band-pass filtering from 0.01 to 0.08 Hz. To further reduce signal noise, the fMRI time series were regressed out head motion parameters and the white matter, cerebrospinal fluid (CSF), and the whole-brain signals.

Calculating FC. We apply the 160 regions of interest (ROIs) defined by Dosenbach et al. [21]. By meaning all voxels signal within the ROI (radius = 6 mm), we extracted 160 ROIs-based time series. Pearson's correlation coefficient between pairs of ROI time course were calculated and normalized to z score using Fisher transformation, resulting in a 160 × 160 symmetric connectivity matrix for each session of each subject.

2.2 Decomposing Subjects' Connectivity into General Substrate and Specific Traits

Previously, FC substrate was usually estimated by averaging FC across subjects, yet few studies could decompose substrate and traits from FC for a subject. Here, we applied a LRMR method [16] to extract substrate and traits from FC for each session of each subject. We performed LRMR on a m × n (472 × 12720) connectivity matrix A, resulting in FC substrate (a low rank matrix L) and FC traits (a sparse matrix X) of connectivity matrix. It is easy to know that each row of the L and X matrices represents the FC substrate and traits of a subject, respectively.

More details in calculating FC substrate and traits are as follow. We used an approach called Principal Component Pursuit (PCP) to perfectly recover the FC substrate and traits [16]. The connectivity matrix A could be represented as:

$$A = L + X \tag{1}$$

Where L is a low rank matrix (FC substrate) and X is a sparse matrix (FC traits) with a small fraction of nonzero entries. The straightforward formulation is to use l0-norm to minimize the energy function:

$$\min_{L,S} \; rank(L) + \gamma \|X\|_0 \quad \text{s.t } A - L - X = 0 \tag{2}$$

Where $\gamma > 0$ is an arbitrary balanced parameter. But since this problem is NP-hard, a typical solution would involve a search with combinatorial complexity. Cand, S [16] seek to solve for L with the following optimization problem:

$$\min_{L,S} \|L\|_* + \lambda \|X\|_1 \quad \text{s.t } A - L - X = 0 \tag{3}$$

Where $\|.\|_*$ and $\|.\|_1$ are the nuclear norm (which is the l1-norm of singular value) and l1-norm, respectively, and $\gamma > 0$ is an arbitrary balanced parameter. Usually, $\gamma = \frac{1}{\sqrt{\max(m,n)}}$. Under these minimal assumptions, this PCP solution perfectly recovers FC substrate and traits.

2.3 Sparse Representation of Specific Traits of Functional Connectivity

The sparse representation was widely used in machine learning field [19] and recently introduced into fMRI researches [20]. Its hypothesis only requests the identified patterns are sparse, which is more reasonable than that of other pattern recognition methods, such as independent component analysis. Based on the sparse hypotheses, we applied effective dictionary learning and sparse representation to FC traits X. We calculated a dictionary D and sparse coefficient weight matrix α as follow. For sparse representation of connectivity matrix X, we aimed to learn an effective over-complete dictionary D which satisfies the constraint that $K > m$ and $K \ll n$ [17]. Specially, the empirical cost function $f_n(D)$ of X considering the average loss of regression to all n connectivity vector using D is

$$f_n(D) = \frac{1}{n} \sum_{i=1}^{n} l(x_i, D) = \frac{1}{n} \sum_{i=1}^{n} \min_{\alpha \in R^k} \frac{1}{2} |x_i - D\alpha_i|_2^2 + \lambda |\alpha_i|_1 \tag{4}$$

Where the loss function $l(D)$ is defined as the optimal value of sparse representation:

$$l(x, D) = \min_{\alpha \in R^k} \frac{1}{2} |x - D\alpha|_2^2 + \lambda |\alpha|_1. \tag{5}$$

Note that the value of $l(D)$ should be small if X is reasonably well sparse represented by D. The l_1 regularization makes sure the sparsity of the resolution α. λ is a regularization parameter between regression residual and sparsity level. Moreover, we have prevented the elements of D from being arbitrarily large using the constraint:

$$C = \{\{D \in R^{m \times k} \; s.t. \; \forall i = 1, \ldots k, \; d_i^T d_i \leq 1\}\} \tag{6}$$

So the problem of minimizing Eq. (4) is rewritten as a matrix factorization problem:

$$\min_{D \in C,\, \alpha \in R^{kxn}} \frac{1}{2} |X - D\alpha|_F^2 + \lambda |\alpha|_{1,1} \tag{7}$$

Dictionary matrix D was calculated via an effective online dictionary learning algorithm performed using a publicly released online dictionary learning toolbox [17]. Once we gained the matrix D, the sparse coefficient weight matrix α could be learning via solving a linear least-square regression problem.

2.4 The Correlations Between the Sparse Traits of Functional Connectivity and the Cognitive Behaviors

The FC traits represent the varying aspect of FC and are proper to account for the cognitive behaviors. Applying sparse representation on FC traits, we gained individual difference patterns in multiple dimensions (dictionary matrix D) and the corresponding connectivity coefficient (sparse coefficient weight matrix α). Then, we calculated the correlation coefficient between fluid intelligence and individual difference patterns in each dimension (each d_i). The significant individual difference patterns ($P < 0.05$, FDR corrected) were selected as the features and regressed to the corresponding behavior via SVR. We performed this algorithm using a publicly released toolbox svm-gun provided by Gunn [18]. Specially, each cognitive behavior was represented by the linear combination of significant individual difference patterns. To avoid over fitting, we apply the LOOCV during feature selecting and regression. The commonly selected individual difference patterns for each loop of LOOCV were regarded as behavior-related patterns D^*. Then its corresponding connectivity patterns was reconstruction by multiply these behavior-related patterns to the corresponding sparse coefficient weight matrix α^*, i.e., connectivity patterns $= D^* \times \alpha^*$.

3   Results

3.1 Intrinsic Functional Connectivity Consists of General Substrate and Specific Traits

Applying LRMR algorithm, we decomposed the FC matrix into substrate and traits for each session of each subject. The similarity in original FC matrices and FC traits matrices were quantified to take a comparison. We found that FC substrate shows stronger similarity both across subjects and sessions ($R = 0.88, 0.89$) than original FC matrices ($R = 0.44, 0.70$). FC traits show extensively differences across subjects, yet its similarity across sessions still remains strong. This demonstrates that FC substrate and traits decomposed from original FC matrices well captured the fundamental and varying aspects of functional brain networks.

3.2 Inter-subject Variability of Specific Traits of Functional Connectivity

To investigating the network-level distributions of inter-subject differences, we calculated the mean of inter-subject differences in FC traits within and between networks

(Shown in Fig. 2A). Moreover, the mean of inter-subject differences in FC traits associated with each network were calculated (shown in Fig. 2B). In general, the connections within FPN show largest individual differences in FC traits, followed by the connections between FPN and DMN, and the connections within DMN. In contrast, cerebellum shows smallest individual differences in FC traits. The connections related to CON, OCC and SMN also exhibit small individual differences in FC traits. These results were consistent with previous studies of individual differences [6–8], which demonstrating the validity of our methods in experimental aspect.

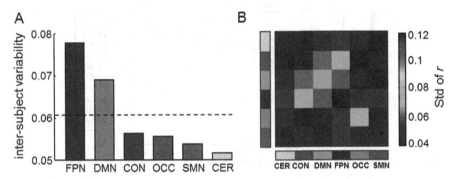

Fig. 2. Inter-subject variability of functional network. (A) Inter-subject variability of FC traits quantified between and within networks. (B) Inter-subject variability of FC traits quantified across brain networks. The dotted line indicates the global mean of inter-subject variability of FC traits. The blue and cyan spheres shown in brain maps represent the nodes affiliated to FPN and DMN, respectively. CER, cerebellum; CON, cingulo-opercular network; DMN, default mode network; FPN, fronto-parietal network; OCC, occipital network; SMN, sensorimotor network. (Color figure online)

3.3   The Sparse Patterns of FC Traits Significantly Correlate to Fluid Intelligence

Based on sparse hypotheses, subjects' FC traits were factorized into a dictionary matrix D and a sparse weight coefficient matrix α via the sparse representation. We selected the regularization parameter λ and dictionary size K via the experimental results based on the criterion of correlations between individual-different patterns in dictionary D and the cognitive behaviors (Shown in Fig. 3; Parameters setting: $\lambda = 0.1$, $K = 1500$). By regressing behavior-correlated patterns to fluid intelligence using SVR, we found that the FC traits were significantly correlated to fluid intelligence (The LOOCV results are shown in Fig. 4). In contrast, the FC substrate was not correlated to the cognitive behaviors, i.e., no behavior-related features of FC substrate were selected during feature selections. This reflects that FC traits represent the functional characteristic underlying the cognitive behaviors.

Further, we sought to identify the FC patterns underlying fluid intelligence. By multiplying the behavior-related patterns to the corresponding sparse coefficient matrix, we can obtain the connectivity patterns underlying the cognitive behavior for all the subjects. For fluid intelligence, we found that FPN and DMN exhibited largest weights in connectivity patterns associated with fluid intelligence. This indicates that FPN and DMN play essential roles in functional mechanism underlying human fluid intelligence.

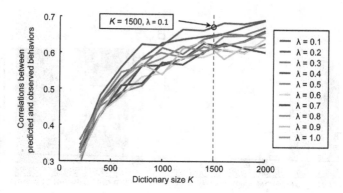

Fig. 3. Selecting the parameters of sparse representation. We repeated the analysis procedure shown in Fig. 1 with the dictionary size K from 250 to 2000 and the regularization parameter λ from 0.1 to 1. The influences of dictionary size K and the regularization parameter λ on the correlation between estimated and observed behaviour were exhibited.

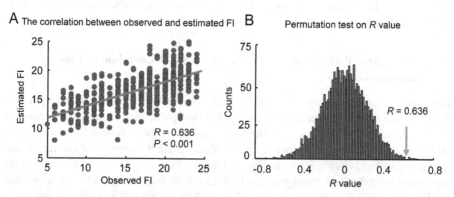

Fig. 4. FC traits are significantly associated with the cognitive behaviour. (A) The correlations between predicted and observed fluid intelligence (FI). (B) The results of 1000 times permutation test. The P value of permutation test is under 0.05. Given that the individual difference units resulted from the sparse representation of FC traits, we calculated the estimated behaviour using Support Vector Regression and leave one out cross validation methods. There is a significant and reliable correlation between estimated behaviour based on individual connectivity traits and observed behaviour.

4 Discussion

Applying LRMR methods, we decomposed FC matrices into general substrate and specific traits for each session of each subject. FC substrate was very consistent across both sessions and subjects ($R = 0.89$ and 0.88, $P < 0.0001$). This similarity was higher than that of original FC matrices ($R = 0.44$ and 0.70, $P < 0.0001$). This demonstrated that FC substrate and traits well captured the fundamental and varying aspects of functional brain networks. Then we investigated the relationships between FC traits and the fluid intelligence. We found that FC traits were significantly associated with the cognitive behaviors. Our findings would enhance our understanding of fundamental and varying aspects of functional brain networks and the cognitive behaviors.

Previously, group-level FC was generally calculated by averaging the FC across many subjects [5]. This is based on a potential hypothesis that subjects' FC contains both group general features and individual specific features [4], which were denoted as FC substrate and traits in this study. Yet few studies focus on the characteristic and relationship of FC substrate and traits. Here, we extracted substrate and traits from each subject's FC via the LRMR method [16]. FC substrate was very consistent across both scan sessions and subjects. In contrast, FC traits exhibited less inter-subject similarity, but remained consistent across scan sessions of the same subject. Once FC substrate and traits were decomposed from original FC of each subject, it is convenient to explore the fundamental and varying aspects of functional networks.

Exploring the functional brain mechanisms underlying the cognitive behaviors is an essential goal for neuroscience researches. It has been suggested that individual FC were associated with cognitive behaviors, such as fluid intelligence [9], emotion, motor and sensor. Yet the subjects' functional networks contained both general and specific features. Here we decomposed functional networks into the substrate and traits via the LRMR method. We found that FC traits were associated with the fluid intelligence, but FC substrate was not. This indicated that FC traits could mainly represent the functional characteristic underlying the cognitive behaviors. Moreover, we reconstructed the connectivity patterns associated with fluid intelligence. FPN and DMN exhibited huge weight behavior related connectivity patterns, reflecting that FPN and DMN play essential roles in human fluid intelligence.

5 Conclusion

We introduced the LRMR approach to decompose subjects' functional networks into the general substrate and specific traits. FC traits were demonstrated to mainly represent the functional characteristic underlying the cognitive behaviors. Our findings indicate that FC substrate shapes a similar functional brain frame reflected in and FC traits could mainly account for inter-subject variability of the cognition and behaviors. This could advance our understanding of fundamental and varying aspects of brain function.

Acknowledgments. This work was supported by the National Science Foundation of China (61420106001 91420302 and 61773391).

References

1. Poldrack, R.A., Farah, M.J.: Progress and challenges in probing the human brain. Nature **526** (7573), 371–379 (2015)
2. Petersen, S.E., Sporns, O.: Brain networks and cognitive architectures. Neuron **88**(1), 207–219 (2015)
3. Van Dijk, K.R., Hedden, T., Venkataraman, A.: Intrinsic functional connectivity as a tool for human connectomics: theory, properties, and optimization. J. Neurophysiol. **103**(1), 297–321 (2010)
4. Gordon, E.M., Laumann, T.O., Adeyemo, B.: Individual-specific features of brain systems identified with resting state functional correlations. Neuroimage **146**, 918–939 (2017)
5. Dubois, J., Adolphs, R.: Building a science of individual differences from fMRI. Trends Cogn. Sci. **20**(6), 425–443 (2016)
6. Mueller, S., Wang, D., Fox, M.D.: Individual variability in functional connectivity architecture of the human brain. Neuron **77**(3), 586–595 (2013)
7. Laumann, T.O., Gordon, E.M., Adeyemo, B.: Functional system and areal organization of a highly sampled individual human brain. Neuron **87**(3), 657–670 (2015)
8. Gordon, E.M., Laumann, T.O., Adeyemo, B.: Individual variability of the system-level organization of the human brain. Cereb. Cortex **27**(1), 386–399 (2017)
9. Finn, E.S., Shen, X., Scheinost, D.: Functional connectome fingerprinting: identifying individuals using patterns of brain connectivity. Nat. Neurosci. **18**(11), 1664–1671 (2015)
10. Tavor, I., Jones, O.P., Mars, R.B.: Task-free MRI predicts individual differences in brain activity during task performance. Science **352**(6282), 216–220 (2016)
11. Glasser, M.F., Sotiropoulos, S.N., Wilson, J.A.: The minimal preprocessing pipelines for the human connectome project. NeuroImage **80**, 105–124 (2013)
12. Smith, S.M., Andersson, J., Auerbach, E.J.: Resting-state fMRI in the human connectome project. Neuroimage **80**(20), 144–168 (2013)
13. Van Essen, D.C., Smith, S.M., Barch, D.M.: The WU-Minn human connectome project: an overview. Neuroimage **80**(8), 62–79 (2013)
14. Barch, D.M., Burgess, G.C., Harms, M.P.: Function in the human connectome: task-fMRI and individual differences in behavior. Neuroimage **80**(8), 169–189 (2013)
15. Bouwmans, T., Zahzah, E.H.: Robust PCA via principal component pursuit: a review for a comparative evaluation in video surveillance. Comput. Vis. Image Underst. **122**(4), 22–34 (2014)
16. Cand, S., Emmanuel, J., Li, X.: Robust principal component analysis? J. ACM **58**(3), 1–73 (2009)
17. Mairal, J., Bach, F., Ponce, J.: Online learning for matrix factorization and sparse coding. J. Mach. Learn. Res. **11**(1), 19–60 (2010)
18. Gunn, S.R.: Support vector machines for classification and regression. ISIS Technical Report. 14 p. (1998)
19. Wright, J., Ma, Y., Mairal, J.: Sparse representation for computer vision and pattern recognition. Proc. IEEE **98**(6), 1031–1044 (2010)
20. Abolghasemi, V., Ferdowsi, S., Sanei, S.: Fast and incoherent dictionary learning algorithms with application to fMRI. Sig., Image Video Process. **9**(1), 147–158 (2015)
21. Dosenbach, N.U., Nardos, B., Cohen, A.L.: Prediction of individual brain maturity using fMRI. Science **329**(5997), 1358–1361 (2010)

Lung Nodule Detection Using Combined Traditional and Deep Models and Chest CT

Junjie Zhang[1], Zhaowei Huang[2], Tairan Huang[3], Yong Xia[1(✉)],
and Yanning Zhang[1]

[1] National Engineering Laboratory for Integrated Aero-Space-Ground-Ocean Big
Data Application Technology, School of Computer Science and Engineering,
Northwestern Polytechnical University, Xi'an 710072, China
yxia@nwpu.edu.cn
[2] Biomedical and Multimedia Information Technology (BMIT) Research Group,
School of Information Technologies, University of Sydney,
Sydney, NSW 2006, Australia
[3] School of Software and Microelectronics, Northwestern Polytechnical University,
Xi'an 710072, China

Abstract. Detection of lung nodules in chest CT scans is of great value
to the early diagnosis of lung cancer. In this paper, we jointly use traditional object detection methods and deep learning, and thus propose
a lung nodule detection algorithm for chest CT scans. We first detect
all candidate nodules using multi-scale Laplace of Gaussian (LoG) filters
and shape priors, and finally construct a multi-scale 3D DCNN to differentiate nodules from non-nodule volumes and estimate nodules' diameters simultaneously. This algorithm has been evaluated on the benchmark
LUng Nodule Analysis 2016 (LUNA16) dataset and achieved an average
diameter estimation error of 0.98 mm and a detection score of 0.913. Our
results suggest that the proposed algorithm can effectively detect lung
nodules on chest CT scans and accurately estimate their diameters.

Keywords: Lung nodule detection · Chest CT · Deep learning
Laplacian of Gaussian (LoG)
Deep convolutional neural network (DCNN)

1 Introduction

Lung cancer is the leading cause of all cancer-related deaths for both men and
women [1]. The 5-year survival rate of lung cancer patients is only about 16%
on average; this number, however, can reach approximately 54% if the diagnosis
is made at an early stage of the disease [2]. Since malignant lung nodules may
be primary lung tumors or metastases, early detection of lung nodules is critical
for best patient care [3]. On chest CT scans, a lung nodule usually refers to a
"spot" of less than 3 cm in diameter on the lung [4]. Many automated lung nodule

© Springer Nature Switzerland AG 2018
Y. Peng et al. (Eds.): IScIDE 2018, LNCS 11266, pp. 655–662, 2018.
https://doi.org/10.1007/978-3-030-02698-1_57

detection methods have been proposed in the literature. Most of them extract hand-crafted features in each suspicious lesion and train a classifier to determine if the lesion is a lung nodule or not [5]. Nithila et al. [6] adopted the intensity cluster, rolling ball and active contour model (ACM) based algorithms to detect solitary, juxta-pleural and juxta-vascular nodules, respectively, then calculated statistical and texture features of those nodules, and used a back propagation neural network (BPNN) optimized by the particle swarm optimization (PSO) technique to improve the performance of detection. Wu et al. [7] firstly applied the thresholding, region growing and morphology operations to segment lung nodules, then extracted 34 visual features to describe nodules, and employed a support vector machine (SVM) for false positive detection. Despite their prevalence, these methods may suffer from limited accuracy, due to the intractable optimization of hand-crafted feature extraction and classification.

Recently, deep learning techniques have been widely applied to many medical image analysis tasks, including lung nodule detection. They have distinct advantages in jointly representation learning and pattern classification in a unified network, and hence are able to largely address the drawbacks of traditional methods. Hamidian et al. [8] used a 3D fully convolutional network (FCN) to generate a score map for candidate nodule identification, and employed another 3D deep convolutional neural network (DCNN) for nodule and non-nodule discrimination. Dou et al. [9] also implemented an FCN to produce candidate nodules, and incorporated two residual blocks into a 3D DCNN for nodule detection. However, due to the extremely limited training dataset, deep learning techniques usually detected fewer candidate nodules than traditional methods, which may result in lower sensitivity.

In this paper, we propose an automated lung nodule detection algorithm that jointly uses a traditional object detection method and a deep learning model. We first segment the lung volume using thresholding and morphological operations, then detect all candidate nodules using multi-scale Laplace of Gaussian (LoG) filters with the area and circularity constraints, and finally construct a multi-scale 3D DCNN to differentiate positive nodules from non-nodule lesions and estimate nodules' diameters. We have evaluated the proposed algorithm on the benchmark LUng Nodule Analysis 2016 (LUNA16) dataset [10] against the best-performed methods listed in the LUNA16 Challenge Leaderboard.

2   Method

The proposed algorithm consists of three major procedures: lung segmentation, candidate nodule identification and differentiation of nodules from non-nodules. A diagram that summarizes the algorithm is shown in Fig. 1.

2.1   Lung Segmentation

Lung segmentation consists of four steps: (a) each chest CT scan is re-sliced to a uniform voxel size of $1.0 \times 1.0 \times 1.0 \, \text{mm}^3$; (b) each re-sliced CT scan is binarized by using the OTSU algorithm [11] on a slice-by-slice basis; (c) morphological

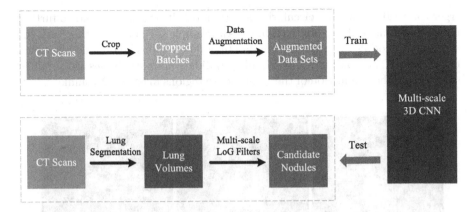

Fig. 1. Diagram of the proposed pulmonary nodule detection algorithm.

closing is used to fill holes, and the morphological dilation with a disk structure element of size 5 is then carried out to produce a lung mask that covers as much lung tissues as possible; and (d) this mask is applied to the re-sliced CT scan to get the volume of lung. An example CT slice and the corresponding results of binarization, mask generation and lung segmentation are shown in Fig. 2.

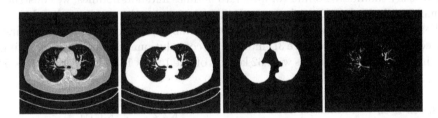

Fig. 2. An example CT slice and the corresponding results of binarization, mask generation and lung segmentation. The left, middle left, middle right and right image is original image, binary image, lung mask and segmented image, respectively.

2.2 Candidate Nodule Identification

Let a CT slice be denoted by $f(x,y)$, and an isotropic LoG filter be denoted by $L(x,y;\delta_i)$, where δ_i is the standard deviation of Gaussian function. Applying the LoG filter to a CT slice, we have a response map $V_i(x,y)$. To detect nodules as different scales, we use 21 LoG filters, whose standard deviation δ_i ranges from 1 to 5 with a step of 0.2.

Then, we apply the OTSU algorithm [11] to the response map of each filter, and calculate the union of 21 binarized response maps, in which each connected region is defined as a candidate nodule region. We calculate the area and circularity of each candidate nodule region. If a region has an area smaller than 9 or larger than 1,000, or has a circularity less than 0.1, it will be removed to further reduce the number of candidate nodules. An example CT slice, together with the true positive nodule and detected candidate nodule regions before and after such region removal, is shown in Fig. 3. Finally, candidate nodule volumes are obtained by analyzing the connection of these candidate regions in each 3D volume.

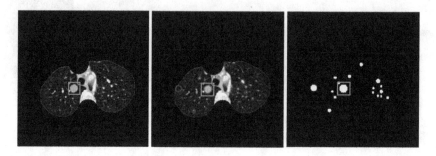

Fig. 3. An example segmented CT slice (left), candidate nodule regions detected by the multi-scale LoG filters (middle), binary image of candidate nodules (right).

2.3 Differentiation of Nodules from Non-nodules

To classify candidate nodule volumes into nodules and non-nodules, we construct a multi-scale 3D DCNN with seven 3D convolutional and four 3D pooling layers. The architecture and parameters of this DCNN model are shown in Fig. 4. The first three convolutional layers are 3D multi-scale convolutional blocks, each consisting of three convolutional layers with 32 kernels of size $1 \times 1 \times 1$, $3 \times 3 \times 3$ and $5 \times 5 \times 5$,

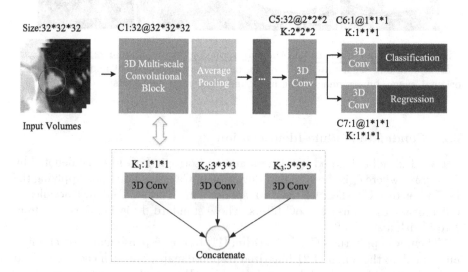

Fig. 4. Structure of the proposed multi-scale 3D DCNN. Both "Conv" and "C" represent conventional layer, and "K" is for kernel size.

respectively, followed by a $2 \times 2 \times 2$ pooling layer. The first pooling layer uses 3D average pooling, while the other three pooling layers use 3D max pooling. The fourth and fifth convolutional layers are traditional 3D conventional layers with 32 kernels of size $3 \times 3 \times 3$ and $2 \times 2 \times 2$, respectively. The last two convolutional layers are two parallel output layers: one for candidate nodule volume classification, and the other for nodule diameter estimation. The classification layer uses the cross-entropy error as the loss function, and the sigmoid function as the activation. The regression layer uses the mean absolute error as the loss function, and replaces the sigmoid function with a linear function. This network takes $32 \times 32 \times 32$ volumes as input. To improve its regularization ability, a dropout with a probability of 0.5 is implemented on the output of subsampling layers.

2.4 Training and Testing

To alleviate the overfitting of our deep model, we performed data augmentation to increase the number of positive nodules in both training and testing. Via rotating each candidate nodule volume along the Z axis with $0°$, $90°$, $180°$ and $270°$ and left-right flipping it, we have five augmented copies for each candidate. At the training stage, we randomly crop negative samples, including false positive nodules, pleura regions, trachea tissues and vessels, from the lung on chest CT scans to ensure the ratio of positive and negative samples to be 1:10. The weights and biases were initialized with standard normal distribution and zero respectively. We choose the mini-batch stochastic gradient descent with a batch size of 16 and a momentum of 0.9 as the optimizer. The learning rate started at 0.01 and was decreased to 0.001 after 10 epochs. Training was stopped when the accuracy on the validation dataset did not improve or the change of the loss function is less than 0.002. At the testing stage, $32 \times 32 \times 32$ patches were cropped according to the central coordinate of identified candidate nodules. For each candidate nodule, five augmented copies were inputted to the trained multi-scale 3D DCNN. The candidate's class label is determined by majority voting and its diameter is estimated as the average of those five results.

3   Experiments and Results

Dataset. The LUNA16 dataset [10] used for this study contains 888 chest CT scans and 1186 positive lung nodules. Each scan has a slice thickness less than 2.5 mm and slice size of 512×512 voxels, and was annotated during a two-phase procedure by four experienced radiologists. Each radiologist marked lesions they identified as non-nodule, nodule <3 mm, and nodules ≥3 mm. The reference standard of the LUNA16 challenge consists of all nodules ≤3 mm accepted by at least 3 out of 4 radiologists.

Results. The proposed multi-scale LoG filter algorithm detected 45,939 candidate nodules, including all 1186 true nodules. We adopted the 10-fold cross-validation to evaluate the candidate nodule classification performance. In each run, 70% of candidate nodules were used for training, 20% for validation and 10% for testing. The free receiver operating characteristic curve (FROC) curve obtained by our nodule detection algorithm was shown in Fig. 5, where the score is calculated as the average sensitivity at seven operating points of the FROC curve: 1/8, 1/4, 1/2, 1, 2, 4, and 8 FPs/scan, and the 95% confidence interval is computed using bootstrapping with 1,000 bootstraps. It shows that our algorithm achieved an average score of 0.913. Figure 6 shows two example cases of estimated results. The results reveal that our algorithm is able to detect small nodules, whose diameter is less than 5 mm, and can estimate nodule diameter accurately. Moreover, the average difference between the detected diameter and true diameter of detection nodules is 0.98 mm.

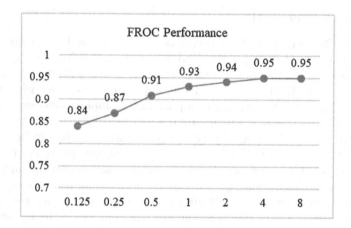

Fig. 5. FROC curve of the score at 1/8, 1/4, 1/2, 1, 2, 4, and 8 FPs/scan.

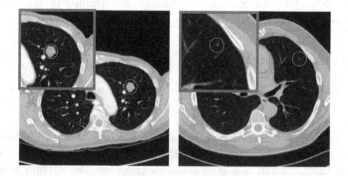

Fig. 6. Sample images of estimated results. Sample images of estimated results. The true probabilities and diameters of the two samples are [1, 18.208 mm] and [1, 4.225 mm], while the estimated values are [0.999, 18.174 mm] and [0.986, 4.228 mm], respectively.

Comparison with Other Methods. We compared the performance of our algorithm to that listed on the LUNA16 Challenge leaderboard in Table 1. These schemes are described briefly as follows:

Model#1: In nodule detection, a 2D DCNN is used to find lung nodule candidates, and 3D DCNNs with different settings to reduce the false positive and then average them.

Model#2: In this system, a fully convolutional network was used to obtain the lung nodule candidates. Then these candidates were further classified by a convolution neural network.

Model#3: A deconvolutional structure was first introduced to Faster Region-based Convolutional Neural Network (Faster R-CNN) for candidate detection on axial slices. Then, a 3D DCNN is presented for the subsequent false positive reduction.

Model#4: Each model is a single stage method based on 3D Region Proposal Network (3D RPN). In total, two different models were trained with different settings, and the results were ensemble by Non-Maximum Suppression (NMS).

Table 1. Comparison with partial results of luna16 challenge leaderboard

Name	Method	Score
Model#1	2D DCNN + 3D DCNN	0.865
Model#2	FCN + CNN	0.882
Model#3	3D Faster R-CNN	0.891
Model#4	3D RPN	0.897
Proposed	**Multi-scale LoG filters + 3D DCNN**	**0.913**

Computational Complexity. Since a multi-scale 3D DCNN has to be trained and tested on an augmented dataset, our algorithm has a high computational complexity. It takes about nine hours to train this algorithm (Intel Xeon E5-2640 V4 CPU, NVIDIA Titan X GPU, 512 GB RAM, Ubuntu 16.04 and MATLAB 2017a). It costs only about four minutes to detect nodules in a chest CT scan, including preprocessing, candidate nodule detection and true nodule identification.

4   Conclusion

In this paper, we propose an algorithm for lung nodule detection using jointly a traditional object detection method and deep learning. Specifically, we use multi-scale LoG filters to detect candidate nodules and a multi-scale 3D DCNN to identify true nodules. When evaluated on the LUNA16 dataset, this algorithm achieved a score of 0.913.

Acknowledgments. This work was supported in part by the National Natural Science Foundation of China under Grants 61471297 and 61771397, in part by the China Postdoctoral Science Foundation under Grant 2017M623245, in part by the Fundamental Research Funds for the Central Universities under Grant 3102018zy031, and in part by the Australian Research Council (ARC) Grants. We appreciate the efforts devoted by LUNA16 challenge organizers to collect and share the data for comparing lung nodule detection algorithms for chest CT scans.

References

1. Ferlay, J., et al.: Cancer incidence and mortality worldwide: sources, methods and major patterns in GLOBOCAN 2012. INT. J. Cancer **136**(5), E359 (2015)
2. Baldwin, D.R.: Prediction of risk of lung cancer in populations and in pulmonary nodules: significant progress to drive changes in paradigms. Lung Cancer **89**(1), 1–3 (2015)
3. Kohan, A.A., et al.: N staging of lung cancer patients with PET/MRI using a three-segment model attenuation correction algorithm: initial experience. Eur. Radiol. **23**(11), 3161–3169 (2013)
4. Callister, M.E., Baldwin, D.R.: How should pulmonary nodules be optimally investigated and managed? Lung Cancer **91**, 48–55 (2016)
5. Valente, I.R., Cortez, P.C., Neto, E.C., Soares, J.M., de Albuquerque, V.H., Tavares, J.M.: Automatic 3D pulmonary nodule detection in CT images: a survey. Comput. Methods Prog. Biomed. **124**, 91–107 (2016)
6. Nithila, E.E., Kumar, S.S.: Automatic detection of solitary pulmonary nodules using swarm intelligence optimized neural networks on CT images. J. Eng. Sci. Technol. **20**(3), 1374–1377 (2016)
7. Wu, P., Xia, K., Yu, H.: Correlation coefficient based supervised locally linear embedding for pulmonary nodule recognition. Comput. Methods Prog. Biomed. **136**, 97–106 (2016)
8. Hamidian S., Sahiner B., Petrick N., Pezeshk, A.: 3D convolutional neural network for automatic detection of lung nodules in chest CT. In: Proceedings of SPIE - The International Society for Optical Engineering, p. 1013409 (2017)
9. Dou, Q., Chen, H., Jin, Y., Lin, H., Qin, J., Heng, P.-A.: Automated pulmonary nodule detection via 3D convnets with online sample filtering and hybrid-loss residual learning. In: Descoteaux, M., Maier-Hein, L., Franz, A., Jannin, P., Collins, D.L., Duchesne, S. (eds.) MICCAI 2017. LNCS, vol. 10435, pp. 630–638. Springer, Cham (2017). https://doi.org/10.1007/978-3-319-66179-7_72
10. Setio, A.A.A., et al.: Validation, comparison, and combination of algorithms for automatic detection of pulmonary nodules in computed tomography images: the LUNA16 challenge. Med. Image Anal. **42**, 1–13 (2017)
11. Otsu, N.: A threshold selection method from gray-level histograms. IEEE Trans. Syst. Man Cybern. **9**(1), 62–66 (1979)

An Enviro-Geno-Pheno State Analysis Framework for Biomarker Study

Hanchen Huang[1], Xianzi Wen[2], Shikui Tu[3], Jiafu Ji[2],
Runsheng Chen[1], and Lei Xu[3(✉)]

[1] Institute of Biophysics, Chinese Academy of Sciences, Beijing, China
huanghanchen12@mails.ucas.ac.cn, crs@sun5.ibp.ac.cn
[2] Peking University Cancer Hospital and Institute, Beijing, China
wenxz@bjmu.edu.cn, jijiafu@hsc.pku.edu.cn
[3] Department of Computer Science and Engineering,
Shanghai Jiao Tong University, Shanghai, China
tushikui@sjtu.edu.cn, lxu@cs.sjtu.edu.cn

Abstract. In this study, we introduce the basic perspectives of E-GPS framework for biomarker study and demonstrate its advantages by a series of experiments on real bio-data in cancer research. Owing to the development of high-throughput technology on gene detection, an increasing number of disease-related biomarkers for diagnosis, subtyping, and prognosis have been discovered. Conventional methods for biomarker discovery mainly focus on detecting the relationship between intrinsic genotypes and extrinsic phenotypes, while recent efforts emphasize the role of a biomarker under particular condition, which extends to the integrated analysis on Enviro-, Geno-, and Pheno-measures. Samples sharing a similar pattern of genotype under certain environment are in a 'state' that is distinctive from other states. Such different states generated by machine learning from samples can further indicate the different outcome of the phenotype of interest, which makes the enviro-geno-pheno state (E-GPS) analysis as an analytic tool for biomarker studies.

Keywords: Biomarker · E-GPS analysis · Prognostic study
Gene expression profile

1 Introduction

A biomarker refers to a marker that is measured to indicate some biological state or event. Since the extraordinary development of molecular biotechnology especially the high-throughput approaches for gene detection, various types of genetic biomarkers have been discovered in a huge amount. In the field of health science and biomedicine, DNA/RNA/protein biomarkers play important roles in the risk assessment, diagnosis, staging and grading, and prognosis prediction of disease [1–4]. Some recent trends on cancer subtype study further emphasize the integrative analysis of gene measures to divide patients into different classes purely by molecular biomarkers. These studies discovered that the patients belonging to the same subtype may share the similar clinical or pathological features, even if those cohorts of patients are from different types of cancers [5].

© Springer Nature Switzerland AG 2018
Y. Peng et al. (Eds.): IScIDE 2018, LNCS 11266, pp. 663–671, 2018.
https://doi.org/10.1007/978-3-030-02698-1_58

Meanwhile, more recent studies revealed the importance of identifying the biological role of a biomarker under certain condition or in some particular environment, represented by the enviro-measures. Recently, Xu proposed a generic tool for examining biomarkers under certain conditions, known as the E-GPS approach [6], considering the joint domain of geno-measures, pheno-measures, and enviro-measures, in which each triple-measured element represents a possible behavior of the bio-system under investigation and a subset of elements that located adjacently and shared a common system status represents a 'state'. Therefore, the system is characterized by several such states learned from samples. As a biomarker, the E-GPS state indicates not only the corresponding system status such as 'health, or good outcome' and 'risk, or bad outcome', but also the closure of the corresponding subset as the condition to specify this state.

For example, CDX2 could predict prognosis of colon cancer patients, which is a well-known clinical fact. A recent study [7] evaluated the efficacy of chemotherapy within the CDX2-positive group and the CDX2-negative group separately, and concluded that CDX2-negative patients with stage II could benefit from chemotherapy significantly with a prolonged survival, while the CDX2-positive patients yielded similar outcome no matter they have received chemotherapy or not. In this case, the geno-measure, CDX2 was investigated under the different therapy conditions that acted as the enviro-measure. Collectively, the CDX2-positive patients had good survival so did the CDX2-negative patients who received chemotherapy, but those CDX2-patients that didn't receive chemotherapy had poor survival. This discovery not only provided a potential chemotherapeutic guidance based on CDX2 value, but also clarified the scope of use of CDX2 as a prognostic biomarker for colon cancer.

Another common presence of the e-measure is the joint analysis of a novel biomarker together with some well-known biomarker. One example is the recent molecular analysis of gastric cancer that identifies four subtypes of gastric cancer by a three-layered binary tree [4], where under the condition of both downergulation of Epithelial–mesenchymal transition (EMT) related genes and microsatellite instability (MSI) related genes, two subtypes namely MSS/TP53+ and MSS/TP53− are identified by an integrated gene expression signature that reflected the TP53 mutation status.

However the referred enviro-measures as listed above were still simply binary or categorical, which degenerated the study into a fractional analysis where we can assess the prognostic feature of the geno-measure separately under a couple of well-defined conditions. As a generalization and extension, Xu proposed a universal framework of E-GPS analysis [6] that is also applicable for the continuous enviro-measure, e.g. the expression value of some specific gene. However, the practical implementation of E-GPS analysis is still lacking, and its application on real bio-data has not been shown.

In this paper, we first introduced the framework of E-GPS analysis, then provided two versions of implementation that were followed by two basic analytical strategies for the prognosis study with geno-enviro-paired 2D data. Via the computational experiments based on the gastric cancer data, we demonstrated the effectiveness of our methods, which indicated that the E-GPS approach could be a promising tool for biomedical study and health science in the future.

2   E-GPS Analysis

2.1   The Basic Concepts of E-GPS Analysis

In the framework E-GPS analysis, whether the survival of a patient is good or not is indicated by one of two states, namely 'good' and 'poor'. Such a state **s** relates to a set **e** of environmental factors and depends on a set **g** of geno-measures. Moreover, a set $\mathbf{p_h}$ of phenotypes of comes from this inner state **s**. Therefore, such a state **s** is named enviro-geno-pheno state.

Featured by a relation among the triple (**e**, **g**, $\mathbf{p_h}$) or a subset in the joint domain of (**e**, **g**, $\mathbf{p_h}$), such an E-GPS state may be approximately learnt from given samples. The E-GPS approach pursues identifying the system status via the E-GPS states, learning and refining the E-GPS states from a given set of samples, and conducting various conditional phenotype analyses based on the E-GPS states [6]. The state that is dominated by samples with either good survival or poor survival is defined as a 'dedicate' state (shortly d-state), while the state with a roughly balanced ratio between good- and poor-survival patients refers to a 'confused' state (shortly c-state).

For each state s, a dedication degree (D-degree), r_s is calculated by Eq. (1) in [6]. Then a cost function J is generated by combining the D-degree of three states, e.g.,

$$J = \sum\nolimits_{s \in S, \text{with } n_s > m \text{ and } r_s > \min(r_s)} \log(1 - r_s) \tag{1}$$

where n_s is the total number of samples in state s, and m is a threshold of minimum sample size, which is introduced to prevent the overfitting.

2.2   E-GPS Analysis on G-E Paired 2D-Data with Prognostic Information

In general cases, a status variate of an E-GPS system is actually a triplet from the space of $\mathbf{G} \times \mathbf{E} \times \boldsymbol{\Phi}$, where $\mathbf{G}$ represents a series of geno-measures (short as g-measures), $\mathbf{E}$ represents a series of enviro-measures (short as e-measures), and $\boldsymbol{\Phi}$ represents the pheono-measures, such as the occurrence, progression, metastasis of cancer, and the death caused by tumor, etc. However, since the characteristics of most current biological and clinical studies, it is still impractical to investigate more than one pair of g-measure and e-measure that are associated with multiple phenotypes. Instead, the common scenario is that, given by one known gene marker acting as e-measure, we aim to observe another gene as the g-measure (usually a newly discovered one that is of interest) under this e-measure, to investigate the clinical features of patients within different states separated by this E-GPS marker. Hence we here investigate the simplest case of only one environmental variable and one genotype variable consisting a G-E pair, which leads to the two hyper-planes degenerating to linear boundaries.

To separate the sample space with linear boundaries, there are two typical implementations: separate the G-E 2D-space by two parallel lines, or, separate the G-E 2D-space by a pair of vertical and horizontal lines.

The first type of separation leads to a trisection with two delicate states and a remaining confused state, and the second type will generate four separated regions in actual by a Boolean logic with two variates.

Parallel Separation. In the case of parallel separation, in total three free parameters should be learned from the samples: the slope factor and two intercepts of the parallel lines. It is complicated to optimize the three factors at the same time so that a two-step implementation could simplify the learning process by first determine a normal direction to project the samples upon, then minimize the J function by calculating the two boundary points based on the projected data. For example, the first step could be done with the help of SVM, so that all the samples are projected to 1D. Then an enumeration to choose the two boundaries generates a minimum J and its corresponding boundary points, which separate the samples by three states.

Notably, in the case of prognostic study, some specific constraints should be introduced to the general E-GPS analysis. First, the c-state should locate between the two parallel margins. Second, the prognostic label of the two d-states should be opposite. In other words, following the normal direction of the parallel lines, the prognostic outcome of the three states should be 'monotonically' arranged. Otherwise, it will be not only complicated to explain how the g-e measures may act as a prognostic biomarker, but also almost impossible to make such E-GPS markers practical for clinical use.

Boolean Logic Separation. In the case of Boolean logic separation, still with two straight lines we may divide the 2D-plane by a pair of vertical and horizontal lines that yield a quardrisection. Obviously, only two free parameters are needed to learn in this scenario, which could be implemented simply by the enumeration to find a minimum J. Different from the case of trisection, there are four subsections here so that two of them will share the same state of good-, poor-, or intermediate-survival outcome, which should be considered additionally when minimizing the J function.

Since there are four states as the result of the Boolean logic separation, it is more flexible to assign the outcome labels. We may also consider the state merging at the same time. Given a Boolean logic separation, to achieve a better performance of prognosis differentiation of the d-states, there are five possible state merging solutions. We can enumerate all of them and generate the one with minimum J as the best merging.

3   Experiments on Gastric Cancer Data

3.1   Datasets

To investigated the performance of E-GPS analysis, we implemented computational experiments on gastric cancer datasets. We first downloaded GSE15459 and GSE62254 from GEO database (https://www.ncbi.nlm.nih.gov/geo/), the normalized gene expression profiling data was directly used. The two datasets were generated by the same microarray platform. We selected the pheno-measure as the five-year overall survival (5-OS). Totally 136 patients from GSE15459 and 283 from GSE62254 were involved in the E-GPS analysis.

3.2   The Selection of E-Measure

In this study we considered the gene *IDH2* as the e-measure. As a well-known bio-marker in glioma, the mutation of which is not only a signature that can indicate the occurrence, but also a predictor for the recurrence and prognosis [8]. In recent years, more extensive studies revealed that the expression of *IDH2* came out to be a useful biomarker in multiple cancers including gastric cancer. The overexpression of *IDH2* was reported to inhibit the cancer invasion in gastric cancer [9]. Moreover, it is also considered as an indicator of prolonged survival in liver cancer [10]. Therefore, we selected *IDH2* as an e-measure of interest.

3.3   The Screening of G-Measure

To select the candidate genes as g-measures for E-GPS analysis, we first performed preliminary screenings for each gene by univariate t-test to test whether there is significant difference between the average expression of patients with good survival and poor survival. The screening process was implemented independently on the two datasets. After the probe-by-probe screening, in total thirty-one genes were selected as a candidate set for g-measures since the p-value of them were considerable small ($p < 0.001$).

Though all the candidate genes could be paired with the e-measure theoretically, we simply chose one of them as an example to demonstrate the characteristics of E-GPS analysis. According to a literature survey, we selected *THBS4* among the thirty-one genes as the g-measure that will be analyzed with *IDH2* together because of several reasons. First, there are no direct reports of the prognostic use of *THBS4* for gastric cancer. Second, on the other hand, the relationship between *THBS4* and the occurrence of gastric cancer has been discovered [11]. Third, a similar gene, *THBS2*, which is in the same family as *THBS4*, has been reported as a prognostic marker of poor survival for gastric cancer [12]. These supporting evidences implied that, a further investigation of *THBS4* is necessary, and its prognostic value for gastric cancer should also be aware of.

3.4   Results of E-GPS Analyses on the *THBS4-IDH2* Marker

The parallel separation and Boolean logic separation were both implemented. For the parallel separation, we first applied Support Vector Machine (SVM) to generate the projection direction as the normal vector of the boundary for the 2D sample data. Then, the 2D data degenerated on such projection direction into 1D values, name as SVM scores. Next, we applied univariate E-GPS analysis on the SVM-scores as follow procedures:

First, all SVM-scores were sorted by increasing order.

Second, every possibility of the two boundaries to separate these 1D data was enumerated, resulting in three groups of samples. The three groups of patients corresponded to the three E-GPS states. The cost function J was calculated by Eq. (1).

Third, the best boundaries were selected when J reached the minimum.

To enhance the robustness of the boundaries, we expanded the line-form boundary to produce a band-form boundary. All points located within the band area (the region of line boundary ±0.1 standard deviation of SVM-scores) were excluded from further calculation and optimization of the cost function J. These samples were also excluded in the stratification for survival outcomes.

As a result, Fig. 1 demonstrated the SVM-based parallel separation. The parallel lines divided the samples into three states: the top-left region possessed more red dots than blue dots so that referred to the 'poor-state'; the bottom-right region possessed more blue dots than red dots so that referred to the 'good-state'; and the patients fall into the banded area were labeled as the 'confused-state' with intermediate survival.

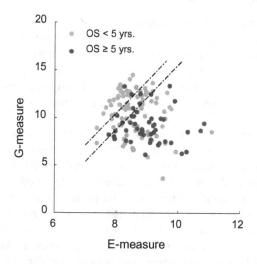

Fig. 1. The scatter plot of E-GPS analysis of *IDH2* (E-measure) and *THBS4* (G-measure) by parallel separation, based on GSE15459. The red dots represent patients with poor outcome (OS < 5 years), and the blue dots represent patients with good outcome (OS ≥ 5 years). The dash-dot line separated all patients into three E-GPS states corresponding to different statuses of prognosis. Patients in the upper left part of the scatter plot are with the poorest OS, whereas the outcome of the patients in the lower right area is the best, and the patients between the two lines have intermediate prognosis (Colour figure online).

Moreover, a K-M plot was generated according to the state separation. As shown in Fig. 2, the prognostic outcome of the three states separated from each other significantly ($p_{\text{log-rank}} = 1.2\text{e-}4$).

We also implemented a Boolean logic separation based on the *THBS4-IDH2* marker. By the enumeration of all possible threshold of the expression of *THBS4* and *IDH2*, the best splitting was generated as shown in Fig. 3a. The corresponding K-M plot referred to Fig. 3b ($p_{\text{log-rank}} = 1.5\text{e-}5$).

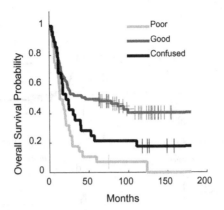

Fig. 2. The K-M plot of the three states defined by *THBS4-IDH2* in GSE15459. The three curves correspond to the E-GPS states defined by Fig. 1.

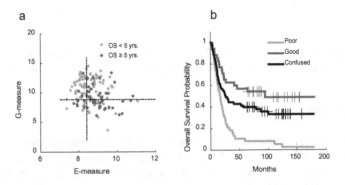

Fig. 3. The Boolean logic separation of the three states defined by *THBS4-IDH2* in GSE15459

The results of Boolean logic demonstrated that, the patients with high expression of *IDH2* and low expression of *THBS4* have the best outcome; the patients with low expression of *IDH2* and high expression of *THBS4* have the worst outcome; and the remaining patients are assigned to the confused state with the intermediate outcome, which is similar to the results of parallel separation.

Similarly, the E-GPS analysis of *THBS4-IDH2* on GSE62254 was implemented in Fig. 4. Figures 4a and 4b demonstrated the parallel separation ($p_{\text{log-rank}} = 1.6e\text{-}06$) while the Figs. 4c and 4d demonstrated the Boolean logic version ($p_{\text{log-rank}} = 2.4e\text{-}04$). Though the distributed shape of the data cloud seems not as same as the one of GSE15459, the separation pattern and the corresponding survival outcomes' differentiation preserve well in this alternative dataset.

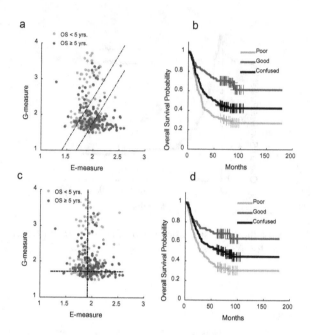

Fig. 4. E-GPS analysis of *THBS4-IDH2* on GSE62254. (a) Parallel separation; (b) K-M plots corresponding to (a); (c) Boolean logic separation; (d) K-M plots corresponding to (c).

4   Discussion

In this paper, we first introduced the framework of E-GPS analysis, then provided two basic implementation by parallel separation and Boolean logic separation. In the experimental analyses on GSE15459 and GSE62254 datasets, we investigated an E-GPS marker, *THBS4-IDH2*, by SVM-based parallel separation and the Boolean logic separation, respectively. The results of the two methods on both datasets showed *THBS4-IDH2* could indicate the prognosis of gastric cancer effectively. Patients with high *IDH2* and low *THBS4* have prolonged overall survival, while the overall survival outcome of the patients with low *IDH2* and high *THBS4* was poor. More importantly, since *IDH2* is a well-known biomarker in cancer, which also has been reported to inhibit the cancer invasion in GC, our study provided an extended observation on how expression of *IDH2* in GC tumor could act as a prognostic marker accompanied by another promising marker, *THBS4*, under certain condition.

In the future, more extensive studies should be done to improve the current work. For example, a combination of the SVM-based parallel separation and Boolean logic separation could be further implemented, by which we may prevent the overfitting problem via SVM and then generate the E-GPS separation by Boolean logic since it is more practicable for generalization in other datasets and the clinical use. Second, in the present study, we still have not investigated how to further separate the confused states effectively. Though there are more conditional enviro-measures could be introduced to help splitting these patients with intermediate outcomes, it is still unknown that how to choose another enviro-measure is reasonable and could lead to a robust result, since the sample sizes of the patients in confused state are usually small and may therefore suffer from the problem of overfitting.

References

1. Ooi, C.H., et al.: Oncogenic pathway combinations predict clinical prognosis in gastric cancer. PLoS Genet. **5**, e1000676 (2009)
2. Lee, J., et al.: Nanostring-based multigene assay to predict recurrence for gastric cancer patients after surgery. PLoS ONE **9**, e90133 (2014)
3. Qian, Z., et al.: Whole genome gene copy number profiling of gastric cancer identifies PAK1 and KRAS gene amplification as therapy targets. Genes Chromosom. Cancer **53**, 883–894 (2014)
4. Cristescu, R., et al.: Molecular analysis of gastric cancer identifies subtypes associated with distinct clinical outcomes. Nat. Med. **21**, 449–456 (2015)
5. Stessman, H.A., Bernier, R., Eichler, E.E.: A genotype-first approach to defining the subtypes of a complex disease. Cell **156**, 872–877 (2014)
6. Xu, L.: Enviro-geno-pheno state approach and state based biomarkers for differentiation, prognosis, subtypes, and staging. Appl. Inform. **3**, 4 (2016)
7. Dalerba, P., et al.: CDX2 as a prognostic biomarker in stage II and stage III colon cancer. New Engl. J. Med. **374**, 211–222 (2016)
8. Ducray, F., Marie, Y., Sanson, M.: IDH1 and IDH2 mutations in gliomas. New Engl. J. Med. **360**, 2248 (2009)
9. Wu, D.: Isocitrate dehydrogenase 2 inhibits gastric cancer cell invasion via matrix metalloproteinase 7. Tumor Biol. **37**, 5225–5230 (2016)
10. Liu, W.-R., et al.: High expression of 5-hydroxymethylcytosine and isocitrate dehydrogenase 2 is associated with favorable prognosis after curative resection of hepatocellular carcinoma. J. Exp. Clin. Cancer Res. **33**, 32 (2014)
11. Lin, X., et al.: Associations of THBS2 and THBS4 polymorphisms to gastric cancer in a Southeast Chinese population. Cancer Genet. **209**, 215–222 (2016)
12. Yasui, W., et al.: Molecular-pathological prognostic factors of gastric cancer: a review. Gastric Cancer **8**, 86–94 (2005)

A Novel Seizure Prediction Method Based on Generative Features

Lili Liu, Lijuan Duan, Ying Xiao, and Yuanhua Qiao[✉]

Faculty of Information Technology, Computer Science,
Beijing University of Technology, Beijing, China
qiaoyuanhua@bjut.edu.cn

Abstract. The diagnosis of epilepsy in hospital is mostly judged by experienced medical personnel visually observing brain waves combined with some characteristic clinical manifestations. As brain signal's complexity the understanding of EEG signal still remains challenge. In this paper we proposed a novel seizure prediction method based on generative features. Then predict seizures according to the EEG information of the epileptic patients. And we use the Extreme learning machine as the classifier of generative features. Finally, we get the highest accuracy score of 98%.

Keywords: Epilepsy · Seizure prediction · Deep learning
Neural network

1 Introduction

Research of brain science is one of the most challenging frontier science problem. In recent years, brain research programs are constantly launched and pushed forward around the world. Epilepsy is a chronic neurologic disease of the brain caused by sudden abnormal discharge of the brain neurons. As a common disease, there are about 0.6%–0.8% population in the world suffer from epilepsy [1]. There are about six to nine million epileptic patients in china, and they are growing at the rate about four hundred thousand per year [2]. The epileptic seizures' onset time is uncertain. When it occurs, if the patient can't get appropriate treatment in time, it may be very dangerous for patients. Even a short seizure may cause severe physiological and psychological damage. If we can predict epileptic seizures automatically and in time, it will help patients to avoid many times damage caused by seizures. The research of epileptic seizures is firstly carried out in 1970s [3]. In recent years, the method of predicting epilepsy based on artificial intelligence attracts a lot of attention.

EEG-based epilepsy prediction mainly includes two steps: feature extraction and classification. When working on feature extraction, researchers can select the common method from time-domain analysis [9–11], frequency domain analysis [12–15], time-frequency domain analysis [16,17] and non-linear analysis [18,19] methods etc. The time-domain analysis is to extract waveform characteristics

© Springer Nature Switzerland AG 2018
Y. Peng et al. (Eds.): IScIDE 2018, LNCS 11266, pp. 672–682, 2018.
https://doi.org/10.1007/978-3-030-02698-1_59

through the analysis of time-domain waveform EEG signals, such as cycle and rhythm [20]. Wavelet Transform, and Hilbert transform [16] are widely used in time frequency analysis. Hybrid dynamics method is proposed in 1990s, which includes maximum lyapunov Index [18], density correlation and kinetic similarity index [21].

Pattern recognition and machine learning methods are frequency used with classification mission. The common used method includes Support Vector Machine (SVM), Decision Tree (DT), Bayesian, etc. Subasi proposed a general EEG processing and analysis system [10]. In this work PCA, ICA and LDA is used to extract features and SVM is used as classification. In 2000, Petrosian combined the original EEG signal and the wavelet decomposition coefficient as the input to RNN [25], which was the first time for RNN be used in EEG processing jobs. In 2015, Achilles [27] used two CNN structures to determine the seizure period in video. Although these studies have achieved some results, the data of individual cases are insufficient because of the obvious difference in body size and the lack of pathological data. How to effectively use a small amount of epileptic data to fully excavate the discriminant characteristics of epileptic EEG data is the key problem to solve the epilepsy prediction.

In this paper, the generation of network model GAN is introduced into the prediction of epilepsy. We propose a new EEG analysis method based on generation features, and practice the training and testing job on it. The main work of this paper are: (1) Transfer the raw EEG data into spectrum pictures (2) Augment the feature data by DCGAN model. (3) Classify the data with ELM. The remainder of this paper is structured as follows: Sect. 2 describes the framework we use. And the results is shown in Sects. 3. Finally, followed by acknowlegments and possible future works.

1.1 Seizure Prediction Method Based on Generative Features

1.2 Experiment Process

EEG signals are characterized by non-stationary and non-linear. Tt is difficult to carry out the data analysis directly on the raw data. Epileptic EEG signal expresses continuity in the time domain and the frequency increases in the frequency domain when seizures onset. Therefore, both the time and frequency domain information should be considered in the feature processing. The epileptic form discharge shows the typical output signal during the seizure period which reflects the energy change. This change can be used as a significant feature of epilepsy detection and prediction. In this paper, we use the short-time Fourier transform method to deal with the original EEG data result in spectrum patterns. Then take advantage of the DCGAN's function of generate new data with the real data. Finally, the ELM is used to classify the combined generated data and real data. The experimental results show that this method can effectively improve the prediction accuracy of seizures. The framework diagram is given in Fig. 1. It consists of five parts: Epilepsy brain electrical signal acquisition includes

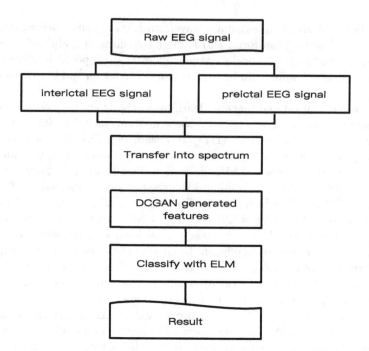

Fig. 1. Framework of epileptic seizure prediction method using generative features

preictal and interictal EEG data; The spectrum data from FFT; Features from DCGAN and classification using ELM.

From the point of feature expression, deep learning has excellent ability in image feature learning and extraction. DCGAN combines CNN with GAN, and combines the advantages of convolution neural network and anti neural network to learn multidimensional high level vector features from input data. DCGAN network has very high application value in the field of image generation. Therefore, when we do epileptic seizure prediction, we transform the EEG mode to maximize the ability of using the neural network. As an unsupervised learning method, DCGAN has more autonomy and depth in the feature extraction of epileptic brain signals, and the characteristics of epileptic data in higher dimensions are extracted from the training of its own structure. ELM is a unique classification method in the field of deep learning. It not only has the high performance of the convolution neural network for data classification, but also has the great learning and computing speed. As a single hidden layer feedforward neural network, ELM abandons the training mode of the traditional neural network, and the weights and biases between the input layer and the hidden layer are generated randomly. It only needs to calculate the weight and bias of the hidden layer to the output layer. To a great extent, the learning ability and learning speed of the network have been improved. At the same time, the generalized inverse theory is introduced, and the result can be obtained by a single calculation through the least square solution method without iteration. Therefore, after DCGAN feature extraction and generation, ELM is used as a classifier to classify epileptic EEG data.

Finally, we can get the classification accuracy of different periods of EEG data which were introduced as prictal and interictal stage. The result can show the great performance of our proposed method applied on this epilepsy prediction mission (Table 1).

Table 1. The experiment process Algorithm Framework description

---

Algorithm Framework

INPUT
Raw epilepsy EEG data, corresponding Label.
OUTPUT
Classification accuracy
BEGIN

for data **in** { Patient_1,Patient_2}
let
    NFFT ; (The number of data points used in each block for the FFT)
    overlap ; (The number of points of overlap between blocks)
    Fs ; (The sampling frequency)
    window; (A function or a vector of length NFFT, the default is $window_h anning().$)
ComputationalFormula :
    $G_f(\beta, u) = \int_R f(t)g(t - u)e^{-j\beta t}dt$
for number **in** training steps
    Sample n training images, X
    Compute n generated images, G

    Compute discriminator probabilities for X and G
    Label training images 1 and generated images 0
        Cost = (1/n)sum[log(D(X)) + log(1 - D(G))]
    Update discriminator weights, hold generator weights constant

    Label generated images 1
        Cost = (1/n)sum[log(D(G))]
    Update generator weights, hold discriminator weights constant

    Classify with Extreme Machine Learning.
    Get the results.
END

---

2   Result

2.1   Dataset

This article is being used on data created by the University of Pennsylvania and Mayo Clinic Intracranial EEG data of two groups of epileptic patients with epilepsy collected from intracranial electroencephalogram. The intracranial electroencephalogram is common way used to determine the stage of epilepsy to prevent future seizures.

The collection of data using different numbers of electrodes, the sampling frequency of 5000 Hz, the reference electrode is the extra-cranial electrodes. The data is divided into a period of ten minutes, the data categories are divided into preictal stage of seizure and interictal stage of seizure. Data Sections were chronologically numbered and the test data were randomly assigned. The pre-seizure consisted of data for one hour before the onset (1: 05-0: 05 pre-seizure). Data taken five minutes prior to seizure (1) Ensure adequate operation Time to patients, family members, medical staff to respond, promptly issued a warning and medication treatment; (2) to ensure that any guarantee data integrity to prevent the pre-attack warning information is missing the same episode interval data is divided into ten minutes for a total of one hour. Duration of the interictal data from the complete data records randomly selected and effectively avoid the pre-seizure and post-seizure signal effects in order to avoid epilepsy Of the pre-, interphase, exacerbation and post-exacerbation of the data interfered with, requiring data collection before and after the week can not have other seizures.

Human subjects were collected during the monitoring process is shorter than the time of the dog subjects were collected, the entire monitoring. The procedure should be less than one week and the interphase data collection time should be strictly controlled at least four hours before or after the epileptic seizure. The interval data should be randomly selected from the canine subjects and the human subjects when the conditions are satisfied (Table 2).

Table 2. The number of used data clips of each subject

Subject	$Patient_1$	$Patient_2$	Channels	Sample frequency
Train interictal	50	42	24	5000 Hz
Train preictal	18	18	24	5000 Hz
Test data	195	150	24	5000 Hz

2.2   Evaluation Index

The evaluation index used in this experiment is shown in the following table. Including sensitivity, specificity, accuracy, precision, AUC etc. In the Table 3, TP represents the number of positive samples correctly classified, TN represents the number of negative samples correctly classified, FP represents the number of samples with negative classification errors, FN represents Positive examples are misclassified samples. If there are M positive samples and N negative samples, the classifier will get a score of samples judged to be positive after prediction, sort the scores from big to small, In this ordered sequence, the positive score is greater than the negative score and the AUC score is obtained.

Table 3. Table of evaluating indicators

Evaluation index	Definition	Function definition
Sensitivity	Positive sample classified correctly	$Sensitivity = \frac{TP}{TP+FN} \times 100\%$
Specificity	Negative samples classified correctly	$Specificity = \times 100\%$
Precision	Correctly classified proportion	$Precision = \frac{TP}{TP+FP} \times 100\%$
Accuracy	All samples classified correctly	$Accuracy = \frac{TP+TN}{TP+FN+TN+FP} \times 100\%$
AUC	Area under the curve	$AUC = \frac{\sum_{ins_i \in positiveclass} rank_{ins_i} - \frac{M \times (M+1)}{2}}{M \times N}$

2.3 Experiment Result

In this experiment we use the training and testing dataset ratio of 3:1. Using FFT change the labeled data into spectrum patterns. All spectrum will be stratified sampling to obtain multiple images, wherein each picture set contains 6000 pictures, the proportion of generated positive samples and negative samples remain the same ratio, after dividing the training set and test set for each group of pictures. The training set data according to the true annotation is divided into two image sets were put into DCGAN for training, generating 6400 pictures of each class, the size normalized images generated for 28 × 28 and the original training samples as ELM classifier training samples, and then to predict the feature map of the test set, the result of prediction the average values are shown in the following table.

Table 4. The classification results of patient1 with DCGAN method

Subjects	Sensitivity	Specificity	Precision	Accuracy	AUC
Rchannel	84.10%	97.02%	90.86%	93.66%	0.96
Gchannel	74.34%	94.69%	83.12%	89.40%	0.94
Bchannel	61.54%	94.94%	81.06%	86.25%	0.90
Grey pic	70.52%	93.39%	78.97%	87.44%	0.91

It can be seen from Table 4 that the color spectrum generated by patient1 has the best classification effect on the R channel, and all the indexes are higher than 2.00%compared with the G channel with the second best performance. In particular, Precision the index is higher than the 7.74%, which reflects the R channel of the R channel with different categories of color maps has the largest distinguishability among the three channels. As can also be seen from Fig. 2, the ROC curve obtained by processing the R channel. The largest area under the same time, the added grayscale contrast experimental results, reflecting the direct use of grayscale classification results worse than the separate use of R channel and G channel classification results, which also shows that different types of grayscale in the distinction Degree has decreased (Table 5).

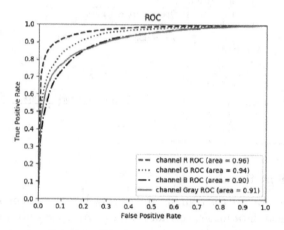

Fig. 2. ROC curve of patient1 feature of DCGAN method

Table 5. The classification results of patient2 with DCGAN method

Subjects	Sensitivity	Specificity	Precision	Accuracy	AUC
Rchannel	91.34%	97.34%	93.83%	95.50%	0.98
Gchannel	81.21%	93.47%	84.62%	89.71%	0.94
Bchannel	88.05%	95.97%	90.63%	93.54%	0.97
Grey pic	75.68%	91.19%	79.17%	86.43%	0.92

As with patient1, patient2 generated the best classification on the R channel, at least % above the B channel, and 3.83% above the Precision index, and at the same time the grayscale performance Still not as good as the R channel classification performance. In order to verify the validity of the experiment, a comparative experiment [33] based on coherence average (CM), spectral entropy (SE), spectral correlation power (SRP) and three feature fusion was added. The classification results obtained are shown in the following Table 6.

Table 6. The classification results of patient1 with DCGAN method

Subjects	Sensitivity	Specificity	Precision	Accuracy	AUC
CM	71.61%	57.34%	37.13%	61.05%	0.66
SE	80.66%	52.71%	37.50%	59.98%	0.69
SRP	75.97%	63.43%	42.22%	66.69%	0.74
(CM, SE, SRP)	84.99%	72.39%	51.84%	75.54%	0.81

The table above reflects the classification results of CM, SE, SRP and patient-specific feature fusion of patient1. It can be seen that the classification result obtained by using the single feature is poor with the maximum Sensitivity of 80.66%, which is derived from the classification result of SE feature The convergence of the 3 features led to a significant increase in classification performance, as can be seen in Fig. 3. For Sensitivity, the value increased by 4.3% after feature fusion, and Accuracy also gained at least 8.85% However, there is still a certain gap between the two methods as a whole (Table 7).

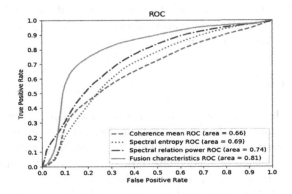

Fig. 3. ROC curve of patient 1 with CM, SE, SRP features

Table 7. The classification results of patient2 with DCGAN method

Subjects	Sensitivity	Specificity	Precision	Accuracy	AUC
CM	56.11%	45.71%	31.39%	48.90%	0.52
SE	69.38%	47.61%	36.96%	54.29%	0.62
SRP	62.66%	46.96%	34.34%	51.78%	0.59
(CM, SE, SRP)	72.58%	54.46%	41.47%	60.16%	0.71

For patient2, the three extracted features do not reflect the differences between samples well. For the single-feature classification, the SE feature obtained the highest Sensitivity, 6.72% higher than the SRP, but the Accuracy

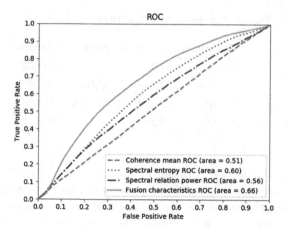

Fig. 4. ROC curve of patient 2 with CM, SE, SRP features

Only 54.29%, while the fusion feature achieved a classification result of at least 3.20% improved on Sensitivity and an increase of 5.87% on Accuracy. As can be seen from Fig. 4, the area under the ROC curve obtained for classification of the fusion features is the largest, which also reflects the improvement in classification performance over the single feature.

3   Conclusion

The research on the prediction of seizures helps to broaden the cognition of brain science in human and also has important physiological significance for patients with epilepsy. However, the study of epilepsy is facing the problem of poor accuracy of prediction of seizures, inadequate data and diseases? Personal information security and other issues. In this paper, the deep convolutional generative adversarial networks (DCGAN) are used to extract and generate the features of the original EEG features, and finally use the extended learning machine (Extreme learning machine, ELM) to classify the different stages of seizures, as early as possible at an earlier warning point to help patients and their families to timely detection and avoid seizures. For medical data will always encounter inadequate experimental data and patients. The information desensitization problem, using DCGAN model generation, can solve these two problems well. The method proposed in this paper can predict the epileptic seizures accurately and play an enlightening role in the study of other diseases such as depression, Parkinson's disease and ADHD.

Acknowledgments. This research is partially sponsored by Natural Science Foundation of China (Nos. 61672070, 81471770, 61572004), the Beijing Municipal Natural Science Foundation (grant number 4182005).

References

1. Mormann, F., Andrzejak, R.G., Elger, C.E.: Seizure prediction: the long and winding road. Brain A J. Neurol. **130**(Pt 2), 314 (2007)
2. Sun, T.: Neurosurgery and Epilepsy, 1st edn. People's Military Medical Press, Beijing (2015)
3. Kesler, J.C.C., Martin, W.B., Ordon, V.A.: Epileptic seizure warning system, U.S. Patent 3863625 (1975)
4. Jia, W.-Y., Gao, S.-K., Gao, X.-R.: The progress in epileptic seizure prediction. J. Biomed. Eng. **21**(2), 325–328 (2004)
5. Wei-Dong, L.S.F.Z., Dong-Mei, Y.Q.C.: Seizure prediction algorithm based on spike rate in EEG. Chin. J. Biomed. Eng. **30**(6), 829–833 (2011)
6. Xu, Y.-H., Cui, J., Hong, W.-X., Liang, H.-J.: Epileptic EEG signal classification based on improved multivariate multiscale entropy. Master dissertation, Yanshan University, China (2015)
7. Li, Y., Yang, C.-J., Ye, M.-N., Zhang, R.: A novel fusion feature extraction method for epileptic EEG. J. Northwest Univ. (Nat. Sci. Edn.) **46**(6), 801–808 (2016)
8. Ma, L., Du, Y.-M., Huang, G., Wang, Y.: A Preliminary study on epileptic seizure prediction using sample entropy and artificial neural network. Chin. J. Biomed. Eng. **32**(2), 243–247 (2013)
9. Ghosh-Dastidar, S., Adeli, H., Dadmehr, N.: Principal component analysis-enhanced cosine radial basis function neural network for robust epilepsy and seizure detection. IEEE Trans. Bio-Med. Eng. **55**(2Pt1), 512 (2008)
10. Subasi, A., Gursoy, M.I.: EEG signal classification using PCA, ICA, LDA and support vector machines. Expert Syst. Appl. Int. J. **37**(12), 8659–8666 (2010)
11. Acharya, U.R., Subbhuraam, V.S., Suri, J.S.: Use of principal component analysis for automatic detection of epileptic EEG activities. Expert Syst. Appl. **39**(10), 9072–9078 (2012)
12. Acharya, U.R., Sree, S.V., Alvin, A.P., et al.: Application of non-linear and wavelet based features for the automated identification of epileptic EEG signals. Int. J. Neural Syst. **22**(02), 37–600 (2012)
13. Welch, P.D.: The use of fast Fourier transform for the estimation of power spectra: a method based on time averaging over short, modified periodograms. IEEE Trans. Audio Electroacoustics **15**(2), 70–73 (1967)
14. Faust, O., Acharya, R.U., Allen, A.R., et al.: Analysis of EEG signals during epileptic and alcoholic states using AR modeling techniques. IRBM **29**(1), 44–52 (2008)
15. Chisci, L., Mavino, A., Perferi, G., et al.: Real-time epileptic seizure prediction using AR models and support vector machines. IEEE Trans. Biomed. Eng. **57**(5), 1124–1132 (2010)
16. Daou, H., Labeau, F.: Dynamic dictionary for combined EEG compression and seizure detection. IEEE J. Biomed. Health Inform. **18**(1), 247 (2014)
17. Costa, R.P., Oliveira, P., Rodrigues, G., Leitão, B., Dourado, A.: Epileptic seizure classification using neural networks with 14 features. In: Lovrek, I., Howlett, R.J., Jain, L.C. (eds.) KES 2008. LNCS (LNAI), vol. 5178, pp. 281–288. Springer, Heidelberg (2008). https://doi.org/10.1007/978-3-540-85565-1_35
18. Iasemidis, L.D., Sackellares, J.C., Zaveri, H.P., et al.: Phase space topography and the Lyapunov exponent of electrocorticograms in partial seizures. Brain Topography **2**(3), 187–201 (1990)

19. Mirowski, P.W., Lecun, Y., Madhavan, D., et al.: Comparing SVM and convolutional networks for epileptic seizure prediction from intracranial EEG. In: IEEE Workshop on Machine Learning for Signal Processing, MLSP 2008, pp. 244–249. IEEE (2008)

20. Zhang, Y.-L.: Research on fractal analysis of epileptic EEG and automatic seizure detection methods. Ph.D. dissertation, Shandong University, China (2016)

21. Le Van Quyen, M., Martinerie, J., Baulac, M., et al.: Anticipating epileptic seizure in real time by a nonlinear analysis of similarity between EEG recordings. Neuroreport 10(10), 2149–2155 (1999)

22. Zheng, Y., Wang, G., Li, K., et al.: Epileptic seizure prediction using phase synchronization based on bivariate empirical mode decomposition. Clin. Neurophysiol. 125(6), 1104–1111 (2014)

23. Kumar, Y., Dewal, M.L., Anand, R.S.: Epileptic seizures detection in EEG using DWT-based ApEn and artificial neural network. Sig. Image Video Process. 8(7), 1323–1334 (2014)

24. Shiao, H.T., Cherkassky, V., Lee, J., et al.: SVM-based system for prediction of epileptic seizures from iEEG signal. IEEE Trans. Bio-Med. Eng. 64(5), 1011–1022 (2016)

25. Petrosian, A., Prokhorov, D., Homan, R., et al.: Recurrent neural network based prediction of epileptic seizures in intra- and extracranial EEG. Neurocomputing 30(1–4), 201–218 (2000)

26. Mirowski, P., et al.: Classification of patterns of EEG synchronization for seizure prediction. Clin. Neurophysiol. 120(11), 1927–1940 (2009)

27. Achilles, F., Belagiannis, V., Tombari, F., et al.: Deep convolutional neural networks for automatic identification of epileptic seizures in infrared and depth images. J. Neurol. Sci. 357, e436 (2015)

28. Deng, C.W., Huang, G.B., Xu, J., et al.: Extreme learning machines: new trends and applications. Sci. China Inf. Sci. 58(2), 1–16 (2015)

29. Goodfellow, I.J., Pouget-Abadie, J., Mirza, M., et al.: Generative adversarial nets. In: International Conference on Neural Information Processing Systems, pp. 2672–2680. MIT Press, Cambridge (2014)

30. Radford, A., Metz, L., Chintala, S.: Unsupervised representation learning with deep convolutional generative adversarial networks. Comput. Sci. (2015)

31. Huang, G.B., Zhu, Q.Y., Siew, C.K.: Extreme learning machine: a new learning scheme of feedforward neural networks. In: Proceedings of International Joint Conference Neural Network, vol. 2, pp. 985–990 (2004)

32. Deng, W., Zheng, Q., Chen, L.: Regularized extreme learning machine. In: IEEE Symposium on Computational Intelligence and Data Mining, CIDM 2009, pp. 389–395. IEEE (2009)

33. Toole, J.M.O., Boylan, G.B.: NEURAL: quantitative features for newborn EEG using Matlab. arXiv preprint arXiv:1704.05694 (2017)

Author Index